Student Solutions Manual

Chemistry
Principles and Reactions

EIGHTH EDITION

William L. Masterton

Cecile N. Hurley

Prepared by

Maria Cecilia D. de Mesa

Thomas D. McGrath

CENGAGE

Australia • Brazil • Mexico • Singapore • United Kingdom • United States

ISBN: 978-1-305-09523-6

Cengage
20 Channel Street
Boston, MA 02210
USA

Cengage is a leading provider of customized learning solutions with employees residing in nearly 40 different countries and sales in more than 125 countries around the world. Find your local representative at:
www.cengage.com.

Cengage products are represented in Canada by Nelson Education, Ltd.

To learn more about Cengage platforms and services, register or access your online learning solution, or purchase materials for your course, visit
www.cengage.com.

Printed at CLDPC, USA, 11-20

TABLE OF CONTENTS

Chapter 17

Chapter 18

Chapter 19

Chapter 20

Chapter 21

Chapter 22

Chapter 23

MATTER AND MEASUREMENTS

1.1 MATTER AND ITS CLASSIFICATIONS

2. Use the flow chart in Figure 1.1 as guide.
   (a) Gold is an **element**.
   (b) Milk is a **mixture**.
   (c) Sugar is a **compound**.
   (d) Vinaigrette dressing with herbs is a **mixture**.

4. Use the flow chart in Figure 1.1 as guide.
   (a) Iron ore is a **heterogeneous mixture**.
   (b) Chicken noodle soup is a **heterogeneous mixture**.
   (c) Tears are **homogeneous mixture or solution**.

6. (a) A mixture of the volatile gases propane, butane and isopropane can be separated by **chromatography**.

   (b) Rubbing alcohol is a homogeneous mixture of isopropyl alcohol and water. These two components can be separated either by **distillation** or by **chromatography**.

8. (a) copper – **Cu**
   (b) carbon – **C**
   (c) bromine – **Br**
   (d) aluminum – **Al**

10. (a) Mn – **manganese**
    (b) Au – **gold**
    (c) Pb – **lead**
    (d) Ag – **silver**

1.2 MEASUREMENTS: Instruments and Units

12. (a) Use a **meter or a yardstick** to measure the length of your bed.

    (b) Use a **graduated cylinder, a buret or a pipet** to measure the amount or volume of the acid delivered by a beaker.

    (c) Use a **thermometer** to determine your temperature. If your temperature is higher than normal body temperature, then you have a fever.

14. Use Example 1.1 as guide. Strategy: $°F \rightarrow °C \rightarrow K$
    Convert 350°F to °C:
    $$t_{°F} = 1.8\, t_{°C} + 32°$$
    $$350°F = 1.8\, t_{°C} + 32°$$
    $$1.8\, t_{°C} = 350°F - 32°$$
    $$t_{°C} = 318°F \div 1.8 = \underline{\textbf{177°C}}$$

The number 1.8 is exact, hence it does not limit the number of significant figures. The final operation is division and the least number of significant figures among the quantities is 3 thus the final answer should have three (3) significant figures.

Convert from °C to K:
$$T_K = t_{°C} + 273.15$$
$$T_K = 177°C + 273.15 = 450\ K = \underline{\textbf{4.50} \times \textbf{10}^2\ \textbf{K}}$$

The operation is addition so the answer should have the same number of decimal places (or decimal digits) as the measurement with the least number of decimal places. The measurement, 177°C has no decimal digit while 273.15 have 2; the answer should have no decimal digit.

$$\underline{\textbf{350°F} = \textbf{177°C} = \textbf{4.50} \times \textbf{10}^2\ \textbf{K}}$$

16. Use Example 1.1 as guide.
    Strategy: K → °C → °F

    Convert 308 K to °C:
    $$T_K = t_{°C} + 273.15$$
    $$308\ K = t_{°C} + 273.15$$
    $$t_{°C} = 308\ K - 273.15 = \underline{\textbf{35°C}}$$

    Convert 35°C to °F:
    $$t_{°F} = 1.8 t_{°C} + 32°$$
    $$t_{°F} = 1.8(35°C) + 32° = \underline{\textbf{95°F}}$$

    $$\underline{\textbf{308K} = \textbf{35°C} = \textbf{95°F}}$$

1.2 MEASUREMENTS: Uncertainties in Measurements – Significant Figures

18. (a) 0.136 m

    Three (3) significant figures. Zeros before a decimal point are not significant.

    (b) 0.0001050 g

    Four (4) significant figures. Zeros after the decimal point but before a non-zero digit are not significant because they just indicate the position of the decimal point. Zeros between two non-zero digits are significant.

    (c) 2.700×10^3 nm

    Four (4) significant figures. Zeros to the right of the decimal point and after a non-zero digit are significant.

    (d) 6×10^{-4} L

    One (1) significant figure. Non-zero digits are significant.

    (e) 56003 cm³

    Five (5) significant figures. Zeros between two non-zero digits are significant.

20. (a) 17.25 cm   (c) 5.00 × 10² °C

    (b) 169 lb   (d) 198 oz

22.

	Scientific Notation
(a) 4020.6 mL	**4.0206 × 10³ mL**
(b) 1.006 g	**1.006 g is already in proper scientific notation**
(c) 100.1°C	**1.001 × 10² °C**

24. (a) Temperature is a measured using a thermometer which reading carries with it a degree of uncertainty. The temperature reading, 72°F is therefore **not exact**.

    (b) Values or numbers obtained by counting are exact numbers. Since the numbers in **6** eggs, **2** cookies and **5** tomatoes are obtained by counting, these numbers are **exact**.

    (c) In the statement "There are 1 × 10⁹ nanometers in 1 meter.", the values are not measured quantities but are from definition of the prefix "nano". Numbers obtained from definition are exact, therefore these numbers are **exact**.

26. The quantity, **5 bedrooms** is a result of counting and is therefore **exact**.

The area of **4000 ft² is ambiguous**.

The quantity, **17 ft has 2 significant figures** and the length **18.5 has 3 significant figures** because they are both measured quantities and carry a degree of uncertainty.

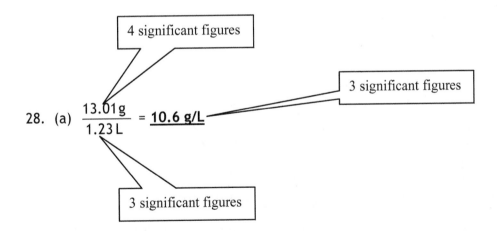

28. (a) $\dfrac{13.01\,g}{1.23\,L} = \underline{10.6\ g/L}$

"When measured quantities are multiplied or divided, the number of significant figures in the result is the same as that in the quantity with the smallest number of significant figures". Since the least number of significant figures is 3, the answer should have only 3 significant figures.

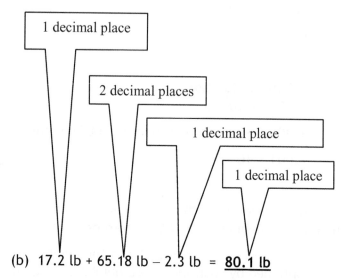

(b) 17.2 lb + 65.18 lb – 2.3 lb = **80.1 lb**

"When measured quantities are added or subtracted, the number of decimal places in the result is the same as that in the quantity with the smallest number of decimal places". Since the least number of decimal places is 1, the answer should have only 1 decimal place.

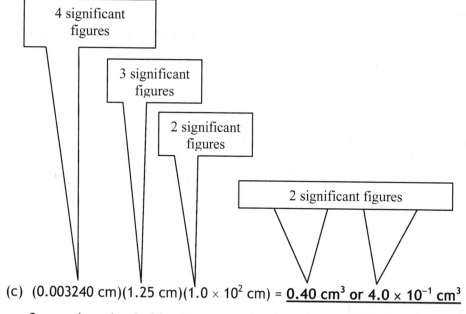

(c) (0.003240 cm)(1.25 cm)(1.0 × 10² cm) = **0.40 cm³ or 4.0 × 10⁻¹ cm³**

See explanation in 28a. The operation is multiplication and the least number of significant figures among the 3 given factors is 2, thus the answer should have only 2 significant figures. To have two significant figures in the calculation result of 0.405 cm³, the number 5 is to be discarded by applying the rules for "rounding off". The number 0.405 is round off such that the last digit of the resulting number is an even number. Since the last digit (0) is already an even number, 0 is left unchanged.

In performing the following mathematical operation:

$$\sqrt{(0.024)^2 + 4(6.3\times10^{-3})} = 0.16\underline{0}549058$$

the result should have 3 decimal places only. Therefore in the result of the calculation, 0.160549058, the last significant figure is the 3rd decimal digit "0" (between 6 and 5).

0 is the last significant figure thus it has 3 decimal places

The 3rd decimal digit is the last significant figure thus, it has 3 significant figures

3 decimal places

(d) $\dfrac{0.024 + \sqrt{(0.024)^2 + 4(6.3\times10^{-3})}}{2} = \dfrac{0.024 + 0.160549058}{2} = \dfrac{0.184549058}{2} = \underline{0.0923}$

3 significant figures

It is given that the numbers 4 and 2 are exact numbers therefore they do not limit the number of significant figures in the calculation. The operation applied to the rest of the numbers in the denominator is addition which follows the least number of decimal places for the result which is 3. The last operation is division which follows the least number of significant figures and since the resulting numerator should have only 3 significant figures, and the denominator is exact, the final answer should have 3 significant figures.

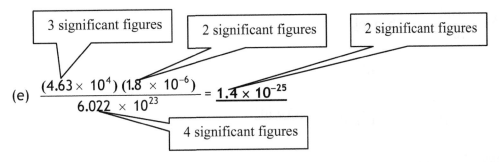

3 significant figures

2 significant figures

2 significant figures

(e) $\dfrac{(4.63\times10^{4})(1.8\times10^{-6})}{6.022\times10^{23}} = \underline{1.4\times10^{-25}}$

4 significant figures

The operations involved are multiplication and division. Assuming all quantities are measurements, the least number of significant figures is 2; therefore the final answer should have 2 significant figures.

30. volume of square pyramid $= \dfrac{B^2 h}{3}$ where   B = length of a side of the base

                                                         h = height of the pyramid

Student 1:   volume of square pyramid $= \dfrac{B^2 h}{3} = \dfrac{(1.33 \text{ m})^2 \,(2.79 \text{ m})}{3} = 1.65 \text{ m}^3$

Student 2:   volume of square pyramid $= \dfrac{B^2 h}{3} = \dfrac{(1.31 \text{ m})^2 \,(2.81 \text{ m})}{3} = 1.61 \text{ m}^3$

difference in the calculated volumes $= 1.65 \text{ m}^3 - 1.61 \text{ m}^3 = \underline{\textbf{0.04 m}^3}$

1.2 MEASUREMENTS: Conversion of Units

32. To compare the given quantities, they should have the same unit of measurements. For each pair, convert one of the measured values to the unit of the other. Once the numbers are expressed in common units, they can be compared directly.

(a) $37.12 \text{ g} \times \dfrac{1 \text{ kg}}{10^3 \text{ g}} = 0.03712 \text{ kg}$;   thus, $\underline{\textbf{37.12 g} < \textbf{0.3712 kg}}$

(b) $28 \text{ m}^3 \times \left(\dfrac{1 \text{ cm}}{10^{-2} \text{ m}}\right)^3 = 28 \times 10^6 \text{ cm}^3$;   thus, $\underline{\textbf{28 m}^3 > \textbf{28} \times \textbf{10}^2 \textbf{ cm}^3}$

(c) $525 \text{ mm} \times \dfrac{10^{-3} \text{ m}}{1 \text{ mm}} \times \dfrac{1 \text{ nm}}{10^{-9} \text{ m}} = 525 \times 10^6 \text{ nm}$;   thus, $\underline{\textbf{525 mm} = \textbf{525} \times \textbf{10}^6 \textbf{ nm}}$

34. (a) Strategy: L → mL

     $0.2156 \text{ L} \times \dfrac{1 \text{ mL}}{10^{-3} \text{ L}} = \underline{\textbf{215.6 mL}}$

(b) Strategy: L → cm^3 → in^3

     $0.2156 \text{ L} \times \dfrac{10^3 \text{ cm}^3}{1 \text{ L}} \times \dfrac{(1 \text{ in})^3}{(2.54 \text{ cm})^3} = \underline{\textbf{13.16 in}^3}$

(c) Strategy: L → qt

     $0.2156 \text{ L} \times \dfrac{1.057 \text{ qt}}{1 \text{ L}} = \underline{\textbf{0.2279 qt}}$

36. (a) Strategy: nautical mi → ft → mi

$$1 \text{ nautical mi} \times \frac{6076.12 \text{ ft}}{1 \text{ nautical mi}} \times \frac{1 \text{ mi}}{5280 \text{ ft}} = \underline{\textbf{1.15078 mi}}$$

The answer should have 6 significant figures. The distance is given as an exact number and 5280 is an exact number. The number of significant figures is constrained by the first conversion factor, 6076.12 ft (6 significant figures).

(b) Strategy: nautical mi → ft → mi → km → m

$$1 \text{ nautical mi} \times \frac{6076.12 \text{ ft}}{1 \text{ nautical mi}} \times \frac{1 \text{ mi}}{5280 \text{ ft}} \times \frac{1.609 \text{ km}}{1 \text{ mi}} \times \frac{10^3 \text{ m}}{1 \text{ km}} = \underline{\textbf{1852 m}}$$

The answer should have 4 significant figures. The distance is given as an exact number and 5280 is an exact number. The number of significant figures is constrained by the first conversion factor, 6076.12 ft (6 significant figures) and by the third conversion factor, 1.609 (4 significant figures), the less of which is 4 significant figures. Thus the answer should have 4 significant figures.

(c) Strategy: knots → nautical mi/h → mi/h

$$22 \text{ knots} \times \frac{1 \text{ nautical mi/h}}{1 \text{ knot}} \times \frac{1.15078 \text{ mi}}{1 \text{ nautical mi}} = 25 \text{ mi/h} = \underline{\textbf{25 mph}}$$

The answer should have 2 significant figures. The given, 22 knots has 2 significant figures. The first conversion factor is an exact number, while the second (from part 36(a) above) has 6 significant figures. Since the operation involved is multiplication and division, the answer should have the least number of significant figures among the quantities which is 2.

38. (a) cost of unleaded gasoline in USD per gallon:

Strategy: gal gasoline → qt gasoline → L gasoline → pesos → USD

$$1 \text{ gal} \times \frac{4 \text{ qt}}{1 \text{ gal}} \times \frac{1 \text{ L}}{1.057 \text{ qt}} \times \frac{38.46 \text{ pesos}}{1 \text{ L}} \times \frac{1 \text{ USD}}{47.15 \text{ pesos}} = \underline{\textbf{3.09 USD}}$$

Therefore, **1 gallon of gasoline costs $3.09.**

(b) cost of unleaded gasoline in USD for a 14 US gallon tank:

Strategy: 14 gal gasoline → qt gasoline → L gasoline → pesos → USD

$$14 \text{ gal} \times \frac{4 \text{ qt}}{1 \text{ gal}} \times \frac{1 \text{ L}}{1.057 \text{ qt}} \times \frac{38.46 \text{ pesos}}{1 \text{ L}} \times \frac{1 \text{ USD}}{47.15 \text{ pesos}} = \underline{\textbf{43 USD}}$$

(c) miles driven by 1255 pesos worth of unleaded gasoline if the car runs 24 miles per gallon:

Strategy: pesos → L gasoline → qt gasoline → gal gasoline → mi

$$1255 \text{ pesos} \times \frac{1 \text{ L}}{38.46 \text{ pesos}} \times \frac{1.057 \text{ qt}}{1 \text{ L}} \times \frac{1 \text{ gal}}{4 \text{ qt}} \times \frac{24 \text{ mi}}{1 \text{ gal}} = \underline{\textbf{2.1} \times \textbf{10}^2 \textbf{ mi}}$$

40. g cholesterol/mL blood:

$$\frac{185 \text{ mg cholesterol}}{1 \text{ dL blood}} \times \frac{10^{-3} \text{ g}}{1 \text{ mg}} \times \frac{10 \text{ dL}}{1 \text{ L}} \times \frac{10^{-3} \text{ L}}{1 \text{ mL}} = \underline{\textbf{1.85} \times \textbf{10}^{-3} \textbf{ g/mL}}$$

42. To find the volume (L) that would cover the area of 3.02×10^6 mi^2 by 2 inches of rainfall, simply multiply the area with the height of the rainfall. However, the units should cancel out to give a final unit of liters.

Strategy: find volume using the formula, V = area × height

area: 3.02×10^6 mi^2 → ft^2

height: 2 in → ft

volume: ft^3 → L

V = area × height

$$= \left(3.02 \times 10^6 \text{ mi}^2 \times \frac{(5280 \text{ ft})^2}{(1 \text{ mi})^2} \right) \times \left(2 \text{ in} \times \frac{1 \text{ ft}}{12 \text{ in}} \right) \times \frac{28.32 \text{ L}}{1 \text{ ft}^3} = \underline{\textbf{3.97} \times \textbf{10}^{14} \textbf{ L}}$$

44. Strategy: tablet → grains → lb → g → mg

$$\text{mg active ingredient} = 1 \text{ tablet} \times \frac{5.000 \text{ grains}}{1 \text{ tablet}} \times \frac{1 \text{ lb}}{5.760 \times 10^3 \text{ grains}} \times \frac{453.6 \text{ g}}{1 \text{ lb}} \times \frac{1 \text{ mg}}{10^{-3} \text{ g}}$$

$$= \underline{\textbf{393.8 mg}}$$

1.3 PROPERTIES OF SUBSTANCES

46. Using equation 1.3 from the textbook the average density of an egg white from one "large" egg can be calculated as follows:

$$d_{egg\ white} = \frac{mass_{egg\ white}}{V_{egg\ white}} = \frac{1.20 \times 10^2\ g}{112\ mL} = \underline{\textbf{1.07 g/mL}}$$

48. The volume of irregularly shaped water–insoluble objects is determined by water displacement. The volume of water displaced corresponds to the volume of the object. As illustrated below, addition of the 11.33 g solid to the graduated cylinder containing water changed the volume from 35.0 mL to 42.3 corresponding to a change of 7.3 mL. This means that the volume of the object is 7.3 mL.

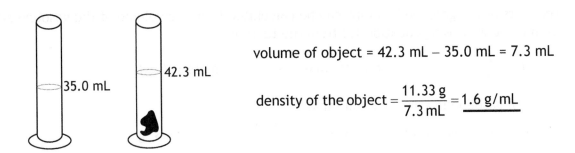

volume of object = 42.3 mL – 35.0 mL = 7.3 mL

$$density\ of\ the\ object = \frac{11.33\ g}{7.3\ mL} = \underline{1.6\ g/mL}$$

50. The first step in solving this problem is to find the volume of the aluminum wire in m^3. The volume can be determined by converting the given mass of the wire to volume using the density equation 1.3 from the textbook. The calculated volume of the wire will be used to determine the length of the 10-lb spool aluminum wire.

volume of the aluminum wire in m^3:

Strategy: $lb/g/cm^3 \rightarrow g/g/cm^3 \rightarrow cm^3 \rightarrow m^3$

$$d = \frac{mass}{V} \Rightarrow solve\ for\ V \Rightarrow \qquad V = \frac{mass}{d}$$

$$V_{Al} = \frac{mass}{d_{Al}} = \frac{10\ lb}{2.70\ g/cm^3} \times \frac{453.6\ g}{1\ lb} \times \frac{1\ m^3}{10^6\ cm^3} = 1.7 \times 10^{-3}\ m^3$$

radius of the wire in m:

Strategy: in $\rightarrow$ cm $\rightarrow$ m

Note: radius $= \dfrac{\text{diameter}}{2}$

radius $= \dfrac{0.0808 \text{ in}}{2} \times \dfrac{2.54 \text{ cm}}{1 \text{ in}} \times \dfrac{10^{-2} \text{ m}}{1 \text{ cm}} = 1.03 \times 10^{-3} \text{ m}$

length of the wire:

$\text{volume}_{\text{cylinder}} = \pi r^2 \ell$ where ℓ is the length and r is the radius of the wire

length $= \dfrac{\text{volume}}{\pi r^2} = \dfrac{1.7 \times 10^{-3} \text{ m}^3}{3.1416 \,(1.03 \times 10^{-3} \text{ m})^2} = \underline{\mathbf{5.10 \times 10^2 \text{ m}}}$

52. The mass of the gasoline in a tank can be calculated from the volume of the gasoline and from its density using equation 1.3 from the textbook.

$d = \dfrac{\text{mass}}{V} \qquad \Rightarrow \text{solve for mass} \Rightarrow \qquad \text{mass} = V \times d$

grams of gasoline in a 14.0 gal full tank:

Strategy: gal $\rightarrow$ qt $\rightarrow$ L $\rightarrow$ m^3 $\xrightarrow{\quad \text{density} \quad}$ kg $\rightarrow$ g

mass $= V \times d = 14.0 \text{ gal} \times \dfrac{4 \text{ qt}}{1 \text{ gal}} \times \dfrac{1 \text{ L}}{1.057 \text{ qt}} \times \dfrac{1 \text{ m}^3}{10^3 \text{ L}} \times \dfrac{732.22 \text{ kg}}{1 \text{ m}^3} \times \dfrac{10^3 \text{ g}}{1 \text{ kg}} = \underline{\mathbf{3.88 \times 10^4 \text{ g}}}$

pounds of gasoline in a 14.0 gal full tank:

Strategy: gal $\rightarrow$ qt $\rightarrow$ L $\rightarrow$ m^3 $\xrightarrow{\quad \text{density} \quad}$ kg $\rightarrow$ lb

mass $= V \times d = 14.0 \text{ gal} \times \dfrac{4 \text{ qt}}{1 \text{ gal}} \times \dfrac{1 \text{ L}}{1.057 \text{ qt}} \times \dfrac{1 \text{ m}^3}{10^3 \text{ L}} \times \dfrac{732.22 \text{ kg}}{1 \text{ m}^3} \times \dfrac{1 \text{ lb}}{0.4536 \text{ kg}} = \underline{\mathbf{85.5 \text{ lb}}}$

54. The amount of potassium sulfate dissolved in a saturated solution is equal to its solubility. If the amount is less than its solubility, it is an unsaturated solution and if the amount is greater than its solubility, it is a supersaturated solution.

Using the solubility of potassium sulfate as conversion factor, the mass of potassium sulfate that would dissolve in 225 g water at 40°C can be obtained.

$225 \text{ g water} \times \dfrac{15 \text{ g potassium sulfate}}{100 \text{ g water}} = 34 \text{ g potassium sulfate}$

Since 225 g water will ONLY dissolve 34 g potassium sulfate, yet the solution contains 39.0 potassium sulfate, **the solution is a supersaturated solution** because it contains an amount greater than the solubility.

The difference between the amount of potassium sulfate dissolved in the supersaturated solution and the maximum amount that dissolves (solubility) is expected to crystallize out of solution.

    39 g – 34 g = 5 g

Therefore, **5 g is expected to crystallize out of the solution**.

56. (a) The mixture is homogeneous if the amount of magnesium chloride dissolved is less than or equal to the solubility. It is a heterogeneous mixture if the amount of magnesium chloride is greater than its solubility in water.

    Determine the maximum amount of magnesium chloride that will dissolve in 38.2 g water using solubility as conversion factor. The solubility of magnesium chloride at 20°C is 54.6 g/100 g water.

$$38.2 \text{ g water} \times \frac{54.6 \text{ g magnesium chloride}}{100 \text{ g water}} = 20.9 \text{ g magnesium chloride}$$

    Since the maximum amount of magnesium chloride that the 38.2 g water can dissolve is 20.9 g and only 16.2 g was placed in it, all 16.2 g is completely dissolved. **It is a homogeneous mixture**.

    To make a saturated solution at 20°C, 4.7 g more of magnesium chloride should be added.

        20.9 g – 16.2 g = **4.7 g more**

(b) Find the maximum amount of magnesium chloride that will dissolve in 38.2 g water at 80°C using solubility as conversion factor. Solubility of magnesium chloride at 80°C = 66.1 g/100 g water

$$38.2 \text{ g water} \times \frac{66.1 \text{ g magnesium chloride}}{100 \text{ g water}} = 25.3 \text{ g magnesium chloride}$$

    To make a saturated solution at 80°C, the mixture should have 25.3 g magnesium chloride. Thus, 9.1 g more of magnesium chloride should be added.

        25.3 g – 16.2 g = **9.1 g more**

UNCLASSIFIED PROBLEMS

58. The amount of rems absorbed by a person for 35 minutes when exposed to radiation of 8217 mSv/h can be determined using the conversion below.

Strategy: min $\rightarrow$ hour $\rightarrow$ mSv $\rightarrow$ Sv $\rightarrow$ rem

$$\text{rems absorbed} = 35 \text{ min} \times \frac{1\text{h}}{60 \text{ min}} \times \frac{8217 \text{ mSv}}{1\text{h}} \times \frac{10^{-3} \text{ Sv}}{1 \text{ mSv}} \times \frac{1 \text{ rem}}{0.0100 \text{ Sv}} = \underline{\textbf{4.8} \times \textbf{10}^2 \textbf{ rems}}$$

number of mammograms that would give a radiation exposure of 4.8×10^2 rems:

$$\text{number of mammograms} = 4.8 \times 10^2 \text{ rems} \times \frac{1 \text{mammogram}}{0.30 \text{ rems}} = \underline{\textbf{1.6} \times \textbf{10}^3 \textbf{ mammograms}}$$

60. To find the grams of solid obtained when the sugar solution reaches saturation, the maximum amount of sugar that the water can dissolve should be determined first. Any amount exceeding this maximum will go out of the solution to form sugar crystals.

The supersaturated sugar solution contained 650.0 g sugar in 150.0 g water. The solubility of sugar at 25°C is only 220.0 g/100 g water therefore the maximum amount of sugar that the 150.0 g water can dissolve at 25°C is only 330 g sugar as shown in the calculation below:

$$150.0 \text{ g water} \times \frac{220.0 \text{ g sugar}}{100 \text{ g water}} = 330 \text{ g sugar}$$

$$650.0 \text{ g} - 330 \text{ g} = \underline{\textbf{3.20} \times \textbf{10}^2 \textbf{ g sugar crystals will be formed}}$$

62. To calculate the thickness or height of the foil in inches: i) find the volume, V of aluminum foil using the density formula (equation 1.3 from the textbook); ii) determine the area, A of the foil and iii) use the calculated V and A to find the thickness of the aluminum foil.

Volume of the aluminum foil:

$$d = \frac{\text{mass}}{V} \quad \Rightarrow \text{solve for } V \Rightarrow \quad V = \frac{\text{mass}}{d}$$

$$V_{\text{Al foil}} = \frac{\text{mass}}{d_{\text{Al foil}}} = \frac{0.83 \text{ kg}}{2.70 \text{ g/cm}^3} \times \frac{10^3 \text{ g}}{1 \text{kg}} \times \frac{(1 \text{ in})^3}{(2.54 \text{ cm})^3} = 19 \text{ in}^3$$

Area (*A*) of the aluminum foil:

dimension of the foil: width of the foil = 12 inches

length of the foil = 66 ⅔ or 66.67 yd

$$A = 12 \text{ in} \times 66.67 \text{ yd} \times \frac{3 \text{ ft}}{1 \text{ yd}} \times \frac{12 \text{ in}}{1 \text{ ft}} = 2.9 \times 10^4 \text{ in}^2$$

Thickness or height (*h*) of the aluminum foil derived from the formula:

$$V = A \times h \implies \text{solve for h} \implies \quad h = \frac{V}{A} = \frac{19 \text{ in}^3}{2.9 \times 10^4 \text{ in}^2} = \underline{\mathbf{6.6 \times 10^{-4} \text{ in}}}$$

64. Use the following steps to calculate the density of the alloy:

   i) Find volume of pycnometer = volume of water that fills the pycnometer

      mass of water = (mass pycnometer + water) – (mass of pycnometer)

$$= 31.486 \text{g} - 20.455 \text{ g}$$

$$= 11.031 \text{ g water}$$

      volume of pycnometer = $11.031 \text{ g} \times \dfrac{1 \text{mL}}{1.00 \text{ g}} = 11.0 \text{ mL}$

   ii) Find volume of alloy = volume of water displaced by the alloy

      mass of water = (mass pycnometer + water + alloy) – (mass of pycnometer + alloy)

$$= 38.689 \text{ g} - 28.695 \text{ g}$$

$$= 9.994 \text{ g water in pycnometer not displaced by the alloy}$$

      volume of water not displaced = $9.994 \text{ g} \times \dfrac{1 \text{mL}}{1.00 \text{ g}} = 9.99 \text{ mL}$

      volume of alloy = volume water displaced = 11.0 mL – 9.99 mL = 1.0 mL

   iii) Calculate the mass of the alloy.

      mass of alloy = (mass pycnometer + alloy) – (mass of pycnometer)

$$= 28.695 \text{g} - 20.455 \text{ g}$$

$$= 8.240 \text{ g alloy}$$

   iv) Find the density of alloy (*d*ₐₗₗₒᵧ) by substituting the calculated volume and the calculated mass of the alloy to equation 1.3 from the textbook.

$$d_{alloy} = \frac{mass_{alloy}}{V_{alloy}} = \frac{8.240 \text{ g}}{1.0 \text{ mL}} = \underline{\mathbf{8.2 \text{ g/mL}}}$$

CONCEPTUAL QUESTIONS

66. Property of Iodine   Intensive or Extensive Property

    (a) Its density is 4.93 g/mL.   <u>intensive</u>

    (b) It is purple.   <u>intensive</u>

    (c) It melts at 114°C.   <u>intensive</u>

    (d) 100g of melted iodine has a volume of 122 mL.   <u>extensive</u>

    (e) Some iodine crystals are large.   <u>extensive</u>

The first 3 properties: density, color and melting point are intensive properties because they do not depend on the amount of iodine present. The volume of melted iodine and the size of iodine crystals are both extensive properties because they depend on the amount of iodine. Greater mass of melted iodine would mean greater volume. Likewise, the larger the amount of iodine present in a crystal, the larger the size of the crystal.

68. (a) Difference between mass and density:

        Mass is an extensive property whereas density is an intensive property.

    (b) Difference between extensive and intensive property:

        Extensive property does not depend on the nature but on the amount of the substance. On the other hand, intensive property is dependent on the nature and not on the amount of the substance.

    (c) Difference between a solvent and a solution:

        A solvent is just a component of a solution. A solution is the homogeneous mixture of a solvent (the dissolving medium) and one or more solutes (the substance(s) dissolved).

70. **Three (3)** significant figures. The length of the line is: $26.15 - 24.25 = 1.90$

72. (a) **A**

    (b) at about **23°C**

    (c) **No**, an increase in temperature does not always increase the solubility of a compound. As shown in the graph for A, the solubility of A is higher at lower temperature.

CHALLENGE PROBLEMS

73. (a)

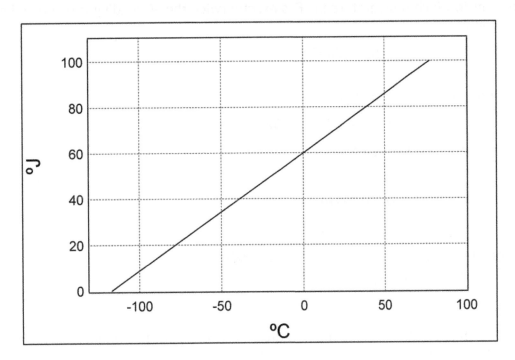

(b) slope $= \dfrac{(100 - 0)°J}{[78 - (-117)]°C} = \dfrac{100 \ °J}{195°C} = \underline{\textbf{0.513 °J/°C}}$

(c) The y-intercept of a line is the value of y when x is zero. In this case, it is the value of °J when °C is zero. As shown in the graph, when °C is zero, °J is 60.

y-intercept = **60°J**

(d) Since the relationship between °J and °C is linear, it follows the equation of the line, y = mx + b, where m is the slope and b is the y-intercept. Therefore, the equation that would convert °J (y) to °C(x) and vice versa is the following.

°J = 0.513 (°C) + 60 $°C = \dfrac{(°J) - 60}{0.513}$

74. The general relation between °F and °C is expressed by the following equation:

$$t_{°F} = 1.8(t_{°C}) + 32 \quad \text{(equation 1)}$$

the point in which temperature in °F is exactly twice the °C reading is expressed by equation 2:

$$t_{°F} = 2(t_{°C}) \quad \text{(equation 2)}$$

substitute equation 2 into equation 1:

$$2(t_{°C}) = 1.8(t_{°C}) + 32$$

$$2(t_{°C}) - 1.8(t_{°C}) = 32$$

$$0.2(t_{°C}) = 32$$

$$(t_{°C}) = 32/0.2$$

$$t_{°C} = 160°C$$

when $t_{°C} = 160°C$, $t_{°F} = 1.8(160°) + 32 = 320°F$

160 °C = 320 °F

75. The first step in solving this problem requires conversion of the volume (gal) to km^3. Use this volume to find the area covered knowing that the thickness or depth of the film formed by oil is 100 nm.

Volume: convert gal to km^3.

Strategy: gal $\rightarrow$ qt $\rightarrow$ in^3 $\rightarrow$ m^3 $\rightarrow$ km^3

$$31.5 \, \text{gal} \times \frac{4 \, \text{qt}}{1 \, \text{gal}} \times \frac{57.75 \, in^3}{1 \, \text{qt}} \times \frac{(1 \, m)^3}{(39.37 \, in)^3} \times \frac{(1 \, km)^3}{(10^3 \, m)^3} = 1.19 \times 10^{-10} \, km^3$$

Thickness or depth: convert nm to km.

Strategy: nm $\rightarrow$ m $\rightarrow$ km

$$100 \, \text{nm} \times \frac{1 \, m}{10^9 \, \text{nm}} \times \frac{1 \, km}{10^3 \, m} = 1.0 \times 10^{-10} \, km$$

Find the area of the oil spill derived from the formula:

volume = area × height

$$\text{Area} = \frac{\text{volume}}{\text{height}} = \frac{1.19 \times 10^{-10} \, km^3}{1.0 \times 10^{-10} \, km} = \underline{\mathbf{1.2 \, km^2}}$$

76. To determine the length of the wire needed in the experiment, follow the following steps: i) calculate the volume of Al wire needed using the density equation 1.3 (from textbook); ii) find the radius of the wire and iii) using the calculated volume and radius substituted in the formula for the volume of a cylinder, determine the length of the wire.

i) volume of Al wire

$$d = \frac{mass}{V} \qquad \Rightarrow \text{solve for } V \Rightarrow \qquad V = \frac{mass}{d}$$

$$V_{Al} = \frac{mass}{d_{Al}} = \frac{12.0 \text{ g}}{2.70 \text{ g/cm}^3} = 4.44 \text{ cm}^3$$

ii) radius of the Al wire in cm

$$r = \frac{0.200 \text{ in}}{2} \times \frac{2.54 \text{ cm}}{1 \text{ in}} = 0.254 \text{ cm}$$

iii) length (ℓ) of the Al wire in cm

$$V = \pi r^2 \ell \quad \Rightarrow \text{solve for } \ell \Rightarrow \qquad \ell = \frac{V}{\pi r^2} = \frac{4.44 \text{ cm}^3}{(3.1416)(0.254 \text{ cm})^2} = \underline{\mathbf{21.9 \text{ cm}}}$$

77. The mass of lead absorbed in 1 year can be determined using the following steps (i) calculate the volume of air that an average adult breathes in 1 year; (ii) find the mass of lead in air if accumulated in 1 year; and (iii) determine the mass of lead absorbed in the lungs assuming that only 50% of the 75% of the lead in the air is absorbed.

i) volume of air an average adult breathes in 1 year

$$\frac{8.50 \times 10^3 \text{ L}}{1 \text{ day}} \times \frac{365 \text{ days}}{1 \text{ year}} = 3.10 \times 10^6 \text{ L air/year}$$

ii) mass of lead in air in 1 year

$$\frac{3.10 \times 10^6 \text{ L air}}{1 \text{ year}} \times \frac{1 \text{ m}^3}{10^3 \text{ L}} \times \frac{7.0 \times 10^{-6} \text{ g Pb}}{1 \text{ m}^3 \text{ air}} = 2.17 \times 10^{-2} \text{ g Pb/year}$$

iii) mass of lead in the lungs assuming that only 50% of the 75% of lead in the air is absorbed

$$\frac{2.17 \times 10^{-2} \text{ g}}{1 \text{ year}} \times 0.75 \times 0.5 = \underline{\mathbf{8.1 \times 10^{-3} \text{ g Pb/year}}}$$

78. To answer this problem, let's visualize it. Mercury is a liquid metal so it can completely fill a cylinder. Both cylinders A and B are completely filled with Hg but cylinder A also contains a metal aside from Hg. The combined weight for the Hg and the metal in cylinder A is 92.60 g. On the other hand the weight of Hg in cylinder B is 52.6 g more than the content of cylinder A resulting in a total Hg weight of 145.2 g. The difference in mass between the contents of the two cylinders is attributed to the mass of the metal.

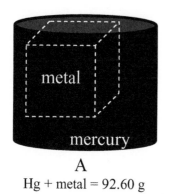

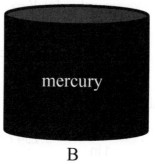

A

Hg + metal = 92.60 g

B

Hg = 92.60 g + 52.6 g = 145.2 g

Masses:

mass of metal = 145.2 g – 92.60 g = 52.6 g metal

mass of mercury in cylinder A = 92.60 g – 52.6 g = 40.0 g Hg

mass of mercury in cylinder B = 92.60 g + 52.6 g = 145.2 g Hg

Using equation 1.3:

$$d = \frac{mass}{V} \qquad \Rightarrow solve\ for\ V \Rightarrow \qquad V = \frac{mass}{d}$$

Since the two cylinders have equal volume then,

volume cylinder A = volume cylinder B

$$(V_{Hg})_{cylinder\ A} = (V_{Hg} + V_{metal})_{cylinder\ B}$$

$$\left(\frac{mass_{Hg}}{d_{Hg}}\right)_{cylinderA} = \left(\frac{mass_{Hg}}{d_{Hg}} + \frac{mass_{metal}}{d_{metal}}\right)_{cylinder\ B}$$

$$\frac{145.2\ g}{13.6\ g/cm^3} = \frac{40.0\ g}{13.6\ g/cm^3} + \frac{52.6\ g}{d_{metal}}$$

solving for the density of the metal,

$$d_{metal} = \frac{52.6\ g}{\left(\frac{145.2\ g - 40.0\ g}{13.6\ g/cm^3}\right)} = \underline{\mathbf{6.80\ g/cm^3}}$$

ATOMS, MOLECULES, AND IONS

2.1 ATOMS AND THE ATOMIC THEORY: Atomic Theory and Laws

2. The law of constant composition states that the ratio of the masses of elements in a compound is constant. For example, different samples of water will have the same mass ratio of hydrogen to oxygen.

4. (a) Neither of the two laws is illustrated by this statement

   (b) Law of Conservation of Mass

   (c) Neither of the two laws is illustrated by this statement

2.2 COMPONENTS OF THE ATOM

6. Rutherford discovered the nucleus. His experiment involved alpha (α) particles which are massive and positively charged. He fired α-particles at a thin sheet of gold foil. He expected all the α-particles to pass through the foil unhindered. Indeed, most of the particles pass through the foil without any change in direction, but a few of them were deflected at acute angles and a very small percentage bounced back. From this observation he concluded that most of the atom is empty space because most of the alpha particles pass through undeflected. He also concluded that the atom has a tiny nucleus which is highly massive and positively charged which caused a few of the α-particles to be deflected and to bounce back.

2.3 QUANTITATIVE PROPERTIES OF THE ATOM: Nuclear Symbols and Isotopes

8. These problems are similar to Example 2.1

   Mass number (A) = number of protons (Z) + number of neutrons

   Rn-220: Atomic number (Z) = 86;   Mass Number (A) = 220

   (a) atomic number = number of protons = **86**

   (b) number of neutrons = A – Z = 220 – 86 = **134**

10. These problems are similar to Example 2.1

    mass number (A) = number of protons (Z) + number of neutrons

    Fe: Atomic number (Z) = 26 = 26 protons

    (a) nuclear symbol for Fe-54: $^{54}_{26}$**Fe**
        nuclear symbol for Fe-56: $^{56}_{26}$**Fe**

    (b) The two isotopes differ in the number of neutrons. Fe-54 has 28 neutrons whereas Fe-56 has 30 neutrons.

        Fe-54: number of neutrons = A − Z = 54 − 26 = **28 neutrons**
        Fe-56: number of neutrons = A − Z = 56 − 26 = **30 neutrons**

12. Mass number (A) = number of protons (Z) + number of neutrons

    Am: Atomic number (Z) = 95 = 95 protons

    (a) Since the Am atom given in this problem is neutral (no charge), then the number of positive particles (protons) is equal to the number of negative particles (electrons).

        atomic number = number of protons = number of electrons = **95**

    (b) Mass number (A) = number of protons + number of neutrons = 95 + 146 = **241**

    (c) nuclear symbol: $^{241}_{95}$**Am**

14. Elements are identified by their atomic number, Z.

Element Representation		Identity of the Element	Number of protons (Z)	Number of neutrons (A − Z)	Number of electrons
A	$^{75}_{33}$A	**A = arsenic, As**	**33 p^+**	75 − 33 = **42 n**	**33 e^-**
L	$^{51}_{23}$L	**L = vanadium, V**	**23 p^+**	51 − 23 = **28 n**	**23 e^-**
Z	$^{131}_{54}$Z	**Z = xenon, Xe**	**54 p^+**	131 − 54 = **77 n**	**54 e^-**

16. Isobars are nuclei which have the same mass number but different atomic numbers. On the other hand isotopes are nuclei which have the same number of protons but different number of neutrons.

(a) <u>isobars: Cr-54 and Fe-54; Fe-58 and Ni-58</u>
<u>isotopes: Fe-54 and Fe-58</u>
Cr-54 and Fe-54 are isobars because they have the same mass number of 54. Fe-58 and Ni-58 are also isobars. Fe-54 and Fe-58 are isotopes because they have the same number of protons, 26.

(b) Fe-54 and Fe-58 have the same atomic number and therefore the <u>**same number of protons**</u>.

(c) <u>**Cr-54 and and Ni-58**</u> have the same number of neutrons, 30.

Cr-54: number of neutrons = mass number (A) – number of protons = 54 – 24 = <u>**30**</u>

Fe-54: number of neutrons = mass number (A) – number of protons = 54 – 26 = 28

Fe-58: number of neutrons = mass number (A) – number of protons = 58 – 26 = 32

Ni-58: number of neutrons = mass number (A) – number of protons = 58 – 28 = <u>**30**</u>

2.3 QUANTITATIVE PROPERTIES OF THE ATOM:
Atomic Masses and Isotopic Abundances

18. The average mass of each can be obtained from the periodic table.

(a) potassium ion: 39.10 amu

(b) phosphorus molecule, P_4: 4 × 30.97 = 123.88 amu

(c) potassium atom: 39.10 amu

(d) platinum atom: 195.1 amu

In terms of increasing mass:

(a) potassium ion < (c) potassium atom < (b) phosphorus molecule, P_4 < (d) platinum atom

ave mass   ~39.10   <   39.10   <   123.88   <   195.1
(amu)

The potassium ion, K^+ is formed when an electron is lost by a potassium atom, K. Since an electron is lost, K^+ is lighter than K. However, the difference in the masses of the potassium ion and the potassium atom is very small because an electron is very light and its mass is insignificant compared to the total mass of the atom.

20. To answer this problem, compare the masses of the three isotopes of O with the atomic mass of oxygen listed in the periodic table (inside front cover of the textbook). Among the three isotopes: O-16 (15.9949 amu), O-17 (16.9993 amu), and O-18 (17.9992 amu), the closest to the atomic mass of oxygen, 16.00 amu, is that of O-16. This information suggests that <u>among the three isotopes, the most abundant is O-16</u>.

22. The percent abundance of ^{87}Rb can be estimated to be **(b) 25%**.

    To approximate the % abundance:

    Compare the difference between the atomic mass of each isotope with the average atomic mass of rubidium, 85.47 amu (see periodic table in the inside front cover of the textbook). ^{85}Rb (84.9118 amu) is only 0.56 amu away from the average atomic mass of Rb (85.47). On the other hand, ^{87}Rb (86.9092 amu) is 1.44 amu away from the average atomic mass of Rb. A comparison of the difference suggests that ^{85}Rb is about 75% closer to the average atomic mass while ^{87}Rb is about 25% closer.

24. The average atomic mass of Si can be calculated using the formula presented the textbook, Equation 2.1.

    $$\text{ave atomic mass of Ne} = 19.9924 \text{ amu} \left(\frac{90.51}{100}\right) + 20.9938 \text{ amu} \left(\frac{0.27}{100}\right) + 21.9914 \text{ amu} \left(\frac{9.22}{100}\right)$$

    $$= 18.10 \text{ amu} + 0.057 \text{ amu} + 2.03 \text{ amu}$$

    $$= \underline{\textbf{20.18 amu}}$$

26. This problem is similar to Example 2.2.
    The atomic mass of the second isotope can be calculated using Equation 2.1. The atomic mass of Cu is 63.55 amu (see inside periodic table in the front cover of the textbook).

    Since there are only 2 Cu isotopes and one of them, ^{63}Cu has a percent abundance of 69.17%, then the other isotope has a percent abundance of 30.83%:

    % abundance of 2nd isotope = 100 − 69.17 = 30.83%

    Let x = atomic mass of 2nd isotope

    $$62.9296 \text{ amu} \left(\frac{69.17}{100}\right) + \left(\frac{30.83}{100}\right) x = 63.55 \text{ amu}$$

    $$43.5284 \text{ amu} + 0.3083x = 63.55 \text{ amu}$$

    $$x = 20.02/0.3083 = \underline{\textbf{64.94 amu}}$$

    atomic mass of second isotope = **64.94 amu**
    The nuclear symbol of the second isotope is $^{65}_{29}$Cu

28. The percent abundances of the other two isotopes of magnesium can be calculated using the formula presented in the textbook, Equation 2.1. The atomic mass of Mg (24.30 amu) can be obtained from the periodic table (see inside front cover of the textbook).

    Since there are 3 isotopes and one of them (24.9858 amu) has a percent abundance of 10.00%, then the sum of the abundances the other two is equal to 90.00%

    $$100 - 10.00 = 90.00\%$$

    Let x = percent abundance of lightest isotope

    90.00 – x = percent abundance of heaviest isotope

    $$23.9850 \text{ amu} \left(\frac{x}{100}\right) + 24.9858 \text{ amu} \left(\frac{10.00}{100}\right) + 25.9826 \text{ amu} \left(\frac{90.00 - x}{100}\right) = 24.30 \text{ amu}$$

    $$0.239850x + 2.499 + 23.38 - 0.2598x = 24.30$$

    $$0.239850x - 0.2598x = 24.30 - 2.499 - 23.38$$

    $$-0.01995x = -1.58$$

    $$x = -1.58/-0.019976 = 79.0\%$$

    percent abundance of lightest isotope = x = **79.0%**

    percent abundance of heaviest isotope = 90.00 – x = 90.00 – 79.0 = **11.0%**

30. Assuming there is only one H isotope, and that there are two Cl isotopes: Cl-35 and Cl-37,

    (a) there are 2 types of HCl molecules: **HCl-35 and HCl-37**

    (b) sum of the mass numbers of the two atoms in each molecule:

    HCl-35: 1 + 35 = **36**   HCl-37: 1 + 37 = **38**

    (c) mass spectrum for HCl

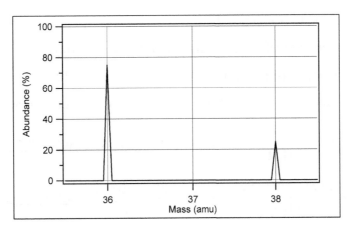

2.3 QUANTITATIVE PROPERTIES OF THE ATOM:
Masses of Individual Atoms; Avogadro's Number

32. This problem is similar to Example 2.3 in the textbook. As described in the textbook, if a sample has a mass in grams numerically equal to the atomic mass of the element, it would have Avogadro's number (6.022×10^{23}) of atoms.

Ag has an atomic mass of 107.87 amu. Thus, 107.87 g Ag contain Avogadro's number of Ag atoms (6.022×10^{23}).

allowed exposure to Ag in air in a 40-hr week: 1×10^{-8} g Ag/L

allowed exposure to Ag in air in a 40-hr week in terms of atoms Ag/L:

$$\frac{1 \times 10^{-8} \text{g Ag}}{1 \text{L}} \times \frac{6.022 \times 10^{23} \text{ atoms Ag}}{107.87 \text{ g Ag}} = \underline{\textbf{6} \times \textbf{10}^{13} \textbf{ atoms Ag/L}}$$

34. This problem is similar to Example 2.3 in the textbook.

Bismuth: atomic number = 83 = number of protons

1 Bi atom has 83 protons.

(a) As described in the textbook, if a sample has a mass in grams numerically equal to the atomic mass of the element, it would have Avogadro's number (6.022×10^{23}) of atoms.

$$0.243 \text{ g Bi} \times \frac{6.022 \times 10^{23} \text{ Bi atoms}}{209.0 \text{ g Bi}} = \underline{\textbf{7.00} \times \textbf{10}^{20} \textbf{ Bi atoms}}$$

(b) mass of 139 protons: $139 \text{ protons} \times \dfrac{1.6726 \times 10^{-24} \text{ g}}{1 \text{ proton}} = \underline{\textbf{2.32} \times \textbf{10}^{-22} \textbf{ g}}$

36. thallium, Tl-201: Atomic Number = 81

mass number (A) = number of protons (Z) + number of neutrons

1 atom of Tl-201: number of neutrons = A − Z = 201 − 81 = 120

Note that atomic, mass, proton, neutron and electron numbers are exact numbers.

(a) $70 \text{ atoms } ^{201}\text{Tl} \times \dfrac{120 \text{ neutrons}}{1 \, ^{201}\text{Tl atom}} = \underline{\textbf{8400 neutrons}} \text{ or } \underline{\textbf{8.400} \times \textbf{10}^3 \textbf{ neutrons}}$

(b) $10^{-12} \text{ g } ^{201}\text{Tl} \times \dfrac{6.022 \times 10^{23} \text{ atoms } ^{201}\text{Tl}}{204.4 \text{ g } ^{201}\text{Tl}} \times \dfrac{120 \text{ neutrons}}{1 \, ^{201}\text{Tl}} = \underline{\textbf{3.535} \times \textbf{10}^{11} \textbf{ neutrons}}$

38. Follow the following steps to solve this problem: (i) convert all dimension to a common unit, cm; (ii) find the volume of the cylinder using the formula $V = \pi r^2 h$; (iii) convert the calculated volume to mass (g) using density as conversion factor then convert the calculated mass to number of atoms using Avogadro's number as conversion factor.

   (i) convert all dimension to a common unit, cm:

   diameter = 1.15 cm

   $$\text{radius} = \frac{\text{diameter}}{2} = \frac{1.15 \text{ cm}}{2} = 0.575 \text{ cm}$$

   $$\text{height} = 4.00 \text{ in} \times \frac{2.54 \text{ cm}}{1 \text{ in}} = 10.2 \text{ cm}$$

   (ii) find the volume of the cylinder using the formula $V = \pi r^2 h$

   $$V = \pi r^2 h = 3.1416 \times (0.575 \text{ cm})^2 \times 10.2 \text{ cm} = 10.6 \text{ cm}^3$$

   (iii) calculate the number of atoms:

   Strategy: volume $\rightarrow$ g $\rightarrow$ number of atoms

   $$10.6 \text{ cm}^3 \times \frac{8.92 \text{ g}}{1 \text{ cm}^3} \times \frac{6.022 \times 10^{23} \text{ Cu atoms}}{63.55 \text{ g}} = \underline{\mathbf{8.96 \times 10^{23} \text{ Cu atoms}}}$$

2.4 INTRODUCTION TO THE PERIODIC TABLE: Elements and the Periodic Table

40. You can find the symbol and the name of the elements in the table presented in the inside back cover of the textbook.

Symbol	name of element	symbol	name of element
(a) S	<u>sulfur</u>	(d) Si	<u>silicon</u>
(b) Sc	<u>scandium</u>	(e) Sr	<u>strontium</u>
(c) Se	<u>selenium</u>		

42. For the classification of the elements, see Section 2.4 and Figure 2.8.

Symbol	classification	symbol	classification
(a) S	<u>nonmetal</u>	(d) Si	<u>metalloid</u>
(b) Sc	transition <u>metal</u>	(e) Sr	<u>metal</u>
(c) Se	<u>nonmetal</u>		

44. See the periodic table in Figure 2.8. The horizontal rows in the periodic table are called periods. The periods are numbered from top to bottom. The first period consists of hydrogen and helium.

(a) The second period is the horizontal row beginning with Li and ending with Ne. B is a metalloid. The two elements to the left of B are metals; the **five** to the right of B (C, N, O, F, and Ne) are nonmetals. Therefore, **there are 5 nonmetals** in period 2.

(b) The fourth period is the horizontal row beginning with K and ending with Kr. Ge and As are metalloids. The 13 elements to the left of the metalloids are metals while the **three** to the right (Se, Br, and Kr) are nonmetals. Therefore, **there are 3 nonmetals in period 4.**

(c) The sixth period is the horizontal row beginning with Cs and ending with Rn. The 30 elements to the left of the thick black line (Cs to Po) are metals, the **two** to the right (At and Rn) are the nonmetals. Therefore, **there are 2 nonmetals** in period 6.

46. See the periodic table in the inside front cover of the textbook. A period is a horizontal row, numbered from top to bottom.

(a) **Period 1** (**first period**) has no metals; it consists of hydrogen and helium.

(b) **Period 7,** which begins with Fr, has no nonmetal (if element 118 is not considered).

(c) **Period 4** has one post-transition metal, Ga, and two metalloids, Ge and As.

2.5 MOLECULES AND IONS

48. For this problem you may also see Example 2.4 for molecular formula.
As discussed in the textbook, a condensed structural formula suggests both the bonding pattern and the reactive group of atoms (also known as functional group) present in a molecule.

To write the condensed structural formula, write the formula of the highlighted portion (reactive groups) separate from the formula of the rest of the molecule.

To write the molecular formula, calculate the total number of atoms of each element and use the sum as the subscript for the symbol of each corresponding element.

(a) condensed structural formula: CH_3COOH

molecular formula: $C_2H_4O_2$

(b) condensed structural formula is also the molecular formula: CH_3Cl

50. Number of protons = sum of the atomic numbers

    Number of electrons = number of protons − charge

    (a) S_8 molecule: Atomic number: S = 16

    Total number of protons = 8×16 = **128 p^+**

    This molecule has no charge thus,

    number of electrons = number of protons − charge = 128 − 0 = **128 e^-**

    (b) SO_4^{2-} ion: Atomic numbers: S = 16, O = 8

    Total number of protons = 1(16)+ 4(8) = **48 p^+**

    This molecule has a charge of −2 thus,

    number of electrons = number of protons − charge = 48 − (−2) = **50 e^-**

    (c) H_2S molecule: Atomic numbers: H = 1, S = 16

    Total number of protons = 2(1) + 1(16) = **18 p^+**

    This molecule has no charge thus,

    number of electrons = number of protons − charge = 18 − 0 = **18 e^-**

    (d) S^{2-} ion: Atomic numbers: S = 16

    Total number of protons = 1(16) = **16 p^+**

    This molecule has a charge of −2 thus,

    number of electrons = number of protons − charge = 16 − (−2) = **18 e^-**

52. Mass number (A) = number of protons (Z) + number of neutrons

Nuclear Symbol	Metal, Nonmetal, Metalloid	Group	Period	Number of Neutrons
Al-27	**metal**	13	3	14
Te-128	metalloid	16	**5**	76
Xe-134	**nonmetal**	**18**	**5**	**80**
C-14	nonmetal	14	**2**	8

54. Molecular compounds like water are nonelectrolytes. In the choices given, (b) calcium nitrate and (c) ammonium carbonate are both ionic compounds and are strong electrolytes. Both **(a) citric acid and (d) iodine tribromide are molecular compounds and are nonelectrolytes**.

2.6 – 2.7 Names and Formulas of Ionic and Molecular Compounds

56. Phosphine is included in the list of binary compounds with common names. The prefixes used are listed in Table 2.4. Use Example 2.9 as guide

     Molecule   __Formula__

  (a) silicon disulfide SiS_2

     The "di" prefix indicates that there are 2 sulfur atoms.

  (b) phosphine PH_3

  (c) diphosphorus pentoxide P_2O_5

     The "di" prefix indicates that there are 2 P atoms and "penta" indicates 5 O atoms.

  (d) radon tetrafluoride RnF_4

     The "tetra" prefix indicates that there are 4 fluorine atoms.

  (e) nitrogen trichloride NCl_3

     The "tri" prefix indicates that there are 3 chlorine atoms.

58. In naming binary molecular compounds, prefixes are used to designate the number of atoms of each nonmetal and the name of the last atom ends in –*ide*. The prefixes used are listed in Table 2.4. Use Example 2.9 as guide.

Formula	Name	Formula	Name
(a) $TeCl_2$	tellurium dichloride	(c) XeO_3	xenon trioxide
(b) I_4O_9	tetraiodine nanoxide	(d) S_4N_4	tetrasulfur tetranitride

60. Use Example 2.7 as guide.
Although ionic compounds are composed of ions or charged species, ionic compounds are neutral substances. The total positive charge must be equal to the total negative charge. Thus, the sum of the charges must be zero.

  (a) Ba^{2+} and I^- form **BaI_2**

    For the charges to add up to zero, for 1 Ba^{2+} there should be 2 I^- ions, thus Ba has no subscript (indicating a subscript of 1) while I has a subscript of 2 in the formula of the compound, **BaI_2**. The total charge = $1(+2) + 2(-1) = 0$

    Ba^{2+} and N^{3-} form **Ba_3N_2**

    For the charges to add up to zero, there should be 3 Ba^{2+} for 2 N^{3-} ions, thus Ba has a subscript of 3 while N has a subscript of 2 in the formula of the compound, **Ba_3N_2**. The total charge = $3(+2) + 2(-3) = 0$

(b) O^{2-} and Fe^{2+} form **FeO**

The charges of O^{2-} and Fe^{2+} add up to zero, so 1 O^{2-} is required for each Fe^{2+}. In the formula FeO, both Fe and O have no subscript (indicating a subscript of 1). The total charge = $1(+2)$ + $1(-2) = 0$

O^{2-} and Fe^{3+} form **Fe$_2$O$_3$**

For the charges to add up to zero, there should be 2 Fe^{3+} for 3 O^{2-} ions, thus Fe has a subscript of 2 while O has a subscript of 3 in the formula of the compound, Fe_2O_3. The total charge = $2(+3)$ + $3(-2) = 0$

62. Use Example 2.7 as guide. The list of some common ions can be found in Tables 2.2 and 2.3 of the textbook.

In order to write the formula of ionic compounds, the formula of the ions that make up the compound must be identified first. The ions are combined such that the total positive charge is equal to the total negative charge. Thus, the sum of the charges must be equal to zero.

(a) The formula for potassium hydrogen phosphate is **K$_2$HPO$_4$**

Cation: K^+ Anion: HPO_4^{2-}

For the charges to add up to zero, there should be 2 K^+ ions for 1 HPO_4^{2-} ion, thus K has a subscript of 2 and HPO_4 has no subscript (indicating a subscript of 1). The total charge = $1(+2)$ + $1(-2) = 0$

(b) The formula for magnesium nitride is **Mg$_3$N$_2$**

Cation: Mg^{2+} Anion: N^{3-}

For the charges to add up to zero, there should be 3 Mg^{2+} ions and 2 N^{3-} ions, thus Mg has a subscript of 3 and N has a subscript of 2. The total charge = $3(+2)$ + $2(-3) = 0$

(c) The formula for lead(IV) bromide is **PbBr$_4$**
The Roman numeral (IV) indicates that Pb is a +4 ion (Pb^{4+}).

Cation: Pb^{4+} Anion: Br^-

For the charges to add up to zero, there should be 4 Br^- ions for 1 Pb^{4+} ion, thus Br has a subscript of 4 and Pb has no subscript (indicating a subscript of 1). The total charge = $1(+4)$ + $4(-1) = 0$

(d) The formula for scandium(III) chloride is $ScCl_3$
The Roman numeral (III) indicates that Sc is a +3 ion (Sc^{3+}).

Cation: Sc^{3+} Anion: Cl^-

For the charges to add up to zero, there should be 3 Cl^- ions for 1 Sc^{3+} ion, thus Cl has a subscript of 3 and Sc has no subscript (indicating a subscript of 1).
The total charge = $1(+3) + 3(-1) = 0$

(e) The formula for barium acetate is $Ba(C_2H_3O_2)_2$

Cation: Ba^{2+} Anion: $C_2H_3O_2^-$

For the charges to add up to zero, there should be 2 $C_2H_3O_2^-$ ions for 1 Ba^{2+} ion, thus $C_2H_3O_2$ has a subscript of 2 and Ba has no subscript (indicating a subscript of 1).
The total charge = $1(+2) + 2(-1) = 0$

64. Use the flowchart for naming binary ionic compounds (Figure 2.18) and Example 2.8 as guide. The list of some common ions can be found in Tables 2.2 and 2.3 of the textbook.

Name ionic compounds by naming first the cation (positive ion) followed by the anion (negative ion). Some transition and post-transition metals form more than one type of cations. To differentiate the different types of ions they form, indicate the charge of the metal with Roman numerals after its chemical symbol.

Formula   Name

(a) $Pb(C_2H_3O_2)_2$ **lead(II) acetate**

Pb forms more than one type of cation so the charge of Pb must be identified by a Roman numeral. Since each acetate has a −1 charge and there are 2 acetates, the total charge from the two acetates is $2(-1)=$ −2. Pb must have a +2 charge for the charges of the cation and the anion to add up to zero.

(b) $Al(OH)_3$ **aluminum hydroxide**

(c) $Sr(HCO_3)_2$ **strontium hydrogen carbonate or strontium bicarbonate**

(d) $Cu_3(PO_4)_2$ **copper(II) phosphate**

Cu forms more than one type of cation so the charge of Cu must be identified by a Roman numeral. Since each phosphate has a −3 charge and there are 2 phosphates, the total charge from the two phosphates is $2(-3)=$ −6. For the charges of the cation and the anion to add up to zero, the total charge from the three Cu cations must be +6 charge, which means that each Cu must have a charge of +2, thus the cation is copper(II).

(e) Rb_2O **rubidium oxide**

66. Use Example 2.7 as guide. The list of some common ions can be found in Table 2.2 and 2.3 of the textbook.

    In order to write the formula of ionic compounds, the formula of the ions that make up the compound must be identified first. The ions are combined such that the total positive charge is equal to the total negative charge. Thus, the sum of the charges must be equal to zero.

    (a) The formula for nitric acid is $HNO_3(aq)$

    Oxoanion: NO_3^-

    The suffix "ic" in the name "nitric" acid suggests that the oxoanion ends in "–ate". The acid is formed from nit<u>rate</u> (NO_3^-). Since the charge of nitrate is –1, it requires 1 H^+ to add up to zero charge. The total charge = 1(+1) + 1(–1) = 0.

    (b) The formula for potassium sulfate is K_2SO_4

    Cation: K^+ Anion: SO_4^{2-}

    For the charges to add up to zero, there should be 2 K^+ ions for 1 SO_4^{2-} ion, thus K has a subscript of 2 and SO_4 has no subscript (indicating a subscript of 1).
    The total charge = 2(+1) + 1(–2) = 0

    (c) The formula for iron(III) perchlorate is $Fe(ClO_4)_3$

    Cation: Fe^{3+} Anion: ClO_4^-

    For the charges to add up to zero, there should be 1 Fe^{3+} ion for 3 ClO_4^- ions, thus Fe has no subscript (indicating a subscript of 1) and ClO_4 has a subscript of 3.
    The total charge = 1(+3) + 3(–1) = 0

    (d) The formula for aluminum iodate is $Al(IO_3)_3$

    Cation: Al^{3+} Anion: IO_3^-

    For the charges to add up to zero, there should be 1 Al^{3+} ion for 3 IO_3^- ions, thus Al has no subscript (indicating a subscript of 1) and IO_3 has a subscript of 3.
    The total charge = 1(+3) + 3(–1) = 0

    (e) The formula for sulfurous acid is $H_2SO_3(aq)$

    Oxoanion: SO_3^{2-}

    The suffix "ous" in the name "sulfurous" acid suggests that the oxoanion ends in "–ite". This implies that the acid is formed from sulf<u>ite</u> for which the formula is SO_3^{2-}. Since the charge of this anion is –2, it requires 2 H^+ to add up to zero charge. Therefore, in the formula for sulfurous acid, H has a subscript of 2 and SO_3 has no subscript (indicating a subscript of 1).
    The total charge = 2(+1) + 1(–2) = 0

68. A review of Examples 2.7, 2.8, and 2.9 will help you solve this problem. You can also use the flowcharts in Figures 2.18 and 2.19 in naming the formulas. A list of some of the ions can be found in Tables 2.2 and 2.3 in the textbook.

Name	Formula	
bromous acid	$HBrO_2(aq)$	(1)
cobalt(III) periodate	$Co(IO_4)_3$	(2)
platinum(IV) chloride	$PtCl_4$	(3)
nitrous acid	$HNO_2(aq)$	(4)
dinitrogen trioxide	N_2O_3	(5)

(1) The name of bromine oxoanions is similar to the naming of chlorine oxoanions. Bromous came from bromite, BrO_2^-.

(2) Periodate has a −1 charge (IO_4^-). Since there are 3 periodates, the total charge of the anion is -3, to balance the charges cobalt should be a +3.

(3) Four (4) chlorides are required to balance the +4 charge of Pt.

(4) The name of NO_2^- is nitrite. The name of the acid formed by nitrite is nitrous acid (*-ite* is change to *-ous*).

(5) The prefix "di" and "tri" denote that there are 2 nitrogens and 3 oxygen, respectively.

UNCLASSIFIED PROBLEMS

70. See Sections 2.2 and 2.3 and Figure 2.8.

   (a) **Ge**   Se and Ge both belong to period 4. Se has an atomic number of 34 while Ge has 32. Since atomic number = proton number, Ge has 2 protons less than Se.

   (b) **W**   W is a group 6 transition metal found in period 6.

   (c) **Sr**   Group 2 elements are called the alkaline earth metals. Sr belongs to this group and it has an atomic number or proton number of 38.

   (d) **Bi**   As stated in the textbook, Ga, In, Tl, Sn, Pb, and Bi are often referred to as post-transition metals. Among these six elements, only Bi is in group 15.

72. See Section 2.5 and Example 2.4.

   (a) C_2H_7N To write the molecular formula, use the total number of atoms of each element as the subscript for the symbol of that corresponding element.

   (b) $C_2H_5NH_2$ To write the condensed structural formula, write the formula of the reactive group separately from the rest of the molecule.

74. See Sections 2.5, 2.6, and 2.7

    (a) "Compounds containing chlorine can be either molecular or ionic." is **always true**. Chlorine can combine with other nonmetals to form molecular compounds e.g. CCl_4. Chlorine can also form ionic compounds by combining with metals e.g. NaCl.

    (b) "An ionic compound always has at least one metal." is **usually true.** Most ionic compounds are formed by a metal and a nonmetal e.g. KBr. However ionic compounds are also formed by polyatomic cations combined with anions. Such ionic compounds do not contain a metal e.g. ammonium chloride, NH_4Cl, and ammonium nitrate, NH_4NO_3.

    (c) "When an element in a molecule has a "di" prefix, it means that the element has a +2 charge" is **never true**. As discussed in the naming of binary compounds (Section 2.7), the Greek prefix like "di" indicates the number of the element in the formula. It does not indicate the charge.

76. See Section 2.3 and Example 2.3.

    To find the volume of CCl_4, convert molecules of CCl_4 to g CCl_4 using Avogadro's number as conversion factor then using the density of CCl_4 as conversion factor convert g CCl_4 to volume.

    Strategy: molecules CCl_4 → g CCl_4 → volume CCl_4

    $$6.00 \times 10^{25} \text{ molecules} \times \frac{153.81 \text{ g } CCl_4}{6.022 \times 10^{23} \text{ molecules}} \times \frac{1 \text{ cm}^3}{1.589 \text{ g}} = \underline{\mathbf{9.64 \times 10^3 \text{ cm}^3}}$$

CONCEPTUAL PROBLEMS

78. See Section 2.2. Unknown element, $^{23}_{11}X$
    Only (a) is true.

    (a) **True**. The atomic number or proton number identifies the element. The unknown element is sodium because the symbol indicates an atomic number of 11 which is the atomic number of Na.

    (b) False. The unknown element cannot be vanadium because V has an atomic number of 23. The number 23 in the symbol indicates the mass number and not the atomic number.

    (c) False. The unknown element has 12 neutrons and not 23.
        number of neutrons = mass number – atomic number = 23 – 11 = 12

    (d) False. X^{2+} would have 9 electrons and not 13 electrons.
        number of electrons = number of protons – charge = 11 – (+2) = 9

    (e) False. The proton/neutron ratio, 11/12, is about 0.9 and not 1.1

CHALLENGE PROBLEMS

88. **Only (c)** follows the law of multiple proportions.

    See Section 2.1. As stated in the textbook, according to the law of multiple proportions, "the masses of one element that combine with a fixed mass of the second element are in a ratio of small whole numbers".

    compound A: $\dfrac{\text{mass H}}{\text{mass C}} = \dfrac{2.39\text{ g}}{28.5\text{ g}} = \dfrac{0.0839\text{ g H}}{1\text{ g C}}$

    compound B: $\dfrac{\text{mass H}}{\text{mass C}} = \dfrac{11.6\text{ g}}{34.7\text{ g}} = \dfrac{0.334\text{ g H}}{1\text{ g C}}$

    To illustrate the law of multiple proportions, let's apply it to the given compounds A and B. With the mass of the second element, C fixed to 1g we will show that the ratio of the masses of H (the first element) that combine with a fixed mass of the second element (1 g C) are in a ratio of small whole numbers.

    $\dfrac{\text{compound B}:\ \text{mass H/1 g C}}{\text{compound A}:\ \text{mass H/1 g C}} = \dfrac{0.334\text{ g H/1 g C}}{0.0839\text{ g H/1 g C}} = \dfrac{4}{1}$

    As illustrated, the ratio of the mass of H (from compounds B and A) that combined with 1g C is a small whole number ratio, 4:1.

    Now, let's try the results given in the problem for the mass of hydrogen in compound C, if they follow the law of multiple proportions:

    (a) compound C: $\dfrac{\text{mass H}}{\text{mass C}} = \dfrac{5.84\text{ g}}{16.2\text{ g}} = \dfrac{0.360\text{ g H}}{1\text{ g C}}$

    $\dfrac{\text{compound C}:\ \text{mass H/1 g C}}{\text{compound A}:\ \text{mass H/1 g C}} = \dfrac{0.360\text{ g H/1 g C}}{0.0839\text{ g H/1 g C}} = \dfrac{4.29}{1}$

    This example does **NOT** follow the law of multiple proportions because 4.29:1 is **NOT** a whole number ratio.

(b) compound C: $\dfrac{\text{mass H}}{\text{mass C}} = \dfrac{3.47 \text{ g}}{16.2 \text{ g}} = 0.214$

$\dfrac{\text{compound C}:\ \text{mass H}/1\text{g C}}{\text{compound A}:\ \text{mass H}/1\text{g C}} = \dfrac{0.214 \text{ g H}/1\text{g C}}{0.0839 \text{ g H}/1\text{g C}} = \dfrac{2.55}{1}$

This example does **NOT** follow the law of multiple proportions because 2.55:1 is **NOT** a small whole number ratio.

(c) compound C: $\dfrac{\text{mass H}}{\text{mass C}} = \dfrac{2.72 \text{ g}}{16.2 \text{ g}} = 0.168$

$\dfrac{\text{compound C}:\ \text{mass H}/1\text{g C}}{\text{compound A}:\ \text{mass H}/1\text{g C}} = \dfrac{0.168 \text{ g H}/1\text{g C}}{0.0839 \text{ g H}/1\text{g C}} = \dfrac{2.00}{1}$

<u>**This example follows the law of multiple proportions because the ratio 2:1 is a small whole number.**</u>

89. See Sections 2.1 and 2.5.

As stated in the textbook, according to the law of multiple proportions, "the masses of one element that combine with a fixed mass of the second element are in a ratio of small whole numbers".

(a) The data and the calculation shown below illustrate the law of multiple proportions. The ratio of the masses of H that combined with a fixed mass (1 g) of C in ethane and in ethylene is in the ratio of 3:2, a small whole number ratio.

ethane: $\quad \dfrac{\text{mass H}}{\text{mass C}} = \dfrac{4.53 \text{ g H}}{18.0 \text{ g C}} = \dfrac{0.252 \text{ g H}}{1 \text{ g C}}$

ethylene: $\quad \dfrac{\text{mass H}}{\text{mass C}} = \dfrac{7.25 \text{ g H}}{43.20 \text{ g C}} = \dfrac{0.168 \text{ g H}}{1 \text{ g C}}$

$\dfrac{\text{ethane}\ :\ \text{mass H}/1\text{g C}}{\text{ethylene}:\ \text{mass H}/1\text{g C}} = \dfrac{0.252 \text{ g H}/1\text{g C}}{0.168 \text{ g H}/1\text{g C}} = \dfrac{3}{2} \quad \Leftarrow \text{ratio of small whole numbers}$

(b) Both compounds have only hydrogen and carbon. However the ratios of H to C that they have are different. As calculated in 89(a), the ratio of the masses of H that combined with 1 g C in ethane and ethylene are 3:2. Assuming one C for both compounds, ethane will have 3 H's while ethylene will have 2 H's. Thus, reasonable formulas for the two compounds are:

	reasonable formula	actual formula
Ethane:	CH_3	C_2H_6
Ethylene:	CH_2	C_2H_4

90. See Sections 1.2 and 2.2.

The average density of a single Al-27 atom can be determined by finding the ratio of the mass of an atom of Al-27 to its volume.

The mass of a single Al-27 atom can be obtained by considering the main components of the atom: protons, neutrons and electrons. A single Al-27 atom contains 13 protons, 14 neutrons and 13 electrons. Note that the mass of each of these sub-atomic particles are:

$$\text{mass of a proton} = 1.6726 \times 10^{-24} \text{ g}$$

$$\text{mass of a neutron} = 1.6749 \times 10^{-24} \text{ g}$$

$$\text{mass of an electron} = 9.1094 \times 10^{-28} \text{ g}$$

$$\text{mass of Al-27} = \text{mass of } 13p^+ + \text{mass of } 14n + \text{mass of } 13e^-$$

$$= 13(1.6726 \times 10^{-24} \text{ g}) + 14(1.6749 \times 10^{-24} \text{ g}) + 13(9.1094 \times 10^{-28} \text{ g})$$

$$= 4.5204 \times 10^{-23} \text{ g}$$

volume of an Al-27 atom assuming it is a sphere with a radius, r, of 0.143 nm:

$$\text{volume} = \frac{4}{3}\pi r^3$$

$$= \frac{4}{3}\pi \left(0.143 \text{ nm} \times \frac{10^{-9} \text{ m}}{1 \text{ nm}} \times \frac{1 \text{ cm}}{10^{-2} \text{ m}} \right)^3$$

$$= 1.22 \times 10^{-23} \text{ cm}^3$$

$$\text{density} = \frac{\text{mass}}{\text{volume}}$$

$$= \frac{4.5204 \times 10^{-23}\,\text{g}}{1.22 \times 10^{-23}\,\text{cm}^3}$$

$$= \underline{\mathbf{3.71\ g/cm^3}}$$

The experimentally determined value of the density of Al ($2.70\ \text{g/cm}^3$) is much lower than the value calculated here ($3.71\ \text{g/cm}^3$), thus **there is lots of space between Al atoms in the metal**.

91. See Section 2.2.

The only difference between a N-14 atom and a N-14^{3-} ion is that the latter has 3 electrons more. The mass of a N-14^{3-} ion can be obtained by adding the mass of 3 electrons to the mass of the N-14 atom.

mass of N atom $= 2.3440 \times 10^{-23}\,\text{g}$

mass of an electron $= 9.1094 \times 10^{-28}\,\text{g}$

mass of N-14^{3-} ion $= 2.3440 \times 10^{-23}\,\text{g} + 3(9.1094 \times 10^{-28}\,\text{g}) = \underline{\mathbf{2.3443 \times 10^{-23}\,g}}$

92. See Sections 1.2 and 2.3.

(a) molecules Lincoln took in when he delivered the address
(He inhaled 200 times.):

$$200\ \text{breaths} \times \frac{500\ \text{mL air}}{1\ \text{breath}} \times \frac{2.5 \times 10^{19}\ \text{molecules}}{1\ \text{mL air}} = \underline{\mathbf{2.5 \times 10^{24}\ molecules}}$$

(b) fraction of the molecules in the earth's atmosphere inhaled by Lincoln at Gettysburg
(molecules in the entire earth's atmosphere $\cong 1.1 \times 10^{44}$):

$$\text{fraction of the molecules} = \frac{2.5 \times 10^{24}\ \text{molecules}}{1.1 \times 10^{44}\ \text{molecules}}$$

$$= \underline{\mathbf{2.3 \times 10^{-20}}}$$

(c) in your next breath, the number of molecules you will inhale that were also inhaled by Lincoln when he delivered the Gettysburg Address:

$$1\,\text{breath} \times \frac{500\,\text{mL air}}{1\,\text{breath}} \times \frac{2.5 \times 10^{19}\,\text{molecules}}{1\,\text{mL air}} \times 2.3 \times 10^{-20} = \underline{\mathbf{2.9 \times 10^2\,\text{molecules}}}$$

93. This problem is an application of the Law of Conservation of Mass. The total mass of the starting materials is equal to total mass of the resulting materials

mass of starting materials = mass of resulting materials

$$18.00\,\text{g Al} + \left(25\,\text{mL HCl} \times \frac{1.025\,\text{g HCl}}{1\,\text{mL HCl}}\right) = 12.00\,\text{g Al} + 30.95\,\text{g resulting solution} + \text{mass } H_{2(g)}$$

$$43.625\,\text{g} = 42.95\,\text{g} + \text{mass } H_2 \text{ gas formed}$$
$$\text{mass } H_2 \text{ gas formed} = 43.625\,\text{g} - 42.95\,\text{g} = 0.68\,\text{g } H_2 \text{ gas formed}$$

$$\text{volume } H_2 \text{ gas formed} = 0.68\,\text{g } H_2 \times \frac{1\,\text{L}}{0.0824\,\text{g } H_2} = \underline{\mathbf{8.2\,\text{L}}}$$

MASS RELATIONS IN CHEMISTRY; STOICHIOMETRY

Chapter **3**

3.1 THE MOLE: The Mole, Molar Mass, and Mole-Gram Conversions

2. See Example 3.1.

   (a) mass of 10^{-6} mol of hazelnut meats

   $$10^{-6}\,mol \times \frac{6.022 \times 10^{23}\,hazelnut\ meat}{1\,mol} \times \frac{0.985g}{1\,hazelnut\ meat} = \underline{\mathbf{5.93 \times 10^{17}\ g}}$$

   (b) moles in a pound of hazelnut meats

   $$1\,lb \times \frac{453.6\ g}{1\,lb} \times \frac{1\,hazelnut\ meat}{0.985\ g} \times \frac{1\,mol}{6.022 \times 10^{23}\,hazelnut\ meat} = \underline{\mathbf{7.65 \times 10^{-22}\ mol}}$$

4. See Section 2.3.

   (a) number of electrons in an ion of Sc^{3+}

   number of electrons = atomic number − charge

   = 21− (+3)

   = **18 electrons**

   (b) number of electrons in a mol of Sc^{3+}

   $$1\,mol\ Sc^{3+} \times \frac{6.022 \times 10^{23}\ Sc^{3+}ions}{1\,mol\ Sc^{3+}} \times \frac{18\ electrons}{1\,Sc^{3+}ion} = \underline{\mathbf{1.084 \times 10^{25}\ electrons}}$$

   (c) number of electrons in a gram of Sc^{3+}

   $$1\,g\ Sc^{3+} \times \frac{6.022 \times 10^{23}\ Sc^{3+}ions}{44.96\ g\ Sc^{3+}} \times \frac{18\ electrons}{1\,Sc^{3+}ion} = \underline{\mathbf{2.411 \times 10^{23}\ electrons}}$$

6. (a) moles Pt in the solid circular cone of platinum

   To solve the problem, find the mass of the platinum cone from its volume and density using equation 1.3. Convert the mass of Pt to moles Pt.

   $$d = \frac{mass}{V} \quad \Rightarrow \quad solve\ for\ mass \quad \Rightarrow \quad mass = d \times V$$

The volume of a solid circular cone is given by the following formula:

$$V_{\text{circular cone}} = \frac{1}{3}\pi r^2 h \qquad \text{where } r \text{ and } h \text{ are the radius and height, respectively}$$

$$\text{mass}_{Pt} = d \times V = d \times \left(\frac{1}{3}\pi r^2 h\right)$$

$$\text{mass}_{Pt} = \frac{21.45\text{ g}}{1\text{ cm}^3} \times \left[\frac{1}{3} \times 3.1416 \times \left(\frac{2.75\text{ cm}}{2}\right)^2\left(3.00\text{ in} \times \frac{2.54\text{ cm}}{1\text{ in}}\right)\right] = 324\text{ g Pt}$$

$$\text{moles Pt} = 324\text{ g Pt} \times \frac{1\text{ mol Pt}}{195.1\text{ g Pt}} = \underline{\mathbf{1.66\text{ mol Pt}}}$$

(b) number of electrons in the platinum cone

$$1.66\text{ mol Pt} \times \frac{6.022 \times 10^{23}\text{ Pt atoms}}{1\text{ mol Pt}} \times \frac{78\text{ electrons}}{1\text{ Pt atom}} = \underline{\mathbf{7.80 \times 10^{25}\text{ electrons}}}$$

8. As stated in the textbook, molar mass is numerically equal to the masses of the atoms in the formula expressed in g/mol instead of amu. For the molar masses, use the numerical values in the periodic table (see inside front cover of the textbook).

(a) Osmium metal = **190.2 g/mol**

(b) $NaHCO_3$ molar mass = 84.01 g/mol

$$\begin{aligned}
Na &= 1 \times 22.99 &&= 22.99 \\
H &= 1 \times 1.008 &&= 1.008 \\
C &= 1 \times 12.01 &&= 12.01 \\
O &= 3 \times 16.00 &&= \underline{48.00} \\
&&\text{Sum} &= \mathbf{84.01\text{ g/mol}}
\end{aligned}$$

(c) Vitamin D, $C_{28}H_{44}O$ molar mass = 396.63 g/mol

$$\begin{aligned}
C &= 28 \times 12.01 &&= 336.28 \\
H &= 44 \times 1.008 &&= 44.352 \\
O &= 1 \times 16.00 &&= \underline{16.00} \\
&&\text{Sum} &= \mathbf{396.63\text{ g/mol}}
\end{aligned}$$

10. See Example 3.1. For the molar masses, use the numerical values in the periodic table (see inside front cover of the textbook).

(a) $C_{12}H_{19}O_8Cl_3$ molar mass = 12(12.01) + 19(1.008) + 8(16.00) + 3(35.45) = 397.62 g/mol

$$\text{mole } C_{12}H_{19}O_8Cl_3 = 128.3\text{ g } C_{12}H_{19}O_8Cl_3 \times \frac{1\text{ mol}}{397.62\text{ g}} = \underline{\mathbf{0.3227\text{ mol}}}$$

(b) $C_5H_4N_4O_3$ molar mass = 5(12.01) + 4(1.008) + 4(14.01) + 3(16.00) = 168.12 g/mol

$$\text{mole } C_5H_4N_4O_3 = 0.3066 \text{ g } C_5H_4N_4O_3 \times \frac{1 \text{ mol}}{168.12 \text{ g}} = \underline{\textbf{0.001824 mol}}$$

(c)   CdTe molar mass = 112.4 + 127.6 = 240.0 g/mol

$$\text{mole CdTe} = 2.664 \text{ g CdTe} \times \frac{1 \text{ mol CdTe}}{240.0 \text{ g CdTe}} = \underline{\textbf{0.01110 mol}}$$

12. See Example 3.1. For the molar masses, use the numerical values in the periodic table (see inside front cover of the textbook).

(a) mass (g) of 1.35 mol TiO_2

TiO_2 molar mass = 47.87 + 2(16.00) = 79.87 g/mol

$$\text{g } TiO_2 = 1.35 \text{ mol } TiO_2 \times \frac{79.87 \text{ g } TiO_2}{1 \text{ mol } TiO_2} = \underline{\textbf{108 g}}$$

(b) mass (g) of 1.35 mol $C_{12}H_{19}O_8Cl_3$

$C_{12}H_{19}O_8Cl_3$ molar mass = 12(12.01) + 19(1.008) + 8(16.00) + 3(35.45) = 397.6 g/mol

$$\text{g } C_{12}H_{19}O_8Cl_3 = 1.35 \text{ mol } C_{12}H_{19}O_8Cl_3 \times \frac{397.6 \text{ g } C_{12}H_{19}O_8Cl_3}{1 \text{ mol } C_{12}H_{19}O_8Cl_3} = \underline{\textbf{537 g}}$$

(c) mass (g) of 1.35 mol $C_{21}H_{22}N_2O_2$

$C_{21}H_{22}N_2O_2$ molar mass = 21(12.01) + 22(1.008) + 2(14.01) + 2(16.00) = 334.4 g/mol

$$\text{g } C_{21}H_{22}N_2O_2 = 1.35 \text{ mol } C_{21}H_{22}N_2O_2 \times \frac{334.4 \text{ g } C_{21}H_{22}N_2O_2}{1 \text{ mol } C_{21}H_{22}N_2O_2} = \underline{\textbf{451 g}}$$

14. See Example 3.1. For the molar masses, use the numerical values in the periodic table (see inside front cover of the textbook).

	Number of Grams	Number of Moles	Number of Molecules	Number of O Atoms
(a)	0.1364	$\underline{7.100 \times 10^{-4}}$	$\underline{4.276 \times 10^{20}}$	$\underline{2.993 \times 10^{21}}$
(b)	$\underline{\textbf{239.8}}$	1.248	$\underline{7.515 \times 10^{23}}$	$\underline{5.261 \times 10^{24}}$
(c)	$\underline{\textbf{13.8}}$	$\underline{7.17 \times 10^{-2}}$	4.32×10^{22}	$\underline{3.02 \times 10^{23}}$
(d)	$\underline{\textbf{0.00253}}$	$\underline{1.32 \times 10^{-5}}$	$\underline{7.93 \times 10^{18}}$	5.55×10^{19}

First, calculate the molar mass of citric acid, $C_6H_8O_7$:

molar mass = 6(12.01) + 8(1.008) + 7(16.00) = 192.1 g/mol $C_6H_8O_7$

(a) Given: 0.1364 g $C_6H_8O_7$

$$0.1364 \text{ g } C_6H_8O_7 \times \frac{1 \text{ mol}}{192.1 \text{ g}} = \underline{\mathbf{7.100 \times 10^{-4} \text{ mol}}}$$

$$0.1364 \text{ g } C_6H_8O_7 \times \frac{1 \text{ mol}}{192.1 \text{ g}} \times \frac{6.022 \times 10^{23} \text{ molecules}}{1 \text{ mol}} = \underline{\mathbf{4.276 \times 10^{20} \text{ molecules}}}$$

$$0.1364 \text{ g } C_6H_8O_7 \times \frac{1 \text{ mol}}{192.1 \text{ g}} \times \frac{6.022 \times 10^{23} \text{ molecules}}{1 \text{ mol}} \times \frac{7 \text{ O atoms}}{1 C_6H_8O_7 \text{ molecule}}$$

$$= \underline{\mathbf{2.993 \times 10^{21} \text{ O atoms}}}$$

(b) Given: 1.248 mol $C_6H_8O_7$

$$1.248 \text{ mol} \times \frac{192.1 \text{ g}}{1 \text{ mol}} = \underline{\mathbf{239.8 \text{ g}}}$$

$$1.248 \text{ mol} \times \frac{6.022 \times 10^{23} \text{ molecules}}{1 \text{ mol}} = \underline{\mathbf{7.515 \times 10^{23} \text{ molecules}}}$$

$$1.248 \text{ mol} \times \frac{6.022 \times 10^{23} \text{ molecules}}{1 \text{ mol}} \times \frac{7 \text{ O atoms}}{1 C_6H_8O_7 \text{ molecule}} = \underline{\mathbf{5.261 \times 10^{24} \text{ O atoms}}}$$

(c) Given: 4.32×10^{22} molecules $C_6H_8O_7$

$$4.32 \times 10^{22} \text{ molecules} \times \frac{1 \text{ mol}}{6.022 \times 10^{23} \text{ molecules}} \times \frac{192.1 \text{ g}}{1 \text{ mol}} = \underline{\mathbf{13.8 \text{ g}}}$$

$$4.32 \times 10^{22} \text{ molecules} \times \frac{1 \text{ mol}}{6.022 \times 10^{23} \text{ molecules}} = \underline{\mathbf{7.17 \times 10^{-2} \text{ mol}}}$$

$$4.32 \times 10^{22} \text{ molecules} \times \frac{7 \text{ O atoms}}{1 C_6H_8O_7 \text{ molecule}} = \underline{\mathbf{3.02 \times 10^{23} \text{ O atoms}}}$$

(d) Given: 5.55×10^{19} O atoms

$$5.55 \times 10^{19} \text{ O atoms} \times \frac{1 C_6H_8O_7 \text{ molecule}}{7 \text{ O atoms}} \times \frac{1 \text{ mol}}{6.022 \times 10^{23} \text{ molecules}} \times \frac{192.12 \text{ g}}{1 \text{ mol}}$$

$$= \underline{\mathbf{0.00253 \text{ g}}}$$

$$5.55 \times 10^{19} \text{ O atoms} \times \frac{1 C_6H_8O_7 \text{ molecule}}{7 \text{ O atoms}} \times \frac{1 \text{ mol}}{6.022 \times 10^{23} \text{ molecules}} = \underline{\mathbf{1.32 \times 10^{-5} \text{ mol}}}$$

$$5.55 \times 10^{19} \text{ O atoms} \times \frac{1 C_6H_8O_7 \text{ molecule}}{7 \text{ O atoms}} = \underline{\mathbf{7.93 \times 10^{18} \text{ molecules}}}$$

3.1 THE MOLE: Moles in Solution; Molarity

16. molarity of sodium hypochlorite, NaClO in the household bleach:

$$\text{molarity} = \frac{\text{mol solute}}{\text{L solution}} = \frac{\text{mol NaClO}}{\text{L solution}} = \frac{5.2 \text{ g NaClO} \left(\dfrac{1 \text{ mol NaClO}}{74.44 \text{ g NaClO}} \right)}{100 \text{g water} \times \dfrac{1.00 \text{ mL}}{1.00 \text{ g}} \times \dfrac{1 \text{ L}}{1000 \text{ mL}}} = \underline{\mathbf{0.70 \text{ } M}}$$

18. See Example 3.3.

Molarity of each ion in the prepared solutions:

(a) K_2O solution; ions present: K^+ and O^{2-}

molar mass of K_2O = 2(39.10) + 16.00 = 94.20 g/mol

$$[K^+] = \frac{28.0 \text{ g K}_2O \times \dfrac{1 \text{ mol K}_2O}{94.20 \text{ g K}_2O} \times \dfrac{2 \text{ mol K}^+}{1 \text{ mol K}_2O}}{785 \text{ mL} \times \dfrac{1 \text{ L}}{1000 \text{ mL}}} = \underline{\mathbf{0.757 \text{ } M}}$$

$$[O^{2-}] = \frac{28.0 \text{ g K}_2O \times \dfrac{1 \text{ mol K}_2O}{94.20 \text{ g K}_2O} \times \dfrac{1 \text{ mol O}^{2-}}{1 \text{ mol K}_2O}}{785 \text{ mL} \times \dfrac{1 \text{ L}}{1000 \text{ mL}}} = \underline{\mathbf{0.379 \text{ } M}}$$

(b) $NaHCO_3$ solution; ions present: Na^+ and HCO_3^-

molar mass of $NaHCO_3$ = 1(22.99) + 1(1.008) + 1(12.01) + 3(16.00)

= 84.01 g/mol

$$[Na^+] = \frac{28.0 \text{ g NaHCO}_3 \times \dfrac{1 \text{ mol NaHCO}_3}{84.01 \text{ g NaHCO}_3} \times \dfrac{1 \text{ mol Na}^+}{1 \text{ mol NaHCO}_3}}{785 \text{ mL} \times \dfrac{1 \text{ L}}{1000 \text{ mL}}} = \underline{\mathbf{0.425 \text{ } M}}$$

$$[HCO_3^-] = \frac{28.0 \text{ g NaHCO}_3 \times \dfrac{1 \text{ mol NaHCO}_3}{84.01 \text{ g NaHCO}_3} \times \dfrac{1 \text{ mol HCO}_3^-}{1 \text{ mol NaHCO}_3}}{785 \text{ mL} \times \dfrac{1 \text{ L}}{1000 \text{ mL}}} = \underline{\mathbf{0.425 \text{ } M}}$$

(c) $Sc(IO_2)_3$ solution; ions present: Sc^{3+} and IO_2^-

molar mass of $Sc(IO_2)_3$ = 44.96 + 3(126.9) + 6(16.00) = 521.66 g/mol

$$[Sc^{3+}] = \frac{28.0 \text{ g } Sc(IO_2)_3 \times \frac{1 \text{ mol } Sc(IO_2)_3}{521.66 \text{ g } Sc(IO_2)_3} \times \frac{1 \text{ mol } Sc^{3+}}{1 \text{ mol } Sc(IO_2)_3}}{785 \text{ mL} \times \frac{1 \text{ L}}{1000 \text{ mL}}} = \underline{\textbf{0.0684 M}}$$

$$[IO_2^-] = \frac{28.0 \text{ g } Sc(IO_2)_3 \times \frac{1 \text{ mol } Sc(IO_2)_3}{521.66 \text{ g } Sc(IO_2)_3} \times \frac{3 \text{ mol } IO_2^-}{1 \text{ mol } Sc(IO_2)_3}}{785 \text{ mL} \times \frac{1 \text{ L}}{1000 \text{ mL}}} = \underline{\textbf{0.205 M}}$$

(d) $Mg_3(PO_4)_2$ solution; ions present: Mg^{2+} and PO_4^{3-}

molar mass of $Mg_3(PO_4)_2$ = 3(24.30) + 2(30.97) + 8(16.00) = 262.84 g/mol

$$[Mg^{2+}] = \frac{28.0 \text{ g } Mg_3(PO_4)_2 \times \frac{1 \text{ mol } Mg_3(PO_4)_2}{262.84 \text{ g } Mg_3(PO_4)_2} \times \frac{3 \text{ mol } Mg^{2+}}{1 \text{ mol } Mg_3(PO_4)_2}}{785 \text{ mL} \times \frac{1 \text{ L}}{1000 \text{ mL}}} = \underline{\textbf{0.407 M}}$$

$$[PO_4^{3-}] = \frac{28.0 \text{ g } Mg_3(PO_4)_2 \times \frac{1 \text{ mol } Mg_3(PO_4)_2}{262.84 \text{ g } Mg_3(PO_4)_2} \times \frac{2 \text{ mol } PO_4^{3-}}{1 \text{ mol } Mg_3(PO_4)_2}}{785 \text{ mL} \times \frac{1 \text{ L}}{1000 \text{ mL}}} = \underline{\textbf{0.271 M}}$$

20. See Example 3.2 and Figure 3.3.

Use the given molarity and the molar mass of the solid substance as conversion factors.

Strategy: volume → moles → mass

(a) $2.00 \text{ L} \times \frac{0.685 \text{ mol } Ni(NO_3)_2}{1 \text{ L}} \times \frac{182.7 \text{ g } Ni(NO_3)_2}{1 \text{ mol } Ni(NO_3)_2} = \underline{\textbf{2.50} \times \textbf{10}^2 \textbf{ g}}$

Dissolve **2.50 × 10² g Ni(NO₃)₂** in sufficient water to make 2.00 L of solution.

(b) $2.00 \text{ L} \times \frac{0.685 \text{ mol } CuCl_2}{1 \text{ L}} \times \frac{134.5 \text{ g } CuCl_2}{1 \text{ mol } CuCl_2} = \underline{\textbf{184 g}}$

Dissolve **184 g CuCl₂** in sufficient water to make 2.00 L of solution.

(c) $2.00 \text{ L} \times \frac{0.685 \text{ mol } C_6H_8O_6}{1 \text{ L}} \times \frac{176.1 \text{ g } C_6H_8O_6}{1 \text{ mol } C_6H_8O_6} = \underline{\textbf{241 g}}$

Dissolve **241 g C₆H₈O₆** in sufficient water to make 2.00 L of solution.

22. See Example 3.2.

    (a) To solve for the maximum volume of $Al_2(CO_3)_3$ solution, use the molar mass of $Al_2(CO_3)_3$ and the molarity of the solution as conversion factors.

    molar mass of $Al_2(CO_3)_3$ = 2(26.98) + 3(12.01) + 9(16.00) = 233.99 g/mol

    $$32.00 \text{ g } Al_2(CO_3)_3 \times \frac{1 \text{ mol } Al_2(CO_3)_3}{233.99 \text{ g } Al_2(CO_3)_3} \times \frac{1 \text{ L solution}}{0.4500 \text{ mol } Al_2(CO_3)_3} = \underline{\textbf{0.3039 L}}$$

    (b) To determine the required volume (mL) of the $Al_2(CO_3)_3$ solution, use molarity as conversion factor. Note that volume should be converted to L.

    $$0.1450 \text{ mol } CO_3^{2-} \times \frac{1 \text{ mol } Al_2(CO_3)_3}{3 \text{ mol } CO_3^{2-}} \times \frac{1 \text{ L solution}}{0.4500 \text{ mol } Al_2(CO_3)_3} \times \frac{1000 \text{ mL}}{1 \text{ L}} = \underline{\textbf{107.4 mL}}$$

24. See Example 3.3.

    The molarity of the resulting Na_2SO_4 solution can be obtained from the ratio of the total moles of Na_2SO_4 with the final (total) volume in L assuming volumes are additive.

    moles of Na_2SO_4 from the 0.388 *M* solution

    $$25 \text{ mL} \times \frac{1 \text{ L}}{1000 \text{ mL}} \times \frac{0.388 \text{ mol } Na_2SO_4}{1 \text{ L}} = 0.00970 \text{ mol } Na_2SO_4$$

    moles of Na_2SO_4 from the 0.229 *M* solution

    $$35.3 \text{ mL} \times \frac{1 \text{ L}}{1000 \text{ mL}} \times \frac{0.229 \text{ mol } Na_2SO_4}{1 \text{ L}} = 0.00808 \text{ mol } Na_2SO_4$$

    total moles of Na_2SO_4 from the two solutions:

    moles $_{total}$ = 0.00970 mol + 0.00808 mol = 0.0178 mol Na_2SO_4

    final volume = 25 mL + 35.3 mL = 60.3 mL or 0.0603 L

    $$M_{final} = \frac{0.0178 \text{ mol}}{0.0603 \text{ L}} = \underline{\textbf{0.295 }\textbf{\textit{M}}}$$

3.2 MASS RELATIONS IN CHEMICAL FORMULAS

26. See Example 3.4.

   molar mass of $C_{16}H_{13}N_2OCl$ = 16(12.01) + 13(1.008) + 2(14.01) + 1(16.00) + 1(35.45)

$$= 284.7 \text{ g/mol}$$

mass percent of carbon:

$$1 \text{ mol } C_{16}H_{13}N_2OCl \times \frac{16 \text{ mol C}}{1 \text{ mol } C_{16}H_{13}N_2OCl} \times \frac{12.01 \text{ g}}{1 \text{ mol C}} = 192.2 \text{ g C}$$

$$\text{mass \% C} = \frac{192.2 \text{ g C}}{284.7 \text{ g } C_{16}H_{13}N_2OCl} \times 100 = \underline{\textbf{67.51\% C}}$$

mass percent of hydrogen:

$$1 \text{ mol } C_{16}H_{13}N_2OCl \times \frac{13 \text{ mol H}}{1 \text{ mol } C_{16}H_{13}N_2OCl} \times \frac{1.008 \text{ g}}{1 \text{ mol H}} = 13.10 \text{ g H}$$

$$\text{mass \% H} = \frac{13.10 \text{ g H}}{284.7 \text{ g } C_{16}H_{13}N_2OCl} \times 100 = \underline{\textbf{4.601\% H}}$$

mass percent of nitrogen:

$$1 \text{ mol } C_{16}H_{13}N_2OCl \times \frac{2 \text{ mol N}}{1 \text{ mol } C_{16}H_{13}N_2OCl} \times \frac{14.01 \text{ g}}{1 \text{ mol N}} = 28.02 \text{ g N}$$

$$\text{mass \% N} = \frac{28.02 \text{ g N}}{284.7 \text{ g } C_{16}H_{13}N_2OCl} \times 100 = \underline{\textbf{9.842\% N}}$$

mass percent of oxygen:

$$1 \text{ mol } C_{16}H_{13}N_2OCl \times \frac{1 \text{ mol O}}{1 \text{ mol } C_{16}H_{13}N_2OCl} \times \frac{16.00 \text{ g}}{1 \text{ mol O}} = 16.00 \text{ g O}$$

$$\text{mass \% O} = \frac{16.00 \text{ g O}}{284.7 \text{ g } C_{16}H_{13}N_2OCl} \times 100 = \underline{\textbf{5.620\% O}}$$

mass percent of chlorine:

$$1 \text{ mol } C_{16}H_{13}N_2OCl \times \frac{1 \text{ mol Cl}}{1 \text{ mol } C_{16}H_{13}N_2OCl} \times \frac{35.45 \text{ g}}{1 \text{ mol Cl}} = 35.45 \text{ g Cl}$$

$$\text{mass \% Cl} = \frac{35.45 \text{ g Cl}}{284.7 \text{ g } C_{16}H_{13}N_2OCl} \times 100 = \underline{\textbf{12.45\% Cl}}$$

28. See Example 3.4.

    molar mass of allicin: $C_6H_{10}O_2S$ = 6(12.01) + 10(1.008) + 2(16.00) + 1(32.07) = 146.2 g/mol

    mass of sulfur from 25.0 g of allicin:

    $$25.0 \text{ g } C_6H_{10}O_2S \times \frac{1 \text{ mol } C_6H_{10}O_2S}{146.2 \text{ g } C_6H_{10}O_2S} \times \frac{1 \text{ mol S}}{1 \text{ mol } C_6H_{10}O_2S} \times \frac{32.07 \text{ g S}}{1 \text{ mol S}} = \underline{\textbf{5.48 g S}}$$

30. See Example 3.4. The percent of aluminum chlorohydrate, $Al_2(OH)_5Cl$, present in the antiperspirant can be obtained by using the following steps:

    (i) convert mass of Al to mass of $Al_2(OH)_5Cl$:

    molar mass of $Al_2(OH)_5Cl$ = 2(26.98) + 5(16.00) + 5(1.008) + 1(35.45) = 174.4 g/mol

    mass of $Al_2(OH)_5Cl$ containing 0.334 g Al:

    $$0.334 \text{ g Al} \times \frac{1 \text{ mol Al}}{26.98 \text{ g Al}} \times \frac{1 \text{ mol } Al_2(OH)_5Cl}{2 \text{ mol Al}} \times \frac{174.4 \text{ g } Al_2(OH)_5Cl}{1 \text{ mol } Al_2(OH)_5Cl} = 1.08 \text{ g } Al_2(OH)_5Cl$$

    (ii) use the mass of $Al_2(OH)_5Cl$ to calculate the mass percent of aluminum chlorohydrate:

    $$\text{mass percent } Al_2(OH)_5Cl = \frac{1.08 \text{ g } Al_2(OH)_5Cl}{2.00 \text{ g antiperspirant sample}} \times 100 = \underline{\textbf{54.0\%}}$$

32. See Examples 3.4 and 3.6. As presented in Section 3.2 of the textbook: the masses of CO_2, H_2O, and Cl_2 obtained on combustion can be converted to the masses of C, H and Cl in the sample. The mass of O is determined by difference.

    mass percent of C:

    $$1.407 \text{ g } CO_2 \times \frac{12.01 \text{ g C}}{44.01 \text{ g } CO_2} = 0.3840 \text{ g C}$$

    $$\text{mass \% C} = \frac{0.3840 \text{ g C}}{1.00 \text{ g sample}} \times 100 = \underline{\textbf{38.40\% C}}$$

    mass percent of H:

    $$0.134 \text{ g } H_2O \times \frac{2(1.008) \text{ g H}}{18.02 \text{ g } H_2O} = 0.0150 \text{ g H}$$

    $$\text{mass \% H} = \frac{0.0150 \text{ g C}}{1.00 \text{ g sample}} \times 100 = \underline{\textbf{1.50\% H}}$$

mass percent of Cl:

$$0.523 \text{ g Cl}_2 \times \frac{2\,(35.45) \text{ g Cl}}{70.90 \text{ g Cl}_2} = 0.523 \text{ g Cl}$$

$$\text{mass \% Cl} = \frac{0.523 \text{ g Cl}}{1.00 \text{ g sample}} \times 100 = \underline{\textbf{52.3\% Cl}}$$

mass percent of O (obtained by difference):

$$\text{mass \% O} = 100\% - [\ 38.40\% \text{ C} + 1.50\% \text{ H} + 52.3\% \text{ Cl}\ ] = \underline{\textbf{7.8\% O}}$$

34. Since the compound, R_2O_3 contained 32.0% O and it is composed of only two elements, R and O, then R is 68.0% by mass.

$$100\% - 32.0\% = 68.0\%$$

If we assume 100g of the compound, it would contain 32.0 g O and 68.0 g R.

$$\text{mol O} = 32.0 \text{ g O} \times \frac{1 \text{ mol O}}{16.00 \text{ g O}} = 2.00 \text{ mol O}$$

$$\text{mol R} = 2.00 \text{ mol O} \times \frac{2 \text{ mol R}}{3 \text{ mol O}} = 1.33 \text{ mol R}$$

$$\text{molar mass of R} = \frac{68.0 \text{ g R}}{1.33 \text{ mol R}} = 51.1 \text{ g/mol}$$

$$\text{molar mass of } R_2O_3 = 2(51.1) + 3(16.00) = \underline{\textbf{1.50} \times \textbf{10}^2 \textbf{ g/mol}}$$

The element that has a molar mass closest to 51.1 g/mol is vanadium (MM= 50.94).

Element R represents the element vanadium (V).

36. See Example 3.5.
    Strategy: (i) find the mass of S that reacted with Ni; (ii) convert the masses of Ni and S to numbers of moles; (iii) calculate the mole ratio; (iv) equate the mole ratio to the atom ratio (to get integers for the atom subscript, you may need to multiply the atom ratio by the smallest whole number) to get the simplest formula.

    (i) The mass of the nickel sulfide formed is 5.433 g, 2.986 g of which is Ni. The difference of these two values is the mass of S.

    $$5.433 \text{ g} - 2.986 \text{ g} = 2.447 \text{ g S}$$

    (ii) convert the masses of Ni and S to numbers of moles

    $$2.986 \text{ g Ni} \times \frac{1 \text{ mol Ni}}{58.69 \text{ g Ni}} = 0.05088 \text{ mol Ni}$$

    $$2.447 \text{ g S} \times \frac{1 \text{ mol S}}{32.07 \text{ g S}} = 0.07630 \text{ mol S}$$

(iii) To find the mole ratio, divide by the smaller number, 0.05088 mol Ni

$$\frac{0.07630 \text{ mol S}}{0.05088 \text{ mol Ni}} = \frac{1.500 \text{ mol S}}{1 \text{ mol Ni}}$$

(iv) get the atom ratio from the mole ratio

the mole ratio is 1 mol Ni: 1.500 mol S

Since the atom numbers should be whole numbers, the mole ratio should be multiplied by 2 because it is the smallest integer that will give the simplest formula. After multiplying the mole ratio by 2, the atom ratio is 2 Ni: 3 S

The simplest formula of the sulfide is **Ni_2S_3**.

The name of the sulfide is **nickel(III) sulfide**.

38. See Example 3.5.
   Strategy: (i) assume 100g of the compound to get the mass in g of each element; (ii) convert the masses of the elements to number of moles; (iii) calculate the mole ratios; (iv) equate the mole ratio to the atom ratio (to get whole numbers for the atom subscript, you may need to multiply the atom ratio with the smallest integer) to obtain the simplest formula.

   (a) tetraethyl lead

      (i) Assume 100 g compound. There will be: 29.71 g C, 6.234 g H, and 64.07 g Pb.

      (ii) Convert the masses to number of moles.

      mole C: $29.71 \text{ g C} \times \dfrac{1 \text{ mol C}}{12.01 \text{ g C}} = 2.474 \text{ mol C}$

      mole H: $6.234 \text{ g H} \times \dfrac{1 \text{ mol H}}{1.008 \text{ g H}} = 6.185 \text{ mol H}$

      mole Pb: $64.07 \text{ g Pb} \times \dfrac{1 \text{ mol Pb}}{207.2 \text{ g Pb}} = 0.3092 \text{ mol Pb}$

      (iii) To find the mole ratios, divide by the smaller number, 0.3092 mol Pb.

      $$\frac{2.474 \text{ mol C}}{0.3092 \text{ mol Pb}} = \frac{8.001 \text{ mol C}}{1 \text{ mol Pb}} \qquad \frac{6.185 \text{ mol H}}{0.3092 \text{ mol Pb}} = \frac{20.00 \text{ mol H}}{1 \text{ mol Pb}}$$

      The mole ratio is 1 mol Pb: 8 mol C: 20 mol H.

      (iv) Get the atom ratio from the mole ratio. As pointed out in the textbook, the mole ratio is the same as the atom ratio. The mole ratios can be rounded off to whole numbers.

      The atom ratio is 1 Pb: 8 C: 20 H

      The simplest formula of tetraethyl lead is **$C_8H_{20}Pb$**.

(b) citric acid

    (i) Assume 100 g compound. There will be: 37.51 g C, 4.20 g H, and 58.29 g O.

    (ii) Convert the masses to number of moles.

$$\text{mole C:} \qquad 37.51\,g\,C \times \frac{1\,mol\,C}{12.01\,g\,C} = 3.123\,mol\,C$$

$$\text{mole H:} \qquad 4.20\,g\,H \times \frac{1\,mol\,H}{1.008\,g\,H} = 4.17\,mol\,H$$

$$\text{mole O:} \qquad 58.29\,g\,O \times \frac{1\,mol\,O}{16.00\,g\,O} = 3.643\,mol\,O$$

    (iii) To find the mole ratios, divide by the smaller number, 3.123 mol C.

$$\frac{4.17\,mol\,H}{3.123\,mol\,C} = \frac{1.34\,mol\,H}{1\,mol\,C} \qquad\qquad \frac{3.643\,mol\,O}{3.123\,mol\,C} = \frac{1.167\,mol\,O}{1\,mol\,C}$$

      The mole ratio is 1 mol C: 1.34 mol H: 1.167 mol O.

    (iv) Get the atom ratio from the mole ratio. Since the atoms should have whole numbers, the mole ratio should be multiplied by 6 because it is the smallest integer that will give the simplest formula.

      After multiplying the mole ratio by 6, the atom ratio is 6 C: 8 H: 7 O

      The simplest formula of citric acid is $C_6H_8O_7$.

(c) cisplatin

    (i) Assume 100 g compound. Then there will be: 9.34 g N, 2.02 g H, 23.36 g Cl, and 65.50 g Pt.

    (ii) Convert the masses to number of moles.

$$\text{mole N:} \qquad 9.34\,g\,N \times \frac{1\,mol\,N}{14.01\,g\,N} = 0.667\,mol\,N$$

$$\text{mole H:} \qquad 2.02\,g\,H \times \frac{1\,mol\,H}{1.008\,g\,H} = 2.00\,mol\,H$$

$$\text{mole Cl:} \qquad 23.36\,g\,Cl \times \frac{1\,mol\,Cl}{35.45\,g\,Cl} = 0.6590\,mol\,Cl$$

$$\text{mole Pt:} \qquad 65.50\,g\,Pt \times \frac{1\,mol\,Pt}{195.1\,g\,Pt} = 0.3357\,mol\,Pt$$

    (iii) To find the mole ratios, divide by the smaller number, 0.3357 mol Pt.

$$\frac{0.667\,mol\,N}{0.3357\,mol\,Pt} = \frac{1.99\,mol\,N}{1\,mol\,Pt}$$

$$\frac{2.00\,mol\,H}{0.3357\,mol\,Pt} = \frac{5.96\,mol\,H}{1\,mol\,Pt}$$

$$\frac{0.6590 \text{ mol Cl}}{0.3357 \text{mol Pt}} = \frac{1.963 \text{ mol Cl}}{1 \text{mol Pt}}$$

The mole ratio is 1 mol Pt: 1.99 mole N: 5.96 mol H: 1.963 mol Cl.

(iv) Get the atom ratio from the mole ratio. As pointed out in the textbook, the mole ratio is the same as the atom ratio. The mole ratios can be rounded off to whole numbers.

The atom ratio is 1 Pt: 2 N: 6 H: 2 Cl

The simplest formula of cisplatin is $PtN_2H_6Cl_2$.

40. See Example 3.6.

Strategy: (i) find the mass of C, H and O; (ii) convert the masses of C, H, and O to number of moles; (iii) calculate the mole ratios; (iv) equate the mole ratio to the atom ratio (to get integers for the atom subscript, you may need to multiply the atom ratio with the smallest whole number) to get the simplest formula.

(i) To find the mass of C, H and O, convert grams of CO_2 and H_2O to grams of C and H, respectively. The mass of O can be obtained by difference.

$$\text{mass C:} \quad 12.24 \text{ g CO}_2 \times \frac{1 \text{mol CO}_2}{44.01 \text{g CO}_2} \times \frac{1 \text{mol C}}{1 \text{mol CO}_2} \times \frac{12.01 \text{g C}}{1 \text{mol C}} = 3.340 \text{ g C}$$

$$\text{mass H:} \quad 2.505 \text{ g H}_2\text{O} \times \frac{1 \text{mol H}_2\text{O}}{18.02 \text{ g}} \times \frac{2 \text{ mol H}}{1 \text{mol H}_2\text{O}} \times \frac{1.008 \text{ g H}}{1 \text{mol H}} = 0.2802 \text{ g H}$$

mass O = 5.287 g – 3.340 g – 0.2802 g = 1.667 g O

(ii) Convert the masses of C, H, and O to number of moles.

$$\text{mole C:} \quad 3.340 \text{ g C} \times \frac{1 \text{mol C}}{12.01 \text{g C}} = 0.2781 \text{mol C}$$

$$\text{mole H:} \quad 0.2802 \text{ g H} \times \frac{1 \text{mol H}}{1.008 \text{ g H}} = 0.2780 \text{ mol H}$$

$$\text{mole O:} \quad 1.667 \text{ g O} \times \frac{1 \text{mol O}}{16.00 \text{ g O}} = 0.1042 \text{ mol O}$$

(iii) To find the mole ratios, divide by the smaller number, 0.1042 mol O.

$$\frac{0.2780 \text{ mol H}}{0.1042 \text{ mol O}} = \frac{2.67 \text{ mol H}}{1 \text{mol O}} \qquad \frac{0.2781 \text{mol C}}{0.1042 \text{ mol O}} = \frac{2.67 \text{ mol C}}{1 \text{mol O}}$$

The mole ratio is 1 mol O: 2.67 mol H: 2.67 mol C.

(iv) Get the atom ratio from the mole ratio. Since the atoms should have whole numbers, the mole ratio should be multiplied by 3 because it is the smallest integer that will give the simplest formula.

After multiplying the mole ratio by 3, the atom ratio is 8C: 8H: 3O

The simplest formula for oil of wintergreen is $C_8H_8O_3$.

42. See Example 3.6.

Strategy: (i) find the mass of C, H, S, N and O; (ii) convert the masses of C, H, S, N and O to number of moles; (iii) calculate the mole ratios; (iv) equate the mole ratio to the atoms ratio (to get integers for the atom subscript, you may need to multiply the atom ratio with the smallest whole number) to get the simplest formula.

(i)   To find the mass of C, H, S, N and O, convert grams of CO_2, H_2O and SO_2 to grams of C, H and S, respectively. The mass of N can be obtained from the given that the compound is 7.65% N by mass. The mass of O can be obtained by difference.

mass C: $12.6 \text{ g CO}_2 \times \dfrac{1\,\text{mol CO}_2}{44.01\,\text{g CO}_2} \times \dfrac{1\,\text{mol C}}{1\,\text{mol CO}_2} \times \dfrac{12.01\,\text{g C}}{1\,\text{mol C}} = 3.44 \text{ g C}$

mass H: $1.84 \text{ g H}_2\text{O} \times \dfrac{1\,\text{mol H}_2\text{O}}{18.02\,\text{g}} \times \dfrac{2\,\text{mol H}}{1\,\text{mol H}_2\text{O}} \times \dfrac{1.008\,\text{g H}}{1\,\text{mol H}} = 0.206 \text{ g H}$

mass S: $2.62 \text{ g SO}_2 \times \dfrac{1\,\text{mol SO}_2}{64.07\,\text{g SO}_2} \times \dfrac{1\,\text{mol S}}{1\,\text{mol SO}_2} \times \dfrac{32.07\,\text{g S}}{1\,\text{mol S}} = 1.31 \text{ g S}$

mass N: $7.500 \text{ g saccharin} \times \dfrac{7.65\,\text{g N}}{100\,\text{g saccharin}} = 0.574 \text{ g N}$

mass O = 7.500 g saccharin – [3.44 g C + 0.206 g H + 1.31 g C + 0.574 g N] = 1.97 g O

(ii)   Convert the masses of C, H, S, N and O to number of moles.

mole C: $3.44 \text{ g C} \times \dfrac{1\,\text{mol C}}{12.01\,\text{g C}} = 0.286 \text{ mol C}$

mole H: $0.206 \text{ g H} \times \dfrac{1\,\text{mol H}}{1.008\,\text{g H}} = 0.204 \text{ mol H}$

mole S: $1.31 \text{ g S} \times \dfrac{1\,\text{mol S}}{32.07\,\text{g S}} = 0.0408 \text{ mol S}$

mole N: $0.574 \text{ g N} \times \dfrac{1\,\text{mol N}}{14.01\,\text{g N}} = 0.0410 \text{ mol N}$

mole O: $1.97 \text{ g O} \times \dfrac{1\,\text{mol O}}{16.00\,\text{g O}} = 0.123 \text{ mol O}$

(iii) To find the mole ratios, divide by the smaller number, 0.0410 mol N

$$\frac{0.286 \text{ mol C}}{0.0410 \text{ mol N}} = \frac{6.98 \text{ mol C}}{1 \text{ mol N}} \qquad\qquad \frac{0.204 \text{ mol H}}{0.0410 \text{ mol N}} = \frac{4.98 \text{ mol H}}{1 \text{ mol N}}$$

$$\frac{0.0408 \text{ mol S}}{0.0410 \text{ mol N}} = \frac{0.995 \text{ mol S}}{1 \text{ mol N}} \qquad\qquad \frac{0.123 \text{ mol O}}{0.0410 \text{ mol N}} = \frac{3.00 \text{ mol O}}{1 \text{ mol N}}$$

The mole ratio is 1 mol N: 6.98 mol C: 4.98 mol H: 0.995 mol S: 3.00 mol O

(iv) Get the atom ratio from the mole ratio. As pointed out in the textbook the mole ratio is the same as the atom ratio. The mole ratios can be rounded off to whole numbers.

The atom ratio is 1 N: 7 C: 5 H: 1 S: 3 O

The simplest formula of saccharin is $C_7H_5NO_3S$.

44. See Examples 3.6 and 3.7.
Strategy: (i) find the mass of C, H, and N; (ii) convert the masses of C, H, and N to number of moles; (iii) calculate the mole ratios; (iv) equate the mole ratio to the atom ratio (to get integers for the atom subscript, you may need to multiply the atom ratio with the smallest whole number) to get the simplest formula; (v) calculate the molar mass of the simplest formula; (vi) find the ratio of the molar masses of the actual (molecular) formula and the simplest formula; and (vii) get the molecular formula by multiplying all the subscripts in the simplest formula with the ratio.

(i)   To find the mass of C, H, and N convert grams of CO_2 and H_2O to grams of C and H, respectively. The mass of N can be obtained by difference.

$$\text{mass C:} \quad 4.190 \text{ g } CO_2 \times \frac{1 \text{ mol } CO_2}{44.01 \text{ g } CO_2} \times \frac{1 \text{ mol C}}{1 \text{ mol } CO_2} \times \frac{12.01 \text{ g C}}{1 \text{ mol C}} = 1.143 \text{ g C}$$

$$\text{mass H:} \quad 3.428 \text{ g } H_2O \times \frac{1 \text{ mol } H_2O}{18.02 \text{ g}} \times \frac{2 \text{ mol H}}{1 \text{ mol } H_2O} \times \frac{1.008 \text{ g H}}{1 \text{ mol H}} = 0.3835 \text{ g H}$$

mass N: 2.859 g dimethylhydrazine – [1.143 g C + 0.3835 g H] = 1.332 g N

(ii) Convert the masses of C, H, and N to number of moles.

$$\text{mole C:} \quad 1.143 \text{ g C} \times \frac{1 \text{ mol C}}{12.01 \text{ g C}} = 0.09517 \text{ mol C}$$

$$\text{mole H:} \quad 0.3835 \text{ g H} \times \frac{1 \text{ mol H}}{1.008 \text{ g H}} = 0.3805 \text{ mol H}$$

$$\text{mole N:} \quad 1.332 \text{ g N} \times \frac{1 \text{ mol N}}{14.01 \text{ g N}} = 0.09507 \text{ mol N}$$

(iii) To find the mole ratios, divide by the smaller number, 0.09507 mol N.

$$\frac{0.09517 \text{ mol C}}{0.09507 \text{ mol N}} = \frac{1.001 \text{ mol C}}{1 \text{ mol N}} \qquad \frac{0.3805 \text{ mol H}}{0.09507 \text{ mol N}} = \frac{4.002 \text{ mol H}}{1 \text{ mol N}}$$

The mole ratio is 1 mol N: 1.001 mol C: 4.002 mol H

(iv) Get the atom ratio from the mole ratio. As pointed out in the textbook, the mole ratio is the same as the atom ratio. The mole ratios can be rounded off to whole numbers.

The atom ratio is 1 N: 1 C: 4 H

The simplest formula of dimethylhydrazine is **CH_4N**.

(v) Calculate the molar mass of the simplest formula.

molar mass of **CH_4N** = 1(12.01) + 4(1.008) + 1(14.01) = 30.05 g/mol

(vi) Find the ratio of the molar masses of the actual (molecular) formula and the simplest formula. (Given: actual molar mass 60.10 g/mol)

$$\frac{\text{actual molar mass}}{\text{simplest formula mass}} = \frac{60.10}{30.05} = 2$$

(vii) Get the molecular formula by multiplying all the subscripts in the simplest formula by the ratio obtained above.

$$CH_4N \xrightarrow{\text{multiply all subscripts by 2}} C_2H_8N_2$$
$$\text{simplest formula} \qquad\qquad\qquad\qquad \text{molecular formula}$$

46. Molar mass of $Na_2B_4O_7 \cdot 10H_2O$ = 2(22.99) + 4(10.81) + 17(16.00) + 20(1.008)
$$= 381.4 \text{ g/mole}$$

a) % water in borax = $\dfrac{\text{mass of water}}{\text{molar mass of borax}} \times 100 = \dfrac{10(18.02 \text{ g/mol})}{381.4 \text{ g/mol}} \times 100 = \underline{\textbf{47.25\%}}$

b) mass of anhydrous sodium borate, $Na_2B_4O_7$:

% sodium borate = 100 − 47.25 = 52.75%

$$15.86 \text{ g Na}_2\text{B}_4\text{O}_7 \cdot 10\text{H}_2\text{O} \times \frac{52.75 \text{ g Na}_2\text{B}_4\text{O}_7}{100 \text{ g Na}_2\text{B}_4\text{O}_7 \cdot 10\text{H}_2\text{O}} = \underline{\textbf{8.366 g Na}_2\textbf{B}_4\textbf{O}_7}$$

3.3 MASS RELATIONS IN REACTIONS: Writing and Balancing Equations

48. See Example 3.8.

    Balance the elements by adjusting the coefficients (*never change the subscripts!*).

    (a) $C_6H_{12}O_6(s) + 6O_2(g) \rightarrow 6CO_2(g) + 6H_2O(l)$

    (b) $XeF_4(g) + 2H_2O(l) \rightarrow Xe(g) + O_2(g) + 4HF(g)$

    (c) $4NaCl(s) + 2H_2O(g) + 2SO_2(g) + O_2(g) \rightarrow 2Na_2SO_4(s) + 4HCl(g)$

50. See Example 3.8.

    Balance the elements by adjusting the coefficients (*never change the subscripts!*).

    (a) $2Sc(s) + 3S(s) \rightarrow Sc_2S_3(s)$ (c) $2Sc(s) + N_2(g) \rightarrow 2ScN(s)$

    (b) $2Sc(s) + 3Cl_2(g) \rightarrow 2ScCl_3(s)$ (d) $4Sc(s) + 3O_2(g) \rightarrow 2Sc_2O_3(s)$

52. See Example 3.8.

    Balance the elements by adjusting the coefficients (*never change the subscripts!*).

    (a) $2H_2S(g) + SO_2(g) \rightarrow 3S(s) + 2H_2O(g)$

    (b) $2CH_4(g) + 2NH_3(g) + 3O_2(g) \rightarrow 2HCN(g) + 6H_2O(g)$

    (c) $Fe_2O_3(s) + 3H_2(g) \rightarrow 2Fe(s) + 3H_2O(g)$

    (d) $UO_2(s) + 4HF(g) \rightarrow UF_4(s) + 2H_2O(g)$

    (e) $C_2H_5OH(l) + 3O_2(g) \rightarrow 2CO_2(g) + 3H_2O(l)$

3.3 MASS RELATIONS IN REACTIONS: Mole–Mass Relations in Reactions

54. See Example 3.9.

    The stoichiometric ratios from the following balanced reaction are used in the solution below:

    $$4NH_3(g) + 5O_2(g) \rightarrow 4NO(g) + 6H_2O(l)$$

    (a) moles of NO obtained:

    $$3.914 \text{ mol } O_2 \times \frac{4 \text{ mol NO}}{5 \text{ mol } O_2} = \underline{\textbf{3.131 mol}} \text{ NO}$$

(b) moles of O_2 required:

$$2.611 \text{ mol NH}_3 \times \frac{5 \text{ mol O}_2}{4 \text{ mol NH}_3} = \textbf{3.264 mol } O_2$$

(c) moles of H_2O obtained:

$$0.8144 \text{ mol NH}_3 \times \frac{6 \text{ mol H}_2O}{4 \text{ mol NH}_3} = \textbf{1.222 mol } H_2O$$

(d) moles of O_2 required:

$$0.2179 \text{ mol H}_2O \times \frac{5 \text{ mol O}_2}{6 \text{ mol H}_2O} = \textbf{0.1816 mol } O_2$$

56. See Example 3.9.

The stoichiometric ratios from the following balanced reaction are used in the solution below:

$$4PH_3(g) + 8O_2(g) \rightarrow P_4O_{10}(s) + 6H_2O(g)$$

(a) mass of P_4O_{10} produced (P_4O_{10} molar mass = 283.88 g/mol)

Strategy: mol $PH_3 \rightarrow$ mol $P_4O_{10} \rightarrow$ g P_4O_{10}

$$12.43 \text{ mol PH}_3 \times \frac{1 \text{ mol P}_4O_{10}}{4 \text{ mol PH}_3} \times \frac{283.88 \text{ g P}_4O_{10}}{1 \text{ mol P}_4O_{10}} = \textbf{882.2 g } P_4O_{10}$$

(b) mass of PH_3 required (PH_3 molar mass = 33.99 g/mol)

Strategy: mol $H_2O \rightarrow$ mol $PH_3 \rightarrow$ g PH_3

$$0.739 \text{ mol H}_2O \times \frac{4 \text{ mol PH}_3}{6 \text{ mol H}_2O} \times \frac{33.99 \text{ g PH}_3}{1 \text{ mol PH}_3} = \textbf{16.7 g } PH_3$$

(c) mass of O_2 required (O_2 molar mass = 32.00 g/mol)

Strategy: g $H_2O \rightarrow$ mol $H_2O \rightarrow$ mol $O_2 \rightarrow$ g O_2

$$1.000 \text{ g H}_2O \times \frac{1 \text{ mol H}_2O}{18.02 \text{ g H}_2O} \times \frac{8 \text{ mol O}_2}{6 \text{ mol H}_2O} \times \frac{32.00 \text{ g O}_2}{1 \text{ mol O}_2} = \textbf{2.368 g } O_2$$

(d) mass of O_2 required (O_2 molar mass = 32.00 g/mol)

Strategy: g $PH_3 \rightarrow$ mol $PH_3 \rightarrow$ mol $O_2 \rightarrow$ g O_2

$$20.50 \text{ g PH}_3 \times \frac{1 \text{ mol PH}_3}{33.99 \text{ g PH}_3} \times \frac{8 \text{ mol O}_2}{4 \text{ mol PH}_3} \times \frac{32.00 \text{ g O}_2}{1 \text{ mol O}_2} = \textbf{38.60 g } O_2$$

58. See Example 3.9.

    (a) balanced equation for the reaction: $SiO_2(s) + 2C(s) \rightarrow Si(s) + 2CO(g)$

    The stoichiometric ratios from the balanced reaction above are used in the solution below.

    (b) moles of SiO_2 required: (Strategy: g Si $\rightarrow$ mol Si $\rightarrow$ mol SiO_2)

    $$20.00 \text{ g Si} \times \frac{1 \text{ mol Si}}{28.09 \text{ g Si}} \times \frac{1 \text{ mol } SiO_2}{1 \text{ mol Si}} = \underline{\textbf{0.7120 mol}} \; SiO_2$$

    (c) mass of CO formed: (Strategy: g Si $\rightarrow$ mol Si $\rightarrow$ mol CO $\rightarrow$ g CO)

    $$98.76 \text{ g Si} \times \frac{1 \text{ mol Si}}{28.09 \text{ g Si}} \times \frac{2 \text{ mol CO}}{1 \text{ mol Si}} \times \frac{28.01 \text{ g CO}}{1 \text{ mol CO}} = \underline{\textbf{197.0 g}} \; CO$$

60. See Example 3.9.

    The stoichiometric ratios from the following balanced reaction are used in the solution below:

    $$Sn(s) + O_2(g) \rightarrow SnO_2(s)$$

    (a) mass of oxidized Sn foil (assuming all Sn reacted):

    Strategy: (i) determine the volume of the Sn foil; (ii) use this volume and the density of Sn to find the mass of Sn; (iii) convert the mass of Sn to mass of SnO_2

    (i)   volume of the Sn foil

    $$0.600 \text{ mm} \times \frac{1 \text{ cm}}{10 \text{ mm}} = 0.0600 \text{ cm}$$

    (ii) mass of Sn using the calculated volume and the density of Sn

    $$(8.25 \text{ cm} \times 21.5 \text{ cm} \times 0.0600 \text{ cm}) \times \frac{7.28 \text{ g Sn}}{1 \text{ cm}^3} = 77.5 \text{ g Sn}$$

    (iii) convert the mass of Sn to mass of SnO_2

    Strategy: g Sn $\rightarrow$ mol Sn $\rightarrow$ mol SnO_2 $\rightarrow$ g SnO_2

    $$77.5 \text{ g Sn} \times \frac{1 \text{ mol Sn}}{118.7 \text{ g Sn}} \times \frac{1 \text{ mol } SnO_2}{1 \text{ mol Sn}} \times \frac{150.7 \text{ g } SnO_2}{1 \text{ mol } SnO_2} = \underline{\textbf{98.4 g}} \; SnO_2$$

    (b) liters of air required:

    Strategy: g SnO_2 $\rightarrow$ mol SnO_2 $\rightarrow$ mol O_2 $\rightarrow$ g O_2 $\rightarrow$ L O_2 $\rightarrow$ L air

    $$98.4 \text{ g } SnO_2 \times \frac{1 \text{ mol } SnO_2}{150.7 \text{ g } SnO_2} \times \frac{1 \text{ mol } O_2}{1 \text{ mol } SnO_2} \times \frac{32.00 \text{ g } O_2}{1 \text{ mol } O_2} \times \frac{1 \text{ L } O_2}{1.309 \text{ g } O_2} \times \frac{100 \text{ L air}}{21 \text{ L } O_2} = \underline{\textbf{76 L}} \text{ air}$$

62. See Example 3.9.

The stoichiometric ratio from the following balanced reaction is used in the solution below:

$$C_6H_{12}O_6(aq) \rightarrow 2C_2H_5OH(l) + 2CO_2(g)$$

(a) volume of ethyl alcohol produced:

Strategy:

$$\text{lb } C_6H_{12}O_6 \rightarrow \text{g } C_6H_{12}O_6 \rightarrow \text{mol } C_6H_{12}O_6 \rightarrow \text{mol } C_2H_5OH \rightarrow \text{g } C_2H_5OH \rightarrow \text{mL } C_2H_5OH$$

$$1\text{lb } C_6H_{12}O_6 \times \frac{453.6 \text{ g}}{1 \text{ lb}} \times \frac{1 \text{ mol } C_6H_{12}O_6}{180.16 \text{ g } C_6H_{12}O_6} \times \frac{2 \text{ mol } C_2H_5OH}{1 \text{ mol } C_6H_{12}O_6} \times \frac{46.07 \text{ g } C_2H_5OH}{1 \text{ mol } C_2H_5OH} \times \frac{1 \text{ mL } C_2H_5OH}{0.789 \text{ g } C_2H_5OH}$$

$$= \underline{\textbf{294 mL}} \text{ } C_2H_5OH$$

(b) grams of fructose required:

Strategy: $\text{gal gasohol} \rightarrow \text{mL gasohol} \rightarrow \text{mL } C_2H_5OH \rightarrow \text{g } C_2H_5OH \rightarrow \text{mol } C_2H_5OH;$

$$\text{mol } C_2H_5OH \rightarrow \text{mol } C_6H_{12}O_6 \rightarrow \text{g } C_6H_{12}O_6$$

(Note: 1 gal = 3785.4 mL)

$$1 \text{ gal gasohol} \times \frac{3785.4 \text{ mL gasohol}}{1 \text{ gal gasohol}} \times \frac{10 \text{ mL } C_2H_5OH}{100 \text{ mL gasohol}} \times \frac{0.789 \text{ g } C_2H_5OH}{1 \text{ mL } C_2H_5OH} \times$$

$$\frac{1 \text{ mol } C_2H_5OH}{46.07 \text{ g } C_2H_5OH} \times \frac{1 \text{ mol } C_6H_{12}O_6}{2 \text{ mol } C_2H_5OH} \times \frac{180.16 \text{ g } C_6H_{12}O_6}{1 \text{ mol } C_6H_{12}O_6} = \underline{\textbf{5.8} \times \textbf{10}^2 \text{ } \textbf{g}} \text{ } C_6H_{12}O_6$$

64. The stoichiometric ratios from the following balanced hypothetical reaction are used in the solution below:

$$2R_2X_5 + 2Z_8 \rightarrow 5X_2 + 4RZ_4$$

(a) moles X_2 produced:

Strategy: $\text{g } Z_8 \rightarrow \text{mol } Z_8 \rightarrow \text{mol } X_2$

$$25.00 \text{ g } Z_8 \times \frac{1 \text{ mol } Z_8}{197.4 \text{ g } Z_8} \times \frac{5 \text{ mol } X_2}{2 \text{ mol } Z_8} = \underline{\textbf{0.3166 mol}} \text{ } X_2$$

(b) molar mass of X_2 = mass $X_2 \div$ mol X_2

$$\text{molar mass of } X_2 = \frac{21.72 \text{ g } X_2}{0.3166 \text{ mol } X_2} = \underline{\textbf{68.60 g/mol}}$$

66. See Example 3.10.

    (a) balanced equation: $Cl_2(g) + 3F_2(g) \rightarrow 2ClF_3(g)$

    (b) Find the number of moles of ClF_3 that each of the given amounts of the reactants will produce. The reactant that will produce the smaller amount of ClF_3 is the limiting reactant.

    mol $Cl_2 \rightarrow$ mol ClF_3

    $$1.75 \text{ mol } Cl_2 \times \frac{2 \text{ mol } ClF_3}{1 \text{ mol } Cl_2} = 3.50 \text{ mol } ClF_3$$

    mol $F_2 \rightarrow$ mol ClF_3

    $$3.68 \text{ mol } F_2 \times \frac{2 \text{ mol } ClF_3}{3 \text{ mol } F_2} = \underline{\textbf{2.45 mol}} \text{ } ClF_3$$

    Since fluorine gas will produce less moles of ClF_3, **F_2 is the limiting reactant**.

    (c) The theoretical yield is **2.45 mol** ClF_3.

    (d) The excess reactant is Cl_2. The number of moles of Cl_2 that remain when the reaction is complete can be obtained from the difference of the moles of Cl_2 initially present and used up.

    $$\text{mol } Cl_2 \text{ used up} = 3.68 \text{ mol } F_2 \times \frac{1 \text{ mol } Cl_2}{3 \text{ mol } F_2} = 1.23 \text{ mol } Cl_2$$

    mol excess reactant unreacted = moles present initially – moles used up

    $$= 1.75 \text{ mol } Cl_2 - 1.23 \text{ mol } Cl_2$$

    $$= \underline{\textbf{0.52 mol}} \text{ } Cl_2$$

    The amount of excess reactant that remained when the reaction is complete is **0.52 mole** Cl_2.

68. See Example 3.11.

    $$3Fe(s) + 4H_2O(g) \rightarrow Fe_3O_4(s) + 4H_2(g)$$

    Strategy: (i) find the theoretical yield of Fe_3O_4

    (ii) use the theoretical yield of Fe_3O_4 to calculate the mass of Fe needed.

    theoretical yield of Fe_3O_4:

    $$\% \text{ yield} = \frac{\text{actual yield}}{\text{theoretical yield}} \times 100\%$$

    $$69\% = \frac{897 \text{ g Fe}_3O_4}{\text{theoretical yield}} \times 100\%$$

    theoretical yield = 1.3×10^3 g Fe_3O_4

mass of iron required:

$$1.3 \times 10^3 \text{ g Fe}_3\text{O}_4 \times \frac{1 \text{ mol Fe}_3\text{O}_4}{231.55 \text{ g Fe}_3\text{O}_4} \times \frac{3 \text{ mol Fe}}{1 \text{ mol Fe}_3\text{O}_4} \times \frac{55.85 \text{ g Fe}}{1 \text{ mol Fe}} = \underline{\textbf{9.4} \times \textbf{10}^2 \textbf{ g Fe}}$$

70. See Examples 3.10 and 3.11.

(a) balanced equation: $4NH_3(g) + 5O_2(g) \rightarrow 4NO(g) + 6H_2O(g)$

(b) Find the mass of NO that each of the given amounts of the reactants produced. The smaller amount of NO produced is the theoretical yield in grams.

mol $NH_3 \rightarrow$ mol NO

$$7.60 \text{ g NH}_3 \times \frac{1 \text{ mol NH}_3}{17.03 \text{ g NH}_3} \times \frac{4 \text{ mol NO}}{4 \text{ mol NH}_3} \times \frac{30.01 \text{ g NO}}{1 \text{ mol NO}} = 13.4 \text{ g NO}$$

mol $O_2 \rightarrow$ mol NO

$$10.00 \text{ g O}_2 \times \frac{1 \text{ mol O}_2}{32.00 \text{ g O}_2} \times \frac{4 \text{ mol NO}}{5 \text{ mol O}_2} \times \frac{30.01 \text{ g NO}}{1 \text{ mol NO}} = \underline{\textbf{7.502 g NO}}$$

Since oxygen gas will produce less moles of NO, O_2 is the limiting reactant; **7.502 g NO can be obtained theoretically.**

(c) The excess reactant is NH_3. The grams of NH_3 that remained when the reaction is complete can be obtained from the difference of the grams of NH_3 initially present and used up.

$$\text{mass NH}_3 \text{ used up} = 10.00 \text{ g O}_2 \times \frac{1 \text{ mol O}_2}{32.00 \text{ g O}_2} \times \frac{4 \text{ mol NH}_3}{5 \text{ mol O}_2} \times \frac{17.03 \text{ g NH}_3}{1 \text{ mol NH}_3} = 4.258 \text{ g NH}_3$$

mass excess reactant unreacted = mass present initially – mass used up

mass excess reactant unreacted = 7.60 g NH_3 – 4.258 g NH_3 = **3.34 g NH_3**

(d) % yield $= \dfrac{\text{actual yield}}{\text{theoretical yield}} \times 100\% = \dfrac{6.22 \text{ g NO}}{7.502 \text{ g NO}} \times 100 = \underline{\textbf{82.9\%}}$

72. To determine the mass of PI_3 needed, follow the following steps: (i) calculate the theoretical yield of H_3PO_3 given that the reaction has a yield of 75% and the actual yield required is 0.250 L H_3PO_3; (ii) convert the volume of H_3PO_3 to mass of H_3PO_3 using density as conversion factor; (iii) calculate the mass of H_3PO_3 to mass of PI_3.

(i) theoretical yield of H_3PO_3:

$$\% \text{ yield} = \frac{\text{actual yield}}{\text{theoretical yield}} \times 100\%$$

$$75.0\% = \frac{0.250\,\text{L}\,H_3PO_3}{\text{theoretical yield}} \times 100\%$$

$$\text{theoretical yield} = \frac{0.250\,\text{L}\,H_3PO_3}{75.0\%} \times 100\% = 0.333\,\text{L}\,H_3PO_4$$

(ii) mass of H_3PO_3

$$0.333\,\text{L}\,H_3PO_3 \times \frac{1000\,\text{mL}}{1\,\text{L}} \times \frac{1\,\text{cm}^3}{1\,\text{mL}} \times \frac{1.651\,\text{g}\,H_3PO_3}{1\,\text{cm}^3\,H_3PO_3} = 5.50 \times 10^2\,\text{g}\,H_3PO_3$$

(iii) calculate the mass of H_3PO_3 to produce this mass of PI_3

$$PI_3(s) + 3H_2O(l) \rightarrow H_3PO_3(l) + 3HI(g)$$

$$5.50 \times 10^2\,\text{g}\,H_3PO_3 \times \frac{1\,\text{mol}\,H_3PO_3}{81.99\,\text{g}\,H_3PO_3} \times \frac{1\,\text{mol}\,PI_3}{1\,\text{mol}\,H_3PO_3} \times \frac{411.67\,\text{g}\,PI_3}{1\,\text{mol}\,PI_3} = \underline{\mathbf{2.76 \times 10^3\,\text{g}\,PI_3}}$$

Follow the following steps to calculate the volume of water that should be used considering that the procedure calls for 45% more water:

(i) find the mass of water theoretically needed; (ii) convert the obtained mass to volume using the density of water; and finally (iii) increase the volume by 45.0%.

$$5.50 \times 10^2\,\text{g}\,H_3PO_3 \times \frac{1\,\text{mol}\,H_3PO_3}{81.99\,\text{g}\,H_3PO_3} \times \frac{3\,\text{mol}\,H_2O}{1\,\text{mol}\,H_3PO_3} \times \frac{18.02\,\text{g}\,H_2O}{1\,\text{mol}\,H_2O} = 363\,\text{g}\,H_2O$$

$$363\,\text{g}\,H_2O \times \frac{1\,\text{cm}^3\,H_2O}{1.00\,\text{g}\,H_2O} \times \frac{1\,\text{mL}}{1\,\text{cm}^3} = 363\,\text{mL}\,H_2O$$

Since 45% excess water is required, increase the amount of water by 45%.

$$363\,\text{mL} + (0.45 \times 363\,\text{mL}) = \underline{\mathbf{526\,\text{mL}\,H_2O}}$$

Therefore, the mass of PI_3 that should be weighed out is $\underline{\mathbf{2.76 \times 10^3\,\text{g}}}$ and the volume of water that should be used is $\underline{\mathbf{526\,\text{mL}}}$.

74. (a)

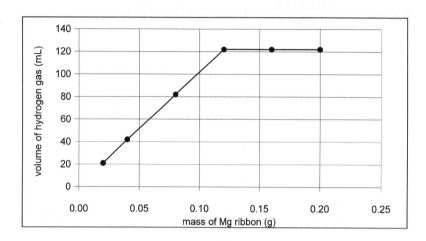

(b) limiting reactant in experiment 1: **Mg**

(c) limiting reactant in experiment 3: **Mg**

(d) limiting reactant in experiment 6: **acid**

(e) **experiment 4** uses stoichiometric amounts of both Mg and acid

(f) if 0.300 g Mg is used, the amount of gas that would be obtained is **122 mL**; if 0.010 g Mg is used, only **11 mL** of gas would be obtained.

76. Balanced chemical equation involved:

$$C_2H_5OH(aq) + O_2(g) \rightarrow HC_2H_3O_2(aq) + H_2O$$

mL of 12.5% (by volume) solution of ethanol, C_2H_5OH:

$$175 \text{ mL } HC_2H_3O_2 \text{ solution} \times \frac{10^{-3}L}{1\text{ mL}} \times \frac{0.664 \text{ mol } HC_2H_3O_2}{1\text{ L } HC_2H_3O_2 \text{ solution}} \times \frac{1 \text{ mol } C_2H_5OH}{1 \text{ mol } HC_2H_3O_2} \times \frac{46.07 \text{ g } C_2H_5OH}{1 \text{ mol } C_2H_5OH}$$

$$\times \frac{1\text{ mL } C_2H_5OH}{0.789\text{g } C_2H_5OH} \times \frac{100 \text{ mL } C_2H_5OH \text{ solution}}{12.5 \text{ mL } C_2H_5OH} = \textbf{54.3 mL } C_2H_5OH \text{ solution}$$

CONCEPTUAL PROBLEMS

78. See Section 3.3 and Example 3.8.

$$2AB_3 + 3C \rightarrow 3CB_2 + 2A$$

80. Representation: ☐ N atom   ⬤ H atom

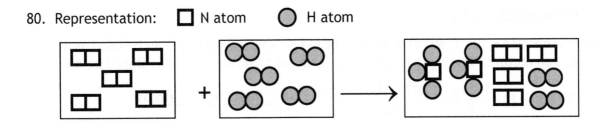

This pictorial representation depicts the following balanced equation:

$$N_2 + 3H_2 \rightarrow 2NH_3$$

82. Representation: ☐ As atom   ⬤ O atom

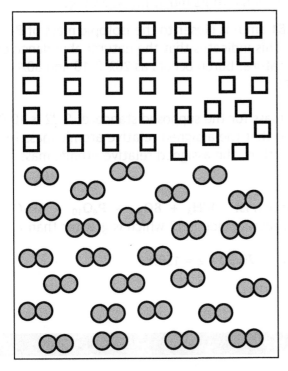 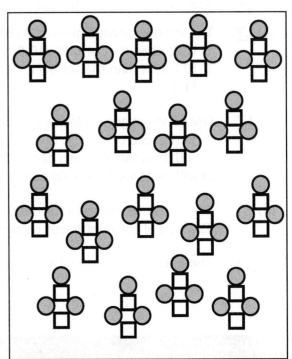

Box A: 36 As atoms ☐

27 O_2 molecules ⬤⬤

Box B: 18 As_2O_3 molecules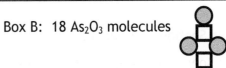

(a) Box A and B have the same number of As and O atoms. In each box there are 36 As atoms and 54 O atoms.

(b) Box A has more discrete particles (63) than Box B (18).

(c) Box A and B have the same mass because they have the same number of As and O atoms.

84. See Sections 2.3 and 3.1.

There are 6.022×10^{23} C atoms in 12.00 g of C-12. This information can be used as a conversion factor in finding the number of atoms in 5.000 kg of C-12 as shown below.

$$5.000 \text{ kg C-12} \times \frac{1000 \text{ g}}{1 \text{ kg}} \times \frac{6.022 \times 10^{23} \text{ C-12 atoms}}{12.00 \text{ g C-12}} = \underline{\textbf{2.509} \times \textbf{10}^{\textbf{26}} \textbf{ C atoms}}$$

Under the condition given in the problem, a mole is defined as the number of atoms in 5.000 kg of C-12. Since 5.000 kg of C-12 has 2.509×10^{26} C atoms then **1 mole under this condition has 2.509×10^{26} atoms**.

86. See Sections 2.3, 3.1, and 3.2.

(a) **EQ.** The mass of 6.022×10^{23} Na atoms is equal to its molar mass, 23.0 g.

(b) **LT.** The average atomic mass of B (10.81 amu) is closer to the isotopic mass of B-11 (11.01) than that of B-10 (10.01). This indicates that the natural abundance of B-10 is less than that of B-11. (In nature, boron occurs as 20% B-10 and 80% B-11.)

(c) **LT.** Since the assigned relative atomic mass for the supposed standard, S-32 is 10.00 amu (instead of ~32), it is expected that the assigned relative atomic mass for all other atoms will also decrease. Thus, the assigned relative atomic mass for H will be lower than 1.00 amu.

(d) **GT.** The balanced equation for the reaction is: $4PH_3 + 8O_2 \rightarrow P_4O_{10} + 6H_2O$
The sum of the coefficient on the reactant side is 12 which is greater than 7.

(e) **GT.** The mass (in grams) of 1 mole of Br_2 is $79.90 \times 2 = 159.8$.

CHALLENGE PROBLEMS

88. See Sections 3.1 and 3.2.

To solve this problem assume 1 mole of chlorophyll. Since each chlorophyll molecule has 1 Mg atom, then 1 mole of chlorophyll has 1 mole Mg atom.

Strategy: 1 mol chlorophyll $\longrightarrow$ mol Mg $\longrightarrow$ g Mg $\longrightarrow$ g chlorophyll

$$1 \text{ mol chlorophyll} \times \frac{1 \text{ mol Mg}}{1 \text{ mol chlorophyll}} \times \frac{24.30 \text{ g Mg}}{1 \text{ mol Mg}} \times \frac{100 \text{ g chlorophyll}}{2.72 \text{ g Mg}} = \underline{\textbf{893 g chlorophyll}}$$

The molar mass of chlorophyll is **893 g/mol**.

89. See Sections 1.2, 2.3, and 3.1.

Use the following steps to solve the problem:

(i) Find the volume of the cube in cm^3.

(ii) Find the number of atoms in 1 mole of silver using density as conversion factor and the given information that 4 atoms occupy the cube.

length of the edge of the cube (s) occupied by 4 atoms of Ag (in cm):

$$s = 0.409 \, nm \times \frac{1 \, m}{10^9 \, nm} \times \frac{100 \, cm}{1 \, m} = 4.09 \times 10^{-8} \, cm$$

volume of the cube occupied by the 4 atoms:

$$V = s^3 = (4.09 \times 10^{-8} \, cm)^3 = 6.84 \times 10^{-23} \, cm^3$$

number of atoms in 1 mole of Ag:

$$1 \, mol \, Ag \times \frac{107.87 \, g \, Ag}{1 \, mol \, Ag} \times \frac{1 \, cm^3 \, Ag}{10.5 \, g \, Ag} \times \frac{4 \, atoms \, Ag}{6.84 \times 10^{-23} \, cm^3} = \underline{\mathbf{6.01 \times 10^{23} \, atoms \, Ag}}$$

90. See Sections 3.2 and 3.3.

When the calcium sample is burned in air, the following balanced chemical reactions occur:

$$2Ca + O_2 \rightarrow 2CaO$$

$$3Ca + N_2 \rightarrow Ca_3N_2$$

The amount of CaO formed can be calculated from the amount of $Ca(OH)_2$ produced when water was added to the mixture of CaO and Ca_3N_2. The following reaction occurred:

$$CaO + H_2O \rightarrow Ca(OH)_2$$

mass of CaO produced when the original sample was burned in air:

$$4.832 \, g \, Ca(OH)_2 \times \frac{1 \, mol \, Ca(OH)_2}{74.10 \, g \, Ca(OH)_2} \times \frac{1 \, mol \, CaO}{1 \, mol \, Ca(OH)_2} \times \frac{56.08 \, g \, CaO}{1 \, mol \, CaO} = \underline{\mathbf{3.657 \, g \, CaO}}$$

mass of Ca from the original sample that formed CaO:

$$3.657 \, g \, CaO \times \frac{1 \, mol \, CaO}{56.08 \, g \, CaO} \times \frac{2 \, mol \, Ca}{2 \, mol \, CaO} \times \frac{40.08 \, g \, Ca}{1 \, mol \, Ca} = 2.614 \, g \, Ca$$

mass of Ca that formed Ca_3N_2 = mass of original Ca sample − mass of Ca (formed CaO)

$$= 5.025 \, g - 2.614 \, g$$

$$= 2.411 \, g \, Ca$$

mass of Ca_3N_2 produced when the original sample was burned in air:

$$2.411\,g\,Ca \times \frac{1\,mol\,Ca}{40.08\,g\,Ca} \times \frac{1\,mole\,Ca_3N_2}{3\,mol\,Ca} \times \frac{148.26\,g\,Ca_3N_2}{1\,mol\,Ca_3N_2} = \underline{\textbf{2.973 g Ca}_3\textbf{N}_2}$$

The mass of CaO and Ca_3N_2 formed are **3.657g and 2.973 g**, respectively.

91. See Sections 3.2 and 3.3.

Since all the KBr in the mixture was converted to KCl, it is expected that the mass of the mixture will decrease because Br has greater mass than Cl. The mass lost is accounted for by the replacement of Br with Cl.

Assume 1 mol Br is converted to 1 mol Cl (1 mol Br has mass of 79.90 g and 1 mol of Cl has mass of 35.45 g).

mass lost in the conversion of 1 mol Br = 79.90 g – 35.45 g = 44.45 g

Therefore, 44.45 g is lost per mole of Br converted to a mole of Cl.

mass lost when all sample is converted to Cl:

mass lost = 3.595 g – 3.129 g = 0.466 g

moles of Br converted to Cl:

$$0.466\,g\,lost \times \frac{1\,mol\,Br\,converted\,to\,Cl}{44.45\,g\,lost} = 0.0105\,mol\,Br\,is\,converted\,to\,Cl$$

mass of KBr in the original mixture:

$$0.0105\,mol\,Br \times \frac{1\,mol\,KBr}{1\,mol\,Br} \times \frac{119.00\,g\,KBr}{1\,mol\,KBr} = 1.25\,g\,KBr\,in\,the\,original\,mixture$$

mass percent of KBr in the original mixture:

$$mass\,\% = \frac{1.25\,g}{3.595\,g} \times 100\% = \underline{\textbf{34.8\% KBr}}$$

92. See Section 3.2.

To visualize the problem follow the schematic diagram below:

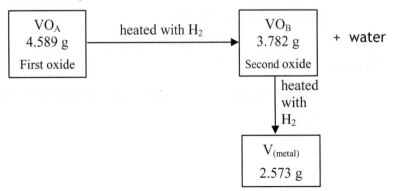

(a) simplest formula of the two oxides

both oxides have 2.573 g V metal or 0.05051 mol V:

$$2.573 \text{ g V} \times \frac{1 \text{ mol V}}{50.94 \text{ g V}} = 0.05051 \text{ mol V}$$

First oxide (VO_A)

mass O = mass VO_A − mass V

mass O = 4.589 g VO_A − 2.573 g V = 2.016 g O

$$\text{mole O} = 2.016 \text{ g O} \times \frac{1 \text{ mol O}}{16.00 \text{ g O}} = 0.1260 \text{ mol O}$$

mol ratio of O and V:

$$\frac{0.1260 \text{ mol O}}{0.05051 \text{ mol V}} = \frac{2.495 \text{ mol O}}{1 \text{ mol V}}$$

mol ratio of V and O: 1 mol V: 2.495 mol O:

Get the atom ratio from the mole ratio. Since the atoms should have whole numbers, the mole ratio should be multiplied by 2 because it is the smallest integer that will give the simplest formula.

After multiplying the mole ratio by 2, the atom ratio is 2 V: 5 O

The simplest formula of the first oxide is V_2O_5.

Second oxide (VO_B)

mass O = mass VO_B − mass V

mass O = 3.782 − 2.573 g = 1.209 g O

$$\text{mole O} = 1.209 \text{ g O} \times \frac{1 \text{ mol O}}{16.00 \text{ g O}} = 0.07556 \text{ mol O}$$

mol ratio of O and V: $\dfrac{0.07556 \text{ mol O}}{0.05051 \text{ mol V}} = \dfrac{1.496 \text{ mol O}}{1 \text{ mol V}}$

mol ratio of O and V: 1.496 mol O : 1 mol V

Get the atom ratio from the mole ratio. Since the atoms should have whole numbers, the mole ratio should be multiplied by 2 because it is the smallest integer that will give the simplest formula.

After multiplying the mole ratio by 2, the atom ratio is 2 V: 3 O

The simplest formula of the second oxide is V_2O_3.

The two oxides are V_2O_3 and V_2O_5.

(b) The two successive heatings with hydrogen removed O and converted it to water. The difference in the mass of the original oxide of vanadium and the final vanadium metal is the mass of O. This O is converted to water.

$$\text{mass O} = 4.589 - 2.573 \text{ g} = 2.016 \text{ g O}$$

$$2.016 \text{ g O} \times \frac{1 \text{ mol O}}{16.00 \text{ g O}} \times \frac{1 \text{ mol H}_2\text{O}}{1 \text{ mol O}} \times \frac{18.02 \text{ g H}_2\text{O}}{1 \text{ mol H}_2\text{O}} = \underline{\textbf{2.271 g H}_2\text{O}}$$

93. See Sections 3.2 and 3.3.

When cocaine ($C_{17}H_{21}O_4N$) and sugar ($C_{12}H_{22}O_{11}$) are burned, the C in these two compounds is converted to CO_2.

$$\text{mass of sample (in g)} = 1.00 \text{ mg} \times \frac{1 \text{ g}}{1000 \text{ mg}} = 0.00100 \text{ g cocaine-sugar mixture}$$

let mass of cocaine in the sample = Y g

let mass of sugar = (0.00100 − Y) g

mass of CO_2 from cocaine ($C_{17}H_{21}O_4N$, molar mass = 303.35 g/mol):

$$Y \text{ g } C_{17}H_{21}O_4N \times \frac{1 \text{ mol } C_{17}H_{21}O_4N}{303.35 \text{ g } C_{17}H_{21}O_4N} \times \frac{17 \text{ mol C}}{1 \text{ mol } C_{17}H_{21}O_4N} \times \frac{1 \text{ mol } CO_2}{1 \text{ mol C}} \times \frac{44.01 \text{ g } CO_2}{1 \text{ mol } CO_2} = 2.466Y \text{ g } CO_2$$

mass of CO_2 from sugar ($C_{12}H_{22}O_{11}$, molar mass = 342.30):

$$(0.00100 - Y) \text{ g } C_{12}H_{22}O_{11} \times \frac{1 \text{ mol } C_{12}H_{22}O_{11}}{342.30 \text{ g } C_{12}H_{22}O_{11}} \times \frac{12 \text{ mol C}}{1 \text{ mol } C_{12}H_{22}O_{11}} \times \frac{1 \text{ mol } CO_2}{1 \text{ mol C}} \times \frac{44.01 \text{ g } CO_2}{1 \text{ mol } CO_2}$$

$$= (0.00154 - 1.54Y) \text{ g } CO_2$$

total mass of CO_2 produced by burning the mixture containing cocaine and sugar is determined from the volume of CO_2 produced:

$$\text{mass of } CO_2 = 1.00 \text{ mL} \times \frac{1 \text{ L}}{1000 \text{ mL}} \times \frac{1.80 \text{ g}}{1 \text{ L}} = 0.00180 \text{ g}$$

Since this total CO_2 came from both cocaine and sucrose then,

total CO_2 = 0.00180 g = CO_2 from cocaine + CO_2 from sucrose

As previously shown,

CO_2 from cocaine = 2.466Y g

CO_2 from sugar = (0.00154 − 1.54Y) g

Then,

0.00180 g = CO_2 from cocaine + CO_2 from sucrose

0.00180 g = 2.466Y g + (0.00154 − 1.54Y) g

Solving for Y,

$$2.466Y \text{ g} - 1.54Y \text{ g} = 0.00180 \text{ g} - 0.00154 \text{ g}$$

$$0.926 \text{ Y} = 0.00026 \text{ g}$$

$$\text{Y} = 2.8 \times 10^{-4} \text{ g cocaine}$$

mass percent cocaine in the sample:

$$\text{mass \% cocaine} = \frac{2.8 \times 10^{-4} \text{ g}}{1.00 \times 10^{-3} \text{ g}} \times 100\% = \underline{\textbf{28\%}}$$

Therefore the mass percent of cocaine in the sample is **28%**.

94. See Sections 3.2 and 3.3.

mass of $NaClO_3$ in the 100.0-g sample:

The mass of $NaClO_3$ in 100.0-g sample of the mixture can be obtained from the 5.95 g O_2 produced by following reaction:

$$2 \text{ NaClO}_3(s) \quad \rightarrow \quad 2\text{NaCl}(s) + 3O_2(g)$$

Strategy: g $O_2 \rightarrow$ mol $O_2 \rightarrow$ mol $NaClO_3 \rightarrow$ g $NaClO_3$

$$5.95 \text{ g O}_2 \times \frac{1 \text{ mol O}_2}{32.00 \text{ g O}_2} \times \frac{2 \text{ mol NaClO}_3}{3 \text{ mol O}_2} \times \frac{106.44 \text{ g NaClO}_3}{1 \text{ mol NaClO}_3} = \underline{\textbf{13.2 g NaClO}_3}$$

mass of $NaHCO_3$ in the 100.0-g sample:

The mass of $NaHCO_3$ in 100.0-g sample of the mixture can be obtained from the 1.67 g H_2O produced by following reaction:

$$2 \text{ NaHCO}_3(s) \quad \rightarrow \quad Na_2O(s) + 2CO_2(g) + H_2O$$

Strategy: g $H_2O \rightarrow$ mol $H_2O \rightarrow$ mol $NaHCO_3 \rightarrow$ g $NaHCO_3$

$$1.67 \text{ g H}_2\text{O} \times \frac{1 \text{ mol H}_2\text{O}}{18.02 \text{ g H}_2\text{O}} \times \frac{2 \text{ mol NaHCO}_3}{1 \text{ mol H}_2\text{O}} \times \frac{84.01 \text{ g NaHCO}_3}{1 \text{ mol NaHCO}_3} = \underline{\textbf{15.6 g NaHCO}_3}$$

mass of Na_2CO_3 in the 100.0-g sample:

The mass of Na_2CO_3 in 100.0-g sample of the mixture can be obtained from the amount of CO_2 produced. However, the total CO_2 formed, 14.5 g comes from two sources: $NaHCO_3$ and Na_2CO_3. The mass of CO_2 from $NaHCO_3$ can be calculated from the amount of H_2O produced by the following reaction:

$$2 \text{ NaHCO}_3(s) \quad \rightarrow \quad Na_2O(s) + 2CO_2(g) + H_2O$$

Strategy: g H$_2$O → mol H$_2$O → mol CO$_2$ → g CO$_2$

$$1.67 \text{ g H}_2\text{O} \times \frac{1 \text{ mol H}_2\text{O}}{18.02 \text{ g H}_2\text{O}} \times \frac{2 \text{ mol CO}_2}{1 \text{ mol H}_2\text{O}} \times \frac{44.01 \text{ g CO}_2}{1 \text{ mol CO}_2} = 8.16 \text{ g CO}_2 \text{ from NaHCO}_3$$

total mass CO$_2$ = mass CO$_2$ from Na$_2$CO$_3$ + mass CO$_2$ from NaHCO$_3$
mass CO$_2$ from Na$_2$CO$_3$ = total mass CO$_2$ − mass CO$_2$ from NaHCO$_3$
= 14.5 g − 8.16 g
= 6.34 g CO$_2$ from Na$_2$CO$_3$

The mass of CO$_2$ from Na$_2$CO$_3$ calculated above is formed by the following reaction and can be converted to the mass of Na$_2$CO$_3$.

Na$_2$CO$_3$(s)   →   Na$_2$O(s) + CO$_2$(g)

Strategy: g CO$_2$ → mol CO$_2$ → mol Na$_2$CO$_3$ → g Na$_2$CO$_3$

$$6.34 \text{ g CO}_2 \times \frac{1 \text{ mol CO}_2}{44.01 \text{ g CO}_2} \times \frac{1 \text{ mol Na}_2\text{CO}_3}{1 \text{ mol CO}_2} \times \frac{105.99 \text{ g Na}_2\text{CO}_3}{1 \text{ mol Na}_2\text{CO}_3} = \underline{\textbf{15.3 g Na}_2\textbf{CO}_3}$$

mass of NaCl in the 100.0-g sample:

The mass of NaCl in 100.0-g sample of the mixture can be obtained by difference as shown below.

g sample = g NaClO$_3$ + g Na$_2$CO$_3$ + g NaHCO$_3$ + g NaCl

g NaCl = 100.0 g sample − [13.2 g NaClO$_3$ + 15.3 g Na$_2$CO$_3$ + 15.6 g NaHCO$_3$]

= **55.9 g NaCl**

The 100.0 g sample is composed of 13.2 g NaClO$_3$, 15.6 g NaHCO$_3$, 15.3 g Na$_2$CO$_3$, and 55.9 g NaCl.

95. See Sections 1.3 and 3.1.

V$_{alloy}$ = 13.0 cm × 22.0 cm × 17.5 cm = 5005 cm^3

Strategy: V$_{alloy}$ → mass of alloy → mass of Mo → number of Mo atoms

$$\text{Mo atoms} = 5005 \text{ cm}^3 \text{ alloy} \times \frac{7.68 \text{ g alloy}}{1 \text{ cm}^3 \text{alloy}} \times \frac{1.34 \text{ g Mo}}{100 \text{ g alloy}} \times \frac{1 \text{ mol Mo}}{95.94 \text{ g Mo}} \times \frac{6.022 \times 10^{23} \text{ Mo atoms}}{1 \text{ mol Mo}}$$

= **3.23 × 10^{24} Mo atoms**

REACTIONS IN AQUEOUS SOLUTION

4.1 PRECIPITATION REACTIONS

2. Use solubility rules (Figure 4.2).

    (a) $BaCl_2$ <u>soluble</u>

    (b) $Mg(OH)_2$ <u>insoluble</u>

    (c) $Cr_2(CO_3)_3$ <u>insoluble</u>

    (d) K_3PO_4 <u>soluble</u>

4. (a) To precipitate $Fe(OH)_3$, <u>add a soluble hydroxide such as a solution of sodium hydroxide</u> **(NaOH)**.

    (b) To precipitate $Fe_2(CO_3)_3$, <u>add a soluble carbonate such as a solution of sodium carbonate</u> **(Na_2CO_3)**.

    (c) To precipitate $FePO_4$, <u>add a soluble phosphate such as a solution of sodium phosphate</u> **(Na_3PO_4)**.

6. This problem is similar to Example 4.2. Using Figure 4.3 in the textbook as your guide, split up the soluble ionic compounds into their respective ions: these ions are present in solution. From these ions, combine a cation with an anion in solution. Two possible combinations of salts are obtained. Check solubility rules (Figure 4.2) to determine which of the possible salts precipitates.

    (a) Ions present: Fe^{3+}, NO_3^-, Na^+, OH^-

    Possible precipitates: $Fe(OH)_3$, $NaNO_3$

    Solubility: $Fe(OH)_3$ is insoluble; $NaNO_3$ is soluble and remains in solution.

    Net ionic equation: $\mathbf{Fe^{3+}(aq) + 3OH^-(aq) \rightarrow Fe(OH)_3(s)}$

    (b) Ions present: Mg^{2+}, SO_4^{2-}, Ba^{2+}, OH^-

    Possible precipitates: $Mg(OH)_2$, $BaSO_4$

    Solubility: $Mg(OH)_2$ and $BaSO_4$ are both insoluble and would both precipitate.

    Net ionic equations: $\mathbf{Ba^{2+}(aq) + SO_4^{2-}(aq) \rightarrow BaSO_4(s)}$

                                        $\mathbf{Mg^{2+}(aq) + 2OH^-(aq) \rightarrow Mg(OH)_2(s)}$

    The net ionic equation can also be written as:

    $$\mathbf{Mg^{2+}(aq) + SO_4^{2-}(aq) + Ba^{2+}(aq) + OH^-(aq) \rightarrow BaSO_4(s) + Mg(OH)_2\ (s)}$$

8. This problem is similar to Example 4.2. Using Figure 4.3 in the textbook as your guide, split up the soluble ionic compounds into their respective ions: these ions are present in solution. From these ions, combine a cation with an anion in solution. Two possible combinations of salts are obtained. Check solubility rules (Figure 4.2) to determine which of the possible salts precipitates.

(a) Ions present: Ag^+, NO_3^-, Na^+, Cl^-

Possible precipitates: AgCl, $NaNO_3$

Solubility: $NaNO_3$ is soluble and remains in solution; AgCl is insoluble and will precipitate.

Net ionic equation: $Ag^+(aq) + Cl^-(aq) \rightarrow AgCl(s)$

(b) Ions present: Co^{2+}, NO_3^-, Na^+, OH^-

Possible precipitates: $Co(OH)_2$, $NaNO_3$

Solubility: $NaNO_3$ is soluble and remains in solution; $Co(OH)_2$ is insoluble and will precipitate.

Net ionic equation: $Co^{2+}(aq) + 2OH^-(aq) \rightarrow Co(OH)_2(s)$

(c) Ions present: NH_4^+, PO_4^{3-}, K^+, OH^-,

Possible precipitates: NH_4OH, K_3PO_4

Solubility: **No reaction.** Both compounds are soluble, thus no precipitate forms.

(d) Ions present: Cu^{2+}, SO_4^{2-}, Na^+, CO_3^{2-}

Possible precipitates: $CuCO_3$, Na_2SO_4

Solubility: Na_2SO_4 is soluble and remains in solution; $CuCO_3$ is insoluble and will precipitate.

Net ionic equation: $Cu^{2+}(aq) + CO_3^{2-}(aq) \rightarrow CuCO_3(s)$

(e) Ions present: Li^+, SO_4^{2-}, Ba^{2+}, OH^-

Possible precipitates: LiOH, $BaSO_4$

Solubility: LiOH is soluble and remains in solution; $BaSO_4$ is insoluble and will precipitate.

Net ionic equation: $Ba^{2+}(aq) + SO_4^{2-}(aq) \rightarrow BaSO_4(s)$

10. This problem is similar to Example 4.2. Using Figure 4.3 in the textbook as your guide, split up the soluble ionic compounds into their respective ions: these ions are present in solution. From these ions, combine a cation with an anion in solution. Two possible combinations of salts are obtained. Check solubility rules (Figure 4.2) to determine which of the possible salts precipitates.

(a) Ions present: Na^+, PO_4^{3-}, Ba^{2+}, Cl^-

Possible precipitates: $Ba_3(PO_4)_2$, $NaCl$

Solubility: $NaCl$ is soluble and remains in solution; $Ba_3(PO_4)_2$ is insoluble and will precipitate.

Net ionic equation: $3Ba^{2+}(aq) + 2PO_4^{3-}(aq) \rightarrow Ba_3(PO_4)_2(s)$

(b) Ions present: Zn^{2+}, SO_4^{2-}, K^+, OH^-

Possible precipitates: $Zn(OH)_2$, K_2SO_4

Solubility: K_2SO_4 is soluble and remains in solution; $Zn(OH)_2$ is insoluble and will precipitate.

Net ionic equation: $Zn^{2+}(aq) + 2OH^-(aq) \rightarrow Zn(OH)_2(s)$

(c) Ions present: NH_4^+, SO_4^{2-}, Na^+, Cl^-

Possible precipitates: NH_4Cl, Na_2SO_4

Solubility: **No reaction.** Both compounds are soluble, so no precipitate forms.

(d) Ions present: Co^{3+}, NO_3^-, Na^+, PO_4^{3-}

Possible precipitates: $CoPO_4$, $NaNO_3$

Solubility: $NaNO_3$ is soluble and remains in solution; $CoPO_4$ is insoluble and will precipitate.

Net ionic equation: $Co^{3+}(aq) + PO_4^{3-}(aq) \rightarrow CoPO_4(s)$

4.1 PRECIPITATION REACTIONS: Stoichiometry

12. These problems are similar to Example 4.3.

(a) Strategy (From Figure 4.6): V of parent compound, Na_2CO_3 $\rightarrow$ mol of parent compound, Na_2CO_3 $\rightarrow$ mol of reacting ion, CO_3^{2-} $\rightarrow$ mol of other reacting ion, Fe^{3+} $\rightarrow$ mol of parent compound, $Fe(NO_3)_3$ $\rightarrow$ M of solution of parent compound, $Fe(NO_3)_3$

Net ionic equation: $2Fe^{3+}(aq) + 3CO_3^{2-}(aq) \rightarrow Fe_2(CO_3)_3(s)$

$$\text{mol } Fe(NO_3)_3 = 12.54 \text{ mL } Na_2CO_3 \times \frac{1 \text{ L}}{1000 \text{ mL}} \times \frac{0.1488 \text{ mol } Na_2CO_3}{1 \text{ L } Na_2CO_3} \times \frac{1 \text{ mol } CO_3^{2-}}{1 \text{ mol } Na_2CO_3}$$

$$\times \frac{2 \text{ mol } Fe^{3+}}{3 \text{ mol } CO_3^{2-}} \times \frac{1 \text{ mol } Fe(NO_3)_3}{1 \text{ mol } Fe^{3+}} = 0.001244 \text{ mol } Fe(NO_3)_3$$

The molarity of the $Fe(NO_3)_3$ solution is:

$$M = \frac{\text{moles}}{\text{volume (L)}} = \frac{0.0012444 \text{ mol } Fe(NO_3)_3}{0.02500 \text{ L}} = \underline{\textbf{0.04976 M}}$$

(b) Strategy (From Figure 4.6): mass of parent compound, K_3PO_4 → mol of parent compound, K_3PO_4 → mol of reacting ion, PO_4^{3-} → mol of other reacting ion, Fe^{3+} → mol of parent compound, $Fe(NO_3)_3$ → M of parent compound, $Fe(NO_3)_3$

Net ionic equation: $Fe^{3+}(aq) + PO_4^{3-}(aq) \rightarrow FePO_4(s)$

$$mol\ Fe(NO_3)_3 = 7.58\ g\ K_3PO_4 \times \frac{1\ mol\ K_3PO_4}{212.27\ g\ K_3PO_4} \times \frac{1\ mol\ PO_4^{3-}}{1\ mol\ K_3PO_4}$$

$$\times \frac{1\ mol\ Fe^{3+}}{1\ mol\ PO_4^{3-}} \times \frac{1\ mol\ Fe(NO_3)_3}{1\ mol\ Fe^{3+}} = 0.03571\ mol\ Fe(NO_3)_3$$

The molarity of the $Fe(NO_3)_3$ solution is:

$$M = \frac{moles}{volume\ (L)} = \frac{0.03571\ mol\ Fe(NO_3)_3}{0.02500\ L} = \underline{\mathbf{1.43\ M}}$$

(c) Strategy (From Figure 4.6): V of parent compound, $Sr(OH)_2$ → mol of parent compound, $Sr(OH)_2$ → mol of reacting ion, OH^- → mol of other reacting ion, Fe^{3+} → mol of parent compound, $Fe(NO_3)_3$ → M of parent compound, $Fe(NO_3)_3$

Net ionic equation: $Fe^{3+}(aq) + 3OH^-(aq) \rightarrow Fe(OH)_3(s)$

$$mol\ Fe(NO_3)_3 = 10.00\ mL\ Sr(OH)_2 \times \frac{1\ L}{1000\ mL} \times \frac{0.1573\ mol\ Sr(OH)_2}{1\ L\ Sr(OH)_2} \times \frac{2\ mol\ OH^-}{1\ mol\ Sr(OH)_2}$$

$$\times \frac{1\ mol\ Fe^{3+}}{3\ mol\ OH^-} \times \frac{1\ mol\ Fe(NO_3)_3}{1\ mol\ Fe^{3+}} = 0.001049\ mol\ Fe(NO_3)_3$$

The molarity of the $Fe(NO_3)_3$ solution is:

$$M = \frac{moles}{volume\ (L)} = \frac{0.001049\ mol\ Fe(NO_3)_3}{0.02500\ L} = \underline{\mathbf{0.04196\ M}}$$

14. These problems are similar to Example 4.3.

(a) Net Ionic Equation: $2Al^{3+}(aq) + 3CO_3^{2-}(aq) \rightarrow Al_2(CO_3)_3(s)$

(b) Strategy (From Figure 4.6): V of parent compound, Na_2CO_3 → mol of parent compound, Na_2CO_3 → mol of reacting ion, CO_3^{2-} → mol of other reacting ion, Al^{3+} → mol of parent compound, $AlCl_3$ → M of $AlCl_3$ solution

$$\text{mol AlCl}_3 = 35.5 \text{ mL Na}_2\text{CO}_3 \times \frac{1\text{L}}{1000 \text{ mL}} \times \frac{0.137 \text{ mol Na}_2\text{CO}_3}{1\text{L Na}_2\text{CO}_3} \times \frac{1 \text{ mol CO}_3^{2-}}{1\text{mol Na}_2\text{CO}_3}$$

$$\times \frac{2 \text{ mol Al}^{3+}}{3 \text{ mol CO}_3^{2-}} \times \frac{1\text{mol AlCl}_3}{1 \text{mol Al}^{3+}} = 0.00324 \text{ mol AlCl}_3$$

The volume of AlCl₃ required is 30.0 mL or 0.0300 L

$$M = \frac{moles}{volume \ (L)} = \frac{0.00324 \text{ mol AlCl}_3}{0.0300 \text{ L}} = \underline{\textbf{0.108 M}}$$

(c) Strategy (From Figure 4.6): V of parent compound, Na₂CO₃ → mol of parent compound, Na₂CO₃ → mol of reacting ion, CO₃²⁻ → mol of precipitate, Al₂(CO₃)₃ → mass of precipitate, Al₂(CO₃)₃

$$\text{mass Al}_2(\text{CO}_3)_3 = 35.5 \text{ mL Na}_2\text{CO}_3 \times \frac{1\text{L}}{1000 \text{ mL}} \times \frac{0.137 \text{ mol Na}_2\text{CO}_3}{1\text{L Na}_2\text{CO}_3} \times \frac{1 \text{ mol CO}_3^{2-}}{1\text{mol Na}_2\text{CO}_3}$$

$$\times \frac{1\text{mol Al}_2(\text{CO}_3)_3}{3 \text{ mol CO}_3^{2-}} \times \frac{233.99 \text{ g Al}_2(\text{CO}_3)_3}{1 \text{mol Al}_2(\text{CO}_3)_3} = \underline{\textbf{0.379 g Al}_2(\text{CO}_3)_3}$$

16. This is a limiting reactant problem. A review of Example 4.3c will help you.

(a) Net Ionic Equation: **Al³⁺(aq) + 3OH⁻(aq) → Al(OH)₃(s)**

(b) To find the mass of precipitate formed, determine first the limiting reactant by identifying the reactant that will produce fewer moles of precipitate.

mol precipitate if Al³⁺ is the limiting reactant:

Strategy (from Figure 4.6): mass of parent compound, Al₂(SO₄)₃ → mol of parent compound, Al₂(SO₄)₃ → mol of reacting ion, Al³⁺ → mol of precipitate, Al(OH)₃

$$2.76 \text{ g Al}_2(\text{SO}_4)_3 \times \frac{1\text{mol Al}_2(\text{SO}_4)_3}{342.17 \text{ g Al}_2(\text{SO}_4)_3} \times \frac{2 \text{ mol Al}^{3+}}{1\text{mol Al}_2(\text{SO}_4)_3} \times \frac{1\text{mol Al(OH)}_3}{1\text{mol Al}^{3+}}$$

$$= 0.0161 \text{ mol Al(OH)}_3$$

mol precipitate if OH⁻ is the limiting reactant:

Strategy (from Figure 4.6): V of parent compound, NaOH → mol of parent compound, NaOH → mol of reacting ion, OH⁻ → mol of precipitate, Al(OH)₃

$$85.0 \text{ mL NaOH} \times \frac{1\text{L}}{1000 \text{ mL}} \times \frac{0.2500 \text{ mol NaOH}}{1\text{L NaOH}} \times \frac{1 \text{ mol OH}^-}{1\text{mol NaOH}} \times \frac{1 \text{ mol Al(OH)}_3}{3 \text{ mol OH}^-}$$

$$= 0.00708 \text{ mol Al(OH)}_3$$

Since 0.00708 mol $Al(OH)_3$ is less than 0.0161 mol $Al(OH)_3$, OH^- is the limiting reactant. Hence, 0.00708 mol of precipitate forms. The mass of this precipitate is

$$0.00708 \text{ mol } Al(OH)_3 \times \frac{78.00 \text{ g } Al(OH)_3}{1 \text{ mol } Al(OH)_3} = \underline{\textbf{0.552 g } Al(OH)_3}$$

(c) Since Al^{3+} will form more precipitate, Al^{3+} is the ion in excess.

mol Al^{3+} originally available: $2.76 \text{ g } Al_2(SO_4)_3 \times \dfrac{1 \text{ mol } Al_2(SO_4)_3}{342.17 \text{ g } Al_2(SO_4)_3} \times \dfrac{2 \text{ mol } Al^{3+}}{1 \text{ mol } Al_2(SO_4)_3}$

$$= 0.0161 \text{ mol } Al^{3+} \text{ available}$$

mol Al^{3+} reacted: $85.0 \text{ mL NaOH} \times \dfrac{1 \text{ L}}{1000 \text{ mL}} \times \dfrac{0.2500 \text{ mol NaOH}}{1 \text{ L NaOH}} \times \dfrac{1 \text{ mol } OH^-}{1 \text{ mol NaOH}} \times \dfrac{1 \text{ mol } Al^{3+}}{3 \text{ mol } OH^-}$

$$= 0.00708 \text{ mol } Al^{3+} \text{ reacted or consumed}$$

excess Al^{3+} = 0.0161 mol Al^{3+} available − 0.00708 mol Al^{3+} consumed

$\qquad\qquad$ = 0.0090 mol Al^{3+}

Assuming volume is additive, total solution volume is 0.235 L or 235 mL (125 mL + 85.0 mL = 210 mL)

$$\text{Molarity} = \frac{0.0090 \text{ mol } Al^{3+}}{0.210 \text{ L}} = \underline{\textbf{0.043 } M}$$

4.2 ACID-BASE REACTIONS

18. Acids and bases **NOT** listed in Table 4.1 are weak.

(a) H_2S is a <u>**weak acid**</u>

(b) H_2SO_4 is a <u>**strong acid**</u>

(c) C_5H_5N is a <u>**weak base**</u>

(d) $Al(OH)_3$ is a <u>**weak base**</u>

20. Acids **NOT** listed in Table 4.1 are weak. As shown in Figure 4.10, for strong acids the reacting species is H^+ while for weak acids the reacting species is the acid molecule.

(a) Hypochlorous acid, $HClO$ is a weak acid; the reacting species is **HClO**.

(b) Formic acid, $HCHO_2$ is a weak acid; the reacting species is **HCHO₂**.

(c) Acetic acid, $HC_2H_3O_2$ is a weak acid; the reacting species is **HC₂H₃O₂**.

(d) Hydrobromic acid, HBr is a strong acid; the reacting species is **H⁺**.

(e) Sulfurous acid, H_2SO_3 is a weak acid; the reacting species is **H₂SO₃**.

22. As shown in Table 4.1, the hydroxides of Group I and Group II metals are the only strong bases. As shown in Figure 4.10, for strong bases the reacting species is OH^- while for weak bases the reacting species is the base molecule.

    (a) Toluidine, C_7H_9N is a weak base; the reacting species is C_7H_9N.

    (b) Strontium hydroxide, $Sr(OH)_2$ is a strong base; the reacting species is OH^-.

    (c) Indol, C_8H_6NH is a weak base; the reacting species is C_8H_6NH.

    (d) Aqueous ammonia, NH_3 is a weak base; the reacting species is NH_3.

24. This problem is similar to Example 4.4. Use Tables 4.1 and 4.2 and Figure 4.10 as guide. Classify the acid and the base as weak or strong, and then identify the reacting species. Write the reaction for the two reacting species.

    (a) Acetic acid, $HC_2H_3O_2$ is a **weak acid**; the reacting species is $HC_2H_3O_2$.

    Strontium hydroxide, $Sr(OH)_2$ is a **strong base**; the reacting species is OH^-.

    Net Ionic Equation: $HC_2H_3O_2(aq) + OH^-(aq) \rightarrow H_2O + C_2H_3O_2^-(aq)$

    (b) Diethylamine, $(C_2H_5)_2NH$ is a **weak base**; the reacting species is $(C_2H_5)_2NH$.

    Sulfuric acid, H_2SO_4 is a **strong acid**; the reacting species is H^+.

    Net Ionic Equation: $(C_2H_5)_2NH(aq) + H^+(aq) \rightarrow (C_2H_5)_2NH_2^+(aq)$

    (c) Hydrofluoric acid (HF) is a **weak acid**; the reacting species is HF.

    Sodium hydroxide (NaOH) is a **strong base**; the reacting species is OH^-.

    Net Ionic Equation: $HF(aq) + OH^-(aq) \rightarrow H_2O + F^-(aq)$

26. The following prototype net ionic equation, $OH^-(aq) + HB(aq) \rightarrow B^-(aq) + H_2O$, would only be correct when the reactants are a weak acid and a strong base. Use Tables 4.1 and 4.2 and Figure 4.10 as guide.

    (a) The equation is **NOT CORRECT**.

    Hydrochloric acid (HCl) is a strong acid; the reacting species is H^+.

    Pyridine (C_5H_5N) is a weak base; the reacting species is C_5H_5N.

    Correct equation: $H^+(aq) + C_5H_5N(aq) \rightarrow C_5H_5NH^+(aq)$

    (b) The equation is **NOT CORRECT**.

    Sulfuric acid (H_2SO_4) is a strong acid; the reacting species is H^+.

    Rubidium hydroxide (RbOH) is a strong base; the reacting species is OH^-.

    Correct Equation: $H^+(aq) + OH^-(aq) \rightarrow H_2O$

(c) The equation is **CORRECT**.

Hydrofluoric acid (HF) is a weak acid; the reacting species is HF.

Potassium hydroxide (KOH) is a strong base; the reacting species is OH^-.

Equation: $OH^-(aq) + HF(aq) \rightarrow F^-(aq) + H_2O$

(d) The equation is **NOT CORRECT**.

Hydroiodic acid (HI) is a strong acid; the reacting species is H^+.

Ammonia (NH_3) is a weak base; the reacting species is NH_3.

Correct Equation: $NH_3(aq) + H^+(aq) \rightarrow NH_4^+(aq)$

(e) The equation is **CORRECT**.

Hydrocyanic acid (HCN) is a weak acid; the reacting species is HCN.

Strontium hydroxide ($Sr(OH)_2$) is a strong base; the reacting species is OH^-.

Equation: $OH^-(aq) + HCN(aq) \rightarrow CN^-(aq) + H_2O$

4.2 ACID-BASE REACTIONS: Acid-Base Titrations

28. This problem is similar to Example 4.5. Use the flowchart for solution stoichiometry (Figure 4.6) as a guide.

    Strategy: V and M $NH_3 \rightarrow$ mol $NH_3 \rightarrow$ mol $H^+ \rightarrow$ mol $H_2SO_4 \rightarrow$ V H_2SO_4

    The stoichiometric ratio of reacting species is 1 mol NH_3 to 1 mol H^+.
    Convert mol H^+ to mol H_2SO_4 (the ratio is 1:2). Finally, use molarity of H_2SO_4 to calculate the volume of H_2SO_4.

    Net ionic equation: $NH_3(aq) + H^+(aq) \rightarrow NH_4^+(aq)$

    mol H_2SO_4: $38.00 \text{ mL } NH_3 \times \dfrac{1 L}{1000 \text{ mL}} \times \dfrac{0.189 \text{ mol } NH_3}{1 L} \times \dfrac{1 \text{ mol } H^+}{1 \text{ mol } NH_3} \times \dfrac{1 \text{ mol } H_2SO_4}{2 \text{ mol } H^+}$

    $= 0.00359 \text{ mol } H_2SO_4$

    volume (mL) of H_2SO_4 required:

    $V = n \div M = \dfrac{0.00359 \text{ mol}}{0.2315 \text{ M}}$ = **0.0155 L or 15.5 mL H_2SO_4 solution**

30. These problems are similar to Example 4.5. Use the flowchart for solution stoichiometry (Figure 4.6) as your guide.

    Strategy: mol base $\rightarrow$ mol $H^+ \rightarrow$ mol $HClO_4 \rightarrow$ M $HClO_4$ solution

    Calculate mol of the given base; use the mole ratio of reacting base to calculate mol H^+.
    Convert mol H^+ to mol $HClO_4$ then use the volume of $HClO_4$ to calculate its molarity.

(a) Net ionic equation: $H^+(aq) + C_2H_5NH_2(aq) \rightarrow C_2H_5NH_3^+(aq)$

mol $HClO_4$:

$$17.25 \text{ mL } C_2H_5NH_2 \times \frac{1 \text{ L}}{1000 \text{ mL}} \times \frac{0.3471 \text{ mol } C_2H_5NH_2}{1 \text{ L}} \times \frac{1 \text{ mol } H^+}{1 \text{ mol } C_2H_5NH_2} \times \frac{1 \text{ mol } HClO_4}{1 \text{ mol } H^+}$$

$$= 0.005987 \text{ mol } HClO_4$$

molarity of $HClO_4$ solution: $M = \dfrac{0.005987 \text{ mol } HClO_4}{0.02500 \text{ L}} = \underline{\textbf{0.2395 M}}$

(b) Net ionic equation: $H^+(aq) + OH^-(aq) \rightarrow H_2O$

mol $HClO_4$:

$$14.17 \text{ g } Sr(OH)_2 \times \frac{1 \text{ mol } Sr(OH)_2}{121.64 \text{ g } Sr(OH)_2} \times \frac{2 \text{ mol } OH^-}{1 \text{ mol } Sr(OH)_2} \times \frac{1 \text{ mol } H^+}{1 \text{ mol } OH^-} \times \frac{1 \text{ mol } HClO_4}{1 \text{ mol } H^+}$$

$$= 0.2330 \text{ mol } HClO_4$$

molarity of $HClO_4$ solution: $M = \dfrac{0.2330 \text{ mol } HClO_4}{0.02500 \text{ L}} = \underline{\textbf{9.320 M}}$

(c) Use density to convert the volume of solution to mass then use mass percent to calculate mass of ammonia in the solution. Then proceed as above.

Net ionic equation: $NH_3(aq) + H^+(aq) \rightarrow NH_4^+(aq)$

mass of NH_3:

$$41.73 \text{ mL} \times \frac{0.9295 \text{ g solution}}{1 \text{ mL}} \times \frac{18 \text{ g } NH_3}{100 \text{ g solution}} \times \frac{1 \text{ mol } NH_3}{17.03 \text{ g } NH_3} \times \frac{1 \text{ mol } H^+}{1 \text{ mol } NH_3} \times \frac{1 \text{ mol } HClO_4}{1 \text{ mol } H^+}$$

$$= 0.41 \text{ mol } HClO_4$$

molarity of $HClO_4$ solution: $M = \dfrac{0.41 \text{ mol } HClO_4}{0.02500 \text{ L}} = \underline{\textbf{16 M}}$

32. This problem is similar to Example 4.5c. Use the flowchart in Figure 4.6 as guide. HAsp is a weak acid so it does not break up into ions.

Strategy: V and M of KOH $\rightarrow$ mol KOH $\rightarrow$ mol OH^- $\rightarrow$ mol HAsp $\rightarrow$ MM aspirin

Net ionic equation: $HAsp(aq) + OH^-(aq) \rightarrow Asp^-(aq) + H_2O$

mol HAsp:

$$17.6 \text{ mL KOH} \times \frac{1 \text{ L}}{1000 \text{ mL}} \times \frac{0.315 \text{ mol KOH}}{1 \text{ L}} \times \frac{1 \text{ mol } OH^-}{1 \text{ mol KOH}} \times \frac{1 \text{ mol HAsp}}{1 \text{ mol } OH^-} = 0.00554 \text{ mol Hasp}$$

Molar mass of aspirin:

$$MM_{aspirin} = mass \div n = \frac{1.00 \text{ g HAsp}}{0.00554 \text{ mol HAsp}} = \underline{\mathbf{1.80 \times 10^2 \text{ g/mol}}}$$

34. Use the flowchart for solution stoichiometry (Figure 4.6) as your guide. $HC_2H_3O_2$ is a weak acid and does not break up into ions.

    Strategy: V and M of $Ba(OH)_2 \rightarrow$ mol $Ba(OH)_2 \rightarrow$ mol $OH^- \rightarrow$ mol $HC_2H_3O_2 \rightarrow$ mass $HC_2H_3O_2$

    Net ionic equation: $OH^-(aq) + HC_2H_3O_2(aq) \rightarrow C_2H_3O_2^-(aq) + H_2O$

    mass $HC_2H_3O_2$:

    $$37.50 \text{ mL Ba(OH)}_2 \times \frac{1 \text{ L}}{1000 \text{ mL}} \times \frac{0.1250 \text{ mol Ba(OH)}_2}{1 \text{ L Ba(OH)}_2} \times \frac{2 \text{ mol OH}^-}{1 \text{ mol Ba(OH)}_2}$$

    $$\times \frac{1 \text{ mol HC}_2\text{H}_3\text{O}_2}{1 \text{ mol OH}^-} \times \frac{60.053 \text{ g HC}_2\text{H}_3\text{O}_2}{1 \text{ mol HC}_2\text{H}_3\text{O}_2} = 0.5630 \text{ g HC}_2\text{H}_3\text{O}_2$$

    $$\% \text{ HC}_2\text{H}_3\text{O}_2 = \frac{\text{mass of HC}_2\text{H}_3\text{O}_2}{\text{mass of sample}} \times 100$$

    $$= \frac{0.5630 \text{ g HC}_2\text{H}_3\text{O}_2}{10.00 \text{ g sample}} \times 100$$

    $$= \underline{\mathbf{5.630\%}}$$

    Yes, the sample can be considered vinegar because percent acetic acid is at least 5.0%.

36. Use the flowchart for solution stoichiometry (Figure 4.6) as your guide. Note that $C_6H_8O_6$ is a weak acid and does not break up into ions.

    Strategy: V and M KOH $\rightarrow$ mol KOH $\rightarrow$ mol $OH^- \rightarrow$ mol $C_6H_8O_6 \rightarrow$ mass $C_6H_8O_6$

    Net ionic equation: $C_6H_8O_6 (aq) + OH^-(aq) \rightarrow C_6H_7O_6^-(aq) + H_2O$

    mass vitamin C, $C_6H_8O_6$:

    $$5.94 \text{ mL KOH} \times \frac{1 \text{ L}}{1000 \text{ mL}} \times \frac{0.450 \text{ mol KOH}}{1 \text{ L KOH}} \times \frac{1 \text{ mol OH}^-}{1 \text{ mol KOH}} \times \frac{1 \text{ mol C}_6\text{H}_8\text{O}_6}{1 \text{ mol OH}^-} \times \frac{176.12 \text{ g C}_6\text{H}_8\text{O}_6}{1 \text{ mol C}_6\text{H}_8\text{O}_6}$$

    $$= 0.471 \text{ g C}_6\text{H}_8\text{O}_6 \text{ (vitamin C)}$$

    $$\% \text{ C}_6\text{H}_8\text{O}_6 = \frac{\text{mass of C}_6\text{H}_8\text{O}_6}{\text{mass of sample}} \times 100 = \frac{0.471 \text{ g C}_6\text{H}_8\text{O}_6}{0.634 \text{ g sample}} \times 100 = \underline{\mathbf{74.3\%}}$$

38. To find the number of moles of OH^- required to neutralize 1 mole of lactic acid ($C_3H_6O_3$), calculate the number of moles of hydroxide ions and lactic acid. Determine the mole ratio of lactic acid to NaOH. Use the flowchart for solution stoichiometry (Figure 4.6) as your guide. Note that $C_3H_6O_3$ is a weak acid and does not break up into ions.

Strategy: mass $C_3H_6O_3$ $\rightarrow$ mol $C_3H_6O_3$

V NaOH $\rightarrow$ mol NaOH $\rightarrow$ mol OH^-

mol $C_3H_6O_3$: $0.100 \text{ g } C_3H_6O_3 \times \dfrac{1 \text{ mol } C_3H_6O_3}{90.08 \text{ g } C_3H_6O_3} = 1.11 \times 10^{-3} \text{ mol } C_3H_6O_3$

mol OH^-: $12.95 \text{ mL NaOH} \times \dfrac{1 \text{ L}}{1000 \text{ mL}} \times \dfrac{0.0857 \text{ mol NaOH}}{1 \text{ L NaOH}} \times \dfrac{1 \text{ mol } OH^-}{1 \text{ mol NaOH}} = 1.11 \times 10^{-3} \text{ mol } OH^-$

Since 1.11×10^{-3} mol OH^- is required to neutralize 1.11×10^{-3} mol lactic acid, then
1 mole OH^- is required to neutralize 1 mole lactic acid.

4.3 OXIDATION-REDUCTION REACTIONS: Oxidation Number

40. This problem is similar to Example 4.6. Apply the rules for oxidation numbers.

(a) CH_4 rule #4: oxidation no. **H = +1**
        rule #5: oxidation no. C: $x + 4(+1) = 0$; solve for $x \Rightarrow x = -4$;
            oxidation no. **C = −4**

(b) CO_3^{2-} rule #6: oxidation no. **O = −2**
        rule #5: oxidation no. C: $x + 3(-2) = -2$; solve for $x \Rightarrow x = +4$;
            oxidation no. **C = +4**

(c) IO_4^- rule #6: oxidation no. **O = −2**
        rule #5: oxidation no. I: $x + 4(-2) = -1$; solve for $x \Rightarrow x = +7$;
            oxidation no. **I = +7**

(d) N_2H_4 rule #4: oxidation no. **H = +1**
        rule #5: oxidation no. N: $x(2) + 4(+1) = 0$; solve for $x \Rightarrow x = -2$;
            oxidation no. **N = −2**

42. These problems are similar to Example 4.6. Apply the rules for oxidation numbers.

(a) HIO_3 rule #4: oxidation no. **H = +1**
        rule #5: oxidation no. I: $1(+1) + x + 3(-2) = 0$; solve for $x \Rightarrow x = +5$;
            oxidation no. **I = +5**
        rule #6: oxidation no. **O = −2**

(b) $NaMnO_4$ rule #3: oxidation no. **Na = +1**
 rule #5: oxidation no. Mn: $1(+1)+ x +4(-2) = 0$; solve for x $\Rightarrow$ x = +7;
 oxidation no. **Mn = +7**
 rule #6: oxidation no. **O = -2**

(c) SnO_2 rule #5: oxidation no. Sn: $x +2(-2) = 0$; solve for x $\Rightarrow$ x = +4;
 oxidation no. **Sn = +4**
 rule #6: oxidation no. **O = -2**

(d) NOF rule #5: oxidation no. N: $x + 1(-2) + 1(-1) = 0$; solve for x $\Rightarrow$ x = +3;
 oxidation no. **N = +3**
 rule #6: oxidation no. **O = -2**
 rule #3: oxidation no. **F = -1**

(e) NaO_2 rule #3: oxidation no. **Na = +1**
 rule #5: oxidation no. O: $1(+1)+ 2x = 0$; solve for x $\Rightarrow$ x = -½ ;
 oxidation no. **O = -½**

4.3 OXIDATION-REDUCTION REACTIONS: Balancing Half-Equations

44. To know if a half-equation is an oxidation or a reduction reaction, determine the change in oxidation number. The half-equation is an oxidation reaction if there is an increase in oxidation number; otherwise it is a reduction reaction.

(a) $CH_3OH(aq) \rightarrow CO_2(g)$
 C: oxidation number *increased* from -2 to +4, so this is **oxidation**.

(b) $NO_3^-(aq) \rightarrow NH_4^+(aq)$
 N: oxidation number *decreased* from +5 to -3, so this is **reduction**.

(c) $Fe^{3+}(aq) \rightarrow Fe(s)$
 Fe: oxidation number *decreased* from +3 to 0, so this is **reduction**.

(d) $V^{2+}(aq) \rightarrow VO_3^-(aq)$
 V: oxidation number *increased* from +2 to +5, so this is **oxidation**.

46. As described in #44, the half-reaction is oxidation if there is an increase in the oxidation number in any of the atoms; otherwise it is reduction.

(a) $TiO_2(s) \rightarrow Ti^{3+}(aq)$
 Ti: oxidation number *decreased* from +4 to +3, so this is **reduction**.

(b) $Zn^{2+}(aq) \rightarrow Zn(s)$

Zn: oxidation number *decreased* from +2 to 0, so this is **reduction**.

(c) $NH_4^+(aq) \rightarrow N_2(g)$

N: oxidation number *increased* from –3 to 0, so this is **oxidation**.

(d) $CH_3OH(aq) \rightarrow CH_2O(aq)$

C: oxidation number *increased* from –2 to 0, so this is **oxidation**.

48. Write unbalanced half-reactions. Assign oxidation number to each element. Identify the element oxidized and reduced. If the oxidation number of an element increased, then that element is oxidized. If the oxidation number of an element decreased, then that element is reduced. The species or reactant (ion or molecule) that has the oxidized atom (lost electrons) is called the reducing agent. On the other hand, the species or reactant (ion or molecule) that has the reduced atom (accepted or gained electrons) is called the oxidizing agent.

(a) $As_2O_3(s) + MnO_4^-(aq) \rightarrow H_3AsO_4(aq) + Mn^{2+}(aq)$
unbalanced half-reactions:

Reduction: $MnO_4^-(aq) + 5e^- \rightarrow Mn^{2+}(aq)$

Mn is reduced: oxidation number *decreased* from +7 to +2

MnO_4^- is the species reduced;   **MnO_4^- is the oxidizing agent**

Oxidation: $As_2O_3(s) \rightarrow H_3AsO_4(aq) + 2e^-$

As is oxidized: oxidation number *increased* from +3 to +5

As_2O_3 is the species oxidized;   **As_2O_3 is the reducing agent**

(b) $N_2H_4(l) + CO_3^{2-}(aq) \rightarrow N_2(g) + CO(g)$
unbalanced half-reactions:

Reduction: $CO_3^{2-}(aq) + 2e^- \rightarrow CO(g)$

C is reduced: oxidation number *decreased* from +4 to +2

CO_3^{2-} is the species reduced;   **CO_3^{2-} is the oxidizing agent**

Oxidation: $N_2H_4(l) \rightarrow N_2(g) + 2e^-$

N is oxidized: oxidation number *increased* from –2 to 0

N_2H_4 is the species oxidized;   **N_2H_4 is the reducing agent**

4.3 OXIDATION-REDUCTION REACTIONS: Stoichiometry

50. Use the flowchart for solution stoichiometry (Figure 4.6) as your guide to determine the mass of vanadium. Use the stoichiometric ratio from the following balanced equation in the calculation.

$$5VO^{2+}(aq) + MnO_4^-(aq) + 11H_2O \rightarrow Mn^{2+}(aq) + 5V(OH)_4^+(aq) + 2H^+(aq)$$

Strategy: V and M $KMnO_4 \rightarrow$ mol $KMnO_4 \rightarrow$ mol $MnO_4^- \rightarrow$ mol $VO^{2+} \rightarrow$ mol V $\rightarrow$ mass V

$$\text{mass vanadium} = 26.45 \text{ mL } KMnO_4 \times \frac{1L}{1000 \text{ mL}} \times \frac{0.02250 \text{ mol } KMnO_4}{1L} \times \frac{1 \text{ mol } MnO_4^-}{1 \text{ mol } KMnO_4}$$

$$\times \frac{5 \text{ mol } VO^{2+}}{1 \text{ mol } MnO_4^-} \times \frac{1 \text{ mol } V}{1 \text{ mol } VO^{2+}} \times \frac{50.94 \text{ g V}}{1 \text{ mol } V} = 0.1516 \text{ g vanadium}$$

$$\% \text{ vanadium in the ore} = \frac{\text{mass of vanadium}}{\text{mass of sample}} \times 100 = \frac{0.1516 \text{ g}}{0.5000 \text{ g}} \times 100 = \underline{\textbf{30.32\%}}$$

52. Use the flowchart for solution stoichiometry (Figure 4.6) as your guide to determine the mass of silver. The stoichiometric ratio is obtained from the following balanced equation.

$$Ag(s) + 2H^+(aq) + NO_3^-(aq) \rightarrow Ag^+(aq) + NO_2(g) + H_2O$$

Strategy: V and M $HNO_3 \rightarrow$ mol $HNO_3 \rightarrow$ mol $H^+ \rightarrow$ mol Ag $\rightarrow$ mass Ag

$$\text{mass Ag} = 42.50 \text{ mL } HNO_3 \times \frac{1L}{1000 \text{ mL}} \times \frac{12.0 \text{ mol } HNO_3}{1L} \times \frac{1 \text{ mol } H^+}{1 \text{ mol } HNO_3} \times \frac{1 \text{ mol } Ag}{2 \text{ mol } H^+} \times \frac{107.9 \text{ g}}{1 \text{ mol } Ag}$$

$$= \underline{\textbf{27.5 g Ag}}$$

54. Use the flowchart for solution stoichiometry (Figure 4.6) as your guide using the stoichiometric ratio from the following balanced equation.

$$Sn^{2+}(aq) + 2Fe^{3+}(aq) \rightarrow 2Fe^{2+}(aq) + Sn^{4+}(aq)$$

Strategy: mass sample $\rightarrow$ mass Fe $\rightarrow$ mol Fe $\rightarrow$ mol $Sn^{2+} \rightarrow$ mol $SnCl_2 \rightarrow$ M $SnCl_2$

$$\text{mass of Fe} = 0.250 \text{ g sample} \times \frac{92.50 \text{ g Fe}}{100 \text{ g sample}} = 0.231 \text{ g Fe}$$

$$\text{mol } SnCl_2 = 0.231 \text{ g Fe} \times \frac{1 \text{ mol Fe}}{55.85 \text{ g Fe}} \times \frac{1 \text{ mol } Fe^{3+}}{1 \text{ mol Fe}} \times \frac{1 \text{ mol } Sn^{2+}}{2 \text{ mol } Fe^{3+}} \times \frac{1 \text{ mol } SnCl_2}{1 \text{ mol } Sn^{2+}}$$

$$= 2.07 \times 10^{-3} \text{ mol } SnCl_2$$

vol of solution = 22.0 mL = 0.022 L

$$M \text{ of SnCl}_2 \text{ solution} = \frac{2.07 \times 10^{-3} \text{ mol SnCl}_2}{0.022 \text{ L solution}} = \underline{\textbf{0.0941 M}}$$

56. Use the flowchart for solution stoichiometry (Figure 4.6) as your guide. Use the stoichiometric ratio from the given balanced equation. Determine the mass of C_2H_5OH using the strategy outlined below. Use this mass to calculate the % C_2H_5OH in blood.

$$16H^+(aq) + 2Cr_2O_7^{2-}(aq) + C_2H_5OH(aq) \rightarrow 4Cr^{3+}(aq) + 2CO_2(g) + 11H_2O$$

Strategy: V and M $K_2Cr_2O_7 \rightarrow$ mol $K_2Cr_2O_7 \rightarrow$ mol $Cr_2O_7^{2-} \rightarrow$ mol $C_2H_5OH \rightarrow$ mass C_2H_5OH

$$\text{mass } C_2H_5OH = 38.94 \text{ mL } K_2Cr_2O_7 \times \frac{1 \text{ L}}{1000 \text{ mL}} \times \frac{0.0723 \text{ mol } K_2Cr_2O_7}{1 \text{ L}} \times \frac{1 \text{ mol } Cr_2O_7^{2-}}{1 \text{ mol } K_2Cr_2O_7}$$

$$\times \frac{1 \text{ mol } C_2H_5OH}{2 \text{ mol } Cr_2O_7^{2-}} \times \frac{46.07 \text{ g } C_2H_5OH}{1 \text{ mol } C_2H_5OH} = 0.0649 \text{ g } C_2H_5OH$$

$$\text{mass \%} = \frac{0.0649 \text{ g } C_2H_5OH}{50.0 \text{ g blood}} \times 100 = \textbf{0.130 \%}$$

Yes, the person is legally drunk because the % C_2H_5OH is more than 0.10% by mass in the blood sample.

UNCLASSIFIED PROBLEMS

58. Use the flowchart for solution stoichiometry (Figure 4.6) as your guide. Use the stoichiometric ratio from the given balanced reaction.

$$MnO_4^-(aq) + 8H^+(aq) + 5Fe^{2+}(aq) \rightarrow Mn^{2+}(aq) + 5Fe^{3+}(aq) + 4H_2O$$

Strategy: V and M $KMnO_4 \rightarrow$ mol $KMnO_4 \rightarrow$ mol $MnO_4^- \rightarrow$ mol $Fe^{2+} \rightarrow$ mass Fe^{2+}

$$\text{mass } Fe^{2+} = 32.3 \text{ mL} \times \frac{1 \text{ L}}{1000 \text{ mL}} \times \frac{0.00210 \text{ mol } KMnO_4}{1 \text{ L}} \times \frac{1 \text{ mol } MnO_4^-}{1 \text{ mol } KMnO_4}$$

$$\times \frac{5 \text{ mol } Fe^{2+}}{1 \text{ mol } MnO_4^-} \times \frac{1 \text{ mol Fe}}{1 \text{ mol } Fe^{2+}} \times \frac{55.85 \text{ g Fe}}{1 \text{ mol Fe}} = 0.0189 \text{ g Fe}$$

$$\text{mass \%} = \frac{0.0189 \text{ g Fe}}{5.00 \text{ g hemoglobin}} \times 100 = \underline{\textbf{0.378 \%}}$$

60. (a) Based on the stoichiometry of the balanced equation below, the ratio of Cl^- to NO_3^- is 4:1, so the required ratio is **4 mol HCl : 1 mol HNO$_3$**

$$Au(s) + 4Cl^-(aq) + 4H^+(aq) + NO_3^-(aq) \rightarrow AuCl_4^-(aq) + NO(g) + 2H_2O$$

(b) Use the flowchart for solution stoichiometry (Figure 4.6) as your guide and the stoichiometric ratio from the balanced reaction above.

Strategy: mass Au $\rightarrow$ mol Au $\rightarrow$ mol Cl^- $\rightarrow$ mol HCl $\rightarrow$ vol HCl
mass Au $\rightarrow$ mol Au $\rightarrow$ mol NO_3^- $\rightarrow$ mol HNO$_3$ $\rightarrow$ vol HNO$_3$

$$\text{vol HCl} = 25.0 \text{ g Au} \times \frac{1 \text{ mol Au}}{197 \text{ g Au}} \times \frac{4 \text{ mol Cl}^-}{1 \text{ mol Au}} \times \frac{1 \text{ mol HCl}}{1 \text{ mol Cl}^-} \times \frac{1 \text{ L HCl}}{12 \text{ mol HCl}}$$

= 0.042 L HCl or **42 mL HCl**

$$\text{vol HNO}_3 = 25.0 \text{ g Au} \times \frac{1 \text{ mol Au}}{197 \text{ g Au}} \times \frac{1 \text{ mol NO}_3^-}{1 \text{ mol Au}} \times \frac{1 \text{ mol HNO}_3}{1 \text{ mol NO}_3^-} \times \frac{1 \text{ L HNO}_3}{16 \text{ mol HNO}_3}$$

= 0.0079 L HNO$_3$ or **7.9 mL HNO$_3$**

62. Balanced net ionic equation (given):
$$H_3PO_4(aq) + 3OH^-(aq) \rightarrow 3H_2O + PO_4^{3-}(aq)$$

Use the flowchart for solution stoichiometry (Figure 4.6) as your guide to calculate for the mass of NaOH that has reacted and subtract this calculated mass from the mass of NaOH originally available (20.0 g).

mass of NaOH that has reacted with H_3PO_4

Strategy: mL H_3PO_4 solution $\rightarrow$ mass H_3PO_4 solution $\rightarrow$ mass H_3PO_4 (pure)
mass H_3PO_4 $\rightarrow$ mol H_3PO_4 $\rightarrow$ mol OH^- $\rightarrow$ mol NaOH $\rightarrow$ mass NaOH

$$\text{mass NaOH} = 10 \text{ mL H}_3PO_4 \text{ solution} \times \frac{1.69 \text{ g H}_3PO_4 \text{ solution}}{1 \text{ mL H}_3PO_4 \text{ solution}} \times \frac{91.7 \text{ g H}_3PO_4}{100 \text{ g H}_3PO_4 \text{ solution}}$$

$$\times \frac{1 \text{ mol H}_3PO_4}{97.99 \text{ g H}_3PO_4} \times \frac{3 \text{ mol OH}^-}{1 \text{ mol H}_3PO_4} \times \frac{1 \text{ mol NaOH}}{1 \text{ mol OH}^-} \times \frac{40.02 \text{ g NaOH}}{1 \text{ mol NaOH}} = 19.0 \text{ g NaOH}$$

mass of NaOH left unreacted = 20.0 g NaOH − 19.0 g NaOH = **1.0 g NaOH unreacted**

CONCEPTUAL QUESTIONS

64. (a) Type: **SA/WB**
Hydrochloric acid (HCl) is a strong acid; the reacting species is H^+.
Ethylamine ($CH_3CH_2NH_2$) is a weak base; the reacting species is $CH_3CH_2NH_2$.
Net Ionic Equation: $H^+(aq) + CH_3CH_2NH_2 \ (aq) \rightarrow CH_3CH_2NH_3^+(aq)$

(b) Type: **WA/SB**
Hydrofluoric acid (HF) is a weak acid; the reacting species is HF.
Calcium hydroxide ($Ca(OH)_2$) is a strong base; the reacting species is OH^-.
Net Ionic Equation: $HF(aq) + OH^-(aq) \rightarrow F^-(aq) + H_2O$

(c) Type: **PPT**
Net Ionic Equation: $3Ca^{2+}(aq) + 2PO_4^{3-}(aq) \rightarrow Ca_3(PO_4)_2(s)$

(d) Type: **PPT**
Net Ionic Equation: $2Ag^+(aq) + SO_4^{2-}(aq) + Ba^{2+}(aq) + 2Cl^-(aq) \rightarrow BaSO_4(s) + 2AgCl(s)$

(e) **NR** $\quad Mg(NO_3)_2(aq) + NaCl(aq) \rightarrow Mg^{2+}(aq) + 2NO_3^-(aq) + Na^+(aq) + Cl^-(aq)$
All ions remain in solution and do not combine to form insoluble salts.

66. (a) This picture represents no precipitation, thus equation (**1**) matches.
Equation (1) involves soluble ions.
$$2Na^+(aq) + SO_4^{2-}(aq) \rightarrow \text{no reaction}$$

(b) This picture represents a precipitation reaction where cations and anions combine to form a solid compound. The ratio of cations to anions is 1:1. It matches **equation (3)** because this reaction forms an insoluble compound with a cation to anion ratio of 1:1.
$$Ba^{2+}(aq) + CO_3^{2-}(aq) \rightarrow BaCO_3(s)$$

(c) This picture represents a precipitation reaction where cations and anions combine to form a solid compound. The ratio of cations to anions is 1:2. It matches **equation (2)** because this equation forms an insoluble compound with a cation to anion ratio of 1:2.
$$Mg^{2+}(aq) + 2OH^-(aq) \rightarrow Mg(OH)_2(s)$$

68. (a) **weak electrolyte.** The figure shows that only some of the species dissociate, thus it is a weak electrolyte.

(b) **nonelectrolyte.** The figure shows that none of the species dissociate, thus it is a nonelectrolyte.

(c) **strong electrolyte.** The figure shows that all of the species dissociate, thus it is a strong electrolyte.

(d) **weak electrolyte.** The figure shows that most of the species did not dissociate, thus it is a weak electrolyte.

70. (a) **False**. The calculated molarity is the same because a change in the amount of water will not change the number of moles of HCl.

(b) **True**. Using $Ba(OH)_2$ instead of NaOH will result in a different volume of base added during the titration, but the number of moles of HCl will be the same.

(c) **False**. The number of moles of HCl is the same.

(d) **False**. The number of moles of HCl is the same.

(e) **True**.

CHALLENGE PROBLEMS

71. The calcium oxalate is dissolved in strong acid following the following reaction:

$$CaC_2O_4(s) + 2H^+(aq) \rightarrow H_2C_2O_4(aq) + Ca^{2+}(aq)$$

The $H_2C_2O_4$ formed is reacted with MnO_4^- following the balanced reaction below:

$$6H^+(aq) + 2MnO_4^-(aq) + 5H_2C_2O_4(aq) \rightarrow 10CO_2(g) + 2Mn^{2+}(aq) + 8H_2O$$

Following the flowchart for solution stoichiometry (Figure 4.6) and the strategy below, the amount of CaC_2O_4 present in the urine sample can be calculated. The stoichiometric ratios in the two balanced reactions above are used in the calculation.

Strategy: M and V $KMnO_4$ $\rightarrow$ mol $KMnO_4$ $\rightarrow$ mol MnO_4^- $\rightarrow$ mol $H_2C_2O_4$
mol $H_2C_2O_4$ $\rightarrow$ mol CaC_2O_4 $\rightarrow$ mass CaC_2O_4

$$\text{mass } CaC_2O_4 = 26.2 \text{ mL } KMnO_4 \times \frac{1 L}{1000 \text{ mL}} \times \frac{0.0946 \text{ mol } KMnO_4}{1 L} \times \frac{1 \text{ mol } MnO_4^-}{1 \text{ mol } KMnO_4}$$

$$\times \frac{5 \text{ mol } H_2C_2O_4}{2 \text{ mol } MnO_4^-} \times \frac{1 \text{ mol } CaC_2O_4}{1 \text{ mol } H_2C_2O_4} \times \frac{128.10 \text{ g } CaC_2O_4}{1 \text{ mol } CaC_2O_4} = \underline{0.794 \text{ g } CaC_2O_4}$$

The urine sample contains **0.794 g** CaC_2O_4.

Now, calculate how much Ca^{2+} is present in the urine sample.

$$0.794 \text{ g } CaC_2O_4 \times \frac{1 \text{ mol } CaC_2O_4}{128.10 \text{ g } CaC_2O_4} \times \frac{1 \text{ mole } Ca^{2+}}{1 \text{ mol } CaC_2O_4} \times \frac{40.08 \text{ g } Ca^{2+}}{1 \text{ mole } Ca^{2+}} = \underline{0.248 \text{ g } Ca^{2+}}$$

The urine sample contains 0.248 g Ca^{2+} or 248 mg Ca^{2+}. **Yes, this amount is within the normal range.**

72. Determine the amount of $Mg(OH)_2$ and $NaHCO_3$ present in the antacid tablet and calculate the volume of acid each could neutralize. Add the two acid volumes. (The NaCl in the tablet cannot neutralize any acid.)

Strategy:
mass sample $\rightarrow$ mass $Mg(OH)_2$ $\rightarrow$ mol $Mg(OH)_2$ $\rightarrow$ mol H^+ $\rightarrow$ mol HCl $\rightarrow$ vol HCl
mass sample $\rightarrow$ mass $NaHCO_3$ $\rightarrow$ mol $NaHCO_3$ $\rightarrow$ mol H^+ $\rightarrow$ mol HCl $\rightarrow$ vol HCl

The balanced net ionic equations are:
$$Mg(OH)_2(s) + 2H^+(aq) \rightarrow Mg^{2+}(aq) + 2H_2O$$
$$HCO_3^-(aq) + H^+(aq) \rightarrow CO_2(g) + H_2O$$

The tablet contains 41.0% $Mg(OH)_2$ and 36.2% $NaHCO_3$

$$41.0\% \, Mg(OH)_2 = \frac{41.0 \text{ g } Mg(OH)_2}{100 \text{ g tablet}} \qquad 36.2 \% \, NaHCO_3 = \frac{36.2 \text{ g } NaHCO_3}{100 \text{ g tablet}}$$

Volume of stomach acid (HCl) neutralized by the $Mg(OH)_2$ in the tablet:

$$330 \text{ mg tablet} \times \frac{1 \text{ g}}{1000 \text{ mg}} \times \frac{41.0 \text{ g } Mg(OH)_2}{100 \text{ g tablet}} \times \frac{1 \text{ mol } Mg(OH)_2}{58.32 \text{ g } Mg(OH)_2} \times \frac{2 \text{ mol } H^+}{1 \text{ mol } Mg(OH)_2}$$

$$\times \frac{1 \text{ mol HCl}}{1 \text{ mol } H^+} \times \frac{1 \text{ L HCl}}{0.020 \text{ mol HCl}} = 0.23 \text{ L HCl}$$

Volume of stomach acid (HCl) neutralized by the $NaHCO_3$ in the tablet:

$$330 \text{ mg tablet} \times \frac{1 \text{ g}}{1000 \text{ mg}} \times \frac{36.2 \text{ g } NaHCO_3}{100 \text{ g tablet}} \times \frac{1 \text{ mol } NaHCO_3}{84.00 \text{ g } NaHCO_3} \times \frac{1 \text{ mol } HCO_3^-}{1 \text{ mol } NaHCO_3}$$

$$\times \frac{1 \text{ mol } H^+}{1 \text{ mol } HCO_3^-} \times \frac{1 \text{ mol HCl}}{1 \text{ mol } H^+} \times \frac{1 \text{ L HCl}}{0.020 \text{ mol HCl}} = 0.071 \text{ L HCl}$$

Total volume of stomach acid (HCl) neutralized by the tablet:
volume = 0.23 L + 0.071 L = **0.30 L**

73. Strategy: mass Cu lost $\rightarrow$ mol Cu lost $\rightarrow$ mol Ag coat $\rightarrow$ mass Ag coat

Net Ionic Equation: $2Ag^+(aq) + Cu(s) \rightarrow 2Ag(s) + Cu^{2+}(aq)$

original mass of Cu = 2.00 g
mass of Cu remaining = 2.00 g $-$ Y   (where Y = mass of Cu lost)

$$\text{mass of Ag coat} = Y \text{ g Cu} \times \frac{1 \text{ mol Cu}}{63.55 \text{ g Cu}} \times \frac{2 \text{ mol Ag}}{1 \text{ mol Cu}} \times \frac{107.9 \text{ g Ag}}{1 \text{ mol Ag}} = 3.40Y$$

mass of coated strip = mass of Cu remaining + mass of Ag coat = 4.18 g substitution yields:

$$(2.00 \text{ g} - Y) + (3.40Y) = 4.18 \text{ g}$$
$$2.00 \text{ g} - Y + 3.40Y = 4.18 \text{ g}$$
$$-Y + 3.40Y = 4.18 \text{ g} - 2.00 \text{ g}$$
$$2.40Y = 2.18 \text{ g}$$
$$Y = (2.18/2.40) \text{ g} = 0.908 \text{ g}$$

Therefore: mass of Cu in strip = 2.00 − 0.908 = **1.09 g Cu**

mass of Ag coat in strip = 3.40Y = 3.40(0.908) = **3.09 g Ag**

74. Calculate moles of Fe^{2+} from the first permanganate titration; then calculate the total moles of Fe (both Fe^{2+} and Fe^{3+}) from the second titration. The second titration gives the total moles Fe because all Fe^{3+} ions present are converted to Fe^{2+} according to the following reaction:

$$Fe^{3+}(aq) + Zn(s) \rightarrow Fe^{2+}(aq) + Zn^{2+}(aq)$$

Finally, calculate moles of Fe^{3+} from the difference in moles of Fe^{2+} obtained from the first and second titrations.

Strategy (for each titration):

volume $KMnO_4 \rightarrow$ mol $KMnO_4 \rightarrow$ mol $MnO_4^- \rightarrow$ mol $Fe^{2+} \rightarrow$ M Fe^{2+}

Balanced net ionic equation: (step-by-step method for balancing is shown in problem #68 above)

$$MnO_4^-(aq) + 8H^+(aq) + 5Fe^{2+}(aq) \rightarrow Mn^{2+}(aq) + 4H_2O + 5Fe^{3+}(aq)$$

First titration (mol Fe^{2+} originally present):

$$35.0 \text{ mL KMnO}_4 \times \frac{1 \text{ L}}{1000 \text{ mL}} \times \frac{0.0280 \text{ mol KMnO}_4}{1 \text{ L}} \times \frac{1 \text{ mol MnO}_4^-}{1 \text{ mol KMnO}_4} \times \frac{5 \text{ mol Fe}^{2+}}{1 \text{ mol MnO}_4^-} = 0.00490 \text{ mol Fe}^{2+}$$

Second titration (total mol of Fe present) :

$$48.0 \text{ mL KMnO}_4 \times \frac{1 \text{ L}}{1000 \text{ mL}} \times \frac{0.0280 \text{ mol KMnO}_4}{1 \text{ L}} \times \frac{1 \text{ mol MnO}_4^-}{1 \text{ mol KMnO}_4} \times \frac{5 \text{ mol Fe}^{2+}}{1 \text{ mol MnO}_4^-} = 0.00672 \text{ mol Fe}^{2+}$$

mol Fe^{3+} = total mol Fe − mol Fe^{2+} = 0.00672 − 0.00490 = 0.00182 mol Fe^{3+}

$$\text{concentration of Fe}^{2+} = \frac{0.00490 \text{ mol Fe}^{2+}}{0.05000 \text{ L}} = \underline{\textbf{0.0980 M Fe}^{2+}}$$

$$\text{concentration of Fe}^{3+} = \frac{0.00182 \text{ mol Fe}^{3+}}{0.05000 \text{ L}} = \underline{\textbf{0.0364 M Fe}^{3+}}$$

75. Balanced Net Ionic Equations:

oxalic acid: $H_2C_2O_4(aq) + 2OH^-(aq) \rightarrow 2H_2O + C_2O_4^{2-}(aq)$

citric acid: $H_3C_6H_5O_7(aq) + 3OH^-(aq) \rightarrow 3H_2O + C_6H_5O_7^{3-}(aq)$

Strategy (for each acid):

mass acid $\rightarrow$ mol acid $\rightarrow$ mol OH^- $\rightarrow$ mol NaOH $\rightarrow$ V NaOH

volume of 0.615 M NaOH required to neutralize 0.0930 g oxalic acid:

$$0.930g\ H_2C_2O_4 \times \frac{1\,mol\ H_2C_2O_4}{90.04\ g\ H_2C_2O_4} \times \frac{2\,mol\ OH^-}{1\,mol\ H_2C_2O_4} \times \frac{1\,mol\ NaOH}{1\,mol\ OH^-} \times \frac{1L\ NaOH}{0.615\,mol\ NaOH} = 0.0336\ L\ NaOH$$

volume of 0.615 M NaOH required to neutralize 0.0930 g citric acid:

$$0.930g\ H_3C_6H_5O_7 \times \frac{1\,mol\ H_3C_6H_5O_7}{192.12\ g\ H_3C_6H_5O_7} \times \frac{3\,mol\ OH^-}{1\,mol\ H_3C_6H_5O_7} \times \frac{1\,mol\ NaOH}{1\,mol\ OH^-} \times \frac{1L\ NaOH}{0.615\,mol\ NaOH}$$

$$= 0.0236\ L\ NaOH$$

The unknown is oxalic acid because complete neutralization of the unknown acid required 0.0336 L NaOH or 33.6 mL of 0.615M NaOH, which is also the amount required for 0.930 g oxalic acid.

76. Balanced Net Ionic Equations:

iron (III) hydroxide and excess acid: $Fe(OH)_3(s) + 3H^+(aq) \rightarrow Fe^{3+}(aq) + 3H_2O$

titration of excess HCl: $H^+(aq) + OH^-(aq) \rightarrow H_2O$

total mol H^+ (added to $Fe(OH)_3$):

$$625\,mL\ HCl \times \frac{1L}{1000\ mL} \times \frac{0.280\,mol\ HCl}{1L} \times \frac{1\,mol\ H^+}{1\,mol\ HCl} = 0.175\,mol\ H^+$$

mol H^+ in excess (titrated with NaOH):

$$238.2\,mL\ NaOH \times \frac{1L}{1000\ mL} \times \frac{0.113\,mol\ NaOH}{1L} \times \frac{1\,mol\ OH^-}{1\,mol\ NaOH} \times \frac{1\,mol\ H^+}{1\,mol\ OH^-} = 0.0269\,mol\ H^+$$

Hence, mol of H^+ that reacted with $Fe(OH)_3$

= total mol H^+ − mol H^+ in excess

= 0.175 mol H^+ − 0.0269 mol H^+ = 0.148 mol H^+

mass of iron (III) hydroxide added originally

$$= 0.148\,mol\ H^+ \times \frac{1\,mole\ Fe(OH)_3}{3\,mol\ H^+} \times \frac{106.87g\ Fe(OH)_3}{1\,mol\ Fe(OH)_3} = \underline{\textbf{5.27 g}}\ \textbf{Fe(OH)}_3$$

77. mass of MgCl$_2$:

Among the ions in solution (Na$^+$, NO$_3^-$, Mg^{2+}, Cl$^-$, and Ba^{2+}) only Mg^{2+} forms a precipitate when reacted with KOH (see Figure 4.2) as shown by the reaction below.

$$Mg^{2+}(aq) + 2OH^-(aq) \rightarrow Mg(OH)_2(s)$$

The mass of the precipitate (Mg(OH)$_2$) from 100.0 g sample is given as 13.47 g which can be converted to the mass of MgCl$_2$.

$$13.47 \text{ g Mg(OH)}_2 \times \frac{1 \text{ mol Mg(OH)}_2}{58.316 \text{ g Mg(OH)}_2} \times \frac{1 \text{ mol Mg}^{2+}}{1 \text{ mol Mg(OH)}_2} \times \frac{1 \text{ mol MgCl}_2}{1 \text{ mol Mg}^{2+}} \times \frac{95.2 \text{ MgCl}_2}{1 \text{ mol MgCl}_2} = 21.99 \text{ g MgCl}_2$$

Thus, for the original 300.0 g sample, the **mass of MgCl$_2$ is 65.97 g** as shown:

$$300.0 \text{ g sample} \times \frac{21.99 \text{ g MgCl}_2}{100.0 \text{ g sample}} = 65.97 \text{ g MgCl}_2$$

mass of BaCl$_2$:

When the sample was treated with AgNO$_3$, the Cl$^-$ in the sample reacted and formed 195.8 g of AgCl as described by the following net ionic equation.

$$Ag^+(aq) + Cl^-(aq) \rightarrow AgCl (s)$$

The total Cl$^-$ in the 200.0 g sample can be calculated from the amount of AgCl formed:

$$\text{total Cl}^- = 195.8 \text{ g AgCl} \times \frac{35.45 \text{ g Cl}^-}{143.35 \text{ g AgCl}} = 48.42 \text{ g Cl}^-$$

This 48.42 g Cl$^-$ comes from two sources: MgCl$_2$ and BaCl$_2$.

$$\text{total Cl}^- = \text{Cl}^- \text{ from MgCl}_2 + \text{Cl}^- \text{ from BaCl}_2 = 48.42 \text{ g}$$

mass of Cl$^-$ from MgCl$_2$: $\quad 200 \text{ g sample} \times \dfrac{21.99 \text{ g MgCl}_2}{100 \text{ g sample}} \times \dfrac{70.90 \text{ g Cl}^-}{95.2 \text{ g MgCl}_2} = 32.75 \text{ g Cl}^-$

mass of Cl$^-$ from BaCl$_2$ can be obtained by difference:

$$\text{Cl}^- \text{ from BaCl}_2 = \text{Cl}^- \text{ total} - \text{Cl}^- \text{ from MgCl}_2$$

$$= 48.42 \text{ g} - 32.75 \text{ g} = 15.67 \text{ g Cl}^- \text{ from BaCl}_2$$

mass of BaCl$_2$ in the 200.0 g sample: $\quad 15.67 \text{ g Cl}^- \times \dfrac{208.20 \text{ g BaCl}_2}{70.90 \text{ g Cl}^-} = 46.02 \text{ g BaCl}_2$

mass of BaCl$_2$ in the 300.0 g sample: $\quad 300.0 \text{ g sample} \times \dfrac{46.02 \text{ g BaCl}_2}{200.0 \text{ g sample}} = \textbf{69.03 g BaCl}_2$

mass of NaNO$_3$ in the 300.0 g sample is obtained by difference.

$$\text{mass of sample} = \text{mass of NaNO}_3 + \text{mass of MgCl}_2 + \text{mass of BaCl}_2$$

$$\text{mass of NaNO}_3 = 300.0 \text{ g} - [65.97 \text{ g MgCl}_2 + 69.03 \text{ g BaCl}_2] = \textbf{165.0 g NaNO}_3$$

The 300.0 g sample contained **165.0 g NaNO$_3$, 65.97 g MgCl$_2$ and 69.03 g BaCl$_2$.**

78. This is a limiting reactant problem. A review of Example 4.3c will help you.

Net Ionic Equation: $Al^{3+}(aq) + 3OH^-(aq) \rightarrow Al(OH)_3(s)$

(a) To find the mass of $Al(OH)_3$ formed, determine first the limiting reactant by identifying the reactant that will produce fewer moles of precipitate.

mol precipitate if Al^{3+} is the limiting reactant:

Strategy (From Figure 4.6): M and V of parent compound, $Al(NO_3)_3 \rightarrow$ mol of parent compound, $Al(NO_3)_3 \rightarrow$ mol of reacting ion, $Al^{3+} \rightarrow$ mol of precipitate, $Al(OH)_3$

$$85.00 \text{ mL Al(NO}_3)_3 \times \frac{1L}{1000 \text{ mL}} \times \frac{0.250 \text{ mol Al(NO}_3)_3}{1L \text{ Al(NO}_3)_3} \times \frac{1 \text{ mol Al}^{3+}}{1 \text{ mol Al(NO}_3)_3} \times \frac{1 \text{ mol Al(OH)}_3}{1 \text{ mol Al}^{3+}}$$

$$= 0.0213 \text{ mol Al(OH)}_3$$

mol precipitate if OH^- is the limiting reactant:

Strategy (From Figure 4.6): M and V of parent compound, $Ba(OH)_2 \rightarrow$ mol of parent compound, $Ba(OH)_2 \rightarrow$ mol of reacting ion, $OH^- \rightarrow$ mol of precipitate, $Al(OH)_3$

$$85.0 \text{ mL Ba(OH)}_2 \times \frac{1L}{1000 \text{ mL}} \times \frac{0.250 \text{ mol Ba(OH)}_2}{1L \text{ Ba(OH)}_2} \times \frac{2 \text{ mol OH}^-}{1 \text{ mol Ba(OH)}_2} \times \frac{1 \text{ mol Al(OH)}_3}{3 \text{ mol OH}^-}$$

$$= 0.0142 \text{ mol Al(OH)}_3$$

Since 0.0142 mol is less than 0.0213 mol $Al(OH)_3$, OH^- is the limiting reactant. Hence, 0.0142 mol of precipitate forms. The mass of this precipitate is

$$0.0142 \text{ mol Al(OH)}_3 \times \frac{78.00 \text{ g Al(OH)}_3}{1 \text{ mol Al(OH)}_3} = \underline{\textbf{1.11 g Al(OH)}_3}$$

(b) The ions present in solution before the reaction are Ba^{2+}, OH^-, Al^{3+} and NO_3^-. After the reaction, all OH^- are used up but Ba^{2+}, Al^{3+} and NO_3^- will remain in the solution. Assuming the volume is additive, the final volume is 170.0 mL (see below).

Total volume = $Ba(OH)_2$ solution volume + $Al(OH)_3$ solution volume
= 85.0 mL + 85.00 mL
= 170.0 mL or 0.1700 L

$[OH^-]$ = $\underline{0}$ because OH^- is the limiting reactant, it is completely used up.

[Ba^{2+}] = <u>0.125 M</u> (solution below)

Ba^{2+} is a spectator ion, thus its original number of moles is not changed.

$$[Ba^{2+}] = \frac{\text{moles Ba}^{2+}}{\text{total volume (L)}}$$

$$= \frac{85.0 \text{ mL Ba(OH)}_2 \times \dfrac{1 \text{ L}}{1000 \text{ mL}} \times \dfrac{0.2500 \text{ mol Ba(OH)}_2}{1 \text{ L Ba(OH)}_2} \times \dfrac{1 \text{ mol Ba}^{2+}}{1 \text{ mol Ba(OH)}_2}}{0.1700 \text{ L}} = 0.125 \text{ M}$$

[NO$_3^-$] = <u>0.375 M</u> (solution below)

NO$_3^-$ is also a spectator ion, thus its original number of moles is not changed.

$$[NO_3^-] = \frac{\text{moles NO}_3^-}{\text{total volume (L)}}$$

$$= \frac{85.00 \text{ mL Al(NO}_3)_3 \times \dfrac{1 \text{ L}}{1000 \text{ mL}} \times \dfrac{0.250 \text{ mol Al(NO}_3)_3}{1 \text{ L Al(NO}_3)_3} \times \dfrac{3 \text{ mol NO}_3^-}{1 \text{ mol Al(NO}_3)_3}}{0.1700 \text{ L}} = 0.375 \text{ M}$$

[Al^{3+}] = <u>0.0415 M</u> (solution below)

Al^{3+} is the reactant present in excess. Some Al^{3+} are consumed in the reaction but some remain unreacted. The unreacted Al^{3+} is dissolved in the solution.

mol Al^{3+} originally available:

$$85.00 \text{ mL Al(NO}_3)_3 \times \frac{1 \text{ L}}{1000 \text{ mL}} \times \frac{0.250 \text{ mol Al(NO}_3)_3}{1 \text{ L Al(NO}_3)_3} \times \frac{1 \text{ mol Al}^{3+}}{1 \text{ mol Al(NO}_3)_3} = 0.02125 \text{ mol}$$

$$= 0.02125 \text{ mol Al}^{3+} \text{ originally available}$$

mol Al^{3+} reacted:

$$85.0 \text{ mL Ba(OH)}_2 \times \frac{1 \text{ L}}{1000 \text{ mL}} \times \frac{0.250 \text{ mol Ba(OH)}_2}{1 \text{ L Ba(OH)}_2} \times \frac{2 \text{ mol OH}^-}{1 \text{ mol Ba(OH)}_2} \times \frac{1 \text{ mol Al}^{3+}}{3 \text{ mol OH}^-}$$

$$= 0.0142 \text{ mol Al}^{3+} \text{ reacted or consumed}$$

excess Al^{3+} (unreacted) = 0.02125 mol Al^{3+} available − 0.0142 mol Al^{3+} consumed

= 0.00705 mol Al^{3+}

$$[Al^{3+}] = \frac{\text{moles Al}^{3+}}{\text{total volume (L)}} = \frac{0.00705 \text{ mol Al}^{3+}}{0.1700 \text{ L}} = 0.0415 \text{ M}$$

Ions in solution: [OH$^-$] = 0; [Ba^{2+}] = 0.125 M; [NO$_3^-$] = 0.375 M; [Al^{3+}] = 0.0415 M

GASES

5-1 MEASUREMENTS ON GASES

2. See Example 5.1.

$$6.00 \text{ ft} \times \frac{12 \text{ in}}{1 \text{ ft}} \times \frac{2.54 \text{ cm}}{1 \text{ in}} = 183 \text{ cm}$$

$$26 \text{ in} \times \frac{2.54 \text{ cm}}{1 \text{ in}} = 66 \text{ cm}$$

$V_{cylinder} = \pi r^2 h = \pi \, (66 \text{ cm})^2 \, (183 \text{ cm}) = 2.5 \times 10^6 \text{ cm}^3$

volume of cylinder (L) $= 2.5 \times 10^6 \text{ cm}^3 \times \dfrac{1 \text{ L}}{10^3 \text{ cm}^3} = \underline{\mathbf{2.5 \times 10^3 \text{ L}}}$

mol He $= 189 \text{ lb He} \times \dfrac{453.6 \text{ g}}{1 \text{ lb}} \times \dfrac{1 \text{ mol He}}{4.003 \text{ g}} = \underline{\mathbf{2.14 \times 10^4 \text{ mol}}}$

temperature (K) $= 25^\circ\text{C} + 273.15 \text{ K} = \underline{\mathbf{298 \text{ K}}}$

4. See Example 5.1.   Use the conversion factors given in the textbook:

1.013 bar = 1 atm = 760 mm Hg = 14.7 psi = 101.3 kPa

1 bar = 10^5 Pa

	mm Hg	atm	psi	kPa
(a)	**3.20×10^2**	**0.422**	**6.20**	42.7
(b)	**1.50×10^3**	**1.98**	29.1	**201**
(c)	**599**	0.788	**11.6**	**79.8**
(d)	1216	**1.600**	**23.5**	**162.1**

(a) $42.7 \text{ kPa} \times \dfrac{760 \text{ mm Hg}}{101.3 \text{ kPa}} = \underline{\mathbf{320 \text{ mm Hg}}}$

$42.7 \text{ kPa} \times \dfrac{1 \text{ atm}}{101.3 \text{ kPa}} = \underline{\mathbf{0.422 \text{ atm}}}$

$42.7 \text{ kPa} \times \dfrac{14.7 \text{ psi}}{101.3 \text{ kPa}} = \underline{\mathbf{6.20 \text{ psi}}}$

(b) $29.1\,\text{psi} \times \dfrac{760\,\text{mm Hg}}{14.7\,\text{psi}} = \underline{\mathbf{1.50 \times 10^3\ mm\ Hg}}$

$29.1\,\text{psi} \times \dfrac{1\,\text{atm}}{14.7\,\text{psi}} = \underline{\mathbf{1.98\ atm}}$

$29.1\,\text{psi} \times \dfrac{101.3\,\text{kPa}}{14.7\,\text{psi}} = \underline{\mathbf{201\ kPa}}$

(c) $0.788\,\text{atm} \times \dfrac{760\,\text{mm Hg}}{1\,\text{atm}} = \underline{\mathbf{599\ mm\ Hg}}$

$0.788\,\text{atm} \times \dfrac{14.7\,\text{psi}}{1\,\text{atm}} = \underline{\mathbf{11.6\ psi}}$

$0.788\,\text{atm} \times \dfrac{101.3\,\text{kPa}}{1\,\text{atm}} = \underline{\mathbf{79.8 kPa}}$

(d) $1216\,\text{mm Hg} \times \dfrac{1\,\text{atm}}{760\,\text{mm Hg}} = \underline{\mathbf{1.600\ atm}}$

$1216\,\text{mm Hg} \times \dfrac{14.7\,\text{psi}}{760\,\text{mm Hg}} = \underline{\mathbf{23.5\ psi}}$

$1216\,\text{mm Hg} \times \dfrac{101.3\,\text{kPa}}{760\,\text{mm Hg}} = \underline{\mathbf{162.1 kPa}}$

5-3 GAS LAW CALCULATIONS

6. See Section 1-2 and Example 5-2. This problem has two sets of conditions. Use the formula for initial state-final state.

$P_1 = 875\,\text{mm Hg} \times \dfrac{1\,\text{atm}}{760.0\,\text{mm Hg}} = 1.15\,\text{atm}$

$T_1 = 25^\circ\text{C} = 25 + 273.15 = 298\ \text{K}$

V_1 = volume of original tank

V_2 = volume of final tank = $2V_1$

$T_2 = ?$

$\dfrac{V_1 P_1}{n_1 T_1} = \dfrac{V_2 P_2}{n_2 T_2}$ $\quad\Rightarrow\quad$ at constant P and n the formula simplifies to $\quad\Rightarrow\quad$ $\dfrac{V_1}{T_1} = \dfrac{V_2}{T_2}$

solving for T_2: $\quad T_2 = \dfrac{V_2 T_1}{V_1} = \dfrac{2V_1(298K)}{V_1} = \underline{\textbf{596 K}}$

$$T_2 = 596 \text{ K} - 273.15 = \underline{\textbf{323°C}}$$

For the pressure to remain constant, the tank should be kept at 323°C

8. See Section 1-2 and Example 5.2. This problem has two sets of conditions. Use the formula for initial state–final state.

   P_1 = 1.22 atm; T, n = constant

   P_2 = ?

   (a) To find P_2 when volume is decreased by 38%, assume V_1 = 100 then V_2 = 100 − 38 = 62.

   $\dfrac{V_1 P_1}{n_1 T_1} = \dfrac{V_2 P_2}{n_2 T_2}$ $\quad\Rightarrow\quad$ at constant n and T the formula simplifies to $\quad\Rightarrow\quad V_1 P_1 = V_2 P_2$

   solving for P_2: $\qquad P_2 = \dfrac{V_1 P_1}{V_2} = \dfrac{100\,(1.22\text{ atm})}{62} = \underline{\textbf{1.97 atm}}$

   (b) To find P_2 when volume is decreased to 38% of its original volume, assume V_1 = 100 then V_2 = 100 × 0.38 = 38.

   $P_2 = \dfrac{V_1 P_1}{V_2} = \dfrac{100(1.22\text{ atm})}{38} = \underline{\textbf{3.21 atm}}$

10. See Section 1-2 and Example 5.2.
    First convert the temperatures to Kelvin and calculate the actual pressure in the tire. Then calculate the new pressure using the ratios of pressure to temperature.

    $T_1 = \dfrac{71°\text{F} - 32}{1.8} = 22°\text{C} = 22 + 273 = 295 \text{ K}$

    $T_2 = \dfrac{115°\text{F} - 32}{1.8} = 46°\text{C} = 46 + 273 = 319 \text{ K}$

    $P_{(actual)} = P_{(gauge)} + 14.7 = 28.0 + 14.7 = 42.7 \text{ psi}$

    $\dfrac{V_1 P_1}{n_1 T_1} = \dfrac{V_2 P_2}{n_2 T_2}$ $\quad\Rightarrow\quad$ at constant V and n the formula simplifies to $\quad\Rightarrow\quad \dfrac{P_1}{T_1} = \dfrac{P_2}{T_2}$

    $P_2 = \dfrac{P_1 T_2}{T_1} = \dfrac{(42.7\text{ psi})\,(319\,\text{K})}{(295\,\text{K})} = \underline{\textbf{46.2 psi}}$ (final actual pressure)

    $P_{(gauge)} = P_{(actual)} - 14.7 = 46.2 - 14.7 = \underline{\textbf{31.5 psi}}$ (final gauge pressure)

12. See Section 1-2 and Example 5.2. Since the problem has two sets of conditions, use the formula for initial state–final state conditions to find the final pressure of the gas in the tank.

T_1 = 25°C + 273.15 = 298 K; T_2= 12°C + 273.15 = 285 K

m_1 = 22.0 g CO_2; n_1 = 22.0 g $CO_2 \times \dfrac{1\,mol\,CO_2}{44.01\,g\,CO_2}$ = 0.500 mol CO_2

m_2 = 22.0 g + 10.0 g CO_2; n_2 = 32.0 g $CO_2 \times \dfrac{1\,mol\,CO_2}{44.01\,g\,CO_2}$ = 0.727 mol CO_2

P_1 = 732 mm Hg

P_2 = ?

$\dfrac{V_1 P_1}{n_1 T_1} = \dfrac{V_2 P_2}{n_2 T_2}$ $\Rightarrow$ at constant V the formula simplifies to $\Rightarrow$ $\dfrac{P_1}{n_1 T_1} = \dfrac{P_2}{n_2 T_2}$

solving for P_2, $P_2 = \dfrac{P_1 n_2 T_2}{n_1 T_1} = \dfrac{732\ mm\,Hg\,(0.727\ mol)(285\ K)}{(0.500\ mol)(298\ K)}$

P_2 = **1.02 × 10³ mm Hg**

14. See Example 5.2.

P_1 = 1.05 atm; P_2 = 1.64 atm

n_1 = 1.35 mol H_2; n_2 = mol N_2 + 1.35 mol H_2

mol N_2 = ?

$\dfrac{V_1 P_1}{n_1 T_1} = \dfrac{V_2 P_2}{n_2 T_2}$ $\Rightarrow$ at constant V and T the formula simplifies to $\Rightarrow$ $\dfrac{P_1}{n_1} = \dfrac{P_2}{n_2}$

solving for n_2 $\Rightarrow$ $n_2 = \dfrac{P_2 n_1}{P_1} = \dfrac{1.64\ atm\,(1.35\ mol)}{1.05\ atm}$ = 2.11 mol

n_2 = 2.11 mol = total mol of gases in the final state which is composed of N_2 and H_2

mol total = mol N_2 + mol H_2

2.11 mol = mol N_2 + 1.35 mol H_2

mol N_2 = **0.76 mol**

16. See Section 1-2 and Example 5.3.

   $V = 162$ L

   $P = 1.00$ atm

   $T = 108°C + 273.15 = 381$ K

   To find the amount of gas, use the ideal gas law to calculate the number of moles (n) of steam.

   $$PV = nRT \qquad \Rightarrow n_{steam} = \frac{PV}{RT} = \frac{(1.00\ \text{atm})(162\ \text{L})}{[0.0821\ \text{L}\cdot\text{atm}/(\text{mol}\cdot\text{K})](381\ \text{K})} = 5.18\ \text{mol steam}$$

   $$g\ H_2O = 5.18\ \text{mol} \times \frac{18.02\ g}{\text{mol}} = \textbf{93.3 g } H_2O$$

   $$ml\ H_2O = 93.3\ g\ H_2O \times \frac{1\text{mL}}{1.00\ g} = \textbf{93.3 mL } H_2O$$

18. See Section 1-2 and Example 5.3.

   $V = 4$ L

   $T = 37°C = 37 + 273 = 310$ K

   $P = 2.68$ atm

   $$PV = nRT \Rightarrow \qquad n_{propane} = \frac{PV}{RT} = \frac{(2.68\ \text{atm})(4\ \text{L})}{[0.0821\ \text{L}\cdot\text{atm}/(\text{mol}\cdot\text{K})](310\ \text{K})}$$

   $$= 0.421\ \text{mol propane } (C_3H_8)$$

   $$0.421\ \text{mol } C_3H_8 \times \frac{44.094\ g\ C_3H_8}{1\text{mol } C_3H_8} = 18.6\ g\ C_3H_8$$

   mass of empty tank = mass of tank after it is filled with butane − mass butane

   mass of empty tank = 1236 g − 18.6 g = **1217 g**

20. See Section 1-2 and Example 5.3.
   Start by converting all units to those used in the ideal gas law. Calculated moles using the ideal gas equation if P, V and T are given, otherwise use the molar mass and the given mass. Then apply the ideal gas law to fill in any other missing data.

	Pressure	Volume	Temperature	Moles	Grams
(a)	18.9 psi	0.886 L	22°C	**0.0470**	**2.07**
(b)	633 mm Hg	1.993 L	**−33°C**	0.0844	**3.72**
(c)	1.876 atm	**36.9 L**	75°F	2.842	**125.3**
(d)	**11.2 atm**	2244 mL	13°C	**1.072**	47.25

(a) $T = 22°C + 273.15 = 295$ K

$$P = 18.9 \, \text{psi} \times \frac{1 \, \text{atm}}{14.7 \, \text{psi}} = 1.29 \, \text{atm}$$

$$n = \frac{PV}{RT} = \frac{(1.29 \, \text{atm})(0.886 \, \text{L})}{(0.0821 \, \text{L} \cdot \text{atm/mol} \cdot \text{K})(295 \, \text{K})} = \underline{\textbf{0.0470 mol} \, C_3H_8}$$

$$0.0470 \, \text{mol} \, C_3H_8 \times \frac{44.09 \, \text{g}}{1 \, \text{mol}} = \underline{\textbf{2.07 g} \, C_3H_8}$$

(b) $P = 633 \, \text{mm Hg} \times \frac{1 \, \text{atm}}{760 \, \text{mm Hg}} = 0.833 \, \text{atm}$

$$T = \frac{PV}{nR} = \frac{(0.833 \, \text{atm})(1.993 \, \text{L})}{(0.0844 \, \text{mol})(0.0821 \, \text{L} \cdot \text{atm/mol} \cdot \text{K})} = 240 \, \text{K}$$

$t_{°C} + 273.15 = 240$ K

$t_{°C} = \underline{\textbf{–33 °C}}$

$$0.0844 \, \text{mol} \, C_3H_8 \times \frac{44.09 \, \text{g}}{1 \, \text{mol}} = \underline{\textbf{3.72 g} \, C_3H_8}$$

(c) $t_{°F} = 1.8 t_{°C} + 32$

$75°F = 1.8 t_{°C} + 32$

$t_{°C} = 24 °C$

$T = 24°C + 273.15 = 297$ K

$$V = \frac{nRT}{P} = \frac{(2.842 \, \text{mol})(0.0821 \, \text{L} \cdot \text{atm/mol} \cdot \text{K})(297 \, \text{K})}{(1.876 \, \text{atm})} = \underline{\textbf{36.9 L}}$$

$$2.842 \, \text{mol} \, C_3H_8 \times \frac{44.09 \, \text{g}}{1 \, \text{mol}} = \underline{\textbf{125.3 g} \, C_3H_8}$$

(d) $T = 13°C + 273.15 = 286$ K

$$V = 2244 \, \text{mL} \times \frac{1 \, \text{L}}{1000 \, \text{mL}} = 2.244 \, \text{L}$$

$$47.25 \, \text{g} \, C_3H_8 \times \frac{1 \, \text{mol}}{44.09 \, \text{g}} = \underline{\textbf{1.072 mol} \, C_3H_8}$$

$$P = \frac{nRT}{V} = \frac{(1.072 \, \text{mol})(0.0821 \, \text{L} \cdot \text{atm/mol} \cdot \text{K})(286 \, \text{K})}{(2.244 \, \text{L})} = \underline{\textbf{11.2 atm}}$$

22. See Section 1-2 and Example 5.4. Use Equation 5.2 from the textbook to solve this problem.

    $T = 97°C + 273 = 370$ K

    $P = 755\,\text{mm Hg} \times \dfrac{1\,\text{atm}}{760.0\,\text{mm Hg}} = 0.993$ atm

    $n = \dfrac{PV}{RT} = \dfrac{(0.993\,\text{atm})(1.00\,\text{L})}{(0.0821\,\text{L} \cdot \text{atm/mol} \cdot \text{K})(370\,\text{K})} = 0.0327$ mol gas

    $\text{density} = MM\left(\dfrac{P}{RT}\right)$

    (a) $\text{density}_{HCl} = 36.46\,\text{g/mol}\left(\dfrac{0.993\,\text{atm}}{(0.0821\,\text{L} \cdot \text{atm/mol} \cdot \text{K})(370\,\text{K})}\right) = \underline{\textbf{1.19 g/L}}$

    (b) $\text{density}_{SO_2} = 64.07\,\text{g/mol}\left(\dfrac{0.993\,\text{atm}}{(0.0821\,\text{L} \cdot \text{atm/mol} \cdot \text{K})(370\,\text{K})}\right) = \underline{\textbf{2.09 g/L}}$

    (c) $\text{density}_{C_4H_{10}} = 58.12\,\text{g/mol}\left(\dfrac{0.993\,\text{atm}}{(0.0821\,\text{L} \cdot \text{atm/mol} \cdot \text{K})(370\,\text{K})}\right) = \underline{\textbf{1.90 g/L}}$

24. See Section 1-2 and Example 5.4. The density of CO_2 can be found using Equation 5.2 from the textbook.

    $\text{density} = MM\left(\dfrac{P}{RT}\right)$

    On the surface of Mars:

    $T = -55°C + 273.15 = 218$ K

    $P = 0.00592$ atm

    $\text{density CO}_2 = 44.01\,\text{g/mol}\left(\dfrac{0.00592\,\text{atm}}{(0.0821\,\text{L} \cdot \text{atm/mol} \cdot \text{K})(218\,\text{K})}\right)$

    $= \underline{\textbf{0.0146 g/L on Mars' surface}}$

    On the surface of Earth:

    $T = 25°C + 273.15 = 298$ K

    $P = 1$ atm

    $\text{density CO}_2 = 44.01\,\text{g/mol}\left(\dfrac{1\,\text{atm}}{(0.0821\,\text{L} \cdot \text{atm/mol} \cdot \text{K})(298\,\text{K})}\right)$

    $= \underline{\textbf{1.80 g/L on Earth's surface}}$

    The density of CO_2 on the surface of Earth (1.80 g/L) is **123 times larger** than its density on Mars (0.0146 g/L).

26. See Section 3-2 and Examples 3.5 and 3.6. Manipulate Equation 5.2 (from the textbook) to derive the formula for solving the *MM* of phosgene.

$$density = MM\left(\frac{P}{RT}\right) \quad \Rightarrow \text{ solve for } MM \Rightarrow \quad MM = \frac{(density)\,RT}{P}$$

(a) $T = 25°C + 273.15 = 298$ K

   $P = 1.05$ atm

   density = 4.24 g/L

   $$MM = \frac{(density)\,RT}{P} = \frac{(4.24 \text{ g/L})(0.082\,1\text{L}\cdot\text{atm/mol}\cdot\text{K})(298 \text{ K})}{1.05 \text{ atm}} = 98.8 \text{ g/mol}$$

   $MM = \underline{\textbf{98.8 g/mol}}$

(b) $12.01 \text{ g C} \times \dfrac{1 \text{ mol C}}{12.01 \text{ g}} = 1.000 \text{ mol C}$

   $16.2 \text{ g O} \times \dfrac{1 \text{ mol O}}{16.00 \text{ g}} = 1.012 \text{ mol O}$

   $71.7 \text{ g Cl} \times \dfrac{1 \text{ mol Cl}}{35.45 \text{ g}} = 2.022 \text{ mol Cl}$

   C: $\dfrac{1.000 \text{ mol}}{1.000 \text{ mol}} = 1$ O: $\dfrac{1.012 \text{ mol}}{1.000 \text{ mol}} = 1$ Cl: $\dfrac{2.022 \text{ mol}}{1.000 \text{ mol}} = 2$

   Thus, the empirical or simplest formula is $COCl_2$

   empirical formula mass = 12.01 + 16.00 + 2(35.45) = 98.9 g/mol

   molar mass from (a) = 98.8 g/mol

   empirical formula = $\underline{\textbf{molecular formula} = \textbf{COCl}_2}$

Since the molar mass is equal to the empirical formula mass, the molecular formula and the empirical formula is the same, $\textbf{COCl}_2$.

28. See also Section1-2.

(a) $MM_{mixture}$ = N$_2$ *MM* contribution + O$_2$ *MM* contribution + CO$_2$ *MM* contribution + H$_2$O *MM* contribution

   = (28.02 g/mol)(0.745) + (32.00 g/mol)(0.157) + (44.01 g/mol)(0.036) + (18.016 g/mol)(0.062)

   = $\underline{\textbf{28.6 g/mol}}$

(b) Use Equation 5.2 for density to calculate the density of exhaled and ordinary air.

$$P = 757 \text{ mm Hg} \times \frac{1 \text{ atm}}{760.0 \text{ mm Hg}} = 0.996 \text{ atm}$$

$T_1 = 37°C = 37 + 273.15 = 310 \text{ K}.$

$$density_{exhaled \ air} = MM_{exhaled \ air}\left(\frac{P}{RT}\right) = 28.6 \text{ g/mol} \left(\frac{0.996 \text{ atm}}{(0.0821 \text{ L} \cdot \text{atm/mol} \cdot \text{K})(310 \text{ K})}\right)$$

$$= \underline{\textbf{1.12 g/L}}$$

$$density_{ordinary \ air} = MM_{ordinary \ air}\left(\frac{P}{RT}\right) = 29.0 \text{ g/mol} \left(\frac{0.996 \text{ atm}}{(0.0821 \text{ L} \cdot \text{atm/mol} \cdot \text{K})(310 \text{ K})}\right)$$

$$= \underline{\textbf{1.13 g/L}}$$

The density of exhaled air (1.12 g/L) is slightly lower than ordinary air (1.13 g/L).

30. Use the formula for *MM* derived in problem 26 to find the *MM* of PX_3.

mass PX_3 = 0.750 g

P = 1.00 atm

T = 26°C + 273.15 = 299 K

$$V = 542 \text{ mL} \times \frac{1 \text{ L}}{1000 \text{ mL}} = 0.542 \text{ L}$$

$$MM = \frac{(density) \ RT}{P} = \frac{\left(\dfrac{0.750 \text{ g}}{0.542 \text{ L}}\right)(0.0821 \text{ L} \cdot \text{atm/mol} \cdot \text{K})(299 \text{ K})}{1.00 \text{ atm}} = 34.0 \text{ g/mol}$$

$MM \ PX_3$ = 34.0 g/mol = (1 mol)(30.97 g/mol) + (3 mol)(MM_X)

MM_X = (34.0 g/mol − 30.97 g/mol)

$$MM_X = \frac{34.0 \text{ g/mol} - 30.97 \text{ g/mol}}{3} = 1.01 \text{ g/mol}$$

The only element that has a *MM* of 1.01 g/mol is H. Therefore, **the element is hydrogen**. Thus, the compound is **phosphine (PH_3)**.

5-4 STOICHIOMETRY OF GASEOUS REACTIONS

32. See Section 3-3 and Examples 32.10 and 5.6. Use Figure 5.5 as your guide.

    (a) **2NF₃(g) + 3H₂O(g) → 6HF(g) + NO(g) + NO₂(g)**

    (b) volume of reactants: 5.22 L NF_3 and 5.22 L H_2O
        moles of reactants, NF_3 and H_2O:

    $$\text{mol } NF_3 = n_{NF_3} = \frac{PV}{RT} = 5.22 \text{ L} \left(\frac{P}{RT} \right)$$

    $$\text{mol } H_2O = n_{H_2O} = \frac{PV}{RT} = 5.22 \text{ L} \left(\frac{P}{RT} \right)$$

    Find the limiting reactant. The volume of NO produced by the limiting reactant is the amount of NO theoretically formed.

    moles NO that can be formed from NF_3:

    $$\text{mol NO} = \text{mol } NF_3 \left(\frac{1 \text{ mol NO}}{2 \text{ mol } NF_3} \right) = 5.22 \text{ L} \left(\frac{P}{RT} \right) \left(\frac{1 \text{ mol NO}}{2 \text{ mol } NF_3} \right) = 2.61 \text{ L} \left(\frac{P}{RT} \right)$$

    moles NO that can be formed from steam, H_2O:

    $$\text{mol NO} = \text{mol } H_2O \left(\frac{1 \text{ mol NO}}{3 \text{ mol } H_2O} \right) = 5.22 \text{ L} \left(\frac{P}{RT} \right) \left(\frac{1 \text{ mol NO}}{3 \text{ mol } H_2O} \right) = 1.74 \text{ L} \left(\frac{P}{RT} \right)$$

    volume of NO from the limiting reactant is the volume actually formed:

    $$n_{NO} = 1.74 \text{ L} \left(\frac{P}{RT} \right) ; \quad n = V \left(\frac{P}{RT} \right) \Rightarrow \underline{\textbf{V = 1.74 L}}$$

    H_2O produced less amount of NO therefore, H_2O is the limiting reactant. The amount produced by the limiting reactant is the amount theoretically produced. Since *P*, *V*, and *T* are constant throughout the reaction, the volume of NO formed is **1.74 L**.

34. See Section 3-3 and Example 3.9.

    $$2TiCl_4(g) + H_2(g) \rightarrow 2TiCl_3(s) + 2HCl(g)$$

    (a) volume of reactants: 3.72 L $TiCl_4$ and 4.50 L H_2

    moles of reactants, $TiCl_4$ and H_2:

    $$\text{mol } TiCl_4: \quad n_{TiCl_4} = \frac{PV}{RT} = \left(\frac{P}{RT} \right) 3.72 \text{ L}$$

    $$\text{mol } H_2: \quad n_{H_2} = \frac{PV}{RT} = \left(\frac{P}{RT} \right) 4.50 \text{ L}$$

Find the limiting reactant. The volume of HCl produced by the limiting reactant is the amount of HCl theoretically formed.

moles HCl that can be formed from $TiCl_4$:

$$mol\ HCl = mol\ TiCl_4 \times \left(\frac{2\ mol\ HCl}{2\ mol\ TiCl_4}\right)$$

$$= \left(\frac{P}{RT}\right) 3.72\ L \times \left(\frac{2\ mol\ HCl}{2\ mol\ TiCl_4}\right)$$

$$= \left(\frac{P}{RT}\right) 3.72\ L\ HCl\ from\ TiCl_4$$

moles HCl that can be formed from H_2:

$$mol\ HCl = mol\ H_2 \times \left(\frac{2\ mol\ HCl}{1\ mol\ H_2}\right)$$

$$= \left(\frac{P}{RT}\right) 4.50\ L \times \left(\frac{2\ mol\ HCl}{1\ mol\ H_2}\right)$$

$$= \left(\frac{P}{RT}\right) 9.00\ L\ HCl\ from\ H_2$$

$TiCl_4$ produced less amount of HCl therefore, $TiCl_4$ is the limiting reactant. The amount produced by the limiting reactant is the amount theoretically produced. Since *P*, *V*, and *T* are constant throughout the reaction, the volume of HCl formed is **3.72 L**.

volume of HCl from the limiting reactant ($TiCl_4$) is the volume of HCl actually formed:

$$n_{HCl} = \left(\frac{P}{RT}\right) 3.72\ L\ ; \quad n = \left(\frac{P}{RT}\right) V \Rightarrow \underline{\textbf{V = 3.72 L}}$$

(b) H_2 is the excess reactant

moles of H_2 available (before the reaction):

$$n_{H_2\ available} = \frac{PV}{RT} = \left(\frac{P}{RT}\right) 4.50\ L$$

moles of H_2 that has reacted with the limiting reactant, $TiCl_4$:

$$n_{H_2\ reacted} = mol\ TiCl_4 \times \left(\frac{1\ mol\ H_2}{2\ mol\ TiCl_4}\right) = \left(\frac{P}{RT}\right) 3.72\ L \times \left(\frac{1\ mol\ H_2}{2\ mol\ TiCl_4}\right)$$

$$= \left(\frac{P}{RT}\right) 1.86\ L$$

$$n_{H_2 \text{ unreacted}} = n_{H_2 \text{ available}} - n_{H_2 \text{ reacted}}$$

$$= \left(\frac{P}{RT}\right) 4.50 \text{ L} - \left(\frac{P}{RT}\right) 1.86 \text{ L}$$

$$= \left(\frac{P}{RT}\right) 2.64 \text{ L}$$

$$n_{H_2 \text{ unreacted}} = \left(\frac{P}{RT}\right) 2.64 \text{ L} \; ; \quad n = \left(\frac{P}{RT}\right) V \Rightarrow \underline{\textbf{\textit{V} = 2.64 L}}$$

36. See Example 5.5

$$2K_2O_2(s) \; + \; 2CO_2(g) \; \rightarrow \; 2K_2CO_3 \; (s) \; + \; O_2 \; (g)$$

$T = 25\,°C + 273.15 = 298 \text{ K}$

$$P = 728 \text{ mm Hg} \times \frac{1 \text{ atm}}{760.0 \text{ mm Hg}} = 0.958 \text{ atm}$$

volume of CO_2 exhaled per person in a five-day trip:

$$V = 5 \text{ days} \times \frac{24 \text{ h}}{1 \text{ day}} \times \frac{60 \text{ min}}{1 \text{ h}} \times \frac{3.0 \text{ L}}{1 \text{ min}} \times \frac{3.4 \text{ L}}{100 \text{ L}} = 7.34 \times 10^2 \text{ L}$$

moles of CO_2 exhaled per person in a five-day trip:

$$n_{CO_2} = \frac{PV}{RT} = \frac{(0.958 \text{ atm})(7.34 \times 10^2 \text{ L})}{(0.0821 \text{ L} \cdot \text{atm/mol} \cdot \text{K})(298 \text{ K})} = 28.7 \text{ mol } CO_2$$

mass of K_2O_2 needed by one person for a five-day trip:

$$g \; K_2O_2 = 28.7 \text{ mol } CO_2 \times \frac{2 \text{ mol } K_2O_2}{2 \text{ mol } CO_2} \times \frac{110.2 \text{ g } K_2O_2}{1 \text{ mol } K_2O_2} = \underline{\textbf{3.2} \times \textbf{10}^3 \textbf{ g } K_2O_2}$$

38. See Section 4-1. Follow the following steps to solve this problem: (i) find mol H_2O_2 (ii) convert mol H_2O_2 to mol O_2 using the coefficient of the balanced equation as conversion factor; and (iii) find the volume of O_2 using ideal gas law.

$$\text{mol } H_2O_2 = 25.00 \text{ mL } H_2O_2 \text{ soln} \times \frac{1.05 \text{ g } H_2O_2 \text{ soln}}{1 \text{ mL } H_2O_2 \text{ soln}} \times \frac{30.00 \text{ g } H_2O_2}{100 \text{ g } H_2O_2 \text{ soln}} \times \frac{1 \text{ mol } H_2O_2}{34.02 \text{ g } H_2O_2}$$

$$= 0.2315 \text{ mol } H_2O_2$$

$$2H_2O_2 \ (aq) \rightarrow 2H_2O(l) \ + \ O_2(g)$$

$$n_{O_2} = 0.2315 \ \text{mol} \ H_2O_2 \times \frac{1 \ \text{mol} \ O_2}{2 \ \text{mol} \ H_2O_2} = 0.1158 \ \text{mol} \ O_2$$

$T = 25°C + 273.15 = 298 \ K$

$P = 1 \ atm$

$V_{O_2} = ?$

$$PV = nRT \Rightarrow V_{O_2} = \frac{nRT}{P} = \frac{(0.1158 \ \text{mol})(0.0821 \ L \cdot atm/mol \cdot K)(298 \ K)}{1.00 \ atm} = \underline{\textbf{2.83 L}}$$

40. See Section 3-3, Figure 5.5, and Examples 3.3 and 5.5.

(a) $2C_9H_{18}O_6(s) \ + \ 21O_2(g) \ \rightarrow \ 18H_2O(g) \ + \ 18CO_2(g)$

(b) $V = 2.00 \ L$

mass of $C_9H_{18}O_6 = 5.00 \ g$

$T = 555°C + 273.15 = 828 \ K$

$P = ?$

From the balanced equation, the total number of moles of gases produced from 2 mol $C_9H_{18}O_6$ is 36 mol gas (18 mol H_2O +18 mol CO_2 = 36 mol gas).

Thus, the stoichiometric ratio to be used is, 2 mol $C_9H_{18}O_6$:36 mol gas.

mol gas produced from 5.00 g $C_9H_{18}O_6$:

$$\text{mol gas} = 5.00 \ g \ C_9H_{18}O_6 \times \frac{1 \ \text{mol} \ C_9H_{18}O_6}{222.23 \ g \ C_9H_{18}O_6} \times \frac{36 \ \text{mol gases}}{2 \ \text{mol} \ C_9H_{18}O_6} = 0.405 \ \text{mol gas}$$

$$PV = nRT \Rightarrow P = \frac{nRT}{V} = \frac{(0.405 \ \text{mol})(0.0821 L \cdot atm/K \cdot mol)(828 \ K)}{2.00 \ L} = \underline{\textbf{13.8 atm}}$$

5-5 GAS MIXTURES

42. See Section 5-3.

$$9.00 \ g \ HCl \left(\frac{1 \text{mol}}{36.46 \ g} \right) = 0.247 \ \text{mol} \ HCl$$

$$2.00 \ g \ H_2 \left(\frac{1 \text{mol}}{2.016 \ g} \right) = 0.992 \ \text{mol} \ H_2$$

$$165.0 \ g \ Ne \left(\frac{1 \text{mol}}{20.18 \ g} \right) = 8.176 \ \text{mol} \ Ne$$

n_{total} = total moles of gas = 0.247 mol + 0.992 mol + 8.176 mol = 9.42 mol

V = 75.0 L

T = 22°C + 273.15 = 295 K

P = ?

$$P = \frac{nRT}{V} = \frac{(9.42\,mol)(0.0821\,L \cdot atm/K \cdot mol)(295\,K)}{75.0\,L} = \underline{\mathbf{3.04\ atm}}$$

The gas that has the fewest moles present has the smallest partial pressure. In this problem the **smallest partial pressure is due to HCl**.

44. See Example 5.9.

P_{tot} = 1.20 atm

$P_A = P_{tot}\,X_A$ where $X_A = n_A/n_{total}$

P_{CH_4} = (1.20 atm)(0.886) = **1.06 atm**

$P_{C_2H_6}$ = (1.20 atm)(0.089) = **0.11 atm**

$P_{C_3H_8}$ = (1.20 atm)(0.025) = **0.030 atm**

46. See Example 5.8. To solve this problem follow the steps: (i) Calculate the moles H_2 formed and (ii) using moles H_2 calculated in (i), find the volume (V) of dry hydrogen

i) $P_{total} = P_{H_2} + P_{H_2O}$

$P_{H_2} = P_{total} - P_{H_2O}$ = 769 mm Hg – 23.8 mm Hg = 745 mm Hg

P_{H_2} = 745 mm Hg $\times \dfrac{1\,atm}{760\,mm\,Hg}$ = 0.981 atm

V_{H_2} = 125 mL $\times \dfrac{1\,L}{1000\,mL}$ = 0.125 L

T = 25°C + 273.15 = 298 K

mol dry H_2 formed: $n_{H_2} = \dfrac{PV}{RT} = \dfrac{(0.981\,atm)(0.125\,L)}{(0.0821\,L \cdot atm/mol \cdot K)(298\,K)}$ = 5.01 × 10⁻³ mol H_2

ii) P = 722 mm Hg $\times \dfrac{1\,atm}{760\,mmHg}$ = 0.950 atm

T = 37°C + 273.15 = 310 K

n_{H_2} = 5.01 × 10⁻³ mol H_2

V_{H_2} = ?

$$V_{H_2} = \frac{nRT}{P} = \frac{(5.01 \times 10^{-3} \text{ mol } H_2)(0.0821 \text{ L} \cdot \text{atm/mol} \cdot \text{K})(310 \text{ K})}{(0.950 \text{ atm})}$$

$$= \underline{0.134 \text{ L} \quad \text{or} \quad 134 \text{ mL}}$$

48. Calculate the moles of each gas using ideal gas law. Since T and R are constant, no values will be used to substitute T and R in the following solution.

$$n_{CO_2} = \frac{PV}{RT} = \frac{(1.25 \text{ atm})(3.00 \text{ L})}{RT} = 3.75 \text{ atm} \cdot \text{L}/RT$$

$$n_{CH_4} = \frac{PV}{RT} = \frac{(2.66 \text{ atm})(6.00 \text{ L})}{RT} = 15.96 \text{ atm} \cdot \text{L}/RT$$

$n_{total} = n_{CO_2} + n_{CH_4} = 3.75 \text{ atm} \cdot \text{L}/RT + 15.96 \text{ atm} \cdot \text{L}/RT = 19.71 \text{ atm} \cdot \text{L}/RT$

$V_{total} = 3.00 \text{ L} + 6.00 \text{ L} = 9.00 \text{ L}$

pressure after opening the valve:

$$P = \frac{nRT}{V} = \frac{(19.71 \text{ atm} \cdot \text{L}/RT)RT}{9.00 \text{ L}} = \underline{2.19 \text{ atm}}$$

50. See Example 5.9

(a) $2NH_4NO_3(s) \rightarrow 2N_2(g) + O_2(g) + 4H_2O(g)$

(b) mol $NH_4NO_3 = 25.0 \text{ g } NH_4NO_3 \times \dfrac{1 \text{ mol } NH_4NO_3}{80.05 \text{ g}} = 0.312 \text{ mol } NH_4NO_3$

mol $N_2 = 0.312 \text{ mol } NH_4NO_3 \times \dfrac{2 \text{ mol } N_2}{2 \text{ mol } NH_4NO_3} = 0.312 \text{ mol } N_2$

mol $O_2 = 0.312 \text{ mol } NH_4NO_3 \times \dfrac{1 \text{ mol } O_2}{2 \text{ mol } NH_4NO_3} = 0.156 \text{ mol } O_2$

mol $H_2O = 0.312 \text{ mol } NH_4NO_3 \times \dfrac{4 \text{ mol } H_2O}{2 \text{ mol } NH_4NO_3} = 0.625 \text{ mol } H_2O$

$n_{total} = n_{N_2} + n_{O_2} + n_{H_2O} = 0.312 + 0.156 + 0.625 = 1.093 \text{ mol gas}$

$T = 125°C + 273.15 = 398 \text{ K}$

$V = 15.0 \text{ L}$

$P_T = ?$

$$PV = nRT \quad \Rightarrow \quad P_T = \frac{nRT}{V} = \frac{(1.093 \text{ mol})(0.0821 \text{L} \cdot \text{atm/mol} \cdot \text{K})(398 \text{ K})}{15.0 \text{ L}} = \underline{2.38 \text{ atm}}$$

(c) Use Equation 5.4 from the textbook

$$P_A = X_A P_{tot} \quad \text{where} \quad X_A = \frac{n_A}{n_{tot}}$$

$$P_{N_2} = \left(\frac{0.312 \text{ mol N}_2}{1.092 \text{ mol gas}} \right) 2.38 \text{ atm} = \underline{\textbf{0.680 atm}}$$

$$P_{O_2} = \left(\frac{0.156 \text{ mol O}_2}{1.092 \text{ mol gas}} \right) 2.38 \text{ atm} = \underline{\textbf{0.340 atm}}$$

$$P_{H_2O} = \left(\frac{0.625 \text{ mol H}_2\text{O}}{1.092 \text{ mol gas}} \right) 2.38 \text{ atm} = \underline{\textbf{1.36 atm}}$$

52. (a) $P_{N_2} = P_{tot} - P_{H_2O} = 745 \text{ mm Hg} - 16.48 \text{ mm Hg} = \underline{\textbf{729 mm Hg}}$

(b) $P_{H_2O} = 16.48 \text{ mm Hg} \times \dfrac{1 \text{ atm}}{760 \text{ mmHg}} = 0.0217 \text{ atm}$

$$V = 500 \text{ mL} \times \frac{1 \text{ L}}{1000 \text{ mL}} = 0.500 \text{ L}$$

$$T = 19°\text{C} + 273.15 = 292 \text{ K}$$

$$n = \frac{PV}{RT} = \frac{(0.0217 \text{ atm})(0.500 \text{ L})}{(0.0821 \text{ L} \cdot \text{atm/mol} \cdot \text{K})(292 \text{ K})} = \underline{\textbf{4.52} \times \textbf{10}^{-4} \textbf{ mol H}_2\textbf{O}}$$

(c) $P_{H_2 \text{ dry}} = 729 \text{ mm Hg} \times \dfrac{1 \text{ atm}}{760 \text{ mmHg}} = 0.959 \text{ atm}$

$$n = \frac{PV}{RT} = \frac{(0.959 \text{ atm})(0.500 \text{ L})}{(0.0821 \text{ L} \cdot \text{atm/mol} \cdot \text{K})(292 \text{ K})} = \underline{\textbf{2.00} \times \textbf{10}^{-2} \textbf{ mol H}_2}$$

(d) mol Ne are added to the flask:

$$n_{He} = 0.128 \text{ g Ne} \times \frac{1 \text{ mol}}{20.18 \text{ g}} = 0.00634 \text{ mol Ne}$$

P_{total} after adding Ne: $\quad P_{total} = \dfrac{n_{total} RT}{V}$

$$P_{total} = \frac{(0.000452 + 0.0200 + 0.00634) \text{mol}(0.0821 \text{ L} \cdot \text{atm/mol} \cdot \text{K})(292 \text{ K})}{0.500 \text{ L}} = 1.28 \text{ atm}$$

$$P_{total} = 1.28 \text{ atm} \times \frac{760 \text{ mmHg}}{1 \text{ atm}} = 976 \text{ mm Hg}$$

$$n_{total} = 0.000452 \text{ mol H}_2\text{O} + 0.0200 \text{ mol N}_2 + 0.00465 \text{ mol Ne} = 0.0268 \text{ mol gas}$$

$$P_{Ne} = X_{Ne}P_{tot} = (n_{Ne}/n_{total}) P_{tot}$$

$$P_{He} = \frac{0.00634 \text{ mol}}{0.0268 \text{ mol}} \times 976 \text{ mm Hg} = \underline{\textbf{231 mm Hg}}$$

(e) P_{total} = **1.28 atm or 976 mm Hg** (see calculation presented above #52d)

5-6 KINETIC THEORY OF GASES

54. The average speed of a gas is inversely proportional to the square root of its molar mass. Lighter gases travel faster than heavier ones.

Gas	Xe	CH_4	F_2	CH_2F_2
MM (g/mol)	131.3	16.042	38	52.026

(a) In terms of increasing speed of effusion through a tiny opening:

$$\underset{\text{slow}}{Xe} < CH_2F_2 < \underset{\text{fast}}{F_2 < CH_4}$$

(b) In terms of increasing time of effusion

$$\underset{\substack{\text{shorter} \\ \text{effusion time}}}{CH_4} < F_2 < \underset{\substack{\text{longer} \\ \text{effusion time}}}{CH_2F_2 < Xe}$$

56. Calculate the ratio of the effusion rates of the two gases.

$$\frac{\text{rate of effusion of } H_2}{\text{rate of effusion of } N_2} = \left(\frac{MM_{N_2}}{MM_{H_2}}\right)^{\frac{1}{2}} \Rightarrow \frac{\text{rate of effusion of } H_2}{\text{rate of effusion of } N_2} = \left(\frac{28.02}{2.016}\right)^{\frac{1}{2}} = \underline{\textbf{3.73}}$$

The hydrogen balloon will deflate **3.73 times faster** than the nitrogen balloon.

58. See Example 5.11

(a) **Yes**, the gas (X) is heavier than helium. Compared to lighter gases, heavier ones take longer effusion time.

(b) $$\frac{\text{rate of effusion X}}{\text{rate of effusion He}} = \left(\frac{MM_{He}}{MM_X}\right)^{\frac{1}{2}}$$

$$\text{rate of effusion X} = \frac{1}{5}\text{rate of effusion He}$$

$$\frac{\frac{1}{5}(\text{rate of effusion He})}{\text{rate of effusion He}} = \left(\frac{4.003}{MM_X}\right)^{\frac{1}{2}} \Rightarrow \text{rate of effusion He cancel out}$$

Take the square of both sides of the equation:

$$\frac{1}{5} = \left(\frac{4.003}{MM_A}\right)^{\frac{1}{2}} \qquad \Rightarrow \qquad \left(\frac{1}{5}\right)^2 = \left(\left(\frac{4.003}{MM_X}\right)^{\frac{1}{2}}\right)^2 \Rightarrow \frac{1}{25} = \frac{4.003}{MM_X}$$

$MM_X = 4.003(25) = \underline{\textbf{100.1 g/mol}}$

60. The key to solving this problem is to realize that because *P*, *T*, and *V* are the same in both experiments, the number of moles of NH_3 and PH_3 effusing through the pinhole is the same. Thus, $n_{NH_3} = n_{PH_3} = n = 1.78 \times 10^{-3}$ mol.

Use Graham's law

$$\frac{\text{rate of effusion } NH_3}{\text{rate of effusion } PH_3} = \left(\frac{MM_{PH_3}}{MM_{NH_3}}\right)^{1/2}$$

let x = time it takes for PH_3 to effuse through the same pinhole

$$\frac{1.78 \times 10^{-3} \text{ mol}/11.2 \text{ s}}{1.78 \times 10^{-3} \text{ mol}/x} = \left(\frac{33.994 \text{ g/mol}}{17.034 \text{ g/mol}}\right)^{1/2}$$

$$\frac{x}{11.2 \text{s}} = 1.4127$$

$$\underline{\textbf{x = 15.8 s}}$$

It will take **15.8 s** for the same amount of PH_3 to effuse through the same pinhole.

62. See Example 5.10. Use Equation 5.6 from the textbook.

$$u = \left(\frac{3RT}{MM}\right)^{\frac{1}{2}} \text{ where } R = 8.31 \times 10^3 \frac{\text{g} \cdot \text{m}^2}{\text{s}^2 \text{mol} \cdot \text{K}}$$

(a) $T = -32°C + 273.15 = 241$ K

$MM_{Cl_2} = 70.9$ g/mol

$$u_{Cl_2} = \left[\frac{3(8.31 \times 10^3 \text{g} \cdot \text{m}^2/\text{s}^2 \text{mol} \cdot \text{K})(241 \text{K})}{70.9 \text{ g/mol}}\right]^{\frac{1}{2}} = \underline{\textbf{291 m/s}}$$

(b) $T = 25°C + 273.15 = 298$ K

$MM_{UF_6} = 352$ g/mol

$$u_{UF_6} = \left[\frac{3\,(8.31 \times 10^3\,\text{g} \cdot \text{m}^2/\text{s}^2\text{mol} \cdot \text{K})(298\text{ K})}{352\text{ g/mol}} \right]^{\frac{1}{2}} = \underline{\textbf{145 m/s}}$$

5-7 REAL GASES

64. Deviations from ideal behavior tend to be largest at high pressures and low temperatures.

   (a) If pressure was reduced from 20 atm to 1 atm, CH_4 should behave **more** ideally.

   (b) If temperature was reduced from 50°C to –50°C, CH_4 should behave **less** ideally.

66. See Section 5-3 and Figure 5.15.

   Use Figure 5.15 to estimate the ratio of the real molar volume to the ideal molar volume. Calculate the ideal molar volume. Calculate the density, substituting the ideal molar volume and the ratio for the real molar volume. Note that a different estimate of the V_m/V_m° ratio will result in a slightly different density.

   (a) According to Figure 5.15, at 100 atm, V_m/V_m° is approximately 0.78.

   Thus, $V_m = 0.78\,V_m^\circ$.

   $$V_m^\circ = \frac{RT}{P} = \frac{(0.0821\,\text{L} \cdot \text{atm/mol} \cdot \text{K})(298\text{ K})}{100\text{ atm}} = 0.245\text{ L/mol}$$

   $$\text{density} = \frac{MM}{V_m} = \frac{MM}{(0.78)(V_m^\circ)} = \frac{16.04\text{ g/mol}}{(0.78)(0.245\text{ L/mol})} = \underline{\textbf{84 g/L}}$$

   (b) $\text{density} = \dfrac{(MM)P}{RT} = \dfrac{(16.04\text{ g/mol})(100\text{ atm})}{(0.0821\text{L} \cdot \text{atm/mol} \cdot \text{K})(298\text{ K})} = \underline{\textbf{65.6 g/L}}$

   (c) The densities will be equal when $V_m^\circ = V_m$, which occurs when $\underline{\textbf{P} \cong \textbf{340 atm}}$.

68. See Sections 5-2 and 5-4

(a) $2C_8H_{18}(l) + 25O_2(g) \rightarrow 16CO_2(g) + 18H_2O(l)$

(b) Car fuel efficiency = 22 mi/gal C_8H_{18}

Distance = 75 mi

$T = 25°C + 273.15 = 298$ K

$P = 1$ atm

$V_{CO_2} = ?$

$$C_8H_{18} \text{ burned} = 75 \text{ mi} \times \frac{1 \text{ gal}}{22 \text{ mi}} \times \frac{4 \text{ qt}}{1 \text{ gal}} \times \frac{1 \text{ L}}{1.057 \text{ qt}} = 13 \text{ L}$$

$$\text{mol } C_8H_{18} = 13 \text{ L } C_8H_{18} \times \frac{1000 \text{ mL}}{1 \text{ L}} \times \frac{0.692 \text{ g}}{1 \text{ mL}} \times \frac{1 \text{ mol } C_8H_{18}}{114.22 \text{ g } C_8H_{18}} = 79 \text{ mol}$$

$$n_{CO_2} = 79 \text{ mol } C_8H_{18} \times \frac{16 \text{ mol } CO_2}{2 \text{ mol } C_8H_{18}} = 6.3 \times 10^2 \text{ mol}$$

$$PV = nRT \Rightarrow V = \frac{nRT}{P}$$

$$V_{CO_2} = \frac{nRT}{P} = \frac{(6.3 \times 10^2 \text{ mol})(0.0821 \text{ L} \cdot \text{atm/mol} \cdot \text{K})(298 \text{ K})}{1 \text{ atm}} = \underline{\mathbf{1.5 \times 10^4 \text{ L}}}$$

70. (a) Since R, n, P, and T are constant, then these variables will cancel out. Thus, V will also be constant. Consequently, the volumes are the same and **the ratio is 1:1**.

$$\frac{V_{Ne}}{V_{HCl}} = \frac{\dfrac{nRT}{P}}{\dfrac{nRT}{P}} = \frac{1}{1}$$

(b) 1.00 mol Ne = 20.18 g;   1.00 mol HCl = 36.46 g;   density = mass/volume

as presented in 70(a), volume, V is the same for the two gases.

$$\frac{\text{density Ne}}{\text{density HCl}} = \frac{20.18 \text{ g}/V}{36.46 \text{ g}/V} = \underline{\mathbf{0.5535}}$$

(c) The translational energy depends only on the temperature. Since the molecules are at the same temperature, they have the **same** translational energy also. Thus, the **ratio is 1:1**.

(d) 1 mol of Ne has 6.022×10^{23} atoms

1 mol of HCl has 6.022×10^{23} atoms

Thus, the **ratio is 1:1**.

72. See Sections 5-2 and 5-4

The number of moles of CO_2 prepared can be determined from the following given information and by using ideal gas law.

$$V_{CO_2} = 118.9 \text{ mL} \times \frac{1 \text{ L}}{1000 \text{ mL}} = 0.1189 \text{ L}$$

$$P_{CO_2} = 758 \text{ mm Hg} \times \frac{1 \text{ atm}}{760 \text{ mm Hg}} = 0.997 \text{ atm}$$

$$T = 22°C + 273.15 = 295 \text{ K}$$

$$PV = nRT \implies n = \frac{PV}{RT}$$

$$n_{CO_2} = \frac{PV}{RT} = \frac{(0.997 \text{ atm})(0.1189 \text{ L})}{(0.0821 \text{ L} \cdot \text{atm}/\text{mol} \cdot \text{K})(295 \text{ K})} = 4.89 \times 10^{-3} \text{ mol } CO_2$$

The number of moles of H^+ required to produce 0.00489 mol CO_2 can be calculated from the stoichiometric ratio given in the following balanced equation.

$$Na_2CO_3(s) + 2H^+(aq) \rightarrow 2Na^+(aq) + CO_2(g) + H_2O$$

$$n_{H^+} = 0.00489 \text{ mol } CO_2 \times \frac{2 \text{ mol } H^+}{1 \text{ mol } CO_2} = 9.79 \times 10^{-3} \text{ mol } H^+$$

Assuming that the acid is HCl, find the number of moles of H^+ in 35.47 mL of a 0.1380 M solution of HCl as shown below.

$$35.47 \text{ mL HCl solution} \times \frac{10^{-3} \text{ L}}{1 \text{ mL}} \times \frac{0.1380 \text{ mol HCl}}{1 \text{ L HCl solution}} \times \frac{1 \text{ mol } H^+}{1 \text{ mol HCl}} = 4.89 \times 10^{-3} \text{ mol } H^+$$

Assuming that the acid is H_2SO_4, the number of moles of H^+ in 35.47 mL of a 0.1380 M solution of H_2SO_4 as shown below.

$$35.47 \text{ mL } H_2SO_4 \text{ solution} \times \frac{10^{-3} \text{ L}}{1 \text{ mL}} \times \frac{0.1380 \text{ mol } H_2SO_4}{1 \text{ L } H_2SO_4 \text{ solution}} \times \frac{2 \text{ mol } H^+}{1 \text{ mol } H_2SO_4} = 9.79 \times 10^{-3} \text{ mol } H^+$$

The strong acid used by the student is H_2SO_4 because 35.47 mL of a 0.1380 M solution of H_2SO_4 has 9.79×10^{-3} mol H^+ which is the amount of H^+ needed to form 4.89×10^{-3} mol CO_2.

74. See Sections 3-2, 5-3 and 5-4.

Determination of % C:

$T = 25°C + 273.15 = 298 \text{ K}$

$P = 1 \text{ atm}$

$V_{CO_2} = 132.9 \text{ mL} \times \dfrac{1 \text{ L}}{1000 \text{ mL}} = 0.1329 \text{ L}$

mol CO_2: $\quad n_{CO_2} = \dfrac{PV}{RT} = \dfrac{(1.00 \text{ atm})(0.1329 \text{ L})}{(0.0821 \text{ L} \cdot \text{atm/mol} \cdot \text{K})(298 \text{ K})} = 5.43 \times 10^{-3} \text{ mol } CO_2$

mass of C: $5.43 \times 10^{-3} \text{ mol } CO_2 \times \dfrac{1 \text{ mol C}}{1 \text{ mol } CO_2} \times \dfrac{12.01 \text{ g C}}{1 \text{ mol C}} = 0.0652 \text{ g C}$

$\% \text{ C} = \dfrac{0.0652 \text{ g C}}{0.2036 \text{ g glycine}} \times 100\% = 32.0\% \text{ C}$

Determination of % H:

mass of H: $0.122 \text{ g } H_2O \times \dfrac{1 \text{ mol } H_2O}{18.02 \text{ g } H_2O} \times \dfrac{2 \text{ mol H}}{1 \text{ mol } H_2O} \times \dfrac{1.008 \text{ g H}}{1 \text{ mol H}} = 0.0136 \text{ g H}$

$\% \text{ H} = \dfrac{0.0136 \text{ g H}}{0.2036 \text{ g glycine}} \times 100\% = 6.68\% \text{ H}$

Determination of % N:

$T = 25°C + 273.15 = 298 \text{ K}$

$P = 1 \text{ atm}$

$V_{N_2} = 40.8 \text{ mL} \times \dfrac{1 \text{ L}}{1000 \text{ mL}} = 0.0408 \text{ L}$

mol N_2: $\quad n_{N_2} = \dfrac{PV}{RT} = \dfrac{(1.00 \text{ atm})(0.0408 \text{ L})}{(0.0821 \text{ L} \cdot \text{atm/mol} \cdot \text{K})(298 \text{ K})} = 1.67 \times 10^{-3} \text{ mol } N_2$

mass of N: $1.67 \times 10^{-3} \text{ mol } N_2 \times \dfrac{2 \text{ mol N}}{1 \text{ mol } N_2} \times \dfrac{14.01 \text{ g N}}{1 \text{ mol N}} = 0.0468 \text{ g N}$

$\% \text{ N} = \dfrac{0.0468 \text{ g N}}{0.2500 \text{ g glycine}} \times 100\% = 18.7\% \text{ N}$

Determination of % O:

$\% \text{ O} = 100\% - 32.0\% \text{ C} - 6.68\% \text{ H} - 18.7\% \text{ N} = 42.6\% \text{ O}$

Percent composition by mass of glycine: __32.0% C, 6.68% H, 18.7% N and 42.6% O__

Determination of empirical formula:

$$32.0 \text{ g C} \times \frac{1 \text{ mol C}}{12.01 \text{ g C}} = 2.66 \text{ mol C} \qquad\qquad 6.68 \text{ g H} \times \frac{1 \text{ mol H}}{1.008 \text{ g H}} = 6.63 \text{ mol H}$$

$$18.7 \text{ g N} \times \frac{1 \text{ mol N}}{14.01 \text{ g N}} = 1.33 \text{ mol N} \qquad\qquad 42.6 \text{ g O} \times \frac{1 \text{ mol O}}{16.00 \text{ g O}} = 2.66 \text{ mol O}$$

$$\text{C: } \frac{2.66 \text{ mol}}{1.33 \text{ mol}} = 2 \qquad \text{H: } \frac{6.63 \text{ mol}}{1.33 \text{ mol}} = 5 \qquad \text{N: } \frac{1.33 \text{ mol}}{1.33 \text{ mol}} = 1 \qquad \text{O: } \frac{2.66 \text{ mol}}{1.33 \text{ mol}} = 2$$

Empirical formula = $C_2H_5NO_2$

CONCEPTUAL QUESTIONS

76. See Sections 5-2 and 5-3.

$T = 85°C + 273.15 = 358 \text{ K}$

$P = 1 \text{ atm}$

$$V_{cylinder} = V_{air} = 542 \text{ cm}^3 \times \frac{1 \text{ mL}}{1 \text{ cm}^3} \times \frac{1 \text{ L}}{1000 \text{ mL}} = 0.542 \text{ L}$$

(a) mol O_2 in cylinder is 21% of the mole of air in the cylinder

$$n_{O_2} = n_{air} \times \frac{21 \text{ mole } O_2}{100 \text{ mole air}}$$

$$\text{where } n_{air \text{ in cylinder}} = \frac{PV}{RT} = \frac{(1 \text{ atm})(0.542 \text{ L})}{(0.0821 \text{ L} \cdot \text{atm/mol} \cdot \text{K})(358 \text{ K})}$$

$$n_{O_2} = \frac{(1 \text{ atm})(0.542 \text{ L})}{(0.0821 \text{ L} \cdot \text{atm/mol} \cdot \text{K})(358 \text{ K})} \times \frac{21 \text{ mole } O_2}{100 \text{ mole air}} = \underline{\textbf{0.00387 mol}} \ O_2 \text{ in each cylinder}$$

(b) The mass of gasoline that should be injected in each cylinder to react with the O_2 can be calculated assuming that hydrocarbons in gasoline has an average molar mass of 1.0×10^2 g/mol and that 1 mole of the hydrocarbon reacts with 12 moles of O_2.

$$\text{mass of gasoline} = 0.00387 \text{ mol } O_2 \times \frac{1 \text{ mol hydrocarbon}}{12 \text{ mol } O_2} \times \frac{1.0 \times 10^2 \text{ g hydrocarbon}}{1 \text{ mol hydrocarbon}}$$

$$= \underline{\textbf{0.032 g hydrocarbon in gasoline}}$$

78. See Sections 5-2, 5-3, 5-5, and 5-6.

    The volume, temperature, and pressure of the two gases, CO_2 and H_2 are constant.

    (a) Since $n = \dfrac{PV}{RT}$ and P, V, and T are constant, then **both gases have the same number of moles, n.**

    (b) **Carbon dioxide, CO_2** has higher density because it has larger molar mass. The density of a gas is directly proportional to its *MM*.

$$\text{density} = MM\left(\dfrac{P}{RT}\right)$$

    (c) **Carbon dioxide, CO_2** takes longer time to effuse out of its tank. The effusion of heavier gases takes longer time than that of lighter ones.

    (d) **Both gases have the same average translational energy** because the two gases have the same temperature.

    (e) **Both gases have the same partial pressures** because they have the same mole fraction and the same total pressure.

80. See Sections 5-2, 5-3, 5-5, and 5-6.

    The volume, temperature, and moles of gases in bulbs A and B are the same.

    (a) Since $P = \dfrac{nRT}{V}$ and n, V, and T are constant, then **the pressure in the two bulbs will be the same.**

    (b) **Bulb A has N_2 which has higher density.** N_2 has higher density than NH_3 because N_2 has larger molar mass (*MM*). The density of a gas is directly proportional to its *MM*.

    (c) **The gases in bulbs A and B have the same average translational energy** because the two gases have the same temperature.

    (d) **The gases in bulb B (NH_3) moves faster than the gases in bulb A (N_2)** because NH_3 is lighter. The average speed is inversely proportional to the square of the molar mass.

    (e) As explained in 80a, the pressure in the two bulbs are the same thus, opening of the bulbs results in **no change in pressure.**

    (f) If 2.0 mole of He are added the fraction of the total pressure due to helium is ½.

$$P_{He} = \dfrac{n_{He}}{n_{total}} P_{total} = \dfrac{2.0 \text{ mol He}}{(1.0 \text{ mol } N_2 + 1.0 \text{ mol } NH_3 + 2.0 \text{ mol He})} P_{total} = \text{½ } P_{total}$$

82. See Section 5-2.

(a)

	Tank A	Tank B
gas present	SO_2	O_2
Pressure	2 atm	1 atm
moles gas: tank A has twice as many mol of SO_2 as tank B has of O_2. Assume mole O_2 = 1 then mole SO_2 = 2	2 mol	1 mol (assumed)
Volume of tank A and B is the same. Assume V = 1L	1 L (assumed)	1 L (assumed)

$$PV = nRT \implies T = \frac{PV}{nR}$$

Ratio of temperature of tank A to tank B:

$$\frac{T_A}{T_B} = \frac{\dfrac{2\,\text{atm}\,(1\ \text{L})}{(2\,\text{mol})\,(0.0821\,\text{L}\cdot\text{atm/mol}\cdot\text{K})}}{\dfrac{1\,\text{atm}\,(1\text{L})}{(1\,\text{mol})\,(0.0821\,\text{L}\cdot\text{atm/mol}\cdot\text{K})}} = \frac{1}{1} \implies T_A = T_B$$

If tank A has twice as many moles of SO_2 as tank B has of O_2, **the temperature of the gases of the two tanks is the same**.

(b)

	Tank A	Tank B
gas present	SO_2	O_2
Pressure	2 atm	1 atm
moles gas: tank A and tank B has the same mol of gases. Assume mole O_2 = 1 then mole SO_2 = 1	1 mol (assumed)	1 mol (assumed)
Volume of tank A and B is the same. Assume V = 1L	1 L (assumed)	1 L (assumed)

Ratio of temperature of tank A to tank B:

$$\frac{T_A}{T_B} = \frac{\dfrac{2\,\text{atm}\,(1\ \text{L})}{(1\,\text{mol})\,(0.0821\,\text{L}\cdot\text{atm/mol}\cdot\text{K})}}{\dfrac{1\,\text{atm}\,(1\text{L})}{(1\,\text{mol})\,(0.0821\,\text{L}\cdot\text{atm/mol}\cdot\text{K})}} = \frac{2}{1} \implies T_A = 2T_B$$

If tank A has the same number of moles of SO_2 as tank B has of O_2, **the temperature of Tank A is twice as that of Tank B**.

(c)		Tank A	Tank B
gas present		SO_2	O_2
Pressure		2 atm	1 atm
mass of gas: tank A has twice as many grams of SO_2 as tank B has of O_2. Assume mass O_2 = 32.00 g then mass SO_2 = 64.00 g moles gas: mole O_2 = 32.00 g/32.00g/mol = 1 mol then, mole SO_2 = 2(32.00 g)/64.07 g/mol = 1 mol		1 mol	1 mol
Volume of tank A and B is the same. Assume V = 1L		1 L (assumed)	1 L (assumed)

Ratio of temperature of tank A to tank B:

$$\frac{T_A}{T_B} = \frac{\dfrac{2\,atm\,(1\ L)}{(1\,mol)\,(0.0821L \cdot atm/mol \cdot K)}}{\dfrac{1\,mol\,(1L)}{(1\,mol)\,(0.0821L \cdot atm/mol \cdot K)}} = \frac{2}{1} \Rightarrow T_A = 2T_B$$

If tank A has twice as many grams of SO_2 as tank B has of O_2, the **temperature of Tank A is twice as that of Tank B**.

84. See Section 5-6

(a) Because He is lighter than CO_2, **curve A** represents CO_2 and curve B represents He (faster molecular speed).

(b) **Curve B** (He) represents the gas that would effuse more quickly (faster).

(c) **Curve B** represents He (faster molecular speed).

86. See Sections 5-2 and 5-5.

(a) **Bulb C** (the one with the most number of atoms) would have the highest pressure because pressure is directly proportional to the number of particles.

(b) The relative pressures are directly proportional to the relative number of particles. Bulb C contains five times as many atoms (or moles) as bulb A, so the pressure would be five times that of A. $P_{bulb\ C}$ = 5 × 0.500 atm = **2.50 atm**

(c) It is given that the pressure in bulb A is 0.500 atm. As answered in part (b) above, the pressure in bulb C is 2.50 atm. The pressure in bulb B is [(4/2)(0.500)] = 1.00 atm. Thus, the total pressure = 0.500 + 2.50 + 1.00 = **4.00 atm**.

(d) Bulbs A and B now contain 3 atoms each, so $P_A = P_B = (3/2)(0.500 \text{ atm}) = 0.750 \text{ atm}$.

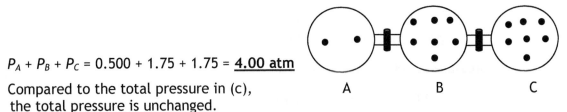

$P_A + P_B + P_C = 0.750 + 0.750 + 2.50 = \underline{\textbf{4.00 atm}}$

Compared to the total pressure in (c), the total pressure is unchanged.

(e) Bulbs B and C now contain 7 atoms each, so $P_B = P_C = (7/2)(0.500 \text{ atm}) = 1.75 \text{ atm}$.

$P_A + P_B + P_C = 0.500 + 1.75 + 1.75 = \underline{\textbf{4.00 atm}}$

Compared to the total pressure in (c), the total pressure is unchanged.

CHALLENGE PROBLEMS

88. Sections 5-3 and 5-7 and Appendix 1.

$$\frac{V_1 P_1}{n_1 T_1} = \frac{V_2 P_2}{n_2 T_2} \Rightarrow \text{ at constant } V \text{ and } n \text{ the formula simplifies to } \Rightarrow \frac{P_1}{T_1} = \frac{P_2}{T_2}$$

$$P_2 = \frac{P_1 T_2}{T_1}$$

$$P_1 = 23.76 \text{ mm Hg} \times \frac{1 \text{ atm}}{760 \text{ mm Hg}} = 0.0313 \text{ atm}$$

$T_1 = 25°C = 25 + 273.15 = 298 \text{ K}$

$T_2 = 40°C = 40 + 273.15 = 313 \text{ K}$

pressure at 40°C: $P_{40°C} = P_2 = ?$

$$P_2 = \frac{P_1 T_2}{T_1} = \frac{0.0313 \text{ atm} (313 \text{ K})}{298 \text{ K}} = 0.0329 \text{ atm}$$

$$P_2 = 0.0329 \text{ atm} \times \frac{760 \text{ mm Hg}}{1 \text{ atm}} = \underline{\textbf{25.0 mm Hg at 40°C}}$$

pressure at 70°C

$P_1 = 0.0313 \text{ atm}$

$T_1 = 25°C = 25 + 273.15 = 298 \text{ K}$

$T_3 = 70°C = 70 + 273.15 = 343 \text{ K}$

pressure at 70°C: $P_{70°C} = P_3 = ?$

$$P_3 = \frac{P_1T_3}{T_1} = \frac{0.0313\,\text{atm}\,(343\,\text{K})}{298\,\text{K}} = 0.0360\,\text{atm}$$

$$P_3 = 0.0360\,\text{atm} \times \frac{760\,\text{mm Hg}}{1\,\text{atm}} = \underline{\textbf{27.4 mm Hg at 70°C}}$$

pressure at 100°C

$P_1 = 0.0313$ atm

$T_1 = 25°C = 25 + 273.15 = 298$ K

$T_4 = 100°C = 100 + 273.15 = 373$ K

pressure at 100°C: $P_{100°C} = P_4 = ?$

$$P_4 = \frac{P_1T_4}{T_1} = \frac{0.0313\,\text{atm}\,(373\,\text{K})}{298\,\text{K}} = 0.0392\,\text{atm}$$

$$P_4 = 0.0392\,\text{atm} \times \frac{760\,\text{mm Hg}}{1\,\text{atm}} = \underline{\textbf{29.8 mm Hg at 100°C}}$$

Comparison of calculated ideal gas pressure and water vapor pressure

Temperature (°C)	ideal gas pressure (mm Hg)	vapor pressure of water (mm Hg)
$T_1 = 25$	23.76	23.76
$T_2 = 40$	25.0	55.3
$T_3 = 70$	27.4	233.7
$T_4 = 100$	29.8	760.0

The calculated ideal gas pressure and water vapor pressure are quite different from each other. Vapor contains not only gas molecules but some liquid molecules as well. As temperature increases, more water is vaporized thus increasing the number of gas molecules (n). As the number of gas molecules increases, gas pressure also increases.

89. See Sections 5-3 and 5-5.

Using mole concept and ideal gas law, the total moles of the gases and the number of moles of each gas, O_2, CO_2 and N_2 can be calculated.

$P = 1.38$ atm

$T = 25°C + 273.15 = 298$ K

$$V = 750\,\text{mL} \times \frac{1\,\text{L}}{1000\,\text{mL}} = 0.750\,\text{L}$$

total moles of gas: $\quad n_{total} = \dfrac{(1.38 \text{ atm})(0.750 \text{ L})}{(0.0821 \text{ L} \cdot \text{atm/mol} \cdot \text{K})(298 \text{ K})} = 0.0423 \text{ mol gas}$

mole of N_2:

$$P_{N_2} = P_{tank \ before \ nitrogen \ is \ removed} - P_{tank \ after \ nitrogen \ is \ removed}$$

$$= 1.38 \text{ atm} - 1.11 \text{ atm} = 0.27 \text{ atm}$$

$$n_{N_2} = \dfrac{P_{N_2} V}{RT} = \dfrac{(0.27 \text{ atm})(0.750 \text{ L})}{(0.0821 \text{ L} \cdot \text{atm/mol} \cdot \text{K})(298 \text{ K})} = 0.0083 \text{ mol } N_2$$

mole of CO_2:

mass CO_2 = mass increase of the CO_2 absorber = 0.114 g

$$n_{CO_2} = 0.114 \text{ g } CO_2 \times \dfrac{1 \text{ mol } CO_2}{44.01 \text{ g } CO_2} = 0.00259 \text{ mol } CO_2$$

mole of O_2:

$$n_{O_2} = n_{total} - \left(n_{N_2} + n_{CO_2} \right)$$

$$= 0.0423 \text{ mol gas} - (0.0083 \text{ mol } N_2 + 0.00259 \text{ mol } CO_2)$$

$$= 0.0314 \text{ mol } O_2$$

mass of each gas:

mass CO_2 = 0.114 g

mass N_2 = $0.0083 \text{ mol } N_2 \times \dfrac{28.02 \text{ g } N_2}{1 \text{ mol } N_2} = 0.23 \text{ g } N_2$

mass O_2 = $0.0314 \text{ mol } O_2 \times \dfrac{32.00 \text{ g } O_2}{1 \text{ mol } O_2} = 1.00 \text{ g } O_2$

total mass of gas = 0.114 g + 0.23 g N_2 + 1.00 g O_2 = 1.344 g

mass percent composition of the gas mixture:

$$\% \ CO_2 = \dfrac{\text{mass } CO_2}{\text{total mass}} \times 100 = \dfrac{0.114 \text{ g } CO_2}{1.344 \text{ g}} \times 100 = \underline{\textbf{8.48 \% } CO_2}$$

$$\% \ N_2 = \dfrac{\text{mass } N_2}{\text{total mass}} \times 100 = \dfrac{0.23 \text{ g } N_2}{1.344 \text{ g}} \times 100 = \underline{\textbf{17.1 \% } N_2}$$

$$\% \ O_2 = \dfrac{\text{mass } O_2}{\text{total mass}} \times 100 = \dfrac{1.00 \text{ g } O_2}{1.344 \text{ g}} \times 100 = \underline{\textbf{74.4 \% } O_2}$$

90. See Section 5-6.

$$\frac{\text{rate of effusion}_{NH_3}}{\text{rate of effusion}_{HCl}} = \left(\frac{MM_{HCl}}{MM_{NH_3}}\right)^{\frac{1}{2}}$$ where rate of effusion = distance travelled/time

$$\frac{(\text{distance/time})_{NH_3}}{(\text{distance/time})_{HCl}} = \left(\frac{MM_{HCl}}{MM_{NH_3}}\right)^{\frac{1}{2}}$$

since the two gases will meet at the same time, then $\Rightarrow$ time$_{NH_3}$ = time$_{HCl}$

Hence, the formula can be simplified as:

$$\frac{\text{distance}_{NH_3}}{\text{distance}_{HCl}} = \left(\frac{MM_{HCl}}{MM_{NH_3}}\right)^{\frac{1}{2}}$$

length of tube = 5.0 ft

let x = distance travelled by NH$_3$

5.0 − x = distance travelled by HCl

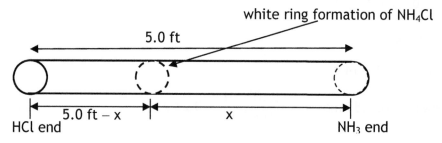

$$\frac{\text{distance}_{NH_3}}{\text{distance}_{HCl}} = \left(\frac{MM_{HCl}}{MM_{NH_3}}\right)^{\frac{1}{2}} \Rightarrow \frac{x}{5.0 - x} = \left(\frac{35.46}{17.03}\right)^{\frac{1}{2}}$$

$$\frac{x}{5.0 - x} = \left(\frac{35.46}{17.03}\right)^{\frac{1}{2}} = 1.45$$

$$x = 7.25 - 1.45x$$

$$x = \frac{7.25}{2.45} = \underline{3.0 \text{ ft}}$$

The white ring will form **3.0 ft** from the end where ammonia was introduced.

91. Sections 5-3, 5-4 and 5-5.

Before the reaction:

mass S = 5.00 g

$T = 25°C + 273.15 = 298$ K

$P = 995 \, mm \, Hg \times \dfrac{1 \, atm}{760.0 \, mm \, Hg} = 1.31 \, atm$

$V_{flask} = V_{O_2} = 5.00$ L

After the reaction:

$T = 138°C + 273.15 = 411$ K

$V_{flask} = V_{gas} = 5.00$ L

(a) To determine the pressure of SO_3 in the flask when the reaction is complete, find the number of moles of SO_3 formed which is dictated by the limiting reactant. Determine the number of moles of SO_3 that each of the given amounts of the reactants will produce. The reactant that will produce the smaller amount of SO_3 is the limiting reactant.

$$2S(s) + 3O_2(g) \rightarrow 2SO_3(g)$$

mol S → mol SO_3

$$5.00 \, g \, S \times \frac{1 \, mol \, S}{32.07 \, g \, S} \times \frac{2 \, mol \, SO_3}{2 \, mol \, S} = 0.156 \, mol \, SO_3$$

mol O_2 → mol SO_3

$$n_{O_2} = \frac{PV}{RT} = \frac{(1.31 \, atm)(5.00 \, L)}{(0.0821 \, L \cdot atm/mol \cdot K)(298 \, K)} = 0.268 \, mol \, O_2 \text{ available}$$

$$0.268 \, mol \, O_2 \times \frac{2 \, mol \, SO_3}{3 \, mol \, O_2} = 0.179 \, mol \, SO_3$$

Since S produce less moles of SO_3, S is the limiting reactant. The theoretical amount of SO_3 that will be produced is 0.156 mol. Thus, the pressure of SO_3 in the flask when the reaction is complete will be 1.05 atm as shown below.

$$P_{SO_3} = \frac{nRT}{V} = \frac{(0.156 \, mol \, SO_3)(0.0821 \, L \cdot atm/K \cdot mol)(411 \, K)}{5.00 \, L} = \underline{\textbf{1.05 atm}}$$

(b) The total pressure in the flask is equal to the sum of the pressures of the gases present inside the flask. Two gases will be present in the flask after the reaction: the product, SO_3 and the excess O_2 which remained unreacted.

$$\text{mole reacted } O_2 = 5.00 \text{ g S} \times \frac{1 \text{ mol S}}{32.07 \text{ g S}} \times \frac{3 \text{ mol } O_2}{2 \text{ mol S}} = 0.234 \text{ mol } O_2$$

$$\text{mole unreacted } O_2 = \text{mole } O_2 \text{ available} - \text{mole } O_2 \text{reacted}$$

$$= 0.268 \text{ mol} - 0.234 \text{ mol}$$

$$= 0.034 \text{ mol}$$

$$P_{unreacted\ O_2} = \frac{nRT}{V} = \frac{(0.034 \text{ mol } O_2)(0.0821 \text{ L} \cdot \text{atm/K} \cdot \text{mol})(411 \text{ K})}{5.00 \text{ L}} = 0.23 \text{ atm}$$

$$P_{total\ in\ flask} = P_{SO_3} + P_{unreacted\ O_2}$$

$$= 1.05 \text{ atm} + 0.23 \text{ atm} = \underline{\textbf{1.28 atm}}$$

(c) When 250.0 mL water is added to SO_3 the following reaction occurs.

$$SO_3(g) + H_2O(l) \rightarrow H_2SO_4(l)$$

Assuming there is enough water such that all SO_3 is converted to H_2SO_4, **the molarity of the H_2SO_4 solution formed is 0.624 M** (as shown below).

$$\text{mole } H_2SO_4 \text{ formed} = 0.156 \text{ mol } SO_3 \times \frac{1 \text{ mol } H_2SO_4}{1 \text{ mol } SO_3} = 0.156 \text{ mol } H_2SO_4$$

$$\text{molarity } H_2SO_4 = \frac{1 \text{ mol } H_2SO_4}{1 \text{ L solution}} = \frac{0.156 \text{ mol } H_2SO_4}{0.2500 \text{ L solution}} = \underline{\textbf{0.624 M}}$$

92. See Sections 3-3 and 5-3.

Solve this problem using the following steps.

i. Calculate the total amount of H_2 produced from Al and Zn using ideal gas law.

$$P = 755 \text{ mm Hg} \times \frac{1 \text{ atm}}{760.0 \text{ mm Hg}} = 0.993 \text{ atm}$$

$V = 0.147 \text{ L}$

$T_1 = 25°C = 25 + 273.15 = 298 \text{ K}$

$n = ?$

$$\text{total moles } H_2: \quad n_{H_2} = \frac{PV}{RT} = \frac{(0.993 \text{ atm})(0.147 \text{ L})}{(0.0821 \text{ L} \cdot \text{atm/mol} \cdot \text{K})(298 \text{ K})} = 0.00597 \text{ mol } H_2$$

Find moles of H_2 produced by each metal and mass of each metal in the sample.

The balanced equations:

$$2Al(s) + 6H^+(aq) \rightarrow 2Al^{3+}(aq) + 3H_2(g)$$

$$Zn(s) + 2H^+(aq) \rightarrow Zn^{2+}(aq) + H_2(g)$$

mass of Al-Zn alloy = 0.2500 g

Let w = mass of Zn

0.2500 g − w = mass of Al

$$\text{mol } H_2 \text{ from Zn} = w\, Zn \times \frac{1\, mol\, Zn}{65.39\, g\, Zn} \times \frac{1\, mol\, H_2}{1\, mol\, Zn} = 0.0153w\, mol\, H_2$$

$$\text{mol } H_2 \text{ from Al} = (0.2500 - w)\, g\, Al \times \frac{1\, mol\, Al}{26.98\, g\, Al} \times \frac{3\, mol\, H_2}{2\, mol\, Al} = (0.0139 - 0.0556w)\, mol\, H_2$$

total mol H_2 = mol H_2 from Zn + mol H_2 from Al

0.00597 mol H_2 = 0.0153w + [0.0139 − 0.0556w]

solving for w:

$$0.0556w - 0.0153w = 0.0139 - 0.00597$$

$$0.0403w = 0.00793$$

$$w = \frac{0.00793}{0.0403}$$

$$w = 0.197\, g$$

mass of Zn = w = 0.197 g

mass of Al = 0.2500 g − w = 0.2500 g − 0.197 g = 0.053 g

$$\text{mass \% Zn} = \frac{mass\, Zn}{total\, mass\, sample} = \frac{0.197\, g\, Zn}{0.2500\, g\, total} \times 100 = \underline{\textbf{78.8\%}}$$

93. In order to get off the ground, the buoyant force of the balloon must exceed the gravitational force holding it down.

buoyant force on the balloon = $mass_{displaced\ air}$

$$mass_{displaced\ air} = (density \times V)_{air}$$

therefore,

buoyant force on the balloon = $(density \times V)_{air}$ (Equation 1) gravitational force on the balloon = $\text{mass}_{(H_2)} + \text{mass}_{(man + balloon)}$

$$\text{mass}_{H_2} = (\text{density} \times V)_{H_2}$$

$$\text{mass}_{(man + balloon)} = 168 \text{ kg} \times \frac{1000 \text{ g}}{1 \text{ kg}} = 1.68 \times 10^5 \text{ g}$$

therefore,

gravitational force on the balloon = $(\text{density} \times V)_{H_2} + 1.68 \times 10^5 \text{ g}$ (Equation 2)

to find $V_{balloon}$ equate the buoyant force with the gravitational force:

buoyant force on the balloon = gravitational force on the balloon   (Equation 3)

substitute Equations 1 and 2 into Equation 3.

$$(\text{density} \times V)_{air} = (\text{density} \times V)_{H_2} + 1.68 \times 10^5 \text{ g}$$

$$(\text{density} \times V)_{air} - (\text{density} \times V)_{H_2} = 1.68 \times 10^5 \text{ g}$$ (Equation 4)

Since gases expand to fill the volume of the container (which in this case is the balloon), $V_{balloon} = V_{air} = V_{H_2}$. Thus, Equation 4 can be rewritten as:

$$(\text{density}_{air} \times V_{balloon}) - (\text{density}_{H_2} \times V_{balloon}) = 1.68 \times 10^5 \text{ g}$$ (Equation 5)

Equation 5 can be simplified as:

$$V_{balloon} (\text{density}_{air} - \text{density}_{H_2}) = 1.68 \times 10^5 \text{ g}$$ (Equation 6)

$$V_{balloon} = \frac{1.68 \times 10^5 \text{ g}}{\text{density}_{air} - \text{density}_{H_2}}$$

$$\text{density} = MM \left(\frac{P}{RT} \right)$$ (Equation 5.2 from the textbook)

Substitution of Equation 5.2 (from the textbook) into Equation 6 above yields:

$$V_{balloon} = \frac{1.68 \times 10^5 \text{ g}}{MM_{air} \left(\dfrac{P}{RT} \right) - MM_{H_2} \left(\dfrac{P}{RT} \right)} = \frac{1.68 \times 10^5 \text{ g}}{\left(MM_{air} - MM_{H_2} \right) \left(\dfrac{P}{RT} \right)}$$

$$P = 758 \text{ mm Hg} \times \frac{1 \text{ atm}}{760.0 \text{ mm Hg}} = 0.997 \text{ atm}$$

$V = 0.147$ L

$T_1 = 22^\circ C = 22 + 273.15 = 295$ K

$MM_{air} = 29.0$ g/mol

$MM_{hydrogen} = 2.016$ g/mol

substitute the values for R, temperature, pressure and molar masses given above:

$$V_{balloon} = \frac{1.68 \times 10^5 g \, (RT)}{(MM_{air} - MM_{H_2})P} = \frac{1.68 \times 10^5 g \, (0.0821 \, L \cdot atm/mol \cdot K)(295 \, K)}{(29.0 \, g/mol - 2.016 \, g/mol) \, 0.997 \, atm} = 1.51 \times 10^5 \, L$$

Convert volume into m³:

$$V \, (m^3) = 1.51 \times 10^5 L \times \frac{1 m^3}{10^3 L} = 1.51 \times 10^2 \, m^3$$

Since the balloon is spherical in shape, the radius, r can be obtained using the formula for the volume of a sphere.

$$V_{balloon} = \frac{4}{3} \pi r^3 = 1.51 \times 10^2 \, m^3$$

$$r_{balloon} = \left(\frac{3 \, (1.51 \times 10^2 \, m^3)}{4\pi} \right)^{1/3} = 3.30 \, m$$

diameter of the balloon = $2r$ = 3.30 m × 2 = **6.60 m**

The diameter of the balloon should be at least 6.60 m to be able to get off the ground.

94. See Sections 3-3 and 5-4.   This problem has two sets of conditions. First the initial and final conditions for P, V, n and T must be determined and finally, use the formula for initial state–final state.

Initial Conditions:

T_i = 25°C = 25 + 273.15 = 298 K

P_i = 0.950 atm

V = constant therefore $V = V_i = V_f$

$n_i = n_R$ = total mol of gas reactants

Final Conditions:

P_f = ?

T_f = 125°C = 25 + 273.15 = 398 K

V = constant therefore $V = V_i = V_f$

n_f = mol gas products + mol unreacted gas reactants

Balanced Equation: $2H_2(g) + O_2(g) \rightarrow 2H_2O(g)$

From the balanced equation, the following mol ratio of gases can be obtained:

$$\frac{\text{total mol gas reactants}}{\text{total mol gas products}} = \frac{2\,\text{mol}\,H_2 + 1\,\text{mol}\,O_2}{2\,\text{moles}\,H_2O} = \frac{3\,\text{mol gas reactants}}{2\,\text{mol gas products}}$$

Let n_R = total mol of gas reactants

Then, total mol gas products = $n_R \times \dfrac{2\,\text{mol gas products}}{3\,\text{mol gas reactants}} = \dfrac{2}{3}\,n_R$

Assuming 88.0% yield, then only 88.0% of the reactants are consumed thus 12.0% is unreacted. Likewise, only 88.0% of the theoretical yield is actually produced.

n_f = mol gas products + mol unreacted gas reactants

$$= \left[\left(\frac{88.0}{100}\right)\left(\frac{2}{3}\,n_R\right)\right] + \left(\frac{12.0}{100}\right)n_R$$

$$= 0.587\,n_R + 0.12\,n_R$$

$$= 0.707\,n_R$$

Use the initial–final state formula.

$$\frac{V_i P_i}{n_i T_i} = \frac{V_f P_f}{n_f T_f} \implies \text{at constant V the formula simplifies to} \implies \frac{P_i}{n_i T_i} = \frac{P_f}{n_f T_f}$$

Solving for P_f, the equation becomes:

$$P_f = \frac{P_i\,n_f\,T_f}{n_i\,T_i}$$

Substitute the values for the initial and final state conditions defined above.

$$P_f = \frac{(0.950\,\text{atm})\,(0.707\,n_R)\,(398\,K)}{n_R\,(298\,K)} = \underline{\mathbf{0.897\ atm}}$$

ELECTRONIC STRUCTURE AND
THE PERIODIC TABLE

6.1 LIGHT, PHOTON ENERGIES, AND ATOMIC SPECTRA

2. See Example 6.1 and 6.2.

Given information: $\lambda = 514$ nm

$h = 6.626 \times 10^{-34}$ J·s

$c = 2.998 \times 10^8$ m/s

(a) $\nu = ?$

$$\lambda = 514 \text{ nm} \times \frac{1 \text{ m}}{10^9 \text{ nm}} = 5.14 \times 10^{-7} \text{ m}$$

Use Equation 6.1 (from the textbook):

$$c = \lambda \nu \implies \text{solve for } \nu \implies \nu = \frac{c}{\lambda}$$

$$\nu = \frac{c}{\lambda} = \frac{2.998 \times 10^8 \text{ m/s}}{5.14 \times 10^{-7} \text{ m}} = \underline{\mathbf{5.83 \times 10^{14} \text{ s}^{-1}}}$$

(b) Use Equation 6.2 (from the textbook) to find the energy in joules per photon:

$$E = \frac{hc}{\lambda} = \frac{(6.626 \times 10^{-34} \text{ J·s})(2.998 \times 10^8 \text{ m/s})}{5.14 \times 10^{-7} \text{ m}} = \underline{\mathbf{3.86 \times 10^{-19} \text{ J/photon}}}$$

(c) Use Avogadro's number to find the energy in kilojoules per mole:

$$\frac{3.86 \times 10^{-19} \text{ J}}{1 \text{ photon}} \times \frac{6.022 \times 10^{23} \text{ photons}}{1 \text{ mol photons}} \times \frac{1 \text{ kJ}}{1000 \text{ J}} = \underline{\mathbf{233 \text{ kJ/mol photons}}}$$

4. See Example 6.1 and 6.2.

   Given information: $\nu = 1.29 \times 10^{15}\ \text{s}^{-1}$

   $c = 2.998 \times 10^8\ \text{m/s}$

   $h = 6.626 \times 10^{-34}\ \text{J·s}$

   (a) $\lambda_{(nm)} = ?$

   Use Equation 6.1 (from the textbook): $c = \lambda\nu \Rightarrow$ solve for $\lambda \Rightarrow \lambda = \dfrac{c}{\nu}$

   $$\lambda = \frac{c}{\nu} = \frac{2.998 \times 10^8\ \text{m/s}}{1.29 \times 10^{15}\ \text{s}^{-1}} \times \frac{10^9\ \text{nm}}{1\ \text{m}} = \underline{\textbf{232 nm}}$$

   (b) **UV range**. Based on Figure 6.3, a frequency of $1.29 \times 10^{15}\ \text{s}^{-1}$ lies in the UV range.

   (c) Use Equation 6.2 (from the textbook) to find the energy absorbed by one photon:

   $$E = h\nu = (6.626 \times 10^{-34}\ \text{J·s})(1.29 \times 10^{15}\ \text{s}^{-1}) = \underline{\textbf{8.55} \times \textbf{10}^{-19}\ \textbf{J per photon}}$$

6. See Example 6.1 and Figure 6.3.

   Given information: $E = 941\ \text{kJ/mol}$

   $c = 2.998 \times 10^8\ \text{m/s}$

   $h = 6.626 \times 10^{-34}\ \text{J·s}$

   (a) $\lambda = ?$

   Use Equation 6.2 to find the wavelength that could break the N–N bond. Note that E in this equation refers to energy per photon, so the given energy should be converted to energy/photon.

   energy per photon: $\dfrac{941\ \text{kJ}}{1\ \text{mol}} \times \dfrac{1000\ \text{J}}{1\ \text{kJ}} \times \dfrac{1\ \text{mol}}{6.022 \times 10^{23}\ \text{photons}} = 1.56 \times 10^{-18}\ \text{J/photon}$

   Use Equation 6.2 (from the textbook): $E = \dfrac{hc}{\lambda} \Rightarrow$ solve for $\lambda \Rightarrow \lambda = \dfrac{hc}{E}$

   $$\lambda = \frac{hc}{E} = \frac{(6.626 \times 10^{-34}\ \text{J·s})(2.998 \times 10^8\ \text{m/s})}{1.56 \times 10^{-18}\ \text{m}} = \underline{\textbf{1.27} \times \textbf{10}^{-7}\ \textbf{m or 127 nm}}$$

   (b) **Ultraviolet or UV range**. Based on Figure 6.3, a radiation with a wavelength of 1.27×10^{-7} m (127 nm) is in the UV range.

8. See Example 6.2.

Given information:
$$c = 2.998 \times 10^8 \text{ m/s}$$
$$h = 6.626 \times 10^{-34} \text{ J·s}$$
$$\text{number of photons emitted} = 0.255 \text{ moles}$$
$$\lambda = 635 \text{ nm} = 635 \text{ nm} \times \frac{1 \text{ m}}{10^9 \text{ nm}} = 6.35 \times 10^{-7} \text{ m}$$

E given off (in kJ) = ?

Use Equation 6.2 (from the textbook) to determine the energy released per photon.

$$E = \frac{hc}{\lambda} = \frac{6.626 \times 10^{-34} \text{ J·s} \, (2.998 \times 10^8 \text{ m/s})}{6.35 \times 10^{-7} \text{ m}} = 3.13 \times 10^{-19} \text{ J}$$

Use Avogadro's number to calculate the energy given off by 0.255 mole photons.

$$E = 0.255 \text{ mole photons} \times \frac{6.022 \times 10^{23} \text{ photons}}{1 \text{ mole photon}} \times \frac{3.13 \times 10^{-19} \text{ J}}{1 \text{ photon}} = \underline{\mathbf{4.81 \times 10^4 \text{ J or } 48.1 \text{ kJ}}}$$

6.2 THE HYDROGEN ATOM

10. See Example 6.3 and Figure 6.3.

Given information:
$$R_H = 2.180 \times 10^{-18} \text{ J}$$
$$h = 6.626 \times 10^{-34} \text{ J·s}$$
$$n_{lo} = 2, \quad n_{hi} = 5$$

(a) Use Equation 6.4 (from the textbook) to find the frequency, ν associated with the transition from the energy level n=4 to n=2.

$$\nu = \frac{R_H}{h} = \left[\frac{1}{(n_{lo})^2} - \frac{1}{(n_{hi})^2} \right] = \frac{2.180 \times 10^{-18} \text{ J}}{6.626 \times 10^{-34} \text{ J·s}} \left[\frac{1}{(2)^2} - \frac{1}{(5)^2} \right] = 6.91 \times 10^{14} \text{ s}^{-1}$$

Convert the calculated ν to wavelength, λ using Equation 6.1 (from the textbook).

$$c = \lambda \nu \quad \Rightarrow \text{solve for } \lambda \Rightarrow \quad \lambda = \frac{c}{\nu} = \frac{2.998 \times 10^8 \text{ m/s}}{6.91 \times 10^{14} \text{ s}^{-1}} \times \frac{10^9 \text{ nm}}{1 \text{ m}} = \underline{\mathbf{434 \text{ nm}}}$$

The wavelength associated with the transition is **434 nm**.

(b) Based on Figure 6.3, the transition 434 nm is at the **visible region**.

(c) **Yes**, energy is absorbed because the transition is from low to high energy level.

12. Use Equation 6.3 (from the textbook) to calculate the energy, E_n for the different n values.

$$E_n = \frac{-R_H}{n^2} \quad \text{(Equation 6.3 where } R_H = 2.180 \times 10^{-18} \text{ J)}$$

$$E_1 = \frac{-R_H}{n^2} = \frac{-2.180 \times 10^{-18} \text{ J}}{1^2} = -2.180 \times 10^{-18} \text{ J}$$

$$E_2 = \frac{-R_H}{n^2} = \frac{-2.180 \times 10^{-18} \text{ J}}{2^2} = -5.450 \times 10^{-19} \text{ J}$$

$$E_3 = \frac{-R_H}{n^2} = \frac{-2.180 \times 10^{-18} \text{ J}}{3^2} = -2.422 \times 10^{-19} \text{ J}$$

$$E_4 = \frac{-R_H}{n^2} = \frac{2.180 \times 10^{-18} \text{ J}}{4^2} = -1.363 \times 10^{-19} \text{ J}$$

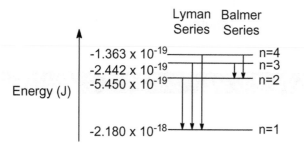

The graph on the left assumes that transitions are from E_{hi} to E_{lo}. If the transitions are in the opposite direction (E_{hi} to E_{lo}), the only change would be in the direction of the arrows but the magnitude of energy change remains the same.

14. See Example 6.3 and Figure 6.3.

Given information: $n_{lo} = 4$ $n_{hi} = 6$ $R_H = 2.180 \times 10^{-18}$ J $h = 6.626 \times 10^{-34}$ J·s

(a) Use Equation 6.4 to calculate the frequency, ν for the transition n = 4 to n = 6. Convert the calculated frequency to wavelength using equation 6.1.

$$\nu = \frac{R_H}{h} = \left[\frac{1}{(n_{lo})^2} - \frac{1}{(n_{hi})^2} \right] = \frac{2.180 \times 10^{-18} \text{ J}}{6.626 \times 10^{-34} \text{ J·s}} \left[\frac{1}{4^2} - \frac{1}{6^2} \right] = 1.142 \times 10^{14} \text{ s}^{-1}$$

Use Equation 6.1 (from the textbook): $c = \lambda \nu \Rightarrow$ solve for $\lambda \Rightarrow \lambda = \frac{c}{\nu}$

$$\lambda = \frac{c}{\nu} = \frac{2.998 \times 10^8 \text{ m/s}}{1.142 \times 10^{14} \text{ s}^{-1}} \times \frac{10^9 \text{ nm}}{1 \text{ m}} = \underline{\textbf{2624 nm or 2.624} \times \textbf{10}^3 \textbf{ nm}}$$

(b) **IR (or infrared) range.** Based on Figure 6.3, transition of $\lambda = 2.624 \times 10^6$ m (2624 nm) occurs in the IR range.

16. See Example 6.3.

    The longest wavelength corresponds to the smallest energy difference which is achieved by a change of one energy level (n_{lo} = 5 to n_{hi} = 6). Use Equation 6.4 (from the textbook) to find the ν associated with the transition from n=5 to n=6. Substitute the calculated ν into Equation 6.1

    Given information:

    $$R_H = 2.180 \times 10^{-18} \text{ J}$$

    $$h = 6.626 \times 10^{-34} \text{ J·s}$$

    $$n_{lo} = 5$$

    $$n_{hi} = 6 \text{ (one energy level different from n = 5)}$$

    $\lambda_{longest}$ = ?

    $$\nu = \frac{R_H}{h} = \left[\frac{1}{(n_{lo})^2} - \frac{1}{(n_{hi})^2} \right] = \frac{2.180 \times 10^{-18} \text{ J}}{6.626 \times 10^{-34} \text{ J·s}} \left[\frac{1}{5^2} - \frac{1}{6^2} \right] = 4.021 \times 10^{13} \text{ s}^{-1}$$

    $$c = \lambda\nu \quad \Rightarrow \text{solve for } \lambda \Rightarrow \quad \lambda = \frac{c}{\nu}$$

    $$\lambda = \frac{c}{\nu} = \frac{2.998 \times 10^8 \text{ m/s}}{4.021 \times 10^{13} \text{ s}^{-1}} \times \frac{10^9 \text{ nm}}{1\text{m}} = \underline{\textbf{7455 nm}}$$

6.3 QUANTUM NUMBERS

18. See Tables 6.3 and 6.4 and Example 6.4.

    Values of m_ℓ vary from $-\ell$ to $+\ell$ (including 0) for any given ℓ value.

    (a) d sublevel: $\ell = 2$ $m_\ell = \underline{-2, -1, 0, +1, +2}$

    (b) s sublevel: $\ell = 0$ $m_\ell = \underline{0}$

    (c) For the n=5 shell, there are 5 values for ℓ = 0, 1, 2, 3, 4

ℓ = 0: s sublevel	$m_\ell = \underline{0}$	
ℓ = 1: p sublevel	$m_\ell = \underline{-1, 0, +1}$	
ℓ = 2: d sublevel	$m_\ell = \underline{-2, -1, 0, +1, +2}$	
ℓ = 3: f sublevel	$m_\ell = \underline{-3, -2, -1, 0, +1, +2, +3}$	
ℓ = 4: g sublevel	$m_\ell = \underline{-4, -3, -2, -1, 0, +1, +2, +3, +4}$	

20. Refer to Figure 6.13 for the order of orbital energy. A more expanded list is shown below:

1s 2s 2p 3s 3p 4s 3d 4p 5s 4d 5p 6s 4f 5d 6p 7s 5f 6d 7p

→

increasing energy

		higher energy orbital
(a)	3s or 2p	**3s**
(b)	4s or 4d	**4d**
(c)	4f or 6s	**4f**
(d)	1s or 2s	**2s**

22. See Example 6.4.

The orbital is characterized by a number and a letter. The number is defined by the n value whereas the letter is defined by the ℓ value e.g. s is $\ell = 0$; p is $\ell = 1$; d is $\ell = 2$; f is $\ell = 3$; g is $\ell = 4$; etc.

(a) n = 3 and $\ell = 1$, the orbital is **3p**

(b) n = 5 and $\ell = 0$, the orbital is **5s**

(c) n = 6 and $\ell = 4$, the orbital is **6g**

24. See Examples 6.4 and 6.8.

The number of orbitals in each sublevel is equal to $2\ell+1$. If there is more than one sublevel, add up the orbitals for each sublevel.

(a) **n = 3** has 3 sublevels: $\ell = 0$, $\ell = 1$ and $\ell = 2$

$\ell = 0$: (s sublevel) $2\ell+1 = 2(0)+1 = 1$ orbital

$\ell = 1$: (p sublevel) $2\ell+1 = 2(1)+1 = 3$ orbitals

$\ell = 2$: (d sublevel) $2\ell+1 = 2(2)+1 = 5$ orbitals

thus the n = 3 main energy shell has a total (1 + 3 + 5) of **9 orbitals**

(b) 4p: $\ell = 1$ $\quad 2\ell+1 = 2(1)+1 = 3$ $\quad$ **3 orbitals**

(c) f: $\ell = 3$ $\quad 2\ell+1 = 2(3)+1 = 7$ $\quad$ **7 orbitals**

(d) d: $\ell = 2$ $\quad 2\ell+1 = 2(2)+1 = 5$ $\quad$ **5 orbitals**

26. See Example 6.5 and Table 6.3.

    Recall that the maximum number of electrons per orbital is 2.

    (a) 1s: n = 1 $\ell = 0$ $m_\ell = 0 \Rightarrow$ 1 orbital

    only one possible value for m_ℓ (0), thus only one orbital

    $$1\,\text{orbital} \times \frac{2\,\text{electrons}}{1\,\text{orbital}} = \textbf{2 electrons}$$

    (b) 4d, $m_\ell = 0$:   n = 4 $\ell = 2$ $m_\ell = 0 \Rightarrow$ 1 orbital

    $$1\,\text{orbital} \times \frac{2\,\text{electrons}}{1\,\text{orbital}} = \textbf{2 electrons}$$

    (c) n = 5, $\ell = 2$:

    for $\ell = 2$ there are 5 possible m_ℓ values ($m_\ell = -2, -1, 0, +1, +2$) thus, there are 5 orbitals

    $$5\,\text{orbitals} \times \frac{2\,\text{electrons}}{1\,\text{orbital}} = \textbf{10 electrons}$$

28. Refer to Table 6.4 for the permissible value of the quantum numbers through n=4. The permissible values for each quantum number is described in Section 6.3 and if a set of quantum numbers is not allowed then the combination of quantum numbers could not occur.

    (a) n = 1 $\ell = 0$ $m_\ell = 0$ $m_s = -\frac{1}{2}$

    The combination of these quantum number values is allowed, so this set **can occur**.

    (b) n = 1 $\ell = 1$ $m_\ell = 0$ $m_s = +\frac{1}{2}$

    This set is not possible because the only allowed value for n=1 is $\ell = 0$ (s subshell). A 1p orbital (n=1 $\ell = 1$) **cannot occur**.

    (c) n = 3 $\ell = 2$ $m_\ell = -2$ $m_s = +\frac{1}{2}$

    The combination of these quantum number values is allowed, so this set **can occur**.

    (d) n = 2 $\ell = 1$ $m_\ell = 2$ $m_s = +\frac{1}{2}$

    The only m_ℓ values allowed when $\ell = 1$ are $m_\ell = -1,\ 0,\ +1$. Since $m_\ell = 2$ is not an allowed value, this set **cannot occur**.

    (e) n = 4 $\ell = 0$ $m_\ell = 2$ $m_s = +\frac{1}{2}$

    The only m_ℓ value allowed when $\ell = 0$ is $m_\ell = 0$. Since $m_\ell = 2$ is not an allowed value when $\ell = 0$, this set **cannot occur**.

6.5 ELECTRON CONFIGURATIONS IN ATOMS

30. See Examples 6.6 and 6.7 and Figure 6.13.

    To write the ground state electron configuration for an element, determine the total number of electrons. If the elements are in their neutral state,

    number of electrons = atomic number

    Fill up the orbitals such that the low energy orbitals are filled up first.

Element	Atomic number	Number of Electrons	Electron Configuration
(a) S	16	16	$1s^2\ 2s^2\ 2p^6\ 3s^2\ 3p^4$
(b) Sc	21	21	$1s^2\ 2s^2\ 2p^6\ 3s^2\ 3p^6\ 4s^2\ 3d^1$
(c) Si	14	14	$1s^2\ 2s^2\ 2p^6\ 3s^2\ 3p^2$
(d) Sr	38	38	$1s^2\ 2s^2\ 2p^6\ 3s^2\ 3p^6\ 4s^2\ 3d^{10}\ 4p^6\ 5s^2$
(e) Sb	51	51	$1s^2\ 2s^2\ 2p^6\ 3s^2\ 3p^6\ 4s^2\ 3d^{10}\ 4p^6\ 5s^2\ 4d^{10}\ 5p^3$

32. See Examples 6.6, 6.7, Table 6.5, and Figure 6.13.

    The abbreviated (or shortened) electron configuration is composed of the symbol of the noble gas (enclosed in brackets) preceding the element followed by the remaining electron configuration that is not part of the noble gas electron configuration. All the elements given in this problem have no change therefore, the atomic number is equal to the number of electrons.

    (a) Mg (atomic number = 12) number of electrons = 12

    electron configuration: $\underbrace{1s^2\ 2s^2\ 2p^6}_{\text{electron configuration of Ne}}\ 3s^2$

    abbreviated electron configuration:

    $$[_{10}\text{Ne}]\ 3s^2 \quad \text{or} \quad \underline{[\text{Ne}]\ 3s^2}$$

    (b) Os (atomic number = 76) number of electrons = 76

    electron configuration: $\underbrace{1s^2\ 2s^2\ 2p^6\ 3s^2\ 3p^6\ 4s^2\ 3d^{10}\ 4p^6\ 5s^2\ 4d^{10}\ 5p^6}_{\text{Electron configuration of Xe}}\ 6s^2\ 4f^{14}\ 5d^6$

    abbreviated electron configuration:

    $$[_{54}\text{Xe}]\ 6s^2\ 4f^{14}\ 5d^6 \quad \text{or} \quad \underline{[\text{Xe}]\ 6s^2\ 4f^{14}\ 5d^6}$$

(c) Ge (atomic number = 32) number of electrons = 32

electron configuration: $\underbrace{1s^2\,2s^2\,2p^6\,3s^2\,3p^6}\,4s^2\,3d^{10}\,4p^2$

Electron configuration of Ar

abbreviated electron configuration: $[_{18}Ar]\,4s^2\,3d^{10}\,4p^2$ or $\underline{[Ar]\,4s^2\,3d^{10}\,4p^2}$

(d) V (atomic number = 23) number of electrons = 23

electron configuration: $\underbrace{1s^2\,2s^2\,2p^6\,3s^2\,3p^6}\,4s^2\,3d^3$

Electron configuration of Ar

abbreviated electron configuration: $[_{18}Ar]\,4s^2\,3d^3$ or $\underline{[Ar]\,4s^2\,3d^3}$

(e) At (atomic number = 85) number of electrons = 85

electron configuration: $\underbrace{1s^2\,2s^2\,2p^6\,3s^2\,3p^6\,4s^2\,3d^{10}\,4p^6\,5s^2\,4d^{10}\,5p^6}\,6s^2\,4f^{14}\,5d^{10}\,6p^5$

electron configuration of Xe

abbreviated electron configuration:

$[_{54}Xe]\,6s^2\,4f^{14}\,5d^{10}\,6p^5$ or $\underline{[Xe]\,6s^2\,4f^{14}\,5d^{10}\,6p^5}$

34. See Figure 6.13.

Assume that all the given elements are neutral or have no charge. Thus, atomic number is equal to the number of electrons. Shown below are the orbitals arranged in order of increasing energy (similar to that of Figure 6.13).

1s 2s 2p 3s 3p 4s 3d 4p 5s 4d 5p 6s 4f 5d 6p 7s 5f 6d 7p

$\longrightarrow$

increasing energy

(a) Write the electron configuration. Fill up the orbitals above starting with the low energy ones first with the maximum number of electrons that the orbital can accommodate (s=2, p=6, d=10, etc). When you reach the first orbital that has an f sublevel, fill that sublevel with 7 electrons only. The electron configuration is:

$1s^2\,2s^2\,2p^6\,3s^2\,3p^6\,4s^2\,3d^{10}\,4p^6\,5s^2\,4d^{10}\,5p^6\,6s^2\,4f^7$

Get the total number of electrons used up, including the 7 electrons in the first f orbital. Since the element is uncharged, number of electrons = atomic number. The atomic number is 63, which corresponds to the element **Europium, Eu**.

(b) The d orbital can only accommodate a total of 10 electrons. For an element to have 12 d electrons, two d orbitals of different n values are filled. Following the same steps in 34a, the last orbital to be filled is the 4d orbital with only 2 electrons in it. The electron configuration is:

$$1s^2\, 2s^2\, 2p^6\, 3s^2\, 3p^6\, 4s^2\, 3d^{10}\, 4p^6\, 5s^2\, 4d^2$$

Total number of electrons = atomic number = 40

The element with an atomic number of 40 is **zirconium, Zr**.

(c) Write the electron configuration that ends up in $3p^3$.

$$1s^2\, 2s^2\, 2p^6\, 3s^2\, 3p^3$$

The total number of electrons = atomic number = 15

The element with an atomic number of 15 is **phosphorus, P**.

(d) The first element with the lowest atomic number that has a completely filled p subshell has the following electron configuration:

$$1s^2\, 2s^2\, 2p^6$$

The total number of electrons = atomic number = 10

The element with an atomic number of 10 is **neon, Ne**.

36. See Example 6.6. Count the total number of electrons in all p subshells.

(a) Mg (atomic number = 12) number of electrons = 12

electron configuration: $1s^2\, 2s^2\, 2p^6\, 3s^2$

electrons in p subshell: $2p^6$ = 6 electrons

fraction of electrons in p subshell = **6/12**

(b) Mn (atomic number = 25) number of electrons = 25

electron configuration: $1s^2\, 2s^2\, 2p^6\, 3s^2\, 3p^6\, 4s^2\, 3d^5$

electrons in p subshells: $2p^6$ and $3p^6$ = 6 + 6 = 12 electrons

fraction of electrons in p subshell = **12/25**

(c) Mo (atomic number = 42) number of electrons = 42

electron configuration: $1s^2\, 2s^2\, 2p^6\, 3s^2\, 3p^6\, 4s^2\, 3d^{10}\, 4p^6\, 5s^2\, 4d^4$

electrons in p subshells: $2p^6$, $3p^6$, and $4p^6$ = 6 + 6 + 6 = 18 electrons

fraction of electrons in p subshell = **18/42**

38. See Section 6.3 and the solution to problem #28 in this Chapter.

    (a) $1s^2 2s^2 1d^1$ **is impossible**

    For n = 1, the possible value of ℓ is only 0 or s subshell, thus $1d^1$ (ℓ = 2) is not allowed.

    (b) $1s^2 2s^2 2p^6 3s^2 3p^4$ is a **ground state** electron configuration because the electrons fill up the orbitals in increasing energy.

    (c) $1s^2 2s^1 2p^7 3s^2$ is **impossible**.

    The p subshell can accommodate a total of 6 electrons only thus, $2p^7$ is not possible.

    (d) $1s^2 2s^2 3s^1 3p^4$ is an **excited state**.

    The ground state electron configuration for this atom is: $1s^2 2s^2 2p^5$. The five 2p electrons in the ground state are in the higher energy levels, 3s and 3p.

    (e) $1s^2 2s^2 3s^2 3p^6 4s^2$ is an **excited state**.

    Its ground state electron configuration is $1s^2 2s^2 2p^6 3s^2 3p^2$. Since there are no electrons in the 2p orbital but there are electrons in the higher energy orbital, 4s, this suggests that the atom is in a higher energy or excited state.

6.6 ORBITAL DIAGRAMS OF ATOMS

40. See Example 6.8 and Figure 6.15.

    (a) Na (atomic number = 11) number of electrons = 11
        orbital diagram:

        1s   2s   2p   3s

        Na: (↑↓) (↑↓) (↑↓)(↑↓)(↑↓) (↑)

    (b) O (atomic number = 8) number of electrons = 8
        orbital diagram:

        1s   2s   2p

        O: (↑↓) (↑↓) (↑↓)(↑)(↑)

    (c) Co (atomic number = 27) number of electrons = 27
        orbital diagram:

        1s 2s   2p   3s   3p   4s   3d

        Co: (↑↓) (↑↓) (↑↓)(↑↓)(↑↓) (↑↓) (↑↓)(↑↓)(↑↓) (↑↓) (↑↓)(↑↓)(↑)(↑)(↑)

(d) Cl (atomic number = 17) number of electrons = 17
   orbital diagram:

$$\begin{array}{ccccc} 1s & 2s & 2p & 3s & 3p \end{array}$$
Cl: (↑↓) (↑↓) (↑↓)(↑↓)(↑↓) (↑↓) (↑↓)(↑↓)(↑)

42. See Figure 6.15 and Example 6.8.

   (a) Electron configuration: $1s^2 2s^1$ Symbol of the atom: <u>Li</u>

   (b) Electron configuration: $1s^2 2s^2 2p^6 3s^2 3p^5$ Symbol of the atom: <u>Cl</u>

   (c) Electron configuration: $1s^2 2s^2 2p^6 3s^1$ Symbol of the atom: <u>Na</u>

44. See Example 6.8.

   (a) <u>Sn, Sb, Te</u>. Sn has 2 half-filled 5p orbitals, Sb has 3 half-filled 5p orbitals, and Te has 2 half-filled 5p orbitals.

   (b) <u>K, Rb, Cs, Fr</u>. All Group 1 elements that have an atomic number greater than that of Ar have completely filled 3p orbitals.

   (c) <u>Ge, As, Sb, Te</u>. All metalloids that have an atomic number greater than that of S have paired 3p electrons.

   (d) <u>None</u>.

46. See Example 6.8 and Figure 6.15

   The number of unpaired electrons can be obtained from the orbital diagram of an atom.

   (a) aluminum (atomic number = 13) number of electrons = 13
      orbital diagram:

$$\begin{array}{ccccc} 1s & 2s & 2p & 3s & 3p \end{array}$$
Al: (↑↓) (↑↓) (↑↓)(↑↓)(↑↓) (↑↓) (↑)( )( )

   As shown in the orbital diagram, the single electron in 3p is unpaired.
   Number of unpaired electrons = 1

   (b) argon (atomic number = 18) number of electrons = 18
      orbital diagram:

$$\begin{array}{ccccc} 1s & 2s & 2p & 3s & 3p \end{array}$$
Ar: (↑↓) (↑↓) (↑↓)(↑↓)(↑↓) (↑↓) (↑↓)(↑↓)(↑↓)

   As shown in the orbital diagram, none of the electrons are unpaired.
   Number of unpaired electrons = 0

(c) arsenic (atomic number = 33) number of electrons = 33
   orbital diagram:

1s	2s	2p	3s	3p	4s	3d	4p

As: (↑↓) (↑↓) (↑↓)(↑↓)(↑↓) (↑↓) (↑↓)(↑↓)(↑↓) (↑↓) (↑↓)(↑↓)(↑↓)(↑↓)(↑↓) (↑)(↑)(↑)

As shown in the orbital diagram, three of the 4p electrons are unpaired.
Number of unpaired electrons = 3

48. There are only three main group metals in the 4th period – K, Ca and Ga, the rest of the elements are transition metals, nonmetals and metalloids. The orbital diagrams of these three metals are shown below:

1s	2s	2p	3s	3p	4s

K: (↑↓) (↑↓) (↑↓)(↑↓)(↑↓) (↑↓) (↑↓)(↑↓)(↑↓) (↑)

1s	2s	2p	3s	3p	4s

Ca: (↑↓) (↑↓) (↑↓)(↑↓)(↑↓) (↑↓) (↑↓)(↑↓)(↑↓) (↑↓)

1s	2s	2p	3s	3p	4s	3d	4p

Ga: (↑↓) (↑↓) (↑↓)(↑↓)(↑↓) (↑↓) (↑↓)(↑↓)(↑↓) (↑↓) (↑↓)(↑↓)(↑↓)(↑↓)(↑↓) (↑)( )( )

(a) **Ca**. Among the three, only Ca has no unpaired electrons.

(b) **K and Ga**. As shown in the orbital diagram, K has 1 unpaired 4s electron while Ga has 1 unpaired 4p electron.

(c) **None**.

(d) **None**.

6.7 ELECTRON ARRANGEMENTS IN MONOATOMIC IONS

50. See Example 6.9.

In writing electron configuration of atoms:

For neutral atoms: number of electrons = atomic number

For anions (negatively charged): number of electrons = atomic number + charge value

For cations (positively charged): number of electrons = atomic number − charge value

Note that the electrons removed are the ones in the highest occupied "n" (main energy level).

(a) F (atomic number = 9) number of electrons = 9

electron configuration: $\underline{\text{F } 1s^2 2s^2 2p^5}$

F^- (atomic number = 9) number of electrons = 9 + 1 = 10

electron configuration: $\underline{\text{F}^- 1s^2 2s^2 2p^6}$

(b) Sc (atomic number = 21) number of electrons = 21

electron configuration: $\underline{\text{Sc } 1s^2 2s^2 2p^6 3s^2 3p^6 4s^2 3d^1}$

Sc^{3+} (atomic number = 21) number of electrons = 21 − 3 = 18

electron configuration: $\underline{\text{Sc}^{3+} 1s^2 2s^2 2p^6 3s^2 3p^6}$

(c) Mn (atomic number = 25) number of electrons = 25

electron configuration: $1s^2 2s^2 2p^6 3s^2 3p^6 4s^2 3d^5$

Mn^{2+} (atomic number = 25) number of electrons = 25 − 2 = 23

electron configuration: $\underline{\text{Mn}^{2+} 1s^2 2s^2 2p^6 3s^2 3p^6 3d^5}$

⇒ The $4s^2$ electrons (highest "n") are removed.

Mn^{5+} (atomic number = 25) number of electrons = 25 − 5 = 20

electron configuration: $\underline{\text{Mn}^{5+} 1s^2 2s^2 2p^6 3s^2 3p^6 3d^2}$

⇒ The $4s^2$ electrons and the 3 electrons in the 3d orbital are removed.

(d) O (atomic number = 8) number of electrons = 8

electron configuration: $1s^2\,2s^2\,2p^4$

O^- (atomic number = 8) number of electrons = 8 + 1 = 9

Electron configuration: <u>O^- $1s^2\,2s^2\,2p^5$</u>

O^{2-}(atomic number = 8) number of electrons = 8 + 2 = 10

Electron configuration: <u>O^{2-} $1s^2\,2s^2\,2p^6$</u>

52. See Examples 6.8 and 6.9.

The number of unpaired electrons can be obtained from the orbital diagram. An electron configuration with completely filled orbitals has no unpaired electrons.

(a) Al^{3+} (atomic number = 13) number of electrons = 13 − 3 = 10

electron configuration: $1s^2\,2s^2\,2p^6$

All orbitals are completely filled, therefore there are no unpaired electrons

Number of unpaired electrons = <u>0</u>

(b) Cl^- (atomic number = 17) number of electrons = 17 + 1 = 18

electron configuration: $1s^2\,2s^2\,2p^6\,3s^2\,3p^6$

All orbitals are completely filled, therefore there are no unpaired electrons

Number of unpaired electrons = <u>0</u>

(c) Sr^{2+} (atomic number = 38) number of electrons = 38 − 2 = 36

electron configuration: $1s^2\,2s^2\,2p^6\,3s^2\,3p^6\,4s^2\,3d^{10}\,4p^6$

All orbitals are completely filled, therefore there are no unpaired electrons

Number of unpaired electrons = <u>0</u>

(d) Zr^{4+} (atomic number = 40) number of electrons = 40 − 4 = 36

Electron configuration: $1s^2\,2s^2\,2p^6\,3s^2\,3p^6\,4s^2\,3d^{10}\,4p^6$

All orbitals are completely filled, therefore there are no unpaired electrons

Number of unpaired electrons = <u>0</u>

6.8 PERIODIC TRENDS IN THE PROPERTIES OF ATOMS

54. See Example 6.10.

    Mg, S, and Cl belong to the same period.

    (a) Atomic radius decreases from left to right across a period therefore in terms of increasing atomic radius:

    $$\underset{\text{smallest}}{Cl} \; < \; S \; < \; \underset{\text{largest}}{Mg}$$

    (b) First ionization energy (IE) increases from left to right across a period therefore in terms of increasing first ionization energy:

    $$\underset{\text{lowest IE}}{Mg} \; < \; S \; < \; \underset{\text{highest IE}}{Cl}$$

    (c) Electronegativity increases from left to right across a period from group 1 to group 17. Therefore, in terms of decreasing electronegativity:

    $$\underset{\substack{\text{most} \\ \text{electronegative}}}{Cl} \; > \; S \; > \; \underset{\text{least}}{Mg}$$

56. See Example 6.10.

    Na, P, and Cl belong to the same period (period 3). K belongs to the 4th period. Na and K belong to the same group.

    (a) **K has the largest atomic radius.** Atomic radius decreases from left to right across a period and increases down a group, thus the largest atom would be the lower leftmost.

    (b) **Cl has the highest ionization energy.** Ionization energy increases from left to right across a period and decreases down a group, thus the atom with the highest ionization energy would be the upper rightmost.

    (c) **Cl has the highest electronegativity.** Electronegativity increases from left to right across a period and decreases down a group, thus the most electronegative atom would be the upper rightmost.

58. See Example 6.10.

    A neutral atom is smaller than its corresponding anion but larger than its corresponding cation.

    (a) **P** is smaller than P^{3-}

    (b) **V^{4+}** is smaller than V^{2+}

    (c) **K^{+}** is smaller than K

    (d) **Co^{3+}** is smaller than Co

60. See Example 6.10 and Figure 6.19.

Atomic radii decrease from left to right across a period and increases down a group. Thus, the ordering of radii from smallest to largest is:

(a) __Kr < K < Rb < Cs__

(b) __Ar < Si < Al < Cs__

UNCLASSIFIED PROBLEMS

62. See Examples 6.1 and 6.2 and Figure 6.3.

Given information: $\nu = 8.429$ GHz

$$\nu = 8.429 \text{ GHz} \times \frac{10^9 \text{ Hz}}{1 \text{ GHz}} = 8.429 \times 10^9 \text{ Hz or } 8.429 \times 10^9 \text{ s}^{-1}$$

$$c = 2.998 \times 10^8 \text{ m/s}$$

$$h = 6.626 \times 10^{-34} \text{ J·s}$$

(a) $\lambda_{(nm)} = ?$

Use Equation 6.1 (from the textbook): $c = \lambda\nu \Rightarrow$ solve for $\lambda \Rightarrow \lambda = \dfrac{c}{\nu}$

$$\lambda = \frac{c}{\nu} = \frac{2.998 \times 10^8 \text{m/s}}{8.429 \times 10^9 \text{s}^{-1}} \times \frac{10^9 \text{nm}}{1\text{m}} = \textbf{3.557} \times \textbf{10}^7 \textbf{ nm}$$

(b) __microwave range.__ Based on Figure 6.3, a frequency of 8.429×10^9 s^{-1} lies in the microwave range.

(c) To find the energy (kJ/mol) associated with this frequency, use Equation 6.2 (from the textbook) to find the energy associated to one photon and convert it to one mole of photons:

$$E_{(per \ photon)} = h\nu = (6.626 \times 10^{-34} \text{ J·s}) (8.429 \times 10^9 \text{ s}^{-1}) = 5.585 \times 10^{-24} \text{ J/photon}$$

$$E_{(per \ mol \ photons)} = \frac{5.585 \times 10^{-24} \text{ J}}{1 \text{photon}} \times \frac{6.022 \times 10^{23} \text{ photons}}{1 \text{mole photons}} \times \frac{1 \text{kJ}}{1000 \text{ J}}$$

$$= \textbf{3.363} \times \textbf{10}^{-3} \textbf{ kJ/mol photons}$$

64. See Sections 6.5, 6.7, and 6.8.

   (a) <u>Si</u>. Based on the given electron configuration, the total number of electrons is 14. Assuming this atom is neutral or uncharged, the number of electrons = atomic number. The element that has an atomic number of 14 is silicon, <u>Si.</u>

   (b) <u>Na</u>. Electronegativity increases from left to right across a period from Group 1 to Group 17. Thus, Group 1 elements in a period will have the least electronegativity.

   (c) <u>Y</u>. A +3 ion is formed by removing 3 electrons. So the atomic number of the element is 3 units more than 36 (the atomic number of Kr). Thus, the atom is element 39 or <u>yttrium, Y</u>.

   (d) <u>At</u>. Atomic radius increases down a group. Thus, for the halogen group (or group 17), At has the largest atomic radius.

   (e) <u>O</u>. Ionization energy decreases down a group. In group 16, the topmost element, O, has the largest ionization energy.

CONCEPTUAL QUESTIONS

66. See Section 6.2.

   (a) Energy is absorbed when the electrons are excited from lower levels to higher ones, thus energy is absorbed for <u>**transitions 2 and 4**</u>.

   (b) Energy is emitted when the electron relaxes from higher level to a lower one, thus energy is emitted for <u>**transitions 1 and 3**</u>.

   (c) We cannot answer this question without knowing what the element is since one must know the number of electrons to determine the ground state configuration. If we assume the atom to be hydrogen, then <u>**transition 1**</u> involves the ground state.

   (d) The transition with largest energy difference (see part (a) and problem 12 above) will absorb the most energy, thus <u>**transition 2**</u>.

   (e) The transition with largest energy difference (see part (b) and problem 12 above) will emit the most energy, thus <u>**transition 1**</u>.

68. See Sections 6.1, 6.3, and 6.6.

   (a) The fourth quantum number, m_s , describes the <u>**spin of the orbital**</u>.

   (b) Photon of lower frequency have lower energy. Thus a photon with a <u>**frequency of 35.8 Hz has less energy than a photon with a frequency of 125 Hz**</u>.

   (c) <u>**Five orbitals**</u> can be associated with the following set of quantum numbers: $n=3$, $\ell=2$. Since $\ell=2$ has 5 permissible values for m_ℓ: -2, -1, 0, $+1$, $+2$, there a 5 possible orbitals.

(d) **Yes**, it is diamagnetic because Zn^{2+} ion is diamagnetic. As shown in the orbital diagram below, Zn^{2+} does not have any unpaired electrons, thus it is diamagnetic.

1s	2s	2p	3s	3p	4s	3d

Zn^{2+}: (↑↓) (↑↓) (↑↓)(↑↓)(↑↓) (↑↓) (↑↓)(↑↓)(↑↓) (↑↓) (↑↓)(↑↓)(↑↓)(↑↓)(↑↓)

70. See Sections 6.1, 6.2, 6.3, and 6.4.

 (a) In contrast with the Bohr model, the following propositions are claimed by the quantum mechanical model (see page 132 of the textbook).

 1. The kinetic energy of an electron is inversely related to the space it occupies.

 2. You cannot know the exact position of an electron at any given instant.

 (b) Frequency and wavelength are inversely related to each other. Unlike frequency which is directly related to energy, wavelength is not.

 (c) The three p orbitals lie along three different axis, x, y, or z. They all have the same shape, like two small balloons with the knot ends joined. This shape is often referred to as dumbbell shaped.

72. See Sections 6.1 to 6.5.

 (a) This statement is **true**. Photons with short wavelengths have high energy.

 (b) This statement is **false**. The energy of an electron is inversely proportional to $\underline{n^2}$, not to the quantum number, ℓ.

 $$E_n = -\frac{R_H}{n^2}$$

 (c) This statement is **false**. Electrons start entering the 5th principal level **before** the fourth is completely filled. The order of filling is [Kr] 5s 4d 5p 6s 4f 5d 6p 7s 5f 6d 7p

74. See Sections 6.6 and 6.7.

 (a) **True**.

 (b) **True**.

 (c) **False**. **Energy is absorbed** when an electron is removed from an atom.

CHALLENGE PROBLEMS

76. See Section 6.2.

For a one electron species, the ground state is n = 1, so the first excited state is n = 2.

$$E = \frac{-BZ^2}{n^2} = \frac{-(2.180 \times 10^{-18}\ J)(3)^2}{2^2} = -4.905 \times 10^{-18}\ J$$

$$E = -4.905 \times 10^{-18}\ J \times \frac{1\,kJ}{1000\,J} \times \frac{6.022 \times 10^{23}}{1\,mol} = -2.954 \times 10^3\ kJ/mol$$

E is the energy of the electron. The energy needed to ionize (or remove) an electron is **2.954×10^3 kJ/mol**.

77. See Section 6.2.

$$\lambda = \frac{hc}{\Delta E} \quad \text{and} \quad \Delta E = -R_H \left[\frac{1}{n_{hi}^2} - \frac{1}{n_{lo}^2} \right]$$

Substituting for ΔE and n_{lo} = 2, we get:

$$\lambda = \frac{hc}{-R_H \left[\frac{1}{n_{hi}^2} - \frac{1}{2^2} \right]} = -\frac{hc}{R_H \left[\frac{1}{n_{hi}^2} - \frac{1}{4} \right]}$$

$$= \frac{(6.626 \times 10^{-34}\ J \cdot s)(2.998 \times 10^8\ m/s)}{-2.180 \times 10^{-18}\ J \left[\frac{1}{n_{hi}^2} - \frac{1}{4} \right]} = \frac{-9.112 \times 10^{-8}\ m}{\left[\frac{1}{n_{hi}^2} - \frac{1}{4} \right]}$$

$$= \frac{9.112 \times 10^{-8}\ m}{\left[\frac{1}{4} - \frac{1}{n_{hi}^2} \right]} = \frac{9.112 \times 10^{-8}\ m}{\left[\frac{n_{hi}^2}{4\,n_{hi}^2} - \frac{4}{4\,n_{hi}^2} \right]} = \frac{9.112 \times 10^{-8}\ m\ (4\,n_{hi}^2)}{(n_{hi}^2 - 4)}$$

$$= \frac{(3.645 \times 10^{-7}\ m)\,n_{hi}^2}{(n_{hi}^2 - 4)}$$

Converting the units from m to nm:

$$\lambda = \frac{(3.645 \times 10^{-7}\ m)\,n_{hi}^2}{(n_{hi}^2 - 4)} \times \frac{10^9\ nm}{1\,m}$$

$$\lambda = \frac{(364.5\ nm)\,n_{hi}^2}{(n_{hi}^2 - 4)}$$

78. See Section 6.3 and Table 6.3.

n	1		2							
ℓ	0	1	0	1			2			
sublevel	1s	1p	2s	2p			2d			
m_ℓ	0	1	0	1	2	0	1	0	1	2	3

electron configuration for an atom with eight electrons: __$1s^4 \, 1p^4$__

79. See Section 6.3 and Table 6.3.

   (a) If there were 3 values for m_s, then each orbital could hold 3 electrons.

   s sublevel: 1 orbital (m_ℓ = 0), 3 electrons.

   p sublevel: 3 orbitals (m_ℓ = –1, 0, 1), 9 electrons.

   d sublevel: 5 orbitals (m_ℓ = –2, –1, 0, 1, 2), 15 electrons.

   (b) The __n = 3 level can hold 27 electrons__; 3e⁻ in the s sublevel(ℓ = 0); 9e⁻ in the p sublevel (ℓ = 1); and 15e⁻ in the d sublevel (ℓ = 2).

   (c) Atomic number = 8   electron configuration: __$1s^3 \, 2s^3 \, 2p^2$__

   Atomic number = 17   electron configuration: __$1s^3 \, 2s^3 \, 2p^9 \, 3s^2$__

80. See Section 6.1.

   Calculate the energy of the light. The difference between that energy and the kinetic energy is the energy needed to eject the electron (E_{min}).

   (a) $E = \dfrac{hc}{\lambda} = \dfrac{(6.626 \times 10^{-34} \text{ J} \cdot \text{s})(2.998 \times 10^8 \text{ m/s})}{5.40 \times 10^{-7} \text{ m}} = 3.68 \times 10^{-19}$ J

   $E_{min} = 3.68 \times 10^{-19}$ J $- 2.60 \times 10^{-20}$ J $= 3.42 \times 10^{-19}$ J

   While the calculations above used the longer wavelength, we also could have used the shorter wavelength and get the same answer as shown below.

   $E = \dfrac{hc}{\lambda} = \dfrac{(6.626 \times 10^{-34} \text{ J} \cdot \text{s})(2.998 \times 10^8 \text{ m/s})}{4.00 \times 10^{-7} \text{ m}} = 4.97 \times 10^{-19}$ J

   $E_{min} = 4.97 \times 10^{-19}$ J $- 1.54 \times 10^{-19}$ J $= $ __3.43×10^{-19} J__

(b) Since longer wavelengths correspond to lower energies, E_{min} would correspond to the longest wavelength.

$$\lambda = \frac{hc}{E_{min}} = \frac{(6.626 \times 10^{-34} \text{ J} \cdot \text{s})(2.998 \times 10^8 \text{ m/s})}{3.42 \times 10^{-19} \text{ J}} = 5.81 \times 10^{-7} \text{ m} = \underline{\textbf{581 nm}}$$

COVALENT BONDING

7.1 LEWIS STRUCTURES

In this section, the solution to the problems follows the same acronyms used in the textbook:

Acronym	Meaning
VE	total valence electrons
NE	total needed electrons
AE	total available electrons

2. See Examples 7.1, 7.2 and 7.4. Use the steps presented on p.158 of the textbook as a guide. If AE ≤ NE follow Figure 7.3 but if AE > NE follow Figure 7.9.

(a) NH₃	Solution
Skeleton	H—N—H | H
VE	3×1 (for H) + 5(for N) = 8
AE	AE = VE − (2 × number of bonds) = 8 − (2 × 3 bonds) = 2
NE	2 (for N to have an octet) = 2
AE = NE ?	Yes. Complete the octet for every atom (except H).
Lewis Structure	H—N̈—H | H

(b) KrF₂	Solution
Skeleton	F——Kr——F
VE	2×7 (for F) + 8(for Kr) = 22
AE	AE = VE − (2 × number of bonds) = 22 − (2 × 2 bonds) = 18
NE	2×6 (for every F to have an octet) + 4 (for Kr) = 16
AE = NE ?	No. AE > NE. Complete the octet for every atom and add the 2 extra electrons to the central atom.
Lewis Structure	:F̈——K̈r——F̈:

(c)	NO$^+$	Solution
	Skeleton	$\left[\,\text{N}\!-\!\text{O}\,\right]^+$
	VE	5 (for N) + 6(for O) − 1(from the +1 charge) = 10
	AE	AE = VE − (2 × number of bonds) = 10 − (2 × 1 bond) = 8
	NE	6 (for N to have an octet) + 6 (for O) = 12
	AE = NE ?	No. AE < NE by 4 electrons ⇒ change the single bond to a triple bond. Complete the octet for every atom.
	Lewis Structure	$\left[\,:\!\text{N}\!\equiv\!\text{O}:\,\right]^+$

(d)	BrO$_2^-$	Solution
	Skeleton	$\left[\,\text{O}\!-\!\text{Br}\!-\!\text{O}\,\right]^-$
	VE	2×6 (for O) + 7(for Br) + 1(from the −1 charge) = 20
	AE	AE = VE − (2 × number of bonds) = 20 − (2 × 2 bonds) = 16
	NE	2×6 (for every O to have an octet) + 4 (for Br) = 16
	AE = NE ?	Yes. Complete the octet for every atom.
	Lewis Structure	$\left[\,:\!\ddot{\text{O}}\!-\!\ddot{\text{Br}}\!-\!\ddot{\text{O}}:\,\right]^-$

4. See Examples 7.1, 7.2 and 7.4. Use the steps presented on p.158 of the textbook as a guide. If AE ≤ NE follow Figure 7.3 but if AE > NE follow Figure 7.9.

(a)	ClF$_4^-$	Solution
	Skeleton	$\left[\begin{array}{c}\text{F}\\|\\\text{F}\!-\!\text{Cl}\!-\!\text{F}\\|\\\text{F}\end{array}\right]^-$
	VE	4×7 (for F) + 7(for Cl) + 1(from the −1 charge) = 36
	AE	AE = VE − (2 × number of bonds) = 36 − (2 × 4 bonds) = 28
	NE	4×6 (for every F to have an octet) = 24
	AE = NE ?	No. AE > NE. Complete the octet for every atom and add the 4 extra electrons to the central atom.
	Lewis Structure	$\left[\begin{array}{c}:\!\ddot{\text{F}}:\\:\!\ddot{\text{F}}\!-\!\ddot{\text{Cl}}\!-\!\ddot{\text{F}}:\\:\!\ddot{\text{F}}:\end{array}\right]^-$

(b) PF_6^-	Solution
Skeleton	
VE	6×7 (for F) + 5(for P) + 1(from the −1 charge) = 48
AE	AE = VE − (2 × number of bonds) = 48 − (2 × 6 bonds) = 36
NE	6×6 (for every F to have an octet) = 36
AE = NE ?	Yes. Complete the octet for all the F atoms.
Lewis Structure	

(c) CNS^-	Solution
Skeleton	[S—C—N]$^-$
VE	6 (for S) + 4(for C) + 5(for N) + 1(from the −1 charge) = 16
AE	AE = VE − (2 × number of bonds) = 16 − (2 × 2 bonds) = 12
NE	6 (for S to have an octet) + 4 (for C) + 6 (for N) = 16
AE = NE ?	No. AE < NE by 4 electrons ⇒ change one of the single bonds to a triple bond. Complete the octet for every atom.
Lewis Structure	$\left[:\ddot{S}—C≡N: \right]^-$

(d) $SnCl_5^-$	Solution
Skeleton	
VE	5×7 (for Cl) + 4(for Sn) + 1(from the −1 charge) = 40
AE	AE = VE − (2 × number of bonds) = 40 − (2 × 5 bonds) = 30

NE	6×5 (for every Cl to have an octet) = 30
AE = NE ?	Yes. Complete the octet for all Cl atoms.
Lewis Structure	

6. See Examples 7.1, 7.2 and 7.4. Use the steps presented on p.158 of the textbook as a guide. If AE ≤ NE follow Figure 7.3 but if AE > NE follow Figure 7.9.

(a) C_2^{2-}	Solution
Skeleton	$\left[C{-}C \right]^{2-}$
VE	2×4 (for C) + 2(from the −2 charge) = 10
AE	AE = VE − (2 × number of bonds) = 10 − (2 × 1 bond) = 8
NE	2×6 (for every C to have an octet) = 12
AE = NE ?	No. AE < NE by 4 electrons ⇒ change the single bond to a triple bond. Complete the octet for every atom.
Lewis Structure	$\left[{:}C{\equiv}C{:} \right]^{2-}$

(b) NFO	Solution
Skeleton	O—N—F
VE	6 (for O) + 5(for N) + 7(for F) = 18
AE	AE = VE − (2 × number of bonds) = 18 − (2 × 2 bonds) = 14
NE	6 (for O to have an octet) + 4 (for N) + 6 (for F) = 16
AE = NE ?	No. AE < NE by 2 electrons ⇒ change the O–N single bond to a double bond. Complete the octet for every atom.
Lewis Structure	:Ö=N̈—F̈ : (Note: halogens like F do not form multiple bonds)

(c) BrF₄⁺	Solution
Skeleton	$$\left[\begin{array}{c} \text{F} \\ | \\ \text{F---Br---F} \\ | \\ \text{F} \end{array}\right]^{+}$$
VE	4×7 (for F) + 7(for Br) − 1(from the +1 charge) = 34
AE	AE = VE − (2 × number of bonds) = 34 − (2 × 4 bonds) = 26
NE	4×6 (for every F to have an octet) = 24
AE = NE ?	No. AE > NE. Complete the octet for every atom and add the 2 extra electrons to the central atom.
Lewis Structure	$$\left[\begin{array}{c} :\ddot{\text{F}}: \\ | \\ :\ddot{\text{F}}\text{---}\dot{\text{Br}}\text{---}\ddot{\text{F}}: \\ | \\ :\ddot{\text{F}}: \end{array}\right]^{+}$$

(d) NI₃	Solution
Skeleton	$$\begin{array}{c} \text{I---N---I} \\ | \\ \text{I} \end{array}$$
VE	3×7 (for I) + 5(for N) = 26
AE	AE = VE − (2 × number of bonds) = 26 − (2 × 3 bonds) = 20
NE	3×6 (for every I to have an octet) + 2 (for N) = 20
AE = NE ?	Yes. Complete the octet for every atom.
Lewis Structure	$$\begin{array}{c} :\ddot{\text{I}}\text{---}\ddot{\text{N}}\text{---}\ddot{\text{I}}: \\ | \\ :\ddot{\text{I}}: \end{array}$$

8. See Examples 7.1, 7.2 and 7.4. Use the steps presented on p.158 of the textbook as a guide. If AE ≤ NE follow Figure 7.3 but if AE > NE follow Figure 7.9.

HOC⁺	Solution
Skeleton	$[\text{H}\!-\!\text{O}\!-\!\text{C}]^{+}$
VE	1 (for H) + 6 (for O) + 4(for C) −1(from the +1 charge) = 10
AE	AE = VE − (2 × number of bonds) = 10 − (2 × 2 bonds) = 6
NE	4 (for O to have an octet) + 6 (for C to have an octet) = 10
AE = NE ?	No. AE < NE by 4 electrons ⇒ change the O–C single bond to a triple bond. Complete the octet for C.
Lewis Structure	$[\text{H}\!-\!\text{O}\!\equiv\!\text{C}\!:\,]^{+}$

10. See Examples 7.1, 7.2 and 7.4. Use the steps presented on p.158 of the textbook as a guide. If AE ≤ NE follow Figure 7.3 but if AE > NE follow Figure 7.9.

(a)	Solution
Skeleton	(skeleton structure with H—C—C backbone, ᵃO above and ᵇO—H below)
VE	4×1 (for H) + 2×4 (for C) + 2×6 (for O) = 24
AE	AE = VE − (2 × number of bonds) = 24 − (2 × 7 bonds) = 10
NE	2 (for 2ND C to have an octet) + 6 (for ᵃO) + 4 (for ᵇO)= 12
AE = NE ?	No. AE < NE by 2 electrons ⇒ change the C–ᵃO single bond to a double bond. Complete the octet for the two oxygen atoms.
Lewis Structure	(Lewis structure with C=O double bond and O—H)

(b) $Br_2C–CF_2$	Solution
Skeleton	
VE	4×7 (for Br and F) + 2×4 (for C) = 36
AE	AE = VE – (2 × number of bonds) = 36 – (2 × 5 bonds) = 26
NE	4×6 (for all Br and F to have an octet) + 2×2 (for each C) = 28
AE = NE ?	No. AE < NE by 2 electrons ⇒ change the C–C single bond to a double bond. Complete the octet for every atom.
Lewis Structure	

(c) $H_3C–NH_2$	Solution
Skeleton	
VE	5×1 (for H) + 4(for C) + 5(for N) = 14
AE	AE = VE – (2 × number of bonds) = 14 – (2 × 6 bonds) = 2
NE	2 (for N to have an octet) = 2
AE = NE ?	Yes. Complete the octet for N.
Lewis Structure	

12. See Examples 7.1, 7.2 and 7.4. Use the steps presented on p.158 of the textbook as a guide. If AE ≤ NE follow Figure 7.3 but if AE > NE follow Figure 7.9.

HCOOH	Solution
Skeleton	
VE	2×1 (for H) + 4 (for C) + 2×6 (for O) = 18
AE	AE = VE – (2 × number of bonds) = 18 – (2 × 4 bonds) = 10
NE	2(for C) + 6 (for every ^{a}O) + 4 (for ^{b}O) = 12
AE = NE ?	No. AE < NE by 2 electrons ⇒ change C–^{a}O to a double bond. Complete the octet for ^{b}O.
Lewis Structure	

14. See Examples 7.1, 7.2 and 7.4. Use the steps presented on p.158 of the textbook as a guide. If AE ≤ NE follow Figure 7.3 but if AE > NE follow Figure 7.9.

One of the two compounds described in the problem with a formula of C_2H_6O:

C_2H_6O	Solution
Skeleton	O bonded to both C
VE	6×1 (for H) + 2×4 (for C) + 6 (for O) = 20
AE	AE = VE – (2 × number of bonds) = 20 – (2 × 8 bonds) = 4
NE	4 (for O to have an octet) = 4
AE = NE ?	Yes. Complete the octet for O.
Lewis Structure	

The second compound described in the problem with a formula of C_2H_6O:

C_2H_6O	Solution
Skeleton	O bonded to only 1 C 2 C's bonded to each other $\begin{array}{cc} H & H \\ | & | \\ H-C-C-O-H \\ | & | \\ H & H \end{array}$
VE	6×1 (for H) + 2×4 (for C) + 6 (for O) = 20
AE	AE = VE − (2 × number of bonds) = 20 − (2 × 8 bonds) = 4
NE	4 (for O to have an octet) = 4
AE = NE ?	Yes. Complete the octet for O.
Lewis Structure	$\begin{array}{cc} H & H \\ | & | \\ H-C-C-\ddot{\underset{\cdot\cdot}{O}}-H \\ | & | \\ H & H \end{array}$

16. See Examples 7.1, 7.2 and 7.4. Use the steps presented on p.158 of the textbook as a guide. If AE ≤ NE follow Figure 7.3 but if AE > NE follow Figure 7.9.

(a) BrO^-	Solution
Skeleton	$\left[Br-O \right]^-$
VE	7 (for Br) + 6 (for O) + 1(from the −1 charge) = 14
AE	AE = VE − (2 × number of bonds) = 14 − (2 × 1 bond) = 12
NE	6 (for Br to have an octet) + 6(for O) = 12
AE = NE ?	Yes. Complete the octet for every atom.
Lewis Structure	$\left[:\ddot{\underset{\cdot\cdot}{Br}}-\ddot{\underset{\cdot\cdot}{O}}: \right]^-$

The Lewis structure of OBr⁻ is similar to that of the halogen molecules, F_2, Cl_2, Br_2, and I_2.

$\left[:\ddot{\underset{\cdot\cdot}{Br}}-\ddot{\underset{\cdot\cdot}{O}}: \right]^-$ $\qquad$ $:\ddot{\underset{\cdot\cdot}{F}}-\ddot{\underset{\cdot\cdot}{F}}:$ $\qquad$ $:\ddot{\underset{\cdot\cdot}{Cl}}-\ddot{\underset{\cdot\cdot}{Cl}}:$ $\qquad$ $:\ddot{\underset{\cdot\cdot}{Br}}-\ddot{\underset{\cdot\cdot}{Br}}:$ $\qquad$ $:\ddot{\underset{\cdot\cdot}{I}}-\ddot{\underset{\cdot\cdot}{I}}:$

BrO^- $\qquad\qquad\qquad$ F_2 $\qquad\qquad$ Cl_2 $\qquad\qquad$ Br_2 $\qquad\qquad$ I_2

(b) NH$_4^+$	Solution
Skeleton	$$\left[\begin{array}{c} H \\ \mid \\ H-N-H \\ \mid \\ H \end{array}\right]^+$$
VE	4×1 (for H) + 5 (for N) − 1(from the +1 charge) = 8
AE	AE = VE − (2 × number of bonds) = 8 − (2 × 4 bonds) = 0
NE	0
AE = NE ?	Yes. Octet rule is already satisfied.
Lewis Structure	$$\left[\begin{array}{c} H \\ \mid \\ H-N-H \\ \mid \\ H \end{array}\right]^+$$

The Lewis structure of NH$_4^+$ is similar to that of CH$_4$.

$$\left[\begin{array}{c} H \\ \mid \\ H-N-H \\ \mid \\ H \end{array}\right]^+ \qquad \begin{array}{c} H \\ \mid \\ H-C-H \\ \mid \\ H \end{array}$$

$$\text{NH}_4^+ \qquad\qquad\qquad \text{CH}_4$$

(c) CN$^-$	Solution
Skeleton	$$\left[\ C-N\ \right]^-$$
VE	4 (for C) + 5 (for N) + 1(from the -1 charge) = 10
AE	AE = VE − (2 × number of bonds) = 10 − (2 × 1 bond) = 8
NE	6 (for C to have an octet) + 6(for N) = 12
AE = NE ?	No. AE < NE by 4 electrons ⇒ change C–N to a triple bond. Complete the octet for both C and N.
Lewis Structure	$$\left[\ :C\!\equiv\!N:\ \right]^-$$

The Lewis structure of CN$^-$ is similar to that of CO.

$$\left[\ :C\!\equiv\!N:\ \right]^- \qquad\qquad :C\!\equiv\!O:$$

$$\text{CN}^- \qquad\qquad\qquad \text{CO}$$

(d) SO_4^{2-}	Solution
Skeleton	
VE	6×4 (for O) + 6 (for S) + 2(from the −2 charge) = 32
AE	AE = VE − (2 × number of bonds) = 32 − (2 × 4bonds) = 24
NE	4×6 (for O to have an octet) = 24
AE = NE ?	Yes. Complete the octet for every atom.
Lewis Structure	

The Lewis structure of SO_4^{2-} is similar to that of CCl_4 and SiF_4.

18. See Examples 7.1, 7.2, and 7.4. Use the steps presented on p.158 of the textbook as a guide. If AE ≤ NE follow Figure 7.3 but if AE > NE follow Figure 7.9.

(a) BCl_4^-	Solution
Skeleton	
VE	4×7 (for Cl) + 3 (for B) + 1(from the −1 charge) = 32
AE	AE = VE − (2 × number of bonds) = 32 − (2 × 4 bonds) = 24
NE	4×6 (for Cl to have an octet) = 24
AE = NE ?	Yes. Complete the octet for all the atoms.
Lewis Structure	

(b) ClO⁻	Solution
Skeleton	$\left[\text{Cl—O} \right]^-$
VE	7 (for Cl) + 6 (for O) + 1(from the −1 charge) = 14
AE	AE = VE − (2 × number of bonds) = 14 − (2 × 1 bond) = 12
NE	6 (for Br to have an octet) + 6(for O) = 12
AE = NE ?	Yes. Complete the octet for every atom.
Lewis Structure	$\left[:\ddot{\text{Cl}}—\ddot{\text{O}}: \right]^-$

(c) S₂O₃²⁻	Solution
Skeleton	$\begin{bmatrix} \text{O} \\ \| \\ \text{S—S—O} \\ \| \\ \text{O} \end{bmatrix}^{2-}$
VE	3×6 (for O) + 2×6 (for S) + 2(from the −2 charge) = 32
AE	AE = VE − (2 × number of bonds) = 32 − (2 × 4 bonds) = 24
NE	4×6 (for O and S to have an octet) = 24
AE = NE ?	Yes. Complete the octet for every atom.
Lewis Structure	$\begin{bmatrix} :\ddot{\text{O}}: \\ \| \\ :\ddot{\text{S}}—\text{S}—\ddot{\text{O}}: \\ \| \\ :\ddot{\text{O}}: \end{bmatrix}^{2-}$

(d) NFCl₂	Solution
Skeleton	$\begin{matrix} \text{F—N—Cl} \\ \| \\ \text{Cl} \end{matrix}$
VE	1×7 (for F) + 2×7 (for Cl) + 5(for N) = 26
AE	AE = VE − (2 × number of bonds) = 26 − (2 × 3 bonds) = 20
NE	3×6 (for F and Cl to have an octet) + 2 (for N) = 20
AE = NE ?	Yes. Complete the octet for every atom.
Lewis Structure	$\begin{matrix} :\ddot{\text{F}}—\ddot{\text{N}}—\ddot{\text{Cl}}: \\ \| \\ :\ddot{\text{Cl}}: \end{matrix}$

20. A number of molecules like BF_3 (see Example 7.5 for the Lewis structure of BF_3) do not follow the octet rule. Some have an expanded octet where the central atom has more than eight electrons while others are electron deficient thus have less than eight.

(a) $BeCl_2$	Solution
Skeleton	Cl—Be—Cl
VE	2×7 (for Cl) + 2 (for Be) = 16
AE	AE = VE − (2 × number of bonds) = 16 − (2 × 2 bonds) = 12
NE	2×6 (for Cl to have an octet) = 12 Be does not follow the octet rule and usually form 2 bonds only
AE = NE ?	Yes. Complete the octet for Cl.
Lewis Structure	$:\overset{\cdot\cdot}{\underset{\cdot\cdot}{Cl}}$—Be—$\overset{\cdot\cdot}{\underset{\cdot\cdot}{Cl}}:$ **$BeCl_2$ does not follow the octet rule.**

(b) SeO_2^-	Solution
Skeleton	$\left[\,O{-}Se{-}O\,\right]^-$
VE	2×6 (for O) + 6 (for Se) + 1(from the −1 charge) = 19
AE	AE = VE − (2 × number of bonds) = 19 − (2 × 2 bonds) = 15
NE	2×6 (for O to have an octet) + 4(for Se) = 16
AE = NE ?	No. AE < NE by 1 electron. Complete the octet for every atom. Since only 15 (not 16) electrons are available for distribution, one of the atoms (Se) does not have an octet.
Lewis Structure	$\left[:\overset{\cdot\cdot}{\underset{\cdot\cdot}{O}}{-}\overset{\cdot}{Se}{-}\overset{\cdot\cdot}{\underset{\cdot\cdot}{O}}:\right]^-$

(c) ClO_3	Solution
Skeleton	O—Cl—O $\quad\quad\mid$ $\quad\quad$O
VE	3×6 (for O) + 1×7 (for Cl) = 25
AE	AE = VE − (2 × number of bonds) = 25 − (2 × 3 bonds) = 19
NE	3×6 (for O to have an octet) + 2(for Cl) = 20
AE = NE ?	No. AE < NE by 1 electron. Complete the octet for every atom. Since only 19 (not 20) electrons are available for distribution, one of the atoms (Cl) does not have an octet.
Lewis Structure	$:\overset{\cdot\cdot}{\underset{\cdot\cdot}{O}}{-}\overset{\cdot}{Cl}{-}\overset{\cdot\cdot}{\underset{\cdot\cdot}{O}}:$ $\quad\quad\mid$ $\quad\quad:\overset{}{\underset{\cdot\cdot}{O}}:$

(d) CH₃	Solution

Skeleton	H—C—H
	      |
	      H

VE	3×1 (for H) + 4 (for C) = 7
AE	AE = VE – (2 × number of bonds) = 7 – (2 × 3 bonds) = 1
NE	2 (for C to have an octet) = 2
AE = NE ?	No. AE < NE by 1 electron. Since only 1 electron (not 2) is available for distribution, C has only 7 electrons around it. CH₃ does not follow the octet rule.

Lewis Structure	H—Ċ—H
	      |
	      H

7.1 LEWIS STRUCTURES: Resonance Forms and Formal Charge

22. See Examples 7.1, 7.2, 7.3 and 7.4. Write the Lewis structures using the steps presented on p.158 of the textbook as a guide. If AE ≤ NE follow Figure 7.3 but if AE > NE follow Figure 7.9. Draw the resonance structures by changing the positions of the multiple bonds and/or the electron pairs without changing the position of the atoms in the skeleton.

(a) SeO₃	Solution

Skeleton	O—Se—O
	      |
	      O

VE	3×6 (for O) + 6 (for Se) = 24
AE	AE = VE – (2 × number of bonds) = 24 – (2 × 3 bonds) = 18
NE	3×6 (for O to have an octet) + 2(for Se) = 20
AE = NE ?	No. AE < NE by 2 electrons ⇒ change one of the Se–O single bond to a double bond. Complete the octet for every atom. Change the position of the double bond and the unshared electron pairs if needed. Remember that each atom should have an octet.
Lewis Resonance Structures	

(b) CS_3^{2-}	Solution
Skeleton	$\left[\begin{array}{c} S-C-S \\ \mid \\ S \end{array}\right]^{2-}$
VE	3×6 (for S) + 4 (for C) + 2(from the –2 charge) = 24
AE	AE = VE – (2 × number of bonds) = 24 – (2 × 3 bonds) = 18
NE	3×6 (for S to have an octet) + 2(for C) = 20
AE = NE ?	No. AE < NE by 2 electrons ⇒ change one of the S–O single bond to a double bond. Complete the octet for every atom. Change the position of the double bond and the unshared electron pairs if needed. Remember that each atom should have an octet.
Lewis Resonance Structures	

(c) CNO^-	Solution
Skeleton	$\left[\begin{array}{c} O-C-N \end{array}\right]^{-}$
VE	6 (for O) + 4 (for C) + 5 (for N) + 1(from the –1 charge) = 16
AE	AE = VE – (2 × number of bonds) = 16 – (2 × 2 bonds) = 12
NE	6 (for O to have an octet) + 4(for C) + 6(for N) = 16
AE = NE ?	No. AE < NE by 4 electrons ⇒ change the two single bonds to two double bonds. Complete the octet for every atom. Change the position of one of the bonded pairs and the unshared electron pairs to have a triple bond and a single bond. Remember that each atom should have an octet and that the total number of electrons should not change.
Lewis/ Resonance Structures	

24. See Examples 7.1, 7.2, 7.3 and 7.4. Write the Lewis structures using the steps presented on p.158 of the textbook as a guide. Follow Figure 7.3 if AE ≤ NE otherwise follow Figure 7.9 if AE > NE. Draw the resonance structures by changing the positions of the multiple bonds and/or the electron pairs without changing the position of the atoms in the skeleton.

$C_2O_4^{2-}$	Solution
Skeleton	
VE	2×4 (for C) + 4×6 (for O) + 2(from the −2 charge) = 34
AE	AE = VE − (2 × number of bonds) = 34 − (2 × 5 bonds) = 24
NE	4×6 (for O to have an octet) + 2×2 (for C) = 28
AE = NE ?	No. AE < NE by 4 electrons ⇒ change two of the C−O single bonds to double bonds. Complete the octet for every atom.
(a) Lewis structure	
(b) Resonance Structures	 Change the position of the multiple bonds and the unshared electron pairs such that an octet for each atom is maintained.

(c) No, the structure drawn is not a resonance structure of the oxalate ion because the position of their atoms are different. They have different skeleton structures.

26. See Examples 7.1, 7.2, 7.3 and 7.4. Draw the Lewis structures using the steps presented on p.158 of the textbook as a guide. If AE ≤ NE follow Figure 7.3 but if AE > NE follow Figure 7.9. Draw the resonance structures by changing the positions of the multiple bonds and/or the electron pairs without changing the position of the atoms in the skeleton.

$B_3N_3H_6$	Solution
Skeleton	
VE	3×3 (for B) + 3×5 (for N) + 6×1 (for H) = 30
AE	AE = VE − (2 × number of bonds) = 30 − (2 × 12 bonds) = 6
NE	2×3 (for N to have an octet) + 2×3 (for B) = 12
AE = NE ?	No. AE < NE by 6 electrons ⇒ change 3 of the B–N single bonds to double bonds. Complete the octet for every atom. Change the position of the multiple bonds and the unshared electron pairs such that an octet for each atom is maintained.
Lewis Resonance Structures	

28. See Table 7.2, the discussion on formal charge and the example given on page 163 on the two possible structures of methanol. As pointed out in the textbook, the formal charge, C_f can be obtained using the equation below or its modified (simplified) version.

C_f = VE − unshared electrons − ½(bonding electrons)

C_f = VE − unshared electrons − number of bonds   (modified equation)

In all problems requiring formal charges, the problems will be solved using the modified equation.

(a) O in HOF

For O: C_f = VE − unshared electrons − number of bonds = 6 − 4 − 2 = 0
For O: C_f = 0

(b) N in NO_2^-

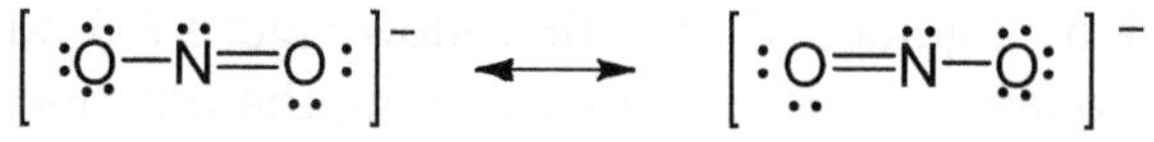

$$\left[:\ddot{O}{-}\ddot{N}{=}\ddot{O}: \right]^- \quad \longleftrightarrow \quad \left[:\ddot{O}{=}\ddot{N}{-}\ddot{O}: \right]^-$$

For N: C_f = VE – unshared electrons – number of bonds = $5 - 2 - 3 = 0$
For N: $C_f = 0$

(c) P in PCl_3

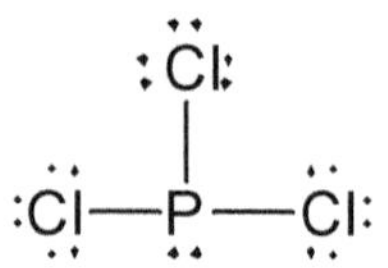

$$:\ddot{Cl}{:}$$
$$|$$
$$:\ddot{Cl}{-}\ddot{P}{-}\ddot{Cl}:$$

For P: C_f = VE – unshared electrons – number of bonds = $5 - 2 - 3 = 0$
For P: $C_f = 0$

30. See Table 7.2, the discussion on formal charge and the example given on page 163 on the two possible structures of methanol. As pointed out in the textbook, the formal charge, C_f can be obtained using the following equation:

$$C_f = \text{VE} - \text{unshared electrons} - \text{number of bonds}$$

Structure I

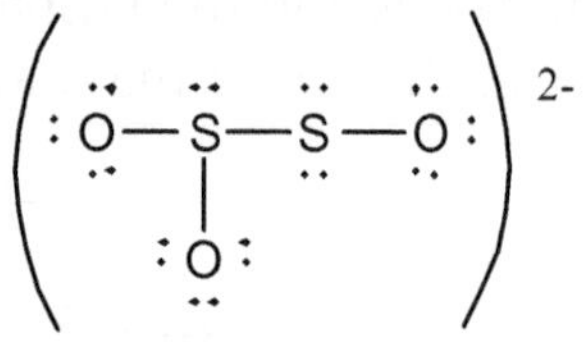

$$\left(:\ddot{O}{-}\ddot{S}{-}\ddot{S}{-}\ddot{O}: \right)^{2-}$$

Structure II

$$\left(\begin{array}{c} :\ddot{O}: \\ | \\ :\ddot{S}{-}\ddot{S}{-}\ddot{O}: \\ | \\ :\ddot{O}: \end{array} \right)^{2-}$$

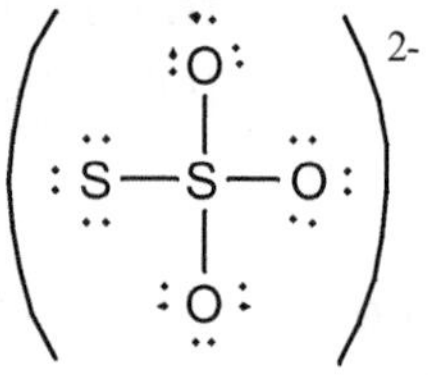

All the O in both structures has a C_f of –1 as shown below:

O: C_f = VE – unshared electrons – number of bonds = $6 - 6 - 1 = -1$

However, the formal charges of the S's in the two structures are different.

Structure I: left S: C_f = VE – unshared electrons – number of bonds = $6 - 2 - 3 = +1$

right S: C_f = VE – unshared electrons – number of bonds = $6 - 4 - 2 = 0$

Structure II: left S: C_f = VE – unshared electrons – number of bonds = $6 - 6 - 1 = -1$

right S: C_f = VE – unshared electrons – number of bonds = $6 - 0 - 4 = +2$

Structure I is the better and the more plausible structure because the formal charges on its sulfur atoms are closer to zero.

7.2 MOLECULAR GEOMETRY

32. See Examples 7.5 and 7.6. Write the Lewis structure and use Table 7.3 to identify the species type (AX_mE_n) and the corresponding geometry.

   (a) O_3

       Lewis structure:

       Species Type: A=O, X=O (2), E=1 $\Rightarrow$ AX_2E

       Geometry: __bent__, 120° bond angle

   (b) OCl_2

       Lewis structure:

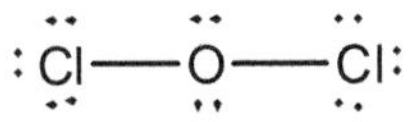

       Species Type: A=O, X=Cl (2), E=2 $\Rightarrow$ AX_2E_2

       Geometry: __bent__, 109.5° bond angle

   (c) $SnCl_3^-$

       Lewis structure:

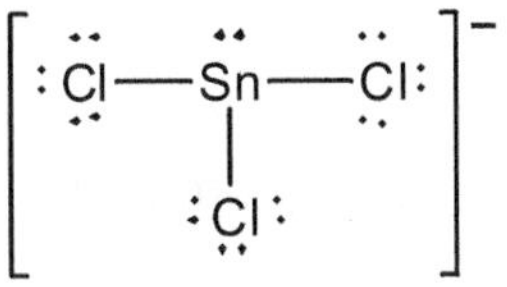

       Species Type: A=Sn, X=Cl (3), E=1 $\Rightarrow$ AX_3E

       Geometry: __trigonal pyramid__, 109.5° bond angle

   (d) CS_2

       Lewis structure:

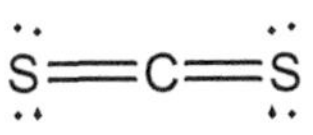

       Species Type: A=C, X=S (2), E=0 $\Rightarrow$ AX_2

       Geometry: __linear__, 180° bond angle

34. See Examples 7.5 and 7.6. Write the Lewis structure and use Table 7.3 to identify the species type and the corresponding geometry.

   (a) NNO

       Lewis structure:

       Species Type: A=N, X=N,O (2), E=0 $\Rightarrow$ AX_2

       Geometry: __linear__, 180° bond angle

(b) ONCl

Lewis structure: $: \overset{..}{O} = N - \overset{..}{\underset{..}{C}} l :$

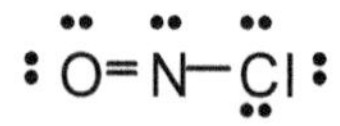

Species Type: A=N, X=O,Cl (2), E=1 ⟹ AX$_2$E
Geometry: **bent**, 120° bond angle

(c) NH$_4^+$

Lewis structure:

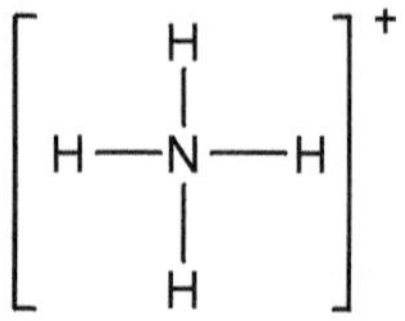

$$\left[\begin{array}{c} H \\ | \\ H - N - H \\ | \\ H \end{array} \right]^+$$

Species Type: A=N, X=H (4), E=0 ⟹ AX$_4$
Geometry: **tetrahedron**, 109.5° bond angle

(d) O$_3$

Lewis structure: $\overset{..}{O} = \overset{..}{O} - \overset{..}{O} : \quad \longleftrightarrow \quad : \overset{..}{O} - \overset{..}{O} = \overset{..}{O}$

Species Type: A=O, X=O (2), E=1 ⟹ AX$_2$E
Geometry: **bent**, 120° bond angle

36. See Examples 7.5 and 7.6. Write the Lewis structure and use Table 7.3 and Figure 7.11 to identify the species type and the corresponding geometry.

(a) ClF$_5$

Lewis structure:

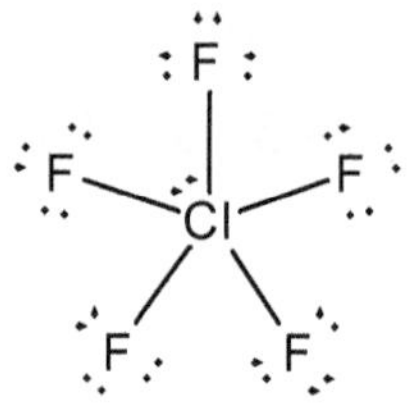

Species Type: A=Cl, X=F (5), E=1 ⟹ AX$_5$E
Geometry: **square pyramid**, 90°, 180° bond angles

(b) XeF$_4$

Lewis structure:

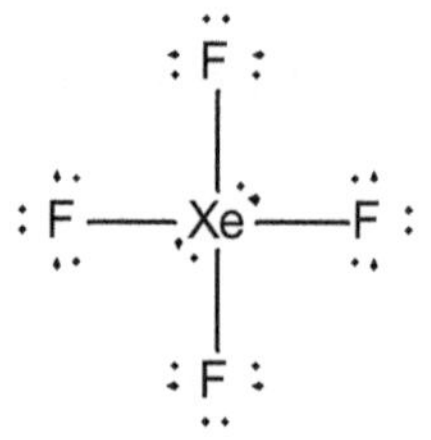

Species Type: A=Xe, X=F (4), E=2 ⟹ AX$_4$E$_2$
Geometry: **square planar**, 90°, 180° bond angles

(c) SiF_6^{2-}

Lewis structure:

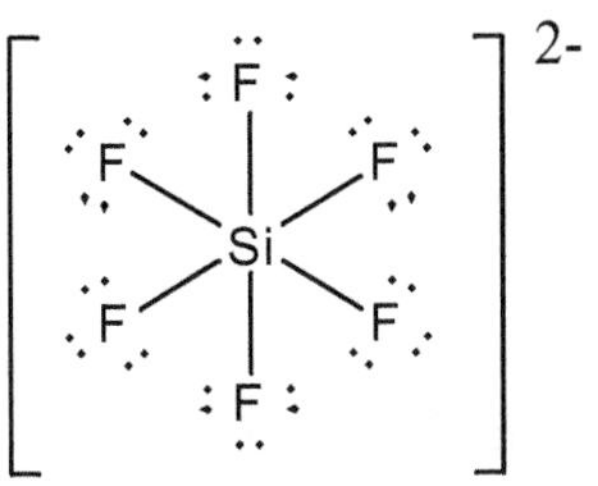

Species Type: A=Si, X=F (6), E=0 $\Rightarrow$ AX$_6$
Geometry: **octahedral**, 90°, 180° bond angles

(d) PCl_5

Lewis structure:

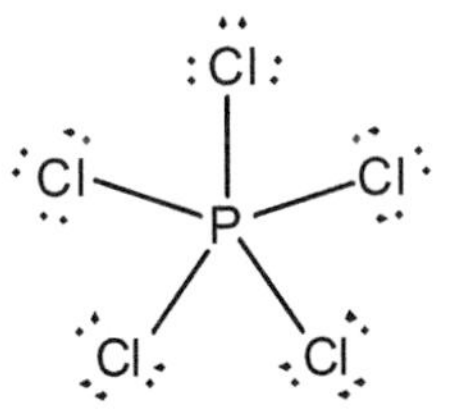

Species Type: A=P, X=Cl (5), E=0 $\Rightarrow$ AX$_5$
Geometry: **trigonal bipyramid**, 90°, 120°, 180° bond angles

38. See Examples 7.5 and 7.6. Write the Lewis structure of the given skeleton structure and use Table 7.3 and Figure 7.11 to identify the species type and the corresponding bond angles.

(a) Lewis structure:

Species Type: A=N, X=O (2), E=1 $\Rightarrow$ AX$_2$E
Geometry: bent, **120° bond angle**

(b) Lewis structure:

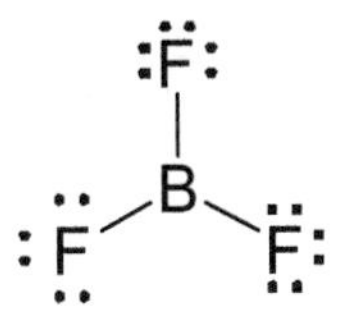

Species Type: A=B, X=F (3), E=0 $\Rightarrow$ AX$_3$
Geometry: trigonal planar, **120° bond angle**

(c) Lewis structure:

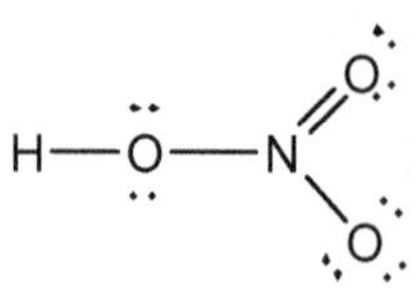

around N:
    Species Type: A=N, X=O (3), E=0 $\Rightarrow$ AX$_3$
    Geometry: trigonal planar, **120° bond angle**

around left O:
    Species Type: A=O, X=H,N (2), E=2 $\Rightarrow$ AX$_2$E$_2$
    Geometry: bent, **109.5° bond angle**

(d) Lewis structure:

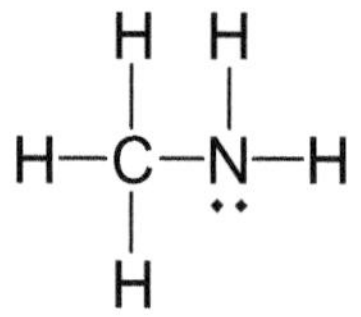

around C:
    Species Type: A=C, X=N,H(3), E=0 $\Rightarrow$ AX$_4$
    Geometry: tetrahedral, **109.5° bond angle**

around N:
    Species Type: A=N, X=C,H(2), E=1 $\Rightarrow$ AX$_3$E
    Geometry: trigonal pyramid, **109.5° bond angle**

40. See Examples 7.5 and 7.6. Write the Lewis structures using the steps presented on p.158 of the textbook as a guide. If AE ≤ NE follow Figure 7.3 but if AE > NE follow Figure 7.9. Use Table 7.3 and Figure 7.11 to identify the species type and the corresponding bond angles.

(a) Lewis structure:

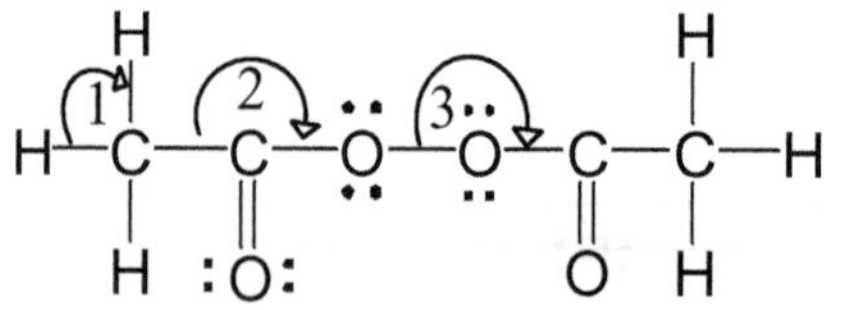

(b) Numbered angle 1 :

    Species Type: A=C, X=H(3),C E=0 $\Rightarrow$ AX$_4$
    Geometry: tetrahedral, **109.5° bond angle**

Numbered angle 2 :

    Species Type: A=C, X=C,O(2) E=0 $\Rightarrow$ AX$_3$
    Geometry: trigonal planar, **120° bond angle**

Numbered angle 3 :

    Species Type: A=O, X=C,O   E=2 $\Rightarrow$ AX_2E_2

    Geometry: bent, **109.5° bond angle**

Numbered angles: **1= 109.5°**   **2= 120°**   **3= 109.5°**

42. See Examples 7.5 and 7.6. Write the Lewis structure of niacin using the steps presented on p.158 of the textbook as a guide. If AE $\leq$ NE follow Figure 7.3 but if AE $>$ NE follow Figure 7.9. Use Table 7.3 and Figure 7.11 to identify the species type and the corresponding values of the indicated bond angles.

Numbered angle 1 :

    Species Type: A=N, X=C (2), E=1 $\Rightarrow$ AX_2E

    Geometry: bent, **120° bond angle**

Numbered angle 2 :

    Species Type: A=O, X=C,H (2), E=2 $\Rightarrow$ AX_2E_2

    Geometry: bent, **109.5° bond angle**

Numbered angle 3 :

    Species Type: A=C, X=C,H,N (3), E=0 $\Rightarrow$ AX_3

    Geometry: trigonal planar, **120° bond angle**

Numbered angles: **1= 120°**   **2= 109.5°**   **3= 120°**

7.3 POLARITY OF MOLECULES

44. See Examples 7.7 and 7.8. Write the Lewis structure. Use Table 7.3 and Figure 7.11 to identify the species type and its geometry. Look at the A–X bonds. The molecule is polar if any of the two conditions apply: (i) the terminal atoms are not identical and (ii) the terminal atoms are identical but they are arranged around the central atom non-symmetrically. If none of these two conditions apply then the molecule is nonpolar.

(a) O_3 – not dipole because the molecule does not have a polar bond.

(b) OCl_2	Solution	
Lewis structure	$:\overset{..}{\underset{..}{Cl}}—\overset{..}{\underset{..}{O}}—\overset{..}{\underset{..}{Cl}}:$	
Species Type	AX_2E_2	
Geometry	bent	
Identical terminal atoms?	yes	dipole or polar
Symmetric A–X bonds?	no	

(c) $SnCl_3^-$	Solution	
Lewis structure	$\left[\,:\overset{..}{\underset{..}{Cl}}—\overset{..}{Sn}—\overset{..}{\underset{..}{Cl}}:\,\right]^{-}$ with $:\overset{}{\underset{..}{Cl}}:$ below	
Species Type	AX_3E	
Geometry	trigonal pyramidal	
Identical terminal atoms?	yes	dipole or polar
Symmetric A–X bonds?	no	

(d) CS_2	Solution	
Lewis structure	$\overset{..}{\underset{..}{S}}{=}C{=}\overset{..}{\underset{..}{S}}$	
Species Type	AX_2	
Geometry	linear	
Identical terminal atoms?	yes	nonpolar
Symmetric A–X bonds?	yes	

Only (b) and (c) are dipoles.

46. See Examples 7.7 and 7.8. Write the Lewis structure. Use Table 7.3 and Figure 7.11 to identify the species type and its geometry. Look at the A–X bonds. The molecule is polar if any of the two conditions apply: (i) the terminal atoms are not identical and (ii) the terminal atoms are identical but they are arranged around the central atom non-symmetrically. If none of these two conditions apply then the molecule is nonpolar.

(a) NNO	Solution
Lewis structure	$:\ddot{N}=N=\ddot{O}:$
Species Type	AX_2
Geometry	linear
Identical terminal atoms?	no $\Rightarrow$ polar

(b) ONCl	Solution
Lewis structure	$:\ddot{O}=\ddot{N}-\ddot{C}l:$
Species Type	AX_2E
Geometry	bent
Identical terminal atoms?	no $\Rightarrow$ polar

(c) NH_4^+	Solution	
Lewis structure	$\begin{bmatrix} & H & \\ & | & \\ H-N-H \\ & | & \\ & H & \end{bmatrix}^+$	
Species Type	AX_4	
Geometry	tetrahedral	
Identical terminal atoms?	yes	nonpolar
Symmetric A–X bonds?	yes	

(d) O_3 – not dipole because the molecule does not have a polar bond.

Only (a) and (b) are dipoles.

48. See Examples 7.7 and 7.8. Write the Lewis structure. Use Table 7.3 and Figure 7.11 to identify the species type and its geometry. Look at the A–X bonds. The molecule is polar if any of the two conditions apply: (i) the terminal atoms are not identical and (ii) the terminal atoms are identical but they are arranged around the central atom non-symmetrically. If none of these two conditions apply then the molecule is nonpolar.

| molecule I | molecule II |

F is more electronegative than N. Thus, the N–F is a polar bond.

In molecule 1 the two dipoles due to N–F do not cancel resulting in a net dipole moment which is indicated by the dipole arrow above. The arrow head is the negative pole.

On the other hand, two dipoles due to N–F in molecule 2 cancel out each other resulting in a zero net dipole moment.

Therefore, of the two molecules **only molecule I is polar**.

7.4 ATOMIC ORBITALS; HYBRIDIZATION

50. See Examples 7.9 and 7.10. Write the Lewis structure. Count bonds (m) and unshared **pairs** (n) around the central atom. Note that for multiple bonds, only one is a hybrid orbital so in counting m for a multiple bond, count it only as 1. Consult the following table for the hybridization.

m + n	hybridization
2	sp
3	sp^2
4	sp^3
5	sp^3d
6	sp^3d^2

Formula	Lewis Structure	m + n (central atom)	Hybridization of the central atom
(a) O_3	$\overset{..}{O}=\overset{..}{O}-\overset{..}{\underset{..}{O}}:$	2 + 1= 3	**sp^2**
(b) OCl_2	$:\overset{..}{\underset{..}{Cl}}-\overset{..}{\underset{..}{O}}-\overset{..}{\underset{..}{Cl}}:$	2 + 2 = 4	**sp^3**

Formula	Lewis Structure	m + n (central atom)	Hybridization of the central atom
(c) $SnCl_3^-$	$\left[\ :\overset{..}{\underset{..}{Cl}}-Sn-\overset{..}{\underset{..}{Cl}}:\ \right]^-$ with $:\overset{}{\underset{..}{Cl}}:$ below	3 + 1 = 4	__sp³__
(d) CS_2	$\overset{..}{S}=C=\overset{..}{S}$	2 + 0 = 2	__sp__

52. See Examples 7.9 and 7.10. Write the Lewis structure. Count bonds (m) and unshared **pairs** (n) around the central atom. Note that for multiple bonds, only one is a hybrid orbital so in counting m for a multiple bond, count it only as 1. Consult the table for the hybridization given in the solution to the previous problem (#50).

Formula	Lewis Structure	m + n (central atom)	Hybridization of the central atom
(a) NNO	$:\overset{..}{N}=N=\overset{..}{O}:$	2 + 0 = 2	__sp__
(b) ONCl	$:\overset{..}{O}=\overset{..}{N}-\overset{..}{\underset{..}{Cl}}:$	2 + 1 = 3	__sp²__
(c) NH_4^+	$\left[\ H-\overset{H}{\underset{H}{N}}-H\ \right]^+$	4 + 0 = 4	__sp³__
(d) O_3	$\overset{..}{\underset{..}{O}}=\overset{..}{O}-\overset{..}{\underset{..}{O}}:$	2 + 1 = 3	__sp²__

54. See Examples 7.9 and 7.10. Write the Lewis structure. Count bonds (m) and unshared **pairs** (n) around the central atom. Note that for multiple bonds, only one is a hybrid orbital so in counting m for a multiple bond, count it only as 1. Consult the table for the hybridization given in the solution to the previous problem (#50).

Formula	Lewis Structure	m + n (central atom)	Hybridization of the central atom
(a) ClF_5	(square pyramidal with central Cl and five F atoms)	5 + 1 = 6	__sp³d²__

Formula	Lewis Structure	m + n (central atom)	Hybridization of the central atom
(b) XeF_4		4 + 2 = 6	**sp^3d^2**
(c) SiF_6^{2-}		6 + 0 = 6	**sp^3d^2**
(d) PCl_5		5 + 0 = 5	**sp^3d**

56. See Examples 7.9 and 7.10. Write the Lewis structure. Count bonds (m) and unshared **pairs** (n) around the central atom. Note that for multiple bonds, only one is a hybrid orbital so in counting m for a multiple bond, count it only as 1. Consult the table for the hybridization given in the solution to the previous problem (#50).

Formula	Lewis Structure	m + n (number of electron pairs around the central atom)	Hybridization of the central atom
(a) ClF_4^-		4 + 2 = 6 **6 electron pairs**	**sp^3d^2**
(b) $GeCl_6^{2-}$		6 + 0 = 6 **6 electron pairs**	**sp^3d^2**

Formula	Lewis Structure	m + n (number of electron pairs around the central atom)	Hybridization of the central atom
(c) $SbCl_4^-$		4 + 1 = 5 **5 electron pairs**	**sp^3d**

58. See Examples 7.9 and 7.10. Count bonds (m) and unshared **pairs** (n) around the central atom. Note that for multiple bonds, only one is a hybrid orbital so in counting m for a multiple bond, count it only as 1. Consult the table for the hybridization given in the solution to the previous problem (#50).

Formula	Lewis Structure	m + n	Hybridization
C_3H_3N		C_1: 3 + 0 = 3 C_2: 3 + 0 = 3 C_3: 2 + 0 = 2 N: 1 + 1 = 2	C_1: **sp^2** C_2: **sp^2** C_3: **sp** N: **sp**

60. See Examples 7.9 and 7.10. Write the Lewis structure. Count bonds (m) and unshared **pairs** (n) around the central atom. Note that for multiple bonds, only one is a hybrid orbital so in counting m for a multiple bond, count it only as 1. Consult the table for the hybridization given in the solution to the previous problem (#50).

Formula	Lewis Structure	m + n (around carbon)	Hybridization of carbon
(a) CH_3Cl		4 + 0 = 4	**sp^3**

	Lewis Structure	m + n (around carbon)	Hybridization of carbon
(b)	$\left[\ :\overset{..}{\underset{..}{O}}-C-\overset{..}{\underset{..}{O}}:\ \right]^{2-}$ (with $:\overset{..}{O}:$ double bonded below C)	3 + 0 = 3	<u>sp</u>^{2}
(c)	$:\overset{..}{O}=C=\overset{..}{O}:$	2 + 0 = 2	<u>sp</u>
(d)	$H-C-\overset{..}{\underset{..}{O}}-H$ (with $:\overset{..}{O}:$ double bonded below C)	3 + 0 = 3	<u>sp</u>^{2}

62. See Examples 7.9 and 7.10. Write the Lewis structure. Count bonds (m) and unshared <u>**pairs**</u> (n) around the central atom. Note that for multiple bonds, only one is a hybrid orbital so in counting m for a multiple bond, count it only as 1. Consult the table for the hybridization given in the solution to the previous problem (#50).

	Lewis Structure	m + n (around Br)	Hybridization of iodine
(a) $HO\underline{Br}O_3$	$\overset{\displaystyle :O:}{\underset{\displaystyle :O:}{H-\overset{..}{O}-\overset{\|}{Br}=O}}$	4 + 0 = 4	<u>sp</u>^{3}

	Lewis Structure	m + n (around O)	Hybridization of oxygen
(b) $Cl_2\underline{O}$	$Cl^{\diagdown}\overset{\overset{..}{O}}{\underset{..}{}}{}^{\diagup}Cl$	2 + 2 = 4	<u>sp</u>^{3}

	Lewis Structure	m + n (around P)	Hybridization of phosphorus
(c) $O\underline{P}Br_2$	$\overset{\displaystyle :\overset{..}{Br}:}{\underset{\displaystyle :\overset{..}{Br}:}{O=\overset{\|}{P}-\overset{..}{Br}:}}$	4 + 0 = 4	<u>sp</u>^{3}

7.4 ATOMIC ORBITALS; *Sigma* and *Pi* Bonds

64. See Example 7.11 particularly the end point. Count all the single bonds and all the multiple bonds in the Lewis structure. Only one bond in a multiple bond is a *sigma* bond. In a double bond, there is 1 *sigma* bond and 1 *pi* bond. In a triple bond, there is 1 *sigma* bond and 2 *pi* bonds.

Lewis structure:

Sigma bonds: 4 (from single bonds) + 1 (from double bond) + 1 (from the triple bond) = 6

Pi bonds: 1 (from double bond) + 2 (from the triple bond) = 3

Acrylontirile has 6 *sigma* bonds and 3 *pi* bonds (6σ, 3π).

66. See Example 7.11. Count all the single bonds and all the multiple bonds in the Lewis structure. Only one bond in a multiple bond is a *sigma* bond. In a double bond, there is 1 *sigma* bond and 1 *pi* bond. In a triple bond, there is 1 *sigma* bond and 2 *pi* bonds.

(a) Lewis structure:

Sigma bonds: 4 (from single bonds) = 4
Pi bonds: none. Since there are no multiple bonds, there are no *pi* bonds.
CH_3Cl has 4 *sigma* bonds and no *pi* bond (4σ).

(b) Lewis structure:

Sigma bonds: 2 (from single bonds) + 1 (from the double bond) = 3
Pi bonds: 1 (from the double bond)
CO_3^{2-} has 3 *sigma* bonds and 1 *pi* bond (3σ, 1π).

(c) Lewis structure:

Sigma bonds: 2 (from the double bond) = 2
Pi bonds: 2 (from the double bond) = 2
CO_2 has 2 *sigma* bonds and 2 *pi* bonds (2σ, 2π).

(d)   Lewis structure:

Sigma bonds: 3 (from single bonds) + 1 (from the double bond)   = 4
Pi bonds: 1 (from the double bond) = 1
CO_2 has **4 *sigma* bonds and 1 *pi* bond (4σ, 1π).**

UNCLASSIFIED PROBLEMS

68. See Section 7.1, Examples 7.1, 7.2 and 7.4. Use the steps presented on p.158 of the textbook as a guide in writing the Lewis structures. If AE ≤ NE follow Figure 7.3 but if AE > NE follow Figure 7.9.

$S_2O_7^{2-}$ Skeleton	
VE	7×6 (for O) + 2×6 (for S) + 2 (for the −2 charge) = 56
AE	AE = VE − (2 × number of bonds) = 56 − (2 × 8 bonds) = 40
NE	1×4 (for central O to have an octet) + 6×6 (for terminal O) = 40
AE = NE ?	Yes. Complete the octet for every atom.
(a) Lewis Structure	
(b) Formal Charge on S	C_f = VE − unshared electrons − number of bonds C_f = 6 − 0 − 4 = +2 Each S has 4 single bonds with no unshared electrons, thus they have the same **formal charge of +2**
(c) Another Lewis Structure	

(d) Formal charges of all atoms in the Lewis structure in (c)	For S, both S have the same C_f: $C_f = \underline{-1}$ C_f = VE – unshared electrons – number of bonds $= 6 - 0 - 7 = -1$ For terminal O's with double bond: $C_f = \underline{0}$ C_f = VE – unshared electrons – number of bonds $= 6 - 4 - 2 = 0$ For central O with single bonds: $C_f = \underline{0}$ C_f = VE – unshared electrons – number of bonds $= 6 - 4 - 2 = 0$

70. See Sections 7.1, 7.2, 7.3, Examples 7.9 and 7.11 Table 7.3 and Figure 7.11.

- The species type (AX_mE_n) which is based on the Lewis structure can be used to determine the geometry of the molecules as shown in Table 7.3 and Figure 7.11. In the formula of the species type (AX_mE_n) m is the number of atoms (X) around the central atom (A) and n is the number of unshared electron pairs around A.

- The hybridization can be obtained from the total number of electron pairs, m+n (bonding and unshared) as shown in Example 7.9 and summarized in a table in the solution to problem #50.

- The polarity of a molecule that has similar X attached to A can be deduced from the geometry and symmetry of the molecule. As discussed in Section 7.3 and in Example 7.7, if all the atoms attached to A are identical, the molecule can only be polar if the X's are not symmetrically arranged around A.

Species	Atoms Around Central Atom A	Unshared Pairs Around A	Geometry	Hybridization	Polarity
AX_2E_2	2	2	bent	sp^3	polar
AX_3	3	0	trigonal planar	sp^2	nonpolar
AX_4E_2	4	2	square planar	sp^3d^2	nonpolar
AX_5	5	0	trigonal bipyramid	sp^3d	nonpolar

CONCEPTUAL QUESTIONS

72. See Section 7.1, Examples 7.1, 7.2 and 7.4. Use the steps presented on p.158 of the textbook as a guide in writing the Lewis structure. If AE ≤ NE follow Figure 7.3 but if AE > NE follow Figure 7.9. See Table 7.2 for the discussion on formal charge. As pointed out in the textbook, the formal charge, C_f can be obtained using the following equation:

C_f = VE – unshared electrons – number of bonds

To determine which atom is central in the structure of the molecules, find the most plausible structure. Draw the possible Lewis structures and select the structure that has formal charges as close to zero as possible. Negative formal charge should be on the most electronegative atom while positive formal charge should be on the least electronegative atom.

(a) HCN	Solution	
Possible Skeleton Structures	H——C——N Structure I	H——N——C Structure II
VE	1 (for H) + 4 (for C) + 5 (for N) = 10	
AE	AE = VE – (2 × number of bonds) = 10 – (2 × 2 bonds) = 6	
NE	Structure I: 4 (for C to have an octet) + 6(for N) = 10 Structure II: 4 (for N to have an octet) + 6(for C) = 10	
AE = NE ?	No. AE < NE by 4 electrons ⇒ change the C–N single bond to a triple bond. Complete the octet for every atom (except H).	
Possible Lewis Structures	H——C≡≡N : Structure I	H——N≡≡C : Structure II
Formal charges	Structure I H: C_f = 1 – 1 – 0 = 0 C: C_f = 4 – 0 – 4 = 0 N: C_f = 5 – 2 – 3 = 0	Structure II H: C_f = 1 – 1 – 0 = 0 C: C_f = 4 – 2 – 3 = –1 N: C_f = 5 – 0 – 4 = +1

Structure I is the better and the more plausible structure for HCN because all formal charges are zero. Therefore, **the central atom in HCN is C.**

(b) NOCl	Solution	
Possible Skeleton Structures	N—O—Cl Structure I	O—N——Cl Structure II
VE	5 (for N) + 6 (for O) + 7 (for Cl) = 18	
AE	AE = VE − (2 × number of bonds) = 18 − (2 × 2 bonds) = 14	
NE	Structure I: 6 (for N) + 4(for O) + 6(for Cl) = 16 Structure II: 6 (for O) + 4(for N) + 6(for Cl) = 16	
AE = NE ?	No. AE < NE by 2 electrons ⇒ change the O–N single bond to a double bond. Complete the octet for every atom.	
Possible Lewis Structures	N̈=Ö—C̈l: Structure I	Ö=N̈—C̈l: Structure II
Formal charges	Structure I N: $C_f = 5 − 4 − 2 = −1$ O: $C_f = 6 − 2 − 3 = +1$ Cl: $C_f = 7 − 6 − 1 = 0$	Structure II N: $C_f = 5 − 2 − 3 = 0$ O: $C_f = 6 − 4 − 2 = 0$ Cl: $C_f = 7 − 6 − 1 = 0$

Structure II is the better and the more plausible structure for NOCl because all formal charges are zero. Therefore, **the central atom in NOCl is N.**

74. See Section 7.2, Examples 7.5 and 7.6. Use the steps presented on p.158 of the textbook as a guide in writing the Lewis structure. If AE ≤ NE follow Figure 7.3 but if AE > NE follow Figure 7.9. Use Table 7.3 and Figure 7.11 to identify the species type and the corresponding bond angles.

SiH_4 Lewis structure:	H | H−Si−H | H
Species Type:	A=Si, X=H (4), E=0 ⇒ AX_4
Geometry:	tetrahedral, **109.5° bond angle**

PH_3 Lewis structure:	•• H−P−H | H
Species Type:	A=P, X=H (3), E=1 ⇒ AX_3E
Geometry:	trigonal pyramid, **109.5° bond angle**

H₂S Lewis structure:	$H-\overset{\cdot\cdot}{\underset{\vert}{S}}\,\overset{\cdot\cdot}{}$ H
Species Type:	A=S, X=H (2), E=2 $\Rightarrow$ AX₂E₂
Geometry:	bent, **109.5° bond angle**

The three molecules, SiH₄, PH₃, and H₂S all have bond angles of about 109.5° because they all have central atoms surrounded by 4 electron groups. However, the electron groups that they have vary. All electron groups in SiH₄ are bonding electron pairs while PH₃ and H₂S contain both bonding pairs and unshared electron pairs (lone pairs). Based on electron pair repulsion theory, lone pair – lone pair repulsion is greater than lone pair – bonding pair repulsion. Bonding pair – lone pair repulsion is greater than bonding pair – bonding pair repulsion or simply:

Repulsion: lone pair – lone pair > lone pair – bonding pair > bonding pair – bonding pair

Consequently, **PH₃, and H₂S have bond angles smaller than 109.5°** due to greater repulsion exerted by the lone (unshared) pair of electrons with a bonding pair or with another lone pair.

76. See Section 7.2, Examples 7.5 and 7.6. Use the steps presented on p.158 of the textbook as a guide in writing the Lewis structure. If AE ≤ NE follow Figure 7.3 but if AE > NE follow Figure 7.9. Use Table 7.3 and Figure 7.11 to identify the species type and the corresponding bond angles.

SnCl₂ Lewis structure:	$:\!\overset{\cdot\cdot}{\underset{\cdot\cdot}{Cl}}\!-\!\overset{\cdot\cdot}{Sn}\!-\!\overset{\cdot\cdot}{\underset{\cdot\cdot}{Cl}}\!:$
Species Type:	A=Sn, X=Cl (2), E=1 $\Rightarrow$ AX₂E
Geometry:	bent, **120° bond angle**
BCl₃ Lewis structure:	$:\!\overset{\cdot\cdot}{\underset{\cdot\cdot}{Cl}}\!-\!\overset{\cdot\cdot}{\underset{\vert}{B}}\!-\!\overset{\cdot\cdot}{\underset{\cdot\cdot}{Cl}}\!:$ $:\!\overset{}{\underset{\cdot\cdot}{Cl}}\!:$
Species Type:	A=B, X=Cl (3), E=0 $\Rightarrow$ AX₃
Geometry:	trigonal planar, **120° bond angle**
SO₂ Lewis structure:	$\overset{\cdot\cdot}{\underset{\cdot\cdot}{O}}\!=\!\overset{\cdot\cdot}{S}\!-\!\overset{\cdot\cdot}{O}\!:$
Species Type:	A=S, X=O (2), E=1 $\Rightarrow$ AX₂E
Geometry:	bent, **120° bond angle**

The three molecules, $SnCl_2$, BCl_3, and SO_2 all have bond angles of about 120° because they all have central atoms surrounded by 3 electron groups. All electron groups in BCl_3 are bonding electron pairs while $SnCl_2$ and SO_2 contain both bonding pairs and unshared electron pairs (lone pairs).

Repulsion: lone pair – lone pair > lone pair – bonding pair > bonding pair – bonding pair

Based on the repulsion trend above, compared to BCl_3, **the bond angles in $SnCl_2$ and SO_2 are <u>smaller than 120°</u>** due to greater repulsion exerted by the lone pair with the bonding which pushes the bonding pair–bonding pair to a smaller bond angle.

CHALLENGE PROBLEMS

78. See chapter 5 and Sections 7.1 to 7.4.

    $ClF_x + U \rightarrow UF_6 + ClF(g)$

    The solution to this problem requires the following steps.

    (i) Calculate mol Cl and F from ClF

    $P = 3.00$ atm

    $T = 75°C + 273.15 = 348$ K

    $V = 457 \text{ mL} \times \dfrac{1 \text{ L}}{1000 \text{ mL}} = 0.457 \text{ L}$

    mol ClF: $\quad n = \dfrac{PV}{RT} = \dfrac{(3.00 \text{ atm})(0.457 \text{ L})}{(0.0821 \text{ L} \cdot \text{atm/mol} \cdot \text{K})(348 \text{ K})} = 0.0480 \text{ mol ClF}$

    mol ClF = mol Cl = mol F (from ClF) = 0.0480 mol

    (ii) Find mol F from UF_6

    mol F (from UF_6) = $5.63 \text{ g UF}_6 \times \dfrac{1 \text{ mol UF}_6}{352 \text{ g UF}_6} \times \dfrac{6 \text{ mol F}}{1 \text{ mol UF}_6} = 0.0960 \text{ mol F}$

    (iii) Determine the Cl to F mol ratio and the value of x in ClF_x

    Total mol F = mol F from ClF + mol F from UF_6

    $\qquad = 0.0480 + 0.0960 = 0.144 \text{ mol F}$

    mol Cl (from ClF) = 0.0480 mol Cl

    $\dfrac{0.144 \text{ mol F}}{0.0480 \text{ mol Cl}} = 3 \text{ mol F/mol Cl} \quad \Rightarrow \textbf{x in ClF}_\textbf{x} \textbf{ is } \underline{\textbf{3}}$

    Since the ratio is 3 mol F for every mol of Cl, then x in ClF_x is 3. The formula of the compound is ClF_3.

ClF$_3$ Lewis structure	
Species Type	A=Cl, X=F (3), E=2 $\Rightarrow$ AX$_3$E$_2$
Geometry	**T-shaped**, **90°**, **180° bond angles**
Polarity	**polar** because T-shaped is not symmetrical
Hybridization of Cl	m + n = 3 + 2 = 5 $\Rightarrow$ **sp^{3}d hybridized**
Number of *sigma* bonds	**3σ**
Number of *pi* bonds	**0** (no multiple bonds therefore no *pi* bonds)

79. See Sections 7.1 to 7.4.

N$_2$H$_4$ Lewis structure	
Species Type	A=N, X=H,H,N (3), E=1 $\Rightarrow$ AX$_3$E
Geometry	**trigonal pyramid**, **109.5° bond angles**
Polarity	**polar**

80. See Sections 7.1 to 7.4.

Lewis structure	
Species Type	A=I, X=O (6), E=0 $\Rightarrow$ AX$_6$
Pairs of electrons around I	**6**
Geometry	**octahedron**
Hybridization of I	m + n = 6 + 0 = 6 $\Rightarrow$ **sp^{3}d^2 hybridized**

81. See Sections 7.1 to 7.4.

SO_4^{2-}	Solution
Skeleton	
VE	6 (for S) + 4×6 (for O) + 2 (for the −2 charge) = 32
AE	AE = VE − (2 × number of bonds) = 32 − (2 × 4 bonds) = 24
NE	4×6 (for O to have an octet) = 24
AE = NE ?	Yes. Complete the octet for every atom.

Possible Lewis Structures	Structure I	Structure II
Species Type	A=S, X=O (4), E=0 $\Rightarrow$ AX$_4$	A=S, X=O (4), E=0 $\Rightarrow$ AX$_4$
Geometry	<u>tetrahedron</u>	<u>tetrahedron</u>
Hybridization of S	m + n = 4 + 0 = 4 $\Rightarrow$ <u>**sp^3**</u>	m + n = 4 + 0 = 4 $\Rightarrow$ <u>**sp^3**</u>
Formal charges	Structure I S: $\quad C_f = 6 - 0 - 4 = $ **+2** All O's: $C_f = 6 - 6 - 1 = $ **−1** **Formal charges:** $\quad$ <u>S = +2; all O's = −1</u>	Structure II S: $\quad C_f = 6 - 0 - 6 = $ **0** Single bonded O's: $\quad C_f = 6 - 6 - 1 = $ **−1** Double bonded O's: $\quad C_f = 6 - 4 - 2 = $ **0** **Formal charges:** $\quad$ <u>S = 0</u> $\quad$ <u>Single bonded O = −1</u> $\quad$ <u>Double bonded O = 0</u>

The formal charges of the atoms in structure II have charges closer to zero, therefore structure II is a more plausible structure.

82. See Sections 7.1 to 7.4.

(a) Lewis structure following the octet rule

POCl$_3$	Solution
Skeleton	$$\begin{array}{c} O \\ \| \\ Cl\!-\!\!-\!P\!-\!\!-\!Cl \\ \| \\ Cl \end{array}$$
VE	5 (for P) + 3×7 (for Cl) + 6 (for O) = 32
AE	AE = VE − (2 × number of bonds) = 32 − (2 × 4 bonds) = 24
NE	4×6(for every O and Cl to have an octet) = 24
AE = NE ?	Yes. Complete the octet for every atom.
Lewis Structure	$$\begin{array}{c} :\!\overset{\cdot\cdot}{O}\!: \\ \| \\ :\!\overset{\cdot\cdot}{Cl}\!-\!P\!-\!\overset{\cdot\cdot}{Cl}\!: \\ \| \\ :\!\overset{\cdot\cdot}{Cl}\!: \end{array}$$
Formal charges	P: $\qquad$ $C_f = 5 - 0 - 4 = +1$ O: $\qquad$ $C_f = 6 - 6 - 1 = -1$ All Cl's: $\quad$ $C_f = 7 - 6 - 1 = 0$ Formal charges: P = +1; O = −1; all Cl = 0

(b) Lewis structure in which all the formal charges are zero:

Lewis Structures	$$\begin{array}{c} :\!\overset{\cdot\cdot}{O}\!: \\ \| \\ :\!\overset{\cdot\cdot}{Cl}\!-\!P\!-\!\overset{\cdot\cdot}{Cl}\!: \\ \| \\ :\!\overset{\cdot\cdot}{Cl}\!: \end{array}$$
Formal charges	P: $C_f = 5 - 0 - 5 = 0$ O: $C_f = 6 - 4 - 2 = 0$ Cl: $C_f = 7 - 6 - 1 = 0$ Formal charges: P = 0; O = 0; all Cl = 0

THERMOCHEMISTRY

8.1 PRINCIPLES OF HEAT FLOW

2. See Example 8.1.

    Given: gold $\Rightarrow$ $q = 1.33$ J; mass $= 5.00$ g; $c = 0.129$ J/g°C

          $\Delta t = ?$

    $q = mass \times c \times \Delta t$ $\Rightarrow$ solve for Δt $\Rightarrow$ $\Delta t = \dfrac{q}{mass \times c}$

    $\Delta t = \dfrac{1.33 \text{ J}}{5.00 \text{ g} \times 0.129 \text{ J/g°C}} = \underline{\mathbf{2.06\,°C}}$

4. See Example 8.1.

    Given: mercury $\Rightarrow$ $q = 46.9$ J; mass $= 100.0$ g

                $\Delta t = t_{final} - t_{initial} = 28.35\,°C - 25.00\,°C = 3.35\,°C$

                $c = ?$

    $q = mass \times c \times \Delta t$ $\Rightarrow$ solve for c $\Rightarrow$ $c = \dfrac{q}{mass \times \Delta t}$

    $c = \dfrac{46.9 \text{ J}}{100.0 \text{ g} \times 3.35\,°C} = \underline{\mathbf{0.140 \text{ J/g°C}}}$

6. See Example 8.1.

    Given: chromium $\Rightarrow$ $c = 0.450$ J/g°C; mass $= 35.0$ g

        $t_{final} = 88\,°F$ $\Rightarrow$ convert to °C $\Rightarrow$ $t_{final(°C)} = \dfrac{(88\,°F - 32)}{1.8} = 31\,°C$

        $t_{initial} = 45\,°F$ $\Rightarrow$ convert to °C $\Rightarrow$ $t_{final(°C)} = \dfrac{(45\,°F - 32)}{1.8} = 7\,°C$

        $\Delta t = t_{final} - t_{initial} = 31\,°C - 7\,°C = 24\,°C$

        $q = ?$

    $q = mass \times c \times \Delta t = 35 \text{ g} \times 0.450 \text{ J/g°C} \times 24\,°C = \underline{\mathbf{3.8 \times 10^2 \text{ J}}}$

8.2 MEASUREMENT OF HEAT FLOW; CALORIMETRY

8. See Example 8.2.

Given: $q_{reaction\ for\ dissolving\ 10.00\ g\ NaCl}$ = 669 J

water $\Rightarrow$ c = 4.18 J/g°C; $t_{initial}$ = 25.0°C

mass = $200.0\ mL \times \dfrac{1.00\ g}{1\ mL} = 200.0\ g$ water

(a) **No.** Heat is absorbed when NaCl is dissolved therefore the solution process is endothermic and not exothermic.

(b) q_{water} = $- q_{reaction}$ = **– 669 J**

(c) t_{final} = ?

$q_{water} = mass \times c \times \Delta t$ $\Rightarrow$ solve for Δt $\Rightarrow$ $\Delta t = \dfrac{q_{water}}{mass_{water} \times c_{water}}$

$\Delta t = \dfrac{-669\ J}{200.0\ g \times 4.18\ J/g°C} = -0.800°C$

$\Delta t = t_{final} - t_{initial}$

$t_{final} = \Delta t + t_{initial} = -0.800°C + 25.0°C = \underline{\textbf{24.2°C}}$

10. See Example 8.2.

Given: hot water $\Rightarrow$ $t_{initial}$ = 35.0°C; t_{final} = 19°C; c = 4.18 J/g°C

mass = $75\ mL \times \dfrac{1.00\ g}{1\ mL} = 75\ g$ hot water

cold water $\Rightarrow$ $t_{initial}$ = 10.0°C; c = 4.18 J/g°C

mL cold water = ?

$q = mass \times c \times \Delta t$ $\Rightarrow$ $q = mass \times c \times (t_{final} - t_{initial})$

$q_{hot\ water} = mass \times c \times (t_{final} - t_{initial}) = 75\ g \times 4.18\ J/g°C \times (19°C - 35.0°C)$

$q_{cold\ water} = mass \times c \times (t_{final} - t_{initial}) = mass_{cold\ water} \times 4.18\ J/g°C \times (10.0°C - 35.0°C)$

$$q_{cold\ water} = - q_{hot\ water}$$

$mass_{cold\ water} \times 4.18\ J/g°C \times (19.0°C - 10.0°C) = - 75\ g \times 4.18\ J/g°C \times (19°C - 35.0°C)$

$$mass_{cold\ water} \times 37.62 = 5016\ g$$

$mass_{cold\ water} = 5016\ g/37.62 = 1.3 \times 10^2\ g$ cold water

mL cold water = $1.3 \times 10^2\ g \times \dfrac{1\ mL}{1.00\ g} = \underline{\textbf{1.3} \times \textbf{10}^2\ \textbf{mL cold water}}$

12. See Section 4.2 and Examples 8.2, 8.4 and 4.5.

   Net reaction involved: $H^+(aq) + OH^-(aq) \rightarrow H_2O$

   (a) Find mL $Sr(OH)_2$ solution used in neutralization:

   Strategy: mol HNO_3 $\rightarrow$ mol H^+ $\rightarrow$ mol OH^- $\rightarrow$ mol $Sr(OH)_2$ $\rightarrow$ V (mL) $Sr(OH)_2$

   mol HNO_3 = $50.00 \text{ mL} \times \dfrac{1L}{1000 \text{ mL}} \times \dfrac{0.743 \text{ mol}}{1L} = 0.0371$ mol HNO_3

   mL $Sr(OH)_2$ = $0.0371 \text{ mol } HNO_3 \times \dfrac{1 \text{ mol } H^+}{1 \text{ mol } HNO_3} \times \dfrac{1 \text{ mol } OH^-}{1 \text{ mol } H^+} \times \dfrac{1 \text{ mol } Sr(OH)_2}{2 \text{ mol } OH^-}$

   $\times \dfrac{1 L\ Sr(OH)_2 \text{ solution}}{1.00 \text{ mol } Sr(OH)_2 \text{ solution}} \times \dfrac{1000 \text{ mL}}{1 L}$ = **18.6 mL** $Sr(OH)_2$ solution

   (b) Find t_{final} of the resulting solution

   solution $\Rightarrow$ c = 4.18 J/g°C; $t_{initial}$ = 27.3°C; t_{final} = ?

   total volume of resulting solution = 50.00 mL + 18.6 mL = 68.6 mL

   mass of resulting solution = $68.6 \text{ mL} \times \dfrac{1 g}{1 \text{ mL}} = 68.6$ g

   $q_{per\ mol\ nitric\ acid}$ $\Rightarrow$ $\Delta H_{neutralization}$ = − 52 kJ

   $q_{reaction}$ = $0.0371 \text{ mol } HNO_3 \times \dfrac{-52 \text{ kJ}}{1 \text{ mol } HNO_3} \times \dfrac{1000 \text{ J}}{1 \text{kJ}} = -1.9 \times 10^3$ J

   $q_{resulting\ solution}$ = $-q_{reaction}$ = $-(-1.9 \times 10^3 \text{ J}) = 1.9 \times 10^3$ J

   $q_{resulting\ solution}$ = *mass* $\times$ *c* $\times$ *Δt* $\Rightarrow$ solve for *Δt* $\Rightarrow$ $\Delta t = \dfrac{q}{mass \times c}$

   $\Delta t = \dfrac{1.9 \times 10^3 \text{ J}}{68.6 \text{ g} \times 4.18 \text{ J/g}°C} = 6.6°C$

   $\Delta t = t_{final} - t_{initial}$

   t_{final} = $\Delta t + t_{initial}$ = 6.6°C + 27.3°C = **33.9°C**

14. See Example 8.3.

   (a) water $\Rightarrow$ $t_{initial}$ = 23.5°C; t_{final} = 39.7°C; c = 4.18 J/g°C

   mass$_{water}$ (g) = $1.200 \text{ kg} \times \dfrac{1000 \text{ g}}{1 \text{kg}} = 1200$ g water

   q_{water} = *mass* $\times$ *c* $\times$ *Δt*

   = 1200 g $\times$ 4.18 J/g°C $\times$ (39.7°C − 23.5°C) = **8.13×10^4 J or 81.3 kJ**

(b) $q_{cal} = C_{cal} \times \Delta t = 5.32 \text{ kJ/}°C \times (39.7°C - 23.5°C) = \underline{\textbf{86.2 kJ}}$

(c) 5.00 mL ethyl ether:

$$q_{reaction} = -(q_{cal} + q_{water}) = -(86.2 \text{ kJ} + 81.3 \text{ kJ}) = \underline{\textbf{-167.5 kJ}}$$

(d) 1 mol ethyl ether: (Note: molar mass of ethyl ether, $C_4H_{10}O = 74.18$ g)

$$q_{reaction} = \frac{-167.5 \text{ kJ}}{5.00 \text{ mL}} \times \frac{1 \text{ mL}}{0.714 \text{ g}} \times \frac{74.18 \text{ g}}{1 \text{ mol}} = \underline{\textbf{-3.48} \times \textbf{10}^3 \textbf{ kJ/mol}}$$

16. See Example 8.3.

(a) $q_{combustion}$ of 7.40 g ethyl alcohol in bomb calorimeter:

$$q_{combustion} = 7.40 \text{ g ethyl alcohol} \times \frac{-29.52 \text{ kJ}}{1.00 \text{ g ethyl alcohol}} = \underline{\textbf{-218 kJ}}$$

(b) $q_{water} = mass \times c \times \Delta t$

$$= \left(0.750 \text{ kg water} \times \frac{1000 \text{g}}{1 \text{ kg}} \right) \times 4.18 \text{ J/g}°C \times (48.04°C - 23.28°C)$$

$$= \underline{\textbf{7.76} \times \textbf{10}^4 \textbf{ J or 77.6 kJ}}$$

(c) $q_{reaction} = q_{combustion} = -(q_{cal} + q_{water})$

solve for q_{cal} $\Rightarrow$ $q_{cal} = -q_{combustion} - q_{water}$

$$= -(-218 \text{ kJ}) - 77.6 \text{ kJ} = \underline{\textbf{1.40} \times \textbf{10}^2 \textbf{ kJ}}$$

(d) $q_{cal} = C_{cal} \Delta t$ (Equation 8.3)

solve for C_{cal} $\Rightarrow$ $C_{cal} = \dfrac{q_{cal}}{\Delta t} = \dfrac{1.40 \times 10^2 \text{ kJ}}{(48.04 - 23.28)°C} = \underline{\textbf{5.65 kJ/}°\textbf{C}}$

18. See Example 8.3.

0.750 g acetylene:

$$q_{reaction} = 0.750 \text{ g acetylene} \times \frac{-48.2 \text{ kJ}}{1 \text{ g acetylene}} = -36.1 \text{ kJ}$$

water $\Rightarrow$ $t_{final} = 32.5°C$; mass = 800.0 g; $c = 4.18 \text{ J/g}°C$

$q_{water} = mass \times c \times \Delta t = 800.0 \text{ g} \times 4.18 \text{ J/g}°C \times \Delta t$

$$= (3.34 \times 10^3 \text{ J/}°C) \Delta t \quad \text{or} \quad (3.34 \text{ kJ/}°C) \Delta t$$

calorimeter:

$$q_{cal} = C_{cal}\,\Delta t = (1.117\ kJ/°C)\,\Delta t$$

$q_{reaction} = -(q_{cal} + q_{water}) \Rightarrow$ substitute the values obtained above for each

$$-36.1\ kJ = -[(1.117\ kJ/°C)\,\Delta t + (3.34\ kJ/°C)\,\Delta t] = -(4.46\ kJ/°C)\,\Delta t$$

$$\Delta t = \frac{-36.1\,kJ/°C}{-4.46\ kJ/°C} = 8.09°C$$

$$\Delta t = t_{final} - t_{initial}$$

$$t_{initial} = t_{final} - \Delta t = 35.2°C + 8.09°C = \underline{\mathbf{27.1°C}}$$

20. See Example 8.2 and 8.3.

$$mL\ CH_3OH\ burned = -71.8\ kJ \times \frac{1\,mole\ CH_3OH}{-1453\ kJ} \times \frac{32.042\ g}{1\,mole\ CH_3OH} \times \frac{1\,mL}{0.791\ g} = \underline{\mathbf{2.00\ mL}}$$

8.3 ENTHALPY
8.4 THERMOCHEMICAL EQUATIONS

22. See Example 8.4 and Figure 8.6.

   (a)   balanced thermochemical equation for the reaction:

$$KClO_3 \rightarrow KCl + \tfrac{3}{2}O_2 \qquad\qquad \Delta H = -44.7\ kJ$$

   (b)   Yes, the reaction is **exothermic** because heat is evolved.

   (c)

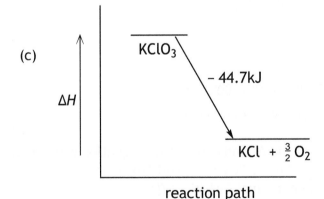

The products (KCl and O_2) have lower enthalpy than the reactant ($KClO_3$). Thus, $\Delta H < 0$ or energy is released to the surroundings in the formation of the products.

   (d)   Given: mass $KClO_3$ = 3.00 g; ΔH = ?

   Strategy: g $KClO_3 \rightarrow$ mol $KClO_3 \rightarrow \Delta H$

$$3.00\ g\ KClO_3 \times \frac{1\,mol\ KClO_3}{122.55\ g\ KClO_3} \times \frac{-44.7\ kJ}{1\,mol\ KClO_3} = \underline{\mathbf{-1.09\ kJ}}$$

(e) Given: $\Delta H = -15$ kJ   g $KClO_3$ = ?

Strategy: $\Delta H \rightarrow$ mol $KClO_3 \rightarrow$ mol C used up $\rightarrow$ g $KClO_3$

$$-15 \text{ kJ} \times \frac{1 \text{ mol } KClO_3}{-44.7 \text{ kJ}} \times \frac{122.55 \text{ g } KClO_3}{1 \text{ mol } KClO_3} = \underline{\textbf{41.1 g } KClO_3}$$

24. See Example 8.5 and 8.7 and the Rules of Thermochemistry.

(a) Precipitation of $BaCl_2$ from aqueous solutions of Ba^{2+} and Cl^- is given by the following:

$$Ba^{2+}(aq) + 2Cl^-(aq) \rightarrow BaCl_2(s) \qquad\qquad \Delta H = \text{?}$$

The ΔH for the precipitation reaction above can be calculated from the $\Delta H°_f$ values given in Table 8.3 using equation 8.4.

$$\Delta H° = \Sigma\Delta H_f° \text{ products} - \Sigma\Delta H_f° \text{ reactants} \qquad \text{(Equation 8.4)}$$

$$= 1 \text{ mol } (\Delta H_f° \ BaCl_2(s)) - [1 \text{ mol } (\Delta H_f° \ Ba^{2+}(aq)) + 2 \text{ mol } (\Delta H_f° \ 2Cl^-(aq))]$$

$$= 1 \text{ mol } (-858.6 \text{ kJ/mol}) - [1 \text{ mol } (-537.6 \text{ kJ/mol}) + 2 \text{ mol } (-167.2 \text{ kJ/mol})]$$

$$= 13.4 \text{ kJ}$$

$$\textbf{Ba}^{2+}\textbf{(aq) + 2Cl}^-\textbf{(aq)} \rightarrow \textbf{BaCl}_2\textbf{(s)} \qquad\qquad \Delta H = \textbf{13.4kJ}$$

(b) Given: mass $BaCl_2$ precipitate = 15.00 g; ΔH = ?

Strategy: g $BaCl_2 \rightarrow$ mol $BaCl_2$ formed $\rightarrow \Delta H$

$$15.00 \text{ g } BaCl_2 \times \frac{1 \text{ mol } BaCl_2}{208.23 \text{ g } BaCl_2} \times \frac{13.4 \text{ kJ}}{1 \text{ mol } BaCl_2} = \underline{\textbf{0.965 kJ}}$$

26. See Example 8.4.

(a) The balanced equation for the reaction of nitroglycerin:

$$2C_3H_5(NO_3)_3(l) \rightarrow 5H_2O(g) + 3N_2(g) + 6CO_2(g) + ½ O_2(g)$$

In writing thermochemical equations, the ΔH indicated in the equation is based on the moles of reactants and products. So, -6.26 kJ/g $C_3H_5(NO_3)_3(l)$ should be converted to its corresponding value per 2 mol $C_3H_5(NO_3)_3(l)$:

$$2 \text{ mol } C_3H_5(NO_3)_3 \times \frac{227.10 \text{ g } C_3H_5(NO_3)_3}{1 \text{ mol } C_3H_5(NO_3)_3} \times \frac{-6.26 \text{ kJ}}{1 \text{ g } C_3H_5(NO_3)_3} = -2.84 \times 10^3 \text{ kJ}$$

The thermochemical equation is:

$$2C_3H_5(NO_3)_3(l) \rightarrow 5H_2O(g) + 3N_2(g) + 6CO_2(g) + ½ O_2(g) \quad \Delta H = -2.84 \times 10^3 \text{ kJ}$$

(b)   Based on the balanced thermochemical equation above, 2 moles of $C_3H_5(NO_3)_3$ yield a total of 14.5 moles of products (total mol product = sum of the coefficient of the products in the balanced equation = 5 + 3 + 6 + ½ = 14.5 mol products).

Given: mol product = 4.65 mol; ΔH = ?

Strategy: mol product → mol $C_3H_5(NO_3)_3$ → ΔH

$$4.65 \text{ mol product} \times \frac{2 \text{ mol } C_3H_5(NO_3)_3}{14.5 \text{ mol product}} \times \frac{-2.84 \times 10^3 \text{ kJ}}{2 \text{ mol } C_3H_5(NO_3)_3} = \underline{-911 \text{ kJ}}$$

28. See question #27 and Example 8.4.

(a) $C_{57}H_{104}O_6(s) + 80 \, O_2(g) \rightarrow 57 \, CO_2(g) + 52 \, H_2O(l)$ $\Delta H = -3.022 \times 10^4$ kJ/mol

(b) kJ lost in one day:

$$425 \text{ nutritional calories} \times \frac{1000 \text{ cal}}{1 \text{ nutritional calorie}} \times \frac{4.184 \text{ J}}{1 \text{ cal}} \times \frac{1 \text{ kJ}}{1000 \text{ J}} = \underline{\textbf{1.78} \times \textbf{10}^3 \textbf{ kJ}}$$

(c) pounds of fat ($C_{57}H_{104}O_6$) lost in one week (7 days):

$$7 \text{ days} \times \frac{1.78 \times 10^3 \text{ kJ}}{1 \text{ day}} \times \frac{1 \text{ mol } C_{57}H_{104}O_6}{-3.022 \times 10^4 \text{ kJ}} \times \frac{885.40 \text{ g } C_{57}H_{104}O_6}{1 \text{ mol } C_{57}H_{104}O_6} \times \frac{2.205 \text{ lbs}}{1000 \text{ g}} = \underline{\textbf{0.805 lb fat}}$$

30. Use the information in Table 8.2 to determine the ΔH for the following phase changes.

$Br_2(l) \rightarrow Br_2(s)$ $\Delta H = -10.8$ kJ/mol

benzene $(l) \rightarrow$ benzene (s) $\Delta H = -40.7$ kJ/mol

ΔH evolved by freezing 100.0 g Br_2:

$$100.0 \text{ g } Br_2 \times \frac{1 \text{ mol } Br_2}{159.80 \text{ g } Br_2} \times \frac{-10.8 \text{ kJ}}{1 \text{ mol } Br_2} = -6.76 \text{ kJ}$$

ΔH evolved by freezing 100.0 g benzene:

$$100.0 \text{ g benzene} \times \frac{1 \text{ mol benzene}}{78.108 \text{ g benzene}} \times \frac{-9.84 \text{ kJ}}{1 \text{ mol benzene}} = -12.6 \text{ kJ}$$

Freezing 100.0 g benzene evolves more heat than freezing same mass of Br_2.

32. See Problem 31 and use the information in Tables 8.1 and 8.2 to determine the ΔH for condensing 100.00 g benzene (C_6H_6) gas at 80.00°C to liquid at 25.00°C. The ΔH for the entire process is equal to the sum of the ΔH of the following steps involved:
(i) conversion of benzene gas to liquid at 80.00°C; and
(ii) cooling of the liquid benzene from 80.00°C to 25.00°C.

   (i) ΔH at the boiling point of benzene (conversion of benzene gas to liquid at 80.00°C)

   $$C_6H_6 \ (g, 80.00°C) \ \rightarrow \ C_6H_6 \ (l, 80.00°C) \qquad\qquad \Delta H = -30.8 \text{ kJ/mol}$$

   For 100.00 g benzene: $\quad \Delta H_{vap} = 100.00 \text{ g } C_6H_6 \times \dfrac{1 \text{ mol } C_6H_6}{78.1 \text{ g } C_6H_6} \times \dfrac{-30.8 \text{ kJ}}{1 \text{ mol } C_6H_6} = -39.4 \text{ kJ}$

   (ii) ΔH for cooling of the liquid benzene from 80.00°C to 25.00°C.

   $$C_6H_6 \ (l, 80.00°C) \ \rightarrow \ C_6H_6 \ (l, 25.00°C) \qquad\qquad \Delta H = ?$$

   Given: mass = 100.00 g; $t_{initial}$ = 80.0°C; t_{final} = 25.0°C; c = 1.72 J/g°C

   $\Delta H_{cooling} = q = mass \times c \times \Delta t = 100.00 \text{ g} \times 1.72 \text{ J/g°C} \times (25.0°C - 80.0°C)$

   $$= -9.46 \times 10^3 \text{ J}$$

   $\Delta H = \Delta H_{condensation} + \Delta H_{cooling} = (-39.4 \text{ kJ}) + (-9.46 \text{ kJ}) = \underline{-48.9 \text{ kJ}}$

   The ΔH for condensing 100.00 g benzene (C_6H_6) gas at 80.00°C to liquid at 25.00°C is $\underline{-48.9 \text{ kJ}}$

34. See Examples 8.5 and 8.6 and the Rules of Thermochemistry.

   The thermochemical equation and the ΔH for the formation of silicon (Si) from sand (SiO_2) can be determined using Hess's law on the following thermochemical equations:

$SiO_2(s) + 2C(s) \rightarrow Si(s) + 2CO(g)$	$\Delta H = 689.6 \text{ kJ}$
$Si(s) + 2Cl_2(g) \rightarrow SiCl_4(g)$	$\Delta H = -657.0 \text{ kJ}$
$SiCl_4(g) + 2Mg(s) \rightarrow 2MgCl_2(s) + Si(s)$	$\Delta H = -625.6 \text{ kJ}$

   Get the sum of the three reactions:

$SiO_2(s) + 2C(s) \rightarrow \cancel{Si(s)} + 2CO(g)$	$\Delta H_1 = 689.6 \text{ kJ}$
$\cancel{Si(s)} + 2Cl_2(g) \rightarrow \cancel{SiCl_4(g)}$	$\Delta H_2 = -657.0 \text{ kJ}$
$\cancel{SiCl_4(g)} + 2Mg(s) \rightarrow 2MgCl_2(s) + Si(s)$	$\Delta H_3 = -625.6 \text{ kJ}$

   (a) $SiO_2(s) + 2C(s) + 2Cl_2(g) + 2Mg(s) \rightarrow 2CO(g) + 2MgCl_2(s) + Si(s)$

(b) $\Delta H = \Delta H_1 + \Delta H_2 + \Delta H_3 =$ 689.6 kJ + (– 657.0 kJ) + (– 625.6 kJ) = **– 593.0 kJ**

(c) **Yes**, the overall reaction is exothermic because the ΔH is negative.

36. See Examples 8.5 and 8.6. See also the solution above for problem #34.
Solve this problem by applying Hess's law and the rules of thermochemistry.

Given thermochemical equations:

$$4B(s) + 3O_2(g) \rightarrow 2B_2O_3(s) \qquad\qquad \Delta H° = -2543.8 \text{ kJ}$$

$$H_2(g) + \tfrac{1}{2}O_2(g) \rightarrow H_2O(g) \qquad\qquad \Delta H° = -241.8 \text{ kJ}$$

$$B_2H_6(s) + 3O_2(g) \rightarrow B_2O_3(s) + 3H_2O(g) \qquad\qquad \Delta H° = -2032.9 \text{ kJ}$$

Desired (overall) reaction is the decomposition of B_2H_6:

$$B_2H_6(s) \rightarrow 2B(s) + 3H_2(g) \qquad\qquad \Delta H = ?$$

Combine the given reactions to get the desired, overall reaction. The ΔH for the reaction will be the sum of the combined reactions.

Remember that based on the rules of thermochemistry, when a reaction is reversed, the sign for ΔH changes and if the reaction is multiplied by a factor, ΔH must also be multiplied by that same factor.

Start by taking the 3rd given equation because it has the starting material, $B_2H_6(s)$. Reverse the 1st reaction. Change the sign of the $\Delta H°$ of the 1st reaction because the equation was reversed (rule 2).

3rd reaction:	$B_2H_6(s) + 3O_2(g) \rightarrow B_2O_3(s) + 3H_2O(g)$	$\Delta H° = -2032.9$ kJ
Reverse of 1st reaction:	$2B_2O_3(s) \rightarrow 4B(s) + 3O_2(g)$	$\Delta H° = +2543.8$ kJ

Multiply the reverse of the 1st reaction by ½ to reduce the coefficient of $B_2O_3(s)$ to 1. The $B_2O_3(s)$ on the right of the 3rd reaction cancels with the $B_2O_3(s)$ on the left of the reverse of the 1st reaction. The 3O_2 on the left of the 3rd reaction reduces to $^3/_2 O_2$ because of cancelation with the $^3/_2 O_2$ on the right of the reverse of the 1st reaction.

3rd reaction:	$B_2H_6(s) + \cancel{3O_2}(g) \rightarrow \cancel{B_2O_3}(s) + 3H_2O(g)$	$\Delta H° = -2032.9$ kJ
Reverse of 1st reaction × ½:	$\cancel{B_2O_3}(s) \rightarrow 2B(s) + {}^3\cancel{/_2 O_2}(g)$	$\Delta H° = \tfrac{1}{2}(2543.8$ kJ$)$

Sum (3rd and 1st):	$B_2H_6(s) + {}^3/_2O_2(g) \rightarrow 3H_2O(g) + 2B(s)$	$\Delta H = -761$ kJ

Take the reverse of the 2nd reaction. Note that the sign of the $\Delta H°$ changed because it was reversed (rule 2):

Reverse of 2nd reaction:	$H_2O(g) \rightarrow H_2(g) + \tfrac{1}{2}O_2(g)$	$\Delta H° = 241.8$ kJ

Multiply the reverse of the 2nd reaction by 3 to get $3H_2O$ and $^3/_2O_2$. Add this modified 2nd equation to the sum of the 3rd reaction and reverse of 1st reaction. The $3H_2O$ and $^3/_2O_2$ cancels out.

Reverse of 2nd reaction × 3: $3H_2O(g) \rightarrow 3H_2(g) + ^3/_2O_2(g)$ $\Delta H° = 3(241.8 \text{ kJ})$

Sum (3rd and 1st) : $B_2H_6(s) + ^3/_2O_2(g) \rightarrow 3H_2O(g) + 2B(s)$ $\Delta H = \quad -761 \text{ kJ}$

Final Equation: $B_2H_6(s) \rightarrow 3H_2(g) + 2B(s)$ $\Delta H = \quad -35.6 \text{ kJ}$

Therefore the ΔH for the decomposition of B_2H_6 (reaction above) is **–35.6 kJ**.

8.5 ENTHALPIES OF FORMATION: ΔH's and Heats of Formation
8.6 BOND ENTHALPY

38. See Example 8.7 and refer to Table 8.3 for $\Delta H_f°$ values.

For each of the following problems, find the $\Delta H_f°$ values from Table 8.3 and substitute them into Equation 8.4 (from textbook).

$\Delta H° = \Sigma\Delta H_f°\text{products} - \Sigma\Delta H_f°\text{reactants}$ Equation 8.4

Remember that the ΔH of elements in their standard states are zero.

(a) $K(s) + ^1/_2Cl_2(g) + ^3/_2O_2(g) \rightarrow KClO_3(s)$ $\Delta H = \Delta H_f°?$

$\Delta H° = \Sigma\Delta H_f°\text{products} - \Sigma\Delta H_f°\text{reactants}$

= 1 mol ($\Delta H_f°$ $KClO_3$) – [1 mol ($\Delta H_f°$ K) + $^1/_2$ mol ($\Delta H_f°$ Cl_2) + $^3/_2$ mol ($\Delta H_f°$ O_2)]

= 1 mol (–397.7 kJ/mol) – [1 mol (0) + $^1/_2$ mol (0) + $^3/_2$ mol (0)]

= –397.7 kJ

$K(s) + ½Cl_2(g) + ^3/_2O_2(g) \rightarrow KClO_3(s)$	$\Delta H = -397.7 \text{ kJ}$

(b) $C(s) + 2Cl_2(g) \rightarrow CCl_4(l)$ $\Delta H = ?$

$\Delta H° = \Sigma\Delta H_f°\text{products} - \Sigma\Delta H_f°\text{reactants}$

= 1 mol ($\Delta H_f°$ CCl_4) – [1 mol ($\Delta H_f°$ C) + 2 mol ($\Delta H_f°$ Cl_2)]

= 1 mol (–135.4 kJ/mol) – [1 mol (0) + 2 mol (0)]

= –135.4 kJ

$C(s) + 2Cl_2(g) \rightarrow CCl_4(l)$	$\Delta H = -135.4 \text{ kJ}$

(c) $\frac{1}{2}H_2(g) + \frac{1}{2}I_2(g) \rightarrow HI(g)$ $\Delta H = ?$

$\Delta H° = \Sigma \Delta H_f° \text{products} - \Sigma \Delta H_f° \text{reactants}$

$\quad = 1 \text{ mol } (\Delta H_f° \text{ HI}) - [\frac{1}{2} \text{ mol } (\Delta H_f° \text{ } H_2) + \frac{1}{2} \text{ mol } (\Delta H_f° I_2)]$

$\quad = 1 \text{ mol } (+26.5 \text{ kJ/mol}) - [\frac{1}{2} \text{ mol } (0) + \frac{1}{2} \text{ mol } (0)]$

$\quad = +26.5 \text{ kJ}$

$\frac{1}{2}H_2(g) + \frac{1}{2}I_2(g) \rightarrow HI(g)$	$\Delta H = 26.5 \text{ kJ}$

(d) $2Ag(s) + \frac{1}{2}O_2(g) \rightarrow Ag_2O(s)$ $\Delta H = ?$

$\Delta H° = \Sigma \Delta H_f° \text{products} - \Sigma \Delta H_f° \text{reactants}$

$\quad = 1 \text{ mol } (\Delta H_f° \text{ } Ag_2O) - [2 \text{ mol } (\Delta H_f° Ag) + \frac{1}{2} \text{ mol } (\Delta H_f° O_2)]$

$\quad = 1 \text{ mol } (-31.0 \text{ kJ/mol}) - [2 \text{ mol } (0) + \frac{1}{2} \text{ mol } (0)]$

$\quad = -31.0 \text{ kJ}$

$2Ag(s) + \frac{1}{2}O_2(g) \rightarrow Ag_2O(s)$	$\Delta H = -31.0 \text{ kJ}$

40. $2Cr_2O_3(s) \rightarrow 4Cr(s) + 3O_2(g)$ $\Delta H° = 2269.4 \text{ kJ}$

(a) The reverse of the given reaction above is the formation of 2 mol Cr_2O_3.

$\quad 4Cr(s) + 3O_2(g) \rightarrow 2Cr_2O_3(s)$ $\Delta H° = -2269.4 \text{ kJ}$

Thus, for 1 mol Cr_2O_3,

$\quad \frac{1}{2} [4Cr(s) + 3O_2(g) \rightarrow 2Cr_2O_3(s)]$ $\Delta H° = -2269.4 \text{ kJ} \times \frac{1}{2}$

$\quad 2Cr(s) + 3/2 \text{ } O_2(g) \rightarrow Cr_2O_3(s)$ $\underline{\Delta H° = -1134.7 \text{ kJ}}$

The heat of formation of Cr_2O_3 is **−1134.7 kJ.**

(b) $\Delta H° = 13.65 \text{ g } Cr_2O_3 \times \dfrac{1 \text{ mol } Cr_2O_3}{152.00 \text{ g } Cr_2O_3} \times \dfrac{-1134.7 \text{ kJ}}{1 \text{ mol } Cr_2O_3} = \underline{-101.9 \text{ kJ}}$

42. See Section 5.3 and Example 8.7.

(a) $N_2H_4(l) + O_2(g) \rightarrow N_2(g) + 2H_2O(g)$ $\Delta H° = ?$

$\Delta H° = \Sigma \Delta H_f° \text{products} - \Sigma \Delta H_f° \text{reactants}$ (Equation 8.4 from textbook)

$\quad = [1 \text{ mol } (\Delta H_f° \text{ } N_2) + 2 \text{ mol } (\Delta H_f° \text{ } H_2O)] - [1 \text{ mol } (\Delta H_f° \text{ } N_2H_4) + 1 \text{ mol } (\Delta H_f° O_2)]$

$\quad = [1 \text{ mol } (0) + 2 \text{ mol } (-241.8 \text{ kJ/mol})] - [1 \text{ mol } (50.6 \text{ kJ/mol}) + 1 \text{ mol } (0)]$

$\quad = -534.2 \text{ kJ}$

thermochemical equation:

$$N_2H_4(l) + O_2(g) \rightarrow N_2(g) + 2H_2O(g) \qquad \Delta H° = -534.2 \text{ kJ}$$

(b) Given: V_{steam} = 1.683 L

T = 125°C + 273.15 = 398 K

$$P = 772 \text{ mm Hg} \times \frac{1 \text{ atm}}{760 \text{ mm Hg}} = 1.02 \text{ atm}$$

heat evolved = ?

$PV = nRT \Rightarrow$ solve for n_{steam}

$$n_{steam} = \frac{PV}{RT} = \frac{(1.02 \text{ atm})(1.683 \text{ L})}{(0.0821 \text{ L} \cdot \text{atm/K} \cdot \text{mol})(398 \text{ K})} = 0.0525 \text{ mol}$$

$$\text{heat} = 0.0525 \text{ mol } H_2O \times \frac{-534.2 \text{ kJ}}{2 \text{ mol } H_2O} = \underline{-14.0 \text{ kJ}}$$

The heat evolved is 14.0 kJ.

44. See Examples 8.7 and 8.8 and refer to Table 8.3 for $\Delta H_f°$ values. For the following problems, use Equation 8.4 (from the textbook) to find the $\Delta H°$ for each.

$$\Delta H° = \Sigma\Delta H_f° \text{products} - \Sigma\Delta H_f° \text{reactants} \qquad \text{(Equation 8.4 from textbook)}$$

(a) $Zn(s) + 2H^+(aq) \rightarrow Zn^{2+}(aq) + H_2(g)$ \qquad $\Delta H° = ?$

$\Delta H° = [\Delta H_f° \text{ } Zn^{2+} + \Delta H_f° \text{ } H_2] - [\Delta H_f° \text{ } Zn + 2 \text{ mol } \Delta H_f° \text{ } H^+]$

$\Delta H° = [(1 \text{ mol})(-153.9 \text{ kJ/mol}) + (1 \text{ mol})(0)] - [(1 \text{ mol})(0) + (2 \text{ mol})(0)]$

$\underline{\Delta H° = -153.9 \text{ kJ}}$

(b) $2H_2S(g) + 3O_2(g) \rightarrow 2SO_2(g) + 2H_2O(g)$ \qquad $\Delta H° = ?$

$\Delta H° = [2\Delta H_f° \text{ } SO_2 + 2\Delta H_f° \text{ } H_2O] - [2\Delta H_f° \text{ } H_2S + 3\Delta H_f° \text{ } O_2]$

$\Delta H° = [(2 \text{ mol})(-296.8 \text{ kJ/mol}) + (2 \text{ mol})(-241.8 \text{ kJ/mol})]$

$\qquad\qquad - [(2 \text{ mol})(-20.6 \text{ kJ/mol}) + (3 \text{ mol})(0)]$

$\underline{\Delta H° = -1036.0 \text{ kJ}}$

(c) $3Ni(s) + 2NO_3^-(aq) + 8H^+(aq) \rightarrow 3Ni^{2+}(aq) + 2NO(g) + 4H_2O(l)$ $\Delta H° = ?$

$\Delta H° = [3\Delta H_f° \ Ni^{2+} + 2\Delta H_f° \ NO + 4\Delta H_f° \ H_2O] - [3\Delta H_f° \ Ni + 2\Delta H_f° \ NO_3^- + 8\Delta H_f° \ H^+]$

$\Delta H° = [(3 \ mol)(-54.0 \ kJ/mol) + (2 \ mol)(+90.2 \ kJ/mol) + (4 \ mol)(-285.8 \ kJ/mol)]$

$- [(3 \ mol)(0 \ kJ/mol) + (2 \ mol)(-205.0 \ kJ/mol) + (8 \ mol)(0 \ kJ/mol)]$

$\Delta H° = -714.8 \ kJ$

46. See Examples 8.7 and 8.8 and refer to Table 8.3 for $\Delta H_f°$ values. For the following problems, use Equation 8.4 (from the textbook) to find the $\Delta H°$ for each.

$\Delta H° = \Sigma \Delta H_f° \ products - \Sigma \Delta H_f° \ reactants$ (Equation 8.4 from textbook)

(a) $MgCO_3(s) + 2H^+(aq) \rightarrow Mg^{2+}(aq) + CO_2(g) + H_2O(l)$ $\Delta H° = ?$

$\Delta H° = [\Delta H_f° \ Mg^{2+} + \Delta H_f° \ CO_2 + \Delta H_f° \ H_2O] - [\Delta H_f° \ MgCO_3 + 2\Delta H_f° \ H^+]$

$\Delta H° = [(1 \ mol)(-466.8 \ kJ/mol) + (1 \ mol)(-393.5 \ kJ/mol) + (1 \ mol)(-285.8 \ kJ/mol)]$

$- [(1 \ mol)(-1095.8 \ kJ/mol) + (2 \ mol)(0)]$

$\Delta H° = -50.3 \ kJ$

(b) $Fe^{3+}(aq) + 3OH^-(aq) \rightarrow Fe(OH)_3(s)$ $\Delta H° = ?$

$\Delta H° = [\Delta H_f° \ Fe(OH)_3] - [\Delta H_f° \ Fe^{3+} + 3\Delta H_f° \ OH^-]$

$\Delta H° = [(1 \ mol)(-823.0 \ kJ/mol)] - [(1 \ mol)(-48.5 \ kJ/mol) + (3 \ mol)(-230.0 \ kJ/mol)]$

$\Delta H° = -84.5 \ kJ$

48. See Examples 8.7 and 8.8 and refer to Table 8.3 for $\Delta H_f°$ values. For the following problems, use Equation 8.4 (from the textbook) to find the $\Delta H°$ for each.

$\Delta H° = \Sigma \Delta H_f° \ products - \Sigma \Delta H_f° \ reactants$ (Equation 8.4 from textbook)

(a) **$CaCO_3(s) + 2NH_3(g) \rightarrow CaCN_2(s) + 3H_2O(l)$**   **$\Delta H° = 90.1 \ kJ$**

(b) $\Delta H° = [\Delta H_f° \ CaCN_2 + 3\Delta H_f° \ H_2O] - [\Delta H_f° \ CaCO_3 + 2\Delta H_f° \ NH_3]$

$90.1 \ kJ = [(1 \ mol)(\Delta H_f° \ CaCN_2) + (3 \ mol)(-285.8 \ kJ/mol)]$

$- [(1 \ mol)(-1206.9 \ kJ/mol) + (2 \ mol)(-46.1 \ kJ/mol)]$

$\Delta H_f° \ CaCN_2 = -351.6 \ kJ/mol$

50. See Examples 8.7 and 8.8 and refer to Table 8.3 for $\Delta H_f°$ values. For the following problem, use Equation 8.4 (from the textbook) to find the $\Delta H°$.

balanced equation:

$$C_2H_4(g) + 6F_2(g) \rightarrow 4HF(g) + 2CF_4(g) \qquad \Delta H° = -2496.7 \text{ kJ}$$

$\Delta H° = \Sigma \Delta H_f° \text{ products} - \Sigma \Delta H_f° \text{ reactants}$ \qquad (Equation 8.4 from textbook)

$\Delta H° = [4 \Delta H_f° \text{ HF} + 2 \Delta H_f° \text{ CF}_4] - [\Delta H_f° \text{ C}_2\text{H}_4 + 6 \Delta H_f° \text{ F}_2]$

$-2496.7 \text{ kJ} = [(4 \text{ mol})(-271.1 \text{ kJ/mol}) + (2 \text{ mol})(\Delta H_f° \text{ CF}_4)]$

$\qquad\qquad - [(1 \text{ mol})(52.3 \text{ kJ/mol}) + (6 \text{ mol})(0]$

$\Delta H_f° \text{ CF}_4 = \underline{-680.0 \text{ kJ/mol}}$

52. See Section 5.3, Examples 8.7 and 8.8 and refer to Table 8.3 for $\Delta H_f°$ values. Employ the following steps to solve this problem: (i) write the balanced thermochemical equation; (2) use Equation 8.4 (from the textbook) to find the $\Delta H°$; (3) use ideal gas law to calculate mol N_2; and (4) convert mol N_2 to heat liberated or absorbed.

balanced equation:

$$2NH_3(g) + 3N_2O(g) \rightarrow 3H_2O(l) + 4N_2(g) \qquad \Delta H° = ?$$

$\Delta H° = \Sigma \Delta H_f° \text{ products} - \Sigma \Delta H_f° \text{ reactants}$ \qquad (Equation 8.4 from textbook)

$\quad = [3 \Delta H_f° \text{ H}_2\text{O} + 4 \Delta H_f° \text{ N}_2] - [2 \Delta H_f° \text{ NH}_3 + 3 \Delta H_f° \text{ N}_2\text{O}]$

$\quad = [(3 \text{ mol})(-285.8 \text{ kJ/mol}) + (4 \text{ mol})(0)]$

$\qquad\qquad - [(2 \text{ mol})(-46.1 \text{ kJ/mol}) + (3 \text{ mol}) (82.05 \text{ kJ/mol})]$

$\quad = -1011.4 \text{ kJ}$

Thus, the balanced thermochemical equation is:

$$2NH_3(g) + 3N_2O(g) \rightarrow 3H_2O(l) + 4N_2(g) \qquad \Delta H° = -1011.4 \text{ kJ}$$

Now using the following given information, the amount of nitrogen gas formed can be determined and can be converted to the amount heat liberated using the balanced thermochemical equation above.

Given: $V_{nitrogen} = 345 \text{ mL} \times \dfrac{1 \text{ L}}{1000 \text{ mL}} = 0.345 \text{ L}$

$\qquad T = 25°C + 273.15 = 298 \text{ K}$

$\qquad P = 717 \text{ mm Hg} \times \dfrac{1 \text{ atm}}{760 \text{ mm Hg}} = 0.943 \text{ atm}$

heat liberated = ?

$$PV = nRT \Rightarrow \text{ solve for n} \Rightarrow n = \frac{PV}{RT} = \frac{(0.943 \text{ atm}) (0.345 \text{ L})}{(0.0821 \text{ L} \cdot \text{atm/K} \cdot \text{mol}) (298 \text{ K})}$$

$$= 0.0133 \text{ mol } N_2 \text{ produced}$$

$$\Delta H = 0.0133 \text{ mol } N_2 \times \frac{-1011.4 \text{ kJ}}{4 \text{ mol } N_2} = \underline{\textbf{-3.36 kJ}}$$

Therefore, the amount of heat liberated is 3.36 kJ.

8.7 THE FIRST LAW OF THERMODYNAMICS

54. As stated in the textbook, "Energy can also be expressed in L·atm". The conversion factor is: 1 L·atm = 0.1013 kJ.

$$? \text{ L} \cdot \text{atm} = 12.2 \text{ kJ} \times \frac{1 \text{ L} \cdot \text{atm}}{0.1013 \text{ kJ}} = \underline{\textbf{1.20} \times \textbf{10}^2 \textbf{ L} \cdot \textbf{atm}}$$

56. See Example 8.9.

(a) Given: $w = -54$ J (negative because work is done by the system)

$\Delta E = -72$ J (negative because the energy of the system decreased or energy was released)

$q = ?$

$\Delta E = q + w \Rightarrow$ solve for $q \Rightarrow q = \Delta E - w = -72$ J $- (-54$J$) = -18$ J

$\underline{\textbf{q} = \textbf{--18 J}}$

(b) Given: $w = +102$ J (positive because work is done on the system)

$q = -38$ J (negative because heat is released)

$\Delta E = ?$

$\Delta E = q + w = -38$ J $+ 102$ J $= +64$ J

$\underline{\Delta \textbf{E} = \textbf{+ 64 J}}$

58. See Example 8.9.

Given: $q = 388$ J   (positive because heat is absorbed)

$\Delta E = 286$ J

$w = ?$

$\Delta E = q + w$ $\Rightarrow$ solve for w $\Rightarrow$ $w = \Delta E - q$ $= 286$ J $- (388$ J$) =$ **−102 J**

$\underline{w = -102 \text{ J}}$

(a) **Yes.** A negative w indicates that work is done by the system.

(b) **−102 J**

60. See also Sections 8.5 and 5.3.

Given: $Br_2(l) \rightarrow Br_2(g)$

1 mole $Br_2(l)$ at 59°C (or 332K)

(a) $\Delta H_{vap} = 29.6$ kJ/mol (from Table 8.2)

$\Delta H = $ 1 mol Br_2 $\times \dfrac{29.6 \text{ kJ}}{1 \text{ mol } Br_2} = 29.6$ kJ

$\Delta H = $ **29.6 kJ**

(b) $\Delta PV = PV_{products} - PV_{reactants}$ (assume $P = 1.00$ atm)

$V_{reactants} = 1 \text{ mol } Br_2 \times \dfrac{159.8 \text{ g } Br_2}{1 \text{ mol } Br_2} \times \dfrac{1 \text{ mL } Br_2}{3.103 \text{ g } Br_2} \times \dfrac{1 \text{ L}}{1000 \text{ mL}} = 0.05150 \text{ L } Br_2$

$PV_{reactants} = 1.00 \text{ atm} \times 0.05150 \text{ L} = 0.05150 \text{ L·atm}$

$PV_{products} = nRT = (1.00 \text{ mol})(0.0821 \text{ L·atm/mol·K})(332 \text{ K}) = 27.3 \text{ L·atm}$

$\Delta PV = 27.3 \text{ L·atm} - 0.05150 \text{ L·atm} = 27.2 \text{ L·atm}$

(Note that PV for the liquid is very small compared to that of the gas thus, it could be ignored.)

$\Delta PV = 27.2 \text{ L·atm} \times \dfrac{0.1013 \text{ kJ}}{1 \text{ L·atm}}$

$\Delta PV = $ **2.76 kJ**

(c) $\Delta H = \Delta E + \Delta PV$ $\Rightarrow$ solve for ΔE $\Rightarrow$ $\Delta E = \Delta H - \Delta PV = 29.6 \text{ kJ} - 2.76 \text{ kJ} = $ **26.8 kJ**

62. See also Example 8.10 and Sections 8.5 and 5.3. Use Table 8.3 to calculate the ΔH for the reaction. Use Equation 8.6 (from the textbook) to find the ΔE.

(a) $CH_3OH\ (l) + {}^3/_2 O_2(g) \rightarrow CO_2(g) + 2H_2O(g)$ $\qquad \Delta H° = ?$

$\Delta H° = [\Delta H_f° \ CO_2 + 2\ \Delta H_f° \ H_2O] - [\Delta H_f° \ CH_3OH(l) + {}^3/_2 \ \Delta H_f° \ O_2(g)]$

$\Delta H° = [(1\ mol)(-393.5\ kJ/mol) + (2\ mol)(-241.8\ kJ/mol)]$

$\qquad - [(1\ mol)(-238.7\ kJ/mol) + ({}^3/_2\ mol)(0)]$

$\underline{\Delta H° = -638.4\ kJ}$

Thermochemical equation:

$\quad CH_3OH\ (l) + {}^3/_2 O_2(g) \rightarrow CO_2(g) + 2H_2O(g)$ $\qquad \Delta H° = -638.4\ kJ$

(b) Based on Equation 8.6:

$\Delta H = \Delta E + \Delta n_g RT$

where Δn_g = change in the number of moles (n) of gas

$\Delta n_g = n_{products} - n_{reactants} = (1 + 2) - (1.5) = 1.5\ mol$

$\Delta n_g RT = 1.5\ mol \times 0.0821\ L \cdot atm/mol \cdot K \times 298K = 36.7\ L \cdot atm$

$\Delta n_g RT = 36.7\ L \cdot atm \times \dfrac{0.1013\ kJ}{1L \cdot atm} = \underline{\mathbf{3.72\ kJ}}$

$\Delta H = \Delta E + \Delta n_g RT \quad \Rightarrow$ solve for $\Delta E \Rightarrow \quad \Delta E = \Delta H - \Delta n_g RT$

Substitute the values for ΔH from (a) and $\Delta n_g RT$

$\Delta E = \Delta H - \Delta n_g RT = -638.4\ kJ - 3.72\ kJ$

$\underline{\Delta E = -642.1\ kJ}$

UNCLASSIFIED PROBLEMS

64. See Section 8.2.

$4.18\ J/g \cdot °C \times \dfrac{9.48 \times 10^{-4}\ BTU}{1\ J} \times \dfrac{453.6\ g}{1\ lb} \times \dfrac{1°\ C}{1.8\ °F} = \underline{\mathbf{1.00\ BTU/lb \cdot °F}}$

66. See Section 8.2.

$120\ nutritional\ cal \times \dfrac{1\ kcal}{1\ nutritional\ cal} \times \dfrac{1\ hr}{250\ kcal} \times \dfrac{60\ min}{1\ hr} = \underline{\mathbf{29\ minutes}}$

68. See Sections 8.2 and 3.3 and Table 8.3.

    $2Al(s) + 3NH_4NO_3(s) \rightarrow 3N_2(g) + 6H_2O(g) + Al_2O_3(s)$

    $\Delta H° = \Sigma \Delta H_f° \, products - \Sigma \Delta H_f° \, reactants$ (Equation 8.4 from textbook)

        = $[3 \, \Delta H_f° \, N_2 + 6 \, \Delta H_f° \, H_2O + \Delta H_f° \, Al_2O_3] - [2 \, \Delta H_f° \, Al + 3 \, \Delta H_f° \, NH_4NO_3]$

        = $[(3 \, mol)(0) + (6 \, mol)(-241.8 \, kJ/mol) + (1 \, mol) \, (-1675.7 \, kJ/mol)]$

            $- [(2 \, mol)(0) + (3 \, mol) \, (-365.6 \, kJ/mol)]$

        = $-2029.7 \, kJ$

    Find the amount of heat that each of the 10.00 kg samples of ammonium nitrate and powdered aluminum will evolve. The fewer mol is the theoretical yield in energy.

    kJ energy released by 10.00 kg Al

    $$10.00 \, kg \, Al \times \frac{1000 \, g}{1 \, kg} \times \frac{1 \, mol \, Al}{26.98 \, g} \times \frac{-2029.7 \, kJ}{2 \, mol \, Al} = -3.761 \times 10^5 \, kJ$$

    kJ energy released by 10.00 kg NH_4NO_3

    $$10.00 \, kg \, NH_4NO_3 \times \frac{1000 \, g}{1 \, kg} \times \frac{1 \, mol \, NH_4NO_3}{80.05 \, g} \times \frac{-2029.7 \, kJ}{3 \, mol \, NH_4NO_3} = \mathbf{-8.452 \times 10^4 \, kJ}$$

    The limiting reactant is NH_4NO_3 and will release **8.452×10^4 kJ** of energy.

70. See Section 8.3 and Table 8.3 for standard enthalpy of formation values.

    $CH_4(g) + 2O_2(g) \rightarrow CO_2(g) + 2H_2O(l)$ $\Delta H° = ?$

    Given: $q = -890 \, kJ$; $w = +5 \, kJ$ $\Delta E = ?$ $\Delta H° = ?$

    $\Delta E = q + w = -890 \, kJ + 5 \, kJ = \underline{\mathbf{-885 \, kJ}}$

    $\Delta H = \Delta E + \Delta n_g RT$ where Δn_g = change in the number of moles (n) of gas

            $\Delta n_g = n_{products} - n_{reactants} = (1) - (1 + 2) = -2 \, mol$

        $\Delta n_g RT = -2 \, mol \times 0.0821 \, L \cdot atm/mol \cdot K \times 298K = -48.9 \, L \cdot atm$

    $$\Delta n_g RT \, (kJ) = -48.9 \, L \cdot atm \times \frac{0.1013 \, kJ}{1 \, L \cdot atm} = -4.95 \, kJ$$

    Substitute the values for ΔE and $\Delta n_g RT$

    $\Delta H = \Delta E + \Delta n_g RT = -885 \, kJ - 4.95 \, kJ = \underline{\mathbf{-890 \, kJ}}$

72. See Sections 8.4, 8.5, and 5.3. Refer to Table 8.3 for standard enthalpy of formation values. To solve this problem first, write the balanced thermochemical equation for the complete combustion of ethane and propane. Use Equation 8.4 from the textbook to find the $\Delta H°$ involved in each combustion reaction.

$C_2H_6(g) + {}^7/_2 O_2(g) \rightarrow 2CO_2(g) + 3H_2O(l)$ $\Delta H° = ?$

$\Delta H° = [2 \Delta H_f° CO_2 + 3 \Delta H_f° H_2O] - [\Delta H_f° C_2H_6 + {}^7/_2 \Delta H_f° O_2]$

$\Delta H° = [(2 \text{ mol})(-393.5 \text{ kJ/mol}) + (3 \text{ mol})(-285.8 \text{ kJ/mol})]$

$- [(1 \text{ mol})(-84.7 \text{ kJ/mol}) + ({}^7/_2 \text{ mol})(0)]$

$\Delta H° = -1559.7 \text{ kJ}$

$C_3H_8(g) + 5O_2(g) \rightarrow 3CO_2(g) + 4H_2O(l)$ $\Delta H° = ?$

$\Delta H° = [3 \Delta H_f° CO_2 + 4 \Delta H_f° H_2O] - [\Delta H_f° C_3H_8 + 5 \Delta H_f° O_2]$

$\Delta H° = [(3 \text{ mol})(-393.5 \text{ kJ/mol}) + (4 \text{ mol})(-285.8 \text{ kJ/mol})]$

$- [(1 \text{ mol})(-103.8 \text{ kJ/mol}) + (5 \text{ mol})(0)]$

$\Delta H° = -2219.9 \text{ kJ}$

Now find the total moles of gas mixture (ethane and propane) using the ideal gas law and the given P, V, and T values.

Given: $V_{gas} = 1.00 \text{ L};$ $P = 1.00 \text{ atm};$ $T = 0°C + 273.15 = 273 \text{ K}$

$PV = nRT \Rightarrow$ solve for $n \Rightarrow n_{total} = \dfrac{PV}{RT} = \dfrac{(1.00 \text{ atm}) (1.00 \text{ L})}{(0.0821 \text{ L} \cdot \text{atm/K} \cdot \text{mol}) (273 \text{ K})} = 0.0446 \text{ mol}$

Therefore, total moles of gas = 0.0446 mol = mol C_2H_6 + mol C_3H_8

let x = number of mol of C_2H_6

0.0446 − x = number of mol of C_3H_8

The total heat evolved, 75.65 kJ is from the combustion of C_2H_6 and C_3H_8. Therefore,

heat evolved by C_2H_6 + heat evolved by C_3H_8 = −75.65 kJ

x (−1559.7 kJ) + [(0.0446 − x) (−2219.9 kJ)] = −75.65 kJ

solve for x: −1559.7x − 99.0 + 2219.9x = −75.65

−1559.7x + 2219.9x = −75.65 + 99.0

660.2x = 23.4

x = 23.4/660.2

x = 0.0354

thus, x, mol C_2H_6 = 0.0354 mol

mol fraction of C_2H_6 = 0.0354 mol/0.0446 mol = <u>0.794</u>

---

74. Given: 3.30×10^9 gal gasoline (mainly n-octane, C_8H_{18})

density n-octane = 0.7028 g/mL

$C_{earth's\ surface\ region}$ = 2.6×10^{23} J/K

Increase in temperature on the eath's surface region due to gasoline consumption = ?

$C_8H_{18}(l)$ + 12 ½ $O_2(g)$ → 8 $CO_2(g)$ + 9 $H_2O(g)$ ΔH = −5564.2 kJ/mol

Heat released by 3.30×10^9 gal gasoline (mainly n-octane, C_8H_{18})

=

$$3.30 \times 10^9 \text{ gal } C_8H_{18} \times \frac{3.786 \text{ L}}{1 \text{ gal}} \times \frac{1000 \text{ mL}}{1 \text{ L}} \times \frac{0.7028 \text{ g } C_8H_{18}}{1 \text{ ml } C_8H_{18}} \times \frac{1 \text{ mol } C_8H_{18}}{114.23 \text{ g } C_8H_{18}} \times \frac{-5564.2 \text{ kJ}}{1 \text{ mol } C_8H_{18}}$$

= $- 4.28 \times 10^{14}$ kJ or $- 4.28 \times 10^{17}$ J

$q_{reaction}$ = $- 4.28 \times 10^{17}$ J

$q_{reaction}$ = $- q_{earth's\ surface}$

$q_{earth's\ surface}$ = $(C \times \Delta t)_{earth's\ surface}$

$q_{reaction}$ = $- (C \times \Delta t)_{earth's\ surface}$ ⇒ solve for Δt ⇒ $\Delta t_{earth's\ surface} = \dfrac{q_{reaction}}{-C}$

$$\Delta t_{earth's\ surface} = \frac{- 4.28 \times 10^{17} \text{ J}}{- 2.6 \times 10^{23} \text{ J/K}}$$

$\Delta t_{earth's\ surface}$ = **1.6 × 10^{-6} K**

Note: A unit temperature change in K is equal to a unit change in °C. The temperature increase on the earth's surface caused by gasoline combustion in the US in 2010 is **1.6 × 10^{-6} °C**

CONCEPTUAL QUESTIONS

76. When NaOH is dissolved in water, the following occurs:

$NaOH(s)$ → $Na^+(aq)$ + $OH^-(aq)$

The ΔH for this dissolution process can be calculated as follows:

ΔH° = $\Sigma \Delta H_f°$ products − $\Sigma \Delta H_f°$ reactants   Equation 8.4

ΔH° = (−240.1 kJ/mol + −230.0 kJ/mol) − (−425.6 kJ/mol)

ΔH° = −44.5 kJ/mol

(a) **Yes.** The temperature of the solution will increase because the process releases energy.

(b) **Negative** $(-)$.

(c) **No.** Comparatively, the Δt for dissolving 10.0 g NaOH is greater than the Δt for dissolving 5.00 g NaOH.

(d) **Yes.** One mole of NaOH is equal to 40.0 g NaOH. Dissolving 40.0 g NaOH causes a larger Δt than that for dissolving 5.00 g NaOH. Thus, dissolving 40.0 g NaOH, results in a higher t_f.

78. See Section 8.7.

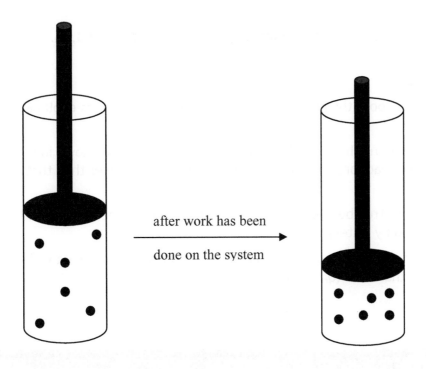

after work has been

done on the system

80. See Section 8.2.

Given: $mass_A = mass_B$

$t_{final} = 80°C$

liquid A: $t_{initial} = 100°C$

liquid B: $t_{initial} = 50°C$

Liquid A: $\qquad q_A = mass_A \times c_A \times \Delta t_A \qquad$ where $\Delta t_A = 80°C - 100°C = -20°C$

thus, $\qquad q_A = mass_A \times c_A \times (-20°C)$

Liquid B: $q_B = mass_B \times c_B \times \Delta t_B$ where $\Delta t_B = 80°C - 50°C = 30°C$

thus, $q_B = mass_B \times c_B \times (30°C)$

since $q_A = -q_B$

then, $mass_A \times c_A \times (-20°C) = -mass_B \times c_B \times (30°C)$

but $mass_A = mass_B$

so the equation simplifies to: $(-20°C) \times c_A = -(30°C) \times c_B$

solving for c_A: $c_A = \dfrac{-(30°C)\,c_B}{-20°C} = \dfrac{30\,c_B}{20}$

$$c_A = 1.5\,c_B$$

The **heat capacity of liquid A, c_A is 1.5 times larger than that of liquid B, c_B.**
Thus, **$c_A > c_B$.**

82. All the statements given about the process are true EXCEPT c, d, and e as explained below (See Section 8.1)

    (c) This statement is not true because the reaction is exothermic. As shown in Figure 8.6, for exothermic reactions, the enthalpy of the products is lower than that of the reactants.

    (d) This statement is not true because $q_{water} = -q_{rxn}$. (assuming all heat given off by the reaction is absorbed by water).

    (e) This statement is not true because the reaction is exothermic which means that q_{rxn} is negative or less than zero ($q_{rxn} < 0$).

CHALLENGE PROBLEMS

84. See Sections 8.2 and 6.1.

    $\lambda = 12.5\ cm \times \dfrac{1m}{100\ cm} = 0.125\ m$

    Energy per photon from the 12.5 cm microwave radiation:

    $E = \dfrac{hc}{\lambda} = \dfrac{(6.626 \times 10^{-34}\ J \cdot s)(2.998 \times 10^{8}\ m/s)}{0.125\ m} = 1.59 \times 10^{-24}\ J/photon$

    $q = mass \times c \times \Delta t$

Given: water $\Rightarrow$ $t_{initial} = 20°C$; $t_{final} = 100°C$; $c = 4.18 \text{ J/g}°C$

$$\text{mass} = 1.00 \times 10^2 \text{ mL water} \times \frac{1.0 \text{ g water}}{1 \text{mL water}} = 1.0 \times 10^2 \text{ g water}$$

$$q_{water} = \text{mass} \times c \times \Delta t = (1.0 \times 10^2 \text{ g}) \times 4.18 \text{ J/g}°C \times (100°C - 20°C) = 3.3 \times 10^4 \text{ J}$$

$$\text{photons} = \frac{3.3 \times 10^4 \text{ J}}{1.59 \times 10^{-24} \text{ J/photon}} = \mathbf{2.1 \times 10^{28} \text{ photons}}$$

85. See Sections 8.2 and 8.3 and Table 8.2.

   Given: 6-pack soda; 12.0 oz soda per can; $c_{soda} = 4.10 \text{ J/g}·°C$

   Al soda can $\Rightarrow$ mass empty can = 12.5 g; $c_{Al} = 0.902 \text{ J/g}°C$

   $t_{initial} = 25.0°C$; $t_{final} = 5.0°C$

   (a) $\Delta t = t_{final} - t_{initial} = 5.0°C - 25.0°C = -20.0°C$

   soda:

   $$\text{mass of soda} = 6 \text{ cans} \times \frac{12.0 \text{ oz}}{1 \text{ can}} \times \frac{1 \text{ lb}}{16 \text{ oz}} \times \frac{454 \text{ g}}{1 \text{ lb}} = 2.04 \times 10^3 \text{ g}$$

   $$q_{soda} = \text{mass} \times c \times \Delta t = 2.04 \times 10^3 \text{ g} \times 4.10 \text{ J/g}°C \times -20.0°C = -1.68 \times 10^5 \text{ J}$$

   Al soda can:

   $$\text{mass of Al cans} = 6 \text{ cans} \times \frac{12.5 \text{ g}}{1 \text{ can}} = 75.0 \text{ g}$$

   $$q_{Al\ can} = \text{mass} \times c \times \Delta t = 75.0 \text{ g} \times 0.902 \text{ J/g}°C \times (-20.0°C) = -1.35 \times 10^3 \text{ J}$$

   $$q_{total} = q_{soda} + q_{Al} = (-1.68 \times 10^5 \text{ J}) + (-1.35 \times 10^3 \text{ J}) = \mathbf{\underline{-1.69 \times 10^5 \text{ J}}}$$

   Therefore, **1.69×10^5 J** of energy must be absorbed by the ice from the six-pack to lower its temperature from 25.0°C to 5.0°C.

   (b) Ice: $\Delta H_{fus} = 6.00 \text{ kJ/mol}$

   mass of ice that must be melted to absorb 1.69×10^5 J of heat from the six-pack:

   $$1.69 \times 10^5 \text{ J} \times \frac{1 \text{ kJ}}{1000 \text{ J}} \times \frac{1 \text{ mol ice}}{6.00 \text{ kJ}} \times \frac{18.02 \text{ g ice}}{1 \text{ mol ice}} = \mathbf{\underline{508 \text{ g ice}}}$$

86. See Section 8.3.

Assume the following:

density of tea = density of H_2O = 1.00 g/mL

density of ice = density of H_2O = 1.00 g/mL

(Note: ice floats in liquid water which indicates that ice has a density < 1.00 g/mL. However, to work on this problem the density of ice is assumed to be similar to that of water)

room temperature = 25°C

volume of the glass = 480 mL (16 fl oz)

volume tea + ice = volume of the glass = 480 mL

mass of ice + tea = $480 \, mL \times \dfrac{1.00 \, g}{1 \, mL}$ = 480 g

Let w = mass of ice

480 g − w = mass of tea

Ice:

q_{ice} = heat to melt ice + heat to warm ice from 0°C to 25°C

$q_{ice} = q_{melting} + q_{warming}$

$$q_{melting} = w \times \dfrac{1 \, mol \; ice}{18.02 \, g} \times \dfrac{6.00 \, kJ}{1 \, mol \; ice} \times \dfrac{1000 \, J}{1 \, kJ} = (333 \, J/g) \, w$$

$$q_{warming} = mass \times c \times \Delta t = w \times 4.18 \, J/g°C \times (25°C - 0°C) = (105 \, J/g) \, w$$

$q_{ice} = q_{melting} + q_{warming} = [(333 \, J/g) \, w] + [(105 \, J/g) \, w] = (438 \, J/g) \, w$

Tea:

$q_{tea} = mass \times c \times \Delta t$

$= (480 \, g - w) \times (4.18 \, J/g°C) \times (0°C - 25°C)$

$= (480 \, g - w) \times (- 104.5 \, J/g)$

$q_{tea} = - 50160 \, J/g + (104.5 \, J/g) \, w$

Ice and Tea: $q_{ice} = - q_{tea}$

$(438 \, J/g) \, w = - [- 50160 \, J/g + (104.5 \, J/g) \, w]$

solve for w:

$$(438 \text{ J/g}) \text{ w} + (104.5 \text{ J/g}) \text{ w} = 50160 \text{ J/g}$$

$$(542.5 \text{ J/g}) \text{ w} = 50160 \text{ J/g}$$

$$\text{w} = \frac{50160 \text{ J/g}}{542.5 \text{ J/g}} = 92$$

The amount of ice required to have a final temperature of 0°C is 92 g.

volume of ice: $92 \text{ g} \times \dfrac{1 \text{mL}}{1.00 \text{ g}} = 92 \text{ mL}$

volume of glass = 480 mL

fraction of the total volume of the glass that should be left empty:

$$\frac{\text{volume of ice}}{\text{volume of glass}} = \frac{92 \text{ mL}}{480 \text{ mL}} \times 100 = \underline{\textbf{19 \%}}$$

Nineteen percent (19%) of the volume of the glass should be left empty for the ice.

87. See Sections 8.3 and 8.5 and Table 8.2.

$$2\text{Al}(s) + \text{Fe}_2\text{O}_3(s) \rightarrow \text{Al}_2\text{O}_3(s) + 2\text{Fe}(s)$$

(a) $\Delta H° = \Sigma \Delta H_f°\text{ products} - \Sigma \Delta H_f°\text{ reactants}$ Equation 8.4

$\Delta H° = [\Delta H_f°\text{ Al}_2\text{O}_3 + 2\ \Delta H_f°\text{ Fe}] - [2\ \Delta H_f°\text{ Al} + \Delta H_f°\text{ Fe}_2\text{O}_3]$

$\Delta H° = [(1 \text{ mol})(-1675.7 \text{ kJ/mol}) + (2 \text{ mol})(0)] - [(2 \text{ mol})(0) + (1 \text{ mol})(-824.2 \text{ kJ/mol})]$

$\Delta H° = \underline{\textbf{-851.5 kJ}}$

(b) specific heats: $c_{Al_2O_3} = 0.77 \text{ J/g°C};$ $c_{Fe} = 0.45 \text{ J/g°C}$

assume: 1 mole $\text{Al}_2\text{O}_3(s)$ and 2 moles $\text{Fe}(s)$

$t_{initial}$ = room temperature = 25°C

$$1 \text{mol Al}_2\text{O}_3 \times \frac{101.96 \text{ g Al}_2\text{O}_3}{1 \text{mol Al}_2\text{O}_3} = 101.96 \text{ g Al}_2\text{O}_3$$

$$2 \text{ mol Fe} \times \frac{55.85 \text{ g Fe}}{1 \text{mol Fe}} = 111.7 \text{ g Fe}$$

$$q_{reaction} = \Delta H° = -851.5 \text{ kJ} = -851.5 \text{ kJ} \times \frac{1000 \text{ J}}{1 \text{ kJ}} = 8.51 \times 10^5 \text{ J}$$

Al_2O_3:

$$q_{Al_2O_3} = mass \times c \times \Delta t = 101.96 \text{ g} \times 0.77 \text{ J/g}°C \times \Delta t = 79 \text{ J/}°C \times \Delta t$$

Fe:

$$q_{Fe} = mass \times c \times \Delta t = 111.7 \text{ g} \times 0.45 \text{ J/g}°C \times \Delta t = 50 \text{ J/}°C \times \Delta t$$

substitute $q_{reaction}$, $q_{Al_2O_3}$ and q_{Fe} to the following equation to find Δt:

$$q_{reaction} = -(q_{Al_2O_3} + q_{Fe})$$

$$-851.5 \text{ kJ} = -[(79 \text{ J/}°C \times \Delta t) + (50 \text{ J/}°C \times \Delta t)]$$

$$-851.5 \text{ kJ} = -129 \text{ J/}°C \times \Delta t$$

$$\Delta t = \frac{-8.515 \times 10^5 \text{ J}}{-129 \text{ J/}°C} = 6.6 \times 10^3 \ °C$$

$$t_{final} = \Delta t + t_{initial} = 6.60 \times 10^3 \ °C + 25.0°C = \underline{\mathbf{6.6 \times 10^3 \ °C}}$$

(c) **Yes.** Assuming 2 mol Fe is produced, melting these amount of Fe requires only 30.2 kJ of energy which can be provided for by the energy released by the reaction (8.51×10^5 J).

$$111.7 \text{ g Fe} \times \frac{270 \text{ J}}{1 \text{ g Fe}} \times \frac{1 \text{kJ}}{1000 \text{ J}} = 30.2 \text{ kJ}$$

Also, the final temperature (6600°C) is much higher than the melting point of iron (1535°C) therefore it is expected that the reaction will produce Fe in the molten state.

88. See Section 8.2.

$$q_{reaction} = -C_{cal}\Delta t = -(22.51 \text{ kJ/}°C)(1.67°C) = -37.6 \text{ kJ}$$

sucrose burned in the reaction based on this amount of heat released (−37.6 kJ):

$$-37.6 \text{ kJ} \times \frac{1 \text{mol sucrose}}{-5.64 \times 10^3 \text{ kJ}} \times \frac{342.30 \text{ g sucrose}}{1 \text{mol sucrose}} = 2.28 \text{ g sucrose}$$

Percentage of sucrose in the sample:

$$\text{mass \% sucrose} = \frac{2.28 \text{ g}}{3.000 \text{ g}} \times 100 = \underline{\mathbf{76.0\%}}$$

89. See Sections 8.2 and 6.1.

$$\lambda_{microwave} = 13.5\ cm \times \frac{1\,m}{100\ cm} = 0.135\ m$$

Energy per photon from the 13.5 cm microwave radiation:

$$E = \frac{hc}{\lambda} = \frac{(6.626 \times 10^{-34}\,J \cdot s)(2.998 \times 10^{8}\,m/s)}{0.135\ m} = 1.47 \times 10^{-24}\ J/photon$$

Rate at which microwave radiation is produced = 925 mol photons/s:

$$photons\ generated = 1.55\ s \times \frac{925\ mol\ photons}{1\,s} \times \frac{6.022 \times 10^{23}\ photons}{1\,mol\ photons}$$

$$= 8.63 \times 10^{26}\ photons$$

Energy from 8.63×10^{26} photons:

$$8.63 \times 10^{26}\ photons \times \frac{1.47 \times 10^{-24}\ J}{1\,photon} = 1.27 \times 10^{3}\ J$$

Mass of steel wool:

$$q = mass_{steel\ wool} \times c \times \Delta t \quad \Rightarrow \text{solve for mass} \Rightarrow \quad mass_{steel\ wool} = \frac{q}{c \times \Delta t}$$

$$q = 1.27 \times 10^{3}\ J$$

$$c = 0.45\ J/g°C$$

$$\Delta t = 400.0°C - 25°C = 375°C$$

$$mass_{steel\ wool} = \frac{q}{c \times \Delta t} = \frac{1.27 \times 10^{3}\,J}{0.45\ J/g \cdot °C \times 375°C} = \underline{\mathbf{7.5\ g}}$$

Therefore, **7.5 g** of steel wool were accidentally placed in the oven.

90. Given: 35.00 mL of 0.217 M $A(NO_3)_2$

25.00 mL of 0.195 M NaOH

$t_{initial} = 24.8°C$

$t_{final} = 28.2°C$

Assumptions: density of the solution = 1.00 g/mL

volumes are additive

specific heat of solution: $c_{solution} = 4.18\ J/g°C$

$$A(NO_3)_2 + 2NaOH \rightarrow A(OH)_2 + 2Na^+ + 2NO_3^- \qquad \Delta H° = ?$$

mole $A(OH)_2$ from 35.00 mL of 0.217 M $A(NO_3)_2$:

$$= 35.00 \text{ mL} \times \frac{1L}{1000 \text{ mL}} \times \frac{0.217 \text{ mol } A(NO_3)_2}{1L} \times \frac{1 \text{ mol } A(OH)_2}{1 \text{ mol } A(NO_3)_2}$$

$$= 7.60 \times 10^{-3} \text{ mol } A(OH)_2$$

mole $A(OH)_2$ from 25.00 mL of 0.195 M NaOH:

$$= 25.00 \text{ mL} \times \frac{1L}{1000 \text{ mL}} \times \frac{0.195 \text{ mol NaOH}}{1L} \times \frac{1 \text{ mol } A(OH)_2}{2 \text{ mol NaOH}}$$

$$= 2.44 \times 10^{-3} \text{ mol } A(OH)_2$$

mole $A(OH)_2$ formed $= 2.44 \times 10^{-3}$ mol

assuming volume is additive:

total volume = 35.00 mL + 25.00 mL = 60.00 mL solution

mass of solution $= V_{solution} \times density_{solution}$

$$= 60.00 \text{ mL solution} \times \frac{1.00 \text{ g solution}}{1 \text{ mL solution}} = 60.00 \text{ g solution}$$

$q_{reaction} = -q_{solution}$

$$= -(mass \times c \times \Delta t)_{solution}$$

$$= -(60.00 \text{ g} \times 4.18 \text{ J/g°C} \times [28.2°C - 24.8°C])$$

$$= -853 \text{ J}$$

$$\Delta H_{precipitation} = \frac{-853 \text{ J}}{2.44 \times 10^{-3} \text{ mol } A(OH)_2} = \underline{-3.50 \times 10^5 \text{ J/mol}} \text{ or } \underline{-3.50 \times 10^2 \text{ kJ/mol}}$$

$$A(NO_3)_{2(aq)} + 2NaOH_{(aq)} \rightarrow A(OH)_{2(s)} + 2Na^+_{(aq)} + 2NO_3^-_{(aq)} \qquad \Delta H = \underline{-3.50 \times 10^2 \text{ kJ/mol}}$$

The ΔH for the precipitation of $A(OH)_2$ is $\underline{-3.50 \times 10^2 \text{ kJ/mol}}$.

LIQUIDS AND SOLIDS

2. See Example 9.1 and Chapter 5.

$$P_1 = 325 \text{ mm Hg} \times \frac{1 \text{ atm}}{760 \text{ mm Hg}} = 0.428 \text{ atm}$$

$T_1 = 80°C + 273.15 = 353 \text{ K}$

(a) at 50°C:

$T_2 = 50°C + 273.15 = 323 \text{ K}; \quad P_2 = ?$

$$\frac{V_1 P_1}{n_1 T_1} = \frac{V_2 P_2}{n_2 T_2}$$

since volume (V) and amount of gas (n) is constant, the formula simplifies to:

$$\frac{P_1}{T_1} = \frac{P_2}{T_2} \qquad \Rightarrow \quad \text{solve for } P_2 \quad \Rightarrow \qquad P_2 = \frac{P_1 T_2}{T_1}$$

$$P_2 = \frac{P_1 T_2}{T_1} = \frac{(0.428 \text{ atm})(323 \text{ K})}{353 \text{ K}} = 0.392 \text{ atm}$$

pressure of the vapor at 50°C (in mm Hg) = $0.392 \text{ atm} \times \dfrac{760 \text{ mm Hg}}{1 \text{ atm}}$ = **298 mm Hg**

at 60°C:

$T_2 = 60°C + 273.15 = 333 \text{ K}; \quad P_2 = ?$

$$P_2 = \frac{P_1 T_2}{T_1} = \frac{(0.428 \text{ atm})(333 \text{ K})}{353 \text{ K}} = 0.404 \text{ atm}$$

pressure of the vapor at 60°C (in mm Hg) = $0.404 \text{ atm} \times \dfrac{760 \text{ mm Hg}}{1 \text{ atm}}$ = **307 mm Hg**

(b) 50°C: **298 mm Hg > 269 mm Hg**

At 50°C, the calculated vapor pressure (298 mm Hg) is **greater than** the equilibrium vapor pressure (269 mm Hg) of benzene.

60°C: <u>307 mm Hg < 389 mm Hg</u>

At 60°C, the calculated vapor pressure (307 mmHg) is <u>less than</u> the equilibrium vapor pressure (389 mm Hg) of benzene.

(c) The pressure exerted by the benzene vapor will never exceed the vapor pressure, therefore, <u>at 50°C, P = 269 mm Hg, and at 60°C, P = 307 mm Hg.</u>

4. See Example 9.1 and Chapter 5.

(a) Use ideal gas law to find mol camphor ($C_{10}H_{16}O$). Convert mol camphor to mg.

$$P = 0.18 \text{ mm Hg} \times \frac{1 \text{ atm}}{760 \text{ mm Hg}} = 2.4 \times 10^{-4} \text{ atm}$$

$T = 20°C + 273.15 = 293 \text{ K}$

$V = 0.500 \text{ L}$

$$PV = nRT \quad \Rightarrow \quad \text{solve for } n \quad \Rightarrow \quad n = \frac{PV}{RT}$$

$$n = \frac{PV}{RT} = \frac{(2.4 \times 10^{-4} \text{ atm})(0.500 \text{ L})}{(0.0821 \text{ L} \cdot \text{atm/K} \cdot \text{mol})(293 \text{ K})} = 5.0 \times 10^{-6} \text{ mol } C_{10}H_{16}O \text{ vapor}$$

$$\text{mg } C_{10}H_{16}O = 5.0 \times 10^{-6} \text{ mol } C_{10}H_{16}O \times \frac{152.23 \text{ g}}{1 \text{ mol } C_{10}H_{16}O} \times \frac{1000 \text{ mg}}{1 \text{ g}} = \underline{\textbf{0.76 mg}}$$

(b) Strategy: Use gas law to calculate P assuming that all the camphor has been converted to vapor. If $P_{calculated} > P_{vapor}$ at 20°C, P_{vapor} is the pressure in the flask. If $P_{calculated} < P_{vapor}$ at 20°C, $P_{calculated}$ is the pressure in the flask.

$T = 20°C + 273.15 = 293 \text{ K}$

$$V = 125 \text{ mL} \times \frac{1 \text{ L}}{1000 \text{ mL}} = 0.125 \text{ L}$$

$$n = 0.15 \text{ mg} \times \frac{1 \text{ g}}{1000 \text{ mg}} \times \frac{1 \text{ mol}}{152.23 \text{ g}} = 9.9 \times 10^{-7} \text{ mol}$$

$$PV = nRT \quad \Rightarrow \quad \text{solve for } P \quad \Rightarrow \quad P = \frac{nRT}{V}$$

$$P = \frac{nRT}{V} = \frac{(9.9 \times 10^{-7} \text{ mol})(0.0821 \text{ L} \cdot \text{atm/K} \cdot \text{mol})(293 \text{ K})}{0.125 \text{ L}} = 1.9 \times 10^{-4} \text{ atm}$$

$$P_{calculated} = 1.9 \times 10^{-4} \text{ atm} \times \frac{760 \text{ mm Hg}}{1 \text{ atm}} = \underline{\textbf{0.14 mm Hg}}$$

$P_{calculated} = 0.14 \text{ mm Hg}$ while $P_{vapor\ 20°C} = 0.18 \text{ mm Hg}$

Since, $P_{calculated} < P_{vapor}$, $P_{calculated}$ (<u>**0.14 mm Hg**</u>) is the pressure in the flask.

6. See Example 9.1 and Chapter 5.

   (a) Use ideal gas law to find mol p–dichlorobenzene ($C_6H_4Cl_2$). Convert mol $C_6H_4Cl_2$ to mg.

   $$P = 0.40 \text{ mm Hg} \times \frac{1 \text{ atm}}{760 \text{ mm Hg}} = 5.3 \times 10^{-4} \text{ atm}$$

   $$T = 20°C + 273.15 = 293 \text{ K}$$

   $$V = 750 \text{ mL} \times \frac{1 \text{ L}}{1000 \text{ mL}} = 0.750 \text{ L}$$

   $$PV = nRT \quad \Rightarrow \quad \text{solve for } n \quad \Rightarrow \quad n = \frac{PV}{RT}$$

   $$n = \frac{PV}{RT} = \frac{(5.3 \times 10^{-4} \text{ atm})(0.750 \text{ L})}{(0.0821 \text{ L} \cdot \text{atm/K} \cdot \text{mol})(293 \text{ K})} = 1.7 \times 10^{-5} \text{ mol } C_6H_4Cl_2 \text{ vapor}$$

   $$\text{mg } C_6H_4Cl_2 = 1.7 \times 10^{-5} \text{ mol } C_6H_4Cl_2 \times \frac{146.99 \text{ g } C_6H_4Cl_2}{1 \text{ mol } C_6H_4Cl_2} \times \frac{1000 \text{ mg}}{1 \text{ g}} = \underline{\textbf{2.5 mg}}$$

   (b) 5.0 mg – 2.5 mg = **2.5 mg of p–dichlorobenzene remaining**.

   Since only 2.5 mg p–dichlorobenzene should vaporize in order to attain the vapor pressure, the other **2.5 mg remains in the solid phase**.

   (c) Strategy: Use gas law to calculate P assuming that all the p–dichlorobenzene has been converted to vapor. If $P_{calculated} > P_{vapor}$ at 20°C, P_{vapor} is the pressure in the flask. If $P_{calculated} < P_{vapor}$ at 20°C, $P_{calculated}$ is the pressure in the flask.

   $$T = 20°C + 273.15 = 293 \text{ K}$$

   $$V = 500 \text{ mL} \times \frac{1 \text{ L}}{1000 \text{ mL}} = 0.500 \text{ L}$$

   $$n = 2.00 \text{ mg} \times \frac{1 \text{ g}}{1000 \text{ mg}} \times \frac{1 \text{ mol}}{146.99 \text{ g}} = 1.36 \times 10^{-5} \text{ mol}$$

   $$PV = nRT \quad \Rightarrow \quad \text{solve for } n \quad \Rightarrow \quad n = \frac{PV}{RT}$$

   $$P = \frac{nRT}{V} = \frac{(1.36 \times 10^{-5} \text{ mol})(0.0821 \text{ L} \cdot \text{atm/K} \cdot \text{mol})(293 \text{ K})}{0.500 \text{ L}} = 6.54 \times 10^{-4} \text{ atm}$$

   $$P_{calculated} = 6.54 \times 10^{-4} \text{ atm} \times \frac{760 \text{ mm Hg}}{1 \text{ atm}} = \underline{\textbf{0.50 mm Hg}}$$

   Since $P_{calculated}$ (0.50 mm Hg) is greater than P_{vapor} (0.40 mm Hg) at 20°C, P_{vapor} is the pressure in the flask.

   <u>**Therefore, the pressure in the flask is 0.40 mm Hg.**</u>
   <u>**Yes, some of the p–dichlorobenzene will remain in the solid phase.**</u>

8. See Example 9.2.

P_{vapor} = 381.0 mm Hg at T = 21.9°C

P_{vapor} = 465.8 mm Hg at T = 26.9°C

R = 8.31 J/mol·K

(a) ΔH_{vap} = ?

P_1 = 381.0 mm Hg; T_1 = 21.9°C $\Rightarrow$ T_1 = 21.9°C + 273.15 = 295.05 K

P_2 = 465.8 mm Hg; T_2 = 26.9°C $\Rightarrow$ T_2 = 26.9°C + 273.15 = 300.05 K

$$\ln \frac{P_2}{P_1} = \frac{\Delta H_{vap}}{R} \left[\frac{1}{T_1} - \frac{1}{T_2} \right]$$

$$\ln \frac{465.8 \text{ mm Hg}}{381.0 \text{ mm Hg}} = \frac{\Delta H_{vap}}{8.31 \text{ J/mol} \cdot \text{K}} \left[\frac{1}{295.05 \text{ K}} - \frac{1}{300.05 \text{ K}} \right]$$

$0.2010 = \Delta H_{vap} [6.80 \times 10^{-6} \text{ J/mol}]$

$\Delta H_{vap} = 0.2010 / [6.80 \times 10^{-6} \text{ J/mol}]$

ΔH_{vap} = 2.96 × 10⁴ J/mol or 29.6 kJ/mol

(b) P_1 = 381.0 mm Hg; T_1 = 21.9°C $\Rightarrow$ T_1 = 21.9°C + 273.15 = 295.05 K

P_2 = 760 mm Hg

T_2 = normal boiling point = ?

ΔH_{vap} = 2.96 × 10⁴ J/mol (obtained from 8a above)

$$\ln \frac{P_2}{P_1} = \frac{\Delta H_{vap}}{R} \left[\frac{1}{T_1} - \frac{1}{T_2} \right]$$

$$\ln \frac{760}{381.0} = \frac{2.96 \times 10^4 \text{ J/mol}}{8.31 \text{ J/mol} \cdot \text{K}} \left[\frac{1}{295.05 \text{ K}} - \frac{1}{T_2} \right]$$

$$0.6905 = 3561.97 \text{ K} \left[\frac{1}{295.05 \text{ K}} - \frac{1}{T_2} \right]$$

$$19.39 \times 10^{-4} \text{ K}^{-1} = \left[\frac{1}{295.05 \text{ K}} - \frac{1}{T_2} \right]$$

T_2 = 312.95 K

T_2 (in °C):   312 K = °C + 273.15

            312.95 K − 273.15 = 39.8 °C

normal boiling point = 39.8 °C

10. See Example 9.2.

water: ΔH_{vap} = 40.7 kJ/mol = $40.7 \text{ kJ/mol} \times \dfrac{1000 \text{ J}}{1 \text{ kJ}}$ = 4.07×10^4 J/mol

normal boiling point of water = 100°C at 760 mm Hg:

P_1 = 760 mm Hg; $\quad T_1$ = 100°C $\Rightarrow$ $\quad T_1$ = 100°C + 273.15 = 373 K

P_2 = 681.0 mm Hg; $\quad T_2$ = ?

$$\ln \frac{P_2}{P_1} = \frac{\Delta H_{vap}}{R}\left[\frac{1}{T_1} - \frac{1}{T_2}\right]$$

$$\ln \frac{681.0 \text{ mm Hg}}{760 \text{ mm Hg}} = \frac{4.07 \times 10^4 \text{ J/mol}}{8.31 \text{ J/mol} \cdot \text{K}}\left[\frac{1}{373 \text{ K}} - \frac{1}{T_2}\right]$$

$$-0.1098 = 4898 \text{ K}\left[\frac{1}{373 \text{ K}} - \frac{1}{T_2}\right]$$

$$-2.241 \times 10^{-5} \text{ K}^{-1} = \left[\frac{1}{373 \text{ K}} - \frac{1}{T_2}\right]$$

T_2 = 370 K

T_2 (in °C): $\qquad$ 370 K = °C + 273.15

$\qquad\qquad$ 370 K − 273.15 = **97°C**

boiling point of water in Glacier National Park (4100 ft above sea level) = **97°C**

12. See Example 9.2.

water: ΔH_{vap} = 40.7 kJ/mol = $40.7 \text{ kJ/mol} \times \dfrac{1000 \text{ J}}{1 \text{ kJ}}$ = 4.07×10^4 J/mol

normal boiling point of water = 100°C at 760 mm Hg:

P_1 = 760 mm Hg = 1 atm; $\qquad T_1$ = 100°C $\quad \Rightarrow \quad T_1$ = 100°C + 273.15 = 373 K

P_2 = 1.75 atm; $\qquad\qquad\qquad T_2$ = ?

$$\ln \frac{P_2}{P_1} = \frac{\Delta H_{vap}}{R}\left[\frac{1}{T_1} - \frac{1}{T_2}\right]$$

$$\ln \frac{1.75 \text{ atm}}{1 \text{ atm}} = \frac{4.07 \times 10^4 \text{ J/mol}}{8.31 \text{ J/mol} \cdot \text{K}}\left[\frac{1}{373 \text{ K}} - \frac{1}{T_2}\right]$$

$$0.5596 = 4898 \text{ K}\left[\frac{1}{373 \text{ K}} - \frac{1}{T_2}\right]$$

$$1.143 \times 10^{-4} \text{ K}^{-1} = \left[\frac{1}{373 \text{ K}} - \frac{1}{T_2} \right]$$

T_2 = 389 K

T_2 (in °C):   390 K = °C + 273.15

390 K − 273.15 = **117°C**

boiling point of water in the pressure cooker set at 1.75 atm = 117°C

normal boiling point of water = 100°C

difference in boiling point: 117°C − 100°C = **17°C**

14. See Figure 9.4 as guide

vp(mm Hg)	ln (vp)	T (°C)	T(K)	1/T (1/K)
146	4.98	−5	268	0.00373
231	5.44	5	278	0.00359
355	5.87	15	288	0.00347
531	6.27	25	298	0.00335

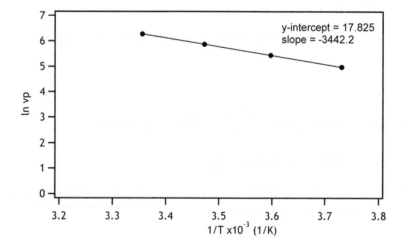

$$\text{slope} = -3442.2 = \frac{-\Delta H_{vap}}{8.31 \text{ J/mol} \cdot \text{K}}$$

$$\Delta H_{vap} = 28605 \text{ J/mol} = 28605 \text{ J/mol} \times \frac{1 \text{ kJ}}{1000 \text{ J}} = 29 \text{ kJ/mol}$$

ΔH_{vap} of diethyl ether = 29 kJ/mol

9.3 PHASE DIAGRAMS

16. See Figure 9.7.

    (a) **liquid** phase   (b) **vapor** phase   (c) **liquid** phase

18. See Example 9.3.

    (a) **vapor**

    (b) **20 °C**

    (c) **critical point**

    (d) **triple point**

    (e) **normal boiling point**

    (f) **deposition**

    (g) **No.** The solid-liquid curve has a positive (+) slope which indicates that the solid phase has a higher density that the liquid phase. Thus, the solid will not float on the liquid.

    (h) **Yes**

20. Use Figure 9.7 and the discussion on page 223 as guide.

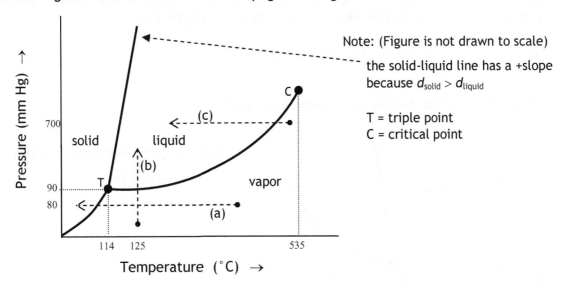

    (a) Iodine vapor at 80 mm Hg condenses to the **<u>solid</u>** when cooled sufficiently. Since this is below the triple point of 90 mm Hg, the liquid cannot form. See arrow marked (a) in the phase diagram above.

(b) Iodine vapor at 125°C condenses to the **liquid** when enough pressure is applied. Since the temperature (125°C) is higher than that of the triple point (114°C), condensation will be to the liquid state.

(c) Iodine vapor at 700 mm Hg condenses to the **liquid** when cooled above the triple point temperature. The pressure is above that of the triple point, so condensation will be to the liquid state.

22. Use Figure 9.7 and the discussion on page 223 and 226 as guide.

(a)

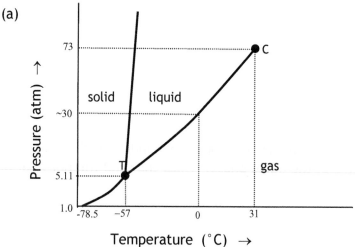

Note: (Figure is not drawn to scale)

T = triple point
C = critical point

Since the solid is the denser phase, the solid-liquid line is inclined to the right.

(b) **About 30 atm**. The value one gets depend on the curve drawn for the vapor–liquid boundary, but the estimated vapor pressure is about **30-40 atm**.

(c) **Yes**. At 10 atm and -50°C, CO_2 is in the **liquid** state.

24. Use Figure 9.7 and the discussion on page 223 as guide.

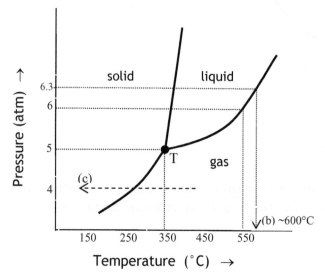

Note: (Figure is not drawn to scale)

T = triple point

(a) The phase diagram for substance A is drawn on page 230 with the appropriate labels. Note that the slope of the solid-liquid line is positive (inclined to the right) since the solid is the denser phase.

(b) The estimated boiling point when 6.3 atm pressure is applied is **600 °C**.

(c) The substance will sublime i.e. **gas will depositing as solid**, as noted with dashed arrow in the phase diagram marked (c).

9.4 MOLECULAR SUBSTANCES; INTERMOLECULAR FORCES

26. Compare the relative strength of the intermolecular forces of the given compounds. The stronger the intermolecular forces, the higher the boiling point of the substance.

    The only intermolecular force present in noble gases is dispersion (London) forces. The strength of dispersion forces depends on two factors: (1) the number of electrons and (2) ease of electron dispersion to form temporary dipoles. Both of these factors increase with molecular size. Thus, the higher the molar mass, the stronger the dispersion forces. The noble gas with the largest molar mass will have the highest boiling point. Therefore in terms of increasing boiling points:

    $$\text{He} \; < \; \text{Ne} \; < \; \text{Ar} \; < \; \text{Xe}$$
    lowest bp   highest bp

28. See Chapter 7 and Example 9.6.

    All substances have dispersion (London) forces. Polar molecules exhibit dipole forces in addition to dispersion forces.

    (a) F_2 **dispersion forces**. This molecule is non–polar thus only dispersion forces are present.

    (b) CO **dispersion and dipole forces**. This molecule is polar thus in addition to dispersion forces, it has dipole forces.

    (c) CO_2 **dispersion forces**. This molecule is non–polar thus only dispersion forces are present.

    (d) H_2CO **dispersion and dipole forces**. This molecule is polar thus in addition to dispersion forces, it has dipole forces.

30. See Example 9.5.

   Molecules that contain an H atom that is bonded to any of the following atoms: N, O and F will show H–bonding.   Draw the Lewis structure of the compound and determine if any of the following bond is presents: H–N, H–O, and H–F. If any of the three bonds is present then the compound shows H–bonding.

   (a)

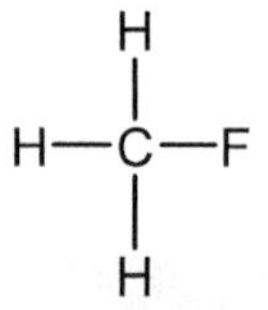

   H
   |
   H—C—F
   |
   H

   This molecule has no H–F bond, thus, **there is no H–bonding**.

   (b) 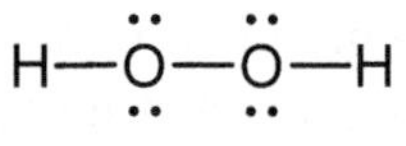 H—Ö—Ö—H

   This molecule has H–O bonds, thus, **there is H–bonding**.

   (c) 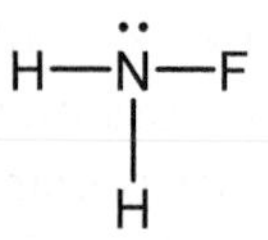 H—N̈—F
   |
   H

   This molecule has H–N bonds, thus, **there is H–bonding**.

   (d)

   H   H
   | |
   H—C—Ö—C—H
   | |
   H   H

   Although, there is an O atom, it is not bonded to H. This molecule has no H–O bond, thus, **there is no H–bonding**.

32. See Example 9.4.

   (a) In boiling ionic compounds like Na_2CO_3, ionic bonds are broken while in molecular compounds like CO_2 intermolecular forces are destroyed. Since ionic bonds are stronger than intermolecular forces, boiling Na_2CO_3 would require higher temperature than CO_2. Thus, **CO_2 has a lower boiling point than Na_2CO_3.**

   (b)

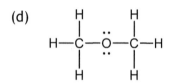

   N_2H_4 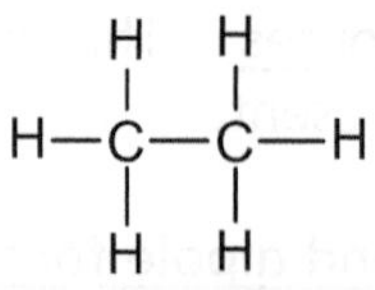 C_2H_6

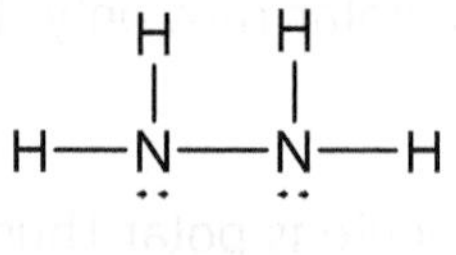

   N_2H_4 has dispersion forces, dipole forces, and H–bonding, while C_2H_6 has only dispersion forces. **Since the intermolecular force of N_2H_4 is stronger, it has a higher boiling point compared to C_2H_6.**

(c)

formic acid

benzoic acid

Both formic acid and benzoic acid have dispersion forces, dipole forces and H–bonding. Since formic acid has a lower molar mass, the strength of its dispersion forces is weaker than that of benzoic acid. Thus, **formic acid has a lower boiling point than benzoic acid.**

(d) CO and N_2 has the same molar mass thus has comparative dispersion forces. However, CO is a polar molecule thus in addition to dispersion forces, CO has dipole forces which is absent in the non–polar N_2 molecules. Consequently, **CO has higher boiling point than N_2.**

34. Processes a and b requires breaking of chemical bonds whereas c and d requires overcoming intermolecular forces only.

(a) **breaking covalent bonds.** Decomposing HCl to the elements composing it requires the breaking of the covalent bond between H and Cl (H–Cl) to form H_2 and Cl_2.

(b) **breaking ionic bonds.** When NaCl is dissolved in water, the ionic bonds between the Na^+ ions and the Cl^- ions must be broken in order for both ions to be surrounded by water.

36. (a) LiCl or **CCl$_4$**. LiCl is an ionic compound. In boiling ionic compounds like LiCl, ionic bonds are broken while in molecular compounds like CCl_4 only intermolecular forces are destroyed. Since ionic bonds are stronger than intermolecular forces, LiCl would require higher boiling temperature than CCl_4. Thus, **CCl_4 has a lower boiling point than LiCl.**

(b) CH_3OH or **CH$_3$F**. Both are polar molecules thu they both have dispersion and dipole forces. However, CH_3OH shows H–bonding which is not present in CH_3F. Consequently, **CH_3F has weaker intermolecular forces than CH_3OH resulting in a lower boiling point for CH_3F.**

(c) H_2O or **SO$_2$**. Both are polar molecules thus, they both have dispersion and dipole forces. However, H_2O shows H–bonding which is not present in SO_2. Thus, **SO_2 has weaker intermolecular forces than H_2O resulting in a lower boiling point for SO_2.**

(d) **N$_2$** or Cl_2. Both molecules are nonpolar and have dispersion forces only. Since **N_2 has a lower molar mass, it has weaker dispersion forces which results in lower boiling point compared to Cl_2.**

38. (a) **H–bonding.** Melting ice or $H_2O(s)$ requires the breaking of intermolecular forces present which are: dispersion forces, dipole forces and H–bonding. Of these three, H–bonding is the strongest.

    (b) **Dispersion forces.** Subliming Br_2 requires the breaking of intermolecular forces. Since Br_2 is a non–polar molecule, the only intermolecular force that should be broken is dispersion forces.

    (c) **Dipole forces.** Boiling $CHCl_3$ requires the breaking of intermolecular forces present which are: dispersion forces and dipole forces. Of these two, dipole force is stronger.

    (d) **Dispersion forces.** Since benzene (C_6H_6) is a non–polar substance, the only intermolecular forces that it has is dispersion forces. This dispersion forces should be broken in vaporizing benzene.

9.5 TYPES OF SUBSTANCES: NETWORK COVALENT, IONIC, AND METALLIC SOLIDS

40. See Example 9.7 and Table 9.6.

    (a) **Molecular solid.** Molecular solids are characterized by low melting point.

    (b) **Ionic solid.** Electrical conductivity in the molten state only is a unique characteristic of ionic compounds.

    (c) **Metallic solid.** Electrical conductivity and insolubility in water are unique characteristics of metallic solids.

42. See Example 9.7 and Table 9.6.

    (a) The three types of solids that are generally insoluble in water are **metallic, network covalent and nonpolar molecular solids**.

    (b) **Network covalent and ionic compounds** generally have very high melting points. Many of the metallic solids have high melting points, but several (notably Hg) have quite low melting points.

    (c) **Metallic compounds** conduct electricity as solids. Most network covalent compounds do not conduct electricity, but a few (such as graphite) do. Ionic compounds conduct electricity only in the molten state or when they are dissolved in water.

44. See Example 9.7 and Table 9.6.

    (a) **Metallic.** W is a transition metal.

    (b) **Molecular.** NO_2 is a molecular compound because it is composed of two nonmetals.

(c) **Network covalent.** Diamond is made only of carbon, but is a 3–dimensional network of C–C bonds (see page 233).

(d) **Ionic.** $FeCl_2$ is an ionic compound because it is composed of ions: a metal ion, Fe^{2+} and a nonmetal ion Cl^-.

(e) **Molecular.** C_2H_2 is a molecular compound because it is composed of two nonmetals.

46. There is a vast number of oxygen–containing solids that would fit into the description given in this problem.

    (a) Polar molecule: **sugar, $NO, C_{12}H_{22}O_{11}$**

    (b) Ionic solid: **$CaCO_3$ (found in TUMS®), CaO**

    (c) Network covalent solid: **SiO_2**

    (d) Non–polar molecule: **dry ice, CO_2, O_2**

48. See Table 9.6.

    (a) **Atoms.** Graphite is composed of C atoms. Each C atom is attached to 3 other C atoms in a three–dimensional array.

    (b) **Atoms.** SiC is composed of Si and C atoms.

    (c) **Ions.** $FeCl_2$ is composed of a metal ion, Fe^{2+} and a nonmetal ion Cl^-.

    (d) **Molecule.** C_2H_2 is a molecular compound therefore its structural unit is molecule.

9.6 CRYSTAL STRUCTURES

50. See Example 9.8 and Table 9.7. Determine which of the equations relating the sides of the unit cell to the given atomic radius is valid. The geometry corresponding to the equation is the geometry of the unit cell.

    atomic radius, r = 0.162 nm

    side of cell, s = 0.458 nm

    - The geometry is not simple cubic because $2r \neq s$ as shown below:

        $2r = 2(0.162) = 0.324$ nm

        $2r = 0.324$ nm $\neq s = 0.458$ nm

- The geometry is not body centered cubic because as shown below, $4r \neq s\sqrt{3}$

$$4r = s\sqrt{3} \qquad \frac{4(0.162)}{\sqrt{3}} = s = 0.374 \neq 0.458$$

- The geometry is face–centered cubic because as shown below, $4r = s\sqrt{2}$

$$4r = s\sqrt{2} \qquad \frac{4(0.162)}{\sqrt{2}} = s = 0.458$$

Thus, the unit cell is **face-centered cubic**.

52. See Example 9.8 and Table 9.7.

A unit cell is a cube. For a cube, volume = s^3

volume = s^3 = $(0.3833 \text{ nm})^3$ = **0.05631 nm³**

For a face–centered cubic structure, $4r = s\sqrt{2}$ solve for r $\Rightarrow$ $r = \dfrac{s\sqrt{2}}{4}$

$$r = \frac{0.3833 \text{ nm } \sqrt{2}}{4}$$

$$r = \underline{\textbf{0.1355 nm}}$$

54. See Example 9.9, Table 9.7 and Figure 9.21.

Ionic radius: K^+: $r = 0.133$ nm; I^-: $r = 0.216$ nm

(a) one side of a cube

K^+ and I^- ions touch along a side

length of one side of the cube, s = 0.216 nm + 2(0.133 nm) + 0.216 nm = **0.698 nm**

(b) face diagonal of the cube

length of the face diagonal of the cube, $d = s\sqrt{2}$ = $0.698\sqrt{2}$ = **0.987 nm**

56. See Example 9.9, Table 9.7 and Figure 9.21.

Ionic radius: Cs^+: $r = 0.169$ nm; Cl^-: $r = 0.181$ nm

Figure 9.21 indicates that CsCl forms a body–centered cubic structure.

(a) length of body diagonal = $4r = 2(r_{Cr^+}) + 2(r_{Cl^-})$ = 2(0.169 nm) + 2(0.181 nm) = **0.700 nm**

(b) length of the side of the cell, s:

For a BCC: $4r = s\sqrt{3}$ $\Rightarrow$ 0.700 nm = $s\sqrt{3}$ $\Rightarrow$ **s = 0.404 nm**

58. See Figure 9.21.

    Visualize the unit cell three dimensionally with ions coming up towards you, and going back away from you.

    The unit cell has in the center one whole cesium ion. At each corner, $^1/_8$ of a chloride ion is inside the cell. There are 8 corners, thus, the number of chloride ions is $8 \times ^1/_8 = 1$ chloride ion. Thus, **there is 1 Cs^+ ion and 1 Cl^- ion**.

UNCLASSIFIED PROBLEMS

60. See Section 9.6, Example 9.8 and Table 9.7.

    (a) atomic radius of aluminum in cm = ?

    A unit cell is a cube. For a cube, volume = s^3

    $$\text{volume} = 0.0662 \text{ nm}^3 = s^3 \implies s = \sqrt[3]{0.0662 \text{ nm}^3} = 0.405 \text{ nm}$$

    For face–centered cubic structure,

    $$4r = s\sqrt{2} \quad \text{solve for r} \implies \quad r = \frac{s\sqrt{2}}{4} \qquad r = \frac{0.405 \text{ nm} \sqrt{2}}{4} = 0.143 \text{ nm}$$

    $$\text{radius (in cm)} = 0.143 \text{ nm} \times \frac{1 \text{m}}{10^9 \text{ nm}} \times \frac{100 \text{ cm}}{1 \text{m}} = \underline{\mathbf{1.43 \times 10^{-8} \text{ cm}}}$$

    (b) volume of a single aluminum atom in cm = ?

    $$V = \frac{4\pi r^3}{3} = \frac{4\pi (1.43 \times 10^{-8})^3}{3} = \underline{\mathbf{1.22 \times 10^{-23} \text{ cm}^3}}$$

    (c) density of a single aluminum atom = ?

    $$\text{mass (g) of a aluminum atom} = 1 \text{ Al atom} \times \frac{26.98 \text{ g Al}}{6.022 \times 10^{23} \text{ Al atoms}} = 4.48 \times 10^{-23} \text{ g}$$

    $$\text{density} = \frac{\text{mass}}{\text{volume}} = \frac{4.48 \times 10^{-23} \text{ g}}{1.22 \times 10^{-23} \text{ cm}^3} = \underline{\mathbf{3.67 \text{ g/cm}^3}}$$

    (d) calculated density of aluminum after factoring in the 26.0% empty space = ?

    $0.260 \times 3.67 \text{ g/cm}^3 = 0.954 \text{ g/cm}^3$

    calculated density = $3.67 \text{ g/cm}^3 - 0.954 \text{ g/cm}^3 = \underline{\mathbf{2.72 \text{ g/cm}^3}}$

62. See Section 9.6 and Table 9.7.

The formula of packing efficiency is

$$\text{packing efficiency} = \frac{\text{volume of the atom in the cell}}{\text{volume of the unit cell}} \times 100$$

(a)   For a **simple cubic** unit cell (1 atom/cell):

volume of the unit cell: $2r = s$ $(2r)^3 = s^3 = 8r^3$

$$\text{volume of 1 atom} = \frac{4\pi\, r^3}{3}$$

$$\text{packing efficiency} = \frac{4\pi\, r^3/3}{8\, r^3} \times 100 = \frac{\pi/3}{2} \times 100 = \underline{\mathbf{52\%}}$$

(b)   For a **face-centered cubic** unit cell (4 atoms/cell):

Volume of the unit cell : $4\,r = s\sqrt{2}$ $\left(\dfrac{4\,r}{\sqrt{2}}\right)^3 = s^3 = \dfrac{32\, r^3}{\sqrt{2}}$

$$\text{Volume of 4 atoms} = \frac{16\pi\, r^3}{3}$$

$$\text{packing efficiency} = \frac{\left(\dfrac{16\pi\, r^3}{3}\right)}{\left(\dfrac{32\, r^3}{\sqrt{2}}\right)} \times 100 = \frac{\left(\dfrac{\pi}{3}\right)}{\left(\dfrac{2}{\sqrt{2}}\right)} \times 100 = \underline{\mathbf{74\%}}$$

(c)   For a **body–centered cubic** unit cell (2 atoms/cell):

Volume of the unit cell : $4\,r = s\sqrt{3}$ $\left(\dfrac{4\,r}{\sqrt{3}}\right)^3 = s^3 = \dfrac{64\, r^3}{3\sqrt{3}}$

$$\text{Volume of 2 atoms} = \frac{8\pi\, r^3}{3}$$

$$\text{packing efficiency} = \frac{\left(\dfrac{8\pi\, r^3}{3}\right)}{\left(\dfrac{64\, r^3}{3\sqrt{3}}\right)} \times 100 = \frac{\pi}{\left(\dfrac{8}{\sqrt{3}}\right)} \times 100 = \underline{\mathbf{68\%}}$$

To check the answers for accuracy, compare these values to the values for % empty space found in Table 9.7

For simple cubic :   100% – 52 % = 48% empty space ✓

For face–centered cubic: 100% – 74% = 26% empty space ✓

For body–centered cubic: 100% – 68% = 32% empty space ✓

64. See Section 9.1 and Chapter 5.

$V = 30.0$ L

$T = 20.0°C + 273.15 = 293.1$ K

amount of HCOOH vaporized = 10.00 g – 7.50 g = 2.50 g

$n = 2.50 \text{ g HCOOH} \times \dfrac{1\,\text{mol HCOOH}}{46.03\,\text{g HCOOH}} = 5.43 \times 10^{-2} \text{ mol HCOOH}$

$PV = nRT$

$P_{vapor} = \dfrac{nRT}{V} = \dfrac{(5.43 \times 10^{-2} \text{ mol HCOOH})(0.0821\,\text{L} \cdot \text{atm}/\text{mol} \cdot \text{K})(293.1\,\text{K})}{30.0\,\text{L}} = \underline{\textbf{0.0436 atm}}$

CONCEPTUAL PROBLEMS

66. See Section 9.2.

(a) **False**   Even if the critical temperature is reached but the proper pressure is not, liquid will not change into gas

(b) **True**

(c) **False** CHF_3 does not exhibit H–bonding.

(d) **True**

68. See Section 9.2.

(a) **GT** C_3H_7OH exhibits H–bonding thus will have higher boiling point.

(b) **MI**   The pressure in the flask depends on the amount of gas and this information is not given.

(c) **LT**   See discussion presented on page 226 on the two types of behavior shown in Figure 9.9.

(d) **EQ**   At normal boiling point, the vapor pressure is 1 atm or 760 mm Hg.

(e) **MI**   Pressure information is needed.

70. See Section 9.3.

    (a) <u>I</u>   Pressure produced when the liquid vaporizes inside the flask cannot exceed 159 mm Hg, the vapor pressure of the liquid at 20°C.

    (b) <u>G</u>   Since the registered pressure is less than the vapor pressure of the liquid, the substance will contain only gas because all the liquid will vaporize.

    (c) <u>G/L</u>   The vapor pressure, 165 mm Hg, was registered at 30°C. When it cools down to 20°C, some of the gas molecules will condense to liquid to lower the vapor pressure to 159 mm Hg. Thus, cooling to 20°C yields a system that has both gas and liquid present.

72. See Section 9.1, 9.2, 9.3, and 9.4.

    (a) A covalent bond is a chemical bond formed by the sharing of electrons between two nonmetal atoms. On the other hand, an ionic bond is a chemical bond formed by the electrostatic attraction between an anion and a cation.

    (b) Boiling point is the temperature at which the vapor pressure equals pressure above the surface of the liquid. On the other hand, critical point is the last point where liquid and vapor can coexist in equilibrium condition and beyond this point the liquid phase cannot exist.

    (c) Deposition is the process where gas changes to solid without passing the liquid state. Sublimation is the opposite of deposition, which is the phase change from a solid directly to a gas.

    (d) Freezing point is the temperature and pressure where liquid and solid coexist while critical point is the temperature and pressure where all the three states: solid, liquid, and gas states are present.

74. See Sections 9.1, 9.2, 9.3, and 9.4.

    (a) <u>A</u>, the liquid with the lowest boiling point will have the weakest intermolecular forces.

    (b) <u>A</u>, the "normal" boiling point, at 760 mm Hg, from the chart for A is about 15°C.

    (c) <u>≈ 34°C</u>, line B crosses the 500 mm Hg line around 34°C.

    (d) <u>gas</u>, from the graph, 25°C is higher than the normal boiling point of A, so it is a gas at 25°C.

    (e) <u>≈ 200 mm Hg</u>, the line for liquid C crosses the 40°C line at 200 mm Hg.

CHALLENGE PROBLEMS

75. See Section 9.2.

$P_1 = 110$ mm Hg; $\quad T_1 = 25\,°C \quad \Rightarrow \quad T_1 = 25\,°C + 273.15 = 298$ K

$P_2 = 760$ mm Hg; $\quad T_2 = 77\,°C \quad \Rightarrow \quad T_2 = 77\,°C + 273.15 = 350$ K

$$\ln \frac{P_2}{P_1} = \frac{\Delta H_{vap}}{R} \left[\frac{1}{T_1} - \frac{1}{T_2} \right]$$

$$\ln \frac{760 \text{ mm Hg}}{110 \text{ mm Hg}} = \frac{\Delta H_{vap}}{8.31 \text{ J/mol} \cdot \text{K}} \left[\frac{1}{298 \text{ K}} - \frac{1}{350 \text{ K}} \right]$$

$$\Delta H_{vap} = 3.22 \times 10^4 \text{ J/mol}$$

$$20.0 \text{ L} \times \frac{1000 \text{ mL}}{1 \text{L}} \times \frac{1.59 \text{ g}}{1 \text{mL}} \times \frac{1 \text{mol}}{153.8 \text{ g}} = 206.8 \text{ mol}$$

$$\text{heat required} = 206.8 \text{ mol} \times \frac{3.22 \times 10^4 \text{ J}}{1 \text{mol}} \times \frac{1 \text{kJ}}{1000 \text{ J}} = \underline{\textbf{6.66} \times \textbf{10}^3 \textbf{ kJ}}$$

76. See Section 9.6.

First calculate the volume of one cell. Using density as a conversion factor, calculate the mass per cell and mass per atom of Fe. Using the mass of a single Fe atom and the mass of a mole of Fe (atomic mass), calculate the number of atoms in a mole.

For a body–centered cube: $4r = s\sqrt{3}$. Thus,

$$s = \frac{4(0.124 \text{ nm})}{\sqrt{3}} = 0.286 \text{ nm}$$

$$s = 0.286 \text{ nm} \times \frac{1 \text{m}}{1 \times 10^9 \text{ nm}} \times \frac{100 \text{ cm}}{1 \text{m}} = 2.86 \times 10^{-8} \text{ cm}$$

$$V = s^3 = (2.86 \times 10^{-8} \text{ cm})^3 = 2.34 \times 10^{-23} \text{ cm}^3$$

$$\frac{2.34 \times 10^{-23} \text{ cm}^3}{1 \text{cell}} \times \frac{7.86 \text{ g}}{1 \text{cm}^3} \times \frac{1 \text{cell}}{2 \text{ atoms}} = 9.19 \times 10^{-23} \text{ g/atom}$$

$$\frac{55.85 \text{ g}}{1 \text{mol}} \times \frac{1 \text{atom}}{9.19 \times 10^{-23} \text{ g}} = \underline{\textbf{6.08} \times \textbf{10}^{23} \textbf{ atoms/mol}}$$

77. See Section 9.2, Chapters 3 and 5 and Appendix 1.

$$2H_2(g) + O_2(g) \rightarrow 2H_2O$$

(a) First, determine the limiting reactant in the reaction. Determine the moles of water that would be produced. Assume that all the water produced will be in the gas state and calculate its corresponding pressure using ideal gas law. Compare the calculated pressure to the vapor pressure of water at 27°C:

If $P_{calculated} > P_{water\ vapor\ (27°C)} \Rightarrow$ both water in the liquid and vapor will be present

If $P_{calculated} < P_{water\ vapor\ (27°C)} \Rightarrow$ only water in the vapor state will be present

Find the limiting reactant:

$$0.400\ g\ H_2 \times \frac{1\ mol\ H_2}{2.016\ g\ H_2} \times \frac{2\ mol\ H_2O}{2\ mol\ H_2} = 0.198\ mol\ H_2O$$

$$3.20\ g\ O_2 \times \frac{1\ mol\ O_2}{32.00\ g\ O_2} \times \frac{2\ mol\ H_2O}{1\ mol\ O_2} = 0.200\ mol\ H_2O$$

Hydrogen gas is the limiting reactant because it formed less (or limited) amount of water, so our calculations will be based on H_2.

$P_{calculated}$ from the water formed by the reaction:

$$n = 0.198\ mol\ H_2O(g);\quad V = 10.0\ L;\quad T = 27°C + 273 = 300\ K$$

$$P_{H_2O(g)} = \frac{nRT}{V} = \frac{(0.198\ mol)(0.0821\ L \cdot atm/mol \cdot K)(300\ K)}{10.0\ L} = 0.488\ atm$$

From Appendix 1, the vapor pressure of water at 27°C is 26.74 mm Hg:

$$P_{H_2O\ (27°C)} = 26.74\ mm\ Hg \times \frac{1\ atm}{760\ mm\ Hg} = 0.0352\ atm$$

The calculated P (0.488 atm) is greater than the vapor pressure of water at 27°C. Since $P_{calculated} > P_{water\ vapor\ (27°C)} \Rightarrow$ both water in the **liquid and vapor** state are present.

(b) **26.7 mm Hg.** The final pressure in the flask will be 0.0352 atm or 26.7 mm Hg, the vapor pressure of the H_2O at 27°C. Although the calculated gas pressure is greater than the vapor pressure, the final pressure cannot be more than the vapor pressure of water at 27°C.

(c) $$3.2\ g\ H_2 \times \frac{1\ mol\ H_2}{2.016\ g\ H_2} \times \frac{2\ mol\ H_2O}{2\ mol\ H_2} = 1.6\ mol\ H_2O$$

$$3.2\ g\ O_2 \times \frac{1\ mol\ O_2}{32.00\ g\ O_2} \times \frac{2\ mol\ H_2O}{1\ mol\ O_2} = 0.20\ mol\ H_2O$$

Oxygen gas is the limiting reactant because it formed less (or limited) amount of water. Since O_2 is the limiting reactant it will be used up completely in the reaction. So at the end of the reaction only two gases will be present: $H_2O(g)$ and the unreacted $H_2(g)$ and the sum of the pressures of these two gases comprise the total pressure in the flask.

Pressure from $H_2O(g)$ = vapor pressure of water at $27°C$
$\qquad\qquad\qquad\quad$ = 0.0352 atm (see part a for calculation and explanation)

Pressure due to unreacted hydrogen:

$$\text{mol } H_2 \text{ available} = 3.2 \text{ g } H_2 \times \frac{1 \text{ mol } H_2}{2.016 \text{ g}} = 1.6 \text{ mol } H_2$$

$$\text{mol } H_2 \text{ reacted} = 3.2 \text{ g } O_2 \times \frac{1 \text{ mol } O_2}{32.00 \text{ g}} \times \frac{2 \text{ mol } H_2}{1 \text{ mol } O_2} = 0.20 \text{ mol } H_2$$

$$\text{mol unreacted } H_2 = \text{mol hydrogen available} - \text{mol reacted hydrogen}$$

$$= 1.6 \text{ mol } H_2 - 0.20 \text{ mol } H_2$$

$$= 1.4 \text{ mol } H_2$$

$$P_{unreacted\ hydrogen} = \frac{nRT}{V} = \frac{(1.4 \text{ mol})(0.0821 \text{ L} \cdot \text{atm/mol} \cdot \text{K})(300 \text{ K})}{10.0 \text{ L}} = 3.4 \text{ atm}$$

$$P_{total} = P_{water\ vapor} + P_{unreacted\ hydrogen}$$

$$= 3.4 \text{ atm} + 0.0352 \text{ atm} = \textbf{3.4 atm}$$

Therefore, the total pressure is **3.4 atm**.

78. See Section 9.2 and Chapter 5.

The following steps are used to solve this problem:
(i) Calculate the moles of trichloroethane and the pressure that would result assuming all $C_2H_3Cl_3$ vaporized.

$$n = 1 \text{ cup} \times \frac{1 \text{ qt}}{4 \text{ cups}} \times \frac{1 \text{ L}}{1.057 \text{ qt}} \times \frac{1000 \text{ mL}}{1 \text{ L}} \times \frac{1.325 \text{ g } C_2H_3Cl_3}{1 \text{ mL } C_2H_3Cl_3} \times \frac{1 \text{ mol } C_2H_3Cl_3}{133.39 \text{ g } C_2H_3Cl_3}$$

n = 2.35 mol $C_2H_3Cl_3$

$$V: \quad 18 \text{ ft}^3 \times \frac{28.32 \text{ L}}{1 \text{ ft}^3} = 510 \text{ L}$$

$T:$ Convert $39°F$ to K:
$$t_{°F} = 1.8 t_{°C} + 32$$
$$39°F = 1.8\ t_{°C} + 32$$
$$t_{°C} = 3.9°C$$

T = 3.9°C + 273.15 = 277.1 K

$$P = \frac{nRT}{V} = \frac{(2.35 \text{ mol})(0.0821 \text{L} \cdot \text{atm/mol} \cdot \text{K})(277.1 \text{K})}{510 \text{L}} = 0.105 \text{ atm}$$

(ii) Calculate the ΔH_{vap} of trichloroethane using the following given information:

$P_1 = 100.0$ mm Hg; $T_1 = 20.0°C$ $\Rightarrow$ $T_1 = 20.0°C + 273.15 = 293.2$ K

$P_2 = 760$ mm Hg; $T_2 = 74.1°C$ $\Rightarrow$ $T_2 = 74.1°C + 273.15 = 347.3$ K

$$\ln \frac{P_2}{P_1} = \frac{\Delta H_{vap}}{R} \left[\frac{1}{T_1} - \frac{1}{T_2} \right]$$

$$\ln \frac{760 \text{ mm Hg}}{100.0 \text{ mm Hg}} = \frac{\Delta H_{vap}}{8.31 \text{ J/mol} \cdot \text{K}} \left[\frac{1}{293.2 \text{ K}} - \frac{1}{347.3 \text{ K}} \right]$$

$$\Delta H_{vap} = 3.17 \times 10^4 \text{ J/mol}$$

(iii) Using the ΔH_{vap}, calculate the vapor pressure of trichloroethane at $3.9°C$ and compare the vapor pressure to the pressure calculated in (i) (0.105 atm).

$P_1 = ?$ $T_1 = 3.9°C$ $\Rightarrow$ $T_1 = 3.9°C + 273.15 = 277.1$ K

$P_2 = 760$ mm Hg; $T_2 = 74.1°C$ $\Rightarrow$ $T_2 = 74.1°C + 273.15 = 347.3$ K

$$\Delta H_{vap} = 3.17 \times 10^4 \text{ J/mol}$$

$$\ln \frac{P_2}{P_1} = \frac{\Delta H_{vap}}{R} \left[\frac{1}{T_1} - \frac{1}{T_2} \right]$$

$$\ln \frac{760 \text{ mm Hg}}{P_1} = \frac{3.17 \times 10^4 \text{ J/mol}}{8.31 \text{ J/mol} \cdot \text{K}} \left[\frac{1}{277.1 \text{K}} - \frac{1}{347.3 \text{K}} \right] = 2.78$$

$$\frac{760 \text{ mm Hg}}{P_1} = e^{2.78}$$

$$\frac{760 \text{ mm Hg}}{P_1} = 16.2$$

$$P_1 = \frac{760 \text{ mm Hg}}{16.2} = 46.9 \text{ mm Hg}$$

$$P_1 = 46.9 \text{ mm Hg} \times \frac{1 \text{ atm}}{760 \text{ mm Hg}} = 0.0617 \text{ atm}$$

Since the pressure that would result from complete vaporization is greater than the vapor pressure of trichloroethane, not all the $C_2H_3Cl_3$ would evaporate. Only the amount required to obtain 0.0617 atm (vapor pressure of trichloroethane at $3.9°C$ or $39°F$) will vaporize.

(iv) Use ideal gas law to determine the amount that would evaporate to produce a pressure of 0.0617 atm. The rest would remain as liquid. Calculate the percentage that remains as liquid.

$$n = \frac{PV}{RT} = \frac{(0.0617 \text{ atm})(510 \text{ L})}{(0.0821 \text{ L} \cdot \text{atm/mol} \cdot \text{K})(277 \text{ K})} = 1.38 \text{ mol}   \text{(would vaporize)}$$

mol liquid remaining in the cup = mol $C_2H_3Cl_3$ available − mol $C_2H_3Cl_3$ vaporized

$$= 2.35 \text{ mol} - 1.38 \text{ mol}$$

$$= 0.97 \text{ mol}$$

$$\text{Percent remaining} = \frac{0.97}{2.35} \times 100\ \% = \underline{\textbf{41 \%}}$$

The percentage of trichloroethane left as a liquid when equilibrium is established is **41 %**.

79. See Section 9.2.

The pressure exerted by a skater is:

$$\text{Pressure} = \frac{120 \text{ lb}}{0.10 \text{ in}^2} \times \frac{1 \text{ atm}}{15 \text{ lbs/in}^2} = \underline{\textbf{80 atm}}$$

$$\text{Decrease in melting point} = 80 \text{ atm} \times \frac{1°\text{C}}{134 \text{ atm}} = \underline{\textbf{0.60°C}}$$

As stated in the textbook, "An increase in pressure of 134 atm is required to lower the melting point of ice by 1°C." Thus, unless the ice temperature is only half a degree below freezing, **the concept that pressure melts the ice is not plausible**. Another possible scenario is heat conduction, given that metals are good conductors of heat. This explanation also falls short, however, since skating is frequently enjoyed when ambient temperatures are well below the freezing point of water. Friction also fails as an explanation since skating is relatively frictionless. Another possible explanation for the ease at which a skater glides over ice is a low coefficient of friction between ice and steel.   John S. Wettlaufer and J. Greg Dash published an article, "Melting Below Zero" in the February 2000 issue of Scientific American Magazine. In this article, the authors talked about "surface melting". When applied to water, surface melting is characterized by a very thin film of liquid water coating ice. The temperature of this thin coat of water is lower than the melting point of ice. Additional information on this phenomenon can be found in the article.

80. See Section 9.6.

LiCl is a face–centered cube, thus, $4r = s\sqrt{2}$

The length of one side: $s = 2(r_{cation} + r_{anion})$

The diagonal: $4r_{anion} = s\sqrt{2}$

Substitution yields: $4r_{anion} = 2(r_{cation} + r_{anion})\sqrt{2}$

$4r_{anion} = 2\sqrt{2}\,r_{cation} + 2\sqrt{2}\,r_{anion}$

Divide both sides of the equation above by $2\sqrt{2}\,r_{anion}$

$$\frac{4\,r_{anion}}{2\sqrt{2}\,r_{anion}} = \frac{2\sqrt{2}\,r_{cation}}{2\sqrt{2}\,r_{anion}} + \frac{2\sqrt{2}\,r_{anion}}{2\sqrt{2}\,r_{anion}}$$

after cancelling common factors, the equation becomes:

$$\frac{4}{2\sqrt{2}} = \frac{r_{cation}}{r_{anion}} + 1$$

$$\frac{r_{cation}}{r_{anion}} = \frac{4}{2\sqrt{2}} - 1$$

$$\frac{r_{cation}}{r_{anion}} = \mathbf{0.414}$$

81. See Section 9.2, Table 9.2 and Chapter 5.

The critical temperature of N_2 is $-147°C$, thus N_2 exists entirely as a gas at both $20°C$ and $10°C$. Propane, on the other hand, has a critical temperature of $97°C$ and thus is present as an equilibrium between the vapor and liquid phases. The change in pressure observed with the N_2 is due to a "contraction" of the gas, or a reduction in the thermal motion of the N_2 molecules, while the pressure change with the propane is due to a change in vapor pressure. The effect of temperature change in gas pressure is linear as governed by gas laws. On the other hand, the effect of temperature change in vapor pressure as shown by Clausius–Clapeyron equation (see equation 9.1) is exponential. Consequently, temperature changes result in larger vapor pressure change compared to changes in gas pressure.

SOLUTIONS

10-1 CONCENTRATION UNITS: Conversions Between Concentration Units

2. (a) mass percent of acetone in the solution: (Note: ethyl alcohol is also called ethanol)

mass of acetone: $35.0 \, \text{mL} \times \dfrac{0.790 \, \text{g acetone}}{1 \, \text{mL acetone}} = 27.6 \, \text{g}$

mass of ethanol: $50.0 \, \text{mL} \times \dfrac{0.789 \, \text{g ethanol}}{1 \, \text{mL ethanol}} = 39.4 \, \text{g}$

mass of solution = mass$_{\text{acetone}}$ + mass$_{\text{ethanol}}$ = 27.6 g + 39.4 g = 67.0 g solution

mass percent of acetone = $\dfrac{\text{mass}_{\text{acetone}}}{\text{mass}_{\text{solution}}} \times 100 \, \% = \dfrac{27.6 \, \text{g}}{67.0 \, \text{g}} \times 100 \, \% = \underline{\textbf{41.2\%}}$

(b) volume percent of ethyl alcohol (ethanol)

volume of acetone = 35.0 mL

volume of ethanol = 50.0 mL

assume volumes are additive: volume of solution = 35.0 mL + 50.0 mL = 85.0 mL

volume percent of ethanol = $\dfrac{\text{volume}_{\text{ethanol}}}{\text{volume}_{\text{solution}}} \times 100 \, \% = \dfrac{50.0 \, \text{mL}}{85.0 \, \text{mL}} \times 100 \, \% = \underline{\textbf{58.8\%}}$

(c) mole fraction of acetone in the solution (This problem is similar to Example 10.2).

mol acetone (C_3H_6O) = $35.0 \, \text{mL} \times \dfrac{0.790 \, \text{g } C_3H_6O}{\text{mL } C_3H_6O} \times \dfrac{1 \, \text{mol } C_3H_6O}{58.08 \, \text{g } C_3H_6O} = 0.476 \, \text{mol } C_3H_6O$

mol ethanol (C_2H_6O) = $50.0 \, \text{mL} \times \dfrac{0.789 \, \text{g } C_2H_6O}{\text{mL } C_2H_6O} \times \dfrac{1 \, \text{mol } C_2H_6O}{46.07 \, \text{g } C_2H_6O} = 0.856 \, \text{mol } C_2H_6O$

total moles = mol acetone + mol ethanol = 0.476 mol + 0.856 mol = 1.332 mol

mole fraction acetone = $\dfrac{\text{mol}_{\text{acetone}}}{\text{mol}_{\text{solution}}} \times 100 \, \% = \dfrac{0.476 \, \text{mol}}{1.332 \, \text{mol}} = \underline{\textbf{0.357}}$

4. The solution to this problem is similar to that of Example 10.4c. To calculate the molarity of the solution, begin by constructing a table similar to those on pages 253–254. Use the given data to fill the appropriate spaces in the table and shade the spots required for molarity, namely moles of solute and volume of solution.

We are given that the solution of NaCl(aq) is 0.90 % by mass. Since we are given mass percent, start with 100 g of solution. The mass of solute is

mass of NaCl = 0.90 % × 100 g = 0.90 g

	moles	$\xleftarrow{MM}$	mass	$\xrightarrow{density}$	volume
Solute			0.90 g		
Solvent					
Solution			100 g		

Convert mass of solute (NaCl) to mol of solute using the molar mass of NaCl:

$$\text{mol of solute} = 0.90 \text{ g} \times \frac{1\,\text{mol}}{58.442 \text{ g}} = 0.015 \text{ mol}$$

Use the given density to obtain the volume of 100 g of solution:

$$\text{vol of solution (L)} = \frac{\text{mass}}{\text{density}} = \frac{100 \text{ g}}{1.00 \text{ g/mL}} \times \frac{1\,\text{L}}{1000 \text{ mL}} = 0.100 \text{ L}$$

	moles	$\xleftarrow{MM}$	mass	$\xrightarrow{density}$	volume
Solute	0.015		0.90 g		
Solvent					
Solution			100 g		0.100

Hence the molarity of the solution is:

$$\text{molarity, } M = \frac{\text{mol solute}}{V_{\text{solution}} \text{ (L)}} = \frac{0.15 \text{ mol}}{0.100 \text{ L}} = 0.15 \text{ M}$$

Finally, note that we are required to find the molarity of Na^+ ions. Hence, using the concentration of NaCl in the solution:

$$\text{molarity of } Na^+ = \frac{0.15 \text{ mol NaCl}}{1\text{L solution}} \times \frac{1\,\text{mol } Na^+}{1 \text{ mol NaCl}} = \underline{\textbf{0.15 M}}$$

6. This problem is quite similar (in part) to Example 10.3b. We are given a concentration of 100 µg Pb/L of blood and the density of the solution (1.00 g/mL), and first asked to determine the corresponding number of mol per liter.

$$\text{mol of Pb in 1 L} = 100 \text{ µg Pb} \times \frac{1 \text{ g}}{10^6 \text{ µg}} \times \frac{1 \text{ mol Pb}}{207.2 \text{ g Pb}} = \underline{\mathbf{4.8 \times 10^{-7} \text{ mol Pb}}}$$

The expression for parts per million (ppm) is similar to that of mass percent except that instead of multiplying the ratio of the mass of solute to mass of solution by 100, the ratio is multiplied by 10^6 as shown below:

$$\text{parts per million:} \quad \text{ppm} = \frac{\text{mass of solute}}{\text{mass of solution}} \times 10^6$$

We know the mass of solute (Pb) per liter of solution; we must obtain the corresponding mass of solution using the density (given).

$$\text{mass of 1 L of solution} = 1 \text{ L} \times \frac{1000 \text{ mL}}{1 \text{ L}} \times \frac{1.00 \text{ g}}{1 \text{ mL}} = 1000 \text{ g solution}$$

Now use mass of solute and mass of solution to determine the concentration in ppm. Note that it is logical to use grams for both mass of solute and mass of solution:

$$\text{ppm} = \frac{\text{mass of solute}}{\text{mass of solution}} \times 10^6 = \frac{100 \text{ µg} \times \dfrac{1 \text{ g}}{10^6 \text{ µg}}}{1000 \text{ g}} \times 10^6 = \underline{\mathbf{0.10 \text{ ppm}}}$$

8. Use the definition of molarity in calculating the missing values in the table.

$$\text{molarity } (M) = \frac{\text{mol solute}}{\text{liter solution}}$$

	Mass of Solute	Volume of Solution	Molarity
(a)	1.672 g	145.0 mL	**0.05413 M**
(b)	2.544 g	**7.073 mL**	1.688 M
(c)	**139 g**	894 mL	0.729 M

(a) $$\text{mol Al(NO}_3)_3 = 1.672 \text{ g} \times \frac{1 \text{ mol Al(NO}_3)_3}{213.01 \text{ g Al(NO}_3)_3} = 7.849 \times 10^{-3} \text{ mol}$$

$$\text{V solution (L)} = 145.0 \text{ mL} \times \frac{1 \text{ L}}{1000 \text{ mL}} = 0.1450 \text{ L}$$

$$\text{molarity } (M) = \frac{\text{moles solute}}{\text{liter solution}} = \frac{7.860 \times 10^{-3} \text{mol}}{0.1450 \text{ L}} = \underline{\textbf{0.05413 } M}$$

(b) $\quad \text{mol Al(NO}_3)_3 = 2.544 \text{ g} \times \dfrac{1 \text{ mol Al(NO}_3)_3}{213.01 \text{ g Al(NO}_3)_3} = 0.01194 \text{ mol}$

solve for liter solution:

$$\text{liter solution} = \frac{\text{moles solute}}{\text{molarity}} = \frac{0.01194 \text{ mol}}{1.688 \text{ mol/L}} = 7.073 \times 10^{-3} \text{ L}$$

$$V \text{ solution (mL)} = 7.073 \times 10^{-3} \text{ L} \times \frac{1000 \text{ mL}}{1 \text{ L}} = \underline{\textbf{7.073 mL solution}}$$

(c) $\quad V \text{ solution (L)} = 894 \text{ mL} \times \dfrac{1 \text{ L}}{1000 \text{ mL}} = 0.894 \text{ L}$

solve for moles solute from M:

moles solute = molarity × liters solution = 0.729 M × 0.894 L = 0.652 mol Al(NO$_3$)$_3$

finally, convert moles solute to mass solute:

$$\text{mass of Al(NO}_3)_3 = 0.652 \text{ mol Al(NO}_3)_3 \times \frac{213.01 \text{ g Al(NO}_3)_3}{1 \text{ mol Al(NO}_3)_3} = \underline{\textbf{139 g Al(NO}_3)_3}$$

10. The completed table (final answer) for this problem is shown below. The solution showing how each value is obtained is given after the table.

	Molality	Mass Percent Solute	Ppm Solvent	Mole Fraction Solute
(a)	<u>3.18</u>	<u>36.4</u>	<u>6.358 × 10^5</u>	0.0542
(b)	<u>4.4116</u>	<u>44.3</u>	557169	<u>0.07364</u>
(c)	<u>1.17</u>	17.4	<u>8.26 × 10^5</u>	<u>0.0206</u>
(d)	0.8341	<u>13.07</u>	<u>8.693 × 10^5</u>	<u>0.01481</u>

As suggested in the textbook, conversion from one concentration unit to another can easily be done by using a table similar to those on pages 253–254 and in Example 10.4. Refer also to the table at the bottom of page 251. For each row given in the question, begin by constructing a table and enter known quantities of solute, solvent and solution.

(a) For the first row (a) the given is "mole fraction solute". Assume 1.000 mol aqueous theobromine solution. From the given information, "mole fraction solute = 0.0542", we know that 1 mol aqueous theobromine solution contains 0.0542 mol solute (theobromine) and the remainder of the solution is 0.9458 mol solvent (water; recall that the solution contains a total of 1.000 mol of solute + solvent). Enter these values in the appropriate spaces in the table (shown below) and shade the spots needed for calculating molality, mass % solute, and ppm solvent.

	moles	$\xleftarrow{\ MM\ }$	mass	$\xrightarrow{\ density\ }$	volume
Solute	0.0542				
Solvent	0.9458				
Solution	1.000				

Complete the table by filling it up with calculated values required in the concentration unit conversions. Note that the volume column need not be completed because none of the required concentration units needs these values.

$$\text{mass of solute} = 0.0542 \text{ mol } C_7H_8N_4O_2 \times \frac{180.16 \text{ g } C_7H_8N_4O_2}{1 \text{ mol } C_7H_8N_4O_2} = 9.76 \text{ g solute}$$

$$\text{mass of solvent} = 0.9458 \text{ mol } H_2O \times \frac{18.02 \text{ g } H_2O}{1 \text{ mol } H_2O} = 17.04 \text{ g solvent}$$

$$\text{mass solution} = 9.76 \text{ g solute} + 17.04 \text{ g solvent} = 26.80 \text{ g solution}$$

Fill up the second column of the table with the calculated values above.

	moles	$\xleftarrow{\ MM\ }$	mass	$\xrightarrow{\ density\ }$	volume
Solute	0.0542		9.76 g		
Solvent	0.9458		17.04 g = 0.01704 kg		
Solution	1.000		26.80 g		

Use the values from the table to convert the given "0.0542 solute mole fraction" to the other concentration units.

$$\text{molality} = \frac{\text{moles solute}}{\text{mass solvent (kg)}} = \frac{0.0542 \text{ mol}}{0.01704 \text{ kg}} = \underline{\textbf{3.18 } \textit{\textbf{m}}}$$

$$\text{mass percent solute} = \frac{\text{mass solvent}}{\text{mass solution}} \times 100 \% = \frac{9.76 \text{ g}}{26.80 \text{ g}} \times 100 \% = \underline{\textbf{36.4\%}}$$

$$\text{ppm solvent} = \frac{\text{mass of solvent}}{\text{mass of solution}} \times 10^6 = \frac{17.04 \text{ g}}{26.80 \text{ g}} \times 10^6 = \underline{\textbf{6.358} \times \textbf{10}^5 \textbf{ ppm}}$$

(b) For the second row (b) the given is "ppm solvent". Assume 10^6 g aqueous theobromine solution. From the given information, "ppm solvent = 557169", we know that 10^6 g aqueous theobromine solution contains 557169 g solvent (water) and $(10^6$ g – 557169 g) = 442831 g theobromine. Enter these three masses in the appropriate spaces in the second column of the table (below). Shade the spots needed for calculating molality, mass solute %, and mole fraction of solute.

	moles	$\xleftarrow{\quad MM \quad}$	mass	$\xrightarrow{\quad density \quad}$	volume
Solute			442831 g		
Solvent			557169 g or 557.169 kg		
Solution			10^6 g or 1000000 g		

Complete the table by filling it up with the calculated values required in the concentration unit conversions. Note that the volume column need not be completed because none of the required concentration units needs these values.

$$\text{moles of solute} = 442831 \text{ g C}_7\text{H}_8\text{N}_4\text{O}_2 \times \frac{1\,\text{mol C}_7\text{H}_8\text{N}_4\text{O}_2}{180.16 \text{ g C}_7\text{H}_8\text{N}_4\text{O}_2} = 2458.0 \text{ mol solute}$$

$$\text{moles of solvent} = 557169 \text{ g H}_2\text{O} \times \frac{1\,\text{mol H}_2\text{O}}{18.02 \text{ g H}_2\text{O}} = 30920 \text{ mol solvent}$$

mol solution = 2458.0 mol solute + 30920 mol solvent = 33378 mol solution

Fill up the first column of the table with the calculated values above.

	moles	$\xleftarrow{\quad MM \quad}$	mass	$\xrightarrow{\quad density \quad}$	volume
Solute	2458.0		442831 g		
Solvent	30920		557169 g or 557.169 kg		
Solution	33378		10^6 g or 1000000 g		

Use the values from the table to convert "557169 ppm solvent" to the different concentration units.

$$\text{molality} = \frac{\text{moles solute}}{\text{mass solvent (kg)}} = \frac{2458.0 \text{ mol}}{557.169 \text{ kg}} = \underline{\textbf{4.4116 } \textbf{\textit{m}}}$$

$$\text{mass percent solute} = \frac{\text{mass solute}}{\text{mass solution}} \times 100\,\% = \frac{442831\,\text{g}}{10^6 \text{ g}} \times 100\,\% = \underline{\textbf{44.3 \%}}$$

$$\text{mole fraction solute} = \frac{\text{mol solute}}{\text{mol solution}} = \frac{2458.0 \text{ mol}}{33378 \text{ mol}} = \underline{\textbf{0.07364}}$$

(c) For the third row (c) the given is "mass percent solute". Assume 100 g aqueous theobromine solution. From the given "17.4 % solute by mass", we know that 100 g aqueous theobromine solution contains 17.4 g theobromine and (100 g – 17.4 g) = 82.6 g solvent (water).   Enter these values in the appropriate spaces in the table to complete the second column. Shade the spots for the values needed for calculating molality, ppm solute, and mole fraction of solvent.

	moles	$\xleftarrow{MM}$	mass	$\xrightarrow{density}$	volume
Solute			17.4 g		
Solvent			82.6 g or 0.0826 kg		
Solution			100.0 g		

Complete the table by filling it up with calculated values required in the concentration unit conversions. Note that the last column need not be completed because none of the required concentration units needs these values.

$$\text{moles of solute} = 17.4 \text{ g C}_7\text{H}_8\text{N}_4\text{O}_2 \times \frac{1 \text{ mol C}_7\text{H}_8\text{N}_4\text{O}_2}{180.16 \text{ g C}_7\text{H}_8\text{N}_4\text{O}_2} = 0.0966 \text{ mol solute}$$

$$\text{moles of solvent} = 82.6 \text{ g H}_2\text{O} \times \frac{1 \text{ mol H}_2\text{O}}{18.02 \text{ g H}_2\text{O}} = 4.58 \text{ mol solvent}$$

mol solution = 0.0966 mol solute + 4.58 mol solvent = 4.68 mol solution

Fill up the first column of the table with the calculated values above.

	moles	$\xleftarrow{MM}$	mass	$\xrightarrow{density}$	volume
Solute	0.0966		17.4 g		
Solvent	4.58		82.6 g or 0.0826 kg		
Solution	4.68		100.0 g		

Use the values from the table to convert "17.4 % solute by mass" to the different concentration units.

$$\text{molality} = \frac{\text{moles solute}}{\text{mass solvent (kg)}} = \frac{0.0966 \text{ mol}}{0.0826 \text{ kg}} = \underline{\textbf{1.17 } \textbf{\textit{m}}}$$

$$\text{ppm solvent} = \frac{\text{mass of solvent}}{\text{mass of solution}} \times 10^6 = \frac{82.6 \text{ g}}{100.0 \text{ g}} \times 10^6 = \underline{\textbf{8.26} \times \textbf{10}^5 \textbf{ ppm}}$$

$$\text{mole fraction solute} = \frac{\text{mol solute}}{\text{mol solution}} = \frac{0.0966 \text{ mol}}{4.68 \text{ mol}} = \underline{\textbf{0.0206}}$$

(d) For row (d) the given is molality which is defined as moles solute per kg solvent. From the given "0.8341 m", we know that there are 0.8341 mol theobromine in 1 kg of water (the solution is aqueous, so the solvent is water). Enter these values in the appropriate spaces in the table. Shade the spots for the values needed for calculating mass % solute, ppm solvent, and mole fraction of solute.

	moles	$\xleftarrow{\quad MM \quad}$	mass	$\xrightarrow{\quad density \quad}$	volume
Solute	0.8341				
Solvent			1 kg or 1000 g		
Solution					

Complete the table by filling it up with calculated values required in the concentration unit conversions. Note that the last column need not be completed because none of the required concentration units needs these values.

$$\text{mass of solute} = 0.8341 \text{ mol } C_7H_8N_4O_2 \times \frac{180.16 \text{ g } C_7H_8N_4O_2}{1 \text{ mol } C_7H_8N_4O_2} = 150.3 \text{ g solute}$$

$$\text{mole of solvent} = 1 \text{ kg } H_2O \times \frac{1000 \text{ g}}{1 \text{ kg}} \times \frac{1 \text{ mol } H_2O}{18.02 \text{ g } H_2O} = 55.49 \text{ mol solvent}$$

mol solution = 0.8341 mol solute + 55.49 mol solvent = 56.32 mol solution

mass solution = 150.3 g solute + 1000 g solvent = 1150.3 g solution

Fill up the first 2 columns of the table with the calculated values above. The third column is not needed in the problem.

	moles	$\xleftarrow{\quad MM \quad}$	mass	$\xrightarrow{\quad density \quad}$	volume
Solute	0.8341		150.3		
Solvent	55.49		1 kg or 1000 g		
Solution	56.32		1150.3 g		

Finally, use the values from the table to convert "0.8341 m" to the different concentration units.

$$\text{mass percent solute} = \frac{\text{mass solute}}{\text{mass solution}} \times 100 \% = \frac{150.3 \text{ g}}{1150.3 \text{ g}} \times 100 \% = \underline{\mathbf{13.07 \%}}$$

$$\text{ppm solvent} = \frac{\text{mass of solvent}}{\text{mass of solution}} \times 10^6 = \frac{1000 \text{ g}}{1150.3 \text{ g}} \times 10^6 = \underline{\mathbf{8.693 \times 10^5 \text{ ppm}}}$$

$$\text{mole fraction solute} = \frac{\text{mol solute}}{\text{mol solution}} = \frac{0.8341 \text{ mol}}{56.32 \text{ mol}} = \underline{\mathbf{0.01481}}$$

12. The solution to this problem is similar (in part) to Example 10.1. Molarity is defined as moles solute per liter solution. In the problem the solute is barium hydroxide, $Ba(OH)_2$ and the solvent is water. The final volume and concentration are 750.0 mL and 0.362 M, respectively.

(a) To prepare the solution from solid barium hydroxide, calculate the mol $Ba(OH)_2$ needed; convert the calculated mol to mass $Ba(OH)_2$ using MM $Ba(OH)_2$ as a conversion factor.

$$\text{mol } Ba(OH)_2 = 750.0 \text{ mL} \times \frac{1 \text{ L}}{1000 \text{ mL}} \times \frac{0.362 \text{ mol}}{1 \text{ L}} = 0.272 \text{ mol } Ba(OH)_2$$

$$\text{mass of solute} = 0.272 \text{ mol } Ba(OH)_2 \times \frac{171.34 \text{ g } Ba(OH)_2}{1 \text{ mol } Ba(OH)_2} = \underline{\textbf{46.6 g}} \text{ } Ba(OH)_2$$

Dissolve 46.6 g solid barium hydroxide in water and dilute with enough water to make 750.0 mL solution.

(b) To prepare the mixture from a stock solution of 4.93 M $Ba(OH)_2$, calculate the volume of the stock solution needed using Equation 10.1.

$$M_C V_C = M_D V_D \quad \text{where C is concentrated and D is diluted}$$

$$M_D = 0.362 \text{ M}; \quad V_D = 750.0 \text{ mL}; \quad M_C = 4.93 \text{ M}; \quad V_C = ?$$

$$V_C = \frac{M_D V_D}{M_C} = \frac{0.362 \text{ M} \times 750.0 \text{ mL}}{4.93 \text{ M}} = \underline{\textbf{55.1 mL}}$$

Take 55.1 M of 4.93 M aqueous barium hydroxide and dilute with enough water to make 750.0 mL solution.

14. Refer to Example 10.1, Section 3.1 and Example 3.3a.

(a) To solve this problem, determine the moles of K_2S dissolve in 0.7850 L of the original solution, convert moles to mass using MM of K_2S as conversion factor.

$$\text{mass } K_2S \text{ dissolved} = 0.7850 \text{ L} \times \frac{1.262 \text{ mol } K_2S}{L} \times \frac{110.27 \text{ g } K_2S}{1 \text{ mol } K_2S} = \underline{\textbf{109.2 g}} \text{ } K_2S$$

(b) Determination of the molarity of K_2S, K^+, and S^{2-} in the diluted solution:

$$\text{molarity of } K_2S = \frac{\text{moles } K_2S}{L \text{ solution}} = \frac{109.2 \text{ g } K_2S \times \frac{1 \text{ mol } K_2S}{110.27 \text{ g } K_2S}}{2.000 \text{ L}} = \underline{\textbf{0.4951 M}}$$

Since K_2S is soluble in water, it dissociates in solution as shown:

$$K_2S(s) \rightarrow 2K^+(aq) + S^{2-}(aq)$$

For every 1 mole of K_2S, 2 mole K^+ and 1 mole S^{2-} are formed. The molarity of the K^+ and S^{2-} ions can be obtained from this relationship.

$$\text{molarity of } K^+ = \frac{\text{moles } K^+}{\text{L solution}} = \frac{109.2 \text{ g } K_2S \times \dfrac{1 \text{ mol } K_2S}{110.27 \text{ g } K_2S} \times \dfrac{2 \text{ mol } K^+}{1 \text{ mol } K_2S}}{2.000 \text{ L}} = \underline{\mathbf{0.9903 \ M}}$$

$$\text{molarity of } S^{2-} = \frac{\text{moles } S^{2-}}{\text{L solution}} = \frac{109.2 \text{ g } K_2S \times \dfrac{1 \text{ mol } K_2S}{110.27 \text{ g } K_2S} \times \dfrac{1 \text{ mol } S^{2-}}{1 \text{ mol } K_2S}}{2.000 \text{ L}} = \underline{\mathbf{0.4951 \ M}}$$

16. As suggested in the textbook, conversion from one concentration unit to another can easily be done by using a table similar to those on pages 253–254 and in Example 10.4. (Refer also to the table at the bottom of page 251.)

(a) We are given a solution of aqueous ammonia whose concentration is 29.8 % NH_3 by mass. Begin by constructing the table. Assume 100.0 g ammonia solution. From the given "29.8 % NH_3 by mass", we know that 100.0 g ammonia solution contains 29.8 g NH_3 and (100.0 g – 29.8 g) = 70.2 g water. Enter these values in the appropriate spaces in the table to complete column 2; since we are asked to find the molarity of the solution, shade the spots needed for calculating molarity — namely moles of solute and volume of solution (liters).

	moles	←——— MM ———→ mass	density ——→ volume	
Solute		29.8 g		
Solvent		70.2 g or 0.0702 kg		
Solution		100.0 g		

Complete the table by calculating the required quantities; note that the density of the solution is given in the question.

$$\text{mole of solute} = 29.8 \text{ g } NH_3 \times \frac{1 \text{ mol } NH_3}{17.031 \text{ g } NH_3} = 1.75 \text{ mol } NH_3$$

$$\text{volume of solution} = 100.0 \text{ g solution} \times \frac{1 \text{ mL solution}}{0.8960 \text{ g solution}} \times \frac{1 \text{ L}}{1000 \text{ mL}} = 0.1116 \text{ L}$$

Enter the two values just calculated into the table:

	moles	←——— MM ———→ mass	density ——→ volume	
Solute	1.75	29.8 g		
Solvent		70.2 g or 0.0702 kg		
Solution		100.0 g	0.1116 L	

Use the values from the table to convert "29.8 % NH_3 by mass" to molarity.

$$\text{molarity, } M = \frac{\text{mol solute}}{V_{\text{solution}} \text{ (L)}} = \frac{1.75 \text{ mol}}{0.1116 \text{ L}} = \underline{\textbf{15.7 M}}$$

(b) Use Equation 10.1 to determine the concentration of the solution obtained by diluting the (concentrated) commercial NH_3 solution.

$$M_C V_C = M_D V_D \quad \text{where C is concentrated and D is diluted}$$

$$M_D = ?; \quad V_D = 2.50 \text{ L}; \quad M_C = 15.7 \text{ M}; \quad V_C = 300 \text{ mL} = 0.300 \text{ L}$$

$$M_D = \frac{M_C V_C}{V_D} = \frac{15.7 \text{ M} \times 0.300 \text{ L}}{2.50 \text{ L}} = \underline{\textbf{1.88 M}}$$

18. The completed table (final answer) for this problem is shown below. The solution showing how each value is obtained is given after the table.

	Density g/mL	Molarity (M)	Molality (m)	Mass percent of Solute
(a)	1.02	0.632	**0.677**	**8.56**
(b)	1.16	**2.39**	**2.88**	28.5
(c)	1.20	**3.60**	5.11	**41.4**

As suggested in the textbook, conversion from one concentration unit to another can easily be done by using a table similar to those on pages 253–254 and in Example 10.4. (Refer also to the table at the bottom of page 251.)
Begin by constructing a table for each row.

(a) For the first row (a) the given is "molarity", which is defined as moles solute per liter solution. From the given "0.632 M", we know that there are 0.632 mol of the solute, K_2CO_3 in 1 liter of solution. Enter these values in the appropriate spaces in the table and shade the spots required for calculating molality and mass percent solute.

	moles	$\xleftarrow{MM}$ mass	$\xrightarrow{\text{density}}$ volume
Solute	0.632		
Solvent			
Solution			1 L = 1000 mL

Calculate the missing values required in the concentration unit conversions and use them to complete the table.

$$\text{mass of solute} = 0.632 \text{ mol } K_2CO_3 \times \frac{138.205 \text{ g } K_2CO_3}{1 \text{ mol } K_2CO_3} = 87.3 \text{ g solute}$$

$$\text{mass of solution} = 1000 \text{ mL solution} \times \frac{1.02 \text{ g}}{1 \text{ mL}} = 1020 \text{ g solution}$$

$$\text{mass of solvent} = \text{mass of solution} - \text{mass of solute} = 1020 \text{ g} - 87.3 \text{ g}$$

$$= 933 \text{ g solvent}$$

	moles ← MM	mass	density →	volume
Solute	0.632	87.3 g		
Solvent		933 g = 0.933 kg		
Solution		1020 g		1 L = 1000 mL

Use the values from the table to convert "0.632 *M*" to the different concentration units.

$$\text{molality} = \frac{\text{moles solute}}{\text{mass solvent (kg)}} = \frac{0.632 \text{ mol}}{0.933 \text{ kg}} = \textbf{\textit{0.677 m}}$$

$$\text{mass percent solute} = \frac{\text{mass of solute}}{\text{mass of solution}} \times 100\% = \frac{87.3 \text{ g}}{1020 \text{ g}} \times 100\% = \textbf{8.56 \%}$$

(b) For the second row (b) the given is "mass percent solute". Assume 100.0 g solution. Since the given is "28.5 % solute by mass", we know that there are 28.5 g K_2CO_3 and 100.0 g – 28.5 g = 71.5 g solvent (water). Enter these values in the appropriate spaces in the table and shade the spots required for calculating molarity and molality.

	moles ← MM	mass	density →	volume
Solute		28.5 g		
Solvent		71.5 g = 0.0715 kg		
Solution		100.0 g		

Calculate the missing values required in the concentration unit conversions and use them to complete the table.

$$\text{mole of solute} = 28.5 \text{ g } K_2CO_3 \times \frac{1 \text{ mol } K_2CO_3}{138.205 \text{ g } K_2CO_3} = 0.206 \text{ mol solute}$$

$$\text{volume of solution} = 100.0 \text{ g solution} \times \frac{1 \text{ mL}}{1.16 \text{ g}} = 86.2 \text{ mL solution}$$

	moles ← MM	mass	density →	volume
Solute	0.206	28.5 g		
Solvent		71.5 g = 0.0715 kg		
Solution		100.0 g		86.2 mL = 0.0862 L

Use the values from the table to convert "28.5% solute by mass" to the different concentration units.

$$\text{molarity} = \frac{\text{moles solute}}{\text{volume solution (L)}} = \frac{0.206 \text{ mol}}{0.0862 \text{ L}} = \underline{\textbf{2.39 } \textbf{\textit{M}}}$$

$$\text{molality} = \frac{\text{moles solute}}{\text{mass solvent (kg)}} = \frac{0.206 \text{ mol}}{0.0715 \text{ kg}} = \underline{\textbf{2.88 } \textbf{\textit{m}}}$$

(c) For the third row (c) the given is **molality** which is defined as moles solute per kg solvent. Assume 1 kg solvent. Since the given is "5.11 *m*", there are 5.11 moles K_2CO_3 in 1 kg solvent. Enter these values in the appropriate spaces in the table and shade the spots required to calculate molarity and mass percent solute.

	moles	⟵ MM	mass	density ⟶	volume
Solute	5.11				
Solvent			1000 g or 1 kg		
Solution					

Calculate the missing values required in the concentration unit conversions and use them to complete the table.

$$\text{mass of solute} = 5.11 \text{ mol } K_2CO_3 \times \frac{138.205 \text{ g } K_2CO_3}{1 \text{ mol } K_2CO_3} = 706 \text{ g solute}$$

$$\text{mass of solution} = \text{mass of solute} + \text{mass of solvent} = 706 \text{ g} + 1000 \text{ g} = 1706 \text{ g}$$

$$\text{volume of solution} = 1706 \text{ g solution} \times \frac{1 \text{ mL}}{1.20 \text{ g}} = 1.42 \times 10^3 \text{ mL solution}$$

	moles	⟵ MM	mass	density ⟶	volume
Solute	5.11		706 g		
Solvent			1000 g or 1 kg		
Solution			1706 g		1.42×10^3 mL = 1.42 L

Use the values from the table to convert "5.11 *m*" to the different concentration units.

$$\text{molarity} = \frac{\text{moles solute}}{\text{volume solution (L)}} = \frac{5.11 \text{ mol}}{1.42 \text{ L}} = \underline{\textbf{3.60 } \textbf{\textit{M}}}$$

$$\text{mass percent solute} = \frac{\text{mass of solute}}{\text{mass of solution}} \times 100 \% = \frac{706 \text{ g}}{1706 \text{ g}} \times 100 \% = \underline{\textbf{41.4 \%}}$$

20. This problem requires conversion from "percent sucrose by mass" to molality. As suggested by the textbook, conversion from one concentration unit to another can be made easy with the use of a table similar to those on pages 253–254 and in Example 10.4. Begin by constructing a table; complete the table with calculated values required in the concentration unit conversions.

	moles $\xleftarrow{\quad MM \quad}$ mass $\xrightarrow{\ density\ }$ volume		
Solute			
Solvent			
Solution			15.0 L or 15000 mL

We are told to assume 15.0 L of grape juice. From this volume, the mass of the juice can be obtained using density as conversion factor.

$$\text{mass of solution} = 15000 \text{ mL juice} \times \frac{1.0 \text{ g}}{1 \text{ mL}} = 1.5 \times 10^4 \text{ g juice}$$

Since the original juice is 24% sucrose by mass (24 g sucrose in 100 g of juice), the mass of sucrose in 1.5×10^4 g juice can be obtained:

$$\text{mass of solute} = 1.5 \times 10^4 \text{ g juice} \times \frac{24 \text{ g sucrose}}{100 \text{ g juice}} = 3.6 \times 10^3 \text{ g}$$

mass of solvent = mass of solution – mass of solute

$$= 1.5 \times 10^4 \text{ g} - 3.6 \times 10^3 \text{ g} = 1.1 \times 10^4 \text{ g}$$

$$\text{mole of solute} = 3.6 \times 10^3 \text{g} \times \frac{1 \text{ mol } C_{12}H_{22}O_{11}}{342.296 \text{ g } C_{12}H_{22}O_{11}} = 11 \text{ mol solute}$$

	moles $\xleftarrow{\quad MM \quad}$ mass $\xrightarrow{\ density\ }$ volume		
Solute	11	3.6×10^3 g	
Solvent		1.1×10^4 g	
Solution		1.5×10^4 g	15.0 L or 15000 mL

Reconstruct the table to accommodate the change due to the removal of 25% of the water by mass.

After removing 25% water by mass, the mass of the water is reduced to 75% of its original amount:

$$\text{mass of solvent} = 1.1 \times 10^4 \text{ g} \times 0.75 = 8.2 \times 10^3 \text{ g water remain}$$

Note that only the water content is reduced. The mass of sucrose remains the same but the total mass of the solution is also changed.

	moles	← MM ―	mass	density →	volume
Solute	11		3.6×10^3 g		
Solvent			8.2×10^3 g or 8.2 kg		
Solution			1.2×10^4 g		

Finally, calculate the molality of this solution:

$$\text{molality } (m) = \frac{\text{moles solute}}{\text{mass solvent (kg)}} = \frac{11\,\text{mol}}{8.2\,\text{kg}} = \underline{\textbf{1.3 } m}$$

10-2 PRINCIPLES OF SOLUBILITY

22. The rule of thumb is "like dissolves like" which means that if two substances have similar type and strength of intermolecular forces, the two substances will be soluble in one another. CCl_4 is a nonpolar molecule thus it is expected to dissolve nonpolar substances.

   (a) **Benzene**   Benzene (C_6H_6) is more soluble than KCl because benzene is nonpolar while KCl is an ionic compound.

   (b) **Octane**   Octane (C_8H_{18}) is nonpolar while glycerol [$HOCH_2$-CH(OH)-CH_2OH] is polar and more similar to H_2O. Since octane and CCl_4 will have a similar type of intermolecular forces, they will be expected to form a solution.

   (c) **CHCl₃**   Both $CHCl_3$ and $C_6H_{11}Cl_3$ are polar but $CHCl_3$ is a more similar molecule to CCl_4, with similar intermolecular dispersion forces, and therefore $CHCl_3$ will be more soluble in CCl_4. In contrast $C_6H_{11}Cl_3$ bears less resemblance to CCl_4 and therefore will be less soluble.

   (d) **CBr₄** CBr_4 is a nonpolar compound, very similar to CCl_4, and thus it will dissolve in the nonpolar CCl_4. $CHBr_3$ is polar and is less likely to dissolve in nonpolar CCl_4.

24. Water is a polar compound and forms hydrogen bonds. Based on the rule of thumb "like dissolves like", water is expected to dissolve polar substances and those substances which can form hydrogen bonds with water. Some ionic compounds are also soluble in water because there are strong forces of attraction between the polar water molecules and the ions.

   (a) **CH₃CH₂OH**   Compared with CH_3CH_2Cl, CH_3CH_2OH is more soluble in water because CH_3CH_2OH and water have similar type and strength of intermolecular forces: dispersion forces, dipole forces, and hydrogen bonding. CH_3CH_2OH forms hydrogen bonds with water, thus they dissolve each other.

(b) **NH₃**   Between NH_3 and PH_3, the first is expected to be more soluble in water. Although both are polar compounds, NH_3 is more polar and can hydrogen bond with water, and so will be more soluble in H_2O.

(c) **Propionic acid**   Between $CH_3(CH_2)_5COOH$ and CH_3CH_2COOH, the second is expected to be more soluble in water. Both are polar compounds, and can hydrogen bond, but heptanoic acid contains a much longer, non-polar, carbon-hydrogen chain than propionic acid. Thus, the non-polar portion of heptanoic acid is more significant than the polar portion of the molecule, making hydrogen bonding less important; hence heptanoic acid has less similarity to H_2O than propionic acid and so will be less soluble than propionic acid.

(d) **NaI**   Between ICl_3 and NaI, the second is expected to be more soluble in water because sodium iodide is an ionic compound and the force of attraction between the ions (Na^+ and I^-) and polar water molecules is strong. ICl_3 is a covalent molecule and, although polar, it will not interact so strongly with H_2O molecules.

26. See Sections 8.5 and 8.6, Example 8.7 and refer to Table 8.3 for $\Delta H_f°$ values.

(a) $CaCO_3(s) \rightarrow Ca^{2+}(aq) + CO_3^{2-}(aq)$ $\Delta H = ?$

$\Delta H° = \Sigma \Delta H_f° \text{products} - \Sigma \Delta H_f° \text{reactants}$

$= [1 \text{ mol } (\Delta H_f° Ca^{2+}) + 1 \text{ mol } (\Delta H_f° CO_3^{2-})] - [1 \text{ mol } (\Delta H_f° CaCO_3)]$

$= [1 \text{ mol } (-542.8 \text{ kJ/mol}) + 1 \text{ mol } (-677.1 \text{ kJ/mol})] - [1 \text{ mol } (-1206.9 \text{ kJ/mol})]$

$= \underline{-13.0 \text{ kJ}}$

$CaCO_3(s) \rightarrow Ca^{2+}(aq) + CO_3^{2-}(aq)$	$\Delta H = -13.0 \text{ kJ}$

(b) **No.** Based on the thermodynamic data, the solubility of $CaCO_3(s)$ is not expected to increase when the temperature is increased because the dissolution process is exothermic.

28. (a) Recall that 1 atm = 760 mm Hg:

$$k = \frac{1.0 \times 10^{-3} \text{ M}}{1 \text{ atm}} \times \frac{1 \text{ atm}}{760 \text{ mm Hg}} = \underline{1.3 \times 10^{-6} \text{ M/mm Hg}}$$

(b) This part of the problem is similar to Example 10.5. Use Henry's Law (Equation 10.2) with the value of k calculated in part (a) above to determine the solubility of the gas:

$$C_{gas} = kP_{gas} = (1.3 \times 10^{-6} \text{ M/mm Hg})(693 \text{ mm Hg}) = \underline{9.1 \times 10^{-4} \text{ M}}$$

(c) The result from part (b) gives the molarity of the gas at this temperature and pressure. Starting from the given volume of solution, use the molarity of the solution and the molar mass of the gas as conversion factors to calculate the corresponding mass of gas:

$$\text{mass of Ar in 29 L} = 29 \text{ L} \times (9.1 \times 10^{-4} \text{ mol/L}) \times \frac{39.948 \text{ g Ar}}{1 \text{ mol Ar}} = \underline{\textbf{1.1 g Ar}}$$

30. The first two parts of this problem are related to Example 10.3. For both (a) and (b), the partial pressure of N_2 in air can be calculated using Equation 5.4: if the air is 78 % N_2 this means that $X_{N_2} = 0.78$. Thus for an air pressure of 1.38 atm

$$P_{N_2} = X_{N_2} P_{total} = (0.78)(1.38 \text{ atm}) = 1.08 \text{ atm.}$$

(a) Now use Henry's Law (Equation 10.2) with the appropriate value of k to calculate the solubility of the gas, and hence (using the molar mass) the corresponding mass of gas dissolved in 3.0 L.

$$C_{N_2} = kP_{N_2} = (6.8 \times 10^{-4} \text{ M/atm})(1.08 \text{ atm}) = 7.3 \times 10^{-4} \text{ M} = 7.3 \times 10^{-4} \text{ mol } N_2/\text{L}$$

$$\text{mass } N_2 \text{ in 3.0 L} = 3.0 \text{ L} \times (7.3 \times 10^{-4} \text{ mol } N_2/\text{L}) \times \frac{28.01 \text{ g } N_2}{1 \text{ mol } N_2} = \underline{\textbf{0.061 g or 61 mg}}$$

(b) Use Henry's Law (Equation 10.2) with the appropriate value of k to calculate the solubility of the gas, and hence (using the molar mass) the corresponding mass of gas dissolved in 3.0 L.

$$C_{N_2} = kP_{N_2} = (6.2 \times 10^{-4} \text{ M/atm})(1.08 \text{ atm}) = 6.7 \times 10^{-4} \text{ M} = 6.7 \times 10^{-4} \text{ mol } N_2/\text{L}$$

$$\text{mass } N_2 \text{ in 3.0 L} = 3.0 \text{ L} \times (6.7 \times 10^{-4} \text{ mol } N_2/\text{L}) \times \frac{28.01 \text{ g } N_2}{1 \text{ mol } N_2} = \underline{\textbf{0.056 g or 56 mg}}$$

(c) The decrease in dissolved N_2 is 0.061 g − 0.056 g = 0.005 g. As a percentage of the original quantity of dissolved N_2, the decrease is

$$\frac{0.005 \text{ g } N_2}{0.061 \text{ g } N_2} \times 100\% = \underline{\textbf{8 \%}}$$

10-3 COLLIGATIVE PROPERTIES OF NONELECTROLYTES

32. This problem is related to Example 10.7. In order to determine the freezing point we must first calculate the molality of the solution. Begin by constructing a table similar to those on page 252 and in Example 10.4 and include the given concentration (66% sucrose) as 66 g solute in 100 g solution.

	moles	← MM —	mass	density →	volume
Solute			66 g		
Solvent					
Solution			100 g		

Convert mass of solute to mol of solute using the molar mass of sucrose, $C_{12}H_{22}O_{11}$:

$$\text{mol of solute} = 66 \text{ g} \times \frac{1\,\text{mol}}{342.296\,\text{g}} = 0.19 \text{ mol}$$

and calculate the mass of solvent from mass of solution and mass of solute:

$$\text{mass of solvent} = (\text{mass of solution}) - (\text{mass of solute}) = 100 \text{ g} - 66 \text{ g}$$

$$= 34 \text{ g} \quad \text{or} \quad 0.034 \text{ kg}$$

	moles $\xleftarrow{\quad MM \quad}$	mass	$\xrightarrow{\quad density \quad}$ volume
Solute	0.19 mol	66 g	
Solvent		34 g = 0.034 kg	
Solution		100 g	

Hence the molality of the solution is:

$$\text{molality} = \frac{\text{mol solute}}{\text{kg solvent}} = \frac{0.19\,\text{mol}}{0.034\,\text{kg}} = 5.6 \text{ } m$$

Now substitute this into the following equation (see page 263 of the textbook and Example 10.7) to calculate ΔT_f:

$$\Delta T_f = k_f m = (1.86 \text{ °C}/m)(5.6 \text{ } m) = 10 \text{ °C}$$

From page 262 and Example 10.7 the freezing point of the pure solution is given by:

$$T_f = T_f° - \Delta T_f = 0\text{°C} - 10\text{°C} = \underline{\mathbf{-10 \text{ °C}}} \quad \text{or} \quad \underline{\mathbf{-1.0 \times 10^1 \text{ °C}}}$$

34. This problem is similar to Example 10.6. In each case, use Equation 10.3 to calculate the vapor pressure over the solution. In order to do this, we must first calculate X_{water} for each given solution.

   (a) Set up a table as in Example 10.4: the solution is 30.6 % glycerol by mass; thus there are 30.6 g glycerol in 100 g solution and therefore the mass of solvent is $(100 - 30.6)$ g = 69.4 g. Shade the spots needed to calculate X_{water}:

	moles $\xleftarrow{\quad MM \quad}$	mass	$\xrightarrow{\quad density \quad}$ volume
Solute		30.6 g	
Solvent		69.4 g	
Solution		100 g	

Convert the masses of solute and solvent to moles of each using their molar masses:

$$\text{mol of } C_3H_8O_3 \text{ solute} = 30.6 \text{ g} \times \frac{1\,\text{mol}}{92.094\,\text{g}} = 0.332 \text{ mol}$$

$$\text{mol of H}_2\text{O solvent} = 69.4 \text{ g} \times \frac{1\,\text{mol}}{18.02\,\text{g}} = 3.85 \text{ mol}$$

	moles	$\xleftarrow{\text{\textit{MM}}}$ mass	$\xrightarrow{\text{density}}$ volume
Solute	0.332	30.6 g	
Solvent	3.85	69.4 g	
Solution		100 g	

Now calculate the mole fraction of H_2O and then the required vapor pressure:

$$X_{\text{water}} = \frac{n_{\text{H}_2\text{O}}}{n_{\text{H}_2\text{O}} + n_{\text{C}_3\text{H}_8\text{O}_3}} = \frac{3.85 \text{ mol}}{(3.85 + 0.332) \text{ mol}} = 0.921$$

$$P_{\text{water}} = (X_{\text{H2O}})(P^{\circ}_{\text{H2O}}) = (0.921)(28.35 \text{ mm Hg}) = \underline{\textbf{26.1 mm Hg}}$$

(b) This part of the problem is similar to part (a) above. Again, use Equation 10.3 to calculate the vapor pressure over the solution having first calculated X_{H2O}. Set up a table as in Example 10.4. The solution is 2.74 *m* glycerol, so there are 2.74 mol glycerol in 1 kg solvent = 1000 g solvent:

	moles	$\xleftarrow{\text{\textit{MM}}}$ mass	$\xrightarrow{\text{density}}$ volume
Solute	2.74		
Solvent		1000 g	
Solution			

Convert the mass of solvent to moles of solvent using its molar mass:

$$\text{mol of H}_2\text{O solvent} = 1000 \text{ g} \times \frac{1\,\text{mol}}{18.02\,\text{g}} = 55.49 \text{ mol}$$

	moles	$\xleftarrow{\text{\textit{MM}}}$ mass	$\xrightarrow{\text{density}}$ volume
Solute	2.74		
Solvent	55.49	1000 g	
Solution			

Now calculate the mole fraction of H_2O and then the required vapor pressure:

$$X_{\text{H2O}} = \frac{n_{\text{H}_2\text{O}}}{n_{\text{H}_2\text{O}} + n_{\text{C}_3\text{H}_8\text{O}_3}} = \frac{55.49 \text{ mol}}{(55.49 + 2.74) \text{ mol}} = 0.953$$

$$P_{\text{H2O}} = (X_{\text{H2O}})(P^{\circ}_{\text{H2O}}) = (0.953)(28.35 \text{ mm Hg}) = \underline{\textbf{27.0 mm Hg}}$$

(c) This part of the problem is similar to parts (a) and (b) above. Again, use Equation 10.3 to calculate the vapor pressure over the solution having first calculated X_{H_2O}. Recall (page 249) that $X_{water} + X_{glycerol} = 1$ and so $X_{water} = 1 - 0.188 = 0.812$.

$$P_{H_2O} = (X_{water})(P°_{water}) = (0.812)(28.35 \text{ mm Hg}) = \underline{\textbf{23.0 mm Hg}}$$

36. This problem is related to Example 10.6. Use Equation 10.3 and the given vapor pressure to calculate X_{H_2O} and hence X_{urea} in the required solution.

$$P_{H_2O} = (X_{H_2O})(P°_{H_2O})$$

$$21.15 \text{ mm Hg} = (X_{H_2O})(25.21 \text{ mm Hg}) \qquad \rightarrow \quad X_{H_2O} = \frac{21.15 \text{ mm Hg}}{25.21 \text{ mm Hg}} = 0.8390$$

Recall (page 249) that $X_{H_2O} + X_{urea} = 1$ and so $X_{urea} = 1 - 0.8390 = 0.1610$. Thus (referring to the table on page 251), assuming 1 mol of solution ($n_{total} = 1$ mol) the number of moles of each component in such a solution (having this vapor pressure) would be:

$$n_{H_2O} = X_{H_2O}n_{total} = 0.8390 \text{ mol} \qquad n_{glycerol} = X_{urea}n_{total} = 0.1610 \text{ mol}$$

Now set up a table as in Example 10.4 and include the moles of the two components:

	moles	← MM	mass	density →	volume
Solute	0.1610				
Solvent	0.8390				
Solution					

Convert moles of solute and solvent to the mass of each using their molar masses, then calculate the total mass of the solution and hence (using its density) the corresponding volume of solution:

$$\text{mass of urea solute} = 0.1610 \text{ mol} \times \frac{60.06 \text{ g}}{1 \text{ mol}} = 9.670 \text{ g}$$

$$\text{mass of H}_2\text{O solvent} = 0.8390 \text{ mol} \times \frac{18.02 \text{ g}}{1 \text{ mol}} = 15.12 \text{ g}$$

$$\text{mass of solution} = (9.670 + 15.12)\text{g} = 24.79 \text{ g}$$

$$\text{volume of solution} = 24.79 \text{ g} \times \frac{1 \text{ mL}}{1.06 \text{ g}} = 23.4 \text{ mL}$$

	moles	← MM	mass	density →	volume
Solute	0.1610		9.670 g		
Solvent	0.8390		15.12 g		
Solution			24.79 g		23.4 mL

Therefore a solution containing 9.670 g urea and 15.12 g of water would have a volume of 23.4 mL of solution and have the required vapor pressure. The corresponding masses of urea and water needed to make 250.0 mL of solution are therefore:

$$250.0 \text{ mL solution} \times \frac{9.670 \text{ g urea}}{23.4 \text{ mL solution}} = 103 \text{ g glycerol}$$

$$250.0 \text{ mL solution} \times \frac{15.12 \text{ g water}}{23.4 \text{ mL solution}} = 162 \text{ g water}$$

To prepare the required solution, **dissolve 103 g urea in 162 g water**.

38. This problem is similar to Example 10.8. Determine the molarity of the solution using the following plan:

$$\text{mass of solute} \xrightarrow{\ MM\ } \text{moles of solute} \xrightarrow{\ V \text{ of solution}\ } \text{molarity}$$

then use Equation 10.5 to calculate the osmotic pressure, remembering to convert into the required units (mm Hg).

$$\text{mol solute} = 0.250 \text{ g} \times \frac{1 \text{ mol}}{3.50 \times 10^4 \text{ g}} = 7.15 \times 10^{-6} \text{ mol}$$

$$\text{molarity} = \frac{\text{mol solute}}{\text{L solution}} = \frac{7.15 \times 10^{-6} \text{ mol}}{0.0550 \text{ L}} = 1.30 \times 10^{-4} \text{ M}$$

$$\pi = MRT = (1.30 \times 10^{-4} \text{ mol/L}) \times (0.0821 \text{ L·atm/mol·K}) \times (303 \text{ K}) \times \frac{760 \text{ mmHg}}{1 \text{ atm}}$$

$$= \mathbf{2.46 \text{ mmHg}}$$

40. For each solute, the number of moles required is the same, since colligative properties depend only on the concentration of particles in solution and not on their identity. An increase of 3.0°C in the boiling point corresponds to $\Delta T_b = 3.0°C$; and a decrease of 2.0°C in the freezing point corresponds to $\Delta T_f = 2.0°C$. For each of these, use the equation on page 263 of the textbook to calculate the molality needed to achieve this value of ΔT_b or ΔT_f and hence how many moles of solute are required. The values of k_b and k_f for *p*-dichlorobenzene are given in Table 10.2.

Boiling point elevation, $\Delta T_b = 3.0°C$:

$$\Delta T_b = k_b(\text{molality})$$

$$3.0°C = (6.2 \text{ °C}/m)(\text{molality}) \qquad \rightarrow \qquad \text{molality} = 0.48 \ m$$

The mass of solvent is 100.0 g = 0.1000 kg. Find the number of moles of solute using the definition of molality; $m = n/(\text{kg solvent})$ so moles of solute, $n = m(\text{kg solvent})$:

$$\text{moles of solute} = n = m(\text{kg solvent}) = (0.48 \ m)(0.1000 \text{ kg}) = 0.048 \text{ mol}$$

<u>Freezing point depression, ΔT_f = 2.0°C:</u>

$$\Delta T_f = k_f(\text{molality})$$

$$2.0°C = (7.1 °C/m)(\text{molality}) \qquad \rightarrow \qquad \text{molality} = 0.28\ m$$

The mass of solvent is 100.0 g = 0.1000 kg. Find the number of moles of solute using the definition of molality; $m = n/(\text{kg solvent})$ \qquad so moles of solute, $n = m(\text{kg solvent})$:

moles of solute $= n = m(\text{kg solvent}) = (0.28\ m)(0.1000\ \text{kg}) = 0.028\ \text{mol}$

(a) To achieve ΔT_b = 3.0°C, 0.048 mol of solute are required. Use the molar mass to calculate the corresponding mass of succinic acid that must be used:

mass succinic acid required $= 0.048\ \text{mol} \times \dfrac{118.09\ \text{g}}{1\,\text{mol}} = $ **5.7 g**

To achieve ΔT_b = 2.0°C, 0.028 mol of solute are required. Use the molar mass to calculate the corresponding mass of succinic acid that must be used:

mass succinic acid required $= 0.028\ \text{mol} \times \dfrac{118.09\ \text{g}}{1\,\text{mol}} = $ **3.3 g**

(b) To achieve ΔT_b = 3.0°C, 0.048 mol of solute are required. Use the molar mass to calculate the corresponding mass of caffeine that must be used:

mass caffeine required $= 0.048\ \text{mol} \times \dfrac{194.19\ \text{g}}{1\,\text{mol}} = $ **9.3 g**

To achieve ΔT_b = 2.0°C, 0.028 mol of solute are required. Use the molar mass to calculate the corresponding mass of caffeine that must be used:

mass caffeine required $= 0.028\ \text{mol} \times \dfrac{194.19\ \text{g}}{1\,\text{mol}} = $ **5.4 g**

42. This problem is related to Example 10.7.

(a) In order to calculate the amount by which the freezing point of water must be depressed (ΔT_f) we must first convert −20°F to °C (using Equation 1.1). Then use the equations on pages 262-263 to calculate ΔT_f, and hence the molality needed to achieve this value of ΔT_f.

$$t_{°F} = 1.8t_{°C} + 32° \qquad \rightarrow \qquad t_{°C} = (t_{°F} - 32°)/1.8 = (-20°F - 32°)/1.8 = -29°C$$

$$\Delta T_f = T_f° - T_f = 0°C - (-29°C) = 29°C$$

$$\Delta T_f = k_f(\text{molality})$$

$$29°C = (1.86 °C/m)(\text{molality}) \qquad \rightarrow \qquad \text{molality} = \underline{16\ m}$$

Thus, the minimum molality of antifreeze needed to protect the engine at −20°F is **16m**.

(b) Calculate the mass of solvent (in kg) using the density, then find the number of moles of solute required using the definition of molality:

$m = n/(\text{kg solvent})$ so $n = m(\text{kg solvent})$

$$\text{mass of solvent} = 250 \text{ mL} \times \frac{1.00 \text{ g}}{1 \text{ mL}} \times \frac{1 \text{ kg}}{1000 \text{ g}} = 0.250 \text{ kg}$$

$$\text{moles of solute} = n = m(\text{kg solvent}) = (16 \, m)(0.250 \text{ kg}) = 4.0 \text{ mol}$$

Finally convert moles of solute to grams using the molar mass, and then convert mass to volume using the density of the solute given in the question:

$$\text{vol of solute} = 4.0 \text{ mol} \times \frac{62.07 \text{ g}}{1 \text{ mol}} \times \frac{1 \text{ mL}}{1.12 \text{ g}} = \underline{2.2 \times 10^2 \text{ mL}}$$

44. This problem is related to Examples 10.9 and 10.10. First calculate the molality of the solution: begin by constructing a table similar to those on page 252 and in Example 10.4. The given concentration is 4.50 g solute (propylene glycol) in 40.5 mL solvent (t-butyl alcohol); the corresponding mass of the solvent is

$$40.5 \text{ mL solvent} \times \frac{0.780 \text{ g}}{1 \text{ mL}} = 31.6 \text{ g}$$

	moles	$\xleftarrow{MM}$	mass	$\xrightarrow{density}$	volume
Solute			4.50 g		
Solvent			31.6 g = 0.0316 kg		
Solution					

Convert mass of solute to mol of solute using the molar mass:

$$\text{mol of solute} = 4.50 \text{ g} \times \frac{1 \text{ mol C}_3\text{H}_8\text{O}_2}{76.094 \text{ g C}_3\text{H}_8\text{O}_2} = 0.0591 \text{ mol}$$

	moles	$\xleftarrow{MM}$	mass	$\xrightarrow{density}$	volume
Solute	0.0591		4.50 g		
Solvent			31.6 g = 0.0316 kg		
Solution					

Hence the molality of the solution is:

$$\text{molality} = \frac{\text{mol solute}}{\text{kg solvent}} = \frac{0.0591 \text{ mol}}{0.0316 \text{ kg}} = 1.87 \, m$$

From page 262 the freezing point depression ΔT_f is given by the equation below; the freezing point ($T_f^\circ = 25.5 \, ^\circ\text{C}$) of the pure solvent and of the solution ($T_f = 8.5 \, ^\circ\text{C}$) are given in the question:

$$\Delta T_f = T_f^\circ - T_f = 25.5 \, ^\circ\text{C} - 8.5 \, ^\circ\text{C} = 17.0 \, ^\circ\text{C}$$

Now substitute this value for ΔT_f and the molality calculated above into the following equation (page 263) relating ΔT_f, molality, and k_f:

$$\Delta T_f = k_f(\text{molality}) \quad \rightarrow \quad 17.0°C = k_f(1.87\ m) \quad \rightarrow \quad k_f = \textbf{9.09 °C/m}$$

46. This problem is related to Example 10.10. First find the molarity M by substituting into Equation 10.5. The osmotic pressure is given and the temperature is 295 K:

$$\pi = MRT \quad \rightarrow \quad M = \frac{\pi}{RT} = \frac{(71.8\ \text{mm Hg}) \times \left(\dfrac{1\ \text{atm}}{760\ \text{mm Hg}}\right)}{(0.0821\ \text{L·atm/mol·K})(295\ \text{K})} = 3.90 \times 10^{-3}\ M$$

The volume of the solution is 0.350 L. Find the number of moles of solute using the definition of molarity, $M = n/V$ so moles of solute, $n = MV$:

$$\text{moles of solute} = n = MV = (3.90 \times 10^{-3}\ \text{mol/L})(0.350\ \text{L}) = 1.36 \times 10^{-3}\ \text{mol}$$

Lastly, knowing the mass of solute that corresponds to this number of moles, calculate the molar mass of epinephrine:

$$\text{molar mass} = \frac{\text{mass solute}}{\text{mol solute}} = \frac{0.250\ \text{g}}{1.36 \times 10^{-3}\ \text{mol}} = \textbf{184 g/mol}$$

48. This problem is related to Example 10.9. First find the freezing point depression ΔT_f; the freezing point ($T_f°$) of the pure solvent and of the solution (T_f) are given in the question:

$$\Delta T_f = T_f° - T_f = 178.40°C - 173.44°C = 4.96°C$$

Now substitute this, along with the given value of k_f, into the equation below (page 263) and determine the molality:

$$\Delta T_f = k_f \times (\text{molality}) \quad \rightarrow \quad m = \frac{\Delta T_f}{k_f} = \frac{4.96°\,C}{40.0°\,C/m} = 0.124\ m$$

The mass of solvent is 50.00 g = 0.05000 kg. Find the number of moles of solute using the definition of molality, $m = (\text{mol solute})/(\text{kg solvent})$ hence:

$$\text{mol solute} = n = m \times (\text{kg solvent}) = (0.124\ m)(0.05000\ \text{kg}) = 6.20 \times 10^{-3}\ \text{mol}$$

Lastly, knowing the (given) mass of solute that corresponds to this number of moles, calculate the molar mass of cortisone acetate:

$$\text{molar mass} = \frac{\text{mass solute}}{\text{mol solute}} = \frac{2.50\ \text{g}}{6.20 \times 10^{-3}\ \text{mol}} = \textbf{403 g/mol}$$

50. The first part of this problem is similar to Example 10.9(a) and problem 48 above. Find the freezing point depression ΔT_f and then use ΔT_f to determine the molality; the freezing point ($T_f°$) and k_f for cyclohexane are given in Table 10.2:

$$\Delta T_f = T_f° - T_f = 6.50°C - (0.0°C) = 6.5°C$$

Now substitute this value into the following equation and determine the molality:

$$\Delta T_f = k_f \times (\text{molality}) \quad \rightarrow \quad m = \frac{\Delta T_f}{k_f} = \frac{6.5°C}{20.2°C/m} = 0.32 \, m$$

Now proceed as in Example 10.9(a): find the mass of solvent in kg using the density; then use the molality calculated above, along with the mass of solvent, to determine the number of moles of solute present. Finally, use moles of solute and mass of solute to calculate molar mass.

$$\text{mass of solvent} = 75.0 \, \text{mL} \times \frac{0.779 \, g}{1 \, \text{mL}} \times \frac{1 \, \text{kg}}{1000 \, g} = 0.0584 \, \text{kg}$$

$$\text{moles solute} = n = m(\text{kg solvent}) = (0.32 \, m)(0.0584 \, \text{kg}) = 0.0187 \, \text{mol}$$

$$\text{molar mass of solute} = \frac{\text{mass solute}}{\text{mol solute}} = \frac{3.16 \, g}{0.0187 \, \text{mol}} = 169 \, g/\text{mol}$$

Now refer to Section 3.2. First determine the simplest formula for the compound based on the % of each element present. This is similar to Example 3.5. For simplicity, assume 100.0 g of compound, then the % masses of each element will be masses in grams. Use these to calculate the number of moles of each element present then follow the flowchart in Figure 3.5.

C: $\dfrac{42.9 \, g \, C}{12.01 \, g/\text{mol}} = 3.57 \, \text{mol}$ H: $\dfrac{2.4 \, g \, H}{1.008 \, g/\text{mol}} = 2.4 \, \text{mol}$

N: $\dfrac{16.6 \, g \, N}{14.01 \, g/\text{mol}} = 1.18 \, \text{mol}$ O: $\dfrac{38.1 \, g \, O}{16.00 \, g/\text{mol}} = 2.38 \, \text{mol}$

Now divide by the smallest number of moles (1.18) to get the mol ratios:

C: 3.57 mol/1.18 = 3 mol   H: 2.4 mol/1.18 = 2 mol

N: 1.18 mol/1.18 = 1 mol   O: 2.38 mol/1.18 = 2 mol

Thus, the simplest formula for the compound is $C_3H_2NO_2$.

Finally, determine the compound's molecular formula. This is similar to Example 3.7. First determine the molar mass of the simplest formula $C_3H_2NO_2$, then calculate the ratio

$$\frac{\text{actual molar mass}}{\text{simplest formula molar mass}}$$

and then multiply the subscripts in the simplest formula to get the molecular formula of the compound.

For $C_3H_2NO_2$ the molar mass:

$$3(12.01 \text{ g}) \text{ C} + 2(1.008 \text{ g H}) + (14.01 \text{ g N}) + 2(16.00 \text{ g O}) = 84.06 \text{ g/mol}$$

$$\frac{\text{actual molar mass}}{\text{simplest formula molar mass}} = \frac{169 \text{ g/mol}}{84.06 \text{ g/mol}} = 2.01 \approx 2$$

Thus, the formula for the compound is $C_{(3 \times 2)}H_{(2 \times 2)}N_{(1 \times 2)}O_{(2 \times 2)}$; that is, $C_6H_4N_2O_4$.

52. This problem is related to Example 10.10. First find the molarity M by substituting into Equation 10.5:

$$\pi = MRT \quad \rightarrow \quad M = \frac{\pi}{RT} = \frac{(1.14 \text{ mm Hg}) \times \left(\frac{1 \text{ atm}}{760 \text{ mm Hg}}\right)}{(0.0821 \text{ L·atm/mol·K})(298 \text{ K})} = 6.13 \times 10^{-5} M$$

The volume of solution is 0.850 L. Find the number of moles of solute using the definition of molarity; $M = n/V$ so moles of solute, $n = MV$:

$$\text{moles of solute} = n = MV = (6.13 \times 10^{-5} \text{ mol/L})(0.850 \text{ L}) = 5.21 \times 10^{-5} \text{ mol}$$

Finally, knowing the (given) mass of solute that corresponds to this number of moles, calculate the molar mass of the enzyme:

$$\text{molar mass} = \frac{\text{mass solute}}{\text{mol solute}} = \frac{7.89 \text{ g}}{5.21 \times 10^{-5} \text{ mol}} = \underline{1.51 \times 10^5 \text{ g/mol}}$$

10-4 COLLIGATIVE PROPERTIES OF ELECTROLYTES

54. For all solutions, the amount of freezing point depression or boiling point elevation depends on the concentration of particles present; when electrolytes are present, we must remember that such substances split up into ions when in solution and that this can affect the concentration of particles in solution.

Referring to Equations 10.6 and 10.7, the magnitude of ΔT_f or ΔT_b depends on i, k, and molality. However, since k_f and k_b are constants for a particular solvent, and in this problem the molality of each solution is the same, only the value of i for each solution need be considered in order to differentiate their freezing and boiling points. The larger the value of i, the larger will be ΔT_f or ΔT_b.

For each solution, consider what happens when each solute dissolves: i is the number of moles of particles produced in solution per mole of solute.

(a) $Al(ClO_3)_3(s) \rightarrow Al^{3+}(aq) + 3ClO_3^-(aq)$ 4 mol particles in solution: $i = 4$

(b) $CH_3OH(l) \rightarrow CH_3OH(aq)$ 1 mol particles in solution: $i = 1$

(c) $(NH_4)_2Cr_2O_7(s) \rightarrow 2NH_4^+(aq) + Cr_2O_7^{2-}(aq)$ 3 mol particles in solution: $i = 3$

(d) $MgSO_4(s) \rightarrow Mg^{2+}(aq) + SO_4^{2-}(aq)$ 2 mol particles in solution: $i = 2$

Thus the magnitude of ΔT_f or ΔT_b will increase in the order (b) < (d) < (c) < (a) and recall that ΔT_f corresponds to a *lowering* of the freezing point whereas ΔT_b corresponds to a *raising* of the boiling point. Therefore, the solutions should be arranged in order of *decreasing* freezing point and boiling point as follows:

freezing point: CH_3OH > $MgSO_4$ > $(NH_4)_2Cr_2O_7$ > $Al(ClO_3)_3$

boiling point: $Al(ClO_3)_3$ > $(NH_4)_2Cr_2O_7$ > $MgSO_4$ > CH_3OH

56. This problem is related to Example 10.8. However, since the solute Na_2SO_4 is an electrolyte, we must determine the osmotic pressure using the equation at the bottom of page 269 (immediately below Equation 10.7), with the molarity and temperature given in the question. To determine the value of i, consider what happens when the solute dissolves: i is the number of moles of ions produced in solution per mole of solute:

$Na_2SO_4(s) \rightarrow 2Na^+(aq) + SO_4^{2-}(aq)$ 3 mol ions in solution, so $i = 3$

$\pi = i \times \text{molarity} \times RT$

$\pi = 3 \times (0.135 \text{ mol/L}) \times (0.0821 \text{ L·atm/mol·K})(293 \text{ K})$

$\pi = \underline{\textbf{9.74 atm}}$

58. This problem is related to Example 10.11.

(a) Find ΔT_f and then use Equation 10.6 to calculate the value of i in this solution. Use the equation below (from page 262) to calculate ΔT_f. Recall that the freezing point $T_f°$ of pure water is 0 °C.

$\Delta T_f = T_f° - T_f = 0°C - (-0.796°C) = 0.796°C$

Now substitute this into Equation 10.6 to determine the value of i:

$\Delta T_f = i \times 1.86°C/m \times \text{molality}$

$0.796°C = i \times 1.86°C/m \times 0.21 \ m$

$i = \underline{\textbf{2.0}}$

(b) When H_2SO_4 dissolves in water, there are three possible reactions that could be written to represent the process; these correspond to the three combinations (i)–(iii) given in the question:

(i) $H_2SO_4(l) \rightarrow H_2SO_4(aq)$

(ii) $H_2SO_4(l) \rightarrow H^+(aq) + HSO_4^-(aq)$

(iii) $H_2SO_4(l) \rightarrow 2H^+(aq) + SO_4^{2-}(aq)$

For these three equations, the value of i equals the number of moles of particles that are produced in solution when 1 mol H_2SO_4 dissolves. Thus the value of i is 1 for equation (i), 2 for equation (ii), and 3 for equation (iii). The value of i determined above is 2.0, so equation (ii) best describes what happens when H_2SO_4 dissolves in water. Thus, **the solution of (ii) H_2SO_4 is primarily made up of H^+ and HSO_4^- ions.**

60. This problem is related to Example 10.9. The solute in this case, M_2O, will dissociate in water as follows:

$$M_2O(s) \rightarrow 2M^+(aq) + O^{2-}(aq)$$

thus M_2O produces 3 mol of ions ($2M^+$, O^{2-}) per mol of electrolyte, so $i = 3$ (assuming complete dissociation of the salt in solution).

We know the freezing point of the solution, so we can calculate the freezing point elevation ΔT_f; then use Equation 10.6 with ΔT_f to calculate the molality of the solution. Recall that the freezing point of pure H_2O is 0°C; thus (see page 262):

$$\Delta T_f = T_f^° - T_f = 0°C - (-4.68°C) = 4.68°C$$

Now substitute this and the value of i into Equation 10.6 to determine the molality:

$$\Delta T_f = i \times 1.86°C/m \times molality$$

$$4.68°C = 3 \times 1.86°C/m \times molality \qquad \rightarrow \qquad molality = 0.839\ m$$

This molality corresponds to 0.839 mol solute in 1 kg solvent.

In order to identify M^+ we must determine the molar mass of atoms of M. We know the mass of solute present and the volume of solvent. Proceed as in Example 10.9: find the mass of solvent in kg using the density; then use the molality calculated above, along with the mass of solvent, to determine the number of moles of solute present.

$$\text{mass of solvent} = 435\ mL \times \frac{1.00\ g}{1\ mL} \times \frac{1\ kg}{1000\ g} = 0.435\ kg$$

$$\text{moles solute in 0.435 kg solvent} = 0.435\ kg\ solvent \times \frac{0.839\ mol\ solute}{1\ kg\ solvent} = 0.365\ mol$$

Finally, use moles of solute and mass of solute to calculate molar mass.

$$\text{molar mass of solute} = \frac{\text{mass solute}}{\text{mol solute}} = \frac{10.91\ g}{0.365\ mol} = 29.9\ g/mol$$

This is the molar mass of the solute M_2O, which can be used to determine the molar mass of M:

let x = molar mass of M (or the mass of 1 mol of M)

assume 1 mole M_2O: $29.9 \text{ g } M_2O = \left(2 \text{ mol M} \times \dfrac{x \text{ g M}}{1 \text{ mol M}} \right) + \left(1 \text{ mol O} \times \dfrac{16.00 \text{ g O}}{1 \text{ mol O}} \right)$,

$\rightarrow \quad 2x = 13.9 \text{ g/mol} \quad \rightarrow \quad x = 6.95 \text{ g/mol}$

The element with molar mass closest to 6.95 g/mol is lithium (MM = 6.94 g/mol). Thus, the element represented by M is <u>**lithium (Li)**</u>.

UNCLASSIFIED PROBLEMS

62. This problem is similar (in part) to Example 10.4. Begin by constructing a table such as those on page 252 and in Example 10.4 and include the given concentration as 32.47 g solute in 100.0 mL of solution:

	moles	$\xleftarrow{\ MM\ }$	mass	$\xrightarrow{\ density\ }$	volume
Solute			32.47 g		
Solvent					
Solution					100.0 mL

(a) Recall (page 247) that molarity is defined as moles of solute per liter of solution. We must therefore convert mass of solute to moles using the molar mass, and express the volume of solution in liters (100.0 mL = 0.1000 L):

$$\text{moles solute} = 32.47 \text{ g FeCl}_3 \times \dfrac{1 \text{ mol FeCl}_3}{162.20 \text{ g FeCl}_3} = 0.2002 \text{ mol}$$

	moles	$\xleftarrow{\ MM\ }$	mass	$\xrightarrow{\ density\ }$	volume
Solute	0.2002		32.47 g		
Solvent					
Solution					0.1000 L

Thus the molarity of the solution is

$$\text{molarity} = \dfrac{\text{mol solute}}{\text{L solution}} = \dfrac{0.2002 \text{ mol}}{0.1000 \text{ L}} = \underline{\textbf{2.002 M}}$$

(b) Recall (page 250) that molality is defined as moles of solute per kilogram of solvent. We must therefore convert volume of solution to mass of solution using the density and hence calculate mass of solvent:

$$\text{mass solution} = 100.0 \text{ mL solution} \times \frac{1.249 \text{ g solution}}{1 \text{mL solution}} = 124.9 \text{ g}$$

$$\text{mass solvent} = (124.9 \text{ g solution}) - (32.47 \text{ g solute}) = 92.4 \text{ g solvent}$$

	moles $\xleftarrow{\quad MM \quad}$	mass $\xrightarrow{\quad density \quad}$	volume
Solute	0.2002	32.47 g	
Solvent		92.4 g	
Solution		124.9 g	100.0 mL

The mass of solvent is 92.4 g = 0.0924 kg; hence the molality of the solution is

$$\text{molality} = \frac{\text{mol solute}}{\text{kg solvent}} = \frac{0.2002 \text{ mol}}{0.0924 \text{ kg}} = \underline{\textbf{2.17 m}}$$

(c) Use the equation for osmotic pressure immediately below Equation 10.7 with the molarity calculated in part (a) above and the value of *i* given in the question:

$$\pi = i \times \text{molarity} \times RT$$

$$\pi = 4 \times (2.002 \text{ mol/L}) \times (0.0821 \text{ L·atm/mol·K})(298 \text{ K}) = \underline{\textbf{196 atm}}$$

(d) Use Equation 10.6 with the molality calculated in part (b) above and the value of *i* given in the question to calculate ΔT_f:

$$\Delta T_f = i \times 1.86°C/m \times \text{molality}$$

$$\Delta T_f = 4 \times 1.86°C/m \times 2.17 \ m = 16.1 \ °C$$

Use the equations from page 262 to calculate T_f; recall that the freezing point $T_f°$ of pure water is 0 °C.

$$T_f = T_f° - \Delta T_f = 0 °C - 16.1°C = \underline{\textbf{−16.1 °C}}$$

64. (a) Since this problem involves boiling point elevation, we must first calculate the molality of the solution. Begin by constructing a table similar to those on page 252 and in Example 10.4 and include the given concentration as 0.287 mol solute in 255 mL CCl_4. Note that here we know the volume of solvent rather than volume of solution:

	moles $\xleftarrow{\quad MM \quad}$	mass $\xrightarrow{\quad density \quad}$	volume
Solute	0.287		
Solvent			255 mL
Solution			

Recall (page 250) that molality is defined as moles of solute per kilogram of solvent. We must therefore convert volume of solvent to mass of solvent using the density:

$$\text{mass solvent} = 255 \text{ mL CCl}_4 \times \frac{1.59 \text{ g}}{1 \text{ mL}} = 405 \text{ g}$$

	moles $\xleftarrow{\hspace{0.3cm} MM \hspace{0.3cm}}$	mass $\xrightarrow{\hspace{0.3cm} \text{density} \hspace{0.3cm}}$	volume
Solute	0.287		
Solvent		405 g	255 mL
Solution			

The mass of solvent is 405 g = 0.405 kg; thus the molality of the solution is

$$\text{molality} = \frac{\text{mol solute}}{\text{kg solvent}} = \frac{0.287 \text{ mol}}{0.405 \text{ kg}} = 0.709 \, m$$

From page 261 the boiling point elevation ΔT_b for a nonelectrolyte solution is given by:

$$\Delta T_b = T_b - T_b^\circ = 80.3°C - 76.8°C = 3.5°C$$

Now substitute this value for ΔT_b and the molality calculated above into the following equation from page 263 of the textbook that relates ΔT_b, molality, and k_b:

$$\Delta T_b = k_b(\text{molality})$$
$$3.5°C = k_b \times 0.709 \, m$$
$$k_b = \underline{\mathbf{4.9 \, °C/m}}$$

(b) As in part (a) above, begin by calculating the molality of the solution. We have 37.1 g solute in 244 mL CCl$_4$.

	moles $\xleftarrow{\hspace{0.3cm} MM \hspace{0.3cm}}$	mass $\xrightarrow{\hspace{0.3cm} \text{density} \hspace{0.3cm}}$	volume
Solute		37.1 g	
Solvent			244 mL
Solution			

We need to know moles of solute and kilograms of solvent: convert mass of solute to moles of solute using the molar mass, and convert volume of solvent to mass of solvent using the density:

$$\text{moles solute} = 37.1 \text{ g solute} \times \frac{1 \text{ mol}}{167 \text{ g}} = 0.222 \text{ mol}$$

$$\text{mass solvent} = 244 \text{ mL CCl}_4 \times \frac{1.59 \text{ g}}{1 \text{ mL}} = 388 \text{ g}$$

	moles	← MM —	mass	density →	volume
Solute	0.222		37.1 g		
Solvent			388 g		244 mL
Solution					

The mass of solvent is 388 g = 0.388 kg; thus the molality of the solution is

$$\text{molality} = \frac{\text{mol solute}}{\text{kg solvent}} = \frac{0.222 \text{ mol}}{0.388 \text{ kg}} = 0.572 \ m$$

From page 261 the boiling point elevation ΔT_b is given by:

$$\Delta T_b = T_b - T_b^\circ = 85.2°C - 76.8°C = 8.4°C$$

Now substitute this value for ΔT_b, the molality calculated above, and the value for k_b calculated in part (a) into the equation below and solve for i. Note that this equation is analogous to Equation 10.7 (which is the form we must use, since the solute is an electrolyte), but instead of using k_b = 0.52 °C/m (the value of k_b for H_2O), we must use the value of k_b for CCl_4.

$$\Delta T_b = i \times k_b \times (\text{molality})$$

$$8.4°C = i \times (4.9 \text{ °C/}m) \times 0.572 \ m \qquad \rightarrow \qquad i = \underline{\mathbf{3.0}}$$

66. Recall (Equation 5.4) that the partial pressure of a gas in a mixture equals its mole fraction multiplied by the total pressure. Thus the partial pressure of radon in this mixture of gases is

$$P_{Rn} = X_{Rn}P_{tot} = (2.7 \times 10^{-6})(28 \text{ atm}) = 7.6 \times 10^{-5} \text{ atm}$$

(a) Henry's Law (Equation 10.2) relates the solubility (concentration, C_{gas}) of a gas to the partial pressure of the gas, P_{gas}, above the solution. Use Henry's Law with P_{Rn} calculated above and the value of k given in the question to calculate the concentration of radon dissolved in the water. (Recall that 1 atm = 760 mm Hg.)

$$C_{Rn} = kP_{Rn} = (9.57 \times 10^{-6} \text{ M/mm Hg}) \times \frac{760 \text{ mm Hg}}{1 \text{ atm}} \times (7.6 \times 10^{-5} \text{ atm})$$

$$C_{Rn} = \underline{\mathbf{5.5 \times 10^{-7} \text{ M}}}$$

(b) Recall the definition of ppm (see page 250):

$$\text{ppm} = \frac{\text{mass solute}}{\text{mass solution}} \times 10^6$$

Set up a table like those on page 252 and in Example 10.4 and include the molarity calculated in part (a) as 5.5×10^{-7} mol solute in 1L = 1000 mL of solution. Based on the definition of ppm, we must convert moles of solute to mass of solute using the molar mass, and convert volume of solution to mass of solution using the density, and hence calculate the mass of solution.

	moles $\xleftarrow{\quad MM \quad}$	mass $\xrightarrow{\quad density \quad}$	volume
Solute	5.5×10^{-7}		
Solvent			
Solution			1000 mL

$$\text{mass solute} = 5.5 \times 10^{-7} \text{ mol Rn} \times \frac{222.0 \text{ g Rn}}{1 \text{ mol Rn}} = 1.2 \times 10^{-4} \text{ g Rn}$$

$$\text{mass solution} = 1000 \text{ mL solution} \times \frac{1 \text{ g solution}}{1 \text{ mL solution}} = 1000 \text{ g}$$

$$\text{mass solvent} = (1000 \text{ g solution}) - (1.2 \times 10^{-4} \text{ g solute}) = 1000 \text{ g solvent}$$

	moles $\xleftarrow{\quad MM \quad}$	mass $\xrightarrow{\quad density \quad}$	volume
Solute	5.5×10^{-7}	1.2×10^{-4} g	
Solvent		1000 g	
Solution		1000 g	1000 mL

Thus the concentration of Rn expressed in ppm is:

$$\text{ppm} = \frac{\text{mass solute}}{\text{mass solution}} \times 10^6 = \frac{1.2 \times 10^{-4} \text{ g}}{1000 \text{ g}} \times 10^6 = \underline{\textbf{0.12 ppm}}$$

68. Osmotic pressure is a colligative property and therefore (page 260) depends only on the concentration of particles in solution, not on their identity. To find the osmotic pressure of the mixture, first determine the concentration of particles in solutions X and Y, and hence the concentration of particles in the mixture. Use Equation 10.5 to determine M_X and M_Y, the molarity of solutions X and Y, respectively:

$$\pi(\text{solution X}) = M_X RT \qquad \rightarrow \qquad 1.8 \text{ atm} = M_X RT \qquad \rightarrow \qquad M_X = 1.8/RT$$

$$\pi(\text{solution Y}) = M_Y RT \qquad \rightarrow \qquad 4.2 \text{ atm} = M_Y RT \qquad \rightarrow \qquad M_Y = 4.2/RT$$

When equal volumes of the two solutions are mixed, the concentration of the mixture is an average of the two: suppose we mix z liters of the two solutions, then the numbers of mol of solute provided by z L of solution X and z L of solution Y are given by MV in each case:

mol of solute from solution X $= (z \text{ L}) \times (1.8/RT \text{ mol/L}) = 1.8z/RT$ mol

mol of solute from solution Y $= (z \text{ L}) \times (4.2/RT \text{ mol/L}) = 4.2z/RT$ mol

So the concentration of the mixture is:

$$n_{total}/V_{total} = \frac{(1.8z/RT + 4.2z/RT)\,mol}{2z\,L} = 3.0/RT \text{ mol/L}$$

which is indeed equal to the average $\frac{1}{2}(M_X + M_Y)$. Note also that this is independent of the value of z, in other words, it does not matter how much of each solution is mixed, so long as equal volumes are used. Now use the molarity of the mixture to calculate osmotic pressure of the mixture, again using Equation 10.5:

$$\pi(\text{mixture}) = MRT = (3.0/RT) \times RT = \underline{\textbf{3.0 atm}}$$

CONCEPTUAL QUESTIONS

70. The solute $CaCl_2$ is a strong electrolyte that dissociates in water as follows:

$$CaCl_2(s) \rightarrow Ca^{2+}(aq) + 2Cl^-(aq)$$

thus each mol of $CaCl_2$ that dissolves produces 1 mol of Ca^{2+} ions and 2 mol of Cl^- ions. The picture in the question shows four moles of $CaCl_2$ to which is being added 1.00 L of H_2O. The molarities of the two ions will be:

$$[Ca^{2+}] = \frac{4\,mol\,CaCl_2}{1.00\,L} \times \frac{1\,mol\,Ca^{2+}}{1\,mol\,CaCl_2} = \underline{\textbf{4 M } Ca^{2+}}$$

$$[Cl^-] = \frac{4\,mol\,CaCl_2}{1.00\,L} \times \frac{2\,mol\,Cl^-}{1\,mol\,CaCl_2} = \underline{\textbf{8 M } Cl^-}$$

therefore there will be 4 mol Ca^{2+} ions and 8 mol Cl^- ions in 1.00 L solution, so the complete reaction may be represented as:

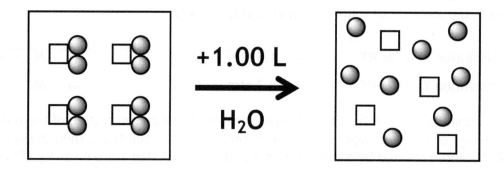

72. Since the molar mass of X is greater than that of Y, 25 g of X will contain fewer moles than 25 g of Y. Each solution has a total mass of (25 g solute + 100 g solvent) = 125 g; and since both solutions have the same density, their volumes will also be equal.

(a) Recall (page 247) that molarity = (mol solute)/(vol solution). Solution 1 contains fewer moles of solute than solution 2 and since the volumes of the two solutions are equal **solution 2 has a higher molarity than solution 1**.

(b) Recall (page 249) that mass % = (mass solute)/(mass solution) × 100. Since mass of solute and mass of solution is the same for both solutions, **the mass percent of solution 1 equals the mass percent of solution 2**.

(c) Recall (page 250) that molality = (mol solute)/(kg solvent). Solution 1 contains fewer moles of solute than solution 2 and since the mass of solvent in the two solutions is the same **solution 2 has a higher molality than solution 1**.

(d) Both X and Y are nonelectrolytes. For nonelectrolytes (see page 269, below Equation 10.6), the value of i is 1, so **the value of i for solution 1 equals that for solution 2**.

(e) Recall (page 249) that mole fraction of solvent = (mol solvent)/(total mol of solution). Both solutions contain the same mass (and hence the same number of moles) of solvent, but solution 1 contains fewer moles of solute than solution 2. Hence, the total number of moles (mol solute + mol solvent) in solution 1 is *less* than that in solution 2, and therefore the ratio (mol solvent)/(total mol of solution) is *greater* for solution 1 than for solution 2. In other words, **solution 1 has a higher mole fraction of solvent than solution 2**.

74. This problem is related to Example 10.11. To determine whether HF is dissociated in this solution calculate the value of i, which is the number of moles of particles in solution when 1 mol of HF molecules dissolves. Use the equation below (from page 262) to calculate ΔT_f. Recall that the freezing point $T_f°$ of pure water is 0 °C.

$$\Delta T_f = T_f° - T_f = 0°C - (-0.38°C) = 0.38°C$$

Now substitute this into Equation 10.6 to determine the value of i:

$$\Delta T_f = i \times 1.86°C/m \times \text{molality}$$

$$0.38°C = i \times 1.86°C/m \times 0.20\ m \qquad \rightarrow \qquad i = 1.0$$

When HF dissolves in water it may remain as HF molecules (1) or it may dissociate into H^+ and F^- ions (2):

$$HF(g) \rightarrow HF(aq) \qquad\qquad ...(1)$$

$$HF(g) \rightarrow H^+(aq) + F^-(aq) \qquad\qquad ...(2)$$

The value of i calculated above was found to be approximately 1.0: in other words, 1 mol HF(g) produces 1 mol particles in solution when it dissolves; this is consistent with reaction (1), so the **HF primarily exists in solution as molecules**.

(This is to be expected: HF is a weak acid, so we expect only a small fraction of its molecules to ionize in solution.)

76. This problem is related to Example 10.11 and to problem 74 above. Find ΔT_f and then use Equation 10.6 to calculate the value of i in this solution.

Use the equation below (from page 262) to calculate ΔT_f. Recall that the freezing point T_f° of pure water is 0 °C.

$$\Delta T_f = T_f^\circ - T_f = 0°C - (-0.38°C) = 0.38°C$$

Now substitute this into Equation 10.6 to determine the value of i:

$$\Delta T_f = i \times 1.86°C/m \times \text{molality}$$

$$0.38°C = i \times 1.86°C/m \times 0.10\ m \qquad \rightarrow \qquad i = 2.0$$

For the three equations given in the question, the value of i equals the number of moles of particles (on the product side of the reaction) produced by 1 mol $KHSO_3$. Thus the value of i should be 1 for equation (a), 2 for equation (b), and 3 for equation (c). The value of i determined above is 2.0, so **equation (b) best describes what happens when $KHSO_3$ dissolves in water.**

(This is to be expected: the anion HSO_3^- is actually a very weak acid, which scarcely dissociates in solution. This behavior is discussed in more detail in Chapter 13.)

78. Refer to Section 10.2; in particular, pages 256–259. Note that we are told that in this case the gas dissolves *exothermically*, so it should follow the rule that gases become less soluble in water as temperature increases.

(a) The temperature has not changed, but pressure is lower. Henry's Law tells us that gases become less soluble as pressure decreases, so **we expect solubility to be less under these conditions**.

(b) The temperature has decreased, but pressure is unchanged. Gases become more soluble as temperature decreases, so **we expect solubility to be greater under these conditions**.

(c) The temperature has decreased, and pressure has increased. Gases become more soluble as temperature decreases, and Henry's Law tells us that gases become more soluble as pressure increases, so **we expect solubility to be greater under these conditions**.

(d) The temperature has increased, and pressure has decreased. Gases become less soluble as temperature increases, and Henry's Law tells us that gases become less soluble as pressure decreases, so **we expect solubility to be less under these conditions**.

80. (a) A solution of an electrolyte will conduct electric current whereas a solution of a nonelectrolyte does not.

(b) See pages 257-259: Beer is a solution containing the gas CO_2. As the solution warms, the CO_2 becomes less soluble and bubbles out of the solution, so that the remaining solution is much less carbonated — in other words, it is flat.

(c) Refer to the definitions of mole fraction of solute and molality of solute on pages 249-250: although both definitions have "mol solute" as the numerator, their denominators differ. In general, the number of moles of solvent will vastly exceed the number of kg of solvent, so the number obtained when calculating X_{solute} will be much smaller than that obtained when calculating m_{solute}.

(d) Recall the definition of boiling point (page 221): In the present case, it is the temperature at which the vapor pressure of the solution equals the prevailing pressure above the solution. However, the vapor pressure of a solution is *lower* than that of the corresponding solvent (Raoult's Law — see page 260); hence, to achieve a vapor pressure above the solution that equals the prevailing pressure, the temperature of the solution must be raised more than is required to achieve the same vapor pressure above the pure solvent. This means that the boiling point of the solution is higher than that of the pure solvent.

CHALLENGE PROBLEMS

82. The solute in this case, KNO_3, is a strong electrolyte that dissociates in water as follows:

$$KNO_3(s) \rightarrow K^+(aq) + NO_3^-(aq)$$

thus KNO_3 produces 2 mol of ions (K^+, NO_3^-) per mol of electrolyte, so $i = 2$ (and assume that the ions are independent in solution).

We are told the boiling point of the solution, which implies the amount of boiling point elevation ΔT_b; using Equation 10.7, we can use ΔT_b to calculate the molality and hence the moles of solute in 1 kg of solvent. This also yields the total mass of the solution.

To calculate the density of the solution, its volume is also required and this can be obtained from the osmotic pressure: specifically, the osmotic pressure gives the molarity — that is, the moles of solute in 1 L of solution, and this can be combined with the information obtained from boiling point elevation to calculate the density.

Recall that the normal boiling point of pure H_2O is 100°C; thus (see page 261):

$$\Delta T_b = T_b - T_b° = 103.0°C - 100°C = 3.0°C$$

Now substitute this into Equation 10.7:

$$\Delta T_b = i \times 0.52°C/m \times \text{molality}$$

$$3.0°C = 2 \times 0.52°C/m \times \text{molality} \qquad \rightarrow \qquad \text{molality} = 2.9\ m$$

Therefore, there are 2.9 mol KNO_3 in 1 kg of solvent. The corresponding mass of KNO_3 is:

$$2.9\ \text{mol } KNO_3 \times \frac{101.1\,g\ KNO_3}{1\,mol\ KNO_3} = 2.9 \times 10^2\ g\ KNO_3 \quad \text{or}\quad 0.29\ kg\ KNO_3$$

and total mass of the solution = (mass solvent) + (mass solute) = 1.29 kg or 1.29×10^3 g

Thus, there are **2.9 mol KNO_3 in 1.29×10^3 g of solution.** ....(1)

Now use the equation for osmotic pressure immediately below Equation 10.7:

$$\pi = i \times \text{molarity} \times RT$$

$$122\ \text{atm} = 2 \times \text{molarity} \times (0.0821\ L\cdot atm/mol\cdot K)(298\ K) \quad \rightarrow \quad \text{molarity} = 2.49\ M$$

Therefore, there are **2.49 mol KNO_3 in 1 L of solution** (that is, in 1000 mL of solution).

Finally, use this last result as a conversion factor to convert the result (mass of solution/mol KNO_3) at **(1)** above into a value for the density:

$$\text{density} = \frac{1.29 \times 10^3\ g\ \text{solution}}{2.9\ \text{mol } KNO_3} \times \frac{2.49\ \text{mol } KNO_3}{1000\ mL\ \text{solution}} = \underline{\textbf{1.1 g/mL}}$$

83. In order to compare the concentration of the given solution (158.2 g KOH/L) with the one required by the technician, we must first convert this concentration into units of molality. Begin by constructing a table such as those on page 252 and in Example 10.4 and include the given concentration as 158.2 g solute in 1L = 1000 mL of solution. Recall (page 250) that molality is defined as moles of solute per kilogram of solvent. We must therefore convert mass of solute to moles using the molar mass, and convert volume of solution to mass using the density and hence calculate mass of solvent.

	moles $\xleftarrow{MM}$	mass $\xrightarrow{\text{density}}$	volume
Solute		158.2 g	
Solvent			
Solution			1000 mL

$$\text{moles solute} = 158.2\ g\ KOH \times \frac{1\,mol\ KOH}{56.11\,g\ KOH} = 2.819\ \text{mol}$$

$$\text{mass solution} = 1000 \text{ mL solution} \times \frac{1.13 \text{ g solution}}{1 \text{ mL solution}} = 1.13 \times 10^3 \text{ g}$$

$$\text{mass solvent} = (1.13 \times 10^3 \text{ g solution}) - (158.2 \text{ g solute}) = 9.7 \times 10^2 \text{ g solvent}$$

	moles	$\xleftarrow{MM}$	mass	$\xrightarrow{\text{density}}$	volume
Solute	2.819 mol		158.2 g		
Solvent			9.7×10^2 g $= 0.97$ kg		
Solution			1.13×10^3 g		1000 mL

Thus the molality of the solution is

$$\text{molality} = \frac{\text{mol solute}}{\text{kg solvent}} = \frac{2.819 \text{ mol}}{0.97 \text{ kg}} = 2.9 \text{ } m$$

This solution is much more concentrated than the one required by the lab technician, so it must be diluted by adding more solvent.

As calculated above, there are 2.819 mol KOH and 9.7×10^2 g solvent in 1000 mL of the concentrated solution; thus in the 100.0 mL sample taken by the technician there are:

$$100.0 \text{ mL} \times \frac{2.819 \text{ mol KOH}}{1000 \text{ mL}} = 0.2819 \text{ mol KOH}$$

$$100.0 \text{ mL} \times \frac{9.7 \times 10^2 \text{ g solvent}}{1000 \text{ mL}} = 97 \text{ g solvent} \quad \text{or} \quad 0.097 \text{ kg of solvent}$$

Let the mass of H_2O that must be added by the technician = x kg; then the total mass of solvent is $(0.097 + x)$kg. We know that the molality of this solution is 0.250 m and contains 0.2819 mol KOH (since dilution does not affect the quantity KOH present) thus the molality of the solution is

$$\text{molality} = 0.250 \text{ } m = \frac{\text{mol solute}}{\text{kg solvent}} = \frac{0.2819 \text{ mol}}{(0.097 + x)\text{kg}} \quad \rightarrow \quad 0.250(0.097 + x) = 0.2819$$

$$0.024 + 0.250x = 0.2819 \quad \rightarrow \quad x = 1.03 \text{ kg} \quad \text{or} \quad 1.03 \times 10^3 \text{ g}$$

The technician must add 1.03×10^3 g H_2O to 100.0 mL of the stock solution to prepare 0.250 m KOH.

84. First use the freezing point depression (ΔT_f = 5.5°C) and the equation below (see page 263) to calculate the overall molality of solute in the solution:

$$\Delta T_f = k_f(\text{molality}) \quad \rightarrow \quad 5.5°C = (1.86 \,°C/m)(\text{molality}) \rightarrow \quad \text{molality} = 3.0 \, m$$

Thus, the concentration of this solution (3.0 m) tells us that there are 3.0 mol solute particles in 1 kg solvent. We have only 100.0 g = 0.1000 kg of solvent, so the number of mol solute present is:

$$\text{mol solute in 100.0 g H}_2\text{O} = 0.1000 \, \text{kg solvent} \times \frac{3.0 \, \text{mol solute}}{1 \, \text{kg solvent}} = 0.30 \, \text{mol}$$

Our solute is a mixture of CH_3OH and $CH_3CH_2CH_2OH$, with a total mass of 14.2 g. Let the number of mol CH_3OH present be X; then the remaining $(0.30 - X)$ mol is $CH_3CH_2CH_2OH$. We can use these expressions, along with their respective molar masses, to obtain the mass of each alcohol present in terms of X.

$$\text{mass CH}_3\text{OH present} = X \, \text{mol} \times \frac{32.042 \, \text{g CH}_3\text{OH}}{1 \, \text{mol CH}_3\text{OH}} = 32.042X \, \text{g}$$

$$\text{mass CH}_3\text{CH}_2\text{CH}_2\text{OH present} = (0.30 - X) \, \text{mol} \times \frac{60.095 \, \text{g CH}_3\text{CH}_2\text{CH}_2\text{OH}}{1 \, \text{mol CH}_3\text{CH}_2\text{CH}_2\text{OH}}$$

$$= (18 - 60.095X) \, \text{g}$$

We know that the total mass of the two alcohols is 14.2 g, so we can solve for X:

$$14.2 \, \text{g} = (\text{mass CH}_3\text{OH}) + (\text{mass CH}_3\text{CH}_2\text{CH}_2\text{OH})$$

$$= (32.042X)\text{g} + (18 - 60.095X)\text{g} = (18 - 28.053X)\text{g} \qquad \rightarrow \qquad X = 0.14$$

Thus we have X = 0.14 mol CH_3OH present in a total of 0.30 mol.

$$\text{mol percent CH}_3\text{OH} = \frac{n_{CH_3OH}}{n_{total}} \times 100 \, \% = \frac{0.14 \, \text{mol}}{0.30 \, \text{mol}} \times 100 \, \% = \underline{\textbf{47\%}}$$

85. This problem is related to Example 10.9. Let the mass of X in the 0.100 g mixture = x; then the mass of sugar in the mixture is $(0.100 - x)$g. The number of moles of each component in the mixture is:

$$\text{mol X present} = x \, \text{g} \times \frac{1 \, \text{mol X}}{410 \, \text{g X}} = \frac{x}{410} \, \text{mol}$$

$$\text{mol sugar present} = (0.100 - x)\text{g} \times \frac{1 \, \text{mol sugar}}{342 \, \text{g sugar}} = \frac{0.100 - x}{342} \, \text{mol}$$

Use the equations from pages 262-263 to calculate ΔT_f and hence the molality of the solution. Recall that the freezing point T_f° of pure water is 0 °C.

$$\Delta T_f = T_f^\circ - T_f = 0°C - (-0.500°C) = 0.500°C$$

Now substitute this into the equation relating ΔT_f and molality:

$$\Delta T_f = 1.86°C/m \times \text{molality}$$

$$0.500°C = 1.86°C/m \times \text{molality}$$

$$\text{molality} = 0.269\ m$$

Therefore, there are 0.269 mol solute in 1 kg of solvent. However, in this case the mass of solvent used was 1.00 g = 1.00×10^{-3} kg and hence, the total number of moles of solute in this solution is:

$$1.00 \times 10^{-3} \text{ kg solvent} \times \frac{0.269 \text{ mol solute}}{1 \text{kg solvent}} = 2.69 \times 10^{-4} \text{ mol solute}$$

This total number of moles of solute equals the sum of the number of moles of X and the number of moles of sugar in the solution. Thus:

$$2.69 \times 10^{-4} \text{ mol solute} = (\text{mol X}) + (\text{mol sugar}) = \frac{x}{410} \text{ mol} + \frac{0.100 - x}{342} \text{ mol}$$

$$2.69 \times 10^{-4} \text{ mol} = \frac{x}{410} \text{ mol} + \frac{0.100}{342} \text{ mol} - \frac{x}{342} \text{ mol}$$

$$\frac{0.100}{342} \text{ mol} - 2.69 \times 10^{-4} \text{ mol} = \frac{x}{342} \text{ mol} - \frac{x}{410} \text{ mol} = \frac{410x - 342x}{(342)(410)} \text{ mol}$$

$$2.3 \times 10^{-5} \text{ mol} = (4.85 \times 10^{-4})x \text{ mol} \qquad \rightarrow \qquad x = 4.7 \times 10^{-2}$$

Thus, there are 4.7×10^{-2} g of X in the mixture. The mass percent of X (see page 249) is:

$$\frac{\text{mass of X}}{\text{total mass}} \times 100 = \frac{0.047 \text{ g}}{0.100 \text{ g}} \times 100 = \underline{\textbf{47\%}}$$

86. Recall the definition of mass percent:

$$\frac{\text{mass of solute}}{\text{mass of solution}} \times 100$$

and (referring to the table on page 251) note that 100 g of the martini solution contains 30.0 g alcohol. Thus, the mass of alcohol in each 142 g martini is

$$\text{mass of alcohol} = 142 \text{ g solution} \times \frac{30.0 \text{ g alcohol}}{100 \text{ g solution}} = 42.6 \text{ g}$$

Of this quantity of alcohol, 15% (15 g/100 g alcohol) enters the bloodstream. Thus, the mass of alcohol entering the bloodstream when two martinis are consumed is

$$2 \text{ martinis} \times \frac{42.6 \text{ g alcohol}}{1 \text{ martini}} \times \frac{15 \text{ g absorbed}}{100 \text{ g alcohol}} = 13 \text{ g}$$

Therefore, the concentration of alcohol in the person's bloodstream is:

$$\frac{13 \text{ g alcohol}}{7.0 \text{ L blood}} \times \frac{1 \text{ L}}{1000 \text{ mL}} \times \frac{1 \text{ mL}}{1 \text{ cm}^3} = \underline{\mathbf{0.0019 \text{ g/cm}^3}}$$

On the basis of this calculation, the person is legally intoxicated.

87. (a) The initial solution contains 49.92 g NaOH on 0.600 L of solution. Thus the molarity is:

$$\frac{49.92 \text{ g NaOH}}{0.600 \text{ L}} \times \frac{1 \text{ mol NaOH}}{40.00 \text{ g NaOH}} = \underline{\mathbf{2.08 \text{ M}}}$$

(b) To determine how much H_2 is produced, first we must establish whether Al or OH^- is the limiting reactant.

$$\text{mol } OH^- \text{ initially} = 49.92 \text{ g NaOH} \times \frac{1 \text{ mol NaOH}}{40.00 \text{ g NaOH}} \times \frac{1 \text{ mol } OH^-}{1 \text{ mol NaOH}} = 1.248 \text{ mol } OH^-$$

$$\text{mol Al initially} = 41.28 \text{ g Al} \times \frac{1 \text{ mol Al}}{26.98 \text{ g Al}} = 1.530 \text{ g Al}$$

$$\text{mol } H_2 \text{ produced if } OH^- \text{ is limiting} = 1.248 \text{ mol } OH^- \times \frac{3 \text{ mol } H_2}{2 \text{ mol } OH^-} = 1.872 \text{ mol } H_2$$

$$\text{mol } H_2 \text{ produced if Al is limiting} = 1.530 \text{ g Al} \times \frac{3 \text{ mol } H_2}{2 \text{ mol Al}} = 2.295 \text{ mol } H_2$$

Therefore OH^- is the limiting reactant and **1.872 mol H_2** were formed.

(c) This part of the problem is similar to Example 5.8. Recall that both $H_2(g)$ and $H_2O(g)$ contribute to the total pressure P_{tot} of the gas collected; thus the partial pressure of H_2 is given by:

$$P_{tot} = P_{H_2} + P_{H_2O} \quad \rightarrow \quad P_{H_2} = P_{tot} - P_{H_2O} = 758.6 \text{ mm Hg} - 23.8 \text{ mm Hg} = 734.8 \text{ mm Hg}$$

Now use the Ideal Gas Law to calculate the volume of H_2 produced.

$$V_{H_2} = \frac{nRT}{P} = \frac{(1.872 \text{ mol})(0.0821 \text{ L·atm/mol·K})(298 \text{ K})}{\left(734.8 \text{ mm Hg} \times \dfrac{1 \text{ atm}}{760 \text{ mm Hg}}\right)} = \underline{\mathbf{47.4\ L}}$$

88. We are given an aqueous solution of M_2X with a boiling point of 105.0°C. Recall that pure H_2O boils at 100.0°C and hence, using the equation below (page 261), calculate ΔT_b:

$$\Delta T_b = T_b - T_b° = (105.0 - 100.0) \text{ °C} = 5.0 \text{ °C}$$

Substituting this value for ΔT_b into Equation 10.7, with the appropriate k_b value, we can calculate the molality of the solution. (Our solute, M_2X, is ionic and will be expected to produce 3 mol of ions — two cations M and one anion X — per mol of M_2X formula units; hence $i = 3$.)

$$\Delta T_b = i \times 0.52 \text{ °C/}m \times (\text{molality}) \qquad \rightarrow \quad 5.0 \text{ °C} = 3 \times 0.52 \text{ °C/}m \times (\text{molality})$$

$$\rightarrow \quad \text{molality of the solution} = 3.2 \ m$$

For a solution with this molality, we know that there are 3.2 mol solute in 1 kg of water. To convert to mol fraction, construct a table similar to those on pages 253-254 and in Example 10.4. Enter mol solute and mass solvent in the table, and shade the spots needed for calculating mole fraction of solute.

	moles	$\xleftarrow{MM}$	mass	$\xrightarrow{density}$	volume
Solute	3.2				
Solvent			1 kg or 1000 g		
Solution					

Complete the table by filling it up with the required values:

$$\text{mol of solvent} = 1000 \text{ g } H_2O \times \frac{1 \text{ mol } H_2O}{18.02 \text{ g } H_2O} = 55.49 \text{ mol solvent}$$

$$\text{mol solution} = 3.2 \text{ mol solute} + 55.49 \text{ mol solvent} = 58.7 \text{ mol solution}$$

	moles	← $\xleftarrow{MM}$	mass	$\xrightarrow{density}$	volume
Solute	3.2				
Solvent	55.49		1 kg or 1000 g		
Solution	58.7				

Finally, use the values from the table to convert "3.2 *m*" to mol fraction solute:

$$\text{mole fraction solute} = \frac{\text{mol solute}}{\text{mol solution}} = \frac{3.2 \text{ mol}}{58.7 \text{ mol}} = \underline{\textbf{0.055}}$$

89. Henry's Law (Equation 10.2, $C_{gas} = kP_{gas}$) relates the solubility (concentration, C_{gas}) of a gas to the partial pressure of the gas, P_{gas}, above the solution; and recall that the concentration of the gas in solution is given by

$$C_{gas} = \frac{n_{gas}}{V_{solution}}$$

Thus n_{gas} (the number of moles of gas that dissolve in a solution of volume $V_{solution}$) can be written as:

$$n_{gas} = C_{gas} \times V_{solution} = (kP_{gas}) \times V_{solution}$$

Using the Ideal Gas Law ($P_{gas}V_{gas} = nRT$), n_{gas} can also be written as:

$$n_{gas} = \frac{P_{gas}V_{gas}}{RT}$$

Combining these two equations we obtain:

$$n_{gas} = \frac{P_{gas}V_{gas}}{RT} = (kP_{gas}) \times V_{solution} \qquad \rightarrow \qquad \frac{V_{gas}}{RT} = k \times V_{solution}$$

Therefore the volume of gas dissolving, $\mathbf{V_{gas}} = \mathbf{kRT} \times \mathbf{V_{solution}}$
and so V_{gas} depends only on the temperature and the volume of solution, **NOT on the partial pressure of the gas.**

RATE OF REACTION

11-1 MEANING OF REACTION RATE

2. This example is similar to Example 11.1(a). Use Equation 11.1 as a guide; recall that reaction rate is always positive.

(a) rate $= \dfrac{-\Delta[N_2O]}{2\Delta t}$ 　　　　(b) rate $= \dfrac{\Delta[O_2]}{\Delta t}$

11-1 MEANING OF REACTION RATE: Measurement of Rate

4. (a) At time = 0 min, there are 0.384 mol of AB_2 present. Then recall that, by applying Equation 11.1, for each time period Δt we obtain that the rate is given by

$$\text{rate} = \frac{-\Delta AB_2}{2\Delta t} = \frac{\Delta A_2}{\Delta t} \qquad \text{so} \qquad \Delta AB_2 = -2\Delta A_2$$

Applying this relationship to the given data (moles of A_2) allows the change in moles of AB_2 over the same time period to be calculated.

For example, between t = 0 and 10 min:
$$\Delta A_2 = (0.0541 - 0) \text{ mol} = 0.0541 \text{ mol} \quad \text{and so} \quad \Delta AB_2 = -2\Delta A_2 = -2(0.0541 \text{ mol})$$

thus, at t = 10 min:

$$\text{moles } AB_2 = 0.384 + \Delta AB_2 = 0.384 - 2(0.0541) = 0.276;$$

and so on to obtain the following table:

Time (min)	0	10	20	30	40	50
Moles AB_2	0.384	0.276	0.217	0.140	0.098	0.071

(b) in the second 10-minute interval (between 10 and 20 min):

$$\text{rate of disappearance of } AB_2 = \frac{\Delta AB_2}{\Delta t} = \frac{(0.217 - 0.276) \text{ mol}}{(20 - 10) \text{ min}} = \underline{-5.90 \times 10^{-3} \text{ mol/min}}$$

in the third 10-minute interval (between 20 and 30 min):

$$\text{rate of disappearance of } AB_2 = \frac{\Delta AB_2}{\Delta t} = \frac{(0.140 - 0.217) \text{ mol}}{(30 - 20) \text{ min}} = \underline{-7.70 \times 10^{-3} \text{ mol/min}}$$

over the second and third 10-minute intervals (between 10 and 30 min):

average rate of disappearance of AB_2

$$= \frac{\Delta AB_2}{\Delta t} = \frac{(0.140 - 0.276)\,\text{mol}}{(30 - 10)\,\text{min}} = \underline{-6.80 \times 10^{-3}\,\text{mol/min}}$$

(c) between 30 and 50 min:

$$\text{rate of appearance of } A_2 = \frac{\Delta A_2}{\Delta t} = \frac{(0.1567 - 0.1221)\,\text{mol}}{(50 - 30)\,\text{min}} = \underline{1.73 \times 10^{-3}\,\text{mol/min}}$$

6. Using Equation 11.1 as a guide, write equivalent expressions for the rate in terms of reactant and product concentrations:

$$\text{rate} = \frac{-\Delta[Br^-]}{5\Delta t} = \frac{-\Delta[BrO_3{}^-]}{\Delta t} = \frac{-\Delta[H^+]}{6\Delta t} = \frac{\Delta[Br_2]}{3\Delta t} = \frac{\Delta[H_2O]}{3\Delta t}$$

Since bromine, Br_2, is being formed at a rate of 0.029 mol/L·s, we can say that

$$\frac{\Delta[Br_2]}{\Delta t} = 0.029\,\text{mol/L·s}$$

(a) Use the above expressions for the rate in terms of $\Delta[Br_2]$ and $\Delta[H^+]$:

$$\frac{-\Delta[H^+]}{6\Delta t} = \frac{\Delta[Br_2]}{3\Delta t}$$

so $\dfrac{\Delta[H^+]}{\Delta t} = -\dfrac{6}{3}\dfrac{\Delta[Br_2]}{\Delta t} = -\dfrac{6}{3}(0.029\,\text{mol/L·s}) = -0.058\,\text{mol/L·s}$

H^+ is being *consumed*, and at a rate of **0.058 mol/L·s**

(b) Use the above expressions for the rate in terms of $\Delta[Br_2]$ and $\Delta[H_2O]$:

$$\frac{\Delta[Br_2]}{3\Delta t} = \frac{\Delta[H_2O]}{3\Delta t}$$

so $\dfrac{\Delta[H_2O]}{\Delta t} = \dfrac{\Delta[Br_2]}{\Delta t} = 0.029\,\text{mol/L·s}$

H_2O is being formed at a rate of **0.029 mol/L·s**

(c) Use the above expressions for the rate in terms of $\Delta[Br_2]$ and $\Delta[Br^-]$:

$$\frac{-\Delta[Br^-]}{5\Delta t} = \frac{\Delta[Br_2]}{3\Delta t}$$

so $\quad\dfrac{\Delta[Br^-]}{\Delta t} = -\dfrac{5}{3}\dfrac{\Delta[Br_2]}{\Delta t} = -\dfrac{5}{3}(0.029\ mol/L\cdot s) = -0.048\ mol/L\cdot s$

Br^- is being consumed at a rate of **0.048 mol/L·s**

8. (a) $N_2(g) + 3H_2(g) \rightarrow 2NH_3(g)$

(b) Use Equation 11.1 as a guide:

$$rate = \dfrac{\Delta[NH_3]}{2\Delta t}$$

(c) Use the equation from part (b) above; see also Example 11.1:

$$average\ rate = \dfrac{\Delta[NH_3]}{2\Delta t} = \dfrac{(0.815\ M - 0.257\ M)}{2(15\ min)} = 0.0186\ M/min\ or\ \textbf{0.0186 mol/L·min}$$

10. See Figure 11.3 and adjacent text.

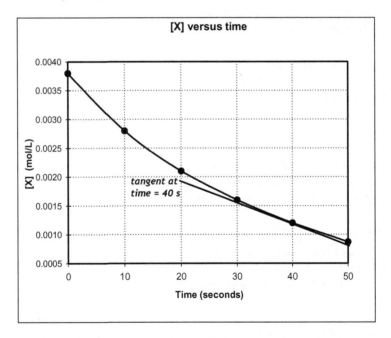

(a) The graph above shows the change in concentration of X over time.

(b) Choose two points that the tangent drawn in the graph above passes through: for example, [X] = 0.0019 *M* and 0.0008 *M* at *t* = 20 s and 50 s, respectively. Therefore the slope of the tangent is $\Delta[X]/\Delta t$ = (0.0008 − 0.0019)*M*/(50 − 20)s = −3.7 × 10^{-5} *M*/s. However, since X is a reactant, the instantaneous rate is −(slope of the tangent) = **3.7 × 10^{-5} *M*/s**.

(c) Using Equation 11.1, the average rate is given by:

$$\text{average rate} = \frac{-\Delta[X]}{\Delta t} = \frac{-(0.00087 - 0.0028)M}{(50 - 10)s} = \underline{\textbf{4.8} \times \textbf{10}^{-5} \textbf{ M/s}}$$

(d) **Instantaneous rate** (at 40 s) **is less than average rate** (over the 40-s interval).

11-2   REACTION RATE AND CONCENTRATION

12. The overall rate expression must be of the form

$$\text{rate} = k[P]^m[Q]^n$$

where: m = order of reaction with respect to P
          n = order of reaction with respect to Q
          $m + n$ = overall order of reaction.

Hence, by inspection:

(a) rate $= k_1 = k_1[P]^0[Q]^0$
   $m = 0$, so the reaction is **zero** order in P
   $n = 0$, so the reaction is **zero** order in Q
   $m + n = 0$, so the reaction is **zero** order overall

(b) rate $= k_2[P]^2[Q] = k_2[P]^2[Q]^1$
   $m = 2$, so the reaction is **second** order in P
   $n = 1$, so the reaction is **first** order in Q
   $m + n = 3$, so the reaction is **third** order overall

(c) rate $= k_3[Q]^2 = k_3[P]^0[Q]^2$
   $m = 0$, so the reaction is **zero** order in P
   $n = 2$, so the reaction is **second** order in Q
   $m + n = 2$, so the reaction is **second** order overall

(d) rate $= k_4[P][Q] = k_4[P]^1[Q]^1$
   $m = 1$, so the reaction is **first** order in P
   $n = 1$, so the reaction is **first** order in Q
   $m + n = 2$, so the reaction is **second** order overall

14. Recall that the units of [P] and [Q] must be mol/L. Since the units of reaction rate are always concentration over time, and given the form of the question, it makes sense to use units of reaction rate as mol/L·min. Substitute all of the units into the given rate expressions and solve for units of k.

(a) rate $= k_1$; or, in terms of units:

   $k_1 = \text{rate} = \underline{\textbf{mol/L·min}}$

(b) rate $= k_2[P]^2[Q]$ so $k_2 = \dfrac{\text{rate}}{[P]^2[Q]}$;   or, in terms of units:

$$k_2 = \dfrac{\text{mol/L}\cdot\text{min}}{(\text{mol/L})^2(\text{mol/L})} = \underline{\text{L}^2/\text{mol}^2\cdot\text{min}}$$

(c) rate $= k_3[Q]^2$ so $k_3 = \dfrac{\text{rate}}{[Q]^2}$;   or, in terms of units:

$$k_3 = \dfrac{\text{mol/L}\cdot\text{min}}{(\text{mol/L})^2} = \underline{\text{L/mol}\cdot\text{min}}$$

(d) rate $= k_4[P][Q]$ so $k_4 = \dfrac{\text{rate}}{[P][Q]}$;   or, in terms of units:

$$k_4 = \dfrac{\text{mol/L}\cdot\text{min}}{(\text{mol/L})(\text{mol/L})} = \underline{\text{L/mol}\cdot\text{min}}$$

16. (a) The graph indicates that the rate is independent of [Y], so it is **zero order with respect to Y.**

(b) **rate $= k[Y]^0$ or rate $= k$**

(c) Reading from the graph, it can be seen that for all [Y], the rate is around 0.045 M/h. Thus, since the rate expression states that rate $= k$, the value of k must be **0.045 M/h.**

18. The given reaction is first order in both X and Y. Thus, the rate expression must be:

rate $= k[X]^1[Y]^1$ or   rate $= k[X][Y]$

Use this expression to solve for the missing quantity in each case.

(a) rate $= k[X][Y] = (1.89 \text{ L/mol}\cdot\text{h})(0.100 \text{ mol/L})(0.400 \text{ mol/L}) = \underline{\textbf{0.0756 mol/L}\cdot\textbf{h}}$

(b) rate $= k[X][Y]$ so $[Y] = \dfrac{\text{rate}}{k[X]} = \dfrac{0.159 \text{ mol/L}\cdot\text{h}}{(0.884 \text{ L/mol}\cdot\text{h})(0.600 \text{ mol/L})} = \underline{\textbf{0.300 M}}$

(c) rate $= k[X][Y]$ so $[X] = \dfrac{\text{rate}}{k[Y]} = \dfrac{0.0479 \text{ mol/L}\cdot\text{h}}{(13.4 \text{ L/mol}\cdot\text{h})(0.250 \text{ mol/L})} = \underline{\textbf{0.0143 M}}$

(d) rate $= k[X][Y]$ so $k = \dfrac{\text{rate}}{[X][Y]} = \dfrac{0.00112 \text{ mol/L}\cdot\text{h}}{(0.600 \text{ mol/L})(0.233 \text{ mol/L})}$

$= \underline{\textbf{0.00801 L/mol}\cdot\textbf{h}}$

20. (a) The rate expression is: rate $= k[NH_3]^0$ or rate $= k$. However, since the value of the rate constant is given (k is 2.5×10^{-4} mol/L·min), the rate expression becomes:

rate = $\underline{2.5 \times 10^{-4} \text{ mol/L·min}}$

(b) The rate is independent of $[NH_3]$; and, as stated in part (a) above:

rate = $\underline{2.5 \times 10^{-4} \text{ mol/L·min}}$

(c) The reaction is zero order in NH_3 and the rate expression can be written as rate $= k$: this is true for all values of $[NH_3]$. Thus, **the rate equals the rate constant for all concentrations of NH_3.**

22. The rate expression is rate $= k[X][Y]^2$.

(a) Solve the rate expression for k and then substitute the given values to obtain the value of k:

$$\text{rate} = k[X][Y]^2 \qquad \text{so} \qquad k = \frac{\text{rate}}{[X][Y]^2}$$

$$k = \frac{0.00389 \text{ mol/L} \cdot \text{min}}{(0.150 \text{ mol/L})(0.0800 \text{ mol/L})^2} = \underline{4.05 \text{ L}^2/\text{mol}^2 \cdot \text{min}}$$

(b) Recall that k is constant, regardless of changes in $[X]$ or $[Y]$. Solve the rate expression for $[Y]$, then substitute the given values to obtain the value of $[Y]$:

$$\text{rate} = k[X][Y]^2 \qquad \text{so} \qquad [Y]^2 = \frac{\text{rate}}{k[X]} \qquad \text{and hence} \qquad [Y] = \left(\frac{\text{rate}}{k[X]}\right)^{1/2}$$

$$[Y] = \left(\frac{\text{rate}}{k[X]}\right)^{1/2} = \left(\frac{0.00948 \text{ mol/L} \cdot \text{min}}{(4.05 \text{ L}^2/\text{mol}^2 \cdot \text{min})(0.0441 \text{ mol/L})}\right)^{1/2} = 0.230 \text{ mol/L or } \underline{0.230 \text{ M}}$$

(c) Given that $[Y] = 2[X]$, substitute this into the rate expression. Solve the rate expression for $[X]$, then substitute the given values to obtain the value of $[X]$:

$$\text{rate} = k[X][Y]^2 = k[X](2[X])^2 = 4k[X]^3$$

$$[X]^3 = \frac{\text{rate}}{4k} \qquad \text{and hence} \qquad [X] = \left(\frac{\text{rate}}{4k}\right)^{1/3}$$

$$[X] = \left(\frac{\text{rate}}{4k}\right)^{1/3} = \left(\frac{0.0124 \text{ mol/L} \cdot \text{min}}{4(4.05 \text{ L}^2/\text{mol}^2 \cdot \text{min})}\right)^{1/3} = 0.0915 \text{ mol/L or } \underline{0.0915 \text{ M}}$$

11-3   REACTION CONCENTRATION AND TIME: Determination of Reaction Order

24. This problem is similar to Example 11.2.

    (a) A reaction involving the decomposition of Y will be of the form

$$Y \rightarrow products$$

Thus, Y is the only reactant and the rate expression will be of the form: rate = $k[Y]^m$. Choose any two experiments — for example the first two — and compare their initial rates and [Y]. For these first two experiments the rate expressions are:

$$rate_1 = k([Y]_1)^m \quad and \quad rate_2 = k([Y]_2)^m$$

Hence:

$$\frac{rate_1}{rate_2} = \frac{k([Y]_1)^m}{k([Y]_2)^m} = \left(\frac{[Y]_1}{[Y]_2}\right)^m$$

Calculate the ratio of rates: and the ratio of concentrations:

$$\frac{rate_1}{rate_2} = \frac{0.288 \, mol/L \cdot min}{0.245 \, mol/L \cdot min} = 1.18 \qquad \left(\frac{[Y]_1}{[Y]_2}\right)^m = \left(\frac{0.200 \, mol/L}{0.170 \, mol/L}\right)^m = 1.18^m$$

Finally, determine the value of m:

$$\frac{rate_1}{rate_2} = \left(\frac{[Y]_1}{[Y]_2}\right)^m \quad \rightarrow \quad 1.18 = 1.18^m \qquad \boldsymbol{m = 1}$$

The reaction is first order (m = 1).

    (b) rate = $k[Y]^1$ or   **rate = $k[Y]$**

    (c) Solve the rate expression for k and then substitute rate and [Y] from any of the given experiments:

$$rate = k[Y] \quad so \quad k = \frac{rate}{[Y]}$$

Using the values from the first experiment:

$$k = \frac{0.288 \, mol/L \cdot min}{0.200 \, mol/L} = \underline{\textbf{1.44 min}^{-1}}$$

26. This problem is similar to Example 11.3. Recall that the rate expression will be of the form:

$$rate = k[NO_2]^m[CO]^n$$

(a) To determine the reaction order with respect to NO_2, select two experiments in which $[NO_2]$ changes while $[CO]$ remains constant (such as experiments 1 and 3 used below).

$$\frac{rate_3}{rate_1} = \frac{k([NO_2]_3)^m([CO]_3)^n}{k([NO_2]_1)^m([CO]_1)^n} = \left(\frac{[NO_2]_3}{[NO_2]_1}\right)^m\left(\frac{[CO]_3}{[CO]_1}\right)^n = \left(\frac{[NO_2]_3}{[NO_2]_1}\right)^m$$

(since $[CO]_1 = [CO]_3$ and $1^n = 1$)

$$\frac{rate_3}{rate_1} = \frac{0.0226 \text{ mol/L·s}}{0.00565 \text{ mol/L·s}} = 4 \qquad \text{and} \qquad \left(\frac{[NO_2]_3}{[NO_2]_1}\right)^m = \left(\frac{0.276 \text{ mol/L}}{0.138 \text{ mol/L}}\right)^m = 2^m$$

thus $\quad 4 = 2^m \qquad$ giving $\quad m = 2$. The reaction is **second order in NO_2**.

To determine the reaction order with respect to CO, select two experiments in which $[CO]$ changes while $[NO_2]$ remains constant (such as experiments 3 and 4 used below).

$$\frac{rate_4}{rate_3} = \frac{k([NO_2]_4)^m([CO]_4)^n}{k([NO_2]_3)^m([CO]_3)^n} = \left(\frac{[NO_2]_4}{[NO_2]_3}\right)^m\left(\frac{[CO]_4}{[CO]_3}\right)^n = \left(\frac{[CO]_4}{[CO]_3}\right)^n$$

(since $[NO_2]_3 = [NO_2]_4$ and $1^m = 1$)

$$\frac{rate_4}{rate_3} = \frac{0.0226 \text{ mol/L·s}}{0.0226 \text{ mol/L·s}} = 1 \qquad \text{and} \qquad \left(\frac{[CO]_4}{[CO]_3}\right)^n = \left(\frac{0.300 \text{ mol/L}}{0.100 \text{ mol/L}}\right)^n = 3^n$$

thus $\quad 1 = 3^n \qquad$ giving $\quad n = 0$. The reaction is **zero order in CO**.

Overall reaction order $= m + n = 2 + 0 = 2$. The reaction is **second order overall**.

(b) **rate $= k[NO_2]^2$**

(c) To calculate k, solve the rate expression for k and then substitute rate and $[NO_2]$ from any of the given experiments:

$$\text{rate} = k[NO_2]^2 \qquad \text{so} \qquad k = \frac{\text{rate}}{[NO_2]^2}$$

Using the values for $[NO_2]$ and initial rate from experiment 1:

$$k = \frac{0.00565 \text{ mol/L·s}}{(0.138 \text{ mol/L})^2} = \underline{0.297 \text{ L/mol·s}}$$

(d) Use the given $[NO_2]$, and the value of k from part (c) above, in the rate expression:

$$\text{rate} = k[NO_2]^2 = (0.297 \text{ L/mol·s})(0.421 \text{ mol/L})^2 = \underline{0.0526 \text{ mol/L·s}}$$

28. This problem is similar to Example 11.3. Recall that the rate expression will be of the form:

$$\text{rate} = k[(C_2H_5)_2(NH)_2]^m[I_2]^n$$

(a) To determine the reaction order with respect to $(C_2H_5)_2(NH)_2$, select two experiments in which $[(C_2H_5)_2(NH)_2]$ changes while $[I_2]$ remains constant (such as experiments 3 and 4 used below).

$$\frac{\text{rate}_4}{\text{rate}_3} = \frac{k([(C_2H_5)_2(NH)_2]_4)^m([I_2]_4)^n}{k([(C_2H_5)_2(NH)_2]_3)^m([I_2]_3)^n} = \left(\frac{[(C_2H_5)_2(NH)_2]_4}{[(C_2H_5)_2(NH)_2]_3}\right)^m \left(\frac{[I_2]_4}{[I_2]_3}\right)^n$$

$$= \left(\frac{[(C_2H_5)_2(NH)_2]_4}{[(C_2H_5)_2(NH)_2]_3}\right)^m \qquad \text{(since } [I_2]_3 = [I_2]_4 \text{ and } 1^n = 1\text{)}$$

$$\frac{\text{rate}_4}{\text{rate}_3} = \frac{3.44 \times 10^{-4} \text{ mol/L} \cdot \text{h}}{2.30 \times 10^{-4} \text{ mol/L} \cdot \text{h}} = 1.50$$

$$\text{and} \quad \left(\frac{[(C_2H_5)_2(NH)_2]_4}{[(C_2H_5)_2(NH)_2]_3}\right)^m = \left(\frac{0.300 \text{ mol/L}}{0.200 \text{ mol/L}}\right)^m = 1.50^m$$

thus $\quad 1.50 = 1.50^m \qquad$ giving $\quad m = 1$. The reaction is **first order in $(C_2H_5)_2(NH)_2$**.

To determine the reaction order with respect to I_2, select two experiments in which $[I_2]$ changes while $[(C_2H_5)_2(NH)_2]$ remains constant (such as experiments 1 and 2 used below).

$$\frac{\text{rate}_2}{\text{rate}_1} = \frac{k([(C_2H_5)_2(NH)_2]_2)^m([I_2]_2)^n}{k([(C_2H_5)_2(NH)_2]_1)^m([I_2]_1)^n} = \left(\frac{[(C_2H_5)_2(NH)_2]_2}{[(C_2H_5)_2(NH)_2]_1}\right)^m \left(\frac{[I_2]_2}{[I_2]_1}\right)^n = \left(\frac{[I_2]_2}{[I_2]_1}\right)^n$$

$$\text{(since } [(C_2H_5)_2(NH)_2]_1 = [(C_2H_5)_2(NH)_2]_2 \text{ and } 1^m = 1\text{)}.$$

$$\frac{\text{rate}_2}{\text{rate}_1} = \frac{1.56 \times 10^{-4} \text{ mol/L} \cdot \text{h}}{1.08 \times 10^{-4} \text{ mol/L} \cdot \text{h}} = 1.44 \qquad \text{and} \quad \left(\frac{[I_2]_2}{[I_2]_1}\right)^n = \left(\frac{0.3620 \text{ mol/L}}{0.250 \text{ mol/L}}\right)^n = 1.45^n$$

thus $\quad 1.44 \approx 1.45 = 1.45^n \qquad$ giving $\quad n = 1$. The reaction is **first order in I_2**.

Overall reaction order $= m + n = 1 + 1 = 2$. The reaction is **second order overall**.

(b) **rate $= k[(C_2H_5)_2(NH)_2][I_2]$**

(c) To calculate k, solve the rate expression for k and then substitute rate and concentration from any of the given experiments:

$$\text{rate} = k[(C_2H_5)_2(NH)_2][I_2] \qquad \text{so} \qquad k = \frac{\text{rate}}{[(C_2H_5)_2(NH)_2][I_2]}$$

Using the values of concentration and initial rate from experiment 1:

$$k = \frac{1.08 \times 10^{-4} \text{ mol/L} \cdot \text{h}}{(0.150 \text{ mol/L})(0.250 \text{ mol/L})} = \underline{\mathbf{2.88 \times 10^{-3} \text{ L/mol} \cdot \text{h}}}$$

(d) Solve the rate expression for $[(C_2H_5)_2(NH)_2]$, then substitute the given $[I_2]$, along with the value of k from part (c) above, into the rate expression:

$$\text{rate} = k[(C_2H_5)_2(NH)_2][I_2] \qquad \text{so} \qquad [(C_2H_5)_2(NH)_2] = \frac{\text{rate}}{k[I_2]}$$

$$[(C_2H_5)_2(NH)_2] = \frac{5.00 \times 10^{-4} \text{ mol/L} \cdot \text{h}}{(2.88 \times 10^{-3} \text{ L/mol} \cdot \text{h})(0.500 \text{ mol/L})} = \mathbf{0.347 \text{ mol/L}} \text{ or } \underline{\mathbf{0.347 \text{ M}}}$$

30. This problem is similar to Example 11.3. Recall that the rate expression will be of the form:

$$\text{rate} = k[NO]^m[O_2]^n$$

(a) To determine the reaction order with respect to NO, select two experiments in which [NO] changes while $[O_2]$ remains constant (experiments 1 and 4 are used below).

$$\frac{\text{rate}_4}{\text{rate}_1} = \frac{k([NO]_4)^m([O_2]_4)^n}{k([NO]_1)^m([O_2]_1)^n} = \left(\frac{[NO]_4}{[NO]_1}\right)^m \left(\frac{[O_2]_4}{[O_2]_1}\right)^n = \left(\frac{[NO]_4}{[NO]_1}\right)^m$$

$$\text{(since } [O_2]_4 = [O_2]_1, \text{ and } 1^n = 1)$$

$$\frac{\text{rate}_4}{\text{rate}_1} = \frac{1.98 \times 10^{-2} \text{ mol/L} \cdot \text{min}}{2.20 \times 10^{-3} \text{ mol/L} \cdot \text{min}} = 9.00 \qquad \text{and} \qquad \left(\frac{[NO]_4}{[NO]_1}\right)^m = \left(\frac{0.0732 \text{ mol/L}}{0.0244 \text{ mol/L}}\right)^m = 3.00^m$$

thus $\quad 9.00 = 3.00^m \qquad$ giving $\quad m = 2$. The reaction is **second order in NO**.

To determine the reaction order with respect to O_2, select two experiments in which $[O_2]$ changes while [NO] remains constant (experiments 2 and 3 are used below).

$$\frac{\text{rate}_3}{\text{rate}_2} = \frac{k([NO]_3)^m([O_2]_3)^n}{k([NO]_2)^m([O_2]_2)^n} = \left(\frac{[NO]_3}{[NO]_2}\right)^m \left(\frac{[O_2]_3}{[O_2]_2}\right)^n = \left(\frac{[O_2]_3}{[O_2]_2}\right)^n$$

$$\text{(since } [NO]_2 = [NO]_3, \text{ and } 1^m = 1)$$

$$\frac{\text{rate}_3}{\text{rate}_2} = \frac{1.55 \times 10^{-3} \text{ mol/L} \cdot \text{min}}{7.23 \times 10^{-4} \text{ mol/L} \cdot \text{min}} = 2.14 \qquad \text{and} \qquad \left(\frac{[O_2]_3}{[O_2]_2}\right)^n = \left(\frac{0.0262 \text{ mol/L}}{0.0122 \text{ mol/L}}\right)^n = 2.15^n$$

thus $\quad 2.14 \approx 2.15 = 2.15^n \qquad$ giving $\quad n = 1$. The reaction is **first order in O_2**.

(b) rate $= k[NO]^2[O_2]^1 \qquad$ or $\quad$ **rate $= k[NO]^2[O_2]$**

(c) To calculate k, solve the rate expression for k and then substitute rate and concentration from any of the given experiments:

$$\text{rate} = k[NO]^2[O_2] \quad \text{so} \quad k = \frac{\text{rate}}{[NO]^2[O_2]}$$

Using the values of concentration and initial rate from experiment 1:

$$k = \frac{2.20 \times 10^{-3} \text{ mol/L} \cdot \text{min}}{(0.0244 \text{ mol/L})^2 (0.0372 \text{ mol/L})} = \textbf{99.3 L}^2\textbf{/mol}^2\textbf{·min}$$

(d) Use the rate expression from part (b) above, and substitute the given concentrations of NO and O_2, along with the value of k from part (c) above:

$$
\begin{aligned}
\text{rate} \quad &= \quad k[NO]^2[O_2] \\
&= \quad (99.3 \text{ L}^2/\text{mol}^2 \cdot \text{min})(0.0100 \text{ mol/L})^2(0.0462 \text{ mol/L}) \\
&= \quad \textbf{4.59} \times \textbf{10}^{-4} \textbf{ mol/L·s}
\end{aligned}
$$

32. This problem is similar to Example 11.3.

(a) Recall that the rate expression will be of the form

$$\text{rate} = k[CH_3CO_2CH_3]^m[OH^-]^n$$

To determine the reaction order with respect to $CH_3CO_2CH_3$, select two experiments in which $[CH_3CO_2CH_3]$ changes while $[OH^-]$ remains constant (experiments 2 and 3 are used below).

$$\frac{\text{rate}_3}{\text{rate}_2} = \frac{k([CH_3CO_2CH_3]_3)^m([OH^-]_3)^n}{k([CH_3CO_2CH_3]_2)^m([OH^-]_2)^n} = \left(\frac{[CH_3CO_2CH_3]_3}{[CH_3CO_2CH_3]_2}\right)^m \left(\frac{[OH^-]_3}{[OH^-]_2}\right)^n$$

$$= \left(\frac{[CH_3CO_2CH_3]_3}{[CH_3CO_2CH_3]_2}\right)^m \quad (\text{since } [OH^-]_2 = [OH^-]_3, \text{ and } 1^n = 1)$$

$$\frac{\text{rate}_3}{\text{rate}_2} = \frac{0.00340 \text{ mol/L} \cdot \text{s}}{0.00203 \text{ mol/L} \cdot \text{s}} = 1.67 \quad \text{and} \quad \left(\frac{[CH_3CO_2CH_3]_3}{[CH_3CO_2CH_3]_2}\right)^m = \left(\frac{0.084 \text{ mol/L}}{0.050 \text{ mol/L}}\right)^m = 1.68^m$$

thus $1.67 \approx 1.68 = 1.68^m$ giving $m = 1$. The reaction is **first order in CH$_3$CO$_2$CH$_3$**.

To determine the reaction order with respect to OH^-, select two experiments in which $[OH^-]$ changes while $[CH_3CO_2CH_3]$ remains constant (experiments 1 and 2 are used below).

$$\frac{\text{rate}_2}{\text{rate}_1} = \frac{k([CH_3CO_2CH_3]_2)^m([OH^-]_2)^n}{k([CH_3CO_2CH_3]_1)^m([OH^-]_1)^n} = \left(\frac{[CH_3CO_2CH_3]_2}{[CH_3CO_2CH_3]_1}\right)^m \left(\frac{[OH^-]_2}{[OH^-]_1}\right)^n = \left(\frac{[OH^-]_2}{[OH^-]_1}\right)^n$$

$$(\text{since } [CH_3CO_2CH_3]_1 = [CH_3CO_2CH_3]_2, \text{ and } 1^m = 1)$$

$$\frac{rate_2}{rate_1} = \frac{0.00203 \text{ mol/L} \cdot s}{0.00158 \text{ mol/L} \cdot s} = 1.28 \quad \text{and} \quad \left(\frac{[OH^-]_2}{[OH^-]_1}\right)^n = \left(\frac{0.154 \text{ mol/L}}{0.120 \text{ mol/L}}\right)^n = 1.28^m$$

thus $\quad 1.28 = 1.28^n \quad$ giving $\quad n = 1$. The reaction is **first order in OH⁻**.

Therefore, the overall rate expression is:

$$rate = k[CH_3CO_2CH_3]^1[OH^-]^1 \quad \text{or} \quad \textbf{rate} = \textbf{\textit{k}[CH}_3\textbf{CO}_2\textbf{CH}_3\textbf{][OH}^-\textbf{]}$$

(b) To calculate k, solve the rate expression for k and then substitute rate and concentration from any of the given experiments:

$$rate = k[CH_3CO_2CH_3][OH^-] \quad \text{so} \quad k = \frac{rate}{[CH_3CO_2CH_3][OH^-]}$$

Using the values of concentration and initial rate from experiment 1:

$$k = \frac{0.00158 \text{ mol/L} \cdot s}{(0.050 \text{ mol/L})(0.120 \text{ mol/L})} = \underline{\textbf{0.26 L/mol·s}}$$

(c) In order to use the rate expression, it is first necessary to calculate $[CH_3CO_2CH_3]$ and $[OH^-]$; then use these concentrations, along with k from part (b) above, to calculate rate. Note that the total volume of the solution is $(75.00 + 10.00) \text{ mL} = 85.00 \text{ mL}$.

(i) concentration of $CH_3CO_2CH_3$:

Strategy: vol $CH_3CO_2CH_3 \rightarrow$ mass $CH_3CO_2CH_3 \rightarrow$ mol $CH_3CO_2CH_3 \rightarrow [CH_3CO_2CH_3]$

$$\text{mol } CH_3CO_2CH_3 = 10.00 \text{ mL } CH_3CO_2CH_3 \times \frac{0.932 \text{ g } CH_3CO_2CH_3}{1 \text{ mL } CH_3CO_2CH_3} \times \frac{1 \text{ mol } CH_3CO_2CH_3}{74.080 \text{ g } CH_3CO_2CH_3}$$

$$= 0.126 \text{ mol } CH_3CO_2CH_3$$

$$[CH_3CO_2CH_3] = \frac{0.126 \text{ mol } CH_3CO_2CH_3}{85.00 \text{ mL} \times \dfrac{1 L}{1000 \text{ mL}}} = 1.48 \text{ mol/L}$$

(ii) concentration of OH^-:

Strategy: mol $OH^- \rightarrow [OH^-]$

$$\text{mol } OH^- \text{ added} = 75.00 \text{ mL solution} \times \frac{1 L}{1000 \text{ mL}} \times \frac{1.50 \text{ mol NaOH}}{1 L \text{ solution}} \times \frac{1 \text{ mol } OH^-}{1 \text{ mol NaOH}}$$

$$= 0.112 \text{ mol } OH^-$$

$$[OH^-] = \frac{0.112 \text{ mol } OH^-}{85.00 \text{ mL} \times \dfrac{1 L}{1000 \text{ mL}}} = 1.32 \text{ mol/L}$$

rate = $k[CH_3CO_2CH_3][OH^-]$ = (0.263 L/mol·s)(1.48 mol/L)(1.32 mol/L)

= **0.51 mol/L·s**

34. This problem is similar to Example 11.6. Prepare a table listing [Q], ln[Q], and 1/[Q] as a function of time from the given data, then make a plot of each concentration—time relationship. Use Table 11.2 in the textbook to correlate order of reaction with the straight-line plot that you obtain.

time (min)	[Q]	ln[Q]	1/[Q]
0	0.334	−1.100	2.99
4	0.25	−1.39	4.0
8	0.20	−1.61	5.0
12	0.167	−1.790	5.99
16	0.143	−1.945	6.99

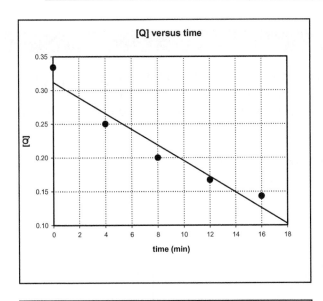

Test for a zero-order reaction: plot of [Q] versus time

These data points (●) do not lie close to the straight line plotted through them. This is not a linear graph, so the reaction is not zero order.

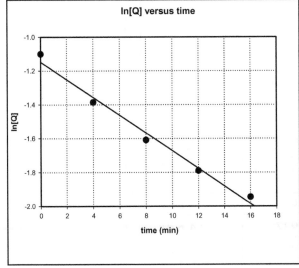

Test for a first-order reaction: plot of ln[Q] versus time

Again, note how the data points curve around the straight line plotted through them. This graph also is not linear, so the reaction is not first order.

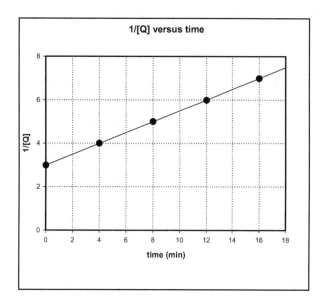

Test for a second-order reaction: plot of 1/[Q] versus time

These data points are well represented by the straight line drawn through them. This graph is linear and the reaction is second order.

The rate law will be:

$$\text{rate} = k[Q]^2$$

11-3   REACTION CONCENTRATION AND TIME: First-Order Reactions

36. (a) See Figure 11.6, Table 11.2, and Examples 11.5 and 11.6: a plot of $\ln[SO_2Cl_2]$ against time will give a straight line if the reaction is first order. Such a plot is shown below, followed by the data used. The graph is clearly linear — the data points all lie on the line drawn through them, so the reaction is indeed first order.

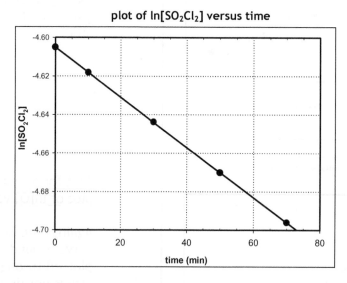

plot of $\ln[SO_2Cl_2]$ versus time

time (min)	$[SO_2Cl_2]$	$\ln[SO_2Cl_2]$
0	0.0100	−4.605
10	0.00987	−4.618
30	0.00962	−4.644
50	0.00937	−4.670
70	0.00913	−4.696

(b) The slope of a plot such as the one in part (a) above is $-k$ (Figure 11.6). Choose two points on the line to determine the slope, such as where it meets the x axis ($y_2 = -4.700$ (to nearest 0.005), $x_2 = 73$ min) and the y-axis ($y_1 = -4.605$ (to nearest 0.005), $x_1 = 0$ min). Thus:

$$k = -(\text{slope}) = -\left(\frac{y_2 - y_1}{x_2 - x_1}\right) = -\left(\frac{-4.700 - (-4.605)}{(73 - 0)\,\text{min}}\right) = \underline{1.3 \times 10^{-3}\,\text{min}^{-1}}$$

(c) Use Equation 11.3, along with the value of k from part (b) above:

$$t_{\frac{1}{2}} = \frac{\ln 2}{k} = \frac{\ln 2}{1.3 \times 10^{-3}\,\text{min}^{-1}} = \underline{5.3 \times 10^2\,\text{min}}$$

(d) Use Equation 11.2a, along with one of the given data pairs — for example, $[SO_2Cl_2] = 0.0100$ M when time = 0 min. Call this value $[SO_2Cl_2]_0$ and then calculate t, the time required for $[SO_2Cl_2]$ to fall from 0.0100 M to 0.00427 M:

$$\ln\frac{[SO_2Cl_2]_0}{[SO_2Cl_2]} = kt$$

so:

$$t = \frac{1}{k}\ln\frac{[SO_2Cl_2]_0}{[SO_2Cl_2]} = \frac{1}{1.3 \times 10^{-3}\,\text{min}^{-1}}\ln\frac{0.0100\,M}{0.00427\,M} = \underline{6.5 \times 10^2\,\text{min}}$$

This value of t is the time required for $[SO_2Cl_2]$ to fall from 0.0100 M (at time = 0 min) to 0.00427 M; therefore, the time when $[SO_2Cl_2] = 0.00427$ M is $(0 + 6.5 \times 10^2)$ min = **6.5 × 10² min**.

(e) Use the rate expression, along with the known value of k and the given $[SO_2Cl_2]$. Recall that we know the reaction to be first order and hence the rate expression is:

$$\text{rate} = k[SO_2Cl_2]$$
$$= (1.3 \times 10^{-3}\,\text{min}^{-1})(0.0153\,M)$$
$$= 2.0 \times 10^{-5}\,M/\text{min} \quad \text{or} \quad \underline{2.0 \times 10^{-5}\,\text{mol/L·min}}$$

38. This problem is related to Example 11.4, except that the value of k must first be calculated.

(a) Use Equation 11.2a, along with the given concentration-time data. The first concentration value will serve as $[COCl_2]_0 = 0.0450$ M, while the second value, $[COCl_2] = 0.0200$ M, occurs after a time interval t.

$$t = (500 - 300)\,s = 200\,s$$

Now substitute the value of t and the two concentrations into Equation 11.2, and solve for k:

$$\ln\frac{[COCl_2]_0}{[COCl_2]} = kt \qquad \text{so:} \qquad k = \frac{1}{t}\ln\frac{[COCl_2]_0}{[COCl_2]} = \frac{1}{200\ s}\ln\frac{0.0450\ M}{0.0200\ M} = \underline{\textbf{4.05} \times \textbf{10}^{-3}\ \textbf{s}^{-1}}$$

(b) Use Equation 11.3, along with the value of k calculated in part (a) above.

$$t_{1/2} = \frac{\ln2}{k} = \frac{\ln2}{4.05\times10^{-3}\ s^{-1}} = \underline{\textbf{171 s}}$$

(c) Use Equation 11.2a, along with the value of k calculated in part (a) above. The data given in the question, $[COCl_2]$ = 0.0450 M at t = 300 s, can be substituted into Equation 11.2a with the equation solved for the initial concentration, $[COCl_2]_0$.

$$\ln\frac{[COCl_2]_0}{[COCl_2]} = kt \qquad \text{so:} \qquad \frac{[COCl_2]_0}{[COCl_2]} = e^{kt}$$

and hence:

$$[COCl_2]_0 = [COCl_2]e^{kt} = (0.0450\ M)e^{(4.05\times10^{-3}\ s^{-1})(300\ s)} = \underline{\textbf{0.152 M}}$$

40. This problem is related to Example 11.4. Note that, since the volume of solution can be assumed to be constant and the molar mass of the drug is constant, the mass of the drug can be used directly as a measure of concentration.

(a) Use Equation 11.2a, with the initial quantity of drug = $[drug]_0$ = 10.0 g:

$$\ln\frac{[drug]_0}{[drug]} = kt \qquad \text{so:} \qquad \frac{[drug]_0}{[drug]} = e^{kt} \qquad \text{and} \qquad [drug] = [drug]_0 e^{-kt}$$

Hence the concentration (mass) of drug remaining after one year (12 months) is:

$$[drug] = [drug]_0 e^{-kt} = (10.0\ g)e^{-(0.215\ month^{-1})(12\ month)} = \underline{\textbf{0.758 g}}$$

(b) Use Equation 11.3, along with the value of k given in the question.

$$t_{1/2} = \frac{\ln2}{k} = \frac{\ln2}{0.215\ month^{-1}} = \underline{\textbf{3.22 month}}$$

(c) Use Equation 11.2a: If 65% of the drug has decomposed, then 35% of the initial quantity must remain. Therefore, if the initial quantity of drug is labeled $[drug]_0$, then the quantity remaining is 0.35 × $[drug]_0$. Substitute both of these values into Equation 11.2a and solve for t:

$$\ln\frac{[drug]_0}{[drug]} = kt \qquad \text{so:} \qquad t = \frac{1}{k}\ln\frac{[drug]_0}{[drug]}$$

Hence:

$$t = \frac{1}{0.215 \text{ month}^{-1}} \ln \frac{[\text{drug}]_0}{0.35[\text{drug}]_0} = \frac{1}{0.215 \text{ month}^{-1}} \ln \frac{1}{0.35} = 4.9 \text{ month}$$

It takes **4.9 months** for 65% of the drug to decompose.

42. Part of this problem is related to Example 11.4.

$$SO_2Cl_2 \rightarrow SO_2 + Cl_2$$

The above decomposition of SO_2Cl_2 is a first order process and therefore the rate expression will be of the form:

rate = $k[SO_2Cl_2]$

(a) Use Equation 11.2a, with the initial concentration $[SO_2Cl_2]_0 = 0.0714$ M, final concentration $[SO_2Cl_2] = 0.0681$ M, and the time interval $t = 1.43$ h; then solve for k:

$$\ln\frac{[SO_2Cl_2]_0}{[SO_2Cl_2]} = kt \qquad \text{so:} \quad k = \frac{1}{t}\ln\frac{[SO_2Cl_2]_0}{[SO_2Cl_2]}$$

Hence the rate constant k is:

$$k = \frac{1}{t}\ln\frac{[SO_2Cl_2]_0}{[SO_2Cl_2]} = \frac{1}{1.43 \text{ h}}\ln\frac{0.0714\ M}{0.0681\ M} = \underline{\mathbf{0.0331\ h^{-1}}}$$

(b) Use the rate expression: substitute the given concentration $[SO_2Cl_2] = 0.0462$ M, along with the value of k calculated in (a) above:

rate = $k[SO_2Cl_2]$
     = $(0.0331 \text{ h}^{-1})(0.0462\ M)$ = 1.53×10^{-3} M/min or **1.53×10^{-3} mol/L·min**

(c) Use Equation 11.2a: If 45% of a sample of SO_2Cl_2 remains during decomposition, then $[SO_2Cl_2] = 0.45 \times [SO_2Cl_2]_0$. Substitute both of these values, plus the value for k, into Equation 11.2a and solve for t:

$$\ln\frac{[SO_2Cl_2]_0}{[SO_2Cl_2]} = kt \qquad \text{so:} \quad t = \frac{1}{k}\ln\frac{[SO_2Cl_2]_0}{[SO_2Cl_2]}$$

Hence:

$$t = \frac{1}{k}\ln\frac{[SO_2Cl_2]_0}{[SO_2Cl_2]} = \frac{1}{k}\ln\frac{[SO_2Cl_2]_0}{0.45 \times [SO_2Cl_2]} = \frac{1}{0.0331 \text{ h}^{-1}}\ln\frac{1}{0.45} = \underline{\mathbf{2.4 \times 10^1 \text{ h}}}$$

44. The hydrolysis of sucrose is a first order process and therefore the half-life can be used to calculate the rate constant k using Equation 11.3:

$$t_{1/2} = \frac{\ln 2}{k} \qquad \text{so} \quad k = \frac{\ln 2}{t_{1/2}} = \frac{\ln 2}{64.2 \text{ min}} = \mathbf{1.08 \times 10^{-2} \text{ min}^{-1}}$$

We are required to find the mass (grams) of sucrose that are hydrolyzed in 1.73 h. Given the initial concentration [sucrose]$_0$ = 0.389 *M*, we can use Equation 11.2a to calculate [sucrose] after 1.73 h, and from this we can determine the change in concentration and hence the mass of sucrose hydrolyzed.

$$t = 1.73 \text{ h} = 1.73 \text{ h} \times \frac{60 \text{ min}}{1 \text{ h}} = 104 \text{ min}$$

$$\ln\frac{[\text{sucrose}]_0}{[\text{sucrose}]} = kt \qquad \text{so:} \qquad \frac{[\text{sucrose}]_0}{[\text{sucrose}]} = e^{kt} \qquad \text{and} \qquad [\text{sucrose}] = [\text{sucrose}]_0 e^{-kt}$$

Hence the concentration of sucrose after 1.73 h (104 min) is:

$$[\text{sucrose}] = (0.389 \text{ } M)e^{-(1.08 \times 10^{-2} \text{ min}^{-1})(104 \text{ min})} = 0.127 \text{ } M$$

Thus, the change in concentration of sucrose is (0.389 − 0.127)*M* = **0.262 *M*** or 0.262 mol/L. Recall that this change in concentration is due to the sucrose being hydrolyzed: therefore, 0.262 mol of sucrose were hydrolyzed per L of solution.

Finally, use the given volume and the change in concentration to determine the mol of sucrose hydrolyzed and hence the mass hydrolyzed.

Strategy: V solution → mol sucrose → mass sucrose

mass of sucrose hydrolyzed =

$$1.25 \text{ L solution} \times \frac{0.262 \text{ mol sucrose}}{1 \text{ L solution}} \times \frac{342.35 \text{ g sucrose}}{1 \text{ mol sucrose}} = \underline{\mathbf{112 \text{ g sucrose}}}$$

46. This problem is similar to Example 11.5(b). The half-life must first be used to calculate *k* using Equation 11.3:

$$t_{1/2} = \frac{\ln 2}{k} \qquad \text{so} \qquad k = \frac{\ln 2}{t_{1/2}} = \frac{\ln 2}{9.7 \text{ d}} = \mathbf{7.1 \times 10^{-2} \text{ d}^{-1}}$$

The initial quantity of [Cs-131]$_0$ is 20.0 mg, and we are required to find *t* such that [Cs-131] has fallen to 33% of its initial value. Thus, [Cs-131] = 33% of [Cs-131]$_0$ or [Cs-131] = 0.33 × 20.0 mg = 6.6 mg. Now use Equation 11.2a and solve for the required *t*:

$$\ln\frac{[\text{Cs-131}]_0}{[\text{Cs-131}]} = kt \qquad \text{so:} \qquad t = \frac{1}{k}\ln\frac{[\text{Cs-131}]_0}{[\text{Cs-131}]} = \frac{1}{7.1 \times 10^{-2} \text{ d}^{-1}}\ln\frac{20.0 \text{ mg}}{6.6 \text{ mg}} = \underline{\mathbf{16 \text{ d}}}$$

48. This problem is similar to Example 11.5(a). The value of k must first be obtained in order to calculate the half-life. Use Equation 11.2a with the given values: $[Na\text{-}24]_0 = 0.050$ mg, $[Na\text{-}24] = 0.016$ mg, and $t = 24.9$ h; and then solve for k.

$$\ln\frac{[Na\text{-}24]_0}{[Na\text{-}24]} = kt \qquad \text{so:} \quad k = \frac{1}{t}\ln\frac{[Na\text{-}24]_0}{[Na\text{-}24]} = \frac{1}{24.9\,\text{h}}\ln\frac{0.050\,\text{mg}}{0.016\,\text{mg}} = 4.6 \times 10^{-2}\,\text{h}^{-1}$$

Now use Equation 11.3 to calculate the half life:

$$t_{\frac{1}{2}} = \frac{\ln 2}{k} = \frac{\ln 2}{4.6 \times 10^{-2}\,\text{h}^{-1}} = \underline{\textbf{15 h}}$$

11-3 REACTION CONCENTRATION AND TIME: Zero- and Second-Order Reactions

50. (a) Use Equation 11.4, with initial concentration $[R]_0 = 739$ mg and $t = 128$ min. Since 41.0% of R has decomposed, $(100 - 41.0)\% = 59.0\%$ remains: thus the concentration at $t = 128$ min is $[R] = 59.0\% \times [R]_0 = 0.590 \times 739$ mg $= 436$ mg.

$$[R] = [R]_0 - kt \qquad \text{so:} \quad k = \frac{1}{t}([R]_0 - [R]) = \frac{1}{128\,\text{min}}(739 - 436)\text{mg} = \underline{\textbf{2.37 mg/min}}$$

   (b) From Table 11.2, the half-life of a zero-order reaction is given by:

$$t_{\frac{1}{2}} = \frac{[A]_0}{2k}$$

Note that for a zero-order reaction, the half life depends on the initial concentration. Use this equation with $[R]_0 = 739$ mg, and the value of k calculated in part (a) above:

$$t_{\frac{1}{2}} = \frac{[A]_0}{2k} = \frac{739\,\text{mg}}{2(2.37\,\text{mg/min})} = \underline{\textbf{156 min}}$$

   (c) The rate expression for a zero-order reaction is of the form

$$\text{Rate} = k[R]^0 = k$$

in other words, the rate equals the rate constant k. Thus, the rate of decomposition of 739 mg is $k = \underline{\textbf{2.37 mg/min}}$.

   (d) Recall that the rate expression for a zero-order reaction is simply **Rate** $= k$ and hence the rate is independent of initial concentration. Thus, the rate of decomposition of 1.25 g is still $k = \underline{\textbf{2.37 mg/min}}$.

---

52. (a) From Table 11.2, the half-life of a zero-order reaction is given by:

$$t_{\frac{1}{2}} = \frac{[A]_0}{2k}$$

Use this equation with $[NH_3]_0 = 0.250\ M$, and the value of k given in the question:

$$t_{\frac{1}{2}} = \frac{[NH_3]_0}{2k} = \frac{0.250\ mol/L}{2(2.08 \times 10^{-4}\ mol/L \cdot s)} = \underline{\mathbf{601\ s}}$$

or:

$$t_{\frac{1}{2}} = 601\ s \times \frac{1\,min}{60\ s} = \underline{\mathbf{10.0\ min}}$$

(b) Use Equation 11.4, with initial concentration $[NH_3]_0 = 1.25\ M$, concentration at time t, $[NH_3] = 0.388\ M$, and the value of k given in the question.

$$[R] = [R]_0 - kt \qquad so: \quad t = \frac{1}{k}([R]_0 - [R])$$

$$t = \frac{1}{(2.08 \times 10^{-4}\ mol/L \cdot s)}(1.25 - 0.388)mol/L = \underline{\mathbf{4.14 \times 10^3\ s}}$$

or:

$$t = 4.14 \times 10^3\ s \times \frac{1\,min}{60\ s} \times \frac{1\,h}{60\ min} = \underline{\mathbf{1.2\ h}}$$

54. (a) Use Equation 11.5, with initial concentration $[C_4H_6]_0 = 0.350\ M$, and $[C_4H_6] = 0.197\ M$ at time $t = 145\ s$.

$$\frac{1}{[A]} - \frac{1}{[A]_0} = kt \qquad so: \quad k = \frac{1}{t}\left(\frac{1}{[A]} - \frac{1}{[A]_0}\right)$$

$$k = \frac{1}{145\ s}\left(\frac{1}{0.197\ M} - \frac{1}{0.350\ M}\right) = \underline{\mathbf{0.0153\ L/mol \cdot s}}$$

(b) From Table 11.2, the half-life of a second-order reaction is given by:

$$t_{\frac{1}{2}} = \frac{1}{k[A]_0}$$

Note that for a second-order reaction, half life depends on initial concentration. Use this equation with $[C_4H_6]_0 = 0.200\ M$, and the value of k calculated in part (a) above:

$$t_{\frac{1}{2}} = \frac{1}{k[A]_0} = \frac{1}{(0.0153\ L/mol \cdot s)(0.200\ mol/L)} = \underline{\mathbf{327\ s}}$$

(c) Use Equation 11.5, with initial concentration $[C_4H_6]_0 = 0.558\ M$ and the value of k calculated above. Since 28.9% of the sample has dimerized, $(100 - 28.9)\% = 71.1\%$ remains: thus the concentration at time t is $[C_4H_6] = 71.1\% \times [C_4H_6]_0 = 0.711 \times 0.558\ M = 0.397\ M$.

$$\frac{1}{[A]} - \frac{1}{[A]_0} = kt \qquad \text{so:} \qquad t = \frac{1}{k}\left(\frac{1}{[A]} - \frac{1}{[A]_0}\right)$$

$$t = \frac{1}{0.0153\ \text{L/mol} \cdot \text{s}}\left(\frac{1}{0.397\ M} - \frac{1}{0.558\ M}\right) = \underline{47.5\ \text{s}}$$

(d) The rate expression for this second-order reaction will be of the form

$$\text{Rate} = k[C_4H_6]^2$$

Use this expression with the value of k calculated in part (a) above and $[C_4H_6] = 0.128\ M$.

$$\text{Rate} = k[C_4H_6]^2 = (0.0153\ \text{L/mol}\cdot\text{s})(0.128\ \text{mol/L})^2 = \underline{2.51 \times 10^{-4}\ \text{mol/L}\cdot\text{s}}$$

56. Use Equation 11.5, with initial concentration $[NOCl]_0 = 0.300\ M$ and $t = 0.20$ min. Since 15.0% of the solution of NOCl has decomposed, $(100 - 15.0)\% = 85.0\%$ remains: thus the concentration at $t = 0.20$ min is $[NOCl] = 85.0\% \times [NOCl]_0 = 0.850 \times 0.300\ M = 0.255\ M$.

$$\frac{1}{[NOCl]} - \frac{1}{[NOCl]_0} = kt \qquad \text{so:} \qquad k = \frac{1}{t}\left(\frac{1}{[NOCl]} - \frac{1}{[NOCl]_0}\right)$$

$$k = \frac{1}{0.20\ \text{min}}\left(\frac{1}{0.255\ M} - \frac{1}{0.300\ M}\right) = \underline{2.9\ \text{L/mol}\cdot\text{min}}$$

11-5   REACTION RATE AND TEMPERATURE

58. Use Equation 11.7b: recall that activation energy must be expressed in J/mol and temperatures in K. Thus $T_1 = 300$ K, $T_2 = 342$ K, and $E_a = 9.13 \times 10^3$ J/mol.

$$\ln\frac{k_2}{k_1} = \frac{E_a}{R}\left[\frac{1}{T_1} - \frac{1}{T_2}\right] = \frac{9.13 \times 10^3\ \text{J/mol}}{8.3145\ \text{J/mol}\cdot\text{K}} \times \left[\frac{1}{300\ \text{K}} - \frac{1}{342\ \text{K}}\right] = 0.450$$

$$\frac{k_2}{k_1} = e^{0.450} = 1.57 \qquad \text{so:}\ k_2 = 1.57 k_1$$

Thus, the value of the rate constant at the higher temperature, k_2, is 1.57 times the rate constant at the lower temperature, k_1. We are asked for the *increase* in the rate constant, which is given by the difference between k_2 and k_1:

$$\text{increase in rate constant} = k_2 - k_1 = 1.57k_1 - k_1 = 0.57k_1$$

thus the fractional increase is $0.57k_1/k_1 = 0.57$ or **57%**

60. See Figure 11.13 and adjacent text: a plot of ln$k$ versus $1/T$ gives a straight line of slope $-E_a/R$. Recall that the units of T should be K.

k (s^{-1})	0.048	2.3	49	590
ln$k$	−3.04	0.833	3.89	6.38
T (K)	773	873	973	1073
$1/T$ (K^{-1})	1.29×10^{-3}	1.15×10^{-3}	1.03×10^{-3}	9.32×10^{-4}

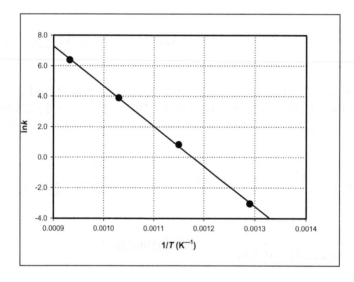

Choose two points on the line to determine the slope, such as where it meets the *x* axis ($y_2 = -4.0$, $x_2 = 1.33 \times 10^{-3}$ K^{-1}) and the *y*-axis ($y_1 = +7.2$, $x_1 = 0.90 \times 10^{-3}$ K^{-1}).

Thus:

$$-\frac{E_a}{R} = \text{slope} = \left(\frac{y_2 - y_1}{x_2 - x_1}\right) = \left(\frac{-4.0 - (7.2)}{(1.33 \times 10^{-3} - 0.90 \times 10^{-3})\,\text{K}^{-1}}\right) = -2.6 \times 10^4 \text{ K}$$

Hence:

$$E_a = -R \times (\text{slope}) = -(8.3145 \text{ J/mol·K})(-2.6 \times 10^4 \text{ K}) = 2.2 \times 10^5 \text{ J/mol}$$

Thus: $E_a = 2.2 \times 10^5 \text{ J/mol} \times \dfrac{1\,\text{kJ}}{1000\,\text{J}} = \underline{\textbf{2.2} \times \textbf{10}^2 \textbf{ kJ/mol}}$

62. See Section 11.4 and in particular Figure 11.10 and the adjacent text. The value of Δ*H* corresponds to the difference in energy between reactants and products; and the activation energy 284 kJ is the energy that must be put into the system in order to form the activated complex. Note that this reaction–energy diagram differs from the one in Figure 11.10 because Δ*H* is positive: the products are *higher* in energy than the reactants.

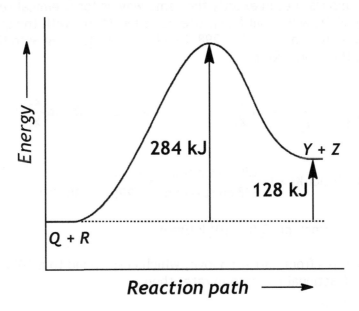

64. This Problem is similar to Problem 58 above. Use Equation 11.7b: recall that activation energy must be expressed in J/mol and temperatures in K. Let the rate constant at 100 °C be k_1 and the rate constant at 92 °C be k_2. The corresponding Kelvin temperatures are T_1 = 373 K, T_2 = 365 K, and the activation energy is E_a = 52.0 kJ/mol = 5.20 × 10^4 J/mol.

$$\ln\frac{k_2}{k_1} = \frac{E_a}{R}\left[\frac{1}{T_1} - \frac{1}{T_2}\right] = \frac{5.2 \times 10^4 \text{ J/mol}}{8.3145 \text{ J/mol} \cdot \text{K}} \times \left[\frac{1}{373 \text{ K}} - \frac{1}{365 \text{ K}}\right] = -0.37$$

$$\frac{k_2}{k_1} = e^{-0.37} = 0.69 \qquad \text{so: } k_2 = 0.69k_1$$

Thus, the rate constant at the lower temperature, k_2, is 0.69 times the rate constant at the higher temperature, k_1. We are asked for the percentage *decrease* in the rate constant, which is given by the difference between k_2 and k_1 expressed as a percentage:

decrease in rate constant = $k_1 - k_2 = k_1 - 0.69k_1 = 0.31k_1$

thus, the fractional decrease is $0.31k_1/k_1 = 0.31$ or **31%**

66. (a) Use the given equation with the temperatures t = 25°C and 35°C:

rate of chirping at t = 25°C: rate of chirping at t = 35°C:

$$X_1 = 7.2(25) - 32 \qquad\qquad X_2 = 7.2(35) - 32$$

$$= \underline{\textbf{148 chirps/min}} \qquad\qquad = \underline{\textbf{220 chirps/min}}$$

(b) Assume that the rate of chirping is proportional to the rate constant k for the process at a given temperature, in exactly the same way as for chemical reactions. If that is so, the ratio X_2/X_1 will equal k_2/k_1. Use equation 11.7b, with the given temperatures t expressed in Kelvin, namely $T_1 = 298$ K and $T_2 = 308$ K, along with the corresponding values of X; then solve for E_a:

$$\ln\frac{k_2}{k_1} = \frac{E_a}{R}\left[\frac{1}{T_1} - \frac{1}{T_2}\right] = \ln\frac{X_2}{X_1} \qquad\qquad \text{so:} \quad E_a = R \times \ln\frac{X_2}{X_1} \times \left[\frac{1}{T_1} - \frac{1}{T_2}\right]^{-1}$$

$$E_a = (8.3145 \text{ J/mol·K}) \times \ln\frac{220 \text{ chirps/min}}{148 \text{ chirps/min}} \times \left[\frac{1}{298 \text{ K}} - \frac{1}{308 \text{ K}}\right]^{-1}$$

$$= 3.0 \times 10^4 \text{ J/mol or } \underline{\textbf{3.0} \times \textbf{10}^1 \textbf{ kJ/mol}}$$

(c) Use the two rates from part (a) above, which correspond to a 10°C temperature rise. Let q be the fractional increase in rate; then:

$$X_2 = X_1 + qX_1$$
$$220 = 148 + q148$$
$$148q = 220 - 148 = 72$$
$$q = 72/148 = 0.49$$

The fractional increase is 0.49, which corresponds to a percentage increase of **49%**.

68. (a) Use Equation 11.7a: recall that activation energy must be expressed in J/mol and temperatures in K. Thus $T_1 = 298$ K, with $k_1 = 0.027$ s^{-1}; and $E_a = 1.74 \times 10^4$ J/mol. At a temperature $T_2 = 338$ K, solve for the corresponding rate constant k_2:

$$\ln k_2 - \ln k_1 = \frac{E_a}{R}\left[\frac{1}{T_1} - \frac{1}{T_2}\right]$$

$$\ln k_2 = \ln k_1 + \frac{E_a}{R}\left[\frac{1}{T_1} - \frac{1}{T_2}\right]$$

$$= \ln(0.027 \text{ s}^{-1}) + \left\{\frac{1.74 \times 10^4 \text{ J/mol}}{8.3145 \text{ J/mol·K}} \times \left[\frac{1}{298 \text{ K}} - \frac{1}{338 \text{ K}}\right]\right\}$$

$$= -2.8$$

$$k_2 = e^{-2.8} = \underline{\textbf{0.062 s}^{-1}}$$

(b) It is already known that $T_1 = 298$ K and $E_a = 1.74 \times 10^4$ J/mol. We require the temperature T_2 at which the rate constant k_2 is 25% greater than k_1. Thus $k_2 = k_1 + (25\% \times k_1) = 1.25k_1$. Substitute all of these into Equation 11.7b and solve for T_2:

$$\ln\frac{k_2}{k_1} = \ln\frac{1.25k_1}{k_1} = \ln(1.25) = \frac{E_a}{R}\left[\frac{1}{T_1} - \frac{1}{T_2}\right] \qquad \text{so:} \qquad \frac{1}{T_2} = \frac{1}{T_1} - \left[\frac{R}{E_a}\ln(1.25)\right]$$

$$\frac{1}{T_2} = \frac{1}{298\,\text{K}} - \left[\frac{8.3145\,\text{J/mol}\cdot\text{K}}{1.74\times10^4\,\text{J/mol}}\ln(1.25)\right] = 3.25 \times 10^{-3}\,\text{K}^{-1}$$

$$T_2 = (3.25 \times 10^{-3}\,\text{K}^{-1})^{-1} = \underline{\textbf{308 K}} = (308 - 273.15)\,^\circ\text{C} = \underline{\textbf{35}\,^\circ\text{C}}$$

70. This problem can be approached in a way similar to Example 11.8(a). Use Equation 11.7b, solve for E_a, and substitute the rate constants at the two given temperatures (the temperatures being expressed in Kelvin). However, only half-lives are given and so these must be used to obtain the corresponding first order rate constants using Equation 11.3.

rate constant k_1 at 25°C:

$$t_{1/2} = \frac{\ln 2}{k} \qquad \text{so:} \qquad k_1 = \frac{\ln 2}{t_{1/2}} = \frac{\ln 2}{2.81\,\text{s}} = 0.247\,\text{s}^{-1}$$

rate constant k_2 at 45°C:

$$k_2 = \frac{\ln 2}{t_{1/2}} = \frac{\ln 2}{0.313\,\text{s}} = 2.21\,\text{s}^{-1}$$

Now use Equation 11.7b, with temperatures $T_1 = 298$ K and $T_2 = 318$ K:

$$\ln\frac{k_2}{k_1} = \frac{E_a}{R}\left[\frac{1}{T_1} - \frac{1}{T_2}\right] \qquad \text{so:} \quad E_a = R \times \ln\frac{k_2}{k_1} \times \left[\frac{1}{T_1} - \frac{1}{T_2}\right]^{-1}$$

$$E_a = (8.3145\,\text{J/mol·K}) \times \ln\frac{2.21\,\text{s}^{-1}}{0.247\,\text{s}^{-1}} \times \left[\frac{1}{298\,\text{K}} - \frac{1}{318\,\text{K}}\right]^{-1}$$

$$= 8.6 \times 10^4\,\text{J/mol} \quad \text{or} \quad \underline{\textbf{86 kJ/mol}}$$

11-6 CATALYSIS

72. The activation energy for the catalyzed reaction is 48% of E_a for the uncatalyzed reaction = 48% × 129 kJ = 62 kJ. Recall that the value of ΔH (−29 kJ) is the same for both the catalyzed (- - - -) and the uncatalyzed (———) reactions. (The graph is not to scale.)

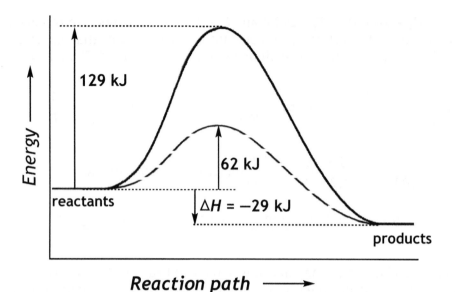

74. Making the same assumptions as suggested for Problem 73, we can use the given equation (Arrhenius Equation, $\ln k = \ln A - E_a/RT$) in the following form (see textbook, page 294):

$$k = Ae^{-E_a/RT}$$

to obtain an expression for the rate constant for the uncatalyzed and catalyzed reactions. For the uncatalyzed reaction, $E_a = 363 \times 10^3$ J at $T = 298$ K, so the rate constant k_1 is:

$$k_1 = A \times e^{-\frac{363 \times 10^3 \text{ J/mol}}{(8.3145 \text{ J/mol·K})(298 \text{ K})}} = A \times 2.36 \times 10^{-64}$$

For the catalyzed reaction, we know that the rate constant k_2 is tenfold larger than k_1, and therefore:

$$k_2 = 10k_1 = A \times 2.36 \times 10^{-63}$$

However, the Arrhenius Equation tells us that for the catalyzed reaction $k_2 = Ae^{-E_a/RT}$ and this will allow us to solve for E_a, the activation energy for the catalyzed reaction:

$$k = Ae^{-E_a/RT} \qquad \text{so: } E_a = -RT \times \ln\frac{k_2}{A}$$

$$E_a = -(8.3145 \text{ J/mol·K})(298 \text{ K}) \times \ln\left(\frac{A \times 2.36 \times 10^{-63}}{A}\right)$$

$$= -(8.3145 \text{ J/mol·K})(298 \text{ K}) \times \ln(2.36 \times 10^{-63})$$

$$= 357 \times 10^3 \text{ J/mol} \qquad \text{or} \quad \underline{\textbf{357 kJ/mol}}$$

11-7   REACTION MECHANISMS

76. Recall that the rate expression for an elementary reaction is equal to a rate constant multiplied by the concentration of each reactant molecule ("molecule" here may mean an atom or a fragment of a molecule).

    (a) **rate = $k[Cl_2]$**

    (b) **rate = $k[N_2O_2][O_2]$**

    (c) **rate = $k[I^-][HClO]$**

78. This problem is similar to Example 11.9. First identify the slow step in the mechanism: this will be the rate-limiting step and its rate expression will determine the overall rate law for the reaction. For the given mechanism, the second step is the slow step, for which the rate expression is:

    rate = $k_2[N_2O_2][H_2]$

    where k_2 is simply the rate constant for the second elementary step.

    However, since N_2O_2 is an intermediate in the reaction its concentration cannot appear in the final form of the experimentally-determined rate expression. In order to eliminate the $[N_2O_2]$ term, use the first (reversible) step in the mechanism. For that step:

    rate of forward reaction = $k_1[NO]^2$

    rate of reverse reaction = $k_{-1}[N_2O_2]$

    and since this first step is in equilibrium, rate of forward reaction = rate of reverse reaction; hence we can write:

    $k_1[NO]^2 = k_{-1}[N_2O_2]$

    Solving for $[N_2O_2]$ then gives

    $$[N_2O_2] = \frac{k_1}{k_{-1}}[NO]^2$$

    This provides an expression for $[N_2O_2]$ in terms of a reactant concentration (i.e., [NO]), and thus we can substitute for $[N_2O_2]$ in the original rate expression:

    rate = $k_2[N_2O_2][H_2]$

    $$= k_2 \times \frac{k_1}{k_{-1}}[NO]^2 \times [H_2] = = \frac{k_2 k_1}{k_{-1}}[NO]^2[H_2]$$

    Finally, combining all of the rate constants into a single constant k gives:

    rate = $k[NO]^2[H_2]$

This is the rate expression given in the problem and so: **YES**, this mechanism IS consistent with the given rate expression.

80. This problem is similar to Example 11.9 and Problem 78 above. For each mechanism, identify the slow step: this will be the rate-limiting step and its rate expression will determine the overall rate law for that mechanism.

    Mechanism 1: The second step is the slow step, for which the rate expression is:

    $$\text{rate} = k_2[NO_3][NO]$$

    where k_2 is simply the rate constant for the second elementary step.

    However, since NO_3 is an intermediate the $[NO_3]$ term must be eliminated in the final form of the rate expression. To do this, use the first (reversible) step in the mechanism and recall that for that step the rate of the forward and reverse reactions are equal, so in this case we can write:

    $$k_1[NO][O_2] = k_{-1}[NO_3]$$

    Solving for $[NO_3]$ then gives

    $$[NO_3] = \frac{k_1}{k_{-1}}[NO][O_2]$$

    Now, substituting for $[NO_3]$ in the rate expression for the rate-determining step gives:

    $$\text{rate} = k_2[NO_3][NO]$$

    $$= k_2 \times \frac{k_1}{k_{-1}}[NO][O_2] \times [NO] = = \frac{k_2 k_1}{k_{-1}}[NO]^2[O_2]$$

    Finally, combining all of the rate constants into a single constant k gives:

    $$\text{rate} = k[NO]^2[O_2]$$

    This is the rate expression given in the problem and so Mechanism 1 is consistent with the observed rate law.

    Mechanism 2: The second step is the slow step, for which the rate expression is:

    $$\text{rate} = k_2[N_2O_2][O_2]$$

    where k_2 is simply the rate constant for the second elementary step.

    However, since N_2O_2 is an intermediate the $[N_2O_2]$ term must be eliminated in the final form of the rate expression. To do this, use the first (reversible) step in the mechanism, and recall that for that step the rate of the forward and reverse reactions are equal, so we can write:

$$k_1[NO]^2 = k_{-1}[N_2O_2]$$

Solving for $[N_2O_2]$ then gives

$$[N_2O_2] = \frac{k_1}{k_{-1}}[NO]^2$$

Now, substituting for $[N_2O_2]$ in the rate expression for the rate-determining step gives:

$$\text{rate} = k_2[N_2O_2][O_2]$$

$$= k_2 \times \frac{k_1}{k_{-1}}[NO]^2 \times [O_2] = = \frac{k_2k_1}{k_{-1}}[NO]^2[O_2]$$

Finally, combining all of the rate constants into a single constant k gives:

$$\text{rate} = k[NO]^2[O_2]$$

This is the rate expression given in the problem and so Mechanism 2 also is consistent with the observed rate law.

UNCLASSIFIED PROBLEMS

82. See Section 11.2. Since the reaction is first order in OH^- and second order in ClO_2, the rate expression for the reaction can be written as:

$$\text{rate} = k[OH^-][ClO_2]^2$$

and, since we are told the initial rate of reaction for certain concentrations of the reactants OH^- and ClO_2, we can use this information to determine the value of the rate constant k. Rearrange the rate expression to solve for k:

$$k = \frac{\text{rate}}{[OH^-][ClO_2]^2} = \frac{6.00 \times 10^{-4}\ \text{mol/L} \cdot \text{s}}{(0.030\ \text{mol/L})(0.010\ \text{mol/L})^2} = 2.0 \times 10^2\ \text{L}^2/\text{mol}^2 \cdot \text{s}$$

This value of k will also apply to the new set of reactant concentrations. However, it is necessary first to calculate $[OH^-]$ and $[ClO_2]$ from the information given. Note that two solutions are mixed and the total volume of the reaction mixture is $(50.0 + 95.0)\text{mL} = 145.0$ mL or 0.145 L.

Strategy (for each reactant concentration):
vol reactant → mol reactant added → molarity of reactant in reaction mixture

$$[ClO_2] = 50.0\ \text{mL ClO}_2 \times \frac{1\ \text{L}}{1000\ \text{mL}} \times \frac{0.200\ \text{mol ClO}_2}{1\ \text{L ClO}_2} \times \frac{1}{0.145\ \text{L mixture}} = 0.0690\ M$$

$$[OH^-] = 95.0 \text{ mL NaOH} \times \frac{1 \text{L}}{1000 \text{ mL}} \times \frac{0.155 \text{ mol NaOH}}{1 \text{L NaOH}} \times \frac{1 \text{ mol OH}^-}{1 \text{ mol NaOH}} \times \frac{1}{0.145 \text{ L mixture}}$$

$$= 0.102 \ M$$

Now substitute these concentrations, along with the value of k already calculated, to determine the initial rate of reaction under these new conditions:

$$
\begin{aligned}
\text{rate} \quad &= k[OH^-][ClO_2]^2 \\[6pt]
&= (2.0 \times 10^2 \text{ L}^2/\text{mol}^2\cdot\text{s})(0.102 \text{ mol/L})(0.0690 \text{ mol/L})^2 \\[6pt]
&= \underline{0.097 \text{ mol/L}\cdot\text{s}}
\end{aligned}
$$

84. See Section 11.6. Use Equation 11.7b: recall that activation energy must be expressed in J/mol and temperatures in K. Thus $T_1 = (28 + 273)\text{K} = 301$ K, $T_2 = (46 + 273)\text{K} = 319$ K, and $E_a = 121$ kJ/mol $= 1.21 \times 10^5$ J/mol.

$$\ln\frac{k_2}{k_1} = \frac{E_a}{R}\left[\frac{1}{T_1} - \frac{1}{T_2}\right] = \frac{1.21\times10^5 \text{ J/mol}}{8.3145 \text{ J/mol}\cdot\text{K}} \times \left[\frac{1}{301\text{K}} - \frac{1}{319\text{K}}\right] = 2.73$$

$$\frac{k_2}{k_1} = e^{2.73} = 15.3 \qquad \text{so: } k_2 = 15.3 k_1$$

Thus, the value of the rate constant at the higher temperature, k_2, is 15.3 times the rate constant at the lower temperature, k_1. Hence **the reaction proceeds around 15 times faster at the higher temperature.**

86. See Section 11.3 and Table 11.2. Calculate the half-life for this first order reaction using Equation 11.3, with the value of k given, and convert the answer obtained into the corresponding value in years:

$$t_{1/2} = \frac{\ln 2}{k} = \frac{\ln 2}{3\times10^{-26} \text{ s}^{-1}} = 2 \times 10^{25} \text{ s}$$

$$= 2 \times 10^{25} \text{ s} \times \frac{1 \text{ min}}{60 \text{ s}} \times \frac{1 \text{ h}}{60 \text{ min}} \times \frac{1 \text{ d}}{24 \text{ h}} \times \frac{1 \text{ y}}{365 \text{ d}} = \underline{7 \times 10^{17} \text{ y}}$$

Since the half-life of the reaction is 7×10^{17} y it is unlikely that it makes any significant contribution to the depletion of the ozone layer.

88. The required equation will be similar in style to Equation 11.7b:

$$\ln\frac{k_2}{k_1} = \frac{E_a}{R}\left[\frac{1}{T_1} - \frac{1}{T_2}\right]$$

which is a two-point form of the Arrhenius Equation:

$$\ln k = \ln A - \frac{E_a}{RT}$$

At a particular temperature T, let's call the rate constant for the uncatalyzed reaction k_1 and for the catalyzed reaction k_2, and label the activation energy for these two reactions $E_{a(uncat)}$ and $E_{a(cat)}$, respectively. The Arrhenius Equation for these two cases will then be written as:

$$\ln k_1 = \ln A - \frac{E_{a(uncat)}}{RT} \qquad \text{and} \qquad \ln k_2 = \ln A - \frac{E_{a(cat)}}{RT}$$

Subtracting the first of these two equations from the second equation, we obtain:

$$\ln k_2 - \ln k_1 = \ln A - \frac{E_{a(cat)}}{RT} - \ln A + \frac{E_{a(uncat)}}{RT}$$

Assuming that A (and hence $\ln A$) is the same for both reactions, this becomes:

$$\ln k_2 - \ln k_1 = \frac{E_{a(uncat)}}{RT} - \frac{E_{a(cat)}}{RT}$$

which simplifies to:

$$\ln \frac{k_2}{k_1} = \frac{1}{RT}\left[E_{a(uncat)} - E_{a(cat)}\right]$$

CONCEPTUAL PROBLEMS

90. (a) Adding water to the reaction mixture will dilute all of the species in solution, so $[CH_3COOCH_3]$ and $[H^+]$ will decrease. Since both of these species appear in the rate law for the reaction, we can say that the rate of the reaction **decreases** when water is added.

(b) See Section 11-5, especially page 293. The value of k **increases** when the temperature is raised.

(c) The activation energy for a reaction does not depend on concentration of reactants, so we would expect that the value of E_a **stays the same** when HCl is added.

(d) Adding NaOH (a strong base) will consume some of the H^+ in the reaction mixture, causing a decrease in $[H^+]$. Since H^+ appears in the rate law, we can say that the rate of the reaction **decreases** when NaOH is added.

(e) See Sections 11-5 and 11-6. Since a catalyst lowers the activation energy (page 296) and decreasing activation energy leads to increasing k (page 291), we can say that the value of k **increases** when a catalyst is added.

92. For the given decomposition reaction, using Equation 11.1 as a guide we obtain the following relationships between the rate of change of concentration for X, Y, and Z:

$$\text{rate} = \frac{-\Delta[X]}{2\Delta t} = \frac{\Delta[Y]}{2\Delta t} = \frac{\Delta[Z]}{\Delta t}$$

<u>Curve 1</u> shows decreasing concentration, which can only apply to a reactant. Since X is the only reactant, this curve must represent **change in [X] with time**.

<u>Curve 2</u> shows increasing concentration, which must apply to a product. Since the increase in concentration is about one half of the amount by which [X] decreases over the same time period, this curve must represent **change in [Z] with time**.

<u>Curve 3</u> shows increasing concentration, which must apply to a product. Since the increase in concentration is about the same as the amount by which [X] decreases over the same time period, this curve must represent **change in [Y] with time**.

94. (a) See Section 11.4. The smaller the activation energy, the larger the rate constant, and hence the faster the reaction rate. Thus, the reaction with the smallest E_a value will be fastest: this is **reaction A**.

    (b) See Sections 11.3 and 11.4. As may be seen from Equation 11.3, half-life is inversely proportional to the rate constant and therefore the reaction with the largest E_a value has the smallest rate constant and hence the longest half-life: this is **reaction C**.

    (c) See Section 11.4. The reaction with the smallest E_a value will be largest rate constant and hence the largest rate of reaction (since reaction rate is proportional to the rate constant): this is **reaction A**.

96. (a) Compare Figure 11.10. The forward activation energy is the energy that must be supplied to convert the reactants (at point A) to the activated complex (point C). Reading from the diagram, this is **+8 kJ**.

    (b) **Point A** on the diagram corresponds to the reactants.

    (c) Compare Figure 11.10. The value of ΔH for the forward reaction is the difference between the energy of the reactants (at point A) and the products (point B). Reading from the diagram, this is **−360 kJ**.

    (d) Compare Figure 11.10. The reverse activation energy is the energy that must be supplied to convert the products (at point B) to the activated complex (point C). Reading from the diagram, this is (360 + 8)kJ = **+368 kJ**.

    (e) **Point B** on the diagram corresponds to the products, so C_2H_6 should certainly be found there.

    (f) Compare Figure 11.10 and adjacent text. Point C correspond to the **activated complex** (transition state).

98. See Sections 11.3 and 11.5. The rate expression for this first order reaction must be of the following form: **rate= k[A]**.

For Expt. 1: k and [A] are known, so the rate = (0.5/min)(0.100 mol/L) = **0.05 mol/L·min**.

For Expt. 2: [A] is known, and both k and E_a must be the same as in Expt. 1 (because the temperature and catalyst are the same), so k = **0.5/min**; the rate = (0.5/min)(0.200 mol/L) = **0.1 mol/L·min**; and E_a = **32 kJ**.

For Expt. 3: k and [A] are known, so the rate = (1/min)(0.100 mol/L) = **0.1 mol/L·min**. The value of E_a **cannot be calculated (_NC_)** without further information.*

For Expt. 4: k and [A] are known, so the rate = (0.6/min)(0.100 mol/L) = **0.06 mol/L·min**. Since the catalyst is the same as in Expts. 1 & 2, E_a must also be the same, so E_a = **32 kJ**.

(*NOTE: Using the assumptions of Problem 88 above, and the equation derived therein, one can estimate E_a for the catalyzed reaction as around 30 kJ. However, we are not told that such assumptions are valid, so NC is the appropriate answer here.)*

CHALLENGE PROBLEMS

99. (a) See Section 8.5 and Example 8.7. The value of $\Delta H°_{(reaction)}$ for the reaction as written can be calculated using Equation 8.4. Note that $\Delta H°_f$ for I_2 is not zero in this case, because it is not in its standard state (solid).

$$\Delta H°_{(reaction)} = \Sigma \Delta H°_{f\ (products)} - \Sigma \Delta H°_{f\ (reactants)}$$
$$= [(2\ mol)(26.48\ kJ/mol) - [(1\ mol)\ (0\ kJ/mol) + (1\ mol)(62.44\ kJ/mol)]$$
$$= -9.48\ kJ$$

By considering Figure 11.10, it can be seen that E_a for the reverse reaction (the energy that must be furnished to convert "products" → "activated complex") is given by:

$$E_{a(reverse)} = E_{a(forward)} - \Delta H_{(forward)}$$
$$= 165\ kJ/mol - (-9.48\ kJ/mol)$$
$$= \underline{\mathbf{174\ kJ/mol}}$$

(b) Use the given equation, $k = Ae^{-Ea/RT}$, to calculate A for the forward reaction:

$$k = Ae^{-Ea/RT} \qquad so \quad A = ke^{Ea/RT}$$
$$A = ke^{Ea/RT} = (138\ L/mol·s)e^{(165000\ J/mol)/(8.3145\ J/mol·K)(973\ K)}$$
$$= 9.94 \times 10^{10}\ L/mol·s$$

Now use this value of A in the given equation, along with the value of $E_{a(reverse)}$ calculated in part (a) to calculate k for the reverse reaction:

$$k = Ae^{-Ea/RT}$$
$$= (9.94 \times 10^{10}\ L/mol·s)e^{-(174000\ J/mol)/(8.3145\ J/mol·K)(973\ K)}$$
$$= \underline{\mathbf{45.3\ L/mol·s}}$$

(c) The reverse reaction is second order in HI, so its rate expression will be of the form

$$\text{rate} = k[\text{HI}]^2$$

Use this rate expression, along with the given [HI] and the value of k from part (b):

$$\text{rate} = (45.3 \text{ L/mol·s})(0.200 \text{ mol/L})^2 = \underline{\textbf{1.81 mol/L·s}}$$

100. See Section 11.6. Let k_1 and k_2 be the rate constants at the two given temperatures, $T_1 = 373$ K and $T_2 = 360$ K, respectively; $E_a = 6.9 \times 10^4$ J/mol. Use Equation 11.7b to determine the ratio k_2/k_1.

$$\ln\frac{k_2}{k_1} = \frac{E_a}{R}\left[\frac{1}{T_1} - \frac{1}{T_2}\right] = \frac{6.9 \times 10^4 \text{ J/mol}}{8.3145 \text{ J/mol·K}} \times \left[\frac{1}{373 \text{ K}} - \frac{1}{360 \text{ K}}\right] = -0.80$$

$$\frac{k_2}{k_1} = e^{-0.80} = 0.45 \qquad \text{so: } k_2 = 0.45k_1$$

Since the rate of reaction is directly proportional to the rate constant, the reaction is clearly slower at the lower temperature T_2 ($\text{rate}_2 = 0.45 \times \text{rate}_1$), as is to be expected. The coagulation takes 5.4 min at the higher temperature, so at the lower temperature

$$\text{time required} = 5.4 \text{ min} \times \frac{\text{rate}_1}{\text{rate}_2} = 5.4 \text{ min} \times \frac{\text{rate}_1}{0.45 \times \text{rate}_1} = \underline{\textbf{12 min}}$$

101. The rate of reaction for a first order reaction is given by

$$\text{rate} = k[\text{A}]$$

and we are also given a second expression for the rate:

$$\text{rate} = -\frac{1}{a}\frac{d[\text{A}]}{dt}$$

therefore: $\qquad \text{rate} = -\frac{1}{a}\frac{d[\text{A}]}{dt} = k[\text{A}] \qquad$ or $\qquad -\frac{d[\text{A}]}{[\text{A}]} = ak\,dt$

integrating both sides of the latter equation gives:

$$-\int_{[\text{A}]}\frac{d[\text{A}]}{[\text{A}]} = ak\int_t dt$$

$$-\{\ln[\text{A}] - \ln[\text{A}]_0\} = ak\{t - 0\}$$

$$\ln[\text{A}]_0 - \ln[\text{A}] = akt \qquad \text{or} \qquad \ln\frac{[\text{A}]_0}{[\text{A}]} = akt$$

102. This problem is similar to Example 11.3. The rate expression will be of the form:

$$\text{rate} = k[A]^m[B]^n[C]^p$$

To determine the reaction order with respect to A, select two experiments in which [A] changes while [B] and [C] remain constant (experiments 1 and 4 are used below).

$$\frac{\text{rate}_4}{\text{rate}_1} = \frac{k([A]_4)^m([B]_4)^n([C]_4)^p}{k([A]_1)^m([B]_1)^n([C]_1)^p} = \left(\frac{[A]_4}{[A]_1}\right)^m\left(\frac{[B]_4}{[B]_1}\right)^n\left(\frac{[C]_4}{[C]_1}\right)^p = \left(\frac{[A]_4}{[A]_1}\right)^m$$

$$\text{(since } [B]_1 = [B]_4, \; [C]_1 = [C]_4, \text{ and } 1^n1^p = 1\text{)}$$

$$\frac{\text{rate}_4}{\text{rate}_1} = \frac{4X}{X} = 4 \quad \text{and} \quad \left(\frac{[A]_4}{[A]_1}\right)^m = \left(\frac{0.40 \text{ mol/L}}{0.20 \text{ mol/L}}\right)^m = 2.0^m$$

thus $4 = 2.0^m$ giving $m = 2$. The reaction is **second order in A.**

Determining the reaction order with respect to B is a little less straightforward, since we are not given two experiments in which [B] changes while [A] and [C] remain constant. Instead use two experiments in which [B] changes but [C] stays constant (experiments 2 and 3 are used below), and substitute $m = 2$ into the rate expressions.

$$\frac{\text{rate}_2}{\text{rate}_3} = \frac{k([A]_2)^m([B]_2)^n([C]_2)^p}{k([A]_3)^m([B]_3)^n([C]_3)^p} = \left(\frac{[A]_2}{[A]_3}\right)^m\left(\frac{[B]_2}{[B]_3}\right)^n\left(\frac{[C]_2}{[C]_3}\right)^p = \left(\frac{[A]_2}{[A]_3}\right)^2\left(\frac{[B]_2}{[B]_3}\right)^n$$

$$\text{(since } [C]_2 = [C]_3 \text{ and } 1^p = 1\text{)}$$

$$\frac{\text{rate}_2}{\text{rate}_3} = \frac{8X}{X} = 8 \quad \text{and} \quad \left(\frac{[A]_2}{[A]_3}\right)^2\left(\frac{[B]_2}{[B]_3}\right)^n = \left(\frac{0.40 \text{ mol/L}}{0.20 \text{ mol/L}}\right)^2\left(\frac{0.40 \text{ mol/L}}{0.20 \text{ mol/L}}\right)^n = 4 \times 2^n$$

thus $8 = 4 \times 2^n$ giving $2 = 2^n$ and hence $n = 1$. The reaction is **first order in B.**

Determining the reaction order with respect to C is similar to the procedure for B above, since we are not given two experiments in which [C] changes while [A] and [B] remain constant. Use any two experiments in which [C] changes (experiments 1 and 2 are used below), and substitute $m = 2$ and $n = 1$ into the rate expressions.

$$\frac{\text{rate}_2}{\text{rate}_1} = \frac{k([A]_2)^m([B]_2)^n([C]_2)^p}{k([A]_1)^m([B]_1)^n([C]_1)^p} = \left(\frac{[A]_2}{[A]_1}\right)^m\left(\frac{[B]_2}{[B]_1}\right)^n\left(\frac{[C]_2}{[C]_1}\right)^p = \left(\frac{[A]_2}{[A]_1}\right)^2\left(\frac{[B]_2}{[B]_1}\right)^1\left(\frac{[C]_2}{[C]_1}\right)^p$$

$$\frac{\text{rate}_2}{\text{rate}_1} = \frac{8X}{X} = 8$$

$$\text{and} \quad \left(\frac{[A]_2}{[A]_1}\right)^2\left(\frac{[B]_2}{[B]_1}\right)^1\left(\frac{[C]_2}{[C]_1}\right)^p = \left(\frac{0.40 \text{ mol/L}}{0.20 \text{ mol/L}}\right)^2\left(\frac{0.40 \text{ mol/L}}{0.40 \text{ mol/L}}\right)^1\left(\frac{0.40 \text{ mol/L}}{0.20 \text{ mol/L}}\right)^p = 4 \times 1 \times 2^p$$

thus $8 = 4 \times 1 \times 2^p$ giving $2 = 2^p$ and hence $p = 1$. The reaction is **first order in C**.

The rate law for the reaction is   **rate = $k[A]^2[B][C]$**.

103. See Section 11.3 and Table 11.2; see also Problem 101 above.

(a) The rate of reaction for a second order reaction is given by

$$\text{rate} = k[A]^2$$

We can also write a second expression for the rate using Equation 11.1 and referring to Problem 101:

$$\text{rate} = -\frac{d[A]}{dt} \qquad \text{(note that for the reaction in Table 11.2, } a = 1)$$

therefore: $\qquad \text{rate} = -\dfrac{d[A]}{dt} = k[A]^2 \qquad$ or $\qquad -\dfrac{d[A]}{[A]^2} = k\,dt$

integrating both sides of the latter equation gives:

$$-\int_{[A]}\frac{d[A]}{[A]^2} = k\int_t dt$$

$$-\left\{-\frac{1}{[A]} - \left[-\frac{1}{[A]_0}\right]\right\} = k\{t - 0\}$$

$$\frac{1}{[A]} - \frac{1}{[A]_0} = kt$$

(b) The rate of reaction for a third order reaction is given by

$$\text{rate} = k[A]^3$$

We can also write a second expression for the rate using Equation 11.1 and referring to Problem 101:

$$\text{rate} = -\frac{d[A]}{dt} \qquad \text{(note again that } a = 1)$$

therefore: $\qquad \text{rate} = -\dfrac{d[A]}{dt} = k[A]^3 \qquad$ or $\qquad -\dfrac{d[A]}{[A]^3} = k\,dt$

integrating both sides of the latter equation gives:

$$-\int_{[A]}\frac{d[A]}{[A]^3} = k\int_t dt$$

$$-\left\{-\frac{1}{2[A]^2}-\left[-\frac{1}{2([A]_0)^2}\right]\right\} = k\{t-0\}$$

$$\frac{1}{2[A]^2}-\frac{1}{2([A]_0)^2} = kt \qquad \text{or:} \qquad \frac{1}{[A]^2}-\frac{1}{([A]_0)^2} = 2kt$$

104. See Section 11.3.

Since the drug is given 3 times per day, Δt must be one third of one day, or $(24/3)$h = 8 h.
The value of $t_{1/2}$ is given as 2.0 days, or 2.0 d × (24 h/1 d) = 48 h.
Therefore, using the second of the two equations given in the question:

$$a = 0.30\frac{\Delta t}{t_{1/2}} = 0.30\frac{8\,\text{h}}{48\,\text{h}} = 0.050$$

Then, using the dose X = 0.100 g, determine the saturation value using the first of the two equations given in the problem, with a = 0.050 as just calculated:

$$\text{saturation value} = \frac{X}{1-10^{-a}} = \frac{0.100\,\text{g}}{1-10^{-0.050}} = \underline{\mathbf{0.92\,g}}$$

The second part of the problem requires us to find the mass X of the drug that gives a maximum saturation value of 0.500 g. Substitute this mass into the first equation given and solve for X (for the same drug, given 3 times per day, a will be 0.050 as before):

$$\text{saturation value} = 0.500\,\text{g} = \frac{X}{1-10^{-0.050}}$$

$$X = (0.500\,\text{g})(1-10^{-0.050}) = \underline{\mathbf{0.054\,g}} \quad \text{or} \quad \underline{\mathbf{54\,mg}}$$

Note that 0.500 g must be the maximum allowable saturation value, and to avoid side effects the patient should receive a dose *no greater than 54 mg* (3 times a day). It can be seen from the first equation given in the problem that the saturation value is directly proportional to the dose of drug (X), so a dose of 54 mg or less will ensure that the patient remains at or below the saturation value that would produce side effects.

105. For a first order reaction, using Equation 11.3, the rate constant k is obtained from the given half-life:

$$k = \frac{\ln 2}{t_{1/2}} = \frac{\ln 2}{19.0 \text{ min}} = 0.0365 \text{ min}^{-1}$$

Since we are given partial pressures, it is convenient to use Equation 11.2a with partial pressures rather than concentrations. Use the Ideal Gas Law (see Chapter 5); and note that the values of V and T are constant, and that n/V is simply molarity:

$$P_A V = n_A(RT) \rightarrow \frac{n_A}{V} = \frac{P_A}{RT} \rightarrow [A] = \frac{P_A}{RT}$$

Substitute for [A] in Equation 11.2a:

$$\ln \frac{[A]_0}{[A]} = kt \rightarrow \ln \frac{(P_A)_0/RT}{P_A/RT} = \ln \frac{(P_A)_0}{P_A} = kt \rightarrow P_A = (P_A)_0\, e^{-kt}$$

Thus P_A after 42 minutes is (using k calculated above):

$$P_A = (622 \text{ mmHg})e^{-(0.0365 \text{ min}^{-1})(42 \text{ min})} = 134 \text{ mmHg}$$

The change in P_A over that 42-minute period is given by:

$$\Delta P_A = (P_A)_{final} - (P_A)_{initial} = (134 - 622) \text{ mmHg} = -488 \text{ mmHg}$$

Now, similar to Equation 11.1, we can develop equivalent expressions that relate ΔP_A to ΔP_X and ΔP_Y:

$$\text{rate} = \frac{-\Delta[A]}{2\Delta t} = \frac{\Delta[X]}{\Delta t} = \frac{\Delta[Y]}{\frac{1}{2}\Delta t}$$

Now, substitute P/RT for concentration of each species:

$$\frac{-\Delta P_A/(RT)}{2\Delta t} = \frac{\Delta P_X/(RT)}{\Delta t} = \frac{\Delta P_Y/(RT)}{\frac{1}{2}\Delta t}$$

Each of the $RT/\Delta t$ terms will cancel, leaving:

$$\frac{-\Delta P_A}{2} = \frac{\Delta P_X}{1} = \frac{\Delta P_Y}{\frac{1}{2}}$$

Hence:

$$\Delta P_X = \frac{-\Delta P_A}{2} \quad \text{and} \quad \Delta P_Y = \frac{-\Delta P_A}{4}$$

$$\Delta P_X = -\tfrac{1}{2}(-488 \text{ mmHg}) = 244 \text{ mmHg}; \quad \Delta P_Y = \tfrac{1}{4}(-488 \text{ mmHg}) = 122 \text{ mmHg}$$

Since P_X and P_Y are zero to begin with, we can say that at $t = 42$ min:

$P_X = \underline{\textbf{244 mmHg}}$; $\quad P_Y = \underline{\textbf{122 mmHg}}$; and recall from above $P_A = \underline{\textbf{134 mmHg}}$

GASEOUS CHEMICAL EQUILIBRIUM

12-1   THE N_2O_4–NO_2 EQUILIBRIUM SYSTEM

2. (a) After equilibrium is established, concentrations (pressures) remain constant (see Section 12-1). This occurs after **75 s** (P_A and P_B both remain constant).

    (b) After 45 s, P_A is still decreasing and P_B is still increasing, so the rate of the forward reaction must be **greater** than the rate of the reverse reaction.

    After 90 s, equilibrium has already been established (and both P_A and P_B remain constant), so the rates of the forward and reverse reactions must be **equal**.

4. Use stoichiometry to relate the partial pressures of each gas: the pressure is proportional to the number of moles of gas present (see Chapter 5).
    Since the P_C is initially zero it must increase, and therefore P_A and P_B must decrease as A and B are being consumed to form C.

Time (min)	0	1	2	3	4	5	6
P_A (atm)	1.000	0.778	**0.580**	**0.415**	**0.355**	0.325	**0.325**
P_B (atm)	0.400	**0.326**	0.260	**0.205**	0.185	**0.175**	0.175
P_C (atm)	0.000	**0.148**	**0.280**	0.390	**0.430**	**0.450**	**0.450**

After 1 min:

Amount of A consumed = 1.000 atm A − 0.778 atm A = 0.222 atm A

Amount of B consumed = 0.222 atm A $\times \dfrac{1\,\text{mol B}}{3\,\text{mol A}}$ = 0.0740 atm B

Amount of B remaining = 0.400 atm B − 0.0740 atm B = **0.326 atm B**

Amount of C formed = 0.222 atm A $\times \dfrac{2\,\text{mol C}}{3\,\text{mol A}}$ = **0.148 atm C**

After 2 min:

Amount of B consumed = 0.400 atm B − 0.260 atm B = 0.140 atm B

Amount of A consumed = 0.140 atm B $\times \dfrac{3\,\text{mol A}}{1\,\text{mol B}}$ = 0.420 atm A

Amount of A remaining = 1.000 atm A − 0.420 atm A = **0.580 atm A**

Amount of C formed = 0.140 atm B $\times \dfrac{2\,\text{mol C}}{1\,\text{mol B}}$ = **0.280 atm C**

After 3 min:

Amount of B consumed = 0.390 atm C $\times \dfrac{1\,\text{mol B}}{2\,\text{mol C}}$ = 0.195 atm B

Amount of B remaining = 0.400 atm B − 0.195 atm B = **0.205 atm B**

Amount of A consumed $= 0.390 \text{ atm C} \times \dfrac{3 \text{ mol A}}{2 \text{ mol C}} = 0.585 \text{ atm A}$

Amount of A remaining $= 1.000 \text{ atm A} - 0.585 \text{ atm A} = \textbf{0.415 atm A}$

__After 4 min:__

Amount of B consumed $= 0.400 \text{ atm B} - 0.185 \text{ atm B} = 0.215 \text{ atm B}$

Amount of A consumed $= 0.215 \text{ atm B} \times \dfrac{3 \text{ mol A}}{1 \text{ mol B}} = 0.645 \text{ atm A}$

Amount of A remaining $= 1.000 \text{ atm A} - 0.645 \text{ atm A} = \textbf{0.355 atm A}$

Amount of C formed $= 0.215 \text{ atm B} \times \dfrac{2 \text{ mol C}}{1 \text{ mol B}} = \textbf{0.430 atm C}$

__After 5 min:__

Amount of A consumed $= 1.000 \text{ atm A} - 0.325 \text{ atm A} = 0.675 \text{ atm A}$

Amount of B consumed $= 0.675 \text{ atm A} \times \dfrac{1 \text{ mol B}}{3 \text{ mol A}} = 0.225 \text{ atm B}$

Amount of B remaining $= 0.400 \text{ atm B} - 0.225 \text{ atm B} = \textbf{0.175 atm B}$

Amount of C formed $= 0.675 \text{ atm A} \times \dfrac{2 \text{ mol C}}{3 \text{ mol A}} = \textbf{0.450 atm C}$

__After 6 min:__

Since P_B has not changed from 5 min, equilibrium has been established and therefore P_A and P_C will also equal the values from 5 min.

12-2   THE EQUILIBRIUM CONSTANT EXPRESSION

6. This problem is similar to Examples 12.1(a) and 12.2. Recall that pure solids and liquids do not appear in equilibrium constant expressions, and that gases are represented by their partial pressures and solutes by their molarities.

   (a) $K = P_{CO_2}$

   (b) $K = \dfrac{(P_{CO})^2 (P_{H_2})^5}{(P_{C_2H_6})}$

   (c) $K = \dfrac{(P_{NH_3})^4 (P_{O_2})^5}{(P_{NO})^4 (P_{H_2O})^6}$

   (d) $K = \dfrac{1}{P_{NH_3}}$

8. Refer to Question 6 above.

   (a) $K = \dfrac{(P_{Cl_2})[I^-]^2}{[Cl^-]^2}$

   (b) $K = \dfrac{[CH_3NH_3^+]}{[CH_3NH_2][H^+]}$

(c) $K = \dfrac{[Au(CN)_4^{2-}]}{[Au^{2+}][CN^-]^4}$

10. This problem is similar to Example 12.2. See also Question 6 above.

(a) $C_3H_6O(l) \rightleftharpoons C_3H_6O(g)$ $\qquad\qquad\qquad$ $K = P_{C_3H_6O}$

(Note: the reactant is a liquid, so it does not appear in the expression for K).

(b) $7H_2(g) + 2NO_2(g) \rightleftharpoons 2NH_3(g) + 4H_2O(g)$ $\qquad$ $K = \dfrac{(P_{NH_3})^2(P_{H_2O})^4}{(P_{H_2})^7(P_{NO_2})^2}$

(c) $H_2S(g) + Pb^{2+}(aq) \rightleftharpoons PbS(s) + 2H^+(aq)$ $\qquad$ $K = \dfrac{[H^+]^2}{(P_{H_2S})[Pb^{2+}]}$

12. Begin with the species that are present in the expression for K — remember that products are on the numerator and reactants on the denominator. (If the partial pressure of a species appears in the expression for K it must be a gas; if it appears as a concentration then it must be a solute in solution.)
Then balance the equation as necessary: pure solids and liquids will be missing from the expression for K but may be required for a balanced equation.

(a) $2SO_2(g) + 2H_2O(g) \rightleftharpoons 2H_2S(g) + 3O_2(g)$ $\qquad\qquad$ (balanced)

(b) $IF(g) \rightleftharpoons \frac{1}{2}F_2(g) + \frac{1}{2}I_2(g)$ $\qquad\qquad$ (balanced)

(c) $Cl_2(g) + 2Br^-(aq) \rightleftharpoons 2Cl^-(aq)$ $\qquad\qquad$ (not balanced)

$\quad$ $Cl_2(g) + 2Br^-(aq) \rightleftharpoons 2Cl^-(aq) + Br_2(l)$ $\qquad\qquad$ (balanced)

(d) $2NO_3^-(aq) + 8H^+(aq) \rightleftharpoons 2NO(g) + 4H_2O(g) + 3Cu^{2+}(aq)$ $\qquad$ (not balanced)

$\quad$ $3Cu(s) + 2NO_3^-(aq) + 8H^+(aq)$
$\qquad\qquad \rightleftharpoons 2NO(g) + 4H_2O(g) + 3Cu^{2+}(aq)$ $\qquad\qquad$ (balanced)

14. (a) $K' = \dfrac{P_{NOCl}}{(P_{NO})(P_{Cl_2})^{1/2}}$

(b) This is related to example 12.1 (c).
Write an equation for the decomposition reaction, noting that *one mole* of $Cl_2(g)$ is the required product. The original equation must be reversed (Reciprocal Rule) and its coefficients multiplied by 2 (Coefficient Rule). Then write the expression for the corresponding equilibrium constant.

$$2NOCl(g) \rightleftharpoons 2NO(g) + Cl_2(g) \qquad K'' = \frac{(P_{NO})^2(P_{Cl_2})}{(P_{NOCl})^2}$$

(c) Apply the two rules:

$$K'' = \left(\frac{1}{K'}\right)^2 = \frac{1}{(K')^2}$$

The answer can be confirmed by applying both rules to the expression for K' given in part (a): the result should be the expression for K'' given in part (b).

12-3   DETERMINATION OF K

16. See pages 310–311. This problem is similar to parts of Example 12.1. Write a balanced reaction for the required equilibrium reaction and relate that to the one given.

(a) The required equation has **one mole** of H_2S as **product** and H_2 and S_2 as **reactants**. Thus the given reaction must be reversed (reciprocal rule) and its coefficients divided by two (coefficient rule).

$$H_2(g) + \tfrac{1}{2}S_2(g) \rightleftharpoons H_2S(g)$$

$$K' = \left(\frac{1}{K}\right)^{\frac{1}{2}} = \left(\frac{1}{2.2 \times 10^{-4}}\right)^{\frac{1}{2}} = 6.7 \times 10^1 \text{ or } \underline{67}$$

(b) The required equation has only **one mole** of H_2S as **reactant**. Thus the given reaction has its coefficients divided by two (coefficient rule).

$$K' = K^{\frac{1}{2}} = \left(2.2 \times 10^{-4}\right)^{\frac{1}{2}} = 1.5 \times 10^{-2} \text{ or } \underline{0.015}$$

18. See Section 12-2. This problem is similar to Example 12.1(d). Formation of one mole of NOBr from its elements proceeds according to the following equation:

$$\tfrac{1}{2}N_2(g) + \tfrac{1}{2}O_2(g) + \tfrac{1}{2}Br_2(g) \rightleftharpoons NOBr(g)$$

Since the elements N, O, and Br exist as diatomic molecules, one half-molecule of each is needed to provide the N, O, and Br atoms present in the product. As with the thermochemical equations in Chapter 8 (see Section 8-4), the equations given in the question must be rearranged so that they sum to the required reaction.
Note that the products (N_2 and O_2) in the first given equation are reactants in the goal equation and their coefficients must be halved. Apply the Reciprocal Rule and the Coefficient Rule (Table 12.3).

The reactant Br_2 and the product NOBr in the second given equation are also reactant and product, respectively, in the goal equation but their coefficients must be halved. Apply the Coefficient Rule.

$$\frac{1}{2}N_2(g) + \frac{1}{2}O_2(g) \rightleftharpoons NO(g) \qquad\qquad K_1 = \left(\frac{1}{1\times 10^{-30}}\right)^{\frac{1}{2}}$$

$$\frac{1}{2}Br_2(g) + NO(g) \rightleftharpoons NOBr(g) \qquad\qquad K_2 = (8 \times 10^1)^{\frac{1}{2}}$$

$$\frac{1}{2}N_2(g) + \frac{1}{2}O_2(g) + \frac{1}{2}Br_2(g) + NO(g)$$
$$\rightleftharpoons NO(g) + NOBr(g)$$

$$\frac{1}{2}N_2(g) + \frac{1}{2}O_2(g) + \frac{1}{2}Br_2(g) \rightleftharpoons NOBr(g)$$

Finally, apply the Rule of Multiple Equilibria to obtain *K* for the target equation:

$$K \quad = K_1 \times K_2 = \left(\frac{1}{1\times 10^{-30}}\right)^{\frac{1}{2}} (8 \times 10^1)^{\frac{1}{2}} = \underline{\textbf{9 x 10}^{\textbf{15}}}$$

20. See Section 12-2 and Problem 18 above.
Rearrange the given equations so that they sum to the required reaction.
The reactant A in the first equation is also the reactant in the goal equation but its coefficient must be divided by 3 (apply the Coefficient Rule).
The reactant D in the second given equation is a product in the goal equation and its coefficient must be divided by 3, so this second equation must be reversed and its coefficients divided by 3 (apply the Reciprocal Rule and the Coefficient Rule).

$$A(g) \rightleftharpoons B(g) + \tfrac{2}{3}C(g) \qquad\qquad K_1' = (K_1)^{\frac{1}{3}} = 0.31^{\frac{1}{3}}$$

$$\tfrac{2}{3}C(g) \rightleftharpoons D(g) + \tfrac{2}{3}B(g) \qquad\qquad K_2' = (K_2)^{-\frac{1}{3}} = \left(\frac{1}{2.8}\right)^{\frac{1}{3}}$$

$$A(g) + \tfrac{2}{3}C(g)$$
$$\rightleftharpoons B(g) + \tfrac{2}{3}C(g) + D(g) + \tfrac{2}{3}B(g)$$

$$A(g) \rightleftharpoons D(g) + \tfrac{5}{3}B(g)$$

Finally, apply the Rule of Multiple Equilibria to obtain *K* for the target equation:

$$K = K_1' \times K_2' = 0.31^{\frac{1}{3}} \times \left(\frac{1}{2.8}\right)^{\frac{1}{3}} = \underline{\textbf{0.48}}$$

22. This problem is similar to Example 12.3. Write an expression for K and substitute the given partial pressures in atm.

$$K = \frac{P_{CH_3OH}}{(P_{CO})(P_{H_2})^2} = \frac{0.0512 \text{ atm}}{(0.814 \text{ atm})(0.274 \text{ atm})^2} = \underline{\textbf{0.838}}$$

24. (a) This is the reverse of the second reaction given in Problem 18 above. Coefficients are halved so that one mol of NOBr is the reactant:

$$NOBr(g) \rightleftharpoons NO(g) + \tfrac{1}{2}Br_2(g)$$

(b) See Section 12-2. This part of the problem is similar to Examples 12.3 and 12.4. Recall that, since the compounds involved in the reaction are gases, the given concentrations must be converted to partial pressures in atm using the Ideal Gas Law (Chapter 5). Write an expression for K and substitute the partial pressures.

Note that the Ideal Gas Law can be written as $P = \dfrac{n}{V}RT$, and that the quantity $\dfrac{n}{V}$ is simply the molarity in mol/L.

$$P_{NOBr} = \frac{n}{V}RT = \frac{0.0162 \text{ mol}}{1\text{L}}(0.0821\text{L}\cdot\text{atm/mol}\cdot\text{K})(298\text{ K}) = 0.396 \text{ atm}$$

$$P_{NO} = \frac{n}{V}RT = \frac{0.0011 \text{ mol}}{1\text{L}}(0.0821\text{L}\cdot\text{atm/mol}\cdot\text{K})(298\text{ K}) = 0.027 \text{ atm}$$

$$P_{Br_2} = \frac{n}{V}RT = \frac{0.072 \text{ mol}}{1\text{L}}(0.0821\text{L}\cdot\text{atm/mol}\cdot\text{K})(298\text{ K}) = 1.76 \text{ atm}$$

$$K = \frac{(P_{NO})(P_{Br_2})^{1/2}}{P_{NOBr}} = \frac{(0.027)(1.76)^{1/2}}{0.396} = \underline{\textbf{0.090}}$$

26. This problem is similar to Example 12.4. We are told that the initial pressure of SO_3 is 0.541 atm and the equilibrium partial pressure of O_2 is 0.216 atm; it is implied that the initial pressures of SO_2 and O_2 are zero. Construct a table and enter this information:

	$2SO_3(g)$ $\rightleftharpoons$	$2SO_2(g)$ +	$O_2(g)$
P_0 (atm)	0.541	0	0
ΔP (atm)			
P_{eq} (atm)			0.216

From this it is seen that ΔP for O_2 is +0.216 atm and thus, according to the stoichiometry of the reaction, ΔP for SO_2 must be +2(0.216 atm) = +0.432 atm and ΔP for SO_3 must be −2(0.216 atm) = −0.432 atm:

	$2SO_3(g)$	$\rightleftharpoons$	$2SO_2(g)$	+	$O_2(g)$
P_0 (atm)	0.541		0		0
ΔP (atm)	−0.432		+0.432		+0.216
P_{eq} (atm)					0.216

Finally, complete the table to obtain the equilibrium concentrations of SO_2 and SO_3:

	$2SO_3(g)$	$\rightleftharpoons$	$2SO_2(g)$	+	$O_2(g)$
P_0 (atm)	0.541		0		0
ΔP (atm)	−0.432		+0.432		+0.216
P_{eq} (atm)	0.109		0.432		0.216

The P_{eq} row gives the equilibrium partial pressures for all three species, which now can be substituted into the expression for K:

$$K = \frac{(P_{SO_2})^2(P_{O_2})}{(P_{SO_3})^2} = \frac{(0.432)^2(0.216)}{(0.109)^2} = \underline{\mathbf{3.39}}$$

12-4   APPLICATIONS OF THE EQUILIBRIUM CONSTANT: K; Direction of the Reaction

28. (a) To determine whether the system is at equilibrium, calculate the value of the reaction quotient Q and compare it with the value of K. Recall that the form of the expression for Q is the same as the expression for K, except that it does not require equilibrium partial pressures.

$$Q = \frac{(P_{Cl_2})(P_{PCl_3})}{(P_{PCl_5})} = \frac{(0.65)(0.33)}{(0.026)} = 8.2$$

Since $Q \neq K$, the system is not at equilibrium.

(b) To determine the direction in which the system must move in order to reach equilibrium, compare Q with K.
Since $Q < K$, the reaction proceeds from left to right ($\rightarrow$).

30. These problems are similar to Example 12.5. To determine the direction in which the system must move in order to reach equilibrium, compare Q with K.

(a) $Q = \dfrac{(P_{NH_3})^2}{(P_{N_2})(P_{H_2})^3} = \dfrac{(0.01)^2}{(0.01)(0.01)^3} = 1 \times 10^4 > K$

$Q > K$, so the reaction proceeds from right to left ($\leftarrow$).

(b) $Q = \dfrac{(P_{NH_3})^2}{(P_{N_2})(P_{H_2})^3} = \dfrac{(0.0045)^2}{(0)(0)^3} = \infty > K$

$Q > K$, so the reaction proceeds from right to left ($\leftarrow$). Note that in this case, since the partial pressure of the reactants is zero, the reaction can *only* proceed to the left and thus calculations were actually unnecessary.

(c) $Q = \dfrac{(P_{NH_3})^2}{(P_{N_2})(P_{H_2})^3} = \dfrac{(0.0058)^2}{(1.2)(1.88)^3} = 4.2 \times 10^{-6} < K$

$Q < K$, so the reaction proceeds from left to right ($\rightarrow$).

32. This problem is similar to Example 12.5. To determine the direction in which the system must move in order to reach equilibrium, compare Q with K. However, in order to calculate Q, it is first necessary to calculate the partial pressure of each gas using the Ideal Gas Law (Chapter 5). Note that R and B will not appear in the expression for Q because they are a solid and a liquid, respectively.

$$P_Q = \frac{nRT}{V} = \frac{(0.30 \text{ mol})(0.0821 \text{L} \cdot \text{atm/mol} \cdot \text{K})(348 \text{ K})}{10.0 \text{ L}} = 0.86 \text{ atm}$$

$$P_A = \frac{nRT}{V} = \frac{(0.50 \text{ mol})(0.0821 \text{L} \cdot \text{atm/mol} \cdot \text{K})(348 \text{ K})}{10.0 \text{ L}} = 1.4 \text{ atm}$$

$$Q = \frac{P_A}{(P_Q)^2} = \frac{1.4}{(0.86)^2} = 1.9 < K$$

$Q < K$, so the reaction proceeds from left to right ($\rightarrow$).

12-4 APPLICATIONS OF THE EQUILIBRIUM CONSTANT: K; Extent of the Reaction

34. Knowing the expression for K and the equilibrium partial pressures of NH_3 and N_2, the partial pressure of H_2 can be calculated directly: write an expression for K, substitute known quantities and solve for partial pressure of H_2.

$$K = \frac{(P_{NH_3})^2}{(P_{N_2})(P_{H_2})^3} \qquad \text{so} \qquad (P_{H_2})^3 = \frac{(P_{NH_3})^2}{(P_{N_2})(K)}$$

Hence:

$$P_{H_2} = \left[\frac{(P_{NH_3})^2}{(P_{N_2})(K)} \right]^{\frac{1}{3}} = \left[\frac{(0.015)^2}{(1.2)(1.5 \times 10^{-5})} \right]^{\frac{1}{3}} = \underline{\mathbf{2.3 \text{ atm}}}$$

36. The balanced equation for the decomposition of 2 moles of ICl_3 is:

$$2ICl_3(s) \rightleftharpoons I_2(g) + 3Cl_2(g) \qquad K = 0.29$$

for which the expression for K is:

$$K = (P_{I_2})(P_{Cl_2})^3$$

(Remember that the solid ICl_3 will not appear in the expression for K.) Given that the equilibrium partial pressure of Cl_2 is three times that of I_2, we can write:

$$P_{Cl_2} = 3 P_{I_2}$$

Substitute for P_{Cl_2} in the expression for K, then solve for P_{I_2} and hence obtain P_{Cl_2}.

$$K = (P_{I_2})(P_{Cl_2})^3 = (P_{I_2})(3P_{I_2})^3 = 27(P_{I_2})^4 = 0.29$$

$$P_{I_2} = \left(\frac{0.29}{27}\right)^{\frac{1}{4}} = \underline{\mathbf{0.32\ atm}}$$

$$P_{Cl_2} = 3P_{I_2} = 3(0.32\ atm) = \underline{\mathbf{0.96\ atm}}$$

38. Recall (Section 5-5) that the total pressure, P_{total} for a mixture of gases is the sum of the partial pressures of the component gases. Thus:

$$P_{total} = P_{NO} + P_{NO_2} + P_{O_2} = 1.25\ atm$$

Since we know the partial pressure of O_2 is 0.515 atm, we can write:

$$P_{NO} + P_{NO_2} + 0.515\ atm = 1.25\ atm$$

$$P_{NO} + P_{NO_2} = (1.25 - 0.515)\ atm = 0.74\ atm \quad \text{and} \quad P_{NO_2} = 0.74 - P_{NO}$$

The expression for K and its value are also given in the problem:

$$K = \frac{(P_{NO})^2(P_{O_2})}{(P_{NO_2})^2} = 0.87$$

Substitute in this equation for the partial pressures of all three gases, then solve for P_{NO}:

$$K = \frac{(P_{NO})^2(P_{O_2})}{(P_{NO_2})^2} = \frac{(P_{NO})^2(0.515)}{(0.74 - P_{NO})^2} = 0.87$$

$$\frac{(P_{NO})^2}{(0.74 - P_{NO})^2} = 0.87/0.515 = 1.7$$

$$\frac{(P_{NO})}{(0.74 - P_{NO})} = (1.7)^{\frac{1}{2}} = 1.3$$

$$P_{NO} = 1.3(0.74 - P_{NO}) = 0.96 - 1.3\,P_{NO}$$

$$2.3\,P_{NO} = 0.96$$

$$P_{NO} = 0.96/2.3 = \underline{\textbf{0.42 atm}} \qquad \text{and hence} \qquad P_{NO_2} = (0.74 - P_{NO})\ \text{atm} = \underline{\textbf{0.32 atm}}$$

40. This problem is similar to Example 12.6. We are told that the initial partial pressure of all three gases is 0.228 atm and that the equilibrium constant is 2.71. First determine the direction in which the system must move in order to reach equilibrium, by comparing Q with K.

$$Q = \frac{(P_Z)^2}{(P_R)(P_Q)} = \frac{(0.228)^2}{(0.228)(0.228)} = 1\ <\ K$$

Since $Q < K$, the reaction proceeds from left to right; in other words, the partial pressures of R and Q will decrease and the partial pressure of Z will increase.

Construct a table and enter the given information:

	R(g) +	Q(g) ⇌	2Z(g)
P_0 (atm)	0.228	0.228	0.228
ΔP (atm)			
P_{eq} (atm)			

Let ΔP for R be $-x$ (since it is decreasing), then by stoichiometry ΔP for Q is also $-x$ and ΔP for Z is $+2x$. Add this information to the table and use it to obtain expressions for P_{eq}:

	R(g) +	Q(g) ⇌	2Z(g)
P_0 (atm)	0.228	0.228	0.228
ΔP (atm)	$-x$	$-x$	$+2x$
P_{eq} (atm)	0.228−x	0.228−x	0.228+2x

Now substitute P_{eq} for each gas into the expression for K:

$$K = \frac{(P_Z)^2}{(P_R)(P_Q)} = \frac{(0.228 + 2x)^2}{(0.228 - x)(0.228 - x)} = \frac{(0.228 + 2x)^2}{(0.228 - x)^2} = 2.71$$

Take the square root of each side of the equation:

$$\frac{(0.228 + 2x)}{(0.228 - x)} = (2.71)^{\frac{1}{2}} = 1.65$$

Now solve for x:

$0.228 + 2x = (1.65)(0.228 - x) = 0.376 - 1.65x$

$1.65x + 2x = 0.376 - 0.228$ or $3.65x = 0.148$ so $x = 0.0405$

Thus the equilibrium partial pressure of each gas is:

$P_R = P_Q = 0.228 - x = 0.228 - 0.0405 = $ **0.188 atm**

$P_Z = 0.228 + 2x = 0.228 + 2(0.0405) = $ **0.309 atm**

42. This problem is similar to Example 12.6.

(a) We are told that the initial partial pressure of all three gases is 1.25 atm and that the equilibrium constant is 84.7. First determine the direction in which the system must move in order to reach equilibrium, by comparing Q with K.

$$Q = \frac{(P_{NO})(P_{SO_3})}{(P_{SO_2})(P_{NO_2})} = \frac{(1.25)(1.25)}{(1.25)(1.25)} = 1 < K$$

Since $Q < K$, the reaction proceeds from left to right to reach equilibrium; in other words, the partial pressures of SO_2 and NO_2 will decrease and the partial pressures of NO and SO_3 will increase.

Construct a table and enter the given information:

	$SO_2(g)$ +	$NO_2(g)$ $\rightleftharpoons$	$NO(g)$ +	$SO_3(g)$
P_0 (atm)	1.25	1.25	1.25	1.25
ΔP (atm)				
P_{eq} (atm)				

Let ΔP for SO_2 be $-x$ (since it is decreasing), then by stoichiometry ΔP for NO_2 is also $-x$ and ΔP for NO and SO_3 is $+x$ in each case. Add this information to the table and use it to obtain expressions for P_{eq}:

	$SO_2(g)$ +	$NO_2(g)$ $\rightleftharpoons$	$NO(g)$ +	$SO_3(g)$
P_0 (atm)	1.25	1.25	1.25	1.25
ΔP (atm)	$-x$	$-x$	$+x$	$+x$
P_{eq} (atm)	$1.25-x$	$1.25-x$	$1.25+x$	$1.25+x$

Now substitute P_{eq} for each gas into the expression for K:

$$K = \frac{(P_{NO})(P_{SO_3})}{(P_{SO_2})(P_{NO_2})} = \frac{(1.25+x)(1.25+x)}{(1.25-x)(1.25-x)} = \frac{(1.25+x)^2}{(1.25-x)^2} = 84.7$$

Take the square root of each side of the equation:

$$\frac{(1.25+x)}{(1.25-x)} = (84.7)^{\frac{1}{2}} = 9.20$$

Now solve for x:

$$1.25 + x = (9.20)(1.25 - x) = 11.5 - 9.20x$$

$$x + 9.20x = 11.5 - 1.25 \quad \text{or} \quad 10.20x = 10.2 \qquad \text{so } x = 1.00$$

Thus the equilibrium partial pressure of each gas is:

$$P_{SO_2} = P_{NO_2} = 1.25 - x = 1.25 - 1.00 = \underline{\textbf{0.25 atm}}$$
$$P_{SO_3} = P_{NO} = 1.25 + x = 1.25 + 1.00 = \underline{\textbf{2.25 atm}}$$

(b) The total pressure is the sum of the partial pressures. Thus the initial total pressure is:

$$P_{total} = P_{SO_2} + P_{NO_2} + P_{SO_3} + P_{NO} = (1.25 + 1.25 + 1.25 + 1.25) \text{ atm} = \underline{\textbf{5.00 atm}}$$

and the total pressure at equilibrium is:

$$P_{total} = P_{SO_2} + P_{NO_2} + P_{SO_3} + P_{NO} = (0.25 + 0.25 + 2.25 + 2.25) \text{ atm} = \underline{\textbf{5.00 atm}}$$

The initial total pressure is the same as the total pressure at equilibrium.
This applies only when $\Delta n = 0$ (that is, when the number of moles of gas is the same on both sides of the reaction equation): recall from the Ideal Gas Law (Section 5-3) that the pressure of a gas is directly proportional to the number of moles of gas present. Any reaction, therefore, in which there is a change in the number of moles of gas as the reaction proceeds will be accompanied by a change in total pressure.

44. This problem is similar (in part) to Example 12.6.

(a) We are told that PH_3BCl_3 is placed in the flask, so the initial pressures of the two gases will be zero. Note also that because PH_3BCl_3 is a solid it will not appear in the equilibrium constant expression.

$$K = \frac{(P_{PH_3})(P_{BCl_3})}{1} = (P_{PH_3})(P_{BCl_3}) = 0.054$$

Set up a table and insert the given information:

	$PH_3BCl_3(s)$	$\rightleftharpoons$	$PH_3(g)$	+	$BCl_3(g)$
P_0 (atm)			0		0
ΔP (atm)					
P_{eq} (atm)					

Let ΔP for PH_3 be $+x$ (since it is initially zero, it can only be increasing), then by stoichiometry ΔP for BCl_3 is also $+x$. Add this information to the table, use it to obtain expressions for P_{eq} in terms of x, and then solve for x.

	$PH_3BCl_3(s)$	$\rightleftharpoons$	$PH_3(g)$	+	$BCl_3(g)$
P_0 (atm)			0		0
ΔP (atm)			$+x$		$+x$
P_{eq} (atm)			x		x

$$K = (P_{PH_3})(P_{BCl_3}) = (x)(x) = x^2 = 0.054 \qquad \text{Hence } x = (0.054)^{1/2} = 0.23.$$

Thus the equilibrium partial pressure of each gas is:

$$P_{PH_3} = P_{BCl_3} = x = \underline{0.23 \text{ atm}}$$

and the total pressure is the sum of the partial pressures:

$$P_{total} = P_{PH_3} + P_{BCl_3} = 0.23 \text{ atm} + 0.23 \text{ atm} = \underline{\textbf{0.46 atm}}$$

(b) In order to determine the mass of PH_3BCl_3 remaining at equilibrium, determine the number of moles of each gas present using the Ideal Gas Law (Section 5-3) and hence the number of moles of PH_3BCl_3 consumed; from this calculate the mass of PH_3BCl_3 consumed and hence the mass remaining.

Strategy: $P\ PH_3 \rightarrow$ mol $PH_3 \rightarrow$ mol PH_3BCl_3 consumed $\rightarrow$ mass PH_3BCl_3 consumed $\rightarrow$ mass PH_3BCl_3 remaining

$$\text{mol } PH_3 \text{ present} = n = \frac{PV}{RT} = \frac{(0.23 \text{ atm})(5.0 \text{ L})}{(0.0821 \text{ L} \cdot \text{atm/mol} \cdot \text{K})(353.15 \text{ K})} = 0.040 \text{ mol } PH_3$$

$$\text{mol } PH_3BCl_3 \text{ consumed} = 0.040 \text{ mol } PH_3 \times \frac{1 \text{ mol } PH_3BCl_3}{1 \text{ mol } PH_3} = 0.040 \text{ mol } PH_3BCl_3$$

$$\text{mass } PH_3BCl_3 \text{ consumed} = 0.040 \text{ mol } PH_3BCl_3 \times \frac{151.167 \text{ g } PH_3BCl_3}{1 \text{ mol } PH_3BCl_3} = 6.0 \text{ g } PH_3BCl_3$$

$$\text{mass } PH_3BCl_3 \text{ remaining} = 20.0 \text{ g} - 6.0 \text{ g} = \underline{\textbf{14.0 g}}$$

46. This problem is similar to Example 12.6.

We are told that only HI is present initially at a partial pressure of 0.200 atm. Thus, the initial partial pressures of H_2 and I_2 are zero.

Set up an equilibrium table and insert the given information:

	2HI(g)	⇌	$H_2(g)$	+	$I_2(g)$
P_0 (atm)	0.200		0		0
ΔP (atm)					
P_{eq} (atm)					

Let ΔP for H_2 be $+x$ (since it is initially zero, it can only be increasing), then by stoichiometry ΔP for I_2 is also $+x$ and ΔP for HI must be $-2x$. Add this information to the table, use it to obtain expressions for P_{eq} in terms of x, and then solve for x.

	2HI(g)	⇌	$H_2(g)$	+	$I_2(g)$
P_0 (atm)	0.200		0		0
ΔP (atm)	$-2x$		$+x$		$+x$
P_{eq} (atm)	$0.200 - 2x$		x		x

$$K = \frac{(P_{H_2})(P_{I_2})}{(P_{HI})^2} = \frac{(x)(x)}{(0.200 - 2x)^2} = \frac{x^2}{(0.200 - 2x)^2} = 0.0169$$

Hence, $\dfrac{x}{0.200 - 2x} = (0.0169)^{1/2} = 0.13$

So: $x = 0.13(0.200 - 2x) = 0.026 - 0.26x$ or $0.026 = x + 0.26x = 1.26x$

Therefore, $x = \dfrac{0.026}{1.26} = 0.021$

Thus the equilibrium partial pressure of each gas is:

$P_{H_2} = P_{I_2} = x = $ **0.021 atm**

12-5   EFFECT OF CHANGES IN CONDITIONS ON AN EQUILIBRIUM SYSTEM
Changing Reaction Conditions; Le Châtelier's Principle

48. See page 323: When a system at equilibrium is expanded (when the volume increases), reaction takes place in the direction that *increases* the total number of moles of gas.

(a) There are three moles of gas on the reactant side of the equation and two moles of gas on the product side, so reaction will take place to increase the number of moles of reactants: since pressure is proportional to number of moles of gas, **the pressure of the reactants will increase.**

(b) There are two moles of gas on the reactant side of the equation and four moles of gas on the product side, so reaction will take place to increase the number of moles of products: since pressure is proportional to number of moles of gas, **the pressure of the products will increase.**

(c) There are two moles of gas on each side of the equation, so reaction in either direction will not change the total number of moles of gas present and therefore changing the volume has no effect on the system at equilibrium: **the pressures of the reactants and products will remain the same.**

50. (a) (1) $O_2(g)$ is a gaseous reactant. When some is removed, the system will shift to the left to restore part of the removed reactant.
This will cause an **increase** in the amount of NH_3.

(2) $N_2(g)$ is a gaseous product. When some is added, the system will shift to the left to consume part of the added product.
This will cause an **increase** in the amount of NH_3.

(3) $H_2O(l)$ is a pure liquid. It will have **no effect** on the position of equilibrium so it will not affect the amount of NH_3.

(4) Expanding the container (increasing the volume) should decrease the total pressure, and so if the expansion takes place at constant pressure then reaction must take place in the direction that increases the total number of moles of gas (since pressure is proportional to number of moles of gas). There are 7 moles of gas on the reactant side of the equation and 2 moles of gas on the product side, so reaction will take place to increase the number of moles of reactants.
This will cause an **increase** in the amount of NH_3.

(5) An increase in temperature causes reaction to occur in the endothermic direction. For this reaction, $\Delta H = -1530.4$ kJ in the forward direction: it is exothermic, so the reverse direction is endothermic. Thus, increasing the temperature causes the reaction to take place in the reverse direction (towards the reactant side).
This will cause an **increase** in the amount of NH_3.

(b) None of factors (1) − (4) will change the value of K; the only factor that will affect K is changing temperature. For factor (5), apply Equation 12.5 (see pages 327–328): for an exothermic reaction, increasing temperature causes K to **decrease**.

52. See pages 325–326 and Example 12.8. When the pressure on a system at equilibrium decreases by expansion, reaction takes place in the direction that *increases* the total number of moles of gas.

(a) There are zero moles of gas on the reactant side of the equation and one mole of gas on the product side, so reaction will take place to increase the number of moles of products: **the system will shift to the right** (→).

(b) There are two moles of gas on the reactant side of the equation and four moles of gas on the product side, so reaction will take place to increase the number of moles of products: **the system will shift to the right** ($\rightarrow$).

(c) There is one mole of gas on the reactant side of the equation and two moles of gas on the product side, so reaction will take place to increase the number of moles of products: **the system will shift to the right** ($\rightarrow$).

54. (a) Write an expression for K and substitute the equilibrium partial pressures given in the question. Recall that the solid R will not appear in the expression for K.

$$K = \frac{P_{X_2R}}{P_{X_2}} = \frac{0.98}{4.3} = \underline{0.23}$$

(b) Since R is a pure solid, it will not affect the position of equilibrium ($Q = K$ after the solid is added), and therefore the partial pressures of the two gases will be unchanged.

$$P_{X_2} = \underline{\text{4.3 atm}} \qquad\qquad P_{X_2R} = \underline{\text{0.98 atm}}$$

(c) This part of the problem is similar to Example 12.7. Set up a table with the given partial pressures of X_2 (2.0 atm) and X_2R (0.98 atm) as initial pressures. Since the equilibrium was disturbed by decreasing the partial pressure of X_2 (a gaseous reactant), we know that the new equilibrium position will be achieved by some X_2R being consumed to replace some of the removed X_2. Let ΔP for X_2 be $+x$ (since we know it will increase), and by stoichiometry ΔP for X_2R is $-x$. Add this information to the table, obtain expressions for P_{eq} in terms of x, substitute into the expression for K, and then solve for x. The value of K, of course, is unchanged.

	$X_2(g)$ + R(s) $\rightleftharpoons$	$X_2R(g)$
P_0 (atm)	2.0	0.98
ΔP (atm)	$+x$	$-x$
P_{eq} (atm)	$2.0 + x$	$0.98 - x$

$$K = \frac{P_{X_2R}}{P_{X_2}} = \frac{0.98 - x}{2.0 + x} = 0.23$$

$$(0.98 - x) = (2.0 + x)0.23 = 0.46 + 0.23x$$

$$0.98 - 0.46 = x + 0.23x \qquad\text{so}\qquad 0.52 = 1.23x \qquad\text{hence:}\qquad x = 0.42$$

Thus the new equilibrium pressures are:

$$P_{X_2} = 2.0 + x = \underline{\text{2.4 atm}} \qquad\qquad P_{X_2R} = 0.98 - x = \underline{\text{0.56 atm}}$$

56. (a) Write an expression for K and substitute the equilibrium partial pressures given in the question.

$$K = \frac{(P_{SO_2})(P_{Cl_2})}{(P_{SO_2Cl_2})} = \frac{(1.88)(0.84)}{(0.27)} = \underline{\mathbf{5.9}}$$

(b) This part of the problem is similar to Example 12.7. Set up a table with the given partial pressures of SO_2 (1.88 atm) and SO_2Cl_2 (0.27 atm), and the new partial pressure of Cl_2 (0.68 atm), as initial pressures. Since the equilibrium was disturbed by removing some Cl_2, a gaseous product, we know that the new equilibrium position will be achieved by some of the reactant SO_2Cl_2 being consumed to partially replace the removed Cl_2; the partial pressures of SO_2 and Cl_2, therefore, will increase. Let ΔP for SO_2 be $+x$ (since we know it will increase), then by stoichiometry ΔP for Cl_2 is $+x$ and for SO_2Cl_2 is $-x$. Add this information to the table, use it to obtain expressions for P_{eq} in terms of x, substitute into the expression for K, and then solve for x. The value of K, of course, is unchanged.

	$SO_2Cl_2(g)$	$\rightleftharpoons$	$SO_2(g)$	$+$	$Cl_2(g)$
P_0 (atm)	0.27		1.88		0.68
ΔP (atm)	$-x$		$+x$		$+x$
P_{eq} (atm)	$0.27 - x$		$1.88 + x$		$0.68 + x$

$$K = \frac{(P_{SO_2})(P_{Cl_2})}{(P_{SO_2Cl_2})} = \frac{(1.88 + x)(0.68 + x)}{(0.27 - x)} = 5.9$$

$$(1.88 + x)(0.68 + x) = (0.27 - x)5.9$$

$$1.3 + 2.56x + x^2 = 1.6 - 5.9x$$

$$x^2 + 8.5x - 0.3 = 0$$

$$x = \frac{-b \pm \sqrt{b^2 - 4ac}}{2a} = \frac{-8.5 \pm \sqrt{(8.5)^2 - 4(1)(-0.3)}}{2(1)}$$

$$x = 0.035 \text{ or } -8.6$$

The negative root of the equation is physically impossible and we already know that the partial pressures of SO_2 and Cl_2 are to increase (so $x > 0$); therefore $x = 0.035$.

Thus the new equilibrium pressures are:

$$P_{SO_2Cl_2} = 0.27 - x = \underline{\mathbf{0.24 \text{ atm}}} \qquad P_{SO_2} = 1.88 + x = \underline{\mathbf{1.92 \text{ atm}}}$$

$$P_{Cl_2} = 0.68 + x = \underline{\mathbf{0.72 \text{ atm}}}$$

58. The variation of K with temperature T is given by the van't Hoff equation (Equation 12.5). See also the example in the text on page 327 (immediately above Example 12.8). In order to use this equation, we must first calculate $\Delta H°$ for the reaction (see also Example 8.7) using $\Delta H_f°$ values (Appendix 1).

$$\Delta H° = \Sigma \Delta H_f°_{(products)} - \Sigma \Delta H_f°_{(reactants)}$$
$$\Delta H° = [(2 \text{ mol})(-296.8 \text{ kJ/mol}) + (1 \text{ mol})(0 \text{ kJ/mol})] - [(2 \text{ mol})(-395.7 \text{ kJ/mol})]$$
$$\Delta H° = 197.8 \text{ kJ} = 197.8 \times 10^3 \text{ J}$$

Now use Equation 12.5 with $\Delta H° = 197.8 \times 10^3$ J and $K_1 = 1.32$ at temperature $T_1 = 900$ K and solve for the equilibrium constant K_2 at temperature $T_2 = 828$ K:

$$\ln \frac{K_2}{K_1} = \frac{\Delta H°}{R} \left[\frac{1}{T_1} - \frac{1}{T_2} \right]$$

$$\ln \frac{K_2}{1.32} = \frac{197.8 \times 10^3 \text{ J/mol}}{8.3145 \text{ J/mol} \cdot \text{K}} \times \left[\frac{1}{900 \text{ K}} - \frac{1}{828 \text{ K}} \right] = -2.30$$

$$\frac{K_2}{1.32} = e^{-2.30} = 0.10 \qquad K_2 = 1.32 \times 0.10 = \underline{\textbf{0.13}}$$

60. The variation of K with temperature T is given by the van't Hoff equation (Equation 12.5). We are told that the equilibrium constant K_1 at a temperature of 37 °C ($T_1 = 310$ K) is 48 % of the equilibrium constant K_2 at a temperature of 27 °C ($T_2 = 300$ K). Thus $K_1 = 0.48K_2$. Use Equation 12.5 with the two values of T and the two expressions for K to solve for $\Delta H°$:

$$\ln \frac{K_2}{K_1} = \frac{\Delta H°}{R} \left[\frac{1}{T_1} - \frac{1}{T_2} \right]$$

$$\ln \frac{K_2}{0.48K_2} = \frac{\Delta H°}{8.3145 \text{ J/mol} \cdot \text{K}} \left[\frac{1}{310 \text{ K}} - \frac{1}{300 \text{ K}} \right]$$

$$\ln \frac{1}{0.48} = 0.73 = \Delta H°(-1.29 \times 10^{-5} \text{ J/mol})$$

$$\Delta H° = -57 \times 10^3 \text{ J/mol} = \underline{\textbf{-57 kJ/mol}}$$

UNCLASSIFIED PROBLEMS

62. (a) Use the Ideal Gas Law (Chapter 5) to calculate the partial pressure of each gas at equilibrium. To do this, however, we need to know the number of moles of each gas present at equilibrium. Set up a table like that in Example 12.6, but using mol (n) instead of partial pressure. Initially, for $S(CH_2CH_2Cl)_2$, zero mol are present and at equilibrium 0.0349 mol are formed, so $\Delta n = +0.0349$ mol. Given the number of moles

of SCl_2 and C_2H_4 present initially, use the stoichiometry of the reaction in order to calculate n_{eq} for these two gases

	$SCl_2(g)$	+	$2C_2H_4(g)$	$\rightleftharpoons$	$S(CH_2CH_2Cl)_2(g)$
n_0 (mol)	0.258		0.592		0
Δn (mol)	−0.0349		−2(0.0349)		+0.0349
n_{eq} (mol)	0.223		0.522		0.0349

$$P_{SCl_2} = \frac{nRT}{V} = \frac{(0.223\ \text{mol})(0.0821\ \text{L} \cdot \text{atm/mol} \cdot \text{K})(293\ \text{K})}{(5.0\ \text{L})} = \underline{\textbf{1.1 atm}}$$

$$P_{C_2H_4} = \frac{nRT}{V} = \frac{(0.522\ \text{mol})(0.0821\ \text{L} \cdot \text{atm/mol} \cdot \text{K})(293\ \text{K})}{(5.0\ \text{L})} = \underline{\textbf{2.5 atm}}$$

$$P_{S(CH_2CH_2Cl)_2} = \frac{nRT}{V} = \frac{(0.0349\ \text{mol})(0.0821\ \text{L} \cdot \text{atm/mol} \cdot \text{K})(293\ \text{K})}{(5.0\ \text{L})} = \underline{\textbf{0.17 atm}}$$

(b) $K = \dfrac{(P_{S(CH_2CH_2Cl)_2})}{(P_{SCl_2})(P_{C_2H_4})^2} = \dfrac{(0.17)}{(1.1)(2.5)^2} = \underline{\textbf{0.025}}$

64. Bearing in mind that solids do not appear in the equilibrium constant expression, K for this reaction is given by

$$K = P_{CO_2} = 1.04$$

Thus a partial pressure of 1.04 atm of CO_2 is present at equilibrium. Use the Ideal Gas Law (Chapter 5) with $V = 5.00$ L and $T = 1173$ K (given) to calculate the corresponding number of moles of CO_2:

$$n = \frac{PV}{RT} = \frac{(1.04\ \text{atm})(5.00\ \text{L})}{(0.0821\ \text{L} \cdot \text{atm/mol} \cdot \text{K})(1173\ \text{K})} = 0.0540\ \text{mol}$$

Use the stoichiometry of the reaction to calculate the number of moles of $CaCO_3$ required to form 0.0540 mol of CO_2 and hence the required mass of $CaCO_3$.
Strategy: mol CO_2 → mol $CaCO_3$ → mass $CaCO_3$

$$\text{mass } CaCO_3 \text{ required} = 0.0540\ \text{mol } CO_2 \times \frac{1\ \text{mol } CaCO_3}{1\ \text{mol } CO_2} \times \frac{100.087\ \text{g } CaCO_3}{1\ \text{mol } CaCO_3} = \underline{\textbf{5.40 g}}$$

66. The equilibrium constant expression for this reaction is given by

$$K = \frac{P_{H_2S}}{P_{H_2}}$$

(remembering that the solid sulfur will not appear in the expression for K). Thus, in order to calculate K we require the equilibrium partial pressure of each gas.

(a) **At 25°C (298 K)**: n for each gas is given so the partial pressures may be calculated using the Ideal Gas Law (Chapter 5) with the given $V = 2.00$ L and $T = 298$ K:

$$P_{H_2} = \frac{nRT}{V} = \frac{(0.120 \text{ mol})(0.0821 \text{L} \cdot \text{atm/mol} \cdot \text{K})(298 \text{ K})}{(2.00 \text{ L})} = 1.47 \text{ atm}$$

$$P_{H_2S} = \frac{nRT}{V} = \frac{(0.034 \text{ mol})(0.0821 \text{L} \cdot \text{atm/mol} \cdot \text{K})(298 \text{ K})}{(2.00 \text{ L})} = 0.4\underline{1}6 \text{ atm}$$

Thus, at 25°C $\qquad K = \dfrac{P_{H_2S}}{P_{H_2}} = \dfrac{0.416}{1.47} = \underline{\mathbf{0.28}}$

(b) **At 35°C (308 K)**: we are told that upon heating to 35°C the pressure of H_2 increases to 1.56 atm. Set up a table as in Example 12.7 with initial partial pressures from part(a) above. Knowing P_0 and P_{eq} for H_2, ΔP for H_2 must be +0.09 atm: use this information to complete the table and hence calculate the new value of K from the new equilibrium partial pressures.

	$H_2(g)$	+	$S(s)$	$\rightleftharpoons$	$H_2S(g)$
P_0 (atm)	1.47				0.416
ΔP (atm)	+0.09				−0.09
P_{eq} (atm)	1.56				0.33

Thus, at 35°C $K = \dfrac{P_{H_2S}}{P_{H_2}} = \dfrac{0.33}{1.56} = \underline{\mathbf{0.21}}$

CONCEPTUAL PROBLEMS

68. (a) Recall (Chapter 5 and the Ideal Gas Law) that the partial pressure of each gas is proportional to the number of moles of that gas present. Thus, in the first time period $(0 \rightarrow 50 \text{ s})$, P_A falls by 0.75 atm while P_B increases by 0.50 atm; in other words, $\Delta P_A = 1.5\Delta P_B$. The same can be observed for any of the other given time intervals. Therefore, it may be concluded that for every 1 mol of B that is formed 1.5 mol of A are consumed, so the stoichiometric ratio in the reaction must be such that we have (1.5 mol A):(1 mol B), or (3 mol A):(2 mol B):

$$3A(g) \rightleftharpoons 2B(g)$$

(b) The pressures of the two gases are still changing, even at time = 250 s, whereas at equilibrium pressures should remain constant. Therefore: **NO, the system has not reached equilibrium**.

70. See Example 12.5; the expression for K is given by

$$K = \frac{(P_{CO})^2}{(P_{CO_2})} = 168$$

To determine the direction in which the system will move in order to reach equilibrium, calculate Q and compare with K. Initially only $C(s)$ and $CO_2(g)$ are present, so $P_{CO} = 0$:

$$Q = \frac{(P_{CO})^2}{(P_{CO_2})} = 0 < K$$

Thus the system must move to the right to reach equilibrium: P_{CO} will increase and both P_{CO_2} and the quantity of $C(s)$ will decrease (this eliminates (d) as a possible answer).

From the above expression for K and the value of K we can say that at equilibrium

$$(P_{CO})^2 = 168 P_{CO_2} \qquad \text{or} \qquad P_{CO} \approx 13(P_{CO_2})^{1/2}$$

from which it is clear that $P_{CO} \gg P_{CO_2}$ at equilibrium. This eliminates statements (a) and (c) and confirms statement (b) as the correct one:

(b) The equilibrium mixture contains mostly CO.

72. See Section 12-5. The equilibrium has been disturbed by increasing the partial pressure of N_2O_4, so the system will respond by shifting in the direction that will consume **_some_** of the added N_2O_4. Thus, after 100 s the partial pressure of N_2O_4 will decrease from 1.0 atm and then level off when the new equilibrium condition is reached. Not all of the added N_2O_4 will be consumed when the new equilibrium is established, so the partial pressure of N_2O_4 at equilibrium will be higher than it was prior to 100 s.

74. See Section 12-5. The box on the left (corresponding to T = 300 K) contains 3 molecules of A_2, 4 molecules of B_2, and 3 molecules of AB_3, whereas the box on the right (corresponding to T = 400 K) contains 2 molecules of A_2, 1 molecule of B_2, and 5 molecules of AB_3. At the higher temperature, therefore, products are favored and reactants are disfavored: in other words, the position of equilibrium lies further to the right and ***K is higher at higher temperatures***. From page 326, this corresponds to a reaction that is **endothermic**.

76. See Sections 12-2 and 12-5. In discussing K, **a balanced chemical equation is required**, otherwise we do not know the form of the expression for K and hence the given value of K would be meaningless. In addition, since K changes with temperature, **temperature must also be specified**.

CHALLENGE PROBLEMS

77. In order to calculate K, we require the equilibrium partial pressure of each gas, but we only know the total pressure at equilibrium and the initial partial pressures of the two reactants. Set up a table as in Example 12.4 and include the information given in the question; let ΔP for $O_2 = -x$.

	$2NO(g)$	$+$	$O_2(g)$	$\rightleftharpoons$	$2NO_2(g)$
P_0 (atm)	0.81		0.70		0
ΔP (atm)	$-2x$		$-x$		$+2x$
P_{eq} (atm)	$0.81 - 2x$		$0.70 - x$		$2x$

We have expressions (in terms of x) for the equilibrium partial pressures of all three gases and we also know that the total pressure (which equals the sum of the partial pressures — see Section 5-5). Thus:

$$P_{total} = 1.20 \text{ atm} = P_{NO} + P_{O_2} + P_{NO_2} = [(0.81 - 2x) + (0.70 - x) + (2x)] \text{ atm}$$

$$1.20 \text{ atm} = (1.51 - x) \text{ atm} \qquad \text{so} \quad x = 0.31$$

$$P_{NO} = 0.81 - 2x = 0.19 \text{ atm}$$
$$P_{O_2} = 0.70 - x = 0.39 \text{ atm}$$
$$P_{NO_2} = 2x = 0.62 \text{ atm}$$

Knowing the equilibrium partial pressures, we can calculate K:

$$K = \frac{(P_{NO_2})^2}{(P_{NO})^2 (P_{O_2})} = \frac{(0.62)^2}{(0.19)^2 (0.39)} = \underline{27}$$

78. See Section 12-2, page 310. For the following general gas-phase reaction,

$$aA(g) + bB(g) + cC(g) + \ldots\ldots \rightleftharpoons xX(g) + yY(g) + zZ(g) + \ldots\ldots$$

the expression for K will be given by:

$$K = \frac{(P_X)^x \times (P_Y)^y \times (P_Z)^z \times \ldots}{(P_A)^a \times (P_B)^b \times (P_C)^c \times \ldots}$$

and the corresponding concentration equilibrium constant will be given by:

$$K_c = \frac{[X]^x \times [Y]^y \times [Z]^z \times \ldots}{[A]^a \times [B]^b \times [C]^c \times \ldots}$$

From the Ideal Gas Law, we know that $PV = nRT$, so for each gas present (assuming ideal behavior) we can write an expression of the form:

$$P_A = \frac{n_A RT}{V} = \frac{n_A}{V} \times RT = [A] \times RT \qquad \text{since } (n_A/V) \text{ is the molarity of A}$$

A similar expression may be written for all of the gases involved in the reaction. Substituting these into the expression for K gives:

$$K = \frac{([X]RT)^x \times ([Y]RT)^y \times ([Z]RT)^z \times \ldots}{([A]RT)^a \times ([B]RT)^b \times ([C]RT)^c \times \ldots}$$

$$= \frac{\{[X]^x[Y]^y[Z]^z \ldots\} \times (RT)^{x+y+z+\ldots}}{\{[A]^a[B]^b[C]^c \ldots\} \times (RT)^{a+b+c+\ldots}}$$

$$= K_c \times (RT)^{\{(x+y+z+\ldots)-(a+b+c+\ldots)\}}$$

$$= K_c \times (RT)^{\Delta n_g}$$

where $\Delta n_g = (x + y + z + \ldots) - (a + b + c + \ldots)$ in other words, Δn_g is the change in number of moles gas as the reaction proceeds from left → right.

79. This problem is similar to Example 12.6. Create a table with $P_0 = 1.00$ atm NH_3 and $P_0 =$ zero for the other two gases. The direction of reaction is implied to be from left → right, since the NH_3 is decomposing; let $\Delta P = x$ for N_2

	$2NH_3(g)$	$\rightleftharpoons$	$N_2(g)$	+	$3H_2(g)$
P_0 (atm)	1.00		0		0
ΔP (atm)	$-2x$		$+x$		$+3x$
P_{eq} (atm)	$1.00 - 2x$		x		$3x$

Now substitute the expressions for P_{eq} into the expression for K:

$$K = \frac{(P_{N_2})(P_{H_2})^3}{(P_{NH_3})^2} = \frac{(x)(3x)^3}{(1.0 - 2x)^2} = \frac{27x^4}{(1.0 - 2x)^2} = 2.5$$

Taking the square root of both sides of the equation then gives

$$\frac{3\sqrt{3}x^2}{(1.0 - 2x)} = \sqrt{2.5} \qquad\qquad 3\sqrt{3}x^2 = \sqrt{2.5}(1.0 - 2x) = \sqrt{2.5} - 2\sqrt{2.5}x)$$

$$3\sqrt{3}x^2 + 2\sqrt{2.5}x - \sqrt{2.5} = 0$$

$$x = \frac{-b \pm \sqrt{b^2 - 4ac}}{2a} = \frac{-2\sqrt{2.5} \pm \sqrt{(2\sqrt{2.5})^2 - 4(3\sqrt{3})(-\sqrt{2.5})}}{2(3\sqrt{3})}$$

$$x = +0.33 \text{ or } -0.93$$

The negative root of the equation is physically impossible therefore $x = 0.33$. Thus the equilibrium partial pressures of each gas are:

$$P_{NH_3} = 1.0 - 2x = \underline{\mathbf{0.34 \text{ atm}}}$$

$$P_{N_2} = x = \underline{\mathbf{0.33 \text{ atm}}}$$

$$P_{H_2} = 3x = \underline{\mathbf{0.99 \text{ atm}}}$$

80. The quantity of I_2 present at equilibrium can be calculated from the titration data. Use the flowchart for solution stoichiometry (Figure 4.6).
    Strategy: vol $Na_2S_2O_3 \rightarrow$ mol $Na_2S_2O_3 \rightarrow$ mol $S_2O_3^{2-} \rightarrow$ mol I_2

    mol I_2 at equilibrium

    $$= 37.0 \text{ mL } Na_2S_2O_3 \times \frac{1L}{1000 \text{ mL}} \times \frac{0.200 \text{ mol } Na_2S_2O_3}{1L} \times \frac{1 \text{ mol } S_2O_3^{2-}}{1 \text{ mol } Na_2S_2O_3} \times \frac{1 \text{ mol } I_2}{2 \text{ mol } S_2O_3^{2-}}$$

    $$= 0.00370 \text{ mol } I_2$$

Use the molar mass to determine the number of moles of HI present initially:

$$\text{mol HI present initially} = 3.20 \text{ g HI} \times \frac{1 \text{ mol HI}}{127.91 \text{ g HI}} = 0.0250 \text{ mol HI}$$

Set up a table like that in Example 12.6, but using mol (n) instead of partial pressure. In addition to the quantities calculated above, we also know that no H_2 or I_2 are present initially:

	2HI(g)	$\rightleftharpoons$	H₂(g)	+	I₂(g)
n_0 (mol)	0.0250		0		0
Δn (mol)					
n_{eq} (mol)					0.00370

Thus $\Delta n = + 0.00370$ mol for I_2 and stoichiometry can be used to complete the table:

	2HI(g)	⇌	H$_2$(g)	+	I$_2$(g)
n_0 (mol)	0.0250		0		0
Δn (mol)	−2(0.00370)		+0.00370		+0.00370
n_{eq} (mol)	0.0176		0.00370		0.00370

For each gas, its equilibrium partial pressure (using the Ideal Gas Law) is given by $\dfrac{nRT}{V}$.

Thus:

$$P_{HI} = 0.0176\frac{RT}{V} \qquad P_{H_2} = 0.00370\frac{RT}{V} \qquad P_{I_2} = 0.00370\frac{RT}{V}$$

Now substitute these into the expression for K:

$$K = \frac{(P_{H_2})(P_{I_2})}{(P_{HI})^2} = \frac{\left(0.00370\dfrac{RT}{V}\right)\left(0.00370\dfrac{RT}{V}\right)}{\left(0.0176\dfrac{RT}{V}\right)^2} = \underline{\mathbf{0.0442}}$$

81. This problem is similar to Example 12.6. Set up a table like that in Example 12.6, with $P_0 = 1.00$ for SO$_3$ and $P_0 = 0$ for the other two gases. Let $\Delta P = +x$ for O$_2$:

	SO$_3$(g)	⇌	SO$_2$(g)	+	½O$_2$(g)
P_0 (atm)	1.00		0		0
ΔP (atm)	−2x		+2x		+x
P_{eq} (atm)	1.00 − 2x		2x		x

Now substitute these partial pressures into the expression for K:

$$K = \frac{(P_{SO_2})(P_{O_2})^{\frac{1}{2}}}{(P_{SO_3})} = \frac{(2x)(x)^{\frac{1}{2}}}{(1.00 - 2x)} = 0.45$$

$$2x^{3/2} = 0.45(1.00 - 2x) = 0.45 - 0.90x$$

This equation may be solved by the method of Successive Approximations: start with an approximate solution, and use that to "refine" the solution and get a new (better) approximate solution, and so on until by successive refinements the approximate solution approaches the true value of x; for example, substitute $x = 0$ (an approximate solution) into the right side of the above equation to obtain

$$2x^{3/2} = 0.45 \qquad \text{so} \quad x = 0.37$$

Now take $x = 0.37$ and substitute that into the right side of the equation to obtain:

$2x^{3/2} = 0.45 - 0.90(0.37)$ so $x = 0.15$

repeating the process gives $x = 0.29$ then 0.21, 0.26, 0.23, 0.25, and ultimately converges at $x = 0.24$ repeatedly. Therefore this is the solution to the above equation. (CHECK: substitute $x = 0.24$ into the above equation and the left and right sides of the equation **DO** equal each other!)

Finally, calculate the partial pressure of SO_3 at equilibrium:

$P_{SO_3} = 1.00 - 2x = 1.00 - 2(0.24) = \underline{\textbf{0.52 atm}}$

82. This problem is similar to Example 12.6. Set up a table like the one in Example 12.6, and let P_0 for $I_2(g)$ be a and $P_0 = 0$ for $I(g)$. Let $\Delta P = -x$ for I_2:

	$I_2(g)$	$\rightleftharpoons$	$2I(g)$
P_0 (atm)	a		0
ΔP (atm)	$-x$		$+2x$
P_{eq} (atm)	$a - x$		$2x$

It is also known that the total pressure at equilibrium, $P_{eq(total)}$, is 40% higher than the initial total pressure, $P_{0(total)}$. Thus:

$P_{eq(total)} = 1.40P_{0(total)}$

$(a - x) + (2x) = 1.40(a + 0)$ $\rightarrow$ $a + x = 1.40a$

$x = 1.40a - a = 0.40a$

Substitute for the equilibrium concentrations in the expression for K:

$K = \dfrac{(P_I)^2}{(P_{I_2})} = \dfrac{(2x)^2}{(a-x)}$ substitute $x = 0.40a$: $K = \dfrac{[2(0.40a)]^2}{(a-0.40a)} = \dfrac{0.64a^2}{0.60a} = 1.1a$

Thus, the equilibrium constant $\underline{\textbf{K = 1.1a}}$ where a is the initial partial pressure of I_2.

83. Since the reaction produces a 50.0% yield of XeF_4, and the stoichiometry of the reaction is such that 1 mol Xe → 1 mol XeF_4, then 50.0% of the initial partial pressure of Xe must be consumed (its partial pressure will decrease by 0.10 atm) and the other partial pressures change accordingly:

	Xe(g)	+	2F$_2$(g)	$\rightleftharpoons$	XeF$_4$(g)
P_0 (atm)	0.20		0.40		0
ΔP (atm)	-0.10		$-2(0.10)$		$+0.10$
P_{eq} (atm)	0.10		0.20		0.10

Substitute the equilibrium partial pressures into the expression for K to calculate the value of K at this temperature:

$$K = \frac{(P_{XeF_4})}{(P_{Xe})(P_{F_2})^2} = \frac{(0.10)}{(0.10)(0.20)^2} = \underline{\textbf{25}}$$

To calculate the initial pressure of F$_2$ required to give a 75.0% yield of XeF$_4$, repeat the above table and denote P_0 for F$_2$ as x. A 75.0% yield means that 75.0% of the initial partial pressure of Xe must be consumed (its partial pressure will decrease by 0.15 atm) and the other partial pressures must change correspondingly:

	Xe(g)	+	2F$_2$(g)	$\rightleftharpoons$	XeF$_4$(g)
P_0 (atm)	0.20		x		0
ΔP (atm)	-0.15		$-2(0.15)$		$+0.15$
P_{eq} (atm)	0.05		$x-0.30$		0.15

Substitute the equilibrium partial pressures into the expression for K; the value of K, of course, is unchanged:

$$K = \frac{(P_{XeF_4})}{(P_{Xe})(P_{F_2})^2} = \frac{(0.15)}{(0.05)(x-0.30)^2} = 25$$

$$(x-0.30)^2 = \frac{(0.15)}{(0.05)(25)} = 0.12$$

Taking the square root of both sides of the equation yields:

$$x-0.30 = \pm 0.35 \qquad \text{so } x = -0.05 \text{ or } 0.65$$

The root $x = -0.05$ gives $P_0 < 0$ and $P_{eq} < 0$ for F$_2$, so it physically impossible; thus $x = 0.65$ and the required initial pressure of F$_2$ = x = **0.65 atm**

84. Use the Ideal Gas Law to calculate the initial partial pressure of benzyl alcohol.
    Strategy: mass C$_6$H$_5$CH$_2$OH $\rightarrow$ mol C$_6$H$_5$CH$_2$OH $\rightarrow$ P C$_6$H$_5$CH$_2$OH

$$\text{mol C}_6\text{H}_5\text{CH}_2\text{OH} = 1.50 \text{ g C}_6\text{H}_5\text{CH}_2\text{OH} \times \frac{1 \text{ mol C}_6\text{H}_5\text{CH}_2\text{OH}}{108.14 \text{ g C}_6\text{H}_5\text{CH}_2\text{OH}} = 0.0139 \text{ mol}$$

$$P_0 \text{ of } C_6H_5CH_2OH = \frac{nRT}{V} = \frac{(0.0139\,\text{mol})(0.0821\,\text{L}\cdot\text{atm/mol}\cdot\text{K})(523\,\text{K})}{(2.0\,\text{L})} = 0.30\,\text{atm}$$

(a) This part of the problem is similar to Example 12.6. Set up an equilibrium table like the one in Example 12.6, with $P_0 = 0.30$ for $C_6H_5CH_2OH$ and $P_0 = 0$ for the other two gases. Let $\Delta P = +x$ for H_2:

	$C_6H_5CH_2OH(g)$	$\rightleftharpoons$	$C_6H_5CHO(g)$	$+$	$H_2(g)$
P_0 (atm)	0.30		0		0
ΔP (atm)	$-x$		$+x$		$+x$
P_{eq} (atm)	$0.30 - x$		x		x

Now substitute these partial pressure expressions into the expression for K:

$$K = \frac{(P_{C_6H_5CHO})(P_{H_2})}{(P_{C_6H_5CH_2OH})} = \frac{(x)(x)}{(0.30 - x)} = \frac{x^2}{(0.30 - x)} = 0.56$$

$$x^2 = 0.56(0.30 - x) = 0.17 - 0.56x \qquad x^2 + 0.56x - 0.17 = 0$$

$$x = \frac{-b \pm \sqrt{b^2 - 4ac}}{2a} = \frac{-(0.56) \pm \sqrt{(0.56)^2 - 4(1)(-0.17)}}{2(1)}$$

$$x = +0.22 \quad \text{or} \quad -0.78$$

The negative root of the equation is physically impossible therefore $x = 0.22$. Thus the equilibrium partial pressure of benzaldehyde is:

$$P_{C_6H_5CHO} = x = \underline{\textbf{0.22 atm}}$$

(b) Use the Ideal Gas Law and the equilibrium partial pressure of benzyl alcohol to calculate the mass of benzyl alcohol remaining at equilibrium.
Strategy: $P\,C_6H_5CH_2OH \rightarrow$ mol $C_6H_5CH_2OH \rightarrow$ mass $C_6H_5CH_2OH$

mol of $C_6H_5CH_2OH$ at equilibrium

$$= \frac{PV}{RT} = \frac{(0.30 - 0.22\,\text{atm})(2.0\,\text{L})}{(0.0821\,\text{L}\cdot\text{atm/mol}\cdot\text{K})(523\,\text{K})} = 0.004\,\text{mol}$$

$$\text{mass } C_6H_5CH_2OH = 0.004\,\text{mol} \times \frac{108.14\,\text{g } C_6H_5CH_2OH}{1\,\text{mol } C_6H_5CH_2OH} = \underline{\textbf{0.4 g}}$$

85. Refer to Sections 5-5 and 10-1. We are given that there is 20.4% NH_3 in the equilibrium mixture: this corresponds to a mol fraction of $X_{NH_3} = 0.204$ and since (page 249) $X_{tot} = 1$, the sum of $X_{N_2} + X_{H_2} = (1 - X_{NH_3}) = (1 - 0.204) = 0.796$.

Assuming that the ratio of $N_2:H_2$ is 1:3 (using the stoichiometry of the reaction), then:

$$X_{H_2} = 3\,X_{N_2} \qquad \rightarrow \qquad X_{N_2} + 3\,X_{N_2} = 0.796$$

$$\rightarrow \quad X_{N_2} = 0.796/4 = 0.199 \qquad\qquad \text{and} \quad X_{H_2} = 3\,X_{N_2} = 0.597$$

Using Equation 5.4, with $P_{tot} = 3.065$ atm (given), the partial pressure of each gas in the mixture can be obtained:

$$P_{NH_3} = X_{NH_3}\,P_{tot} = (0.204)(3.065 \text{ atm}) = 0.625 \text{ atm}$$

$$P_{N_2} = X_{N_2}\,P_{tot} = (0.199)(3.065 \text{ atm}) = 0.610 \text{ atm}$$

$$P_{H_2} = X_{H_2}\,P_{tot} = (0.597)(3.065 \text{ atm}) = 1.83 \text{ atm}$$

Finally, use these partial pressures to calculate K:

$$K = \frac{(P_{NH_3})^2}{(P_{N_2})(P_{H_2})^3} = \frac{(0.625)^2}{(0.610)(1.83)^3} = \underline{\mathbf{0.104}}$$

86. See Section 12-3 and Examples 12.4 and 12.6; see also Section 4-1 and 5-2. Addition of silver(I) ion to the mixture of $H_2O(g)$ and $HCl(g)$ gases results in precipitation of AgCl, with the chloride of the precipitate coming from the HCl:

$$HCl(g) + Ag^+(aq) \rightarrow AgCl(s) + H^+(aq)$$

Use the following strategy, along with the stoichiometry of the above reaction, to deduce mol HCl in the mixture of gases:

$$\text{mass AgCl} \xrightarrow{\text{molar mass}} \text{mol AgCl} \xrightarrow{\text{mol ratio}} \text{mol HCl}$$

$$\text{mol HCl} = 7.29 \text{ g AgCl} \times \frac{1 \text{ mol AgCl}}{143.32 \text{ g AgCl}} \times \frac{1 \text{ mol HCl}}{1 \text{ mol AgCl}} = 0.0509 \text{ mol HCl}$$

Now use the stoichiometry of the reaction given in the question to determine how many mol H_2O are consumed to form 0.0509 mol HCl, and hence mol H_2O remaining at equilibrium:

$$\text{mol } H_2O \text{ consumed} = 0.0509 \text{ mol HCl} \times \frac{1 \text{ mol } H_2O}{2 \text{ mol HCl}} = 0.0254 \text{ mol } H_2O$$

$$\text{mol } H_2O \text{ remaining} = (0.0417 - 0.0254) \text{ mol } H_2O = 0.0163 \text{ mol } H_2O$$

Use the Ideal Gas Law (Equation 5.2) to determine the corresponding partial pressures of H_2O and HCl in the equilibrium mixture, using $V = 2.50$ L (given) and $T = (165 + 273)$ K = 438 K (given):

$$P_{H_2O} = \frac{n}{V}RT = \frac{0.0163 \text{ mol}}{2.50 \text{ L}}(0.0821 \text{ L} \cdot \text{atm/mol} \cdot \text{K})(438 \text{ K}) = 0.234 \text{ atm}$$

$$P_{HCl} = \frac{n}{V}RT = \frac{0.0509 \text{ mol}}{2.50 \text{ L}}(0.0821 \text{ L} \cdot \text{atm/mol} \cdot \text{K})(438 \text{ K}) = 0.732 \text{ atm}$$

Finally calculate the value of K:

$$K = \frac{(P_{HCl})^2}{P_{H_2O}} = \frac{(0.732)^2}{0.234} = \underline{\mathbf{2.29}}$$

ACIDS AND BASES

13-1 BRØNSTED-LOWRY ACID-BASE MODEL

2. This problem is related to Example 13.1. Recall that the acid is a ***proton*** (H⁺ ion) ***donor***, so as the reaction proceeds an acid will lose an H⁺ ion. Likewise, the base is a ***proton*** (H⁺ ion) ***acceptor***, so as the reaction proceeds the base will gain an H⁺ ion. (Note that here we must consider both forward and reverse directions.) A conjugate acid/base pair consists of two species (one acid plus one base) that differ from each other by the presence or absence of a proton (H⁺).

   (a) Brønsted-Lowry acids: H_2O, HCN
   Brønsted-Lowry bases: CN^-, OH^-
   conjugate acid/base pairs: H_2O, OH^-; HCN, CN^-

   (b) Brønsted-Lowry acids: H_3O^+, H_2CO_3
   Brønsted-Lowry bases: HCO_3^-, H_2O
   conjugate acid/base pairs: H_3O^+, H_2O; H_2CO_3, HCO_3^-

   (c) Brønsted-Lowry acids: $HC_2H_3O_2$, H_2S
   Brønsted-Lowry bases: HS^-, $C_2H_3O_2^-$
   conjugate acid/base pairs: $HC_2H_3O_2$, $C_2H_3O_2^-$; H_2S, HS^-

4. (a) CH_3O^- is likely to act as a **base**: since it is negatively charged, it is likely to act as an acceptor of (positively charged) protons.

   (b) CO_3^{2-} is likely to act as a **base**: since it is negatively charged, it is likely to act as an acceptor of (positively charged) protons.

   (c) $HAsO_4^{2-}$ is **amphiprotic**: It is likely to act as a **base**: since it is negatively charged, it is likely to act as an acceptor of (positively charged) protons. However, since it contains a proton (bonded to oxygen), it may also act as an **acid** (compare HSO_4^- in the Table on page 332 and HCO_3^- in Example 13.1(b)).

   Acids: **$HAsO_4^{2-}$**   Bases: **CH_3O^-, CO_3^{2-}, $HAsO_4^{2-}$**

6. (See page 332.) The conjugate base of any acid is the species formed after the acid has donated its proton. Thus, to find the conjugate base of the given species, simply remove a proton (H⁺). In each of the reactions below, the last species given is the required conjugate base:

   (a) $H_2AsO_4^- \rightarrow H^+ + \mathbf{HAsO_4^{2-}}$

   (b) $Fe(H_2O)_5(OH)^+ \rightarrow H^+ + \mathbf{Fe(H_2O)_4(OH)_2}$

   (c) $HClO_3 \rightarrow H^+ + \mathbf{ClO_3^-}$

(d) $NH_4^+ \rightarrow H^+ + NH_3$

(e) $HC_2H_3O_2 \rightarrow H^+ + C_2H_3O_2^-$

8. When the HCO_3^- ion behaves as a Brønsted-Lowry acid, it must donate a proton to a base; in an aqueous solution of HCO_3^-, the base will be H_2O:

$$HCO_3^-(aq) + H_2O \rightleftharpoons CO_3^{2-}(aq) + H_3O^+(aq)$$

Likewise, when the HCO_3^- ion behaves as a Brønsted-Lowry base, it must accept a proton from an acid; in an aqueous solution of HCO_3^-, the acid will be H_2O:

$$HCO_3^-(aq) + H_2O \rightleftharpoons H_2CO_3(aq) + OH^-(aq)$$

10. This problem is related to Example 13.4(a); see also Section 13-4. When these species behave as weak Brønsted-Lowry acids in water, they must donate a proton to a H_2O molecule (which is acting as a base): this makes the solution acidic because hydronium ions are present.

(a) (See page 339.) The $Zn(H_2O)_3OH^+$ cation contains H_2O molecules that are the source of H^+ when this species behaves as an acid: one such H_2O molecule is converted to hydroxide during this process, but remains bonded to Zn:

$$Zn(H_2O)_3(OH)^+(aq) + H_2O \rightleftharpoons Zn(H_2O)_2(OH)_2(aq) + H_3O^+(aq)$$

(b) $HSO_4^-(aq) + H_2O \rightleftharpoons SO_4^{2-}(aq) + H_3O^+(aq)$

(c) $HNO_2(aq) + H_2O \rightleftharpoons NO_2^-(aq) + H_3O^+(aq)$

(d) The $Fe(H_2O)_6^{2+}$ cation behaves similarly to the $Zn(H_2O)_3OH^+$ cation (see part (a) above) or $Al(H_2O)_6^{3+}$ cation (see page 339): one of its H_2O molecules is converted to hydroxide when the cation acts as an acid, but the hydroxide remains bonded to Fe

$$Fe(H_2O)_6^{2+}(aq) + H_2O \rightleftharpoons Fe(H_2O)_5(OH)^+(aq) + H_3O^+(aq)$$

(e) $HC_2H_3O_2(aq) + H_2O \rightleftharpoons C_2H_3O_2^-(aq) + H_3O^+(aq)$

(f) $H_2PO_4^-(aq) + H_2O \rightleftharpoons HPO_4^{2-}(aq) + H_3O^+(aq)$

12. This problem is related to Example 13.10; see also Section 13-5. When these species behave as weak Brønsted-Lowry bases in water, they must accept a proton from a H_2O molecule (which is acting as an acid): this makes the solution basic because hydroxide ions are present in excess.

(a) $(CH_3)_3N(aq) + H_2O \rightleftharpoons (CH_3)_3NH^+(aq) + OH^-(aq)$

(b) $HC_2O_4^-(aq) + H_2O \rightleftharpoons H_2C_2O_4(aq) + OH^-(aq)$

(c) $CN^-(aq) + H_2O \rightleftharpoons HCN\ (aq) + OH^-(aq)$

(d) $S^{2-}(aq) + H_2O \rightleftharpoons HS^-(aq) + OH^-(aq)$

(e) $BrO_3^-(aq) + H_2O \rightleftharpoons HBrO_3(aq) + OH^-(aq)$

(f) $H_2C_6O_5H_7^-(aq) + H_2O \rightleftharpoons H_3C_6O_5H_7(aq) + OH^-(aq)$

13-2 & 13-3 THE ION PRODUCT OF WATER; pH AND pOH: [H⁺], [OH⁻], pH, and pOH

14. To calculate pH, use Equation 13.3; if pH < 7 the solution is acidic and if pH > 7 the solution is basic (see page 333). Note that rules regarding significant figures in logarithms are given in Appendix 3.

    (a) $pH = -log_{10}[H^+] = -log_{10}(2.7 \times 10^{-3}) = \underline{\textbf{2.57}}$ The solution is **acidic**.

    (b) $pH = -log_{10}[H^+] = -log_{10}(1.5) = \underline{\textbf{-0.18}}$ The solution is **acidic**.

    (c) $pH = -log_{10}[H^+] = -log_{10}(1.45 \times 10^{-13}) = \underline{\textbf{12.84}}$ The solution is **basic**.

    (d) $pH = -log_{10}[H^+] = -log_{10}(6.4 \times 10^{-9}) = \underline{\textbf{8.19}}$ The solution is **basic**.

16. This problem is similar to Example 13.2(b). To convert pH to [H⁺], use the second form of Equation 13.3; substitute [H⁺] into Equation 13.1 to obtain [OH⁻]. Note that rules regarding significant figures in inverse logarithms are given in Appendix 3.

    (a) $[H^+] = 10^{-pH}\ 10^{-(-0.76)} = \underline{\textbf{5.8 M}}$

    $[OH^-] = K_w/[H^+] = 1.0 \times 10^{-14}/5.8 = \underline{\textbf{1.7} \times \textbf{10}^{-15}\ \textbf{M}}$

    (b) $[H^+] = 10^{-pH}\ 10^{-9.11} = \underline{\textbf{7.8} \times \textbf{10}^{-10}\ \textbf{M}}$

    $[OH^-] = K_w/[H^+] = 1.0 \times 10^{-14}/7.8 \times 10^{-10} = \underline{\textbf{1.3} \times \textbf{10}^{-5}\ \textbf{M}}$

    (c) $[H^+] = 10^{-pH}\ 10^{-3.81} = \underline{\textbf{1.5} \times \textbf{10}^{-4}\ \textbf{M}}$

    $[OH^-] = K_w/[H^+] = 1.0 \times 10^{-14}/1.5 \times 10^{-4} = \underline{\textbf{6.7} \times \textbf{10}^{-11}\ \textbf{M}}$

    (d) $[H^+] = 10^{-pH}\ 10^{-12.08} = \underline{\textbf{8.3} \times \textbf{10}^{-13}\ \textbf{M}}$

    $[OH^-] = K_w/[H^+] = 1.0 \times 10^{-14}/8.3 \times 10^{-13} = \underline{\textbf{1.2} \times \textbf{10}^{-2}\ \textbf{M}}$ or $\underline{\textbf{0.012 M}}$

18. This problem is related to Example 13.2. Use Equation 13.1 to convert [OH⁻] to [H⁺], or to convert [H⁺] to [OH⁻]; use Equation 13.3 to convert [H⁺] to pH, or to convert pH to [H⁺]; use Equation 13.4 to convert [OH⁻] to pOH, or to convert pOH to [OH⁻]; recall (page 334) that pH + pOH = pK_w = 14.00 so that pOH can be obtained directly from pH and vice versa. The solution is basic if pH > 7 (see page 333).

    (a) $pH + pOH = pH + 8.14 = 14.00$ $pH = 14.00 - 8.14 = \underline{\textbf{5.86}}$

$[OH^-] = 10^{-pOH} = 10^{-8.14} = \underline{\mathbf{7.2 \times 10^{-9}\ M}}$

$[H^+] = 10^{-pH} = 10^{-5.86} = \underline{\mathbf{1.4 \times 10^{-6}\ M}}$

(b) $pH + pOH = 13.28 + pOH = 14.00 \qquad pOH = 14.00 - 13.28 = \underline{\mathbf{0.72}}$

$[H^+] = 10^{-pH} = 10^{-13.28} = \underline{\mathbf{5.2 \times 10^{-14}\ M}}$

$[OH^-] = 10^{-pOH} = 10^{-0.72} = \underline{\mathbf{1.9 \times 10^{-1}\ M}}$

(c) $[H^+] = K_w/[OH^-] = 1.0 \times 10^{-14}/4.3 \times 10^{-3} = \underline{\mathbf{2.3 \times 10^{-12}\ M}}$

$pH = -\log_{10}[H^+] = -\log_{10}(2.3 \times 10^{-12}) = \underline{\mathbf{11.63}}$

$pOH = -\log_{10}[OH^-] = -\log_{10}(4.3 \times 10^{-3}) = \underline{\mathbf{2.37}}$

(d) $[OH^-] = K_w/[H^+] = 1.0 \times 10^{-14}/2.75 = \underline{\mathbf{3.6 \times 10^{-15}\ M}}$

$pH = -\log_{10}[H^+] = -\log_{10}(2.75) = \underline{\mathbf{-0.44}}$

$pOH = -\log_{10}[OH^-] = -\log_{10}(3.6 \times 10^{-15}) = \underline{\mathbf{14.44}}$

	$[H^+]$	$[OH^-]$	pH	pOH	Basic?
(a)	1.4×10^{-6}	7.2×10^{-9}	5.86	8.14	no
(b)	5.2×10^{-14}	1.9×10^{-1}	13.28	0.72	yes
(c)	2.3×10^{-12}	4.3×10^{-3}	11.63	2.37	yes
(d)	2.75	3.6×10^{-15}	−0.44	14.44	no

20. See Example 13.2. For solution R, calculate $[H^+]$ using Equation 13.3; for solution Q, calculate $[H^+]$ using Equation 13.1.

$[H^+]$ in solution R $= 10^{-pH} = 10^{-13.42} = 3.8 \times 10^{-14}\ M$

$[H^+]$ in solution Q $= K_w/[OH^-] = 1.0 \times 10^{-14}/0.16 = 6.2 \times 10^{-14}\ M$

From Figure 13.2, solutions become increasingly basic as pH increases and as $[H^+]$ decreases, so the solution with lower $[H^+]$ is the more basic; and the solution with the higher $[H^+]$ will have the lower pH.

Thus, **solution R is more basic and solution Q has the lower pH.**

22. (a) For solutions A and C, calculate $[H^+]$ using Equation 13.3; for solution B, $[H^+]$ is derived from the $[H^+]$ in solution A. The pH of solution C is ½(pH of solution A) = 6.16.

$[H^+]$ in solution A $= 10^{-pH} = 10^{-12.32} = \underline{\mathbf{4.8 \times 10^{-13}\ M}}$

$[H^+]$ in solution B $= 3 \times ([H^+]$ in solution A$) = \underline{\mathbf{1.4 \times 10^{-12}\ M}}$

$[H^+]$ in solution C $= 10^{-pH} = 10^{-6.16} = \underline{\mathbf{6.9 \times 10^{-7}\ M}}$

(b) Calculate pH using Equation 13.3.

pH of solution B $= -\log_{10}[H^+] = -\log_{10}(1.4 \times 10^{-12}) = \underline{\mathbf{11.85}}$

pH of solution C = ½(pH of solution A) = **6.16**

(c) Recall (page 333) that a solution with $[H^+] > 10^{-7}$ M is acidic and one with $[H^+] < 10^{-7}$ M is basic. Thus, **solutions A and B are basic and solution C is acidic.**

24. See Example 13.2.

(a) For each drink, calculate $[H^+]$ using Equation 13.3:

For cola: $[H^+] = 10^{-pH} = 10^{-3.1} =$ **7.9×10^{-4} *M***

For green tea: $[H^+] = 10^{-pH} = 10^{-5.8} =$ **1.6×10^{-6} *M***

(b) The ratio of H^+ concentration in cola to that of green tea is

7.9×10^{-4} *M* : 1.6×10^{-6} *M*

or, dividing both numbers by the smaller of the two, the ratio becomes:

$$\frac{7.9 \times 10^{-4}}{1.6 \times 10^{-6}} : \frac{1.6 \times 10^{-6}}{1.6 \times 10^{-6}} = \underline{\mathbf{4.9 \times 10^2 : 1}}$$

Thus **the cola is around 490 times more acidic than the green tea.**

26. (a) Since HNO_3 is a strong acid (compare Figure 13.3), it completely ionizes in H_2O and therefore $[HNO_3(aq)] = [H_3O^+(aq)] = [H^+(aq)]$:

$$HNO_3(aq) + H_2O \rightarrow H_3O^+(aq) + NO_3^-(aq)$$

Calculate $[HNO_3]$ in a 12.0 % by mass solution (12.0 g HNO_3 in 100 g solution). Strategy: use the density to convert mass of solution → volume of solution; then mass HNO_3 → mol HNO_3 → $[HNO_3]$

$$\text{volume of solution} = 100 \text{ g} \times \frac{1\,mL}{1.00\,g} \times \frac{1\,L}{1000\,mL} = 0.100 \text{ L}$$

$$\text{mol of } HNO_3 = 12.0 \text{ g } HNO_3 \times \frac{1\,mol\,HNO_3}{63.013\,g\,HNO_3} = 0.190 \text{ mol } HNO_3$$

$$[HNO_3] = \frac{mol\,HNO_3}{L\,solution} = \frac{0.190\,mol\,HNO_3}{0.100\,L} = 1.90 \text{ *M*}$$

$$[H^+] = \frac{1.90\,mol\,HNO_3}{1\,L} \times \frac{1\,mol\,H^+}{1\,mol\,HNO_3} = \underline{\mathbf{1.90 \text{ *M*}}}$$

$$pH = -\log_{10}[H^+] = -\log_{10}(1.90) = \underline{\mathbf{-0.28}}$$

The concentration of the HNO_3 solution does not change, regardless of whether we consider the whole solution or a 45.6 mL or 10.00 mL sample. Since pH depends on concentration, it also does not change when the volume of sample changes.
Thus, **$[H^+]$ in both solutions is 1.90 M and the pH of both solutions is −0.28.**

(b) Since HCl is a strong acid (compare Example 13.3), it completely ionizes in H_2O and therefore $[HCl(aq)] = [H_3O^+(aq)] = [H^+(aq)]$:

$$HCl(aq) + H_2O \rightarrow H_3O^+(aq) + Cl^-(aq)$$

Calculate $[HCl]$. Strategy: mass HCl $\rightarrow$ mol HCl $\rightarrow$ $[HCl]$

$$\text{mol of HCl} = 1.0 \text{ g HCl} \times \frac{1 \text{ mol HCl}}{36.460 \text{ g HCl}} = 0.027 \text{ mol HCl}$$

$$[HCl] = \frac{\text{mol HCl}}{\text{L solution}} = \frac{0.027 \text{ mol HCl}}{1.28 \text{ L}} = 0.021 \text{ M}$$

$$[H^+] = \frac{0.021 \text{ mol HCl}}{1 \text{ L}} \times \frac{1 \text{ mol H}^+}{1 \text{ mol HCl}} = 0.021 \text{ M}$$

$$pH = -\log_{10}[H^+] = -\log_{10}(0.21) = 1.67$$

Thus, in the 1.28 L solution, **[H$^+$] = 0.021 M** and **pH = 1.67**.

Repeat the concentration calculation for the same number of moles of HCl in a solution of volume 128 mL = 0.128 L:

$$[HCl] = \frac{\text{mol HCl}}{\text{L solution}} = \frac{0.027 \text{ mol HCl}}{0.128 \text{ L}} = 0.21 \text{ M} = [H^+]$$

$$pH = -\log_{10}[H^+] = -\log_{10}(0.21) = 0.67$$

Thus, in the 128 mL solution, **[H$^+$] = 0.21 M** and **pH = 0.67**.

28. (a) Since $Sr(OH)_2$ is a strong base (compare Example 13.3), it is completely ionized in H_2O:

$$Sr(OH)_2(aq) \rightarrow Sr^{2+}(aq) + 2OH^-(aq)$$

and therefore in 0.106 M $Sr(OH)_2$ solution:

$$[OH^-] = \frac{0.106 \text{ mol Sr(OH)}_2}{1 \text{ L}} \times \frac{2 \text{ mol OH}^-}{1 \text{ mol Sr(OH)}_2} = 0.212 \text{ M}$$

The number of moles of OH^- in thirty-eight mL of this solution is:

$$\text{mol OH}^- = \frac{0.212 \text{ mol OH}^-}{1 \text{ L}} \times \frac{1 \text{ L}}{1000 \text{ mL}} \times 38 \text{ mL} = 0.00806 \text{ mol}$$

Diluting this solution (to 275 mL) changes the volume but not number of moles of OH^-; thus in the diluted solution:

$$[OH^-] = \frac{\text{mol OH}^-}{\text{L of solution}} = \frac{0.00806 \text{ mol OH}^-}{275 \text{ mL}} \times \frac{1000 \text{ mL}}{1 \text{ L}} = \underline{0.0293 \text{ M}}$$

To obtain [H$^+$], use Equation 13.1, and to obtain pH, use Equation 13.3:

$$[H^+] = K_w/[OH^-] = 1.0 \times 10^{-14}/0.0293 = \underline{\textbf{3.4} \times \textbf{10}^{-13} \textbf{\textit{M}}}$$

$$pH = -\log_{10}[H^+] = -\log_{10}(3.4 \times 10^{-13}) = \underline{\textbf{12.47}}$$

(b) Since KOH is a strong base (compare Example 13.3), it is completely ionized in H$_2$O and therefore [KOH(*aq*)] = [OH$^-$(*aq*)]:

$$KOH(aq) \rightarrow K^+(aq) + OH^-(aq)$$

Calculate [KOH]. Strategy: mass KOH $\rightarrow$ mol KOH $\rightarrow$ [KOH]

$$\text{mol of KOH} = 5.00 \text{ g KOH} \times \frac{1 \text{ mol KOH}}{56.106 \text{ g KOH}} = 0.0891 \text{ mol KOH}$$

$$[KOH] = \frac{\text{mol KOH}}{\text{L of solution}} = \frac{0.0891 \text{ mol KOH}}{447 \text{ mL}} \times \frac{1000 \text{ mL}}{1 \text{ L}} = 0.199 \text{ } M$$

$$[OH^-] = \frac{0.199 \text{ mol KOH}}{1 \text{ L}} \times \frac{1 \text{ mol OH}^-}{1 \text{ mol KOH}} = \underline{\textbf{0.199} \textbf{\textit{M}}}$$

To obtain [H$^+$], use Equation 13.1, and to obtain pH, use Equation 13.3:

$$[H^+] = K_w/[OH^-] = 1.0 \times 10^{-14}/0.199 = \underline{\textbf{5.0} \times \textbf{10}^{-14} \textbf{\textit{M}}}$$

$$pH = -\log_{10}[H^+] = -\log_{10}(5.0 \times 10^{-14}) = \underline{\textbf{13.30}}$$

30. To calculate the pH of the mixture we need to know [H$^+$] in the mixture. First calculate the total number of moles of H$^+$ that the two acids provide; then use that with the total volume to calculate [H$^+$] in the mixture, and hence pH. Recall that both HCl and HNO$_3$ are strong acids and so will completely ionize in solution. (See also problem 26 above.)

mol H$^+$ provided by HCl solution

$$= 145 \text{ mL} \times \frac{1 \text{ L}}{1000 \text{ mL}} \times \frac{0.575 \text{ mol HCl}}{1 \text{ L}} \times \frac{1 \text{ mol H}^+}{1 \text{ mol HCl}} = 0.0834 \text{ mol H}^+$$

Use Equation 13.3 to calculate [H$^+$] in the HNO$_3$ solution (before mixing):

$$[H^+] = 10^{-pH} = 10^{-1.39} = 0.041 \text{ } M$$

mol H$^+$ provided by HNO$_3$ solution

$$= 493 \text{ mL} \times \frac{1 \text{ L}}{1000 \text{ mL}} \times \frac{0.041 \text{ mol HNO}_3}{1 \text{ L}} \times \frac{1 \text{ mol H}^+}{1 \text{ mol HNO}_3} = 0.020 \text{ mol H}^+$$

$$[H^+] \text{ in mixture} = \frac{\text{total mol } H^+}{\text{total volume}} = \frac{(0.0834 + 0.020) \text{ mol } H^+}{(145 + 493) \text{ mL}} \times \frac{1000 \text{ mL}}{1 \text{ L}} = 0.161 \text{ M}$$

$$\text{pH of mixture} = -\log_{10}[H^+] = -\log_{10}(0.161) = \underline{\textbf{0.793}}$$

32. Since we are dealing with bases, it is simpler to work in terms of $[OH^-]$ and pOH. First convert pH to pOH for both solutions and use that to calculate $[OH^-]$ for each solution (use Equation 13.4); from that determine the number of moles of OH^- that each base provides and hence the total number of moles of OH^-. Then use that total, along with the total volume, to calculate $[OH^-]$ in the mixture, and hence pOH and then pH. Recall that both NaOH and $Sr(OH)_2$ are strong bases and so will completely ionize in solution. (See also problem 28 above).

For the NaOH solution: pOH = 14.00 − pH = 14.00 − 11.57 = 2.43.

$$[OH^-] = 10^{-pOH} = 10^{-2.43} = 0.0037 \text{ M}$$

mol OH^- provided by NaOH solution

$$= 235 \text{ mL} \times \frac{1 \text{ L}}{1000 \text{ mL}} \times \frac{0.0037 \text{ mol } OH^-}{1 \text{ L}} = 8.7 \times 10^{-4} \text{ mol } OH^-$$

For the $Sr(OH)_2$ solution: pOH = 14.00 − pH = 14.00 − 12.09 = 1.91.

$$[OH^-] = 10^{-pOH} = 10^{-1.91} = 0.012 \text{ M}$$

mol OH^- provided by $Sr(OH)_2$ solution

$$= 316 \text{ mL} \times \frac{1 \text{ L}}{1000 \text{ mL}} \times \frac{0.012 \text{ mol } OH^-}{1 \text{ L}} = 3.8 \times 10^{-3} \text{ mol } OH^-$$

$$[OH^-] \text{ in mixture} = \frac{\text{total mol } OH^-}{\text{total volume}} = \frac{(8.7 \times 10^{-4} + 3.8 \times 10^{-3}) \text{ mol } OH^-}{(235 + 316) \text{ mL}} \times \frac{1000 \text{ mL}}{1 \text{ L}}$$

$$= 0.0085 \text{ M}$$

$$\text{pOH of mixture} = -\log_{10}[OH^-] = -\log_{10}(0.0085) = 2.07$$
$$\text{pH of mixture} = 14.00 - pOH = 14.00 - 2.07 = \underline{\textbf{11.93}}$$

13-4 WEAK ACIDS AND THEIR EQUILIBRIUM CONSTANTS: Ionization Expressions, Weak Acids

34. (See page 339.) Since these species are weak acids, their ionization in water will be of the following form:

$$HB(aq) + H_2O \rightleftharpoons H_3O^+(aq) + B^-(aq)$$

or, in simplified form:

$$HB(aq) \rightleftharpoons H^+(aq) + B^-(aq)$$

for which the equilibrium constant K_a is:

$$K_a = \frac{[H^+][B^-]}{[HB]} \qquad \text{(Equation 13.5)}$$

(a) $HSO_3^-(aq) \rightleftharpoons H^+(aq) + SO_3^{2-}(aq)$ $\qquad K_a = \dfrac{[H^+][SO_3^{2-}]}{[HSO_3^-]}$

(b) $HPO_4^{2-}(aq) \rightleftharpoons H^+(aq) + PO_4^{3-}(aq)$ $\qquad K_a = \dfrac{[H^+][PO_4^{3-}]}{[HPO_4^{2-}]}$

(c) $HNO_2(aq) \rightleftharpoons H^+(aq) + NO_2^-(aq)$ $\qquad K_a = \dfrac{[H^+][NO_2^-]}{[HNO_2]}$

36. Use Equation 13.6 to calculate pK_a from K_a.

(a) $pK_a = -\log_{10}K_a = -\log_{10}(1.8 \times 10^{-4}) = \underline{\mathbf{3.74}}$

(b) $pK_a = -\log_{10}K_a = -\log_{10}(6.8 \times 10^{-8}) = \underline{\mathbf{7.17}}$

(c) $pK_a = -\log_{10}K_a = -\log_{10}(4.0 \times 10^{-11}) = \underline{\mathbf{10.40}}$

38. (a) Recall (page 340) that the weaker the acid, the smaller the value of K_a. Thus, to arrange the acids in order of ***increasing*** strength, simply arrange them in order of ***increasing*** K_a value:

K_a value: $\quad 2 \times 10^{-6} \quad < \quad 3 \times 10^{-4} \quad < \quad 9 \times 10^{-4} \quad < \quad 1.6 \times 10^{-3}$

acids: $\qquad\quad$ C $\qquad < \qquad$ D $\qquad < \qquad$ B $\qquad < \qquad$ A

(b) Since $pK_a = -\log_{10}K_a$, it follows that ***decreasing*** values of K_a will correspond to ***increasing*** values of pK_a. Thus **acid A will have the smallest pK_a value**. This can be verified by calculating the pK_a values for the four acids using Equation 13.6:

pK_a value: 5.70 $\quad>\quad$ 3.52 $\quad>\quad$ 3.05 $\quad>\quad$ 2.80

acids: $\qquad\quad$ C $\quad>\quad$ D $\quad>\quad$ B $\quad>\quad$ A

40. The solutions all have the same reactant concentration, so $[H^+]$ in each solution will depend on the extent of ionization of the acid component. Recall (page 340) that the smaller the K_a value, the weaker the acid is and hence the lower $[H^+]$ will be.
The acids present are HF, HCl, HCN, and HClO. HCl is a strong acid (see Section 13-3), so it will provide the highest $[H^+]$ since it ionizes completely. Of the others, the K_a values (from Table 13.2) are: HF, 6.9×10^{-4}; HCN, 5.8×10^{-10}; HClO, 2.8×10^{-8}. Thus, ranking HCl as the most acidic and the others in order of decreasing K_a, $[H^+]$ in the four solutions decreases in the order:

HCl > HF > HClO > HCN

42. Recall (Section 13-3 and Figure 13.2) that pH decreases as [H⁺] increases, and vice versa. Thus when the four solutions are ranked in order of **_decreasing_** pH it will correspond to **_increasing_** [H⁺], in other words the reverse of the ordering for Question 40 above:

HCN > HClO > HF > HCl

13-4 WEAK ACIDS AND THEIR EQUILIBRIUM CONSTANTS: Equilibrium Calculations, Weak Acids

---

44. This problem is similar to Example 13.5; see also Example 13.7. Use Equation 13.3 to convert pH to [H⁺]:

$$[H^+] = 10^{-pH} = 10^{-5.32} = 4.8 \times 10^{-6} \, M$$

Draw an equilibrium table as in example 13.5, based on the following equilibrium reaction (see page 339):

$$HB(aq) \rightleftharpoons H^+(aq) + B^-(aq) \qquad K_a = \frac{[H^+][B^-]}{[HB]} \qquad \text{(Equation 13.5)}$$

In completing the following table, we know that $[HB]_0 = 2.642 \, M$ (given) and initially we assume there is no ionization, so $[H^+]_0 = [B^-]_0 = 0$. Let $\Delta[H^+] = x$:

	HB(aq) $\rightleftharpoons$	H⁺(aq) +	B⁻(aq)
[ ]₀ (M)	2.642	0	0
Δ[ ] (M)	−x	+x	+x
[ ]eq (M)	2.642 − x	x	x

Now substitute the equilibrium concentrations of each species into the expression for K_a, and recall that we already know the value of x because $x = [H^+]_{eq} = 4.8 \times 10^{-6} \, M$ (calculated using the pH of the solution):

$$K_a = \frac{[H^+][B^-]}{[HB]} = \frac{(x)(x)}{(2.642 - x)} = \frac{(4.8 \times 10^{-6})(4.8 \times 10^{-6})}{(2.642 - 4.8 \times 10^{-6})} = \underline{8.7 \times 10^{-12}}$$

46. This problem is similar to Example 13.5 and problem 44 above. Use Equation 13.3 to convert pH to [H⁺]:

$$[H^+] = 10^{-pH} = 10^{-5.24} = 5.8 \times 10^{-6} \, M$$

We must also calculate the molarity of the acetaminophen solution:

$$[HC_8H_8NO_2] = \frac{6.54 \text{ g } HC_8H_8NO_2}{250.0 \text{ mL}} \times \frac{1 \text{ mol } HC_8H_8NO_2}{151.17 \text{ g } HC_8H_8NO_2} \times \frac{1000 \text{ mL}}{1 \text{ L}} = 0.173 \text{ } M$$

Draw an equilibrium table as in example 13.5, based on the following equilibrium reaction (see page 339):

$$HC_8H_8NO_2(aq) \rightleftharpoons H^+(aq) + C_8H_8NO_2^-(aq) \qquad K_a = \frac{[H^+][C_8H_8NO_2^-]}{[HC_8H_8NO_2]}$$

In completing the following table, we know that $[HC_8H_8NO_2]_0 = 0.173$ M (calculated above) and initially we assume there is no ionization, so $[H^+]_0 = [C_8H_8NO_2^-]_0 = 0$. Let $\Delta[H^+] = x$:

	$HC_8H_8NO_2(aq) \rightleftharpoons$	$H^+(aq)$ +	$C_8H_8NO_2^-(aq)$
[ ]$_0$ (M)	0.173	0	0
Δ[ ] (M)	$-x$	$+x$	$+x$
[ ]$_{eq}$ (M)	$0.173 - x$	x	x

Now substitute the equilibrium concentrations of each species into the expression for K_a, and recall that we already know the value of x because $x = [H^+]_{eq} = 5.8 \times 10^{-6}$ M (calculated using the pH of the solution):

$$K_a = \frac{[H^+][C_8H_8NO_2^-]}{[HC_8H_8NO_2]} = \frac{(x)(x)}{(0.173 - x)} = \frac{(5.8 \times 10^{-6})(5.8 \times 10^{-6})}{(0.173 - 5.8 \times 10^{-6})} = \underline{1.9 \times 10^{-10}}$$

48. This problem is similar to Example 13.5 and very similar to problem 46 above. Use Equation 13.3 to convert pH to $[H^+]$:

$$[H^+] = 10^{-pH} = 10^{-2.34} = 4.6 \times 10^{-3} \text{ } M$$

We must also calculate the molarity of the barbituric acid solution:

$$[HC_4H_3N_2O_3] = \frac{9.00 \text{ g } HC_4H_3N_2O_3}{325 \text{ mL}} \times \frac{1 \text{ mol } HC_4H_3N_2O_3}{128.09 \text{ g } HC_4H_3N_2O_3} \times \frac{1000 \text{ mL}}{1 \text{ L}} = 0.216 \text{ } M$$

Draw an equilibrium table as in example 13.5, based on the following equilibrium reaction (see page 339):

$$HC_4H_3N_2O_3(aq) \rightleftharpoons H^+(aq) + C_4H_3N_2O_3^-(aq) \qquad K_a = \frac{[H^+][C_4H_3N_2O_3^-]}{[HC_4H_3N_2O_3]}$$

In completing the following table, we know that $[HC_4H_3N_2O_3]_0 = 0.216$ M (calculated above) and initially we assume there is no ionization, so $[H^+]_0 = [C_4H_3N_2O_3^-]_0 = 0$. Let $\Delta[H^+] = x$:

	$HC_4H_3N_2O_3(aq)$	$\rightleftharpoons$	$H^+(aq)$	$+$	$C_4H_3N_2O_3^-(aq)$
$[\ ]_0$ (M)	0.216		0		0
$\Delta[\ ]$ (M)	$-x$		$+x$		$+x$
$[\ ]_{eq}$ (M)	$0.216 - x$		x		x

Now substitute the equilibrium concentrations of each species into the expression for K_a, and recall that we already know the value of x because $x = [H^+]_{eq} = 4.6 \times 10^{-3}$ M (calculated using the pH of the solution):

$$K_a = \frac{[H^+][C_4H_3N_2O_3^-]}{[HC_4H_3N_2O_3]} = \frac{(x)(x)}{(0.216 - x)} = \frac{(4.6 \times 10^{-3})(4.6 \times 10^{-3})}{(0.216 - 4.6 \times 10^{-3})} = \underline{1.0 \times 10^{-4}}$$

50. This problem is similar to Example 13.7. The anion of the salt $NaHSO_3$, namely the HSO_3^- ion, is listed as an acid in Table 13.2 (see also Section 13-6), and K_a is 6.0×10^{-8}. When this ion acts as an acid, the following equilibrium reaction will occur (see page 339):

$$HSO_3^-(aq) \rightleftharpoons H^+(aq) + SO_3^{2-}(aq) \qquad K_a = \frac{[H^+][SO_3^{2-}]}{[HSO_3^-]}$$

The molarity of the HSO_3^- ion in the given solution of $NaHSO_3$ is

$$[HSO_3^-] = \frac{0.47 \text{ mol } NaHSO_3}{1 L} \times \frac{1 \text{ mol } HSO_3^-}{1 \text{ mol } NaHSO_3} = 0.47 \text{ M}$$

Draw an equilibrium table as in Example 13.5, based on the above equilibrium reaction. In completing the table, recall that $[HSO_3^-]_0 = 0.47$ M (calculated above) and initially we assume there is no ionization, so $[H^+]_0 = [SO_3^{2-}]_0 = 0$. Let $\Delta[H^+] = x$:

	$HSO_3^-(aq)$	$\rightleftharpoons$	$H^+(aq)$	$+$	$SO_3^{2-}(aq)$
$[\ ]_0$ (M)	0.47		0		0
$\Delta[\ ]$ (M)	$-x$		$+x$		$+x$
$[\ ]_{eq}$ (M)	$0.47 - x$		x		x

Now substitute the equilibrium concentrations of each species into the expression for K_a:

$$K_a = 6.0 \times 10^{-8} = \frac{[H^+][SO_3^{2-}]}{[HSO_3^-]} = \frac{(x)(x)}{(0.47 - x)}$$

Assume $x \ll 0.47$:

$$6.0 \times 10^{-8} = \frac{x^2}{0.47} \qquad \text{so } x^2 = (0.47)(6.0 \times 10^{-8}) = 2.8 \times 10^{-9} \quad \text{and} \quad x = 1.7 \times 10^{-4} \text{ M}$$

Note that the value of x is indeed much less that 0.47, so the assumption made in solving the above equation was valid; % ionization = $[H^+]_{eq}/[HSO_3^-]_0 \times 100 = (1.7 \times 10^{-4})/(0.47) \times 100 = 0.036\% < 5\%$ (see page 345).

Finally, since $x = [H^+]_{eq}$, we can use the value of x to calculate pH using Equation 13.3:

$$pH = -\log_{10}[H^+] = -\log_{10}(1.7 \times 10^{-4}) = \mathbf{\underline{3.77}}$$

52. This problem is similar to Example 13.8; see also Example 13.7. When penicillin acts as an acid, the following equilibrium reaction will occur (see page 339; label the acid "HB"):

$$HB(aq) \rightleftharpoons H^+(aq) + B^-(aq) \qquad\qquad K_a = \frac{[H^+][B^-]}{[HB]}$$

For each of the two solutions of penicillin, calculate $[HB]_0$ from the given information, then set up an equilibrium table (as in Example 13.7) and use it to obtain $[H^+]_{eq}$.

(a) $[HB]_0 = \dfrac{0.187 \text{ mol HB}}{725 \text{ mL}} \times \dfrac{1000 \text{ mL}}{1 \text{ L}} = 0.258 \text{ M}$

In completing the equilibrium table, use $[HB]_0 = 0.258$ M (calculated above) and initially we assume there is no ionization, so $[H^+]_0 = [B^-]_0 = 0$. Let $\Delta[H^+] = x$:

	HB(aq) $\rightleftharpoons$	H$^+$(aq) +	B$^-$(aq)
[ ]$_0$ (M)	0.258	0	0
Δ[ ] (M)	$-x$	$+x$	$+x$
[ ]$_{eq}$ (M)	$0.258 - x$	x	x

Now substitute equilibrium concentrations of each species into the expression for K_a:

$$K_a = 1.7 \times 10^{-3} = \frac{[H^+][B^-]}{[HB]} = \frac{(x)(x)}{(0.258 - x)}$$

Assume $x \ll 0.258$:

$$1.7 \times 10^{-3} = \frac{x^2}{0.258} \qquad \text{so } x^2 = (0.258)(1.7 \times 10^{-3}) = 4.4 \times 10^{-4} \text{ and } x = 2.1 \times 10^{-2} \text{ M}$$

The value of x is NOT much less that 0.258, so an invalid assumption was made in solving the above equation; % ionization = $[H^+]_{eq}/[HB]_0 \times 100 = (2.1 \times 10^{-2})/(0.258) \times 100 = 8.1\% > 5\%$ (see page 345). We cannot make the simplifying assumption ($x \ll 0.258$) so we must use the quadratic formula to solve the equation above.

$$1.7 \times 10^{-3} = \frac{x^2}{(0.258 - x)}$$

$$x^2 = (0.258 - x)(1.7 \times 10^{-3}) = (4.4 \times 10^{-4}) - (1.7 \times 10^{-3})(x)$$

$$x^2 + (1.7 \times 10^{-3})(x) - (4.4 \times 10^{-4}) = 0$$

$$x = \frac{-b \pm \sqrt{b^2 - 4ac}}{2a} = \frac{-(1.7 \times 10^{-3}) \pm \sqrt{(1.7 \times 10^{-3})^2 - 4(1)(-4.4 \times 10^{-4})}}{2(1)}$$

$$x = 0.020 \text{ or } -0.022$$

The negative root of the equation is physically impossible since x is a concentration; therefore $x = [H^+]_{eq} = \underline{\textbf{0.020 M}}$.

(b) $[HB]_0 = \dfrac{127 \text{ g HB}}{725 \text{ mL}} \times \dfrac{1 \text{ mol HB}}{356 \text{ g HB}} \times \dfrac{1000 \text{ mL}}{1 \text{ L}} = 0.492 \text{ M}$

In completing the equilibrium table, use $[HB]_0 = 0.492$ M (calculated above) and initially we assume there is no ionization, so $[H^+]_0 = [B^-]_0 = 0$. Let $\Delta[H^+] = x$:

	HB(*aq*) ⇌	H⁺(*aq*) +	B⁻(*aq*)
[ ]₀ (M)	0.492	0	0
Δ[ ] (M)	−x	+x	+x
[ ]ₑq (M)	0.492 − x	x	x

Now substitute equilibrium concentrations of each species into the expression for K_a:

$$K_a = 1.7 \times 10^{-3} = \frac{[H^+][B^-]}{[HB]} = \frac{(x)(x)}{(0.492 - x)}$$

Assume $x \ll 0.492$:

$$1.7 \times 10^{-3} = \frac{x^2}{0.492} \quad \text{so } x^2 = (0.492)(1.7 \times 10^{-3}) = 8.4 \times 10^{-4} \quad \text{and} \quad x = 2.9 \times 10^{-2} \text{ M}$$

The value of x is NOT much less that 0.492, so an invalid assumption was made in solving the above equation; % ionization = $[H^+]_{eq}/[HB]_0 \times 100 = (2.9 \times 10^{-2})/(0.492) \times 100 = 5.9\% > 5\%$ (see page 345). We cannot make the simplifying assumption ($x \ll 0.492$) so we must use the quadratic formula to solve the equation above.

$$1.7 \times 10^{-3} = \frac{x^2}{(0.492 - x)}$$

$$x^2 = (0.492 - x)(1.7 \times 10^{-3}) = (8.4 \times 10^{-4}) - (1.7 \times 10^{-3})(x)$$

$$x^2 + (1.7 \times 10^{-3})(x) - (8.4 \times 10^{-4}) = 0$$

$$x = \frac{-b \pm \sqrt{b^2 - 4ac}}{2a} = \frac{-(1.7 \times 10^{-3}) \pm \sqrt{(1.7 \times 10^{-3})^2 - 4(1)(-8.4 \times 10^{-4})}}{2(1)}$$

$$x = 0.028 \quad \text{or} \quad -0.030$$

The negative root of the equation is physically impossible since x is a concentration, therefore $x = [H^+]_{eq} = \underline{\textbf{0.028 M}}$.

54. This problem is similar to Example 13.7. When anisic acid behaves as an acid in aqueous solution, the following equilibrium reaction will occur (see page 339; label the acid "HB"):

$$HB(aq) \rightleftharpoons H^+(aq) + B^-(aq) \qquad\qquad K_a = \frac{[H^+][B^-]}{[HB]}$$

Draw an equilibrium table as in Example 13.5, based on the above equilibrium reaction. In completing the table, recall that $[HB]_0 = 0.279$ M (given) and initially we assume there is no ionization, so $[H^+]_0 = [B^-]_0 = 0$. Let $\Delta[H^+] = x$:

	HB(*aq*) $\rightleftharpoons$	H$^+$(*aq*) +	B$^-$(*aq*)
[ ]$_0$ (M)	0.279	0	0
Δ[ ] (M)	$-x$	$+x$	$+x$
[ ]$_{eq}$ (M)	$0.279 - x$	x	x

Now substitute the equilibrium concentrations of each species into the expression for K_a:

$$K_a = 3.38 \times 10^{-5} = \frac{[H^+][B^-]}{[HB]} = \frac{(x)(x)}{(0.279 - x)}$$

Assume $x \ll 0.279$:

$$3.38 \times 10^{-5} = \frac{x^2}{0.279} \quad \text{so} \quad x^2 = (0.279)(3.38 \times 10^{-5}) = 9.43 \times 10^{-6} \quad \text{and} \quad x = 3.07 \times 10^{-3} \text{ M}$$

Note that the value of x is indeed much less that 0.279, so the assumption made in solving the above equation was valid; % ionization = $[H^+]_{eq}/[HB]_0 \times 100$ = $(3.07 \times 10^{-3})/(0.279) \times 100 = 1.10\% < 5\%$ (see page 345).

(a) Recall that $x = [H^+]_{eq}$ so $[H^+]$ = **3.07×10^{-3} M**

(b) Recall (page 333) that $[H^+][OH^-] = 1.0 \times 10^{-14}$ so $(3.07 \times 10^{-3})[OH^-] = 1.0 \times 10^{-14}$ and hence $[OH^-] = (1.0 \times 10^{-14})/(3.07 \times 10^{-3})$ = **3.3×10^{-12} M**

(c) Calculate pH using Equation 13.3:

$$pH = -\log_{10}[H^+] = -\log_{10}(3.07 \times 10^{-3}) = \mathbf{2.51}$$

(d) Calculate % ionization using Equation 13.7:

$$\% \text{ ionization} = \frac{[H^+]_{eq}}{[HB]_0} \times 100 = \frac{3.07 \times 10^{-3}}{0.279} \times 100 = \underline{\mathbf{1.10\%}}$$

56. This problem is similar to Example 13.7. When benzoic acid behaves as an acid in aqueous solution, the following equilibrium reaction will occur (see page 339; label the acid "HB"):

$$HB(aq) \rightleftharpoons H^+(aq) + B^-(aq) \qquad\qquad K_a = \frac{[H^+][B^-]}{[HB]}$$

Calculate $[HB]_0$ from the given information, then set up an equilibrium table (as in Example 13.7) and use it to obtain $[H^+]_{eq}$.

$$[HB]_0 = \frac{0.288 \text{ mol HB}}{726 \text{ mL}} \times \frac{1000 \text{ mL}}{1 \text{ L}} = 0.397 \text{ M}$$

In completing the following table, use $[HB]_0 = 0.397$ M (calculated above) and recall that initially we assume there is no ionization, so $[H^+]_0 = [B^-]_0 = 0$. Let $\Delta[H^+] = x$:

	HB(aq) $\rightleftharpoons$	H$^+$(aq) +	B$^-$(aq)
[ ]$_0$ (M)	0.397	0	0
Δ[ ] (M)	$-x$	$+x$	$+x$
[ ]$_{eq}$ (M)	$0.397 - x$	x	x

Now substitute the equilibrium concentrations of each species into the expression for K_a:

$$K_a = 6.6 \times 10^{-5} = \frac{[H^+][B^-]}{[HB]} = \frac{(x)(x)}{(0.397 - x)}$$

Assume $x \ll 0.397$:

$$6.6 \times 10^{-5} = \frac{x^2}{0.397} \quad \text{so} \quad x^2 = (0.397)(6.6 \times 10^{-5}) = 2.6 \times 10^{-5} \quad \text{and} \quad x = 5.1 \times 10^{-3}$$

Note that the value of x is indeed much less that 0.397, so the assumption made in solving the above equation was valid; % ionization = $[H^+]_{eq}/[HB]_0 \times 100 = (5.1 \times 10^{-3})/(0.397) \times 100 = 1.3\% < 5\%$ (see page 345).

Recall that $x = [H^+]_{eq}$ so we can use the value of x to calculate pH using Equation 13.3:

$$pH = -\log_{10}[H^+] = -\log_{10}(5.1 \times 10^{-3}) = \underline{\mathbf{2.29}}$$

Finally, calculate % ionization using Equation 13.7:

$$\% \text{ ionization} = \frac{[H^+]_{eq}}{[HB]_0} \times 100 = \frac{5.1 \times 10^{-3}}{0.397} \times 100 = \underline{\mathbf{1.3\%}}$$

13-4   WEAK ACIDS AND THEIR EQUILIBRIUM CONSTANTS: Polyprotic Acids

58. For a polyprotic acid, the ionizations occur in steps, with one H$^+$ ion being produced in each step (see page 346). Thus, for citric acid, three K_a values indicate that there will be *three* successive ionizations:

$H_3C_6H_5O_7(aq) \rightleftharpoons H^+(aq) + H_2C_6H_5O_7^-(aq)$ $K_{a1} = 7.5 \times 10^{-4}$

$H_2C_6H_5O_7^-(aq) \rightleftharpoons H^+(aq) + HC_6H_5O_7^{2-}(aq)$ $K_{a2} = 1.7 \times 10^{-5}$

$HC_6H_5O_7^{2-}(aq) \rightleftharpoons H^+(aq) + C_6H_5O_7^{3-}(aq)$ $K_{a3} = 4.0 \times 10^{-7}$

To obtain the overall net ionic equation, add together the three above equations:

$$H_3C_6H_5O_7(aq) + H_2C_6H_5O_7^-(aq) + HC_6H_5O_7^{2-}(aq)$$
$$\rightleftharpoons 3H^+(aq) + H_2C_6H_5O_7^-(aq) + HC_6H_5O_7^{2-}(aq) + C_6H_5O_7^{3-}(aq)$$

which simplifies to:

$$H_3C_6H_5O_7(aq) \rightleftharpoons 3H^+(aq) + C_6H_5O_7^{3-}(aq)$$

To obtain K for this reaction, apply the Rule of Multiple Equilibria (see Table 12.3):

$$K = K_{a1} \times K_{a2} \times K_{a3} = (7.5 \times 10^{-4})(1.7 \times 10^{-5})(4.0 \times 10^{-7}) = \underline{\textbf{5.1} \times \textbf{10}^{-15}}$$

60. This problem is very similar to Example 13.9. Recall that for polyprotic acids most of the H^+ is obtained from the first ionization:

$$H_2X(aq) \rightleftharpoons H^+(aq) + HX^-(aq) \qquad K_{a1} = \frac{[H^+][HX^-]}{[H_2X]} = 3.3 \times 10^{-4}$$

After the first ionization, $[HX^-]_{eq} = [H^+]_{eq}$ and although these concentrations will decrease and increase, respectively, after the second ionization, they will not do so to a significant extent. Thus the values of these two concentrations (and also the value of pH) can be calculated by considering *only* the first ionization.

Set up an equilibrium table (similar to that in Example 13.7) with $[H_2X]_0 = 0.33\ M$ (given), $[H^+]_0 = [HX^-]_0 = 0$, and letting $\Delta[H^+] = x$:

	$H_2X(aq)$	$\rightleftharpoons$	$H^+(aq)$	$+$	$HX^-(aq)$
[ ]$_0$ (M)	0.33		0		0
Δ[ ] (M)	$-x$		$+x$		$+x$
[ ]$_{eq}$ (M)	$0.33 - x$		x		x

Now substitute the equilibrium concentrations of each species into the expression for K_{a1}:

$$K_{a1} = 3.3 \times 10^{-4} = \frac{[H^+][HX^-]}{[H_2X]} = \frac{(x)(x)}{(0.33 - x)}$$

Assume $x \ll 0.33$:

$$3.3 \times 10^{-4} = \frac{x^2}{0.33} \quad \text{so} \quad x^2 = (0.33)(3.3 \times 10^{-4}) = 1.1 \times 10^{-4} \quad \text{and} \quad x = 1.0 \times 10^{-2}$$

Note that the value of x is indeed much less that 0.33, so the assumption made in solving the above equation was valid; % ionization = $[H^+]_{eq}/[H_2X]_0 \times 100 = (1.0 \times 10^{-2})/(0.33) \times 100 = 3.0\% < 5\%$ (see page 345). Hence $x = [H^+]_{eq} = 1.0 \times 10^{-2}\ M$.
Now use equation 13.3 to calculate pH:

$$pH = -\log_{10}[H^+] = -\log_{10}(1.0 \times 10^{-2}) = \underline{\textbf{2.00}}$$

Note also that $[HX^-]_{eq} = [H^+]_{eq} = x = \underline{1.0 \times 10^{-2} \ M}$.

Now consider the second ionization:

$$HX^-(aq) \rightleftharpoons H^+(aq) + X^{2-}(aq) \qquad K_{a2} = \frac{[H^+][X^{2-}]}{[HX^-]} = 9.7 \times 10^{-8}$$

and recall that $[HX^-]_{eq} = [H^+]_{eq}$ after the first ionization. Since the extent to which HX^- ionizes in the second equation is negligible, these two concentrations remain essentially equal. Thus $[H^+]$ and $[HX^-]$ cancel each other out from the K_{a2} expression (see page 347):

$$K_{a2} = \frac{[H^+][X^{2-}]}{[HX^-]} = [X^{2-}] = \underline{9.7 \times 10^{-8} \ M}$$

62. This problem is very similar to Example 13.9 and problem 60 above. Recall that for polyprotic acids most of the H^+ is obtained from the first ionization:

$$H_2SeO_3(aq) \rightleftharpoons H^+(aq) + HSeO_3^-(aq) \qquad K_{a1} = \frac{[H^+][HSeO_3^-]}{[H_2SeO_3]} = 2.7 \times 10^{-3}$$

After the first ionization, $[HSeO_3^-]_{eq} = [H^+]_{eq}$ and although these concentrations will decrease and increase, respectively, after the second ionization, they do not do so to a significant extent. Thus, the values of these two concentrations (and also the value of pH) can be calculated by considering *only* the first ionization.

Set up an equilibrium table (similar to that in Example 13.7) with $[H_2SeO_3]_0 = 2.89 \ M$ (given), $[H^+]_0 = [HSeO_3^-]_0 = 0$, and letting $\Delta[H^+] = x$:

	H_2SeO_3 (aq)	$\rightleftharpoons$	H^+(aq)	+	$HSeO_3^-$(aq)
[ ]$_0$ (M)	2.89		0		0
Δ[ ] (M)	$-x$		$+x$		$+x$
[ ]$_{eq}$ (M)	$2.89 - x$		x		x

Now substitute the equilibrium concentrations of each species into the expression for K_{a1}:

$$K_{a1} = 2.7 \times 10^{-3} = \frac{[H^+][HSeO_3^-]}{[H_2SeO_3]} = \frac{(x)(x)}{(2.89 - x)}$$

Assume $x \ll 2.89$:

$$2.7 \times 10^{-3} = \frac{x^2}{2.89} \quad \text{so} \quad x^2 = (2.89)(2.7 \times 10^{-3}) = 7.8 \times 10^{-3} \quad \text{and} \quad x = 8.8 \times 10^{-2}$$

Note that the value of x is indeed much less that 2.89, so the assumption made in solving the above equation was valid; % ionization = $[H^+]_{eq}/[H_2SeO_3]_0 \times 100 = (8.8 \times 10^{-2})/(2.89) \times 100 = 3.0\% < 5\%$ (see page 345). Hence $x = [H^+]_{eq} = 8.8 \times 10^{-2} \ M$.
Now use equation 13.3 to calculate pH:

$$pH = -\log_{10}[H^+] = -\log_{10}(8.8 \times 10^{-2}) = \mathbf{1.06}$$

Note also that $[HSeO_3^-]_{eq} = [H^+]_{eq} = x = \mathbf{8.8 \times 10^{-2}\ M}$.

Now consider the second ionization:

$$HSeO_3^-(aq) \rightleftharpoons H^+(aq) + SeO_3^{2-}(aq) \qquad K_{a2} = \frac{[H^+][SeO_3^{2-}]}{[HSeO_3^-]} = 5.0 \times 10^{-8}$$

and recall that $[HSeO_3^-]_{eq} = [H^+]_{eq}$ after the first ionization. Since the extent to which HX^- ionizes in the second equation is negligible, these two concentrations remain essentially equal. Thus $[H^+]$ and $[HSeO_3^-]$ cancel each other out from the K_{a2} expression (see also page 347):

$$K_{a2} = \frac{[H^+][SeO_3^{2-}]}{[HSeO_3^-]} = [SeO_3^{2-}] = \mathbf{5.0 \times 10^{-8}\ M}$$

13-5   WEAK BASES AND THEIR EQUILIBRIUM CONSTANTS: Ionization Expressions, Weak Bases

64. See page 349. Since these species are bases, their ionization in water will be of the form:

$$B^-(aq) + H_2O \rightleftharpoons HB(aq) + OH^-(aq)$$

(where the anion B^- is any weak base). For this reaction the equilibrium constant K_b is:

$$K_b = \frac{[HB][OH^-]}{[B^-]} \qquad \text{(Equation 13.8)}$$

(a) $NH_3(aq) + H_2O \rightleftharpoons NH_4^+(aq) + OH^-(aq)$ 　　　 $K_b = \dfrac{[NH_4^+][OH^-]}{[NH_3]}$

(b) $HS^-(aq) + H_2O \rightleftharpoons H_2S(aq) + OH^-(aq)$ 　　　 $K_b = \dfrac{[H_2S][OH^-]}{[HS^-]}$

(c) $(CH_3)_3N(aq) + H_2O \rightleftharpoons (CH_3)_3NH^+(aq) + OH^-(aq)$ 　　　 $K_b = \dfrac{[(CH_3)_3NH^+][OH^-]}{[(CH_3)_3N]}$

66. Use the values of K_b given in Table 13.2 (right column) directly and arrange the ions in order of increasing K_b. Note that the given concentration has no effect on the value of K_b.

K_b value:	1.7×10^{-11}	<	5.3×10^{-11}	<	1.7×10^{-7}
base:	NO_2^-	<	CHO_2^-	<	SO_3^{2-}

68. Since these species are bases, it is convenient to work with $[OH^-]$ and then convert to pH. The solutions all have the same reactant concentration, so in principle $[OH^-]$ in each solution should depend on the extent of ionization of the base.

The bases present are KOH, CN^-, HCO_3^-, and $Ba(OH)_2$. Of these, KOH and $Ba(OH)_2$ are strong bases (see Section 13-3), so both ionize completely in solution and will have high $[OH^-]$. However, $Ba(OH)_2$ produces **two** mol of OH^- per formula unit, whereas KOH only provides **one** mol of OH^- per formula unit. Thus $[OH^-]$ in these two solutions is:

$$\text{in } Ba(OH)_2: \quad [OH^-] = \frac{0.1\,\text{mol } Ba(OH)_2}{1\,L} \times \frac{2\,\text{mol } OH^-}{1\,\text{mol } Ba(OH)_2} = 0.2\ M$$

$$\text{in KOH:} \quad [OH^-] = \frac{0.1\,\text{mol KOH}}{1\,L} \times \frac{1\,\text{mol } OH^-}{1\,\text{mol KOH}} = 0.1\ M$$

For the other two bases, recall (page 349) that the larger the K_b value, the stronger the base and the higher $[OH^-]$ will be. For the two weak bases, the K_b values (Table 13.2) are: CN^-, 1.7×10^{-5}; HCO_3^-, 2.3×10^{-8}, so $[OH^-]$ in the solution containing CN^- ions will be higher than in the solution in the solution containing HCO_3^- ions. The order of decreasing $[OH^-]$ for the four bases, therefore, is:

$$Ba(OH)_2 > KOH > CN^- > HCO_3^-$$

Recall also (page 333 and Figure 13.1) that decreasing $[OH^-]$ corresponds to increasing $[H^+]$, and (page 334 and Figure 13.2) that increasing $[H^+]$ corresponds to decreasing pH. Thus decreasing $[OH^-]$ also corresponds to decreasing pH and so the above ordering of the four bases is also in order of decreasing pH:

$$Ba(OH)_2 > KOH > NaCN > HCO_3^-$$

13-5 WEAK BASES AND THEIR EQUILIBRIUM CONSTANTS: Equilibrium Calculations, Weak Bases

70. Recall from Equation 13.10 that, for a weak base—conjugate acid pair (at 25 °C), $K_a \times K_b = K_w = 1.0 \times 10^{-14}$. Thus $K_b = K_w/K_a$: substitute the given K_a values to obtain K_b.

(a) $K_b = \dfrac{K_w}{K_a} = \dfrac{1.0 \times 10^{-14}}{1.5 \times 10^{-4}} = \underline{\mathbf{6.7 \times 10^{-11}}}$

(b) $K_b = \dfrac{K_w}{K_a} = \dfrac{1.0 \times 10^{-14}}{1.5 \times 10^{-5}} = \underline{\mathbf{6.7 \times 10^{-10}}}$

72. This problem is related to Example 13.11(a). The anion of the salt $NaHCO_3$ (namely the HCO_3^- ion) is listed as a base in Table 13.2 (see also Section 13-6), with K_b is 2.3×10^{-8}. When this ion acts as a base, the following equilibrium reaction occurs (see page 349):

$$HCO_3^-(aq) + H_2O \rightleftharpoons H_2CO_3(aq) + OH^-(aq) \qquad K_b = \frac{[H_2CO_3][OH^-]}{[HCO_3^-]}$$

The molarity of the HCO_3^- ion in the given solution of $NaHCO_3$ is:

$$[HCO_3^-] = \frac{0.72 \text{ mol } NaHCO_3}{1L} \times \frac{1 \text{ mol } HCO_3^-}{1 \text{ mol } NaHCO_3} = 0.72 \text{ M}$$

Draw an equilibrium table as in examples 13.5 and 13.11, based on the above equilibrium reaction. In completing the table, recall that $[HCO_3^-]_0 = 0.72$ M and initially we assume there is no ionization, so $[H_2CO_3]_0 = [OH^-]_0 = 0$. Let $\Delta[OH^-] = x$:

	$HCO_3^-(aq) + H_2O \rightleftharpoons$	$H_2CO_3(aq)$ +	$OH^-(aq)$
[ ]$_0$ (M)	0.72	0	0
Δ[ ] (M)	$-x$	$+x$	$+x$
[ ]$_{eq}$ (M)	$0.72 - x$	x	x

Now substitute the equilibrium concentrations of each species into the expression for K_b:

$$K_b = 2.3 \times 10^{-8} = \frac{[H_2CO_3][OH^-]}{[HCO_3^-]} = \frac{(x)(x)}{(0.72 - x)}$$

Assume $x \ll 0.72$:

$$2.3 \times 10^{-8} = \frac{x^2}{0.72} \qquad \text{so } x^2 = (0.72)(2.3 \times 10^{-8}) = 1.7 \times 10^{-8} \quad \text{and} \quad x = 1.3 \times 10^{-4} \text{ M}$$

Note that the value of x is indeed much less that 0.72: the % ionization = $[OH^-]_{eq}/[HCO_3^-]_0$ × 100 = $(1.3 \times 10^{-4})/(0.72)$ × 100 = 0.02% < 5% (see page 345), so the assumption made in solving the above equation was valid.

Note that $x = [OH^-]_{eq}$, so $[OH^-] = \underline{\mathbf{1.3 \times 10^{-4} \text{ M}}}$

Use this value of $[OH^-]$ with equation 13.4 to calculate pOH:

$$pOH = -\log_{10}[OH^-] = -\log_{10}(1.3 \times 10^{-4}) = 3.89$$

Finally, since pH + pOH = 14.00 (see page 334):

$$pH = 14.00 - pOH = 14.00 - 3.89 = \underline{\mathbf{10.11}}$$

74. (a) Since pyridine is a weak base, it will react with H_2O in a way similar to that for ammonia (NH_3) to produce OH^- ions in solution (see pages 348–349):

$$C_5H_5N(aq) + H_2O \rightleftharpoons C_5H_5NH^+(aq) + OH^-(aq)$$

(b) Recall (Equation 13.10) that, for a weak base–conjugate acid pair (at 25 °C), $K_a \times K_b = K_w = 1.0 \times 10^{-14}$. Thus $K_b = K_w/K_a$: substitute the given K_a value to obtain K_b.

$$K_b = \frac{K_w}{K_a} = \frac{1.0 \times 10^{-14}}{6.7 \times 10^{-6}} = \underline{\mathbf{1.5 \times 10^{-9}}}$$

(c) This part of the problem is similar to Example 13.11(a). First calculate $[C_5H_5N]_0$, then set up an equilibrium table and determine $[OH^-]$ and hence $[H^+]$ and pH.

$$[C_5H_5N]_0 = \frac{2.74 \text{ g } C_5H_5N}{685 \text{ mL}} \times \frac{1 \text{ mol } C_5H_5N}{79.10 \text{ g } C_5H_5N} \times \frac{1000 \text{ mL}}{1 \text{ L}} = 0.0506 \text{ M}$$

In completing the equilibrium table, recall that $[C_5H_5N]_0 = 0.0506$ M (calculated above) and initially we assume there is no ionization, so $[C_5H_5NH^+]_0 = [OH^-]_0 = 0$. Let $\Delta[OH^-] = x$:

	$C_5H_5N(aq) + H_2O$	$\rightleftharpoons$ $C_5H_5NH^+(aq)$ +	$OH^-(aq)$
$[\ ]_0$ (M)	0.0506	0	0
$\Delta[\ ]$ (M)	$-x$	$+x$	$+x$
$[\ ]_{eq}$ (M)	$0.0506 - x$	x	x

Now substitute the equilibrium concentrations of each species into the expression for K_b:

$$K_b = 1.5 \times 10^{-9} = \frac{[C_5H_5NH^+][OH^-]}{[C_5H_5N]} = \frac{(x)(x)}{(0.0506 - x)}$$

Assume $x \ll 0.0506$:

$$1.5 \times 10^{-9} = \frac{x^2}{0.0506} \quad \text{so } x^2 = (0.0506)(1.5 \times 10^{-9}) = 7.6 \times 10^{-11} \quad \text{and} \quad x = 8.7 \times 10^{-6}$$

Check the assumption that the value of x is indeed much less that 0.84: the % ionization = $[OH^-]_{eq}/[C_5H_5NH^+]_0 \times 100 = (8.7 \times 10^{-6})/(0.0506) \times 100 = 0.017\% < 5\%$ (see page 345), so the assumption made in solving the above equation was valid.

Note that $x = [OH^-]_{eq}$, so we say that $[OH^-] = 8.7 \times 10^{-6}$ M and hence, using Equations 13.1 and 13.2: $K_w = [H^+][OH^-] = 1.0 \times 10^{-14}$. Thus:

$$[H^+] = \frac{K_w}{[OH^-]} = \frac{1.0 \times 10^{-14}}{8.7 \times 10^{-6}} = 1.1 \times 10^{-9} \text{ M}$$

Finally, use Equation 13.3 to calculate pH:

$$pH = -\log_{10}[H^+] = -\log_{10}(1.1 \times 10^{-9}) = \underline{\mathbf{8.94}}$$

76. The cyanide ion acts as a weak base (since it is the conjugate base of a weak acid, HCN):

$$CN^-(aq) + H_2O \rightleftharpoons HCN(aq) + OH^-(aq) \qquad K_b = \frac{[HCN][OH^-]}{[CN^-]} = 1.7 \times 10^{-5} \text{ (Table 13.3)}$$

For a solution with pH = 12.10, $[H^+] = 10^{-12.10}$ (Equation 13.3) and hence from Equations 13.1 and 13.2:

$$[OH^-] = \frac{K_w}{[H^+]} = \frac{1.0 \times 10^{-14}}{10^{-12.10}} = 1.3 \times 10^{-2} \ M$$

Set up an equilibrium table and work in reverse to obtain $[CN^-]_0$. Initially we assume there is no ionization, so $[HCN]_0 = [OH^-]_0 = 0$. Let $\Delta[OH^-] = x$ and $[CN^-]_0 = y$:

	$CN^-(aq) + H_2O \rightleftharpoons$	$HCN(aq) +$	$OH^-(aq)$
[ ]₀ (M)	y	0	0
Δ[ ] (M)	$-x$	$+x$	$+x$
[ ]eq (M)	$y - x$	x	x

However, we know that $[OH^-]_{eq} = 1.3 \times 10^{-2} \ M$ (calculated above), so $x = 1.3 \times 10^{-2}$. Now substitute the equilibrium concentrations of each species into the expression for K_b:

$$K_b = 1.7 \times 10^{-5} = \frac{[HCN][OH^-]}{[CN^-]} = \frac{(x)(x)}{(y-x)} = \frac{(1.3 \times 10^{-2})(1.3 \times 10^{-2})}{(y - 1.3 \times 10^{-2})}$$

$$y - 1.3 \times 10^{-2} = \frac{(1.3 \times 10^{-2})^2}{1.7 \times 10^{-5}} = 9.9 \qquad \text{so} \quad y = 9.9 \qquad \text{and } [CN^-]_0 = y = 9.9 \ M$$

Thus the concentration of the NaCN solution is:

$$[NaCN] = \frac{9.9 \ mol \ CN^-}{1L} \times \frac{1 \ mol \ NaCN}{1 \ mol \ CN^-} = 9.9 \ M$$

Use the flowchart for solution stoichiometry (Figure 4.6) to determine the mass NaCN needed to make the given solution. Strategy: M NaCN → mol NaCN → g NaCN

$$\text{Mass NaCN required} = \frac{9.9 \ mol \ NaCN}{1L} \times \frac{1L}{1000 \ mL} \times 425 \ mL \times \frac{49.01 \ g \ NaCN}{1 \ mol \ NaCN}$$

$$= \underline{2.1 \times 10^2 \ g}$$

13-6 ACID-BASE PROPERTIES OF SALT SOLUTIONS

78. There are many possible correct answers to this problem. See Figure 13.14 and also Example 13.12.

(a) Since NH_4^+ is an acidic cation (see Table 13.5), in order to form the required salt NH_4^+ must be combined with a basic anion. Moreover, note from Figure 13.14 (last row) that K_b for the anion must be greater than K_a for NH_4^+ to give a basic salt. From Table 13.2, K_a for NH_4^+ is 5.6×10^{-10}; so we require an anion with $K_b > 5.6 \times 10^{-10}$. Suitable anions from Table 13.2 are SO_3^{2-} ($K_b = 1.7 \times 10^{-7}$), ClO^- ($K_b = 3.6 \times 10^{-7}$), and CN^- ($K_b = 1.7 \times 10^{-5}$) for which the corresponding salts would be:

(NH₄)₂SO₃, NH₄ClO, NH₄CN

(b) Since CO_3^{2-} is a basic anion (it is the conjugate base of a weak acid; see Table 13.4), each of the *spectator* cations from Table 13.5 will form a basic salt with this anion:

Li₂CO₃, Na₂CO₃, K₂CO₃, CaCO₃, SrCO₃, BaCO₃

(c) Since Br^- is a spectator anion (see Table 13.5), each of the *spectator* cations from Table 13.5 will form a neutral salt with this anion:

LiBr, NaBr, KBr, CaBr₂, SrBr₂, BaBr₂

(d) See Problem 57. The acid H_2CrO_4 is diprotic with $K_{a2} = 3.2 \times 10^{-7}$. This K_{a2} value refers to the second ionization reaction of the acid H_2CrO_4:

$$HCrO_4^-(aq) \rightleftharpoons CrO_4^{2-}(aq) + H^+(aq)$$

— in other words, it is the K_a value for the anion $HCrO_4^-$, for which CrO_4^{2-} is the conjugate base. Using Equation 13.10, K_b for $CrO_4^{2-} = K_w / (K_a$ for $HCrO_4^-) = (1.0 \times 10^{-14}) / (3.2 \times 10^{-7}) = 3.1 \times 10^{-8}$.

Now, since CrO_4^{2-} is basic it must be combined with a more acidic cation in order to form an acidic salt. Note from Figure 13.14 that K_a for the cation must be greater than K_b for CrO_4^{2-} to give an acidic salt. From Table 13.2, cations with $K_a > 3.1 \times 10^{-8}$ are $Fe(H_2O)_6^{3+}$ ($K_a = 6.7 \times 10^{-3}$) and $Al(H_2O)_6^{3+}$ ($K_a = 1.2 \times 10^{-5}$) for which the corresponding salts would be $[Fe(H_2O)_6]_2[CrO_4]_3$ and $[Al(H_2O)_6]_2[CrO_4]_3$; or, more simply:

Fe₂(CrO₄)₃ and Al₂(CrO₄)₃

80. This problem is similar to Example 13.12; see also Figure 13.14 and Table 13.5. Identify the ions present in solution when the salt dissolves, label them as acidic, basic, or spectator using Table 13.5, then follow the flowchart in Figure 13.14 to describe the solution.

	Salt	Cation	Anion	Solution
(a)	FeCl₃	Fe^{3+} (acidic)	Cl^- (spectator)	**acidic**
(b)	BaI₂	Ba^{2+} (spectator)	I^- (spectator)	**neutral**
(c)	NH₄NO₂	NH_4^+ (acidic)	NO_2^- (basic)	**acidic***
(d)	Na₂HPO₄	Na^+ (spectator)	HPO_4^{2-} (amphiprotic)	**basic****
(e)	K₃PO₄	K^+ (spectator)	PO_4^{3-} (basic)	**basic**

* For NH₄NO₂: in this case (acidic cation + basic anion; see Figure 13.14), it is necessary to consider K_a of the acidic species and K_b of the basic species. Referring to Table 13.2, K_a for NH_4^+ is 5.6×10^{-10} and K_b for NO_2^- is 1.7×10^{-11}. Thus:

K_a of the acidic species > K_b of the basic species, so the salt solution is **acidic**.

** For Na₂HPO₄: in this case (spectator cation + amphiprotic anion; see Figure 13.14), it is necessary to consider both K_a and K_b of the amphiprotic species. From Table 13.2, K_a for HPO_4^{2-} is 4.5×10^{-13} and K_b for HPO_4^{2-} is 1.6×10^{-7}. Thus:

$K_a < K_b$ for the amphiprotic anion, so the salt solution is **basic**.

82. The net ionic equations can be written by analogy with those given in the text (pages 352−354); for part (d), see also Section 13-4, Polyprotic Weak Acids (pages 346−347).

   (a) See also page 339. The hydrated $Fe^{3+}(aq)$ ion is the complex ion $Fe(H_2O)_6^{3+}$:

   $$Fe(H_2O)_6^{3+}(aq) + H_2O \rightleftharpoons H_3O^+(aq) + Fe(H_2O)_5(OH)^{2+}(aq)$$

   (b) BaI_2 is neutral

   (c) $NH_4^+(aq) + H_2O \rightleftharpoons H_3O^+(aq) + NH_3(aq)$ $\qquad K_a = 5.6 \times 10^{-10}$
   $NO_2^-(aq) + H_2O \rightleftharpoons HNO_2(aq) + OH^-(aq)$ $\qquad K_b = 1.7 \times 10^{-11}$

   (d) $HPO_4^{2-}(aq) + H_2O \rightleftharpoons H_3O^+(aq) + PO_4^{3-}(aq)$ $\qquad K_a = 4.5 \times 10^{-13}$
   $HPO_4^{2-}(aq) + H_2O \rightleftharpoons H_2PO_4^-(aq) + OH^-(aq)$ $\qquad K_b = 1.6 \times 10^{-7}$

   (e) $PO_4^{3-}(aq) + H_2O \rightleftharpoons HPO_4^{2-}(aq) + OH^-(aq)$

84. Of the compounds listed, the solution of $ZnBr_2$ has the lowest pH since it is the only one that is acidic (see Table 13.5 and Figure 13.14): the Zn^{2+} cation is acidic, but the Br^- anion is a spectator.
The solution of $KClO_4$ will be neutral, since the cation (K^+) and anion (ClO_4^-) are both spectators (see Table 13.5 and Figure 13.14).
The solution of LiF will be basic, since the anion F^- is basic — it is the conjugate base of the weak acid HCN. (See also Table 13.5 and Figure 13.14: spectator cation and basic anion.) However, F^- is only a very weak base ($K_b = 1.4 \times 10^{-11}$ from Table 13.2), so although the solution will have a basic pH (pH > 7) but it will only be weakly basic.
Both NaOH and $Sr(OH)_2$ are strong bases (see page 336), so both of those solutions will be strongly basic, with a high pH (high $[OH^-]$). However, a 0.1 M solution of NaOH has $[OH^-]$ = 0.1 M whereas in 0.1 M $Sr(OH)_2$ $[OH^-]$ = 0.2 M (since every mole of $Sr(OH)_2$ produces *two* mol of OH^- ions; compare KOH/$Ba(OH)_2$ in problem 68). Hence, of these two, the solution of $Sr(OH)_2$ will have the higher $[OH^-]$ and hence the highest pH.

   Thus, combining all of these, the order of increasing pH will be:

   $$ZnBr_2 < KClO_4 < LiF < NaOH < Sr(OH)_2$$

UNCLASSIFIED PROBLEMS

86. Referring to Table 13.5, the Cl^- anion of ammonium chloride and the Na^+ cation of sodium nitrite are spectator ions and so do not undergo acid-base reactions. However, the ammonium ion is acidic and the nitrite ion is basic (see Table 13.2).
   When the ammonium ion behaves as an acid, the following reaction takes place:
   $$NH_4^+(aq) \rightleftharpoons H^+(aq) + NH_3(aq) \qquad K_a = 5.6 \times 10^{-10} \text{ (Table 13.2)}$$
   and the nitrite ion can act as a base, accepting the H^+ produced in the above reaction:
   $$NO_2^-(aq) + H^+(aq) \rightleftharpoons HNO_2(aq) \qquad K = 1/K_{a(HNO_2)} = 1/(6.0 \times 10^{-4})$$

The latter reaction is the reverse of the HNO_2 dissociation reaction: K_a for HNO_2 is 6.0 × 10^{-4} (Table 13.2); then apply the Reciprocal Rule (Table 12.3).

Combining the two above reactions gives:

$NH_4^+(aq) + NO_2^-(aq) + H^+(aq) \rightleftharpoons H^+(aq) + NH_3(aq) + HNO_2(aq)$

or (simplify by cancelling duplications) :

$NH_4^+(aq) + NO_2^-(aq) \rightleftharpoons NH_3(aq) + HNO_2(aq)$

for which (using the Rule for Multiple Equilibria, Table 12.3) the corresponding equilibrium constant K is given by:

$K = K_a \times 1/K_{a(HNO_2)} = (5.6 \times 10^{-10}) \times (6.0 \times 10^{-4})^{-1} = 9.3 \times 10^{-7}$

Since K is so small, the reaction does not happen to any significant extent; so **NO, the reaction is not likely**.

88. Ammonia is a weak base:

$$NH_3(aq) + H_2O \rightleftharpoons NH_4^+(aq) + OH^-(aq) \qquad K_b = \frac{[NH_4^+][OH^-]}{[NH_3]} = 1.8 \times 10^{-5} \text{ (Table 13.2)}$$

For a solution with pH = 11.55, $[H^+] = 10^{-11.55}$ (Equation 13.3) and hence from Equations 13.1 and 13.2:

$$[OH^-] = \frac{K_w}{[H^+]} = \frac{1.0 \times 10^{-14}}{10^{-11.55}} = 3.5 \times 10^{-3} \, M$$

Set up an equilibrium table and work in reverse to obtain $[NH_3]_0$. Initially we assume there is no ionization, so $[NH_4^+]_0 = [OH^-]_0 = 0$. Let $\Delta[OH^-] = x$ and $[NH_3]_0 = y$:

	$NH_3(aq) + H_2O$	$\rightleftharpoons$	$NH_4^+(aq) +$	$OH^-(aq)$
[ ]$_0$ (M)	y		0	0
Δ[ ] (M)	$-x$		$+x$	$+x$
[ ]$_{eq}$ (M)	$y - x$		x	x

However, we know that $[OH^-]_{eq} = 3.5 \times 10^{-3} \, M$ (calculated above), so $x = 3.5 \times 10^{-3}$. Now substitute the equilibrium concentrations of each species into the expression for K_b:

$$K_b = 1.8 \times 10^{-5} = \frac{[NH_4^+][OH^-]}{[NH_3]} = \frac{(x)(x)}{(y - x)} = \frac{(3.5 \times 10^{-3})(3.5 \times 10^{-3})}{(y - 3.5 \times 10^{-3})}$$

$$y - 3.5 \times 10^{-3} = \frac{(3.5 \times 10^{-3})^2}{1.8 \times 10^{-5}} = 0.68 \qquad \text{so} \quad y = 0.68 \quad \text{and} \quad [NH_3]_0 = y = 0.68 \, M$$

Use the flowchart for solution stoichiometry (Figure 4.6) to determine the mol NH_3 needed to form a solution of molarity 0.68 M; then use the Ideal Gas Law (Equation 5.1) to calculate the corresponding volume of NH_3 required. Strategy: M $NH_3 \rightarrow$ mol $NH_3 \rightarrow$ V NH_3

$$\text{mol NH}_3 \text{ required} = \frac{0.68 \text{ mol NH}_3}{1 \text{L}} \times 4.00 \text{ L} = 2.7 \text{ mol}$$

$$\text{vol NH}_3 \text{ required} = V = \frac{nRT}{P} = \frac{(2.7 \text{ mol})(0.0821 \text{L} \cdot \text{atm/mol} \cdot \text{K})(298 \text{ K})}{1 \text{ atm}} = \underline{\textbf{66 L}}$$

90. For a solution with pH = 13.33, $[H^+] = 10^{-13.33}$ (Equation 13.3) and hence from Equations 13.1 and 13.2:

$$[OH^-] = \frac{K_w}{[H^+]} = \frac{1.0 \times 10^{-14}}{10^{-13.33}} = 0.21 \ M$$

Since KOH is a strong base, this gives us directly the concentration of KOH:

$$[KOH] = \frac{0.21 \text{ mol OH}^-}{1 \text{L}} \times \frac{1 \text{ mol KOH}}{1 \text{ mol OH}^-} = 0.21 \ M$$

Use the flowchart for solution stoichiometry (Figure 4.6) to determine the mass KOH needed to make the given solution.   Strategy: M KOH → mol KOH → g KOH

$$\text{Mass KOH required} = \frac{0.21 \text{ mol KOH}}{1 \text{L}} \times \frac{1 \text{L}}{1000 \text{ mL}} \times 455 \text{ mL} \times \frac{56.11 \text{ g KOH}}{1 \text{ mol KOH}} = \underline{\textbf{5.4 g}}$$

92. For a solution with pH = 1.19, $[H^+] = 10^{-1.19} = 0.065 \ M$ (using Equation 13.3).

Note that the reaction given in the question shows that the reaction of Cl_2 with H_2O produces HCl(*aq*) and HOCl(*aq*). HCl is a strong acid (see Section 13-4) whereas HOCl is a rather weak acid ($K_a = 3.2 \times 10^{-8}$, given), so essentially all of the H^+ in solution can be assumed to have come from dissociation of the HCl. Thus, use the stoichiometry of the given reaction to determine the number of mol Cl_2 required to produce the given $[H^+]$, and hence the mass Cl_2 required. (Refer to Figure 4.6.)

Strategy: $[H^+] \rightarrow$ mol $H^+ \rightarrow$ mol $Cl_2 \rightarrow$ mass Cl_2

$$\text{mol H}^+ \text{ present in 1 L solution} = \frac{0.065 \text{ mol H}^+}{1 \text{L}} \times 1 \text{ L solution} = 0.065 \text{ mol H}^+$$

$$\text{mass Cl}_2 \text{ required} = 0.065 \text{ mol H}^+ \times \frac{1 \text{ mol Cl}_2}{1 \text{ mol H}^+} \times \frac{70.91 \text{ g Cl}_2}{1 \text{ mol Cl}_2} = \underline{\textbf{4.6 g Cl}_2}$$

94. Call the weak base B, for which the ionization reaction with be:

$$B(aq) + H_2O \rightleftharpoons BH^+(aq) + OH^-(aq) \qquad K_b = \frac{[BH^+][OH^-]}{[B]}$$

The solution has pH = 8.73, so $[H^+] = 10^{-8.73}$ (Equation 13.3) and hence from Equations 13.1 and 13.2:

$$[OH^-] = \frac{K_w}{[H^+]} = \frac{1.0 \times 10^{-14}}{10^{-8.73}} = 5.4 \times 10^{-6} \ M$$

For the given osmotic pressure, calculate the concentration of particles in solution (see Example 10.9(b)).

$$\pi = MRT \qquad \text{so} \quad M = \frac{\pi}{RT} = \frac{55 \ mm \ Hg \times \dfrac{1 \ atm}{760 \ mm \ Hg}}{(0.0821 \ L \cdot atm/mol \cdot K)(298 \ K)} = 3.0 \times 10^{-3} \ M$$

We can take this to be the molarity of B in solution, specifically $[B]_0$. This is an approximation since the base B further dissociates into BH^+ and OH^-; however, since B is a weak base, $[BH^+]$ and $[OH^-]$ are very small — in fact, $[OH^-]_{eq} = 5.4 \times 10^{-6}$ M (calculated above) and $[BH^+]_{eq} = [OH^-]_{eq}$ by stoichiometry — so the approximation is reasonable. Thus $[B]_{eq} = 3.0 \times 10^{-3}$ M.

Finally substitute the equilibrium concentrations of each species into the K_b expression:

$$K_b = \frac{[BH^+][OH^-]}{[B]} = \frac{(5.4 \times 10^{-6})(5.4 \times 10^{-6})}{(3.0 \times 10^{-3})} = \underline{\mathbf{9.7 \times 10^{-9}}}$$

CONCEPTUAL PROBLEMS

96. (a) $Ba(OH)_2$ is a strong base, which will completely dissociate in aqueous solution:

$$Ba(OH)_2(aq) \rightarrow Ba^{2+}(aq) + 2OH^-(aq)$$

Thus $[OH^-]$ is given by:

$$[OH^-] = \frac{0.1 \ mol \ Ba(OH)_2}{1 \ L} \times \frac{2 \ mol \ OH^-}{1 \ mol \ Ba(OH)_2} = 0.2 \ M$$

Hence, this statement is **false**.

(b) Using the stoichiometry of the reaction given in part (a) above, $[Ba^{2+}]$ is given by:

$$[Ba^{2+}] = \frac{0.1 \ mol \ Ba(OH)_2}{1 \ L} \times \frac{1 \ mol \ Ba^{2+}}{1 \ mol \ Ba(OH)_2} = 0.1 \ M$$

from which it can be seen that $[Ba^{2+}] \neq [OH^-]$. Therefore, this statement is **false**.

(c) Using Equation 13.4 with $[OH^-]$ from part (a) above, pOH = $-\log_{10}(0.2) = 0.7$; and since pH + pOH = 14.00 (page 334), pH = 14.00 − pOH = 14.00 − 0.7 = 13.3. The given statement is **true**.

(d) From Equations 13.1 and 13.2, $[H^+] = \dfrac{K_w}{[OH^-]} = \dfrac{1.0 \times 10^{-14}}{[OH^-]} = \dfrac{1.0 \times 10^{-14}}{0.2} = 5 \times 10^{-14}\,M$; therefore the given statement is **false**.

98. Use Tables 13.2 and 13.4 to identify whether each beaker contains a strong or weak acid or base:

    Beaker A: strong acid (HI)
    Beaker B: weak acid (HNO$_2$; $K_a = 6.0 \times 10^{-4}$)
    Beaker C: strong base (NaOH)
    Beaker D: strong base (Ba(OH)$_2$)
    Beaker E: weak acid (NH$_4^+$; $K_a = 5.6 \times 10^{-10}$)
    Beaker F: weak base (C$_2$H$_5$NH$_2$ is a primary amine: see Table B (page 357))

    (a) Beaker A contains a strong acid and beaker B contains a weak acid. A strong acid will contain more H$^+$ in solution than a weak acid and so will have a lower pH than a weak acid (Figure 13.2):

    LT   The pH in beaker A is less than the pH in beaker B.

    (b) Both beakers contain strong bases, but Ba(OH)$_2$ will produce more OH$^-$ in solution than NaOH: [OH$^-$] will be 0.1 M (beaker C) and 0.2 M (beaker D). Higher [OH$^-$] means lower [H$^+$] (Equation 13.1) and therefore higher pH (Figure 13.2):

    LT   The pH in beaker C is less than the pH in beaker D.

    (c) Beaker A contains a strong acid and beaker C contains a strong base. Since both solutes are strong electrolytes, by definition these are completely ionized in solution and so the % ionization in both beakers is 100 %:

    EQ   The % ionization in beaker A equals the % ionization in beaker C.

    (d) Based on K_a values, the weak acid in beaker B is stronger (more ionized) than the weak acid in beaker E. Since [H$^+$] depends on the extent of ionization, [H$^+$] must be higher in beaker B and so the pH in beaker B is lower than in beaker E (Figure 13.2):

    LT   The pH in beaker B is less than the pH in beaker E.
    The answer "MI" is also possible here: the extra information required is the K_a values.

    (e) Beaker E contains an acid, for which the pH should be < 7, and beaker F contains a base, for which the pH should be > 7. Therefore:

    LT   The pH in beaker E is less than the pH in beaker F.

    (f) Beaker C contains a strong base and beaker F contains a weak base. A strong base will contain more OH$^-$ in solution than a weak base; and so (using Equation 13.1) a strong base will contain *less* H$^+$ in solution than a weak base. Hence, a strong base will have a higher pH than a weak base (Figure 13.2):

    GT   The pH in beaker C is greater than the pH in beaker F.

100. A molecule of the acid will consist of a square and a circle together; and recall that strong acids completely dissociate into H$^+$ and an anion (Section 13-3), whereas weak acids only partially dissociate (Section 13-4). Thus, the box with all of the squares (H$^+$) and circles (anions) *separated* will represent a strong acid (*all of the molecules are*

separated into ions), and the box with some squares and circles together will represent a weak acid (*some molecules not separated into ions*).

Therefore, **box 2** represents a strong acid and **box 1** represents a weak acid.

102. Dissolving some of the solid in water and measuring the pH of the solution will immediately tell you whether the solid is acidic (pH < 7) or basic (pH > 7). However, this by itself will not differentiate between a strong or a weak acid/base. To do that, it is necessary to measure the pH of a solution of known concentration. Since the molar mass of the compound is known, one must prepare (for example) a 0.10 *M* solution by dissolving 0.10 mol of the compound in 1.00 L H_2O; then measure the pH and consider the following:

If the solid is a strong acid then $[H^+]$ will be (at least) 0.10 *M*, since the solute must completely dissociate in solution to produce (at least) 0.10 mol H^+ in solution; and then pH = $-\log_{10}[H^+]$ = 1.00 (or less). If the solid is a weak acid, then $[H^+]_{eq}$ will be much less than 0.10 *M* (because the weak acid only partially dissociates in solution) and the pH will be greater than 1.00 – but still less than 7 for a weak acid.

If the solid is a strong base then $[OH^-]$ will be (at least) 0.10 *M*, since the solute must completely dissociate in solution to produce (at least) 0.10 mol OH^- in solution; and then pOH = $-\log_{10}[OH^-]$ = 1.00 (or less), which would make pH = 13.00 (or higher). If the solid is a weak base, then $[OH^-]$ will be much less than 0.10 *M* (because the weak base only partially dissociates in solution) and then pOH will be greater than 1.00 and pH will be less than 13.00 – but still greater than 7 for a weak base.

To summarize, **for a 0.1 *M* solution of the compound: pH ≤ 1.00, strong acid; 1.00 < pH < 7, weak acid; 7 < pH < 13.00, weak base; pH ≥ 13.00, strong base.**

CHALLENGE PROBLEMS

103. See page 336, and pages 346–347. The first ionization of H_2SO_4 is complete:

$$H_2SO_4(aq) \rightarrow H^+(aq) + HSO_4^-(aq)$$

and so for a solution with nominal concentration $[H_2SO_4]$ = 0.020 *M*, after the first ionization step, $[H^+]$ and $[HSO_4^-]$ are:

$$[H^+] = \frac{0.020 \text{ mol } H_2SO_4}{1 \text{ L}} \times \frac{1 \text{ mol } H^+}{1 \text{ mol } H_2SO_4} = 0.020 \text{ } M$$

$$[HSO_4^-] = \frac{0.020 \text{ mol } H_2SO_4}{1 \text{ L}} \times \frac{1 \text{ mol } HSO_4^-}{1 \text{ mol } H_2SO_4} = 0.020 \text{ } M$$

The second ionization of H_2SO_4 is reversible; K_a = 0.010 (given) :

$$HSO_4^-(aq) \rightleftharpoons H^+(aq) + SO_4^{2-}(aq)$$

To determine pH, we must first find $[H^+]_{eq}$ after this second ionization and then use Equation 13.3. Set up an equilibrium table with $[H^+]_0$ = $[HSO_4^-]_0$ = 0.020 *M* (these are the concentrations (calculated above) after the first ionization); we assume that initially no reaction has occurred so $[SO_4^{2-}]_0$ is zero, and let $\Delta[H^+] = x$:

	$HSO_4^-(aq)$ $\rightleftharpoons$	$H^+(aq)$ +	$SO_4^{2-}(aq)$
$[\ ]_0$ (M)	0.020	0.020	0
$\Delta[\ ]$ (M)	$-x$	$+x$	$+x$
$[\ ]_{eq}$ (M)	$0.020 - x$	$0.020 + x$	x

Now substitute the equilibrium concentrations of each species into the expression for K_a:

$$K_a = 0.010 = \frac{[H^+][SO_4^{2-}]}{[HSO_4^-]} = \frac{(0.020 + x)(x)}{(0.020 - x)} = \frac{0.020x + x^2}{0.020 - x}$$

Therefore

$$0.020x + x^2 = (0.010)(0.020 - x) = 0.0002 - 0.010x$$
$$x^2 + 0.030x - 0.0002 = 0$$

Using the quadratic formula:

$$x = \frac{-b \pm \sqrt{b^2 - 4ac}}{2a} = \frac{-(0.030) \pm \sqrt{(0.030)^2 - 4(1)(-0.0002)}}{2(1)}$$

$$x = +0.006 \text{ or } -0.036$$

The negative root of the equation is physically impossible since x is a concentration; therefore $x = 0.006$ and $[H^+]_{eq} = 0.020 + x = 0.026$

Hence, $\quad pH = -\log_{10}[H^+] = -\log_{10}(0.026) = \underline{\mathbf{1.59}}$

104. For each reaction, write the net ionic equation (see Table 4.2) and then calculate $\Delta H°$ for the net reaction (use Equation 8.4). Note that the quantities specified in the question are not 1 mol of each reactant, whereas $\Delta H°$ is ***per mol*** of reaction, so this must be taken into account. (See Example 8.8(b).)

(a) The reaction is between a strong acid and a strong base so the net ionic equation is:

$$H^+(aq) + OH^-(aq) \rightarrow H_2O$$

$$\Delta H° = \Sigma \Delta H_f°_{(products)} - \Sigma \Delta H_f°_{(reactants)}$$

$$= [(1 \text{ mol})(-285.8 \text{ kJ/mol})] - [(1 \text{ mol})(0 \text{ kJ/mol}) + (1 \text{ mol})(-230.0 \text{ kJ/mol})]$$

$$= -55.8 \text{ kJ}$$

This is the value of $\Delta H°$ per mol of H^+ (or OH^-). The quantity of H^+ present is:

$$\text{mol } H^+ = \frac{0.100 \text{ mol HCl}}{1 L} \times 1.00 \text{ L} \times \frac{1 \text{ mol } H^+}{1 \text{ mol HCl}} = 0.100 \text{ mol } H^+$$

Thus, the required $\Delta H° = \dfrac{-55.8\,\text{kJ}}{1\,\text{mol H}^+} \times 0.100\,\text{mol H}^+ = \underline{\underline{-5.58\,\text{kJ}}}$

(b) The reaction is between a weak acid and a strong base so the net ionic equation is:

$$HF(aq) + OH^-(aq) \rightarrow H_2O + F^-(aq)$$

$$\Delta H° = \Sigma\,\Delta H_f°_{\,(products)} - \Sigma\,\Delta H_f°_{\,(reactants)}$$

$$= [(1\,\text{mol})(-285.8\,\text{kJ/mol}) + (1\,\text{mol})(-332.6\,\text{kJ/mol})]$$
$$- [(1\,\text{mol})(-320.1\,\text{kJ/mol}) + (1\,\text{mol})(-230.0\,\text{kJ/mol})]$$

$$= -68.3\,\text{kJ}$$

This is the value of $\Delta H°$ per mol of HF (or OH$^-$). The quantity of HF present is:

$$\text{mol HF} = \dfrac{0.100\,\text{mol HF}}{1\,\text{L}} \times 1.00\,\text{L} = 0.100\,\text{mol HF}$$

Thus, the required $\Delta H° = \dfrac{-68.3\,\text{kJ}}{1\,\text{mol HF}} \times 0.100\,\text{mol HF} = \underline{\underline{-6.83\,\text{kJ}}}$

105. The reaction is between a strong acid and a strong base so the net ionic equation is:

$$H^+(aq) + OH^-(aq) \rightarrow H_2O$$

Determine the number of mol of H$^+$ and OH$^-$ present: one of them will be in excess and the final concentration of that species (after neutralization) will determine the pH of the solution.

$$\text{mol H}^+ \text{ present} = \dfrac{0.12\,\text{mol HCl}}{1\,\text{L}} \times 0.45\,\text{L} \times \dfrac{1\,\text{mol H}^+}{1\,\text{mol HCl}} = 0.054\,\text{mol H}^+$$

$$\text{mol OH}^- \text{ present} = \dfrac{0.233\,\text{mol Ba(OH)}_2}{1\,\text{L}} \times 0.30\,\text{L} \times \dfrac{2\,\text{mol OH}^-}{1\,\text{mol Ba(OH)}_2} = 0.14\,\text{mol OH}^-$$

Clearly OH$^-$ is in excess: all of the H$^+$ will be neutralized and $(0.14 - 0.054)\text{mol} = 0.086$ mol OH$^-$ will remain. This can also be demonstrated in tabular form, similar to the equilibrium table used in Example 13.5. (See also Example 14.3.) Set up a reaction table based on the net ionic equation and allow all of the limiting reactant to be consumed (n is the number of moles of each species):

	$H^+(aq)$	$+$	$OH^-(aq)$	$\rightarrow$	H_2O
n_0 (mol)	0.054		0.14		0
Δn (mol)	−0.054		−0.054		+0.054
n_{final} (mol)	0		0.086		0.054

Thus after neutralization 0.14 mol of OH⁻ remain, in a total volume of (0.30 + 0.45)L = 0.75 L, and therefore the final concentration of OH⁻ is:

$$[OH^-] = \frac{0.086 \text{ mol OH}^-}{0.75 \text{ L}} = 0.11 \text{ } M$$

Hence, pOH = $-\log_{10}[OH^-]$ = $-\log_{10}(0.11)$ = 0.96   and pH = 14.00 − pOH = **13.04**

106. From Equation 13.3, [H⁺] in the mixture of acids is [H⁺] = 10^{-pH} = $10^{-0.39}$ = 0.4$\underline{0}$7 M. The number of mol H⁺ provided by the HNO₃ is:

$$\text{mol H}^+ \text{ from HNO}_3 = \frac{0.164 \text{ mol HNO}_3}{1 \text{ L}} \times 273 \text{ mL} \times \frac{1 \text{ L}}{1000 \text{ mL}} \times \frac{1 \text{ mol H}^+}{1 \text{ mol HNO}_3} = 0.0448 \text{ mol}$$

Let the volume of HCl solution added be *x* ml; then mol H⁺ provided by HCl is:

$$\text{mol H}^+ \text{ from HCl} = \frac{0.800 \text{ mol HCl}}{1 \text{ L}} \times x \text{ mL} \times \frac{1 \text{ L}}{1000 \text{ mL}} \times \frac{1 \text{ mol H}^+}{1 \text{ mol HCl}} = (8.00 \times 10^{-4})x \text{ mol}$$

Now [H⁺] in the solution is the sum of mol H⁺ from both sources divided by the total V (liters); and we already know [H⁺] = 0.407 M (calculated above using the pH). Thus:

$$[H^+] = 0.407 \text{ } M = \frac{[0.0448 + (8.00 \times 10^{-4})x] \text{ mol H}^+}{(273 + x) \text{ mL} \times \dfrac{1 \text{ L}}{1000 \text{ mL}}}$$

$$0.111 + (4.07 \times 10^{-4})x = 0.0448 + (8.00 \times 10^{-4})x$$

$$(3.93 \times 10^{-4})x = 0.066 \qquad \rightarrow \quad x = 1.7 \times 10^2$$

The volume of HCl solution used was 1.7 × 10² mL.

107. Recall (Sections 10-3 and 10-4) that the freezing point of a solution is lowered relative to the pure solvent, and that this is a colligative property: the amount of freezing point depression depends on the number of solute particles. In this case, the solute (acetic acid) dissociates into H⁺ and C₂H₃O₂⁻ ions and the number of moles of these particles in solution must be determined. From this, calculate the molality and hence the amount of freezing point depression, and thence the freezing point. For simplicity we can assume a one-liter solution.

$$\text{mass of solution} = 1.000 \text{ L} \times \frac{1000 \text{ cm}^3}{1 \text{ L}} \times \frac{1.006 \text{ g}}{1 \text{ cm}^3} = 1006 \text{ g}$$

$$\text{mass of HC}_2\text{H}_3\text{O}_2 \text{ present} = 1006 \text{ g solution} \times \frac{5.00 \text{ g HC}_2\text{H}_3\text{O}_2}{100 \text{ g solution}} = 50.3 \text{ g}$$

$$\text{mol HC}_2\text{H}_3\text{O}_2 \text{ present} = 50.3 \text{ g} \times \frac{1 \text{ mol HC}_2\text{H}_3\text{O}_2}{60.05 \text{ g HC}_2\text{H}_3\text{O}_2} = 0.838 \text{ mol HC}_2\text{H}_3\text{O}_2$$

$$[HC_2H_3O_2] = \frac{0.838 \text{ mol } HC_2H_3O_2}{1.000 \text{ L}} = 0.838 \text{ M}$$

Now set up an equilibrium table as in Example 13.5 to establish the equilibrium concentration of the species present in this solution of $HC_2H_3O_2$:

	$HC_2H_3O_2(aq)$	$\rightleftharpoons$	$H^+(aq)$	+	$C_2H_3O_2^-(aq)$
$[\ ]_0$ (M)	0.838		0		0
$\Delta[\ ]$ (M)	$-x$		$+x$		$+x$
$[\ ]_{eq}$ (M)	$0.838 - x$		x		x

Substitute the equilibrium concentrations of each species into the expression for K_a:

$$K_a = 1.8 \times 10^{-5} = \frac{[H^+][C_2H_3O_2^-]}{[HC_2H_3O_2]} = \frac{(x)(x)}{(0.838 - x)} \approx \frac{x^2}{0.838} \quad \text{(assuming } x << 0.838\text{)}$$

Hence, $x^2 = (0.838)(1.8 \times 10^{-5}) = 1.51 \times 10^{-5}$

so $x = 3.88 \times 10^{-3}$ M $= [H^+]_{eq} = [C_2H_3O_2^-]_{eq}$ and $[HC_2H_3O_2] = 0.838 - x = 0.834$ M

Since we are in 1.000 L of solution, the number of mol of each species present is readily obtained: 3.88×10^{-3} mol H^+, 3.88×10^{-3} mol $C_2H_3O_2^-$ and 0.834 mol $HC_2H_3O_2$; and the total number of mol present is 0.842 mol.

Molality is defined as mol solute per kg solvent; from the calculations above, we know that in 1006 g of solution we have 50.3 g $HC_2H_3O_2$ and so the remainder of the solution must be solvent H_2O: mass H_2O present = 1006 g $-$ 50.3 g = 956 g = 0.956 kg.

$$\text{molality of the solution} = \frac{\text{total mol solute}}{\text{kg solvent}} = \frac{0.842 \text{ mol solute}}{0.956 \text{ kg solvent}} = 0.881 \text{ } m$$

Calculate the freezing point depression using the equation on page 263. (Note: we need not incorporate a Van't Hoff factor (Section 10-4), since we have already calculated the *total* number of particles in solution.) The value of k_f for H_2O (Table 10.2) is 1.86 °C/m:

$$\Delta T_f = k_f m = (1.86 \text{ °C/}m)(0.881 \text{ } m) = 1.64 \text{ °C}$$

Thus, the freezing point of the solution (see page 262; from Table 10.2, $T_f°$ for H_2O is 0 °C):

$$T_f = T_f° - \Delta T_f = (0 - 1.64) \text{ °C} = \underline{-1.64 \text{ °C}}$$

108. For simplicity, assume a one-liter solution and (since it is a strong base) assume also that the $Ca(OH)_2$ dissociates completely in solution. The solubility can be used to calculate $[OH^-]$, from which Equation 13.4 will give pOH and hence pH may be determined since pH + pOH = 14.00 (page 334). Here the solubility is given as 0.153 g in 100 g H_2O, which corresponds to 0.153 g $Ca(OH)_2$ in (100 + 0.153 g) = 100 g solution.

$$[OH^-] = \frac{0.153 \text{ g Ca(OH)}_2}{100 \text{ g solution}} \times \frac{1.00 \text{ g solution}}{1 \text{ mL solution}} \times \frac{1000 \text{ mL}}{1 \text{ L}} \times \frac{1 \text{ mol Ca(OH)}_2}{74.096 \text{ g Ca(OH)}_2}$$

$$\times \frac{2 \text{ mol OH}^-}{1 \text{ mol Ca(OH)}_2} = 0.0413 \text{ } M$$

$$pOH = -\log_{10}[OH^-] = -\log_{10}(0.0413) = 1.384$$

$$pH = 14.00 - pOH = 14.00 - 1.384 = \textbf{12.62}$$

109. From the given information, calculate the nominal concentration of each acid, $[HA]_0$ and $[HB]_0$.

$$[HA]_0 = \frac{11.0 \text{ g HA}}{745 \text{ mL solution}} \times \frac{1000 \text{ mL}}{1 \text{ L}} \times \frac{1 \text{ mol HA}}{138 \text{ g HA}} = 0.107 \text{ } M$$

$$[HB]_0 = \frac{5.00 \text{ g HB}}{525 \text{ mL solution}} \times \frac{1000 \text{ mL}}{1 \text{ L}} \times \frac{1 \text{ mol HB}}{72.0 \text{ g HB}} = 0.132 \text{ } M$$

Both solutions have the same pH (and hence the same $[H^+]$), but the solution of HB is more concentrated in order to achieve this. Thus HB must dissociate to a lesser extent than HA; so **HA is the stronger acid.**

This may be shown by calculation. For both solutions, let $[H^+] = x$. Refer to Example 13.7 and the discussion on pages 345–346; for each acid, K_a is related to a (the original concentration of each acid) by the following relationship, assuming that $x/a \leq 0.05$:

$$K_a \approx \frac{x^2}{a} \qquad \text{and hence} \quad x = (K_a \times a)^{\frac{1}{2}}$$

Recall that x is the same for both solutions. Now substitute for a the values of $[HA]_0$ and $[HB]_0$ calculated above:

$$x = \{K_{a(HA)}(0.107)\}^{\frac{1}{2}} = \{K_{a(HB)}(0.132)\}^{\frac{1}{2}} \text{ or } K_{a(HA)}(0.107) = K_{a(HB)}(0.132)$$

$$\text{Hence } K_{a(HA)} = \frac{0.132}{0.107} K_{a(HB)} = 1.23 K_{a(HB)}$$

Thus $K_{a(HA)} > K_{a(HB)}$ so **HA is stronger than HB.**

110. For the weak base NaB, we can deduce that the anion B^- is the weak base (Na^+ is a spectator cation):

$$B^-(aq) + H_2O \rightleftharpoons HB(aq) + OH^-(aq) \qquad K_b = \frac{[HB][OH^-]}{[B^-]}$$

The solution has pH = 10.54, so $[H^+] = 10^{-10.54}$ (Equation 13.3) and hence from Equations 13.1 and 13.2:

$$[OH^-] = \frac{K_w}{[H^+]} = \frac{1.0 \times 10^{-14}}{10^{-10.54}} = 3.47 \times 10^{-4} \, M$$

From the freezing point of the solution we can calculate the amount of freezing point depression (see page 262; T_f^o for H_2O is 0 °C):

$$\Delta T_f = T_f^o - T_f = (0 - (-0.89)) \, °C = 0.89 \, °C$$

and from this we will obtain the concentration (molality) of particles in solution (see Example 10.9). The solute is an electrolyte so we must use Equation 10.6, incorporating the Van't Hoff factor (see Section 10-4). Use $i = 2$: in other words, assume that 1 mol of NaB produces 2 mol of ions, Na^+ and B^-, in solution. (This is an approximation since the anion B^- further dissociates into HB and OH^-; but B^- is a weak base and so [HB] and [OH⁻] are very small — in fact, $[OH^-] = [HB] = 3.47 \times 10^{-4} \, M$ (calculated above) — so the approximation $i \approx 2$ is reasonable.):

$$\Delta T_f = i \times 1.86 \, °C/m \times m$$

$$\rightarrow m = \frac{\Delta T_f}{i \times 1.86 \, °C/m} = \frac{0.89 °C}{2 \times 1.86 °C/m} = 0.239 \, m$$

This is the molality of NaB in solution, and corresponds to 0.239 mol NaB/kg solvent. To obtain the corresponding molarity of NaB, set up a table as in Example 10.4 and shade the boxes required to calculate molarity.

	moles	⟵ MM	mass	density ⟶	volume
Solute	0.239				
Solvent			1 kg = 1000 g		
Solution					

Calculate the required quantities from those known.

$$\text{mass NaB} = 0.239 \text{ mol NaB} \times \frac{233 \text{ g NaB}}{1 \text{ mol NaB}} = 55.7 \text{ g}$$

total mass of solution = mass (solute + solvent) = (1000 + 55.7)g = 1056 g

$$\text{volume of solution (L)} = 1056 \text{ g solution} \times \frac{1 \text{ mL solution}}{1 \text{ g solution}} \times \frac{1 \text{ L}}{1000 \text{ mL}} = 1.056 \text{ L}$$

Now complete the table:

	moles	⟵ MM	mass	density ⟶	volume
Solute	0.239		55.7 g		
Solvent			1 kg = 1000 g		
Solution			1056 g		1.056 L

$$\text{Molarity of NaB solution} = \frac{0.239 \text{ mol NaB}}{1.056 \text{ L solution}} = 0.226 \text{ M}$$

Hence, $[B^-]_0$ is given by:

$$[B^-]_0 = \frac{0.226 \text{ mol NaB}}{1 \text{ L}} \times \frac{1 \text{ mol B}^-}{1 \text{ mol NaB}} = 0.226 \text{ M}$$

To calculate K_b we must obtain the equilibrium concentrations of the species involved in the dissociation reaction for B^-. Set up an equilibrium table and fill in concentrations we already know. Initially we assume there is no ionization, so $[HB]_0 = [OH^-]_0 = 0$ and recall from above that $[OH^-]_{eq} = 3.47 \times 10^{-4}$ M and $[B^-]_0 = 0.226$ *M*

	$B^-(aq)$	+	H_2O	$\rightleftharpoons$	$HB(aq)$	+	$OH^-(aq)$
[ ]$_0$ (M)	0.226				0		0
Δ[ ] (M)							
[ ]$_{eq}$ (M)							3.47×10^{-4}

From this, clearly, $\Delta[OH^-] = 3.47 \times 10^{-4}$ and by stoichiometry $\Delta[HB] = 3.47 \times 10^{-4}$ and $\Delta[B^-] = -3.47 \times 10^{-4}$ *M*; hence the table can be completed:

	$B^-(aq)$	+	H_2O	$\rightleftharpoons$	$HB(aq)$	+	$OH^-(aq)$
[ ]$_0$ (M)	0.226				0		0
Δ[ ] (M)	-3.47×10^{-4}				$+3.47 \times 10^{-4}$		$+3.47 \times 10^{-4}$
[ ]$_{eq}$ (M)	0.226				3.47×10^{-4}		3.47×10^{-4}

Finally, substitute equilibrium concentrations of each species into the expression for K_b:

$$K_b = \frac{[HB][OH^-]}{[B^-]} = \frac{(3.47 \times 10^{-4})(3.47 \times 10^{-4})}{(0.226)} = \underline{\mathbf{5.3 \times 10^{-7}}}$$

EQUILIBRIA IN ACID-BASE SOLUTIONS

2. Refer to Sections 4-2 and 13-4, and Table 13.2.
   Write the reactions, eliminating the spectator ions. Bear in mind that soluble salts ionize, thus they exist in solution as ions. Weak acids and bases (Table 13.2) do not ionize significantly, thus they exist in solution as the undissociated acid or base.

   (a) $NaC_2H_3O_2(aq) + HNO_3(aq) \rightarrow HC_2H_3O_2(aq) + NaNO_3(aq)$
   $NaC_2H_3O_2$ and $NaNO_3$ are salts and HNO_3 is a strong acid, thus all three dissociate in solution, while $HC_2H_3O_2$ is a weak acid and so remains undissociated.
   $Na^+(aq) + C_2H_3O_2^-(aq) + H^+(aq) + NO_3^-(aq) \rightarrow HC_2H_3O_2(aq) + Na^+(aq) + NO_3^-(aq)$

   $C_2H_3O_2^-(aq) + H^+(aq) \rightarrow HC_2H_3O_2(aq)$

   (b) $2HBr(aq) + Sr(OH)_2(aq) \rightarrow SrBr_2(aq) + 2H_2O$
   HBr and $Sr(OH)_2$ are a strong acid and a strong base, respectively, and thus dissociate completely in solution.
   $2H^+(aq) + 2Br^-(aq) + Sr^{2+}(aq) + 2OH^-(aq) \rightarrow Sr^{2+}(aq) + 2Br^-(aq) + 2H_2O$
   $2H^+(aq) + 2OH^-(aq) \rightarrow 2H_2O$

   $H^+(aq) + OH^-(aq) \rightarrow H_2O$

   (c) $HClO(aq) + NaCN(aq) \rightarrow NaClO(aq) + HCN(aq)$
   HClO and HCN are weak acids and do not dissociate; NaCN and NaClO are salts and completely dissociate in solution.
   $HClO(aq) + Na^+(aq) + CN^-(aq) \rightarrow Na^+(aq) + ClO^-(aq) + HCN(aq)$

   $HClO(aq) + CN^-(aq) \rightarrow ClO^-(aq) + HCN(aq)$

   (d) $HNO_2(aq) + NaOH(aq) \rightarrow NaNO_2(aq) + H_2O$
   HNO_2 is a weak acid and does not dissociate; NaOH is a strong base and $NaNO_2$ is a salt, thus both dissociate in solution.
   $HNO_2(aq) + Na^+(aq) + OH^-(aq) \rightarrow Na^+(aq) + NO_2^-(aq) + H_2O$

   $HNO_2(aq) + OH^-(aq) \rightarrow NO_2^-(aq) + H_2O$

4. See Sections 4-2, 13-3, and 13-4. For each reaction, identify acids and bases as strong/weak (see Table 13.2) and use Table 4.2 to write the net ionic equations, eliminating the spectator ions. In each case, the OH^- ion acts as a strong base.

(a) $Fe(H_2O)_6^{3+}$ is a weak acid so from Table 4.2 the equation will be of the form

$HB(aq) + OH^-(aq) \rightarrow H_2O + B^-(aq)$

$$Fe(H_2O)_6^{3+}(aq) + OH^-(aq) \rightarrow H_2O + Fe(OH)(H_2O)_5^{2+}(aq)$$

(b) Sodium hydrogen carbonate, $NaHCO_3$, is a source of the HCO_3^- anion, which is a weak acid; so from Table 4.2 the equation will be of the form

$HB(aq) + OH^-(aq) \rightarrow H_2O + B^-(aq)$

$$HCO_3^-(aq) + OH^-(aq) \rightarrow H_2O + CO_3^{2-}(aq)$$

(c) Ammonium chloride, NH_4Cl, is a source of the NH_4^+ cation, which is a weak acid; so from Table 4.2 the equation will be of the form

$HB(aq) + OH^-(aq) \rightarrow H_2O + B^-(aq)$

$$NH_4^+(aq) + OH^-(aq) \rightarrow H_2O + NH_3(aq)$$

6. (a) The reaction in Question 2(a) is the reverse of the following reaction:

$HC_2H_3O_2(aq) \rightleftharpoons H^+(aq) + C_2H_3O_2^-(aq)$

which is a weak acid ionization reaction, and for which the equilibrium constant is K_a (in this case, K_a for $HC_2H_3O_2$, which from Table 13.2 is 1.8×10^{-5}). Recall from Table 12.3 that when a reaction is reversed, $K' = 1/K$ (Reciprocal rule). Thus the required equilibrium constant here is $1/(1.8 \times 10^{-5})$ = **5.6×10^4**.

(b) The reaction in Question 2(b) is the reverse of the following reaction:

$H_2O \rightleftharpoons H^+(aq) + OH^-(aq)$

which is the self-ionization of H_2O, and for which the equilibrium constant is K_w (which from Equation 13.2 is 1.0×10^{-14}). Recall from Table 12.3 that when a reaction is reversed, $K' = 1/K$ (Reciprocal rule). Thus the required equilibrium constant here is $1/(1.0 \times 10^{-14})$ = **1.0×10^{14}**.

(c) The reaction in Question 2(c) is the sum of the following two reactions:

$HClO(aq) \rightleftharpoons ClO^-(aq) + H^+(aq)$

$H^+(aq) + CN^-(aq) \rightleftharpoons HCN(aq)$

which are, respectively, the ionization of the weak acid HClO ($K_a = 2.8 \times 10^{-8}$) and the reverse of the ionization of the weak acid HCN ($K_a = 5.8 \times 10^{-10}$; apply the Reciprocal rule (Table 12.3) so $K = 1/(5.8 \times 10^{-10}) = 1.7 \times 10^9$). Recall from Table 12.3 that when two reactions are added together, K for the new reaction is the product of the two individual reactions (Rule of Multiple Equilibria). Thus the required equilibrium constant here is $(2.8 \times 10^{-8})(1.7 \times 10^9)$ = **48**.

(d) The reaction in Question 2(d) is the reverse of the following reaction:

$$NO_2^-(aq) + H_2O \rightleftharpoons HNO_2(aq)\ OH^-(aq)$$

which is a weak base ionization reaction, and for which the equilibrium constant is K_b (in this case, K_a for NO_2^-, which from Table 13.2 is 1.7×10^{-11}). Recall from Table 12.3 that when a reaction is reversed, $K' = 1/K$ (Reciprocal rule). Thus the required equilibrium constant here is $1/(1.7 \times 10^{-11})$ = **5.9×10^{10}**.

8. (a) The reaction in Question 4(a) is the reverse of the following reaction:

$$Fe(OH)(H_2O)_5^{2+}(aq) + H_2O \rightleftharpoons Fe(H_2O)_6^{3+}(aq) + OH^-(aq)$$

which is a weak base ionization reaction, and for which the equilibrium constant is K_b (in this case, K_b for $Fe(OH)(H_2O)_5^{2+}$, which from Table 13.2 is 1.5×10^{-12}). Recall from Table 12.3 that when a reaction is reversed, $K' = 1/K$ (Reciprocal rule). Thus the required equilibrium constant here is $1/(1.5 \times 10^{-12})$ = **6.7×10^{11}**.

(b) The reaction in Question 4(b) is the reverse of the following reaction:

$$CO_3^{2-}(aq) + H_2O \rightleftharpoons HCO_3^-(aq) + OH^-(aq)$$

which is a weak base ionization reaction, and for which the equilibrium constant is K_b (in this case, K_b for CO_3^{2-}, which from Table 13.2 is 2.1×10^{-4}). Recall from Table 12.3 that when a reaction is reversed, $K' = 1/K$ (Reciprocal rule). Thus the required equilibrium constant here is $1/(2.1 \times 10^{-4})$ = **4.8×10^3**.

(c) The reaction in Question 4(c) is the reverse of the following reaction:

$$NH_3(aq) + H_2O \rightleftharpoons NH_4^+(aq) + OH^-(aq)$$

which is a weak base ionization reaction, and for which the equilibrium constant is K_b (in this case, K_b for NH_3, which from Table 13.2 is 1.8×10^{-5}). Recall from Table 12.3 that when a reaction is reversed, $K' = 1/K$ (Reciprocal rule). Thus the required equilibrium constant here is $1/(1.8 \times 10^{-5})$ = **5.6×10^4**.

14-1 BUFFERS: Buffers

10. The HSO_3^- ion is a weak acid ("HB") with $K_a = 6.0 \times 10^{-8}$ (Table 13.2); SO_3^{2-} ion is its conjugate base ("B$^-$"). For each solution, calculate [H$^+$] using Equation 14.1 and, from that, pH and [OH$^-$] (Sections 13-2 and 13-3).

(a) $[H^+] = K_a \times \dfrac{[HB]}{[B^-]} = (6.0 \times 10^{-8}) \times \dfrac{0.429}{0.0249} = 1.03 \times 10^{-6}\ M$

$pH = -\log_{10}[H^+] = -\log(1.03 \times 10^{-6})$ = **5.99**

$[OH^-] = \dfrac{K_w}{[H^+]} = \dfrac{1.0 \times 10^{-14}}{1.03 \times 10^{-6}}$ = **$9.7 \times 10^{-9}\ M$**

(b) $[H^+] = K_a \times \dfrac{[HB]}{[B^-]} = (6.0 \times 10^{-8}) \times \dfrac{0.429}{0.247} = 1.04 \times 10^{-7} \, M$

$pH = -\log_{10}[H^+] = -\log(1.04 \times 10^{-7}) = \underline{\mathbf{6.98}}$

$[OH^-] = \dfrac{K_w}{[H^+]} = \dfrac{1.0 \times 10^{-14}}{1.04 \times 10^{-7}} = \underline{\mathbf{9.6 \times 10^{-8} \, M}}$

(c) $[H^+] = K_a \times \dfrac{[HB]}{[B^-]} = (6.0 \times 10^{-8}) \times \dfrac{0.429}{0.504} = 5.1 \times 10^{-8} \, M$

$pH = -\log_{10}[H^+] = -\log(5.1 \times 10^{-8}) = \underline{\mathbf{7.29}}$

$[OH^-] = \dfrac{K_w}{[H^+]} = \dfrac{1.0 \times 10^{-14}}{5.1 \times 10^{-8}} = \underline{\mathbf{2.0 \times 10^{-7} \, M}}$

(d) $[H^+] = K_a \times \dfrac{[HB]}{[B^-]} = (6.0 \times 10^{-8}) \times \dfrac{0.429}{0.811} = 3.2 \times 10^{-8} \, M$

$pH = -\log_{10}[H^+] = -\log(3.2 \times 10^{-8}) = \underline{\mathbf{7.50}}$

$[OH^-] = \dfrac{K_w}{[H^+]} = \dfrac{1.0 \times 10^{-14}}{3.2 \times 10^{-8}} = \underline{\mathbf{3.2 \times 10^{-7} \, M}}$

(e) $[H^+] = K_a \times \dfrac{[HB]}{[B^-]} = (6.0 \times 10^{-8}) \times \dfrac{0.429}{1.223} = 2.1 \times 10^{-8} \, M$

$pH = -\log_{10}[H^+] = -\log(2.1 \times 10^{-8}) = \underline{\mathbf{7.68}}$

$[OH^-] = \dfrac{K_w}{[H^+]} = \dfrac{1.0 \times 10^{-14}}{2.1 \times 10^{-8}} = \underline{\mathbf{4.8 \times 10^{-7} \, M}}$

12. This problem is similar to Example 14.1. Use Equation 14.1 to calculate $[H^+]$ for the solution and hence calculate pH using Equation 13.3. To use Equation 14.1, we need K_a, [HF] and [F⁻]. From Table 13.2, K_a for HF is 6.9×10^{-4} and [HF] = 0.0399 M (given).

$[F^-] = \dfrac{0.062 \text{ mol NaF}}{127 \text{ mL}} \times \dfrac{1000 \text{ mL}}{1 \text{L}} \times \dfrac{1 \text{ mol F}^-}{1 \text{ mol NaF}} = 0.49 \, M$

$$[H^+] = K_a \times \frac{[HB]}{[B^-]} = 6.9 \times 10^{-4} \times \frac{0.0399}{0.49} = 5.6 \times 10^{-5} \, M$$

$$pH = -\log_{10}[H^+] = -\log(5.6 \times 10^{-5}) = \underline{\textbf{4.25}}$$

14. (a) This problem is similar to Example 14.1.

$$\text{mol NH}_4^+ = 5.50 \text{ g NH}_4\text{Cl} \times \frac{1 \text{ mol NH}_4\text{Cl}}{53.49 \text{ g NH}_4\text{Cl}} \times \frac{1 \text{ mol NH}_4^+}{1 \text{ mol NH}_4\text{Cl}} = 0.103 \text{ mol}$$

Use Equation 14.2 to determine $[H^+]$:

$$[H^+] = K_a \times \frac{n_{\text{NH}_4^+}}{n_{\text{NH}_3}} = (5.6 \times 10^{-10}) \times \frac{0.103 \text{ mol NH}_4^+}{0.0188 \text{ mol NH}_3} = 3.1 \times 10^{-9} \, M$$

$$pH = -\log_{10}[H^+] = -\log_{10}(3.1 \times 10^{-9}) = \underline{\textbf{8.51}}$$

(b) Since volume does not appear in Equation 14.2, it follows that $[H^+]$ is independent of volume. We would therefore expect that **the pH remains the same.**

16. This problem is similar (in part) to Example 14.1(b). Use Equation 13.3 to obtain $[H^+]$:

$$[H^+] = 10^{-pH} = 10^{-4.65} = 2.2 \times 10^{-5} \, M$$

and recall (Section 13-4) that for a weak acid at equilibrium $[H^+] = [X^-]$ and that $[HX]_0 = [HX]_{eq}$. Thus we can calculate K_a for the acid HX:

$$K_a = \frac{[H^+][X^-]}{[HX]} = \frac{(2.2 \times 10^{-5})(2.2 \times 10^{-5})}{(0.057)} = 8.5 \times 10^{-9}$$

One liter of the 0.057 M weak acid HX will contain 0.057 mol HX = n_{HX}; to this, 0.018 mol of KX are added providing 0.018 mol X^- = n_{X^-}. Use Equation 14.2 to calculate $[H^+]$ in the mixture and then use Equation 13.3 to calculate pH:

$$[H^+] = K_a \times \frac{n_{HX}}{n_{X^-}} = (8.5 \times 10^{-9}) \times \frac{0.057 \text{ mol HX}}{0.018 \text{ mol X}^-} = 2.7 \times 10^{-8} \, M$$

$$pH = -\log_{10}[H^+] = -\log_{10}(2.7 \times 10^{-8}) = \underline{\textbf{7.57}}$$

18. This problem is related to Example 14.3. Recall that for a solution to act as a buffer it must contain significant amounts of a weak acid and its conjugate base. For each case consider how the added compound will react with the HCl, then construct a table as in

Example 14.3 to determine the composition of the product mixture and hence establish whether it contains the two required buffer components. In all cases the initial quantity of HCl is:

$$\text{mol HCl} = \frac{0.380 \text{ mol HCl}}{1\text{L}} \times 295 \text{ mL} \times \frac{1\text{L}}{1000 \text{ mL}} = 0.112 \text{ mol}$$

and so the initial quantity of H^+ is

$$\text{mol } H^+ = 0.112 \text{ mol HCl} \times \frac{1 \text{mol } H^+}{1 \text{mol HCl}} = 0.112 \text{ mol}$$

(a) The mixture containing NH_4Cl and HCl is a mixture of a weak acid (NH_4^+) and a strong acid (H^+): there is no weak base present, so **this solution CANNOT act as a buffer.**

(b) The reaction between HCl and KF is of the type strong acid (H^+) + weak base (F^-) (see Table 4.2); the limiting reactant (F^-) is completely consumed:

$$H^+(aq) + F^-(aq) \rightarrow HF(aq)$$

	n_{H^+}	n_{F^-}	n_{HF}
Original	0.112	0.033	0
Change	−0.033	−0.033	+0.033
Final	0.079	0	0.033

Thus a strong acid (H^+) and a weak acid (HF) remain after reaction, so **this solution WILL NOT form a buffer.**

(c) The reaction between HCl and $Sr(OH)_2$ is of the type strong acid + strong base (see Table 4.2) and so there is no way to form a mixture containing a weak acid plus its conjugate base: thus, **this solution CANNOT act as a buffer.**

(d) This is similar to part (b) above, with the weak base in this case being NO_2^-; the limiting reactant (H^+) will be completely consumed:

$$H^+(aq) + NO_2^-(aq) \rightarrow HNO_2(aq)$$

	n_{H^+}	$n_{NO_2^-}$	n_{HNO_2}
Original	0.112	0.279	0
Change	−0.112	−0.112	+0.112
Final	0	0.167	0.112

Thus a weak acid (HNO_2) and its conjugate base (NO_2^-) remain after reaction, so **this solution WILL form a buffer.**

(e) Set this up similarly to part (d) above; the weak base (ClO^-) is completely consumed in this case:

$$H^+(aq) + ClO^-(aq) \rightarrow HClO(aq)$$

	n_{H^+}	n_{ClO^-}	n_{HClO}
Original	0.112	0.112	0
Change	−0.112	−0.112	+0.112
Final	0	0	0.112

Thus all of the weak base OCl^- is consumed by reaction with H^+, so although 0.112 mol of the weak acid HClO remain after reaction, **this solution CANNOT act as a buffer.**

20. This problem is related to Example 14.3. The reaction between aniline and HCl is of the type strong acid + weak base (see Table 4.2):

$$H^+(aq) + C_6H_5NH_2(aq) \rightarrow C_6H_5NH_3^+(aq)$$

From the given information it is first necessary to determine the initial quantity of H^+ and $C_6H_5NH_2$ present:

$$\text{mol } H^+ = \frac{1.67 \text{ mol HCl}}{1 L} \times 35.0 \text{ mL} \times \frac{1 L}{1000 \text{ mL}} \times \frac{1 \text{ mol } H^+}{1 \text{ mol HCl}} = 0.0584 \text{ mol}$$

$$\text{mol } C_6H_5NH_2 = 20.00 \text{ mL } C_6H_5NH_2 \times \frac{1.022 \text{ g}}{1 \text{ mL}} \times \frac{1 \text{ mol } C_6H_5NH_2}{93.126 \text{ g } C_6H_5NH_2} = 0.2195 \text{ mol}$$

Construct a table as in Example 14.3 to determine the composition of the mixture after the two species react; the limiting reactant (H^+) is completely consumed:

	n_{H^+}	$n_{C_6H_5NH_2}$	$n_{C_6H_5NH_3^+}$
Original	0.0584	0.2195	0
Change	−0.0584	−0.0584	+0.0584
Final	0	0.1611	0.0584

Now, given the quantity of weak acid ($C_6H_5NH_3^+$) and conjugate base ($C_6H_5NH_2$) present, we can use Equation 14.2 to calculate $[H^+]$ and hence pH (using Equation 13.3). However, Equation 14.2 uses K_a of the acid component of the buffer whereas in this case K_b is known. To calculate K_a for $C_6H_5NH_3^+$, use Equation 13.10:

$$K_a \text{ for } C_6H_5NH_3^+ = K_w/(K_b \text{ for } C_6H_5NH_2) = (1.0 \times 10^{-14})/(4.3 \times 10^{-10}) = 2.3 \times 10^{-5}$$

$$[H^+] = K_a \times \frac{n_{C_6H_5NH_3^+}}{n_{C_6H_5NH_2}} = 2.3 \times 10^{-5} \times \frac{0.0584}{0.1611} = 8.3 \times 10^{-6} \text{ M}$$

$$pH = -\log_{10}[H^+] = -\log_{10}(8.3 \times 10^{-6}) = \underline{\textbf{5.08}}$$

22. This problem is related to Example 14.3. The reaction between ethanolamine and HCl is of the type strong acid + weak base (see Table 4.2):

$$H^+(aq) + C_2H_5ONH_2(aq) \rightarrow C_2H_5ONH_3^+(aq)$$

From the given information it is first necessary to determine the initial quantity of H^+ and $C_2H_5ONH_2$ present:

$$\text{mol } H^+ = \frac{1.0 \text{ mol HCl}}{1 L} \times 50.0 \text{ mL} \times \frac{1 L}{1000 \text{ mL}} \times \frac{1 \text{ mol } H^+}{1 \text{ mol HCl}} = 0.050 \text{ mol}$$

$$\text{mol } C_2H_5ONH_2 = \frac{1.20 \text{ mol } C_2H_5ONH_2}{1 L} \times 100.00 \text{ mL } C_2H_5ONH_2 \times \frac{1 L}{1000 \text{ mL}} = 0.120 \text{ mol}$$

Construct a table as in Example 14.3 to determine the composition of the mixture after the two species react; the limiting reactant (H^+) is completely consumed:

	n_{H^+}	$n_{C_2H_5ONH_2}$	$n_{C_2H_5ONH_3^+}$
Original	0.050	0.120	0
Change	−0.050	−0.050	+0.050
Final	0	0.070	0.050

Now, given the quantity of weak acid ($C_2H_5ONH_3^+$) and conjugate base ($C_2H_5ONH_2$) present, we can use Equation 14.2 to calculate $[H^+]$ and hence pH (using Equation 13.3); the required K_a value is given in the question:

$$[H^+] = K_a \times \frac{n_{C_2H_5ONH_3^+}}{n_{C_2H_5ONH_2}} = 3.6 \times 10^{-10} \times \frac{0.050}{0.070} = 2.6 \times 10^{-10} \text{ M}$$

$$pH = -\log_{10}[H^+] = -\log_{10}(2.6 \times 10^{-10}) = \mathbf{9.59}$$

24. This problem is related to Example 14.2(a); see also page 364. For each case, choose an acid which has pK_a approximately equal to the required pH (or K_a approximately equal to the required $[H^+]$). Since Table 13.2 lists K_a values, it is convenient to convert the given pH into $[H^+]$ using Equation 13.3.

(a) pH = 4.5 $\rightarrow$ $[H^+]$ = $10^{-4.5}$ = 3.2×10^{-5} M. The acid with K_a closest to this is acetic acid, $HC_2H_3O_2$ ($K_a = 1.8 \times 10^{-5}$), and since a buffer requires both a weak acid and its conjugate base we should use $\mathbf{HC_2H_3O_2/C_2H_3O_2^-}$.

(b) pH = 9.2 $\rightarrow$ $[H^+]$ = $10^{-9.2}$ = 6.3×10^{-10} M. The acid with K_a closest to this is hydrocyanic acid, HCN ($K_a = 5.8 \times 10^{-10}$), and since a buffer requires both a weak acid and its conjugate base we should use $\mathbf{HCN/CN^-}$.

(c) pH = 11.0 → [H⁺] = $10^{-11.0}$ = 1.0×10^{-11} M. The acid with K_a closest to this is the hydrogen carbonate anion, HCO_3^- (K_a = 4.7×10^{-11}), and since a buffer requires both a weak acid and its conjugate base we should use **HCO_3^-/CO_3^{2-}**.

26. This problem is related to Example 14.2(b, c). The given pH may be converted into the corresponding [H⁺] using Equation 13.3:

$$pH = 10.00 \quad \rightarrow \quad [H^+] = 10^{-10.00} = 1.00 \times 10^{-10} \ M$$

(a) Use Equation 14.1: [H⁺] is known and K_a for the acid component of the buffer (NH_4^+) is 5.6×10^{-10} (from Table 13.2):

$$[H^+] = K_a \times \frac{[HB]}{[B^-]} \qquad 1.00 \times 10^{-10} = (5.6 \times 10^{-10}) \times \frac{[NH_4^+]}{[NH_3]}$$

$$\frac{[NH_4^+]}{[NH_3]} = \underline{0.18}$$

(b) Note that since Equations 14.1 and 14.2 are equivalent the ratio $\dfrac{n_{NH_4^+}}{n_{NH_3}}$ is also 0.18.

Find mol of NH_3 present and hence mol of NH_4^+ required; and thus find the corresponding mol of NH_4Cl needed.

$$mol \ NH_3 = \frac{1.24 \ mol \ NH_3}{1 \ L} \times 465 \ mL \ NH_3 \times \frac{1 \ L}{1000 \ mL} = 0.577 \ mol$$

$$\frac{n_{NH_4^+}}{n_{NH_3}} = 0.18 \quad \rightarrow \quad mol \ NH_4^+ \ required = 0.18 \times n_{NH_3} = 0.18 \times 0.577 \ mol = 0.10 \ mol$$

$$mol \ NH_4Cl \ required = 0.10 \ mol \ NH_4^+ \times \frac{1 \ mol \ NH_4Cl}{1 \ mol \ NH_4^+} = \underline{\textbf{0.10 mol}}$$

(c) Find mol of NH_4^+ present and hence mol of NH_3 and vol NH_3 required:

$$mol \ NH_4^+ \ present = 2.08 \ g \ NH_4Cl \times \frac{1 \ mol \ NH_4Cl}{53.49 \ g \ NH_4Cl} \times \frac{1 \ mol \ NH_4^+}{1 \ mol \ NH_4Cl} = 0.0389 \ mol$$

$$\frac{n_{NH_4^+}}{n_{NH_3}} = 0.18 \quad \rightarrow \quad mol \ NH_3 \ required = \frac{n_{NH_4^+}}{0.18} = \frac{0.0389 \ mol}{0.18} = 0.22 \ mol$$

$$vol \ NH_3 \ required = 0.22 \ mol \ NH_3 \times \frac{1 \ L}{0.236 \ mol \ NH_3} \times \frac{1000 \ mL}{1 \ L} = \underline{\textbf{9.3} \times 10^2 \ \textbf{mL}}$$

(d) Find mol of NH_4^+ present and hence mol of NH_3 and vol NH_3 required:

$$\text{mol NH}_4^+ = \frac{0.109 \text{ mol NH}_4Cl}{1L} \times 395 \text{ mL} \times \frac{1L}{1000 \text{ mL}} \times \frac{1 \text{ mol NH}_4^+}{1 \text{ mol NH}_4Cl} = 0.0431 \text{ mol}$$

$$\frac{n_{NH_4^+}}{n_{NH_3}} = 0.18 \quad \rightarrow \quad \text{mol NH}_3 \text{ required} = \frac{n_{NH_4^+}}{0.18} = \frac{0.0431 \text{ mol}}{0.18} = 0.24 \text{ mol}$$

$$\text{vol NH}_3 \text{ required} = 0.24 \text{ mol NH}_3 \times \frac{1L}{0.499 \text{ mol NH}_3} \times \frac{1000 \text{ mL}}{1L} = \textbf{4.8} \times \textbf{10}^2 \textbf{ mL}$$

28. In theory, a buffer will react with added H^+ or OH^- until all of the base or acid component, respectively, of the buffer is consumed. (In practice, as Figure 14.5 shows, the buffer's capacity to maintain pH is exhausted sometime before the buffer components are spent.) For simplicity, assume one liter of buffer solution and then n_{acid} or n_{base} in the buffer is numerically equal to [acid] or [base], respectively. The amount of acid component, HSO_3^-, in 1 L of buffer is the same in each case, 0.429 mol, so this will equal the quantity of base (OH^-) that each solution can absorb; the amount of acid (H^+) that each buffer can absorb will equal the number of moles of base component, SO_3^{2-}, present.

(a) $[HSO_3^-]$ = 0.429 M, so one liter of buffer can absorb 0.429 mol of base; its capacity is **0.429 M OH$^-$**.
    $[SO_3^{2-}]$ = 0.0249 M, so one liter of buffer can absorb 0.0249 mol of acid; its capacity is **0.0249 M H$^+$**.

(b) $[HSO_3^-]$ = 0.429 M, so one liter of buffer can absorb 0.429 mol of base; its capacity is **0.429 M OH$^-$**.
    $[SO_3^{2-}]$ = 0.247 M, so one liter of buffer can absorb 0.247 mol of acid; its capacity is **0.247 M H$^+$**.

(c) $[HSO_3^-]$ = 0.429 M, so one liter of buffer can absorb 0.429 mol of base; its capacity is **0.429 M OH$^-$**.
    $[SO_3^{2-}]$ = 0.504 M, so one liter of buffer can absorb 0.504 mol of acid; its capacity is **0.504 M H$^+$**.

(d) $[HSO_3^-]$ = 0.429 M, so one liter of buffer can absorb 0.429 mol of base; its capacity is **0.429 M OH$^-$**.
    $[SO_3^{2-}]$ = 0.811 M, so one liter of buffer can absorb 0.811 mol of acid; its capacity is **0.811 M H$^+$**.

(e) $[HSO_3^-]$ = 0.429 M, so one liter of buffer can absorb 0.429 mol of base; its capacity is **0.429 M OH$^-$**.
    $[SO_3^{2-}]$ = 1.223 M, so one liter of buffer can absorb 1.223 mol of acid; its capacity is **1.223 M H$^+$**.

30. This problem is similar in part to Example 14.4.

    (a) Calculate mol of the acid ($HC_4H_4O_6^-$) and base ($C_4H_4O_6^{2-}$) components of the buffer and then use Equation 14.2 to calculate $[H^+]$ and hence pH.

    $$\text{mol of } HC_4H_4O_6^- = \frac{0.187 \text{ mol } KHC_4H_4O_6}{1L} \times 239 \text{ mL} \times \frac{1L}{1000 \text{ mL}} \times \frac{1 \text{ mol } HC_4H_4O_6^-}{1 \text{ mol } KHC_4H_4O_6}$$

    $$= 0.0447 \text{ mol}$$

    $$\text{mol of } C_4H_4O_6^{2-} = \frac{0.288 \text{ mol } K_2C_4H_4O_6}{1L} \times 137 \text{ mL} \times \frac{1L}{1000 \text{ mL}} \times \frac{1 \text{ mol } C_4H_4O_6^{2-}}{1 \text{ mol } K_2C_4H_4O_6}$$

    $$= 0.0395 \text{ mol}$$

    $$[H^+] = K_a \times \frac{n_{HC_4H_4O_6^-}}{n_{C_4H_4O_6^{2-}}} = 4.55 \times 10^{-5} \times \frac{0.0447}{0.0395} = 5.15 \times 10^{-5} \text{ M}$$

    $$pH = -\log_{10}[H^+] = -\log_{10}(5.15 \times 10^{-5}) = \underline{\textbf{4.288}}$$

    (b) Added HCl will react with the base component of the buffer; the reaction is of the strong acid + weak base type.

    $$H^+(aq) + C_4H_4O_6^{2-}(aq) \rightarrow HC_4H_4O_6^-(aq)$$

    Construct a table as in Example 14.3 to determine the composition of the mixture after the two species react. Assuming that volumes are additive, 0.376 L of buffer contains all of the buffer components given in the question; the limiting reactant (H^+) is completely consumed.

	n_{H^+}	$n_{C_4H_4O_6^{2-}}$	$n_{HC_4H_4O_6^-}$
Original	0.0250	0.0395	0.0447
Change	−0.0250	−0.0250	+0.0250
Final	0	0.0145	0.0697

    Now, given the quantity of buffer acid and base components present after reaction with H^+, use Equation 14.2 to calculate $[H^+]$ and hence pH (using Equation 13.3):

    $$[H^+] = K_a \times \frac{n_{HC_4H_4O_6^-}}{n_{C_4H_4O_6^{2-}}} = 4.55 \times 10^{-5} \times \frac{0.0697}{0.0145} = 2.19 \times 10^{-4} \text{ M}$$

    $$pH = -\log_{10}[H^+] = -\log_{10}(2.19 \times 10^{-4}) = \underline{\textbf{3.660}}$$

    (c) Added KOH will react with the acid component of the buffer; the reaction is of the weak acid + strong base type.

    $$OH^-(aq) + HC_4H_4O_6^-(aq) \rightarrow C_4H_4O_6^{2-}(aq) + H_2O$$

Construct a table as in Example 14.3 to determine the composition of the mixture after the two species react. The initial quantities of the buffer components were calculated in part (a) above; the limiting reactant (OH^-) is completely consumed.

	n_{OH^-}	$n_{HC_4H_4O_6^-}$	$n_{C_4H_4O_6^{2-}}$
Original	0.0250	0.0447	0.0395
Change	−0.0250	−0.0250	+0.0250
Final	0	0.0197	0.0645

Now, given the quantity of buffer acid and base components present, use Equation 14.2 to calculate [H^+] and hence pH (using Equation 13.3):

$$[H^+] = K_a \times \frac{n_{HC_4H_4O_6^-}}{n_{C_4H_4O_6^{2-}}} = 4.55 \times 10^{-5} \times \frac{0.0197}{0.0645} = 1.39 \times 10^{-5}\ M$$

$$pH = -\log_{10}[H^+] = -\log_{10}(1.39 \times 10^{-5}) = \underline{\mathbf{4.857}}$$

---

32. This problem is similar in part to Example 14.4.

(a) Although the buffer in Question 30 has been diluted, the number of mol of $HC_4H_4O_6^-$ and $C_4H_4O_6^{2-}$ are unaffected by dilution. Therefore, using Equation 14.2, the [H^+] and pH as calculated in Question 30(a) above will be identical:

$$[H^+] = K_a \times \frac{n_{HC_4H_4O_6^-}}{n_{C_4H_4O_6^{2-}}} = 4.55 \times 10^{-5} \times \frac{0.0447}{0.0395} = 5.15 \times 10^{-5}\ M$$

$$pH = -\log_{10}[H^+] = -\log_{10}(5.15 \times 10^{-5}) = \underline{\mathbf{4.288}}$$

(b) In order to determine pH changes in the diluted buffer, it is first necessary to know how much of the buffer components are present in 0.376 L of diluted buffer. The quantities of $HC_4H_4O_6^-$ and $C_4H_4O_6^{2-}$ calculated in Question 30(a) above are now present in 5.00 L of diluted buffer, from which a 0.376 L sample is being taken.

mol of $HC_4H_4O_6^-$ in 0.376 L of diluted buffer

$$= 0.376\ \text{L buffer} \times \frac{0.0447\ \text{mol}\ HC_4H_4O_6^-}{5.00\ \text{L buffer}} = 0.00336\ \text{mol}$$

mol of $C_4H_4O_6^{2-}$ in 0.376 L of diluted buffer

$$= 0.376\ \text{L buffer} \times \frac{0.0395\ \text{mol}\ C_4H_4O_6^{2-}}{5.00\ \text{L buffer}} = 0.00297\ \text{mol}$$

Construct a table as in Example 14.3 to determine the composition of the mixture after the two species react. Added HCl will react with the base component of the buffer; the reaction is of the strong acid + weak base type; the limiting reactant now is $C_4H_4O_6^{2-}$, which is completely consumed.

$$H^+(aq) + C_4H_4O_6^{2-}(aq) \rightarrow HC_4H_4O_6^-(aq)$$

	n_{H^+}	$n_{C_4H_4O_6^{2-}}$	$n_{HC_4H_4O_6^-}$
Original	0.0250	0.00297	0.00336
Change	−0.00297	−0.00297	+0.00297
Final	0.0220	0	0.00633

The excess H^+ will then react with, and consume, the remaining $HC_4H_4O_6^-$, forming $H_2C_4H_4O_6$:

$$H^+(aq) + HC_4H_4O_6^-(aq) \rightarrow H_2C_4H_4O_6(aq)$$

	n_{H^+}	$n_{HC_4H_4O_6^-}$	$n_{H_2C_4H_4O_6}$
Original	0.0220	0.00633	0
Change	−0.00633	−0.00633	+0.00633
Final	0.0157	0	0.00633

Again, excess H^+ will remain. Given the low concentration of the $H_2C_4H_4O_6$ (relative to H^+) and the fact that $H_2C_4H_4O_6$ is a weak acid, it will contribute very few H^+ ions to solution, so pH calculations can be based entirely on the remaining H^+.

$$[H^+] = \frac{0.0157 \text{ mol } H^+}{0.376 \text{ L}} = 0.0418 \; M$$

$$pH = -\log_{10}[H^+] = -\log_{10}(0.0418) = \underline{\textbf{1.379}}$$

Note that if the reaction between H^+ and $HC_4H_4O_6^-$ to form $H_2C_4H_4O_6$ is ignored and pH is calculated based the quantity of H^+ remaining after all of the $C_4H_4O_6^{2-}$ is consumed, we obtain:

$$[H^+] = \frac{0.0220 \text{ mol } H^+}{0.376 \text{ L}} = 0.0585 \; M$$

$$pH = -\log_{10}[H^+] = -\log_{10}(0.0585) = \underline{\textbf{1.233}}$$

(c) Added NaOH will react with the acid component of the buffer; the reaction is of the weak acid + strong base type. Construct a table as in Example 14.3 to determine the composition of the mixture after the two species react the limiting reactant now is $HC_4H_4O_6^-$, which is completely consumed.

$$HC_4H_4O_6^-(aq) + OH^-(aq) \rightarrow C_4H_4O_6^{2-}(aq) + H_2O$$

	$n_{HC_4H_4O_6^-}$	n_{OH^-}	$n_{C_4H_4O_6^{2-}}$
Original	0.00336	0.0250	0.00297
Change	−0.00336	−0.00336	+0.00336
Final	0	0.0216	0.00633

The system is no longer a buffer since the acid component has been exhausted. Use Equations 13.1 and 13.2 to calculate [H$^+$] and hence pH (using Equation 13.3):

$$[OH^-] = \frac{0.0216 \text{ mol } OH^-}{0.376 \text{ L}} = 0.0574 \text{ M}$$

$$[H^+] = \frac{K_w}{[OH^-]} = \frac{1.0 \times 10^{-14}}{0.0574} = 1.7 \times 10^{-13} \text{ M}$$

$$pH = -\log_{10}[H^+] = -\log_{10}(1.7 \times 10^{-13}) = \underline{\textbf{12.77}}$$

(d) The pH of the diluted buffer is unchanged [32(a) compared with 30(a)]. However, when strong acid or strong base is added to the diluted buffer [32(b and c) compared with 30(b and c)] the pH of the solution changes dramatically.

(e) Dilution significantly compromises the buffer's capacity to completely absorb strong acid or base. The quantity of each buffer component in 0.376 L of the diluted buffer is much less than in 0.376 L of the original buffer.

34. This problem is similar in part to Example 14.4.

(a) Use Equation 14.2 with the given K_a value and acid/base ratio ([HB]/[B$^-$] = 2.2):

$$[H^+] = K_a \times \frac{[HB]}{[B^-]} = (1.54 \times 10^{-5}) \times 2.2 = 3.4 \times 10^{-5} \text{ M}$$

$$pH = -\log_{10}[H^+] = -\log_{10}(3.4 \times 10^{-5}) = \underline{\textbf{4.47}}$$

(b) For the buffer in part (a), [HB]/[B$^-$] = 2.2: set [B$^-$] = 1.0 M then [HB] = 2.2 M. Now 15% of the acid component is to be converted to the base component:

new [HB] = 2.2 M − (0.15 × 2.2 M) = 1.9 M

and since [HB] has decreased by 0.3 M. it follows that [B$^-$] must increase by the same amount:

new [B$^-$] = 1.0 M + 0.3 M = 1.3 M

Now use Equation 14.2 with these new acid and base concentrations:

$$[H^+] = K_a \times \frac{[HB]}{[B^-]} = (1.54 \times 10^{-5}) \times \frac{1.9\ M}{1.3\ M} = 2.3 \times 10^{-5}\ M$$

$$pH = -\log_{10}[H^+] = -\log_{10}(2.3 \times 10^{-5}) = \underline{\textbf{4.64}}$$

(c) The pH is to increase by 1 unit; thus the required pH is 5.47, and therefore [H$^+$] = $10^{-5.47}$ = 3.4 × 10^{-6} M. Use Equation 14.2 with the given K_a value and this [H$^+$] to determine the new acid/base ratio:

$$[H^+] = K_a \times \frac{[HB]}{[B^-]}$$

$$3.4 \times 10^{-6} = (1.54 \times 10^{-5}) \times \frac{[HB]}{[B^-]} \qquad \rightarrow \qquad \frac{[HB]}{[B^-]} = \underline{\textbf{0.22}}$$

36. This problem is related to Example 14.4.

(a) Since pH = 7.40, [H$^+$] = $10^{-7.40}$ = 4.0 × 10^{-8} M. Use this [H$^+$] and the given K_a value in Equation 14.2 to determine the acid/base ratio:

$$[H^+] = K_a \times \frac{[H_2PO_4^-]}{[HPO_4^{2-}]}$$

$$4.0 \times 10^{-8} = 6.2 \times 10^{-8} \times \frac{[H_2PO_4^-]}{[HPO_4^{2-}]} \qquad so \qquad \frac{[H_2PO_4^-]}{[HPO_4^{2-}]} = \underline{\textbf{0.65}}$$

(b) To determine the new acid/base ratio, repeat the procedure in part (a) with pH = 7.00. For pH = 7.00, [H$^+$] = $10^{-7.00}$ = 1.0 × 10^{-7} M. Use Equation 14.2 with the given K_a value and this new [H$^+$]:

$$[H^+] = K_a \times \frac{[H_2PO_4^-]}{[HPO_4^{2-}]}$$

$$1.0 \times 10^{-7} = 6.2 \times 10^{-8} \times \frac{[H_2PO_4^-]}{[HPO_4^{2-}]} \qquad so \qquad \frac{[H_2PO_4^-]}{[HPO_4^{2-}]} = 1.6$$

To determine the percentage of HPO$_4^{2-}$ ions converted, set [HPO$_4^{2-}$] = 1.0 in part (a) and then [H$_2$PO$_4^-$] = 0.65. Let x be the decrease in [HPO$_4^{2-}$] then (by stoichiometry) [H$_2$PO$_4^-$] will increase by the same amount. Thus:

$$1.6 = \frac{[H_2PO_4^-]}{[HPO_4^{2-}]} = \frac{0.65 + x}{1.0 - x} \qquad so\ 0.65 + x = 1.6(1.0 - x) = 1.6 - 1.6x$$

$$2.6x = 1.6 - 0.65 = 1.0 \qquad therefore\ x = 0.38\ \ or\ 38\ \%$$

The percentage of HPO$_4^{2-}$ ions converted is **38%**.

(c) To determine the new acid/base ratio, repeat the procedure in part (a) with pH = 8.00. For pH = 8.00, $[H^+] = 10^{-8.00} = 1.0 \times 10^{-8}$ M. Use Equation 14.2 with the given K_a value and this $[H^+]$:

$$[H^+] = K_a \times \frac{[H_2PO_4^-]}{[HPO_4^{2-}]}$$

$$1.0 \times 10^{-8} = 6.2 \times 10^{-8} \times \frac{[H_2PO_4^-]}{[HPO_4^{2-}]} \qquad \text{so} \qquad \frac{[H_2PO_4^-]}{[HPO_4^{2-}]} = 0.16$$

To determine the percentage of $H_2PO_4^-$ ions converted, set $[H_2PO_4^-] = 1.0$ in part (a) and then $[HPO_4^{2-}] = 1.5$. Let x be the decrease in $[H_2PO_4^-]$ then (by stoichiometry) $[HPO_4^{2-}]$ will increase by the same amount. Thus:

$$0.16 = \frac{[H_2PO_4^-]}{[HPO_4^{2-}]} = \frac{1.0 - x}{1.5 + x} \qquad \text{so } 1.0 - x = 0.16(1.5 + x) = 0.24 + 0.16x$$

$$1.16x = 1.0 - 0.24 = 0.8 \qquad \text{therefore } x = 0.69 \text{ or } 69 \%$$

The percentage of $H_2PO_4^-$ ions converted is **69%**.

14-2   ACID-BASE INDICATORS

38. See page 371: For an acid-base indicator to be suitable for a given titration, its color change (***end point***) should coincide with the ***equivalence point*** of the titration. In practice this means that the color change should coincide with the region of rapid pH change that occurs close to the equivalence point: see Figures 14.7, 14.8, 14.10 and 14.12. These figures also indicate that the pH at the equivalence point depends on the nature (weak *versus* strong) of the acid and base involved. See also Table 14.2.

(a) Titration of $NaCHO_2$ with HNO_3 corresponds to addition of a strong acid (HNO_3) to a weak base (CHO_2^-; see Table 13.2 and Section 13-6), similar to the experiment shown in Figure 14.12. The equivalence point will be at a weakly acidic pH (pH = 3—6), so **methyl orange** (color change at pH 4) is suitable.

(b) Titration of $HOCl$ with $Ba(OH)_2$ corresponds to addition of a strong base ($Ba(OH)_2$) to a weak acid ($HOCl$), similar to the experiment shown in Figure 14.10. The equivalence point will be at a weakly basic pH (pH = 8—11), so **phenolphthalein** (color change at pH 9) is suitable.

(c) Titration of HNO_3 with HI corresponds to addition of a strong acid to another strong acid (HNO_3), so the pH will remain strongly acidic and **no indicator is appropriate**.

(d) Titration of HCl with NH_3 corresponds to addition of a weak base (NH_3) to a strong acid (HCl), somewhat similar to the experiment shown in Figure 14.12. (In this case, however, the pH will be strongly acidic at the start of the experiment and then increase to basic pH when NH_3 is in excess: in other words, the reverse of that shown in Figure 14.12.) The equivalence point will be to weakly acidic pH (pH = 3–6), so **methyl orange** (color change at pH 4) is suitable.

40. This question is related to Example 14.6.

    (a) See page 372: A color change occurs when pH ≈ pK_a, in this case when pH = 9.5, so pK_a for the indicator is 9.5. Recall from Equation 13.6 that pK_a = $-\log_{10}K_a$; so K_a = 10^{-pK_a}. Thus:

    $$K_a = 10^{-9.5} = \mathbf{3.2 \times 10^{-10}}$$

    (b) The range over which an indicator changes color is normally around 2 units of pH, which here means one unit of pH above and below 9.5. Thus, the range of the indicator is **pH = 8.5 – 10.5**.

    (c) See pages 371–372: The color observed depends on the ratio [HIn]/[In⁻]. At pH = 9.5, pH = pK_a and therefore [H⁺] = K_a or [H⁺]/K_a = 1. Substituting the latter into Equation 14.3 gives [HIn]/[In⁻] = 1 or [HIn] = [In⁻], so at pH = 9.5 the observed color is a mixture of colorless and blue. When the pH increases to 10.0, we expect there to be less of the colorless acid form of the indicator (HIn) and more of the blue base form of the indicator (In⁻). Thus, at pH = 10.0 the color of a solution containing the indicator is **blue**.

14-3 ACID-BASE TITRATIONS

42. This problem is related to Example 14.7.

    (a) The reaction between HCl and KOH is of the strong acid/strong base type (see Table 4.2):

    $$H^+(aq) + OH^-(aq) \rightarrow H_2O$$

    Determine mol H⁺ added and then find mol KOH using this plan:

    $$\text{mol HCl} \xrightarrow{\text{atom ratio}} \text{mol H}^+ \xrightarrow{\text{stoichiometric ratio}} \text{mol OH}^- \xrightarrow{\text{atom ratio}} \text{mol KOH}$$

    $$\text{mol H}^+ \text{ added} = 27.66 \text{ mL} \times \frac{1 \text{ L}}{1000 \text{ mL}} \times \frac{0.2500 \text{ mol HCl}}{1 \text{ L}} \times \frac{1 \text{ mol H}^+}{1 \text{ mol HCl}}$$

    $$= 6.915 \times 10^{-3} \text{ mol H}^+$$

    $$\text{mol KOH present} = 6.915 \times 10^{-3} \text{ mol H}^+ \times \frac{1 \text{ mol OH}^-}{1 \text{ mol H}^+} \times \frac{1 \text{ mol KOH}}{1 \text{ mol OH}^-} = 6.915 \times 10^{-3} \text{ mol}$$

    Thus we know mol KOH and are asked to determine the volume of KOH solution added: the molarity of KOH is the conversion factor required. Knowing the pH of the solution we can calculate [H⁺] and hence [OH⁻] (using Equation 13.1) and, from that, [KOH].

$$pH = 13.29 \qquad \text{so} \quad [H^+] = 10^{-pH} = 10^{-13.29} = 5.13 \times 10^{-14} \; M$$

$$[OH^-] = \frac{K_w}{[H^+]} = \frac{1.0 \times 10^{-14}}{5.13 \times 10^{-14}} = 0.195 \; M$$

$$[KOH] = \frac{0.195 \; mol \; OH^-}{1 \; L} \times \frac{1 \; mol \; KOH}{1 \; mol \; OH^-} = 0.195 \; M$$

Finally calculate volume of the KOH solution:

$$V = \frac{n}{M} = \frac{6.915 \times 10^{-3} \; mol}{0.195 \; mol/L} = \mathbf{3.55 \times 10^{-2} \; L} \quad \text{or} \quad \mathbf{35.5 \; mL}$$

(b) (See also Figure 14.8.) Since the reactants are a strong acid and a strong base, the equivalence point will be at exactly **pH = 7**.

(c) The total volume of the solution at the equivalence point will be the sum of the volumes of the HCl and KOH solutions = 27.66 mL + 35.5 mL = 63.2 mL = 0.0632 L.

Note also that mol Cl^- added = mol H^+ added and mol OH^- present = mol K^+ present. At the equivalence point mol H^+ added = mol OH^- present, so mol Cl^- = mol K^+ = 6.915 × 10^{-3} mol. Thus:

$$[K^+] = [Cl^-] = \frac{n}{V} = \frac{6.915 \times 10^{-3} \; mol}{0.0632 \; L} = \mathbf{0.109 \; M}$$

44. This problem is related to Examples 14.7 and 14.8.

(a) The reaction between $NaHSO_3$ and KOH is between a weak acid and a strong base. Here the weak acid is the HSO_3^- anion and (see Table 4.2) the net ionic equation is:

$$HSO_3^-(aq) + OH^-(aq) \rightarrow SO_3^{2-}(aq) + H_2O$$

(b) Given M and V for the KOH solution we can calculate n for HSO_3^- and hence, using V for the HSO_3^- solution, determine $[HSO_3^-]$:

$$mol \; OH^- \; added = 22.94 \; mL \times \frac{1 \; L}{1000 \; mL} \times \frac{0.238 \; mol \; KOH}{1 \; L} \times \frac{1 \; mol \; OH^-}{1 \; mol \; KOH}$$
$$= 5.46 \times 10^{-3} \; mol$$

$$mol \; HSO_3^- \; originally \; present = 5.46 \times 10^{-3} \; mol \; OH^- \times \frac{1 \; mol \; HSO_3^-}{1 \; mol \; OH^-} = 5.46 \times 10^{-3} \; mol$$

Thus we know mol HSO_3^- and we are told V = 50.0 mL = 0.0500 L

$$[HSO_3^-] = \frac{n}{V} = \frac{5.46 \times 10^{-3}\,\text{mol}}{0.0500\,\text{L}} = \underline{\textbf{0.109 M}}$$

(c) At the equivalence point all of the HSO_3^- ion initially present and all of the added OH^- have been converted to $SO_3^{2-} + H_2O$ according to the net ionic equation in part (a). Thus mol of SO_3^{2-} formed = mol HSO_3^- originally present = 5.46×10^{-3} mol. Recall that the total volume of solution at the equivalence point is 50.0 mL + 22.94 mL = 72.9 mL or 0.0729 L. Thus a preliminary value for $[SO_3^{2-}]$ is

$$[SO_3^{2-}] = \frac{n}{V} = \frac{5.46 \times 10^{-3}\,\text{mol}}{0.0729\,\text{L}} = 0.0749\,M$$

Now the SO_3^{2-} ion may act as weak base:

$$SO_3^{2-}(aq) + H_2O \rightleftharpoons HSO_3^-(aq) + OH^-(aq)$$

and the position of equilibrium for this reaction will determine $[SO_3^{2-}]$, $[HSO_3^-]$, and $[OH^-]$. The value of K_b for this reaction is 1.7×10^{-7} (Table 13.2). Thus:

$$K_b = \frac{[HSO_3^-][OH^-]}{[SO_3^{2-}]} \qquad \text{so:} \quad 1.7 \times 10^{-7} = \frac{[HSO_3^-][OH^-]}{[SO_3^{2-}]} \approx \frac{[OH^-]^2}{[SO_3^{2-}]}$$

since $[HSO_3^-] \approx [OH^-]$. (See also Example 13.11.) Since SO_3^{2-} is a weak base, its concentration (0.0749 M) will not change due to the weak base dissociation reaction so we can say

$$[SO_3^{2-}] = \underline{\textbf{0.0749 M}}$$

Substituting this value for $[SO_3^{2-}]$ into the K_b relationship gives:

$$1.7 \times 10^{-7} = \frac{[OH^-]^2}{0.0749} \qquad \text{from which } \textbf{[OH}^-\textbf{]} = \underline{\textbf{1.1} \times \textbf{10}^{-4}\textbf{ M}}$$

and therefore also $\textbf{[HSO}_3^-\textbf{]} = \textbf{[OH}^-\textbf{]} = \underline{\textbf{1.1} \times \textbf{10}^{-4}\textbf{ M}}$

Finally, mol K^+ added must equal mol OH^- added = 5.46×10^{-3} mol and likewise mol Na^+ originally present must equal mol HSO_3^- originally present = 5.46×10^{-3} mol (and since both of these are spectators their quantities will not change during the reactions). Thus their concentrations at the equivalence point are:

$$[K^+] = \frac{n}{V} = \frac{5.46 \times 10^{-3}\,\text{mol}}{0.0729\,\text{L}} = \underline{\textbf{0.0749 M}}$$

$$[Na^+] = \frac{n}{V} = \frac{5.46 \times 10^{-3}\,\text{mol}}{0.0729\,\text{L}} = \underline{\textbf{0.0749 M}}$$

(d) At the equivalence point $[OH^-]$ is 1.1×10^{-4} M; therefore (Equations 13.1 and 13.2)

$$[H^+] = \frac{K_w}{[OH^-]} = \frac{1.0 \times 10^{-14}}{1.1 \times 10^{-4}} = 9.1 \times 10^{-11} \text{ M}$$

$$pH = -\log_{10}[H^+] = -\log_{10}(9.1 \times 10^{-11}) = \underline{\textbf{10.04}}$$

46. This problem is related to Examples 14.7 and 14.8.

(a) The reaction is between a strong acid (HNO_3) and the potassium salt of a weak base (BrO^-) for which the net ionic equation (Table 4.2) is:

$$H^+(aq) + BrO^-(aq) \rightarrow HBrO(aq)$$

(b) From the net ionic equation, at the equivalence point mol BrO^- present = mol H^+ added. We can determine the mol BrO^- present, then find mol H^+ and mol HNO_3 required, and finally vol HNO_3 required using this plan:

$$\text{vol KBrO} \xrightarrow{\text{M}} \text{mol KBrO} \xrightarrow{\text{atom ratio}} \text{mol BrO}^- \xrightarrow{\text{stoichiometric ratio}} \text{mol H}^+$$

$$\xrightarrow{\text{atom ratio}} \text{mol HNO}_3 \xrightarrow{\text{M}} \text{vol HNO}_3$$

$$\text{mol BrO}^- \text{ present} = 35.00 \text{ mL KBrO} \times \frac{1 \text{L}}{1000 \text{ mL}} \times \frac{0.487 \text{ mol KBrO}}{1 \text{L}} \times \frac{1 \text{ mol BrO}^-}{1 \text{ mol KBrO}}$$

$$= 1.70 \times 10^{-2} \text{ mol BrO}^-$$

$$\text{vol HNO}_3 \text{ needed} = 1.70 \times 10^{-2} \text{ mol BrO}^- \times \frac{1 \text{ mol H}^+}{1 \text{ mol BrO}^-} \times \frac{1 \text{ mol HNO}_3}{1 \text{ mol H}^+}$$

$$\times \frac{1 \text{L}}{0.264 \text{ mol HNO}_3} \times \frac{1000 \text{ mL}}{1 \text{L}}$$

$$= \underline{\textbf{64.4 mL HNO}_3}$$

(c) At the equivalence point, all of the BrO^- initially present and all of the added H^+ have been converted to HBrO according to the net ionic equation in part (a). Thus mol of HBrO formed = mol BrO^- originally present = 1.70×10^{-2} mol (from part (b)). Recall that the total volume of solution at the equivalence point is 35.00 mL + 64.4 mL = 99.4 mL or 0.0994 L. Thus a preliminary value for [HBrO] is

$$[HBrO] = \frac{n}{V} = \frac{1.70 \times 10^{-2} \text{ mol}}{0.0994 \text{ L}} = 0.171 \text{ M}$$

Now proceed as in Example 13.8. The HBrO may act as weak acid:

$$HBrO(aq) \rightleftharpoons H^+(aq) + BrO^-(aq) \qquad K_a = \frac{[H^+][BrO^-]}{[HBrO]}$$

and the position of equilibrium for this reaction determinea [HBrO], [BrO⁻], and [H⁺]. The value of K_a for this reaction can be obtained from Equation 13.10 since we know that K_b for the conjugate base BrO^- is 4.0×10^{-6} (given). Thus:

$$K_a = \frac{K_w}{K_b} = \frac{1.0 \times 10^{-14}}{4.0 \times 10^{-6}} = 2.5 \times 10^{-9}$$

Use $[HBrO]_0 = 0.171$ M and initially we assume there is no ionization, so $[H^+]_0 = [BrO^-]_0 = 0$. Let $\Delta[H^+] = x$:

	$HBrO(aq)$	$\rightleftharpoons$ $H^+(aq)$ +	$BrO^-(aq)$
[ ]₀ (M)	0.171	0	0
Δ[ ] (M)	$-x$	$+x$	$+x$
[ ]$_{eq}$ (M)	$0.171 - x$	x	x

Now substitute the equilibrium concentrations of each species into the K_a expression:

$$K_a = 2.5 \times 10^{-9} = \frac{[H^+][BrO^-]}{[HBrO]} = \frac{(x)(x)}{(0.171 - x)}$$

Assume $x \ll 0.171$:

$$2.5 \times 10^{-9} = \frac{x^2}{0.171} \qquad \text{so } x^2 = 4.3 \times 10^{-10} \rightarrow \quad x = 2.1 \times 10^{-5} = [H^+]$$

Check the assumption that the value of x is indeed much less that 0.171: the % ionization = $[H^+]_{eq}/[HBrO]_0 \times 100$ = $(2.1 \times 10^{-5})/(0.171) \times 100 = 0.012\% < 5\%$ (see page 345), so the assumption made in solving the above equation was valid.
At the equivalence point $[H^+]$ is 2.1×10^{-5} M; therefore (using Equation 13.3):

$$pH = -\log_{10}[H^+] = -\log_{10}(2.1 \times 10^{-5}) = \underline{\textbf{4.68}}$$

(d) The value of x determined in part (c) above gives us the equilibrium $[H^+]$ and $[BrO^-]$ at the equivalence point. So we can say:

$$\underline{[H^+] = [BrO^-] = 2.1 \times 10^{-5} M}$$

Moreover, from part (c) above the equilibrium [HBrO] is $0.171 - x$. Thus:

$$\underline{[HBrO] = 0.171 M}$$

Finally, we know that mol H^+ added = mol BrO^- present at the equivalence point However, mol H^+ added = mol NO_3^- added (since HNO_3 is the source of each one) and likewise mol K^+ present = mol BrO^- present (since $KBrO$ is the source of each ion). Thus mol K^+ = mol NO_3^- = 1.70×10^{-2} mol (from part (b); and since both are spectators their quantity will not change during the reactions). Hence:

$$[K^+] = [NO_3^-] = \frac{n}{V} = \frac{1.70 \times 10^{-2}\,\text{mol}}{0.0994\,\text{L}} = \underline{\textbf{0.171 M}}$$

48. This problem is related to Example 14.7.

(a) The reaction between HBr and KOH is between a strong acid and a strong base for which the net ionic equation (Table 4.2) is:

$$H^+(aq) + OH^-(aq) \rightarrow H_2O$$

(b) The overall reaction taking place is:

$$HBr(aq) + KOH(aq) \rightarrow H_2O + KBr(aq)$$

or, in terms of the ions present:

$$H^+(aq) + Br^-(aq) + K^+(aq) + OH^-(aq) \rightarrow H_2O + K^+(aq) + Br^-(aq)$$

At the equivalence point all of the H^+ and OH^- have reacted so all that remains are **water** and the spectator ions **K^+ and Br^-**.

(c) First find mol HBr present, then use the following strategy:

$$\text{mol HBr} \xrightarrow{\text{atom ratio}} \text{mol } H^+ \xrightarrow{\text{stoichiometric ratio}} \text{mol } OH^- \xrightarrow{\text{atom ratio}}$$

$$\text{mol KOH} \xrightarrow{M} \text{V KOH}$$

$$\text{mol HBr present} = 0.03000\,\text{L HBr} \times \frac{0.269\,\text{mol HBr}}{1\,\text{L}} = 8.07 \times 10^{-3}\,\text{mol}$$

$$\text{mol KOH added} = 8.07 \times 10^{-3}\,\text{mol HBr} \times \frac{1\,\text{mol } H^+}{1\,\text{mol HBr}} \times \frac{1\,\text{mol } OH^-}{1\,\text{mol } H^+} \times \frac{1\,\text{mol KOH}}{1\,\text{mol } OH^-}$$

$$= 8.07 \times 10^{-3}\,\text{mol KOH}$$

$$\text{vol KOH} = \frac{n}{M} = \frac{8.07 \times 10^{-3}\,\text{mol}}{0.2481\,\text{mol/L}} = \underline{\textbf{0.0325 L}} \quad \text{or} \quad \underline{\textbf{32.5 mL}}$$

(d) 1. Before any KOH is added, pH is determined by the HBr present. Since HBr is a strong acid, $[HBr] = [H^+] = 0.269\,M$; therefore (using Equation 13.3):

$$pH = -\log_{10}[H^+] = -\log_{10}(0.269) = \underline{\textbf{0.570}}$$

2. Halfway to the equivalence point means that half of the HBr has been neutralized so $\frac{1}{2}(8.07 \times 10^{-3}\,\text{mol}) = 4.04 \times 10^{-3}\,\text{mol}$ of HBr remain; and since HBr is a strong acid, mol HBr = mol H^+ = 4.04×10^{-3} mol. In addition, half of the total volume of KOH has been added, that is, $\frac{1}{2}(32.5)\,\text{mL} = 16.2\,\text{mL}$. The total volume at that point will be $(30.00 + 16.2)\text{mL} = 46.2\,\text{mL}$ or $0.0462\,\text{L}$. Thus:

$$[H^+] = \frac{n}{V} = \frac{4.04 \times 10^{-3}\,\text{mol}}{0.0462\,\text{L}} = 0.0874\,M$$

$$pH = -\log_{10}[H^+] = -\log_{10}(0.0874) = \underline{\mathbf{1.06}}$$

3. (See also Figure 14.8.) Since the reactants are a strong acid and a strong base, the equivalence point will be at exactly **pH = 7**.

50. This problem is related to Examples 14.7 and 14.8.

(a) The reaction between HCl and $C_{17}H_{19}O_3N$ is between a strong acid and a weak base for which the net ionic equation (Table 4.2) is:

$$H^+(aq) + C_{17}H_{19}O_3N(aq) \rightleftharpoons C_{17}H_{19}O_3NH^+(aq)$$

(b) The overall reaction taking place is:

$$HCl(aq) + C_{17}H_{19}O_3N(aq) \rightarrow C_{17}H_{19}O_3NHCl(aq)$$

or, in terms of the ions present:

$$H^+(aq) + Cl^-(aq) + C_{17}H_{19}O_3N(aq) \rightarrow C_{17}H_{19}O_3NH^+(aq) + Cl^-(aq)$$

At the equivalence point all of the H^+ and $C_{17}H_{19}O_3N$ have reacted to form $C_{17}H_{19}O_3NH^+$. However, since $C_{17}H_{19}O_3NH^+$ is the conjugate acid of the weak base $C_{17}H_{19}O_3N$ it will act as a weak acid:

$$C_{17}H_{19}O_3NH^+(aq) \rightleftharpoons H^+(aq) + C_{17}H_{19}O_3N(aq)$$

Thus the species present at the equivalence point are **$C_{17}H_{19}O_3NH^+$, H^+, $C_{17}H_{19}O_3N$, and Cl^-.**

(c) First find mol $C_{17}H_{19}O_3N$ originally present, then use the following strategy:

$$\text{mol } C_{17}H_{19}O_3N \xrightarrow{\text{stoichiometric ratio}} \text{mol } H^+ \xrightarrow{\text{atom ratio}} \text{mol HCl} \xrightarrow{M} V\,\text{HCl}$$

$$\text{mol } C_{17}H_{19}O_3N = 0.0500\,\text{L } C_{17}H_{19}O_3N \times \frac{0.1500\,\text{mol } C_{17}H_{19}O_3N}{1\,\text{L}} = 7.50 \times 10^{-3}\,\text{mol}$$

$$\text{mol HCl added} = 7.50 \times 10^{-3}\,\text{mol } C_{17}H_{19}O_3N \times \frac{1\,\text{mol } H^+}{1\,\text{mol } C_{17}H_{19}O_3N} \times \frac{1\,\text{mol } H^+}{1\,\text{mol HCl}}$$

$$= 7.50 \times 10^{-3}\,\text{mol}$$

$$\text{vol HCl} = \frac{n}{M} = \frac{7.50 \times 10^{-3}\,\text{mol}}{0.1045\,\text{mol/L}} = \underline{\mathbf{0.0718\,L}} \quad \text{or} \quad \underline{\mathbf{71.8\,mL}}$$

(d) Before any HCl is added, pH is determined by dissociation of the weak base $C_{17}H_{19}O_3N$: this is similar to Example 13.11(a). Set up an equilibrium table and determine $[OH^-]$ and hence $[H^+]$ and pH. In completing the equilibrium table, recall that $[C_{17}H_{19}O_3N]_0 = 0.1500\ M$ (given) and initially we assume there is no ionization, so $[C_{17}H_{19}O_3NH^+]_0 = [OH^-]_0 = 0$. Let $\Delta[OH^-] = x$:

	$C_{17}H_{19}O_3N(aq) + H_2O$ ⇌	$C_{17}H_{19}O_3NH^+(aq)$ +	$OH^-(aq)$
$[\ ]_0\ (M)$	0.1500	0	0
$\Delta[\ ]\ (M)$	$-x$	$+x$	$+x$
$[\ ]_{eq}\ (M)$	$0.1500 - x$	x	x

Now substitute the equilibrium concentrations of each species into the K_b expression:

$$K_b = 7.4 \times 10^{-7} = \frac{[C_{17}H_{19}O_3NH^+][OH^-]}{[C_{17}H_{19}O_3N]} = \frac{(x)(x)}{(0.1500 - x)}$$

Assume $x \ll 0.1500$:

$$7.4 \times 10^{-7} = \frac{x^2}{0.1500} \qquad \text{so } x^2 = (0.1500)(7.4 \times 10^{-7}) = 1.1 \times 10^{-7} \text{ and } x = 3.3 \times 10^{-4}$$

Check the assumption that the value of x is indeed much less that 0.1500: the % ionization = $[OH^-]_{eq}/[C_{17}H_{19}O_3N]_0 \times 100 = (3.3 \times 10^{-4})/(0.1500) \times 100 = 0.22\% < 5\%$ (see page 416), so the assumption made in solving the above equation was valid.

Note that $x = [OH^-]_{eq}$, so we can say that $[OH^-] = 3.3 \times 10^{-4}\ M$ and hence, using Equations 13.1 and 13.2: $K_w = [H^+][OH^-] = 1.0 \times 10^{-14}$. Thus:

$$[H^+] = \frac{K_w}{[OH^-]} = \frac{1.0 \times 10^{-14}}{3.3 \times 10^{-4}} = 3.0 \times 10^{-11}\ M$$

Finally, use Equation 13.3 to calculate pH:

$$pH = -\log_{10}[H^+] = -\log_{10}(3.0 \times 10^{-11}) = \underline{\mathbf{10.52}}$$

(e) See Figure 14.12: At halfway to the equivalence point, half of the weak base has been converted to the conjugate acid and half of it remains; thus $[C_{17}H_{19}O_3NH^+] = [C_{17}H_{19}O_3N]$. Therefore:

$$K_b = 7.4 \times 10^{-7} = \frac{[C_{17}H_{19}O_3NH^+][OH^-]}{[C_{17}H_{19}O_3N]} = [OH^-]$$

$$[H^+] = \frac{K_w}{[OH^-]} = \frac{1.0 \times 10^{-14}}{7.4 \times 10^{-7}} = 1.4 \times 10^{-8}\ M$$

Finally, use Equation 13.3 to calculate pH:

$$pH = -\log_{10}[H^+] = -\log_{10}(1.4 \times 10^{-8}) = \underline{\textbf{7.87}}$$

Alternatively, note from Figure 14.12 that at this point pH = pK_a: use Equation 13.10 to obtain K_a from the given K_b value, then use Equation 13.6 to obtain pK_a and hence pH.

(f) At the equivalence point all of the $C_{17}H_{19}O_3N$ initially present and all of the added H^+ have been converted to $C_{17}H_{19}O_3NH^+$ according to the net ionic equation in part (a). Thus mol of $C_{17}H_{19}O_3NH^+$ formed = mol $C_{17}H_{19}O_3N$ originally present = 7.50×10^{-3} mol. Recall that the total volume of solution at the equivalence point is 50.0 mL + 71.8 mL = 121.8 mL or 0.1218 L. Thus a preliminary value for $[C_{17}H_{19}O_3NH^+]$ at the equivalence point is

$$[C_{17}H_{19}O_3NH^+] = \frac{n}{V} = \frac{7.50 \times 10^{-3} \, mol}{0.1218 \, L} = 0.0616 \, M$$

Now the $C_{17}H_{19}O_3NH^+$ ion may act as weak acid:

$$C_{17}H_{19}O_3NH^+(aq) \rightleftharpoons C_{17}H_{19}O_3N(aq) + H^+(aq) \qquad K_a = \frac{[C_{17}H_{19}O_3N][H^+]}{[C_{17}H_{19}O_3NH^+]}$$

and the position of equilibrium for this reaction will determine $[H^+]$ and hence pH. The value of K_a for this reaction can be obtained from Equation 13.10 since we know that K_b for the conjugate base $C_{17}H_{19}O_3N$ is 7.4×10^{-7} (given). Thus:

$$K_a = \frac{K_w}{K_b} = \frac{1.0 \times 10^{-14}}{7.4 \times 10^{-7}} = 1.4 \times 10^{-8}$$

Thus: $$1.4 \times 10^{-8} = \frac{[C_{17}H_{19}O_3N][H^+]}{[C_{17}H_{19}O_3NH^+]} \approx \frac{[H^+]^2}{[C_{17}H_{19}O_3NH^+]}$$

since $[C_{17}H_{19}O_3N] \approx [H^+]$. (See also Examples 13.7 and 13.8.) Since $C_{17}H_{19}O_3NH^+$ is a weak acid, its concentration (0.0616 M) will not change due to the weak acid dissociation reaction so we can say $[C_{17}H_{19}O_3NH^+] = 0.0616 \, M$

Substituting this value for $[C_{17}H_{19}O_3NH^+]$ into the K_a relationship gives:

$$1.4 \times 10^{-8} = \frac{[H^+]^2}{0.0616} \qquad \text{from which } [H^+] = 2.9 \times 10^{-5} \, M$$

Finally, use Equation 13.3 to calculate pH:

$$pH = -\log_{10}[H^+] = -\log_{10}(2.9 \times 10^{-5}) = \underline{\textbf{4.54}}$$

52. (a) First determine $[HC_7H_5O_2]$:

$$\text{mol } HC_7H_5O_2 = 0.350 \text{ g } HC_7H_5O_2 \times \frac{1 \text{ mol } HC_7H_5O_2}{122.121 \text{ g } HC_7H_5O_2} = 0.00287 \text{ mol}$$

$$[HC_7H_5O_2] = \frac{0.00287 \text{ mol } HC_7H_5O_2}{100.0 \text{ mL}} \times \frac{1000 \text{ mL}}{1 \text{ L}} = 0.0287 \text{ } M$$

Now proceed as in Example 13.7. Set up an equilibrium table and determine $[H^+]$ and hence pH. In completing the equilibrium table, use $[HC_7H_5O_2]_0 = 0.0287$ M (calculated above) and initially we assume there is no ionization, so $[H^+]_0 = [C_7H_5O_2^-]_0 = 0$. Let $\Delta[H^+] = x$:

	$HC_7H_5O_2(aq)$	$\rightleftharpoons$	$H^+(aq)$	$+$	$C_7H_5O_2^-(aq)$
$[\ ]_0$ (M)	0.0287		0		0
$\Delta[\ ]$ (M)	$-x$		$+x$		$+x$
$[\ ]_{eq}$ (M)	$0.0287 - x$		x		x

From Table 13.2, K_a for benzoic acid is 6.6×10^{-5}. Substitute the equilibrium concentrations of each species into the K_a expression:

$$K_a = 6.6 \times 10^{-5} = \frac{[H^+][C_7H_5O_2^-]}{[HC_7H_5O_2]} = \frac{(x)(x)}{(0.0287 - x)}$$

Assume $x \ll 0.0287$:

$$6.6 \times 10^{-5} \approx \frac{x^2}{0.0287} \qquad \text{so} \quad x^2 = (0.0287)(6.6 \times 10^{-5}) = 1.9 \times 10^{-6}$$

$$x = 1.4 \times 10^{-3}$$

Check the assumption that the value of x is indeed much less that 0.0287: the % ionization $= [H^+]_{eq}/[HC_7H_5O_2]_0 \times 100 = (1.4 \times 10^{-3})/(0.0287) \times 100 = 4.9\% < 5\%$ (see page 345), so the assumption made in solving the above equation was valid.

Note that $x = [H^+]_{eq}$, so we can say that $[H^+] = 1.4 \times 10^{-3}$ M and hence, using Equation 13.3:

$$\text{pH} = -\log_{10}[H^+] = -\log_{10}(1.4 \times 10^{-3}) = \underline{\textbf{2.85}}$$

(b) See Figure 14.10: At halfway to the equivalence point, half of the weak acid has been converted to the conjugate base and half of it remains; thus $[C_7H_5O_2^-] = [HC_7H_5O_2]$. Therefore:

$$K_a = 6.6 \times 10^{-5} = \frac{[H^+][C_7H_5O_2^-]}{[HC_7H_5O_2]} = [H^+]$$

Now use Equation 13.3 to calculate pH:

$$\text{pH} = -\log_{10}[H^+] = -\log_{10}(6.6 \times 10^{-5}) = \underline{\textbf{4.18}}$$

Alternatively, note from Figure 14.10 that at this point pH = pK_a: use Equation 13.6 to obtain pK_a and hence pH.

(c) The 30.00 mL sample used in the titration came from the prepared $HC_7H_5O_2$ solution (total 100.0 mL), and so mol $HC_7H_5O_2$ present in 30.00 mL = (30.00/100.0) × (0.00287 mol) = 8.61×10^{-4} mol. At the equivalence point all of the $HC_7H_5O_2$ initially present in the 30.00 mL sample has reacted with all of the added OH^- to form $C_7H_5O_2^-$ according to the following net ionic equation:

$$HC_7H_5O_2(aq) + OH^-(aq) \rightleftharpoons C_7H_5O_2^-(aq) + H_2O$$

Thus mol $C_7H_5O_2^-$ formed = mol $HC_7H_5O_2$ present in the 30.00 mL sample; so mol $C_7H_5O_2^-$ formed at the equivalence point = 8.61×10^{-4} mol.

To determine the volume of KOH solution added to reach the equivalence point, use the following strategy:

$$\text{mol } HC_7H_5O_2 \xrightarrow{\text{stoichiometric ratio}} \text{mol } OH^- \xrightarrow{\text{atom ratio}} \text{mol KOH} \xrightarrow{M} \text{V KOH}$$

$$\text{mol KOH added} = 8.61 \times 10^{-4} \text{ mol } HC_7H_5O_2 \times \frac{1\,\text{mol } OH^-}{1\,\text{mol } HC_7H_5O_2} \times \frac{1\,\text{mol KOH}}{1\,\text{mol } OH^-}$$

$$= 8.61 \times 10^{-4} \text{ mol}$$

$$\text{vol KOH} = \frac{n}{M} = \frac{8.61 \times 10^{-4} \text{ mol}}{0.272 \text{ mol/L}} = 0.00317 \text{ L} \quad \text{or} \quad 3.17 \text{ mL}$$

Hence the total volume of solution at the equivalence point is 30.00 mL + 3.17 mL = 33.17 mL or 0.03317 L. A preliminary value for $[C_7H_5O_2^-]$ at the equivalence point is:

$$[C_7H_5O_2^-] = \frac{n}{V} = \frac{8.61 \times 10^{-4}\,\text{mol}}{0.03317 \text{ L}} = 0.0260 \text{ M}$$

Now the $C_7H_5O_2^-$ ion may act as weak base:

$$C_7H_5O_2^-(aq) + H_2O \rightleftharpoons HC_7H_5O_2(aq) + OH^-(aq) \qquad K_b = \frac{[HC_7H_5O_2][OH^-]}{[C_7H_5O_2^-]}$$

and the position of equilibrium for this reaction will determine $[OH^-]$ and hence $[H^+]$ and pH. The value of K_b for $C_7H_5O_2^-$ is 1.5×10^{-10} (Table 13.2). Thus:

$$1.5 \times 10^{-10} = \frac{[HC_7H_5O_2][OH^-]}{[C_7H_5O_2^-]} \approx \frac{[OH^-]^2}{[C_7H_5O_2^-]}$$

since $[HC_7H_5O_2] \approx [OH^-]$. (See also Examples 13.7 and 13.8.) Since $C_7H_5O_2^-$ is a very weak base, its concentration (0.0260 M) will not change due to the weak base dissociation reaction so we can say $[C_7H_5O_2^-] = 0.0260$ M

Substituting this value for $[C_7H_5O_2^-]$ into the K_a relationship gives:

$$1.5 \times 10^{-10} = \frac{[OH^-]^2}{0.0260} \qquad \text{from which } [OH^-] = 2.0 \times 10^{-6} \text{ M}$$

Use Equations 13.1 and 13.2 to calculate $[H^+]$ and thence pH (using Equation 13.3):

$$[H^+] = \frac{K_w}{[OH^-]} = \frac{1.0 \times 10^{-14}}{2.0 \times 10^{-6}} = 5.0 \times 10^{-9} \text{ M}$$

$$pH = -\log_{10}[H^+] = -\log_{10}(5.0 \times 10^{-9}) = \underline{\textbf{8.30}}$$

UNCLASSIFIED PROBLEMS

54. See Section 14-1. Calculate mol of NH_4Cl, and hence mol NH_4^+, added:

$$\text{mass } NH_4Cl \xrightarrow{\text{MM}} \text{mol } NH_4Cl \xrightarrow{\text{atom ratio}} \text{mol } NH_4^+$$

$$\text{mol } NH_4^+ = 15.0 \text{ g } NH_4Cl \times \frac{1 \text{ mol } NH_4Cl}{53.491 \text{ g } NH_4Cl} \times \frac{1 \text{ mol } NH_4^+}{1 \text{ mol } NH_4Cl} = 0.280 \text{ mol}$$

We require pH = 9.56, so using Equation 13.3 and then Equations 13.1 and 13.2:

$$[H^+] = 10^{-pH} = 10^{-9.56} = 2.8 \times 10^{-10} \text{ M}$$

$$[OH^-] = \frac{K_w}{[H^+]} = \frac{1.0 \times 10^{-14}}{2.8 \times 10^{-10}} = 3.6 \times 10^{-5} \text{ M}$$

From the expression for K_b for NH_3 we can obtain a relationship between mol NH_4^+ present (calculated above) and mol NH_3 required. Note that K_b for NH_3 is 1.8×10^{-5} (Table 13.2):

$$K_b = \frac{[OH^-][NH_4^+]}{[NH_3]} \qquad \text{thus: } [OH^-] = K_b \times \frac{[NH_3]}{[NH_4^+]} = K_b \times \frac{n_{NH_3}}{n_{NH_4^+}}$$

(compare Equations 14.1 and 14.2)

$$[OH^-] = K_b \times \frac{n_{NH_3}}{n_{NH_4^+}}$$

$$3.6 \times 10^{-5} = 1.8 \times 10^{-5} \times \frac{n_{NH_3}}{0.280 \text{ mol}} \qquad \text{so } n_{NH_3} = 5.6 \times 10^{-1} \text{ mol}$$

Use the Ideal Gas Law to obtain the required volume of NH_3:

$$V = \frac{nRT}{P} = \frac{(5.6 \times 10^{-1} \, \text{mol})(0.0821 \, \text{L} \cdot \text{atm/mol} \cdot \text{K})(298 \, \text{K})}{0.981 \, \text{atm}} = \underline{\textbf{14 L}}$$

56. See Section 14-3. The net ionic equation for the reaction between RNH (weak base) and HCl (strong acid) is:

$$\text{RNH(aq)} + \text{H}^+\text{(aq)} \rightarrow \text{RNH}_2{}^+\text{(aq)}$$

(a) mol RNH present $= 0.05990 \, \text{L} \times \dfrac{0.925 \, \text{mol HCl}}{1 \, \text{L}} \times \dfrac{1 \, \text{mol H}^+}{1 \, \text{mol HCl}} \times \dfrac{1 \, \text{mol RNH}}{1 \, \text{mol H}^+} = 0.0554 \, \text{mol}$

molar mass RNH $= \dfrac{2.500 \, \text{g RNH}}{0.0554 \, \text{mol RNH}}$ $\underline{\textbf{45.1 g/mol}}$

(b) The second titration described in the question will take us halfway to the equivalence point, since the volume of HCl solution added is half that needed to reach the equivalence point in the first titration. As seen in Figure 14.12, at halfway to the equivalence point, pH = pK_a for the conjugate acid $\text{RNH}_2{}^+$ or $[\text{H}^+] = K_a$. Since $K_w = [\text{H}^+][\text{OH}^-]$ (Equation 13.1) and $K_w = K_a \times K_b$ (for an acid–base conjugate pair; Equation 13.10), we can say that at halfway to the equivalence point $[\text{OH}^-] = K_b$ for RNH.

$\text{pOH} = 14.00 - 10.77 = 3.23$

$[\text{OH}^-] = 10^{-3.23} = 5.9 \times 10^{-4} \, M$

$K_b = \underline{\textbf{5.9} \times \textbf{10}^{-4}}$

(c) From Equation 13.10: $K_a = \dfrac{K_w}{K_b} = \dfrac{1.0 \times 10^{-14}}{5.9 \times 10^{-4}} = \underline{\textbf{1.7} \times \textbf{10}^{-11}}$

58. See Section 14-3. The reaction between HCl and NaOH is of the strong acid/strong base type, for which the net ionic equation is:

$$\text{H}^+(aq) + \text{OH}^-(aq) \rightarrow \text{H}_2\text{O}$$

mol OH$^-$added $= 0.0788 \, \text{mL} \times \dfrac{4.85 \, \text{mol NaOH}}{1 \, \text{L}} \times \dfrac{1 \, \text{mol OH}^-}{1 \, \text{mol NaOH}} = 0.382 \, \text{mol}$

mol H$^+$ present $= 0.382 \, \text{mol} \times \dfrac{1 \, \text{mol H}^+}{1 \, \text{mol OH}^-} = 0.382 \, \text{mol}$

$[\text{H}^+] = \dfrac{n}{V} = \dfrac{0.382 \, \text{mol}}{0.07500 \, \text{L}} = 5.10 \, M = [\text{HCl}]$ in the diluted solution

Recall (page 247 and Equation 10.1) that for any dilution problem: (mol solute in concentrated solution) = (mol solute in dilute solution); that is:

$$M_c V_c = M_d V_d$$

$$(6.00\ M)(0.35000\ L) = (5.10\ M) V_d \quad \rightarrow \quad \text{vol dilute solution} = V_d = 0.412\ L = 412\ mL$$

Thus, volume H_2O added = 412 mL − 350.00 mL = **62 mL**

60. See Chapters 8, 12, and 13.

(a) In order to determine the pH of a solution of HF at 100°C, we must know the value of K_a at that temperature; the variation in equilibrium constants with temperature is given by the van't Hoff equation (Equation 12.5), for which a value of $\Delta H°$ is required. Thus we need $\Delta H°$ for the following reaction:

$$HF(aq) \rightleftharpoons H^+(aq) + F^-(aq)$$

$$\Delta H° = \Sigma\,\Delta H_f°\text{ (products)} - \Sigma\,\Delta H_f°\text{ (reactants)}$$

$$= [\Delta H_f°\ H^+(aq) + \Delta H_f°\ F^-(aq)] - [\Delta H_f°\ HF(aq)]$$

$$\Delta H° = [(1\ \text{mol})(0\ kJ/mol) + (1\ \text{mol})(-332.6\ kJ/mol)] - [(1\ \text{mol})(-320.1\ kJ/mol)]$$

$$\Delta H° = -12.5\ kJ = -12.5 \times 10^3\ J$$

From Table 13.2, K_a for HF at 25°C is $6.9 \times 10^{-4} = K_1$

$$\ln\frac{K_2}{K_1} = \frac{\Delta H°}{R}\left[\frac{1}{T_1} - \frac{1}{T_2}\right]$$

$$\ln\frac{K_2}{6.9 \times 10^{-4}} = \frac{-12.5 \times 10^3}{8.3145\ J/mol\cdot K}\left[\frac{1}{298\ K} - \frac{1}{373\ K}\right]$$

$$\ln K_2 - \ln(6.9 \times 10^{-4}) = \ln K_2 - (-7.28) = -1.01$$

$$\ln K_2 = -8.29 \rightarrow K_2 = e^{-8.29} = 2.5 \times 10^{-4} = K_a \text{ at } 100°C$$

Finally use Equation 13.5 (and see also Examples 13.8 and 13.9):

$$K_a = \frac{[H^+][F^-]}{[HF]} \approx \frac{[H^+]^2}{[HF]} \qquad \text{so } [H^+]^2 = K_a[HF] = (2.5 \times 10^{-4})(0.100)$$

$$[H^+] = 5.0 \times 10^{-3}\ M \qquad pH = -\log_{10}[H^+] = -\log_{10}(5.0 \times 10^{-3}) = \underline{\mathbf{2.30}}$$

(b) Repeat the last two lines of part (a) to obtain the pH at 25°C:

$$K_a = \frac{[H^+][F^-]}{[HF]} \approx \frac{[H^+]^2}{[HF]} \qquad \text{so } [H^+]^2 = K_a[HF] = (6.9 \times 10^{-4})(0.100)$$

$$[H^+] = 8.3 \times 10^{-3}\ M \qquad pH = -\log_{10}[H^+] = -\log_{10}(8.3 \times 10^{-3}) = \underline{\mathbf{2.08}}$$

CONCEPTUAL PROBLEMS

62. See Section 14-3.

(a) 2 mol of H^+ will react with 2 mol of X^- to give 2 mol of HX:

$\{4HX \ + \ 5X^-\}$ $+$ $\{2H^+\}$ $\rightarrow$ $\{6HX \ + \ 3X^-\}$

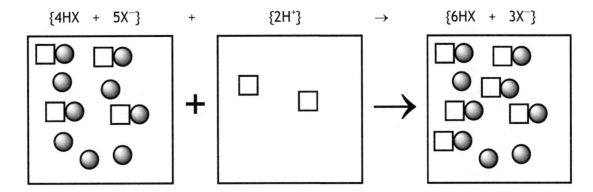

(b) 2 mol of OH^- will react with 2 mol of HX forming 2 mol X^- and 2 mol H_2O (not shown).

$\{4HX \ + \ 5X^-\}$ $+$ $\{2OH^-\}$ $\rightarrow$ $\{2HX \ + \ 7X^-\} + 2H_2O$

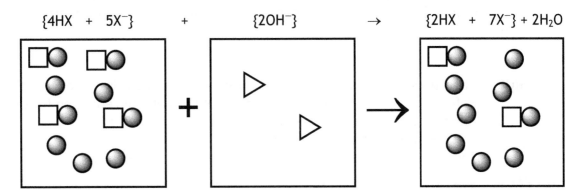

(c) 5 mol of OH^- will react with the 4 mol HX forming 4 mol X^- and 4 mol H_2O (not shown); 1 mol OH^- (excess) will remain.

$\{4HX \ + \ 5X^-\}$ $+$ $\{5OH^-\}$ $\rightarrow$ $\{9X^- \ + \ 1OH^-\} + 4H_2O$

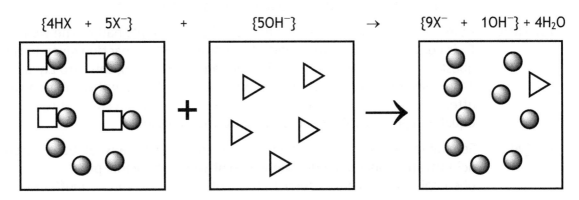

64. See Section 14-1.

Consider the system present in each beaker:

   A   [HA] = [NaA] = [A⁻]. The Henderson-Hasselbalch equation shows that pH = pK_a.

   B   The same system as Beaker A, only more dilute; pH = pK_a.

   C   HCl reacts with NaA forming HA: 0.150 mol HA; 0.050 mol A⁻; 0 mol HCl remain.

   D   NaOH reacts with HA forming NaA: 0 mol HA; 0.200 mol NaA; 0 mol NaOH remain.

   E   HCl and NaOH in equimolar amounts react to form water and NaCl; pH = 7.00.

   (a) **EQ** For both beakers, pH = pK_a for the same acid, HA.

   (b) **GT** For Beaker A, pH = pK_a. For Beaker C, pH = pK_a + log₁₀{[A⁻]/[HA]}. Since [A⁻]/[HA] is less than 1 in Beaker C, log₁₀{[A⁻]/[HA]} < 0; thus pH will be less than pK_a, making the pH in Beaker C less than the pH in Beaker A.

   (c) **GT** Beaker D contains only NaA, the salt of a weak acid. These salts are basic and have a pH > 7.00.

   (d) **MI** The pK_a of weak acids vary (see Table 14.1). One needs to know the value of the pK_a before a comparison to a known pH can be made.

   (e) **LT** Beaker A has pH = pK_a. Beaker D corresponds to Beaker A with some strong base added. It will therefore be more basic than the HA/A⁻ buffer system originally present.

66. (a) **TRUE.** The fluoride ion F⁻ (from NaF) is a weak base and H_2SO_4 is a strong acid. Referring to Table 4.2, the general form of the net ionic equation for the reaction between a strong acid (H⁺) and a weak base (B) is:

   $$H^+(aq) + B(aq) \rightleftharpoons HB^+(aq)$$

   Taking F⁻ as the weak base gives the reaction in the question.

   (b) **FALSE.** See part (a) above: the fluoride ion is a weak base.

   (c) **FALSE.** A buffer requires the presence of both members of a weak acid—weak base conjugate pair. According to the reaction in the question, at the equivalence point, all of the weak base F⁻ will have been consumed and converted to the weak acid HF. Hence the solution cannot act as a buffer.

   (d) **TRUE.** Refer to Figure 14.12. At the half-neutralization point, half of the weak base F⁻ is converted to the weak acid HF according to the reaction given in the question. Thus, at this point [F⁻] = [HF] and substituting this into Equation 14.1 gives pH = pK_a (see discussion on page 362). Diluting this solution will affect [F⁻] and [HF] equally, so the concentration of the two will remain equal and pH will still equal pK_a despite the dilution.

   (e) **FALSE.** See Table 14.3. Methyl red is a preferred indicator for a weak base—strong acid titration since the equivalence point will be to acidic pH; phenolphthalein changes color to basic pH and hence will be ineffective in this titration.

68. One liter of each solution contains:

HCl: 2.0×10^{-3} mol of ions (H^+, Cl^-); since it is a strong acid, it completely dissociates

$HCHO_2$: 6.0×10^{-3} mol of molecules ($HCHO_2$); since it is a weak acid, it hardly dissociates — there will also be some ions present, but very few.

$C_6H_5NH_3^+$: 8.0×10^{-2} mol of ions ($C_6H_5NH_3^+$ cations and the corresponding (unspecified) anion)

Since the three solutions differ in the number of particles present, one could determine the value of a colligative property such as freezing point depression (see Chapter 10): the solution with the most particles would have the lowest freezing point. Likewise, the conductivity of the three solutions will be different: the one with the greatest number of ions will have the highest conductivity.

Alternatively, titration with a strong base such as NaOH would afford solutions with different pH at the equivalence point: for HCl the titration will form $NaCl(aq)$, which is neutral; for $HCHO_2$ the titration will form $NaCHO_2(aq)$, which is a weakly basic salt; for $C_6H_5NH_3^+$ the titration will form $C_6H_5NH_2(aq)$, a weak base (and note that it will also be more concentrated than the solution of $NaCHO_2$). The order of increasing pH at the equivalence point will be:

$$HCl < HCHO_2 < C_6H_5NH_3^+$$

CHALLENGE PROBLEMS

70. See Section 14-3. The reaction between HF and NaOH is of the weak acid + strong base type:

$$HF(aq) + OH^-(aq) \rightleftharpoons H_2O + F^-(aq)$$

At the one-third neutralization point, one-third of the HF has been converted to F^- and two-thirds of the HF remains, so $[F^-] = \frac{1}{2}[HF]$. Substituting this into the expression for K_a gives:

$$K_a = 6.9 \times 10^{-4} = \frac{[H^+][F^-]}{[HF]} \approx \frac{[H^+] \times \frac{1}{2}[HF]}{[HF]} = \frac{1}{2}[H^+] \quad \text{so } [H^+] = 1.4 \times 10^{-3} \, M$$

$$pH = -\log_{10}[H^+] = -\log_{10}(1.4 \times 10^{-3}) = \underline{\mathbf{2.85}}$$

71. See Sections 14-1 and 14-3. Write the net ionic equation for the reaction. Calculate mol of acetate (from the reaction stoichiometry) and $[H^+]$ (from pH). From those two with Equation 14.2, determine mol of acetic acid still in solution, then calculate the mol of acetic acid added and finally the volume of acetic acid.

$$HC_2H_3O_2(aq) + OH^-(aq) \rightarrow C_2H_3O_2^-(aq) + H_2O$$

$$\text{mol } C_2H_3O_2^- = 0.1000 \text{ L NaOH} \times \frac{1.25 \text{ mol NaOH}}{1\text{L}} \times \frac{1 \text{ mol OH}^-}{1 \text{ mol NaOH}} \times \frac{1 \text{ mol } C_2H_3O_2^-}{1 \text{ mol OH}^-}$$

$$= 0.125 \text{ mol}$$

$$[H^+] = 10^{-pH} = 10^{-4.20} = 6.3 \times 10^{-5} \text{ M}$$

$$[H^+] = K_a \times \frac{n_{HC_2H_3O_2}}{n_{C_2H_3O_2^-}} \qquad 6.3 \times 10^{-5} = 1.8 \times 10^{-5} \times \frac{n_{HC_2H_3O_2}}{0.125}$$

$$n_{HC_2H_3O_2} = 0.44 \text{ mol}$$

This is the number of moles of $HC_2H_3O_2$ present after reaction with NaOH has occurred. The total moles of acid that must have been present before reaction is:

$$n_{total} = n_{C_2H_3O_2^-} + n_{HC_2H_3O_2} = 0.125 \text{ mol} + 0.44 \text{ mol} = 0.56 \text{ mol}$$

Note that this solution contains approximately equal amounts of acid ($HC_2H_3O_2$) and base ($C_2H_3O_2^-$), and is therefore a buffer.

Finally calculate the volume of glacial acetic acid required to provide 0.56 mol $HC_2H_3O_2$:

$$0.56 \text{ mol} \times \frac{60.05 \text{ g } HC_2H_3O_2}{1 \text{ mol } HC_2H_3O_2} \times \frac{100 \text{ g solution}}{98 \text{ g } HC_2H_3O_2} \times \frac{1 \text{ mL}}{1.0542 \text{ g solution}} = \underline{\underline{33 \text{ mL}}}$$

72. See Sections 14-1 and 14-3. After solution A is titrated to its equivalence point, all of the acid "HB" will have been converted to the conjugate base "B^-". Also, since solutions A and B were initially identical, when they are combined $[HB] = [B^-]$, and so $[HB]/[B^-] = 1$. Calculate $[H^+]$ and hence pH of this solution, and K_a for the acid. From K_a and pH of the original solution, one can calculate $[HB]$ and thus the molar mass.

For the final solution:

$$[H^+] = 10^{-pH} = 10^{-4.26} = 5.5 \times 10^{-5} \text{ M}$$

$$[H^+] = K_a \times \frac{[HB]}{[B^-]} \qquad 5.5 \times 10^{-5} = K_a \times 1 \quad \text{so } K_a = 5.5 \times 10^{-5}$$

For the initial solution:

$$[H^+] = 10^{-pH} = 10^{-2.56} = 2.8 \times 10^{-3} \text{ M}$$

$$K_a = 5.5 \times 10^{-5} = \frac{[H^+][B^-]}{[HB]} = \frac{(2.8 \times 10^{-3})(2.8 \times 10^{-3})}{[HB]} \qquad \text{so } [HB] = 0.14 \text{ M}$$

$$\text{mol HB} = 0.2500 \text{ L} \times \frac{0.14 \text{ mol HB}}{1\text{L}} = 0.035 \text{ mol HB}$$

$$\text{molar mass of HB} = \frac{\text{mass HB}}{\text{mol HB}} = \frac{4.00 \text{ g}}{0.035 \text{ mol}} = \underline{\mathbf{1.1 \times 10^2 \text{ g/mol}}}$$

73. See Section 14-1. Calculate the required mole ratio of NH_4^+ to NH_3 needed to make a buffer with pH = 6.50. K_a for NH_4^+ is 5.6×10^{-10} (Table 13.2).

    pH = 6.50   so $[H^+] = 10^{-pH} = 10^{-6.50} = 3.2 \times 10^{-7}$ M

    Now substitute $[H^+]$ and K_a into Equation 14.2:

    $$[H^+] = K_a \times \frac{n_{NH_4^+}}{n_{NH_3}} \qquad\qquad 3.2 \times 10^{-7} = 5.6 \times 10^{-10} \times \frac{n_{NH_4^+}}{n_{NH_3}}$$

    $$\frac{n_{NH_4^+}}{n_{NH_3}} = 5.7 \times 10^2$$

    A buffer requires roughly equal amounts of a weak acid and its conjugate base, whereas the ratio $\dfrac{n_{NH_4^+}}{n_{NH_3}}$ required here to achieve a pH of 6.50 is around 570 and therefore this solution will not act as a buffer.

74. See Sections 14-1 and 14-3, and Example 14.6.

    (a) $K_a = 6.0 \times 10^{-4} = \dfrac{[H^+][NO_2^-]}{[HNO_2]} \approx \dfrac{[H^+]^2}{[HNO_2]} = \dfrac{[H^+]^2}{1.000} \qquad \rightarrow [H^+] = 0.024$ M

    pH $= -\log_{10}[H^+] = -\log_{10}(0.024) = \underline{\mathbf{1.62}}$

    (b) See Figure 14.10: At half-neutralization, pH = pK_a:

    pH $= -\log_{10}K_a = -\log_{10}(6.0 \times 10^{-4}) = \underline{\mathbf{3.22}}$

    (c) At the equivalence point, all of the HNO_2 and all of the added OH^- have been consumed according to the following (weak acid/strong base) net ionic equation:

    $$HNO_2(aq) + OH^-(aq) \rightarrow NO_2^-(aq) + H_2O$$

    The NO_2^- anion is weakly basic:

    $$NO_2^-(aq) + H_2O \rightleftharpoons HNO_2(aq) + OH^-(aq)$$

    Calculate the volume of NaOH added and hence the total volume of the solution at the equivalence point, then calculate $[NO_2^-]$. Use this to construct an equilibrium table (see Example 13.11(a)) and determine $[OH^-]$ and finally pH.

$$\text{V HNO}_2 \xrightarrow{\ M\ } \text{mol HNO}_2 \xrightarrow{\text{atom ratio}} \text{mol H}^+ \xrightarrow{\text{stoichiometric ratio}} \text{mol OH}^-$$

$$\xrightarrow{\text{atom ratio}} \text{mol NaOH} \xrightarrow{\ M\ } \text{V NaOH}$$

$$\text{mol NaOH needed } = 0.0500 \text{ L HNO}_2 \times \frac{1.000 \text{ mol HNO}_2}{1 \text{ L}} \times \frac{1 \text{ mol H}^+}{1 \text{ mol HNO}_2} \times \frac{1 \text{ mol OH}^-}{1 \text{ mol H}^+}$$

$$\times \frac{1 \text{ mol NaOH}}{1 \text{ mol OH}^-} = 0.0500 \text{ mol}$$

$$\text{vol NaOH} = \frac{n}{M} = \frac{0.0500 \text{ mol}}{0.850 \text{ mol/L}} = 0.0588 \text{ L} \quad \text{or} \quad 58.8 \text{ mL}$$

$$\text{total volume} = 0.0500 \text{ L HNO}_2 + 0.0588 \text{ L NaOH} = 0.1088 \text{ L}$$

$$[\text{NO}_2^-] \text{ at equivalence point} = \frac{0.0500 \text{ mol NO}_2^-}{0.1088 \text{ L}} = 0.460 \text{ } M$$

Now proceed as in Example 13.11(a). Set up an equilibrium table and determine $[\text{OH}^-]$ and hence $[\text{H}^+]$ and pH. In completing the equilibrium table, use $[\text{NO}_2^-]_0 = 0.460 \text{ } M$ (calculated above) and initially we assume there is no ionization, so $[\text{HNO}_2]_0 = [\text{OH}^-]_0 = 0$. Let $\Delta[\text{OH}^-] = x$:

	$\text{NO}_2^-(aq)$	$+$	H_2O	$\rightleftharpoons$	$\text{HNO}_2(aq)$	$+$	$\text{OH}^-(aq)$
$[\ ]_0$ (M)	0.460				0		0
$\Delta[\ ]$ (M)	$-x$				$+x$		$+x$
$[\ ]_{eq}$ (M)	$0.460 - x$				x		x

Now substitute the equilibrium concentrations of each species into the K_b expression:

$$K_b = 1.7 \times 10^{-11} = \frac{[\text{HNO}_2][\text{OH}^-]}{[\text{NO}_2^-]} = \frac{(x)(x)}{(0.460 - x)}$$

Assume $x \ll 0.460$:

$$1.7 \times 10^{-11} \approx \frac{x^2}{0.460} \quad \text{so} \quad x^2 = (0.460)(1.7 \times 10^{-11}) = 7.8 \times 10^{-12}$$

$$x = 2.8 \times 10^{-6}$$

Check the assumption that the value of x is indeed much less that 0.460: the % ionization = $[\text{OH}^-]_{eq}/[\text{NO}_2^-]_0 \times 100 = (2.8 \times 10^{-6})/(0.460) \times 100 = 0.00061\% < 5\%$ (see page 345), so the assumption made in solving the above equation was valid.

Note that $x = [\text{OH}^-]_{eq}$, so we can say that $[\text{OH}^-] = 2.8 \times 10^{-6} \text{ } M$ and hence, using Equations 13.1 and 13.2:

$$[H^+] = \frac{K_w}{[OH^-]} = \frac{1.0 \times 10^{-14}}{2.8 \times 10^{-6}} = 3.6 \times 10^{-9} \ M$$

Finally, use Equation 13.3 to calculate pH:

$$pH = -\log_{10}[H^+] = -\log_{10}(3.6 \times 10^{-9}) = \underline{\mathbf{8.44}}$$

(d) As calculated in part (c) above, 0.0588 L NaOH solution is needed to reach the equivalence point. When 0.10 mL less than that are added (0.0587 L NaOH solution), and the total volume of the solution is 0.1088 L − 0.00010 L = 0.1087 L

$$\text{mol } OH^- \text{ added} = 0.0587 \ L \ NaOH \times \frac{0.850 \ \text{mol NaOH}}{1 \ L} \times \frac{1 \ \text{mol } OH^-}{1 \ \text{mol NaOH}} = 0.0499 \ \text{mol}$$

mol HNO_2 remaining = 0.05000 mol − 0.0499 mol = 0.0001 mol
mol NO_2^- produced = mol OH^- added = 0.0499 mol

$$[HNO_2] = \frac{n}{V} = \frac{0.0001 \ \text{mol } HNO_2}{0.1087 \ L} = 0.0009 \ M$$

$$[NO_2^-] = \frac{n}{V} = \frac{0.0499 \ \text{mol } NO_2^-}{0.1087 \ L} = 0.459 \ M$$

The pH is determined by the ionization of HNO_2 in the presence of NO_2^-; the composition of the mixture is such that it will not act as a buffer (similar quantities of each species are not present), so the situation must be treated via a full equilibrium calculation. Set up an equilibrium table and determine $[H^+]$ and hence pH. In completing the table, use $[HNO_2]_0 = 0.0009 \ M$ and $[NO_2^-] = 0.0499 \ M$ (calculated above) and initially we assume there is no ionization, so $[H^+] = 0$. Let $\Delta[H^+] = x$:

	$HNO_2(aq)$	$\rightleftharpoons$	$H^+(aq)$	+	$NO_2^-(aq)$
[ ]$_0$ (M)	0.0009		0		0.459
Δ[ ] (M)	−x		+x		+x
[ ]$_{eq}$ (M)	0.0009 − x		x		0.459 + x

Now substitute the equilibrium concentrations of each species into the K_a expression:

$$K_a = 6.0 \times 10^{-4} = \frac{[H^+][NO_2^-]}{[HNO_2]} = \frac{(x)(0.459 + x)}{(0.0009 - x)}$$

Assume $x << 0.0009$:

$$6.0 \times 10^{-4} \approx \frac{x(0.459)}{0.0009} \quad \text{so} \quad (0.459)x = (0.0009)(6.0 \times 10^{-4}) \quad \text{and} \quad x = 1.2 \times 10^{-6}$$

Check the assumption that the value of x is indeed much less that 0.0009: the % ionization = $[H^+]_{eq}/[HNO_2]_0 \times 100 = (1.2 \times 10^{-6})/(0.0009) \times 100 = 0.1\% < 5\%$ (see page

416), so the assumption made in solving the above equation was valid.
Note that $x = [H^+]_{eq}$ so now use Equation 13.3 to calculate pH:

$$pH = -\log_{10}[H^+] = -\log_{10}(1.2 \times 10^{-6}) = \textbf{5.92}$$

(e) Beyond the equivalence point, any further added NaOH solution is in excess. If 0.10 mL excess are added then the total added is 0.0589 L, and the total volume of the solution is 0.1088 L + 0.00010 L = 0.1089 L. Since NaOH is a strong base it will dominate and $[OH^-]$ will dictate $[H^+]$ and hence pH of the solution. Calculate how many mol excess OH^- are present and hence $[OH^-]$, then $[H^+]$ and hence pH

$$\text{mol } OH^- \text{ excess} = 0.00010 \text{ L NaOH} \times \frac{0.850 \text{ mol NaOH}}{1 \text{ L}} \times \frac{1 \text{ mol } OH^-}{1 \text{ mol NaOH}}$$

$$= 8.5 \times 10^{-5} \text{ mol}$$

$$[OH^-] = \frac{n}{V} = \frac{8.5 \times 10^{-5} \text{ mol } OH^-}{0.1089 \text{ L}} = 7.8 \times 10^{-4} \text{ M}$$

$$[H^+] = \frac{K_w}{[OH^-]} = \frac{1.0 \times 10^{-14}}{7.8 \times 10^{-4}} = 1.3 \times 10^{-11} \text{ M}$$

Finally, use Equation 13.3 to calculate pH:

$$pH = -\log_{10}[H^+] = -\log_{10}(1.3 \times 10^{-11}) = \textbf{10.89}$$

(f) To complete the graph, include one more data point, with (say) 10.00 mL NaOH in excess having been added. Repeat the calculation in part (e) above using this value of added NaOH. If 10.00 mL excess is added then the total volume NaOH solution added is 0.0688 L, and the total volume of the solution is 0.1088 L + 0.01000 L = 0.1188 L. Again $[OH^-]$ will dictate $[H^+]$ and hence pH of the solution. Calculate how many mol excess OH^- are present and hence $[OH^-]$, then $[H^+]$ and hence pH

$$\text{mol } OH^- \text{ excess} = 0.01000 \text{ L NaOH} \times \frac{0.850 \text{ mol NaOH}}{1 \text{ L}} \times \frac{1 \text{ mol } OH^-}{1 \text{ mol NaOH}} = 0.0085 \text{ mol}$$

$$[OH^-] = \frac{n}{V} = \frac{0.0085 \text{ mol } OH^-}{0.1188 \text{ L}} = 0.072 \text{ M}$$

$$[H^+] = \frac{K_w}{[OH^-]} = \frac{1.0 \times 10^{-14}}{0.072} = 1.4 \times 10^{-13} \text{ M}$$

Finally, use Equation 13.3 to calculate pH:

$$pH = -\log_{10}[H^+] = -\log_{10}(1.4 \times 10^{-13}) = \textbf{12.85}$$

A plot of pH *versus* volume NaOH added is shown below.

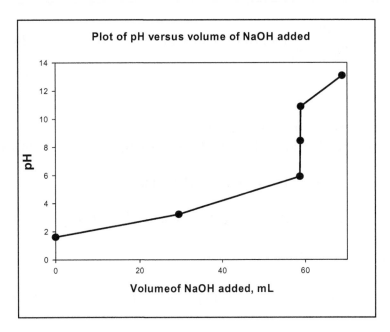

Volume NaOH added (mL)	pH
0.00	1.62
29.4	3.22
58.7	5.92
58.8	8.45
58.9	10.89
68.8	12.85

75. See Figure 14.13 and the discussion around it (pages 380–381). The overall neutralization reaction between H_2B and NaOH will be:

$$H_2B(aq) + 2NaOH(aq) \rightleftharpoons Na_2B(aq) + 2H_2O$$

From the given information we can calculate how many mol H_2B are present in the 25 mL sample that was tirated with NaOH:

$$\text{V NaOH} \xrightarrow{M} \text{mol NaOH} \xrightarrow{\text{stoichiometry}} \text{mol } H_2B$$

$$\text{mol } H_2B \text{ present} = 48.5 \text{ mL NaOH} \times \frac{1L}{1000 \text{ mL}} \times \frac{0.425 \text{ mol NaOH}}{1L} \times \frac{1 \text{ mol } H_2B}{2 \text{ mol NaOH}}$$

$$= 1.03 \times 10^{-2} \text{ mol } H_2B$$

and hence in the original 150.00 mL solution that was prepared there will be

$$(150.00/25) \times (1.03 \times 10^{-2} \text{ mol}) = 6.18 \times 10^{-2} \text{ mol } H_2B.$$

This solution was prepared using 10.00 g of the hydrate $H_2B \cdot xH_2O$. Since the molar mass of H_2B is 126 g/mol (given) and the molar mass of H_2O is 18 g/mol, the molar mass of the hydrate will be $(126 + 18x)$ g/mol. Given this molar mass and the 10.00 g mass of hydrate used, mol of hydrate used to prepare the original solution will be:

$$\text{mol hydrate} = 10.00 \text{ g hydrate} \times \frac{1 \text{ mol hydrate}}{(126 + 18x) \text{ g hydrate}} = \frac{10}{(126 + 18x)} \text{ mol hydrate}$$

Since 1 mol of hydrate contains 1 mol H_2B, we can say that this is also the mol H_2B present in the original solution. Using the results of the titration we calculated (see above) that the original solution contained 6.18×10^{-2} mol H_2B. Thus:

$$\frac{10}{(126+18x)} \text{ mol } H_2B = 6.18 \times 10^{-2} \text{ mol } H_2B \qquad \rightarrow \qquad 126 + 18x = 162$$

$$\rightarrow \qquad 18x = 36 \qquad \rightarrow \qquad \underline{\mathbf{x = 2}}$$

76. See Section 13-4 and Example 13.9.
   (a) In its first ionization, H_2SO_4 is a strong acid which ionizes completely:

$$H_2SO_4(aq) \;\rightarrow\; H^+(aq) + HSO_4^-(aq)$$

Thus $[H^+] = [H_2SO_4] = 0.1500$ M and pH $= -\log_{10}[H^+] = -\log_{10}(0.1500) = \underline{\mathbf{0.8239}}$

   (b) For the second ionization:

$$HSO_4^-(aq) \;\rightleftharpoons\; H^+(aq) + SO_4^{2-}(aq) \qquad\qquad K_a = \frac{[H^+][SO_4^{2-}]}{[HSO_4^-]} = 1.1 \times 10^{-2}$$

After the first ionization $[H^+] = [HSO_4^-] = 0.1500$ M. Set up an equilibrium table with $[H^+]_0 = [HSO_4^-]_0 = 0.1500$ M, and $[SO_4^{2-}]_0 = 0$; and let $\Delta[H^+] = x$:

	$HSO_4^-(aq)$	$\rightleftharpoons$	$H^+(aq)$	+	$SO_4^{2-}(aq)$
[ ]$_0$ (M)	0.1500		0.1500		0
Δ[ ] (M)	$-x$		$+x$		$+x$
[ ]$_{eq}$ (M)	$0.1500 - x$		$0.1500 + x$		x

Now substitute the equilibrium concentrations of each species into the K_a expression:

$$K_a = 1.1 \times 10^{-2} = \frac{[H^+][SO_4^{2-}]}{[HSO_4^-]} = \frac{(0.1500+x)(x)}{(0.1500-x)}$$

$$1.1 \times 10^{-2}(0.1500 - x) = x(0.1500 + x)$$

$$(1.6 \times 10^{-3}) - (1.1 \times 10^{-2})x = 0.1500x + x^2$$

$$x^2 + 0.16x - 1.6 \times 10^{-3} = 0$$

This must be solved using the quadratic formula:

$$x = \frac{-b \pm \sqrt{b^2 - 4ac}}{2a} = \frac{-(0.16) \pm \sqrt{(0.16)^2 - 4(1)(-0.0016)}}{2(1)}$$

$$x = +0.01 \;\text{ or }\; -0.17$$

The negative root of the equation is physically impossible since x is a concentration; therefore $x = 0.01$ and $[H^+]_{eq} = 0.1500 + x = 0.16$ M.

Hence, pH $= -\log_{10}[H^+] = -\log_{10}(0.16) = \underline{\mathbf{0.80}}$

77. The reaction between the weak base (call it B) and the strong acid (a source of H^+) will proceed according to the following net ionic equation:

$$H^+(aq) + B(aq) \rightleftharpoons BH^+(aq)$$

Since 21.0 mL of strong acid is needed for neutralization, the titration with 7.00 mL of strong acid represents one-third neutralization. Thus one third of the base has been converted to the conjugate acid BH^+, and two-thirds of the base remain; so we can say:

$$\frac{n_{HB^+}}{n_B} = \frac{\frac{1}{3}}{\frac{2}{3}} = 0.5$$

Note that we have significant quantities of weak base B and its conjugate acid BH^+ present, so the mixture can be treated as a buffer. At this point we also know that pH = 8.95, so $[H^+] = 10^{-8.95} = 1.1 \times 10^{-9}$ M. Substitute these into Equation 14.2 to obtain K_a for the acid component of the mixture and then use Equation 13.10 to calculate K_b for the base.

$$[H^+] = K_a \times \frac{n_{HB^+}}{n_B} \qquad \rightarrow \qquad 1.1 \times 10^{-9} = K_a \times 0.5 \qquad \rightarrow \qquad K_a = 2.2 \times 10^{-9}$$

$$K_b = \frac{K_w}{K_a} = \frac{1.0 \times 10^{-14}}{2.2 \times 10^{-9}} = \underline{\mathbf{4.5 \times 10^{-6}}}$$

78. See Section 14-1.

For the initial buffer, $[NH_3] = 0.150$ M and $[NH_4^+]$ is:

$$[NH_4^+] = \frac{10.0 \text{ g NH}_4\text{Cl}}{1 \text{ L}} \times \frac{1 \text{ mol NH}_4\text{Cl}}{53.49 \text{ g NH}_4\text{Cl}} \times \frac{1 \text{ mol NH}_4^+}{1 \text{ mol NH}_4\text{Cl}} = 0.187 \text{ M}$$

Thus, using Equation 14.1 $[H^+]$ in the buffer initially is:

$$[H^+] = K_a \times \frac{[NH_4^+]}{[NH_3]} = (5.6 \times 10^{-10}) \times \frac{0.187}{0.150} = 7.0 \times 10^{-10} \text{ M}$$

We are required to determine the quantity of NaOH needed to increase the pH of the buffer by one unit: this corresponds to a decrease in $[H^+]$ of a factor of 10. In other words, we must produce $[H^+] = 7.0 \times 10^{-11}$ M.
Addition of NaOH to the buffer will convert NH_4^+ to NH_3:

$$NH_4^+(aq) + OH^-(aq) \rightarrow NH_3(aq) + H_2O$$

By stoichiometry, if $[NH_4^+]$ decreases by x then $[NH_3]$ must increase by x. Thus, using Equation 14.1:

$$[H^+] = K_a \times \frac{[NH_4^+]}{[NH_3]} \qquad \rightarrow 7.0 \times 10^{-11} = 5.6 \times 10^{-10} \times \frac{0.187 - x}{0.150 + x}$$

$$\frac{0.187 - x}{0.150 + x} = 0.12 \qquad \rightarrow 0.187 - x = 0.12(0.150 + x) = 0.018 + 0.12x$$

$$1.12x = 0.169 \qquad \rightarrow x = 0.151$$

The stoichiometry of the reaction also shows that x is the quantity of OH^- consumed by reaction with NH_4^+. Therefore, $[OH^-] = 0.151$ M prior to reaction with NH_4^+. Use this to determine the corresponding mass NaOH that must be added to the 1.00 L of buffer in order to produce the initial concentration of $[OH^-] = 0.151$ M.

$$\text{mass NaOH} = \frac{0.151\,\text{mol OH}^-}{1\,\text{L}} \times \frac{1\,\text{mol NaOH}}{1\,\text{mol OH}^-} \times \frac{40.00\,\text{g NaOH}}{1\,\text{mol NaOH}} = \underline{\textbf{6.04 g NaOH}}$$

79. The reaction between HNO_3 and NaF is between a strong acid (HNO_3) and a weak base (F^-), for which the net ionic equation (Table 4.2) in this case will be:

$$H^+(aq) + F^-(aq) \rightarrow HF(aq)$$

Recall that a buffer requires both members of a weak acid–weak base conjugate pair to be present, and a buffer can be prepared from HNO_3 and NaF if some F^- is converted to HF according to the above equation, resulting in a mixture of HF (weak acid) and F^- (conjugate base).

Our required buffer has pH = 4.68, for which (using Equation 13.3) is $[H^+] = 10^{-pH} = 10^{-4.68} = 2.1 \times 10^{-5}$ M. The value of K_a for HF is 6.9×10^{-4} (Table 13.2). Thus, using Equation 14.2 the required molar ratio of HF to F^- can be calculated:

$$[H^+] = K_a \times \frac{n_{HF}}{n_{F^-}} \qquad \rightarrow \qquad \frac{n_{HF}}{n_{F^-}} = \frac{[H^+]}{K_a} = \frac{2.1 \times 10^{-5}}{6.9 \times 10^{-4}} = 0.030$$

Now set up a table as in Example 14.3. Initially, mol H^+ present is:

$$\text{mol H}^+ = \frac{0.150\,\text{mol HNO}_3}{1\,\text{L}} \times 70.00\,\text{mL} \times \frac{1\,\text{L}}{1000\,\text{mL}} \times \frac{1\,\text{mol H}^+}{1\,\text{mol HNO}_3} = 0.0105\,\text{mol H}^+$$

and since we know that we must make a buffer, this H^+ from the strong acid must be completely consumed: H^+ is the limiting reactant. Let the original mol F^- be x and at the start HF is not present.

	n_{H^+}	n_{F^-}	n_{HF}
Original	0.0105	x	0
Change	−0.0105	−0.0105	+0.0105
Final	0	$x − 0.0105$	0.0105

From above, $\dfrac{n_{HF}}{n_{F^-}}$ = 0.030; substitute for n_{F^-} and n_{HF} from the table and solve for x.

$$\frac{n_{HF}}{n_{F^-}} = \frac{0.0105}{x − 0.0105} = 0.030 \qquad\qquad \rightarrow 0.0105 = 0.030(x − 0.0105) = 0.030x − 0.00032$$

$$\rightarrow 0.030x = 0.0108 \qquad\qquad \rightarrow x = 0.36$$

Since x = mol F^- present initially, we can use this to find mol NaF present initially and hence mass NaF needed to make the buffer.

$$\text{mass NaF} = 0.36\ \text{mol } F^- \times \frac{1\,\text{mol NaF}}{1\,\text{mol } F^-} \times \frac{41.988\ \text{g NaF}}{1\,\text{mol NaF}} = 15\ \text{g}$$

Thus **15 g NaF should be added to make the required buffer.**

80. (a) See Figure 14.8. Since $HClO_4$ is a strong acid and KOH is a strong base, the equivalence point is at pH = 7. Therefore, ___YES the student went beyond the equivalence point.___

    (b) This part of the problem is related to Example 14.7(c). When the pH has reached 12.39, pOH = 14.00 − 12.39 = 1.61 (see page 334) and hence (using Equation 13.4) $[OH^-] = 10^{-1.61} = 0.0245\ M$. The total volume of solution at that point is (25.00 + 42.35)mL = 67.35 mL. Thus:

$$\text{mol } OH^-\ (\text{at pH 12.39}) = 67.35\ \text{mL} \times \frac{1\,L}{1000\ \text{mL}} \times \frac{0.0245\ \text{mol } OH^-}{1\,L}$$

$$= 1.65 \times 10^{-3}\ \text{mol } OH^-$$

We also know mol OH^- added during the experiment:

total mol OH^- added (from KOH)

$$= 42.35\ \text{mL} \times \frac{1\,L}{1000\ \text{mL}} \times \frac{0.731\ \text{mol KOH}}{1\,L} \times \frac{1\,\text{mol } OH^-}{1\,\text{mol KOH}}$$

$$= 3.10 \times 10^{-2}\ \text{mol } OH^-$$

and hence mol OH^- consumed during the reaction is $(3.10 \times 10^{-2}) − (1.65 \times 10^{-3}) = 2.94 \times 10^{-2}$ mol OH^-. The reaction between a strong acid and strong base proceeds according to the following net ionic equation (Table 4.2):

$$H^+(aq) + OH^-(aq) \rightarrow H_2O$$

Thus mol OH^- consumed by reaction with strong acid equals mol H^+ present initially and therefore 2.94×10^{-2} mol H^+ are contained in the original 25.00 mL of $HClO_4$ solution. Hence, the molarity of the $HClO_4$ solution is:

$$[HClO_4] = 2.94 \times 10^{-2} \text{ mol } H^+ \times \frac{1 \text{ mol } HClO_4}{1 \text{ mol } H^+} \times \frac{1}{25.00 \text{ mL}} \times \frac{1000 \text{ mL}}{1 \text{ L}} = \underline{\textbf{1.18 M}}$$

(c) From part (a) we know that the student did go beyond the equivalence point. To reach the equivalence point, he should have added 2.94×10^{-2} mol OH^- since this is the number of mol of H^+ present initially (see part (b)). Thus the volume KOH solution required to reach the equivalence point is:

$$\text{vol KOH} = 2.94 \times 10^{-2} \text{ mol } OH^- \times \frac{1 \text{ mol KOH}}{1 \text{ mol } OH^-} \times \frac{1 \text{ L}}{0.731 \text{ mol KOH}} \times \frac{1000 \text{ mL}}{1 \text{ L}}$$

$$= 40.2 \text{ mL KOH solution}$$

Thus the excess is $(42.35 - 40.2)$ mL = $\underline{\textbf{2.2 mL}}$

COMPLEX ION AND PRECIPITATION EQUILIBRIA

15-1 COMPLEX ION EQUILIBRIA; FORMATION CONSTANT (K_f)

2. This problem is related to Example 15.1.

(a) For the complex ion $Cd(CN)_4^{2-}$, the reaction describing its formation and the corresponding expression for the formation constant K_f are:

$$Cd^{2+}(aq) + 4CN^-(aq) \rightleftharpoons Cd(CN)_4^{2-}(aq) \qquad K_f = \frac{[Cd(CN)_4^{2-}]}{[Cd^{2+}][CN^-]^4}$$

and from Table 15.1 the value of K_f is 2×10^{18}. Rearrange the expression for K_f to solve for $[CN^-]$ and substitute $[Cd^{2+}] = 10^{-8} \times [Cd(CN)_4^{2-}]$:

$$2 \times 10^{18} = \frac{[Cd(CN)_4^{2-}]}{[Cd^{2+}][CN^-]^4}$$

$$[CN^-]^4 = \frac{[Cd(CN)_4^{2-}]}{10^{-8} \times [Cd(CN)_4^{2-}] \times (2 \times 10^{18})} = 5 \times 10^{-11} \qquad \rightarrow \qquad [CN^-] = \underline{\mathbf{3 \times 10^{-3}\ M}}$$

(b) For the complex ion $Fe(CN)_6^{4-}$, the reaction describing its formation and the corresponding expression for the formation constant K_f are:

$$Fe^{2+}(aq) + 6CN^-(aq) \rightleftharpoons Fe(CN)_6^{4-}(aq) \qquad K_f = \frac{[Fe(CN)_6^{4-}]}{[Fe^{2+}][CN^-]^6}$$

and from Table 15.1 the value of K_f is 4×10^{45}. Rearrange the expression for K_f to solve for $[CN^-]$ and substitute $[Fe^{2+}] = 10^{-20} \times [Fe(CN)_6^{4-}]$:

$$4 \times 10^{45} = \frac{[Fe(CN)_6^{4-}]}{[Fe^{2+}][CN^-]^6}$$

$$[CN^-]^6 = \frac{[Fe(CN)_6^{4-}]}{10^{-20} \times [Fe(CN)_6^{4-}] \times (4 \times 10^{45})} = 2 \times 10^{-26} \qquad \rightarrow \qquad [CN^-] = \underline{\mathbf{5 \times 10^{-5}\ M}}$$

4. For the complex ion $Hg(Cl)_4^{2-}$, the reaction describing its formation and the corresponding expression for the formation constant K_f are:

$$Hg^{2+}(aq) + 4Cl^-(aq) \rightleftharpoons Hg(Cl)_4^{2-}(aq) \qquad K_f = \frac{[Hg(Cl)_4^{2-}]}{[Hg^{2+}][Cl^-]^4} = 1.2 \times 10^{15}$$

If the initial $[Hg^{2+}] = x$, then when 58% of the Hg^{2+} has been converted to $Hg(Cl)_4^{2-}$, 42% of the Hg^{2+} has been consumed, so $[Hg^{2+}] = 0.42x$ and (by stoichiometry) $[Hg(Cl)_4^{2-}] = 0.58x$. Now substitute for $[Hg^{2+}]$ and $[Hg(Cl)_4^{2-}]$ into the K_f expression and solve for $[Cl^-]$. Thus:

$$K_f = \frac{[Hg(Cl)_4^{2-}]}{[Hg^{2+}][Cl^-]^4} = 1.2 \times 10^{15}$$

$$[Cl^-]^4 = \frac{[Hg(Cl)_4^{2-}]}{(1.2 \times 10^{15})[Hg^{2+}]} = \frac{0.58x}{(1.2 \times 10^{15})(0.42x)} = 1.\underline{1}51 \times 10^{-15}$$

$$[Cl^-] = 1.8 \times 10^{-4} \ M$$

Thus, when $[Cl^-] = \underline{1.8 \times 10^{-4}} \ M$, 58% of Hg^{2+} is converted to the complex ion.

15-2   SOLUBILITY; SOLUBILITY PRODUCT CONSTANT (K_{SP}):   Expression for K_{sp}

6. This problem is similar to Example 15.2. In each case, begin by writing the chemical equation for the solution process (solid on the left, ions on the right). Then write the K_{sp} expression following the rules for writing equilibrium constant expressions (Section 12-2).

    (a) $AgCl(s) \rightleftharpoons Ag^+(aq) + Cl^-(aq)$ $\qquad\qquad$ $K_{sp} = [Ag^+][Cl^-]$

    (b) $Al_2(CO_3)_3(s) \rightleftharpoons 2Al^{3+}(aq) + 3CO_3^{2-}(aq)$ $\qquad$ $K_{sp} = [Al^{3+}]^2[CO_3^{2-}]^3$

    (c) $MnS_2(s) \rightleftharpoons Mn^{4+}(aq) + 2S^{2-}(aq)$ $\qquad\quad$ $K_{sp} = [Mn^{4+}][S^{2-}]^2$

    (d) $Mg(OH)_2(s) \rightleftharpoons Mg^{2+}(aq) + 2OH^-(aq)$ $\qquad$ $K_{sp} = [Mg^{2+}][OH^-]^2$

8. This problem is related to Example 15.2, in reverse. In each case, the concentration terms in the K_{sp} expression indicate which ions are present and their exponents indicate their stoichiometry: these must be on the right side of the chemical equation for the solution process and from that the solid (on the left side of the equation) can be deduced.

    (a) $K_{sp} = [Pb^{4+}][O^{2-}]^2$ $\quad \rightarrow$ ions present: $Pb^{4+} + 2O^{2-}$ $\quad \rightarrow$ solid: PbO_2

    Thus the equilibrium reaction is: $\quad PbO_2(s) \rightleftharpoons Pb^{4+}(aq) + 2O^{2-}(aq)$

    (b) $K_{sp} = [Hg^{2+}]^3[PO_4^{3-}]^2$ $\quad \rightarrow$ ions present: $3Hg^{2+} + 2PO_4^{3-}$ $\quad \rightarrow$ solid: $Hg_3(PO_4)_2$

    Thus the equilibrium reaction is: $\quad Hg_3(PO_4)_2(s) \rightleftharpoons 3Hg^{2+}(aq) + 2PO_4^{3-}(aq)$

    (c) $K_{sp} = [Ni^{3+}][OH^-]^3$ $\quad \rightarrow$ ions present: $Ni^{3+} + 3OH^-$ $\quad \rightarrow$ solid: $Ni(OH)_3$

    Thus the equilibrium reaction is: $\quad Ni(OH)_3(s) \rightleftharpoons Ni^{3+}(aq) + 3OH^-(aq)$

    (d) $K_{sp} = [Ag^+]^2[SO_4^{2-}]$ $\quad \rightarrow$ ions present: $2Ag^+ + SO_4^{2-}$ $\quad \rightarrow$ solid: Ag_2SO_4

    Thus the equilibrium reaction is: $\quad Ag_2SO_4(s) \rightleftharpoons 2Ag^+(aq) + SO_4^{2-}(aq)$

15-2 SOLUBILITY; SOLUBILITY PRODUCT CONSTANT (K_{SP}): K_{sp} and the Equilibrium Concentration of Ions

10. This problem is related to Example 15.3. For each salt, write the chemical equation for its dissolution and hence the expression for K_{sp}; then substitute the given K_{sp} value and ion concentration and solve to find the concentration of the other ion.

(a) $Li_3PO_4(s) \rightleftharpoons 3Li^+(aq) + PO_4^{3-}(aq)$ $\qquad K_{sp} = [Li^+]^3[PO_4^{3-}]$

$3.2 \times 10^{-9} = [Li^+]^3(7.5 \times 10^{-4}) \rightarrow [Li^+]^3 = 4.3 \times 10^{-6} \rightarrow [Li^+] = \underline{\mathbf{1.6 \times 10^{-2}\ M}}$

(b) $AgNO_2(s) \rightleftharpoons Ag^+(aq) + NO_2^-(aq)$ $\qquad K_{sp} = [Ag^+][NO_2^-]$

$6.0 \times 10^{-4} = (0.025)[NO_2^-] \rightarrow [NO_2^-] = \underline{\mathbf{2.4 \times 10^{-2}\ M}}$

(c) $Sn(OH)_2(s) \rightleftharpoons Sn^{2+}(aq) + 2OH^-(aq)$ $\qquad K_{sp} = [Sn^{2+}][OH^-]^2$

$pH = 9.35 \rightarrow [H^+] = 10^{-9.35} = 4.5 \times 10^{-10}\ M$

$[OH^-] = \dfrac{K_w}{[H^+]} = \dfrac{1.0 \times 10^{-14}}{4.5 \times 10^{-10}} = 2.2 \times 10^{-5}\ M$

$1.4 \times 10^{-28} = [Sn^{2+}](2.2 \times 10^{-5})^2 \rightarrow [Sn^{2+}] = \underline{\mathbf{2.9 \times 10^{-19}\ M}}$

12. This problem is similar to Example 15.3. For each anion, write the chemical equation for dissolution of the corresponding salt with Ca^{2+} and hence write the expression for K_{sp}. Using K_{sp} values from Table 15.2 and the given $[Ca^{2+}]$, solve to find the concentration of each anion.

(a) $CaCO_3(s) \rightleftharpoons Ca^{2+}(aq) + CO_3^{2-}(aq)$ $\qquad K_{sp} = [Ca^{2+}][CO_3^{2-}] = 4.9 \times 10^{-9}$

$4.9 \times 10^{-9} = (1.24 \times 10^{-4})[CO_3^{2-}] \rightarrow [CO_3^{2-}] = \underline{\mathbf{4.0 \times 10^{-5}\ M}}$

(b) $Ca(OH)_2(s) \rightleftharpoons Ca^{2+}(aq) + 2OH^-(aq)$ $\qquad K_{sp} = [Ca^{2+}][OH^-]^2 = 4.0 \times 10^{-6}$

$4.0 \times 10^{-6} = (1.24 \times 10^{-4})[OH^-]^2 \rightarrow [OH^-]^2 = 0.032 \rightarrow [OH^-] = \underline{\mathbf{0.18\ M}}$

(c) $Ca_3(PO_4)_2(s) \rightleftharpoons 3Ca^{2+}(aq) + 2PO_4^{3-}(aq)$ $\qquad K_{sp} = [Ca^{2+}]^3[PO_4^{3-}]^2 = 1 \times 10^{-33}$

$1 \times 10^{-33} = (1.24 \times 10^{-4})^3[PO_4^{3-}]^2 \rightarrow [PO_4^{3-}]^2 = 5 \times 10^{-22} \rightarrow [PO_4^{3-}] = \underline{\mathbf{2 \times 10^{-11}\ M}}$

14. (a) This part of the problem is related to Example 15.4(a); see also the discussion in the text immediately above that example. Let s = the molar solubility of BaC_2O_4 (that is, s mol of BaC_2O_4 will dissolve in 1 L of H_2O). The equation for the compound dissolving in water is:

$$BaC_2O_4(s) \rightleftharpoons Ba^{2+}(aq) + C_2O_4^{2-}(aq) \qquad\qquad K_{sp} = [Ba^{2+}][C_2O_4^{2-}]$$
$$ \quad s \qquad\qquad s \qquad\quad s$$

The solubility of BaC_2O_4 is equal to the equilibrium concentration of Ba^{2+} and $C_2O_4^{2-}$; in other words $[Ba^{2+}] = [C_2O_4^{2-}] = s$. Or, more explicitly, for Ba^{2+}:

$$[Ba^{2+}] = \frac{s \text{ mol } BaC_2O_4}{1L} \times \frac{1 \text{ mol } Ba^{2+}}{1 \text{ mol } BaC_2O_4} = s \text{ mol/L}$$

and likewise for $[C_2O_4^{2-}]$. Thus:

$$K_{sp} = [Ba^{2+}][C_2O_4^{2-}] = (s)(s) = s^2$$
$$s = [Ba^{2+}] = [C_2O_4^{2-}] = K_{sp}^{\frac{1}{2}} = (1.6 \times 10^{-6})^{\frac{1}{2}} = \underline{\mathbf{1.3 \times 10^{-3} \ M}}$$

The next two parts of the problem are similar to Example 15.3(a). For each salt, write the chemical equation for its dissolution and hence the expression for K_{sp}; then substitute the given K_{sp} value and ion concentration and solve to find the concentration of the other ion.

(b) $Cr(OH)_3 \ (s) \rightleftharpoons Cr^{3+}(aq) + 3OH^-(aq) \qquad\qquad K_{sp} = [Cr^{3+}][OH^-]^3$

$$6.3 \times 10^{-31} = (2.7 \times 10^{-8})[OH^-]^3 \rightarrow [OH^-]^3 = 2.3 \times 10^{-23} \rightarrow [OH^-] = \underline{\mathbf{2.9 \times 10^{-8} \ M}}$$

(c) $Pb_3(PO_4)_2(s) \rightleftharpoons 3Pb^{2+}(aq) + 2PO_4^{3-}(aq) \qquad K_{sp} = [Pb^{2+}]^3[PO_4^{3-}]^2 = 1 \times 10^{-54}$

$$1 \times 10^{-54} = [Pb^{2+}]^3(8 \times 10^{-6})^2 \rightarrow [Pb^{2+}]^3 = 2 \times 10^{-44} \rightarrow [Pb^{2+}] = \underline{\mathbf{3 \times 10^{-15} \ M}}$$

Thus the complete table is:

	Compound	[cation]	[anion]	K_{sp}
(a)	BaC_2O_4	1.3×10^{-3}	1.3×10^{-3}	1.6×10^{-6}
(b)	$Cr(OH)_3$	2.7×10^{-8}	2.9×10^{-8}	6.3×10^{-31}
(c)	$Pb_3(PO_4)_2$	3×10^{-15}	8×10^{-6}	1×10^{-54}

15-2 SOLUBILITY; SOLUBILITY PRODUCT CONSTANT (K_{SP}): K_{sp} and Solubility

16. This problem is similar to Example 15.4(a); see also the discussion in the text immediately before that example. Obtain K_{sp} values from Table 15.2.

(a) Let s = the molar solubility of MgF_2 (that is, s mol of MgF_2 will dissolve in 1 L of H_2O). The equation for the compound dissolving in water is:

$$MgF_2(s) \rightleftharpoons Mg^{2+}(aq) + 2F^-(aq) \qquad\qquad K_{sp} = [Mg^{2+}][F^-]^2$$
$$ \quad s \qquad\quad s \qquad 2s$$

The equilibrium concentration of Mg^{2+} is equal to the solubility of MgF_2 and the equilibrium concentration of F^- is equal to twice the solubility of MgF_2:

$$[Mg^{2+}] = \frac{s \text{ mol } MgF_2}{1L} \times \frac{1 \text{ mol } Mg^{2+}}{1 \text{ mol } MgF_2} = s \text{ mol/L}$$

$$[F^-] = \frac{s \text{ mol } MgF_2}{1L} \times \frac{2 \text{ mol } F^-}{1 \text{ mol } MgF_2} = 2s \text{ mol/L}$$

Thus:

$$K_{sp} = [Mg^{2+}][F^-]^2 = (s)(2s)^2 = 4s^3$$

$$s = \left(\frac{K_{sp}}{4}\right)^{1/3} = \left(\frac{7 \times 10^{-11}}{4}\right)^{1/3} = \underline{3 \times 10^{-4} \; M}$$

(b) Let s = the molar solubility of $Fe(OH)_3$ (that is, s mol of $Fe(OH)_3$ will dissolve in 1 L of H_2O). The equation for the compound dissolving in water is:

$$Fe(OH)_3(s) \rightleftharpoons Fe^{3+}(aq) + 3OH^-(aq) \qquad\qquad K_{sp} = [Fe^{3+}][OH^-]^3$$
$$\;\; s \qquad\quad\; s \qquad\quad 3s$$

The equilibrium concentration of Fe^{3+} is equal to the solubility of $Fe(OH)_3$ and the equilibrium concentration of OH^- is equal to three times the solubility of $Fe(OH)_3$:

$$[Fe^{3+}] = \frac{s \text{ mol } Fe(OH)_3}{1L} \times \frac{1 \text{ mol } Fe^{3+}}{1 \text{ mol } Fe(OH)_3} = s \text{ mol/L}$$

$$[OH^-] = \frac{s \text{ mol } Fe(OH)_3}{1L} \times \frac{3 \text{ mol } OH^-}{1 \text{ mol } Fe(OH)_3} = 3s \text{ mol/L}$$

Thus:

$$K_{sp} = [Fe^{3+}][OH^-]^3 = (s)(3s)^3 = 27s^4$$

$$s = \left(\frac{K_{sp}}{27}\right)^{1/4} = \left(\frac{3 \times 10^{-39}}{27}\right)^{1/4} = \underline{1 \times 10^{-10} \; M}$$

(c) Let s = the molar solubility of $Mg_3(PO_4)_2$ (that is, s mol of $Mg_3(PO_4)_2$ will dissolve in 1 L of H_2O). The equation for the compound dissolving in water is:

$$Mg_3(PO_4)_2(s) \rightleftharpoons 3Mg^{2+}(aq) + 2PO_4^{3-}(aq) \qquad\qquad K_{sp} = [Mg^{2+}]^3[PO_4^{3-}]^2$$
$$\;\; s \qquad\qquad 3s \qquad\quad\; 2s$$

The equilibrium concentration of Mg^{2+} equals three times the solubility of $Mg_3(PO_4)_2$ and the equilibrium concentration of PO_4^{3-} equals twice the solubility of $Mg_3(PO_4)_2$:

$$[Mg^{2+}] = \frac{s \text{ mol } Mg_3(PO_4)_2}{1 L} \times \frac{3 \text{ mol } Mg^{2+}}{1 \text{ mol } Mg_3(PO_4)_2} = 3s \text{ mol/L}$$

$$[PO_4^{3-}] = \frac{s \text{ mol } Mg_3(PO_4)_2}{1 L} \times \frac{2 \text{ mol } PO_4^{3-}}{1 \text{ mol } Mg_3(PO_4)_2} = 2s \text{ mol/L}$$

Thus:

$$K_{sp} = [Mg^{2+}]^3[PO_4^{3-}]^2 = (3s)^3(2s)^2 = 108s^5$$

$$s = \left(\frac{K_{sp}}{108}\right)^{1/5} = \left(\frac{1 \times 10^{-24}}{108}\right)^{1/5} = \underline{\mathbf{6 \times 10^{-6} \; M}}$$

18. This problem is similar to Example 15.4(a), in reverse; see also the discussion in the text immediately before that example.

(a) Let s = the molar solubility of MgC_2O_4 (that is, s mol of MgC_2O_4 will dissolve in 1 L of H_2O). The equation for the compound dissolving in water is:

$$MgC_2O_4(s) \rightleftharpoons Mg^{2+}(aq) + C_2O_4^{2-}(aq) \qquad\qquad K_{sp} = [Mg^{2+}][C_2O_4^{2-}]$$
$$\;\;\;\;s \qquad\qquad\qquad s \qquad\quad s$$

The equilibrium concentration of both Mg^{2+} and $C_2O_4^{2-}$ are equal to the solubility of MgC_2O_4:

$$[Mg^{2+}] = \frac{s \text{ mol } MgC_2O_4}{1 L} \times \frac{1 \text{ mol } Mg^{2+}}{1 \text{ mol } MgC_2O_4} = s \text{ mol/L}$$

$$[C_2O_4^{2-}] = \frac{s \text{ mol } MgC_2O_4}{1 L} \times \frac{1 \text{ mol } C_2O_4^{2-}}{1 \text{ mol } MgC_2O_4} = s \text{ mol/L}$$

Thus: $K_{sp} = [Mg^{2+}][C_2O_4^{2-}] = (s)(s) = s^2 = (9.2 \times 10^{-3})^2 = \underline{\mathbf{8.5 \times 10^{-5}}}$

(b) Let s = the molar solubility of $Mn(OH)_2$ (that is, s mol of $Mn(OH)_2$ will dissolve in 1 L of H_2O). The equation for the compound dissolving in water is:

$$Mn(OH)_2(s) \rightleftharpoons Mn^{2+}(aq) + 2OH^-(aq) \qquad\qquad K_{sp} = [Mn^{2+}][OH^-]^2$$
$$\;\;\;\;s \qquad\qquad\qquad s \qquad\qquad 2s$$

The equilibrium concentration of Mn^{2+} is equal to the solubility of $Mn(OH)_2$ and the equilibrium concentration of OH^- is equal to twice the solubility of $Mn(OH)_2$:

$$[Mn^{2+}] = \frac{s \text{ mol } Mn(OH)_2}{1 L} \times \frac{1 \text{ mol } Mn^{2+}}{1 \text{ mol } Mn(OH)_2} = s \text{ mol/L}$$

$$[OH^-] = \frac{s \text{ mol } Mn(OH)_2}{1 L} \times \frac{2 \text{ mol } OH^-}{1 \text{ mol } Mn(OH)_2} = 2s \text{ mol/L}$$

Thus: $K_{sp} = [Mn^{2+}][OH^-]^2 = (s)(2s)^2 = 4s^3 = 4(3.5 \times 10^{-5})^3 = \underline{\mathbf{1.7 \times 10^{-13}}}$

(c) Let s = the molar solubility of $Cd_3(PO_4)_2$ (that is, s mol of $Cd_3(PO_4)_2$ will dissolve in 1 L of H_2O). The equation for the compound dissolving in water is:

$$Cd_3(PO_4)_2(s) \rightleftharpoons 3Cd^{2+}(aq) + 2PO_4^{3-}(aq) \qquad\qquad K_{sp} = [Cd^{2+}]^3[PO_4^{3-}]^2$$
$$\quad\; s \qquad\qquad\qquad 3s \qquad\quad 2s$$

The equilibrium concentration of Cd^{2+} equals three times the solubility of $Cd_3(PO_4)_2$ and the equilibrium concentration of PO_4^{3-} equals twice the solubility of $Cd_3(PO_4)_2$:

$$[Cd^{2+}] = \frac{s \text{ mol } Cd_3(PO_4)_2}{1 L} \times \frac{3 \text{ mol } Cd^{2+}}{1 \text{ mol } Cd_3(PO_4)_2} = 3s \text{ mol/L}$$

$$[PO_4^{3-}] = \frac{s \text{ mol } Cd_3(PO_4)_2}{1 L} \times \frac{2 \text{ mol } PO_4^{3-}}{1 \text{ mol } Cd_3(PO_4)_2} = 2s \text{ mol/L}$$

Thus: $K_{sp} = [Cd^{2+}]^3[PO_4^{3-}]^2 = (3s)^3(2s)^2 = 108s^5 = 108(1.5 \times 10^{-7})^5 = \underline{\mathbf{8.2 \times 10^{-33}}}$

20. (a) This part of the problem is similar to Example 15.4; see also the discussion in the text immediately before that example. From Table 15.2, $K_{sp} = 6 \times 10^{-12}$. Let s = the molar solubility of $Mg(OH)_2$ (that is, s mol of $Mg(OH)_2$ will dissolve in 1 L of H_2O). The equation for the compound dissolving in water is:

$$Mg(OH)_2(s) \rightleftharpoons Mg^{2+}(aq) + 2OH^-(aq) \qquad\qquad K_{sp} = [Mg^{2+}][OH^-]^2$$
$$\quad\; s \qquad\qquad\quad s \qquad\quad 2s$$

The equilibrium concentration of Mg^{2+} is equal to the solubility of $Mg(OH)_2$ and the equilibrium concentration of OH^- is equal to twice the solubility of $Mg(OH)_2$:

$$[Mg^{2+}] = \frac{s \text{ mol } Mg(OH)_2}{1 L} \times \frac{1 \text{ mol } Mg^{2+}}{1 \text{ mol } Mg(OH)_2} = s \text{ mol/L}$$

$$[OH^-] = \frac{s \text{ mol } Mg(OH)_2}{1 L} \times \frac{2 \text{ mol } OH^-}{1 \text{ mol } Mg(OH)_2} = 2s \text{ mol/L}$$

Thus:

$$K_{sp} = [Mg^{2+}][OH^-]^2 = (s)(2s)^2 = 4s^3 \rightarrow s = \left(\frac{K_{sp}}{4}\right)^{1/3} = \left(\frac{6 \times 10^{-12}}{4}\right)^{1/3} = 1 \times 10^{-4} M$$

$$\text{solubility in g/L} = \frac{1 \times 10^{-4} \text{ mol } Mg(OH)_2}{1 L} \times \frac{58.32 \text{ g } Mg(OH)_2}{1 \text{ mol } Mg(OH)_2} = \underline{\mathbf{6 \times 10^{-3} \text{ g/L}}}$$

The next two parts of the problem are similar to Example 15.6.

(b) Let s = the molar solubility of $Mg(OH)_2$ in the $Ba(OH)_2$ solution. Set up a table as in Example 15.6. Initially $[Mg^{2+}] = 0$ because $Mg(OH)_2$ has not yet dissolved in the solution, and there are no OH^- ions contributed by the $Mg(OH)_2$, but some OH^- ions *are* contributed by the $Ba(OH)_2$:

$$\text{initial } [OH^-] = \frac{0.041 \text{ mol } Ba(OH)_2}{1L} \times \frac{2 \text{ mol } OH^-}{1 \text{ mol } Ba(OH)_2} = 0.082 \ M$$

	$[Mg^{2+}]$	$[OH^-]$
Original	0	0.082
Change	+s	+2s
Equilibrium	s	0.082 + 2s

Substitute the entries in the equilibrium row into the K_{sp} expression:

$$K_{sp} = 6 \times 10^{-12} = [Mg^{2+}][OH^-]^2 = (s)(0.082 + 2s)^2$$

Assume $s \ll 0.082$ (this is reasonable, since we expect that the Common Ion Effect will lower the solubility):

$$6 \times 10^{-12} \approx (s)(0.082)^2 \qquad\qquad s = \ = 9 \times 10^{-10} \ M$$

(The assumption is justified: s is indeed much less than 0.082.)

$$\text{solubility in g/L} = \frac{9 \times 10^{-10} \text{ mol } Mg(OH)_2}{1L} \times \frac{58.32 \text{ g } Mg(OH)_2}{1 \text{ mol } Mg(OH)_2} = \underline{\underline{5 \times 10^{-8} \text{ g/L}}}$$

(c) Let s = the molar solubility of $Mg(OH)_2$ in the $MgCl_2$ solution. Set up a table as in Example 15.6. Initially $[OH^-] = 0$ because $Mg(OH)_2$ has not yet begun to dissolve, and there are no Mg^{2+} ions contributed by the $Mg(OH)_2$, but some Mg^{2+} ions *are* contributed by the $MgCl_2$:

$$\text{initial } [Mg^{2+}] = \frac{0.0050 \text{ mol } MgCl_2}{1L} \times \frac{1 \text{ mol } Mg^{2+}}{1 \text{ mol } MgCl_2} = 0.0050 \ M$$

	$[Mg^{2+}]$	$[OH^-]$
Original	0.0050	0
Change	+s	+2s
Equilibrium	0.0050 + s	2s

Substitute the entries in the equilibrium row into the K_{sp} expression:

$$K_{sp} = 6 \times 10^{-12} = [Mg^{2+}][OH^-]^2 = (0.0050 + s)(2s)^2$$

Assume $s \ll 0.0050$:

$$6 \times 10^{-12} \approx (0.0050)(2s)^2 = (0.0050)(4s^2) \qquad s = \left(\frac{6 \times 10^{-12}}{4(0.0050)} \right)^{\frac{1}{2}} = 2 \times 10^{-5}\ M$$

$$\text{solubility in g/L} = \frac{2 \times 10^{-5}\ \text{mol Mg(OH)}_2}{1\text{L}} \times \frac{58.32\ \text{g Mg(OH)}_2}{1\ \text{mol Mg(OH)}_2} = \underline{\mathbf{1 \times 10^{-3}\ g/L}}$$

22. This problem is similar to Example 15.5 and related to Example 15.4. Use the given information to calculate the molar solubility, s, of $Ni(OH)_2$; then write a net ionic equation to represent $Ni(OH)_2$ dissolving in H_2O, and the expression for K_{sp}; next use stoichiometry to find relationships between $[Ni^{2+}]$ and $[OH^-]$ and the solubility, and finally substitute these into the K_{sp} expression and evaluate it.

$$s = \frac{0.239\ \text{mg Ni(OH)}_2}{500.0\ \text{mL}} \times \frac{1\text{g}}{1000\ \text{mg}} \times \frac{1000\ \text{mL}}{1\text{L}} \times \frac{1\ \text{mol Ni(OH)}_2}{92.71\text{g Ni(OH)}_2} = 5.16 \times 10^{-6}\ M$$

The equation for the compound dissolving in water is:

$$Ni(OH)_2(s) \rightleftharpoons Ni^{2+}(aq) + 2OH^-(aq) \qquad\qquad K_{sp} = [Ni^{2+}][OH^-]^2$$
$$\quad\ s \qquad\qquad\quad s \qquad\quad 2s$$

The equilibrium concentration of Ni^{2+} is equal to the solubility s and the equilibrium concentration of OH^- is equal to twice the solubility:

$$[Ni^{2+}] = \frac{s\ \text{mol Ni(OH)}_2}{1\text{L}} \times \frac{1\ \text{mol Ni}^{2+}}{1\ \text{mol Ni(OH)}_2} = s\ \text{mol/L}$$

$$[OH^-] = \frac{s\ \text{mol Ni(OH)}_2}{1\text{L}} \times \frac{2\ \text{mol OH}^-}{1\ \text{mol Ni(OH)}_2} = 2s\ \text{mol/L}$$

Thus: $K_{sp} = [Ni^{2+}][OH^-]^2 = (s)(2s)^2 = 4s^3 = 4(5.16 \times 10^{-6})^3 = \underline{\mathbf{5.50 \times 10^{-16}}}$

24. This problem is related to Example 15.4. Let s = the molar solubility of $CaSO_4$ at a particular temperature (that is, s mol of $CaSO_4$ will dissolve in 1 L of H_2O at that temperature). The equation for the compound dissolving in water is:

$$CaSO_4(s) \rightleftharpoons Ca^{2+}(aq) + SO_4^{2-}(aq) \qquad\qquad K_{sp} = [Ca^{2+}][SO_4^{2-}]$$
$$\quad s \qquad\qquad\quad s \qquad\qquad s$$

The equilibrium concentration of both Ca^{2+} and SO_4^{2-} are equal to the solubility of $CaSO_4$:

$$[Ca^{2+}] = \frac{s\ \text{mol CaSO}_4}{1\text{L}} \times \frac{1\ \text{mol Ca}^{2+}}{1\ \text{mol CaSO}_4} = s\ \text{mol/L}$$

$$[SO_4^{2-}] = \frac{s \text{ mol CaSO}_4}{1L} \times \frac{1 \text{ mol SO}_4^{2-}}{1 \text{ mol CaSO}_4} = s \text{ mol/L}$$

Thus: $K_{sp} = [Ca^{2+}][SO_4^{2-}] = (s)(s) = s^2$ and this relationship applies at any temperature; only the values of K_{sp} and s will change.

(a) At 25°C, K_{sp} for $CaSO_4$ is 7.1×10^{-5} (Table 15.2)

$$s = K_{sp}^{\frac{1}{2}} = (7.1 \times 10^{-5})^{\frac{1}{2}} = 8.4 \times 10^{-3} M$$

$$\text{solubility in g/L} = \frac{8.4 \times 10^{-3} \text{ mol CaSO}_4}{1L} \times \frac{136.14 \text{ g CaSO}_4}{1 \text{ mol CaSO}_4} = 1.1 \text{ g/L}$$

Thus at 25°C, 1.1 g of $CaSO_4$ can dissolve per liter of solution so: **YES**, 0.915 g $CaSO_4$ will dissolve in one liter at 25°C.

(b) At 100°C, K_{sp} for $CaSO_4$ is 1.6×10^{-5} (given) → $s = K_{sp}^{\frac{1}{2}} = (1.6 \times 10^{-5})^{\frac{1}{2}} = 4.0 \times 10^{-3} M$

$$\text{solubility in g/L} = \frac{4.0 \times 10^{-3} \text{ mol CaSO}_4}{1L} \times \frac{136.14 \text{ g CaSO}_4}{1 \text{ mol CaSO}_4} = 0.54 \text{ g/L}$$

Thus at 100°C, only 0.54 g of $CaSO_4$ can dissolve per liter of solution so: **NO**, 0.915 g $CaSO_4$ will **NOT** dissolve in one liter at this temperature.

26. This problem is related to Examples 15.4 and 15.5. The net ionic equation that represents $Cr(OH)_2$ dissolving in H_2O, and the corresponding K_{sp} expression, are:

$$Cr(OH)_2(s) \rightleftharpoons Cr^{2+}(aq) + 2OH^-(aq) \qquad\qquad K_{sp} = [Cr^{2+}][OH^-]^2$$

For the given mass of $Cr(OH)_2$ and the volume of solution prepared, calculate $[Cr^{2+}]$ in the solution:

$$[Cr^{2+}] = \frac{10.24 \text{ mg Cr(OH)}_2}{125 \text{ mL}} \times \frac{1g}{1000 \text{ mg}} \times \frac{1000 \text{ mL}}{1L} \times \frac{1 \text{ mol Cr(OH)}_2}{86.01g \text{ Cr(OH)}_2} = 9.52 \times 10^{-4} M$$

The stoichiometry of the above reaction indicates that $[OH^-]$ should be $2 \times [Cr^{2+}]$; however, using the given pH to determine $[H^+]$ and hence $[OH^-]$ at equilibrium gives:

$$pH = 8.49 \rightarrow [H^+] = 10^{-pH} = 10^{-8.49} = 3.2 \times 10^{-9} M$$

$$[OH^-] = \frac{K_w}{[H^+]} = \frac{1.0 \times 10^{-14}}{3.2 \times 10^{-9}} = 3.1 \times 10^{-6} M$$

Thus it must be assumed that a reaction such as the one below occurs, whereby *some* of the OH^- ions provided by dissolution of $Cr(OH)_2$ are consumed (compare Example 15.9):

$$Cr(OH)_2(s) + 2H^+(aq) \rightleftharpoons Cr^{2+}(aq) + 2H_2O$$

(In other words, the solution must be slightly *acidic* before $Cr(OH)_2$ is added.) Note that this does not change $[Cr^{2+}]$. Thus $[Cr^{2+}]$ and $[OH^-]$ are known: substitute these into the K_{sp} expression and evaluate it.

$$K_{sp} = [Cr^{2+}][OH^-]^2 = (9.52 \times 10^{-4})(3.1 \times 10^{-6})^2 = \textbf{9.1} \times \textbf{10}^{-15}$$

15-3 PRECIPITATE FORMATION: K_{sp} and Precipitation

28. This problem is related to Example 15.7.
    (a) Use the given pH to calculate the initial $[OH^-]$ before $CdCl_2$ is added:

$$pH = 9.62 \rightarrow [H^+] = 10^{-pH} = 10^{-9.62} = 2.4 \times 10^{-10} \ M$$

$$[OH^-] = \frac{K_w}{[H^+]} = \frac{1.0 \times 10^{-14}}{2.4 \times 10^{-10}} = 4.2 \times 10^{-5} \ M$$

The equation for the dissolution of $Cd(OH)_2$ in water is:

$$Cd(OH)_2(s) \rightleftharpoons Cd^{2+}(aq) + 2OH^-(aq) \qquad\qquad K_{sp} = [Cd^{2+}][OH^-]^2$$

The expression for Q is identical to the one for K_{sp}, except that it does not use equilibrium concentrations:

$$Q = [Cd^{2+}][OH^-]^2$$

Initially (before any $CdCl_2$ is added; $[Cd^{2+}] = 0$), the value of Q is zero. As $[Cd^{2+}]$ increases, so does the value of Q, until the point where $Q = K_{sp}$ and then precipitation commences. Thus, to determine $[Cd^{2+}]$ at this point, use $[OH^-]$ calculated above, substitute into the Q expression and solve to find the corresponding value for $[Cd^{2+}]$. Precipitation begins when:

$$Q = [Cd^{2+}][OH^-]^2 = K_{sp} \qquad \rightarrow \qquad [Cd^{2+}](4.2 \times 10^{-5})^2 = 2.5 \times 10^{-14}$$

$$[Cd^{2+}] = \underline{\textbf{1.4} \times \textbf{10}^{-5} \ \textbf{M}}$$

(b) Beyond the point (part (a)) where $Q = K_{sp}$ and precipitation commences, addition of more $CdCl_2$ causes further $Cd(OH)_2$ to precipitate: applying Le Châtelier's Principle to the equilibrium reaction

$$Cd(OH)_2(s) \rightleftharpoons Cd^{2+}(aq) + 2OH^-(aq)$$

indicates that increasing $[Cd^{2+}]$ will cause the system to shift to the left, thereby producing $Cd(OH)_2$ precipitate and removing OH^- ions from solution; the solution itself remains saturated (equilibrium) and $Q = K_{sp}$.
To determine $[OH^-]$ for the given $[Cd^{2+}]$, substitute into the K_{sp} expression and solve to find the corresponding value for $[OH^-]$.

$$K_{sp} = [Cd^{2+}][OH^-]^2 \qquad \rightarrow \qquad (0.0013)[OH^-]^2 = 2.5 \times 10^{-14}$$

$$[OH^-] = 4.4 \times 10^{-6} \, M$$

To determine pH at this point, first calculate $[H^+]$:

$$[H^+] = \frac{K_w}{[OH^-]} = \frac{1.0 \times 10^{-14}}{4.4 \times 10^{-6}} = 2.3 \times 10^{-9} \, M$$

$$pH = -\log_{10}[H^+] = -\log_{10}(2.3 \times 10^{-9}) = \mathbf{8.64}$$

(c) Originally, from part (a), $[OH^-] = 4.2 \times 10^{-5} \, M$; after precipitation, part (b), $[OH^-] = 4.4 \times 10^{-6}$ M. Thus:

$$\% \, [OH^-] \text{ remaining} = \frac{[OH^-]_{final}}{[OH^-]_{initial}} \times 100 = \frac{4.4 \times 10^{-6}}{4.2 \times 10^{-5}} \times 100 = \underline{\mathbf{1 \times 10^1 \, \%}}$$

30. This problem is related to Example 15.7. The equation for the dissolution of Ag_2SO_4 in water is:

$$Ag_2SO_4(s) \rightleftharpoons 2Ag^+(aq) + SO_4^{2-}(aq) \qquad\qquad K_{sp} = [Ag^+]^2[SO_4^{2-}]$$

To determine whether a precipitate will form, calculate the initial concentrations of the individual ions, use them to calculate Q, and compare the value of Q with K_{sp}.

$$[Ag^+]_{initial} = \frac{15 \text{ g AgNO}_3}{1.0 \text{ L}} \times \frac{1 \text{ mol AgNO}_3}{169.87 \text{ g AgNO}_3} \times \frac{1 \text{ mol Ag}^+}{1 \text{ mol AgNO}_3} = 8.8 \times 10^{-2} \, M$$

$$[SO_4^{2-}]_{initial} = \frac{20 \text{ g K}_2SO_4}{1.0 \text{ L}} \times \frac{1 \text{ mol K}_2SO_4}{174.26 \text{ g K}_2SO_4} \times \frac{1 \text{ mol SO}_4^{2-}}{1 \text{ mol K}_2SO_4} = 1.1 \times 10^{-1} \, M$$

$$Q = [Ag^+]^2[SO_4^{2-}] = (8.8 \times 10^{-2})^2(1.1 \times 10^{-1}) = 8.5 \times 10^{-4}$$

$Q > K_{sp}$, so **yes, a precipitate will form**

Assuming that $[SO_4^{2-}]_{initial} = 1.1 \times 10^{-1}$ M, as calculated above, to determine $[Ag^+]$ such that precipitation just begins, recall that this will occur when $Q = K_{sp}$. Thus, substitute this $[SO_4^{2-}]$ into the K_{sp} expression, and solve for $[Ag^+]$.   Precipitation begins when:

$$Q = [Ag^+]^2[SO_4^{2-}] = K_{sp} \qquad \rightarrow \qquad [Ag^+]^2(1.1 \times 10^{-1}) = 1.2 \times 10^{-5}$$

$$[Ag^+]^2 = 1.1 \times 10^{-4} \qquad \rightarrow \qquad [Ag^+] = \underline{\mathbf{1.0 \times 10^{-2} \, M}}$$

32. This problem is related to Example 15.7. The equation for the dissolution of Ag_2CO_3 in water is:

$$Ag_2CO_3(s) \rightleftharpoons 2Ag^+(aq) + CO_3^{2-}(aq) \qquad\qquad K_{sp} = [Ag^+]^2[CO_3^{2-}]$$

(a) To determine whether a precipitate will form, calculate the initial concentrations of the individual ions, use them to calculate Q, and compare the value of Q with K_{sp}. (From Table 15.2, $K_{sp} = 8 \times 10^{-12}$.) Note that the total volume of solution after the two components are mixed is 13.00 mL + 45.00 mL = 58.0 mL = 0.0580 L.

mol Ag^+ present initially

$$= \frac{0.022 \text{ mol AgNO}_3}{1.0 \text{ L}} \times 45.00 \text{ mL} \times \frac{1 \text{ L}}{1000 \text{ mL}} \times \frac{1 \text{ mol Ag}^+}{1 \text{ mol AgNO}_3} = 9.9 \times 10^{-4} \text{ mol}$$

$$[Ag^+]_{\text{initial}} = \frac{9.9 \times 10^{-4} \text{ mol Ag}^+}{0.0580 \text{ L}} = 1.7 \times 10^{-2} \text{ M}$$

mol CO_3^{2-} present initially

$$= \frac{0.0014 \text{ mol Na}_2\text{CO}_3}{1.0 \text{ L}} \times 13.00 \text{ mL} \times \frac{1 \text{ L}}{1000 \text{ mL}} \times \frac{1 \text{ mol CO}_3^{2-}}{1 \text{ mol Na}_2\text{CO}_3} = 1.8 \times 10^{-5} \text{ mol}$$

$$[CO_3^{2-}]_{\text{initial}} = \frac{1.8 \times 10^{-5} \text{ mol CO}_3^{2-}}{0.0580 \text{ L}} = 3.1 \times 10^{-4} \text{ M}$$

$$Q = [Ag^+]^2[CO_3^{2-}] = (1.7 \times 10^{-2})^2 (3.1 \times 10^{-4}) = 9.0 \times 10^{-8}$$

$Q > K_{sp}$, so **yes, a precipitate will form**

(b) To determine concentrations at equilibrium, set up an equilibrium table with the initial concentrations of Ag^+ and CO_3^{2-} calculated above. Since it is known that $Q > K_{sp}$ and therefore a precipitate will form, the concentrations of both ions must decrease in order to reach equilibrium; let the change in $[CO_3^{2-}] = x$:

	$[Ag^+]$	$[CO_3^{2-}]$
Original	1.7×10^{-2}	3.1×10^{-4}
Change	$-2x$	$-x$
Equilibrium	$1.7 \times 10^{-2} - 2x$	$3.1 \times 10^{-4} - x$

Substitute the entries in the equilibrium row into the K_{sp} expression:

$$K_{sp} = 8 \times 10^{-12} = [Ag^+]^2[CO_3^{2-}] = (1.7 \times 10^{-2} - 2x)^2 (3.1 \times 10^{-4} - x)$$

Solving this equation is not straightforward; however, since $[Ag^+]_{\text{initial}} \gg [CO_3^{2-}]_{\text{initial}}$, it can reasonably be assumed that essentially all of the CO_3^{2-} will be precipitated. Thus, if $[CO_3^{2-}]$ falls to zero then (by referring to the above table) clearly $x = 3.1 \times 10^{-4}$. Use this value of x to determine $[Ag^+]$ at equilibrium:

$$[Ag^+] = 1.7 \times 10^{-2} - 2x = 1.7 \times 10^{-2} - 2(3.1 \times 10^{-4}) = \underline{\mathbf{1.6 \times 10^{-2} \text{ M}}}$$

Substitute this value of $[Ag^+]$ into the K_{sp} expression and solve for the corresponding value of $[CO_3^{2-}]$

$$K_{sp} = 8 \times 10^{-12} = [Ag^+]^2[CO_3^{2-}] = (1.6 \times 10^{-2})^2[CO_3^{2-}] \quad \rightarrow \quad [CO_3^{2-}] = \underline{\mathbf{3 \times 10^{-8} \ M}}$$

(Note that this very small value for $[CO_3^{2-}]$ confirms that our assumption above — namely that essentially all of the CO_3^{2-} will precipitate — was reasonable.)

Finally, note that Na^+ and NO_3^- are spectator ions in this process so their initial concentrations are not affected by precipitation of Ag_2CO_3. There concentrations are related to the values of $[CO_3^{2-}]_{initial}$ and $[Ag^+]_{initial}$, respectively:

$$[Na^+] = \frac{3.1 \times 10^{-4} \ \text{mol} \ CO_3^{2-}}{1 L} \times \frac{2 \ \text{mol} \ Na^+}{1 \ \text{mol} \ Na_2CO_3} = \underline{\mathbf{6.2 \times 10^{-4} \ M}}$$

$$[NO_3^-] = \frac{1.7 \times 10^{-2} \ \text{mol} \ Ag^+}{1 L} \times \frac{1 \ \text{mol} \ NO_3^-}{1 \ \text{mol} \ AgNO_3} = \underline{\mathbf{1.7 \times 10^{-2} \ M}}$$

15-3 PRECIPITATE FORMATION: Selective Precipitation

34. This problem is similar to Example 15.8.

   (a) From Table 15.2, K_{sp} for $PbSO_4$ is 1.8×10^{-8}, and K_{sp} for $Pb(OH)_2$ is 2.8×10^{-16} (given). For each salt, precipitation begins when $Q = K_{sp}$; therefore substitute the given concentrations of SO_4^{2-} and OH^- in the K_{sp} expression for $PbSO_4$ and $Pb(OH)_2$, respectively, and solve to find $[Pb^{2+}]$ at which precipitation begins:

   for $PbSO_4$:
   $$K_{sp} = [Pb^{2+}][SO_4^{2-}] = [Pb^{2+}](0.020) = 1.8 \times 10^{-8} \quad \rightarrow \quad [Pb^{2+}] = 9.0 \times 10^{-7} \ M$$

   for $Pb(OH)_2$:
   $$K_{sp} = [Pb^{2+}][OH^-]^2 = [Pb^{2+}](0.020)^2 = 2.8 \times 10^{-16} \quad \rightarrow \quad [Pb^{2+}] = 7.0 \times 10^{-13} \ M$$

   Therefore, a lower concentration of Pb^{2+} is required to precipitate $Pb(OH)_2$, meaning that **$Pb(OH)_2$ will precipitate first.**

   (b) From part (a), when $PbSO_4$ begins to precipitate, $[Pb^{2+}] = 9.0 \times 10^{-7}$ M. Recall that at that point some $Pb(OH)_2$ has already precipitated and the solution is saturated in $Pb(OH)_2$. Substitute this concentration of Pb^{2+} into the K_{sp} expression for $Pb(OH)_2$, solve to find the corresponding $[OH^-]$ at this point, and use that to calculate the pH:

   $$K_{sp} = [Pb^{2+}][OH^-]^2 = (9.0 \times 10^{-7})[OH^-]^2 = 2.8 \times 10^{-16} \quad \rightarrow \quad [OH^-] = 1.8 \times 10^{-5} \ M$$

   $$[H^+] = \frac{K_w}{[OH^-]} = \frac{1.0 \times 10^{-14}}{1.8 \times 10^{-5}} = 5.6 \times 10^{-10} \ M$$

   $$pH = -\log_{10}[H^+] = -\log_{10}(5.6 \times 10^{-10}) = \underline{\mathbf{9.25}}$$

36. This problem is similar to Example 15.8.

    (a) From Table 15.2, K_{sp} for Ag_2CrO_4 is 1×10^{-12}, and K_{sp} for $PbCrO_4$ is 2×10^{-14}. For each salt, precipitation begins when $Q = K_{sp}$; therefore, use the information given to calculate the initial $[Ag^+]$ and $[Pb^{2+}]$, then substitute these concentrations into the K_{sp} expressions for Ag_2CrO_4 and $PbCrO_4$, respectively, and solve to find $[CrO_4^{2-}]$ at which precipitation begins.

$$[Ag^+]_{initial} = \frac{0.839 \text{ g AgNO}_3}{0.492 \text{ L}} \times \frac{1 \text{ mol AgNO}_3}{161.91 \text{ g AgNO}_3} \times \frac{1 \text{ mol Ag}^+}{1 \text{ mol AgNO}_3} = 1.05 \times 10^{-2} \text{ M}$$

$$[Pb^{2+}]_{initial} = \frac{1.024 \text{ g Pb(NO}_3)_2}{0.492 \text{ L}} \times \frac{1 \text{ mol Pb(NO}_3)_2}{331.20 \text{ g Pb(NO}_3)_2} \times \frac{1 \text{ mol Pb}^{2+}}{1 \text{ mol Pb(NO}_3)_2}$$
$$= 6.28 \times 10^{-3} \text{ M}$$

for Ag_2CrO_4:
$$K_{sp} = [Ag^+]^2[CrO_4^{2-}] = (1.05 \times 10^{-2})^2[CrO_4^{2-}] = 1 \times 10^{-12}$$
$$\rightarrow [CrO_4^{2-}] = 9 \times 10^{-9} \text{ M}$$

for $PbCrO_4$:
$$K_{sp} = [Pb^{2+}][CrO_4^{2-}] = (6.28 \times 10^{-3})[CrO_4^{2-}] = 2 \times 10^{-14}$$
$$\rightarrow [CrO_4^{2-}] = 3 \times 10^{-12} \text{ M}$$

Therefore, a lower concentration of CrO_4^{2-} is required to precipitate $PbCrO_4$, meaning that **$PbCrO_4$ will precipitate first.**

    (b) From part (a), when $PbCrO_4$ begins to precipitate, $[CrO_4^{2-}] = \underline{\textbf{3} \times \textbf{10}^{-12} \textbf{ M}}$.

38. This problem is related to Example 15.8.
    From Table 15.2, K_{sp} for $BaCO_3$ is 2.6×10^{-9}, and K_{sp} for $CaCO_3$ is 4.9×10^{-9}. For the three salts that are dissolved initially, two possible precipitation reactions may occur:

$$Ba(NO_3)_2(aq) + Na_2CO_3(aq) \rightarrow BaCO_3(s) + 2NaNO_3(aq)$$
$$\text{net ionic equation:} \quad Ba^{2+}(aq) + CO_3^{2-}(aq) \rightarrow BaCO_3(s)$$

$$Ca(NO_3)_2(aq) + Na_2CO_3(aq) \rightarrow CaCO_3(s) + 2NaNO_3(aq)$$
$$\text{net ionic equation:} \quad Ca^{2+}(aq) + CO_3^{2-}(aq) \rightarrow CaCO_3(s)$$

For each of the possible precipitates, recall that precipitation begins when $Q = K_{sp}$; therefore, use the information given to calculate the initial $[Ba^{2+}]$, $[Ca^{2+}]$ and $[CO_3^{2-}]$, then substitute these concentrations into the Q expressions for $BaCO_3$ and $CaCO_3$, calculate Q and compare its value with K_{sp} to determine whether precipitation occurs.

$$[Ba^{2+}]_{initial} = \frac{0.095 \text{ g Ba(NO}_3)_2}{0.5000 \text{ L}} \times \frac{1 \text{ mol Ba(NO}_3)_2}{261.35 \text{ g Ba(NO}_3)_2} \times \frac{1 \text{ mol Ba}^{2+}}{1 \text{ mol Ba(NO}_3)_2} = 7.3 \times 10^{-4} \text{ M}$$

$$[\text{Ca}^{2+}]_{\text{initial}} = \frac{0.095 \text{ g Ca(NO}_3)_2}{0.5000 \text{ L}} \times \frac{1 \text{ mol Ca(NO}_3)_2}{164.09 \text{ g Ca(NO}_3)_2} \times \frac{1 \text{ mol Ca}^{2+}}{1 \text{ mol Ca(NO}_3)_2} = 1.2 \times 10^{-3} \text{ M}$$

$$[\text{CO}_3{}^{2-}]_{\text{initial}} = \frac{0.1000 \text{ g Na}_2\text{CO}_3}{0.5000 \text{ L}} \times \frac{1 \text{ mol Na}_2\text{CO}_3}{105.99 \text{ g Na}_2\text{CO}_3} \times \frac{1 \text{ mol CO}_3{}^{2-}}{1 \text{ mol Na}_2\text{CO}_3} = 1.887 \times 10^{-3} \text{ M}$$

For BaCO$_3$: $Q = [\text{Ba}^{2+}][\text{CO}_3{}^{2-}] = (7.3 \times 10^{-4})(1.887 \times 10^{-3}) = 1.4 \times 10^{-6} > K_{\text{sp}}$

For CaCO$_3$: $Q = [\text{Ca}^{2+}][\text{CO}_3{}^{2-}] = (1.2 \times 10^{-3})(1.887 \times 10^{-3}) = 2.3 \times 10^{-6} > K_{\text{sp}}$

For both BaCO$_3$ and CaCO$_3$, $Q > K_{\text{sp}}$, so **a precipitate of both CaCO$_3$ and BaCO$_3$ will form.**

15-4   DISSOLVING PRECIPITATES

40. This problem is related to Example 15.10.
    (a) See pages 399–400: the anion of CaF$_2$ (F$^-$) is a weak base (it is the conjugate base of HF), so F$^-$ from CaF$_2$ reacts with strong acid (H$^+$) to form molecules of HF (equation ② below):

$$\text{CaF}_2(s) \rightleftharpoons \text{Ca}^{2+}(aq) + 2\text{F}^-(aq) \qquad\qquad ①$$
$$\text{H}^+(aq) + \text{F}^-(aq) \rightleftharpoons \text{HF}(aq) \qquad\qquad ②$$

① + (2 × ②): CaF$_2$(s) + 2H$^+$(aq) + 2F$^-$(aq) $\rightleftharpoons$ Ca^{2+}(aq) + 2F$^-$(aq) + 2HF(aq)

CaF$_2$(s) + 2H$^+$(aq) $\rightleftharpoons$ Ca^{2+}(aq) + 2HF(aq)

(b) See pages 399–400; this is very similar to the second part of Example 15.10: the anion of CuCO$_3$ (CO$_3{}^{2-}$) is a weak base (it is the conjugate base of H$_2$CO$_3$):

CuCO$_3$(s) + 2H$^+$(aq) $\rightleftharpoons$ Cu^{2+}(aq) + H$_2$CO$_3$(aq)

(c) See pages 399–400. This is very similar to the first part of Example 15.10: the anion of Ti(OH)$_3$ (OH$^-$) is a weak base (it is the conjugate base of H$_2$O):

$$\text{Ti(OH)}_3(s) \rightleftharpoons \text{Ti}^{3+}(aq) + 3\text{OH}^-(aq) \qquad\qquad ①$$
$$\text{H}^+(aq) + \text{OH}^-(aq) \rightleftharpoons \text{H}_2\text{O} \qquad\qquad ②$$

① + (3 × ②): Ti(OH)$_3$(s) + 3H$^+$(aq) + 3OH$^-$(aq) $\rightleftharpoons$ Ti^{3+}(aq) + 3OH$^-$(aq) + 3H$_2$O

Ti(OH)$_3$(s) + 3H$^+$(aq) $\rightleftharpoons$ Ti^{3+}(aq) + 3H$_2$O

(d) This is somewhat similar to part (c) above: the complex anion Sn(OH)$_6{}^{2-}$ is a source of OH$^-$ ions that will react with strong acid. (Note that reaction ① below is the reverse of the complex ion formation reaction.):

$$\text{Sn(OH)}_6{}^{2-}(aq) \rightleftharpoons \text{Sn}^{4+}(aq) + 6\text{OH}^-(aq) \qquad\qquad ①$$
$$\text{H}^+(aq) + \text{OH}^-(aq) \rightleftharpoons \text{H}_2\text{O} \qquad\qquad ②$$

① + (6 × ②): $Sn(OH)_6^{2-}(aq) + 6H^+(aq) + 6OH^-(aq) \rightleftharpoons Sn^{4+}(aq) + 6OH^-(aq) + 6H_2O$

$Sn(OH)_6^{2-}(aq) + 6H^+(aq) \rightleftharpoons Sn^{4+}(aq) + 6H_2O$

(e) This is related to part (d) above: the complex anion $Cd(NH_3)_4^{2+}$ is a source of NH_3 molecules (a weak base) that will react with strong acid. (Note that reaction ① below is the reverse of the complex ion formation reaction.):

$Cd(NH_3)_4^{2+}(aq) \rightleftharpoons Cd^{2+}(aq) + 4NH_3(aq)$ ①
$H^+(aq) + NH_3(aq) \rightleftharpoons NH_4^+$ ②

① + (4× ②): $Cd(NH_3)_4^{2+}(aq) + 4H^+(aq) + 4NH_3(aq)$
$\rightleftharpoons Cd^{2+}(aq) + 4NH_3(aq) + 4NH_4^+(aq)$

$Cd(NH_3)_4^{2+}(aq) + 4H^+(aq) \rightleftharpoons Cd^{2+}(aq) + 4NH_4^+(aq)$

42. (a) This is almost identical to the set of reactions for $Zn(OH)_2$ shown on page 401 of the textbook. Cu^{2+} ions form the complex ion $Cu(NH_3)_4^{2+}$ with NH_3 (Table 15.3).

$Cu(OH)_2(s) \rightleftharpoons Cu^{2+}(aq) + 2OH^-(aq)$ ①
$Cu^{2+}(aq) + 4NH_3(aq) \rightleftharpoons Cu(NH_3)_4^{2+}(aq)$ ②

① + ②: $Cu(OH)_2(s) + Cu^{2+}(aq) + 4NH_3(aq)$
$\rightleftharpoons Cu^{2+}(aq) + 2OH^-(aq) + Cu(NH_3)_4^{2+}(aq)$

$Cu(OH)_2(s) + 4NH_3(aq) \rightleftharpoons Cu(NH_3)_4^{2+}(aq) + 2OH^-(aq)$

(b) Cd^{2+} ions form the complex ion $Cd(NH_3)_4^{2+}$ with NH_3 (Table 15.3). See also problem 40(e) above.

$Cd^{2+}(aq) + 4NH_3(aq) \rightleftharpoons Cd(NH_3)_4^{2+}(aq)$

(c) Pb^{2+} ions do not form a complex ion with NH_3. In order to form a precipitate, the lead cation must combine with an anion to make an electrically neutral salt. Recall (Chapter 13) that solutions of NH_3 are basic, and so will contain OH^- ions:

$NH_3(aq) + H_2O \rightleftharpoons NH_4^+(aq) + OH^-(aq)$ ①

these can combine with Pb^{2+} to form $Pb(OH)_2$:

$Pb^{2+}(aq) + 2OH^-(aq) \rightleftharpoons Pb(OH)_2(s)$ ②

(2 × ①) + ②: $Pb^{2+}(aq) + 2OH^-(aq) + 2NH_3(aq) + 2H_2O$
$\rightleftharpoons Pb(OH)_2(s) + 2NH_4^+(aq) + 2OH^-(aq)$

$Pb^{2+}(aq) + 2NH_3(aq) + 2H_2O \rightleftharpoons Pb(OH)_2(s) + 2NH_4^+(aq)$

44. (a) From Table 15.3, the complex ion of Al^{3+} and OH^- is $Al(OH)_4^-$: this requires the Al^{3+} ion to combine with four OH^- ions:

$$Al^{3+}(aq) + 4OH^-(aq) \rightleftharpoons Al(OH)_4^-(aq)$$

(b) According to Table 15.2, the salt $AlPO_4$ is a precipitate: this requires the Al^{3+} ion to combine with one PO_4^{3-} ion:

$$Al^{3+}(aq) + PO_4^{3-}(aq) \rightleftharpoons AlPO_4(s)$$

(c) According to Table 15.2, the salt $Al(OH)_3$ is a precipitate: this must be the species that dissolves in strong acid. This is very similar to problem 40(c) above:

$$Al(OH)_3(s) + 3H^+(aq) \rightleftharpoons Al^{3+}(aq) + 3H_2O$$

SOLUTION EQUILIBRIA

46. This is very similar to Example 15.9(a). The equation for the dissolution of $Co(OH)_2$ in water is below. Note that the equilibrium constant for reaction ①, K_1, is simply K_{sp} for dissolution of $Co(OH)_2$ (given in the question):

$$Co(OH)_2(s) \rightleftharpoons Co^{2+}(aq) + 2OH^-(aq) \qquad \text{...①} \qquad K_1 = K_{sp}$$

The OH^- formed in reaction ① must react with H^+ (provided by the strong acid HCl):

$$H^+(aq) + OH^-(aq) \rightleftharpoons H_2O \qquad \text{...②} \qquad K_2 = 1/K_w$$

Reaction ② is the reverse of the reaction $H_2O \rightleftharpoons H^+(aq) + OH^-(aq)$, for which the equilibrium constant is K_w (Section 13-1). Thus, to obtain K_2, the equilibrium constant for reaction ②, apply the Reciprocal Rule (Table 12.3): $K_2 = 1/K_w$.

Since there are two mol OH^- in equation ①, we require 2 mol of H^+ and hence 2 mol of reaction ②; apply the Coefficient Rule (Table 12.3) to get the equilibrium constant K_2' for this reaction:

$$2H^+(aq) + 2OH^-(aq) \rightleftharpoons 2H_2O \qquad 2 \times ② \qquad K_2' = (1/K_w)^2$$

Finally, we obtain the overall equation and hence the net ionic equation by adding together ① + (2 × ②), so we must apply the Product Rule (Table 12.3) to get K for the overall reaction:

①:　　　　$Co(OH)_2(s) \rightleftharpoons Co^{2+}(aq) + 2OH^-(aq)$　　$K_1 = K_{sp} = 2 \times 10^{-16}$

2 × ②:　　$2H^+(aq) + 2OH^-(aq) \rightleftharpoons 2H_2O$　　$K_2' = (1/K_w)^2 = 1.0 \times 10^{28}$

---

① + (2 × ②):　$Co(OH)_2(s) + 2H^+(aq) + 2OH^-(aq)$
　　　　　　　　　$\rightleftharpoons Co^{2+}(aq) + 2OH^-(aq) + 2H_2O$

　　　　$Co(OH)_2(s) + 2H^+(aq) \rightleftharpoons Co^{2+}(aq) + 2H_2O$　　$K = K_{sp} \times (1/K_w)^2$

$$K = K_{sp} \times (1/K_w)^2 = (2 \times 10^{-16}) \times (1.0 \times 10^{28}) = \underline{\mathbf{2 \times 10^{12}}}$$

48. Two precipitates are given in the equation in the question: BaF_2 and $BaSO_4$. The equations for their dissolution in water, and their K_{sp} values From Table 15.2, are:

$$BaF_2(s) \rightleftharpoons Ba^{2+}(aq) + 2F^-(aq) \qquad \text{...①} \qquad K_1 = K_{sp}\,(BaF_2) = 1.8 \times 10^{-7}$$

$$BaSO_4(s) \rightleftharpoons Ba^{2+}(aq) + SO_4^{2-}(aq) \qquad \text{...②} \qquad K_2 = K_{sp}\,(BaSO_4) = 1.1 \times 10^{-10}$$

(a) The equation in the question is obtained by subtracting equation ② from equation ① (i.e., adding ③, the reverse of equation ②). The equilibrium constant, K_3, for equation ③ is obtained by the Reciprocal Rule (Table 12.3):

$$Ba^{2+}(aq) + SO_4^{2-}(aq) \rightleftharpoons BaSO_4(s) \qquad \text{...③} \qquad K_3 = 1/K_2 = 9.1 \times 10^9$$

Finally, add together equations ① and ③; the equilibrium constant K for the overall reaction is obtained by the Product Rule (Table 12.3):

①+③: $BaF_2(s) + Ba^{2+}(aq) + SO_4^{2-}(aq)$
$$\rightleftharpoons Ba^{2+}(aq) + 2F^-(aq) + BaSO_4(s)$$

$$BaF_2(s) + SO_4^{2-}(aq) \rightleftharpoons BaSO_4(s) + 2F^-(aq) \qquad K = K_1 \times K_3$$

$$K = (1.8 \times 10^{-7})(9.1 \times 10^9) = \underline{\mathbf{1.6 \times 10^3}}$$

(b) A saturated solution of BaF_2 contains Ba^{2+} ions, so it would be expected that addition of SO_4^{2-} ions (from Na_2SO_4) could form a precipitate of $BaSO_4$. Moreover, according to the equilibrium reaction given in the question, for which K is around 1600, the position of equilibrium lies towards the product side, so formation of $BaSO_4$ is favored over BaF_2 and addition of SO_4^{2-} ions will encourage this.
Therefore: **YES**, a precipitate of $BaSO_4$ will form.

50. This problem is very similar to Example 15.9.

(a) For the two equilibrium constants given in the question, the corresponding reactions are:

$$Cu(OH)_2(s) \rightleftharpoons Cu^{2+}(aq) + 2OH^-(aq) \qquad K_1 = K_{sp}\,(Cu(OH)_2) = 2 \times 10^{-19}$$

$$Cu^{2+}(aq) + 4NH_3(aq) \rightleftharpoons Cu(NH_3)_4^{2+}(aq) \qquad K_2 = K_f\,(Cu(NH_3)_4^{2+}) = 2 \times 10^{12}$$

The reaction given in the question is obtained by adding the two above equations together, so the corresponding equilibrium constant K is obtained by the Product Rule (Table 12.3):

$Cu(OH)_2(s) + Cu^{2+}(aq) + 4NH_3(aq)$
$$\rightleftharpoons Cu^{2+}(aq) + 2OH^-(aq) + Cu(NH_3)_4^{2+}(aq)$$

$$Cu(OH)_2(s) + 4NH_3(aq) \rightleftharpoons Cu(NH_3)_4^{2+}(aq) + 2OH^-(aq) \qquad K = K_1 \times K_2$$

$$K = K_1 \times K_2 = (2 \times 10^{-19})(2 \times 10^{12}) = \underline{\mathbf{4 \times 10^{-7}}}$$

(b) The expression for K is:

$$K = \frac{[Cu(NH_3)_4^{2+}][OH^-]^2}{[NH_3]^4}$$

and note from the stoichiometry of the reaction that $[OH^-] = 2[Cu(NH_3)_4^{2+}]$. Let s = solubility of $Cu(OH)_2$ in mol/L. Thus:

$$[Cu(NH_3)_4^{2+}] = \frac{s \text{ mol } Cu(OH)_2}{1\,L} \times \frac{1 \text{ mol } Cu(NH_3)_4^{2+}}{1 \text{ mol } Cu(OH)_2} = s \text{ mol/L} \quad \text{and} \quad [OH^-] = 2s \text{ mol/L}$$

Substitute these, and the given $[NH_3]$ into the expression for K:

$$K = \frac{s(2s)^2}{(4.5)^4} = 4 \times 10^{-7} \qquad \rightarrow \quad s^3 = 4 \times 10^{-5} \qquad\qquad \rightarrow \qquad s = \underline{\mathbf{3 \times 10^{-2}\ M}}$$

52. This problem is similar to Example 15.9(a). The equation for the dissolution of $PbCl_2$ in water is:

$$PbCl_2(s) \;\rightleftharpoons\; Pb^{2+}(aq) + 2Cl^-(aq) \qquad\qquad K_1 = K_{sp}\ (PbCl_2) = 1.7 \times 10^{-5}$$

The equation for forming the complex ion $Pb(OH)_3^-$ from its components is:

$$Pb^{2+}(aq) + 3OH^-(aq) \;\rightleftharpoons\; Pb(OH)_3^-(aq) \qquad\qquad K_2 = K_f\ (Pb(OH)_3^-) = 3.8 \times 10^{14}$$

The reaction whereby $PbCl_2$ dissolves in excess OH^- (in this case, 0.2 M NaOH) is obtained by adding together the two above equations, so the corresponding equilibrium constant K is obtained by the Product Rule (Table 12.3):

$$PbCl_2(s) + Pb^{2+}(aq) + 3OH^-(aq)$$
$$\rightleftharpoons Pb^{2+}(aq) + 2Cl^-(aq) + Pb(OH)_3^-(aq)$$

$$PbCl_2(s) + 3OH^-(aq) \;\rightleftharpoons\; Pb(OH)_3^-(aq) + 2Cl^-(aq) \qquad K = K_1 \times K_2$$

$$K = K_1 \times K_2 = (1.7 \times 10^{-5})(3.8 \times 10^{14}) = 6.5 \times 10^9$$

For this reaction, the expression for K is:

$$K = \frac{[Pb(OH)_3^-][Cl^-]^2}{[OH^-]^3}$$

Let s = the solubility of $PbCl_2$ in mol/L. Thus, using the stoichiometry of the overall reaction:

$$[Pb(OH)_3^-] = \frac{s \text{ mol } PbCl_2}{1\,L} \times \frac{1 \text{ mol } Pb(OH)_3^-}{1 \text{ mol } PbCl_2} = s \text{ mol/L}$$

and

$$[Cl^-] = \frac{s \text{ mol PbCl}_2}{1 \text{ L}} \times \frac{2 \text{ mol Cl}^-}{1 \text{ mol PbCl}_2} = 2s \text{ mol/L}$$

Substitute these, plus the given $[OH^-]$, into the expression for K and solve for s

$$K = \frac{(s)(2s)^2}{(0.2)^3} = 6.5 \times 10^9 \qquad \rightarrow \qquad s^3 = 1 \times 10^7 \qquad \rightarrow \qquad s = \underline{\mathbf{2 \times 10^2 \; M}}$$

The solubility of $PbCl_2$ in 0.2 M NaOH is $\underline{\mathbf{2 \times 10^2 \; M}}$.

54. Two reactions are relevant to the equation given in the question: dissolution of $Zn(OH)_2$ in water and formation of the complex ion $Zn(OH)_4^{2-}$. The equations for these reactions and their equilibrium constants (Tables 15.2 and 15.1, respectively) are:

$$Zn(OH)_2(s) \; \rightleftharpoons \; Zn^{2+}(aq) + 2OH^-(aq) \qquad \ldots ① \qquad\qquad K_1 = K_{sp} \, (Zn(OH)_2) = 4 \times 10^{-17}$$

$$Zn^{2+}(aq) + 4OH^-(aq) \; \rightleftharpoons \; Zn(OH)_4^{2-}(aq) \quad \ldots ② \qquad\qquad K_2 = K_f \, (Zn(OH)_4^{2-}) = 3 \times 10^{14}$$

(a) The reaction in the question is obtained by adding together equations ① and ②; use the Product Rule (Table 12.3) to obtain the equilibrium constant K for the overall reaction:

$$Zn(OH)_2(s) + Zn^{2+}(aq) + 4OH^-(aq)$$
$$\rightleftharpoons \; Zn^{2+}(aq) + 2OH^-(aq) + Zn(OH)_4^{2-}(aq)$$

$$Zn(OH)_2(s) + 2OH^-(aq) \; \rightleftharpoons \; Zn(OH)_4^{2-}(aq) \qquad K = K_1 \times K_2$$

$$K = (4 \times 10^{-17})(3 \times 10^{14}) = \underline{\mathbf{1.2 \times 10^{-2}}}$$

(b) The expression for K is:

$$K = \frac{[Zn(OH)_4^{2-}]}{[OH^-]^2}$$

We are told that 10.0 g of $Zn(OH)_2$ are dissolved in 1.0 L of solution. Thus:

$$[Zn(OH)_4^{2-}] = \frac{10.0 \text{ g Zn(OH)}_2}{1.0 \text{ L}} \times \frac{1 \text{ mol Zn(OH)}_2}{99.42 \text{ g Zn(OH)}_2} \times \frac{1 \text{ mol Zn(OH)}_4^{2-}}{1 \text{ mol Zn(OH)}_2} = 0.10 \; M$$

Substitute this concentration into the expression for K and solve for $[OH^-]$:

$$K = \frac{(0.10)}{[OH^-]^2} = 1.2 \times 10^{-2} \quad \rightarrow \quad [OH^-]^2 = 8.3 \qquad \rightarrow [OH^-] = \underline{\mathbf{3 \; M}}$$

UNCLASSIFIED PROBLEMS

56. For the given reaction, the expression for K must be:

$$K = \frac{[\text{hemoglobin·CO}]P_{O_2}}{[\text{hemoglobin·O}_2]P_{CO}} = 2.0 \times 10^2$$

Let the initial [hemoglobin·O$_2$] = x. If 12% of the hemoglobin in the blood is converted to the CO complex then (assuming the remaining hemoglobin is the O$_2$ complex):

[hemoglobin·CO] = 0.12x and   [hemoglobin·O$_2$] = 0.88x

Substitute these concentrations into the expression for K and hence obtain the required ratio of partial pressures:

$$K = \frac{[\text{hemoglobin·CO}]P_{O_2}}{[\text{hemoglobin·O}_2]P_{CO}} = \frac{(0.12x)P_{O_2}}{(0.88x)P_{CO}} = \frac{(0.12)P_{O_2}}{(0.88)P_{CO}} = 2.0 \times 10^2$$

$$\frac{P_{CO}}{P_{O_2}} = \underline{\mathbf{6.8 \times 10^{-4}}}$$

58. See Sections 10-3 and 10-4. Recall that the freezing point of pure H$_2$O is $0°$C. Since the given solution freezes at $-0.33°$C, $\Delta T_f = 0.33°$C. For a salt such as Li$_2$CO$_3$, the value of the van't Hoff factor $i = 3$ (assuming complete dissociation, we expect two Li$^+$ ions and one CO$_3^{2-}$ ion per formula unit of Li$_2$CO$_3$). Substitute these values into Equation 10.6:

$\Delta T_f = i \times 1.86°$C/$m$ × molality   → $0.33°$C = 3 × 1.86$°$C/m × molality

molality = 0.059 m = (0.059 mol solute)/(kg solvent)

To convert the molality to molarity of Li$_2$CO$_3$, construct a table as in Example 10.4 and shade the spots required for molarity, namely moles of solute and volume of solution. The molality gives moles of solute and mass of solvent.

	moles	$\xleftarrow{\text{MM}}$	mass	$\xrightarrow{\text{density}}$	volume
Solute	0.059				
Solvent			1 kg = 1000 g		
Solution					

Convert moles of solute (Li$_2$CO$_3$) to mass of solute using the molar mass:

$$\text{mass of solute} = 0.059 \text{ mol} \times \frac{73.891\,\text{g}}{1\,\text{mol}} = 4.4 \text{ g}$$

Hence the total mass of solution = 4.4 g solute + 1000 g solvent = 1004 g solution

The corresponding volume of solution (using the given density) is

$$\text{vol of solution (L)} = 1004 \text{ g} \times \frac{1 \text{ mL}}{1.0 \text{ g}} \times \frac{1 \text{ L}}{1000 \text{ mL}} = 1.004 \text{ L}$$

Complete the table with the values calculated above:

	moles $\xleftarrow{ MM }$ mass $\xrightarrow{ \text{density} }$ volume		
Solute	0.059	4.4 g	
Solvent		1 kg = 1000 g	
Solution		1004 g	1.004 L

Hence the molarity of the solution is:

$$\text{molarity, } M = \frac{\text{mol solute}}{V_{\text{solution}} \text{ (L)}} = \frac{0.059 \text{ mol}}{1.004 \text{ L}} = 0.059 \text{ } M$$

This is the concentration of Li_2CO_3 in this saturated solution; in other words, this is the compound's solubility, *s*. The equation for dissolution of Li_2CO_3 in water is:

$$Li_2CO_3(s) \rightleftharpoons 2Li^+(aq) + CO_3^{2-}(aq) \qquad K_{sp} = [Li^+]^2[CO_3^{2-}]$$

from which it can be seen that $[Li^+] = 2[Li_2CO_3] = 2s$ and $[CO_3^{2-}] = [Li_2CO_3] = s$. Thus $K_{sp} = 4s^3$. Use the value for *s* calculated above to determine K_{sp}:

$$K_{sp} = 4s^3 = 4(0.059)^3 = \underline{\mathbf{8.1 \times 10^{-4}}}$$

60. See Section 12-5.

(a) Addition of $A(NO_3)_3$ will provide A^{3+} ions, increasing the concentration of a product in this equilibrium reaction. Thus the system will **shift to the left** to consume some of the added product, causing more AB_3 to precipitate.

(b) ΔH is negative (the reaction is exothermic). Increasing temperature will decrease *K* (see pages 326–328) and — since the expression for *K* will involve only the products of this reaction — this must correspond to less AB_3 dissolving (a **shift to the left**) to decrease the products' concentrations.

(c) Added Na^+ may interact with B^- to form an ion pair (see Figure 10.18 and adjacent discussion), thereby effectively removing B^- ions (a product in this equilibrium reaction) from solution. Thus the system will **shift to the right** to replace some of the removed product, causing more AB_3 to dissolve.

62. (a) Use the given information to calculate $[NH_4^+]$. Assume that the volume of solution is constant:

$$[NH_4^+] = \frac{21.0 \text{ g } NH_4Cl}{2.0 \text{ L}} \times \frac{1 \text{ mol } NH_4Cl}{53.49 \text{ g } NH_4Cl} \times \frac{1 \text{ mol } NH_4^+}{1 \text{ mol } NH_4Cl} = 0.20 \text{ M}$$

Now set up an equilibrium table based on the dissociation of NH_3 in water, for which K_b = 1.8×10^{-5} (Table 13.4); this is somewhat similar to Example 13.11 but with the added complication that the common ion NH_4^+ is present. (Note that although we have a mixture of NH_3/NH_4^+, the concentrations of the two buffer components differ significantly, and Equation 14.1 should not be used.)
Use $[NH_3]_0 = 4.17 \text{ M}$ (given) and $[NH_4^+]_0 = 0.20 \text{ M}$ (calculated above); initially we assume there is no ionization, so $[OH^-]_0 = 0$. Let $\Delta[OH^-] = x$:

	$NH_3(aq)$	+	H_2O	$\rightleftharpoons$	$NH_4^+(aq)$	+	$OH^-(aq)$
$[\ ]_0$ (M)	4.17				0.20		0
$\Delta[\ ]$ (M)	$-x$				$+x$		$+x$
$[\ ]_{eq}$ (M)	$4.17 - x$				$0.20 + x$		x

Now substitute the equilibrium concentrations of each species into the K_b expression:

$$K_b = 1.8 \times 10^{-5} = \frac{[NH_4^+][OH^-]}{[NH_3]} = \frac{(0.20 + x)(x)}{(4.17 - x)}$$

Assume $x \ll 0.20$:

$$1.8 \times 10^{-5} \approx \frac{0.20x}{4.17} \qquad \text{so} \quad x = 3.8 \times 10^{-4}$$

Check the assumption that the value of x is indeed much less that 0.20: the % ionization = $[OH^-]_{eq}/[NH_4^+]_0 \times 100$ = $(3.8 \times 10^{-4})/(0.20) \times 100 = 0.19\% < 5\%$ (see page 345), so the assumption made in solving the above equation was valid.

Thus $x = [OH^-]_{eq} = \mathbf{3.8 \times 10^{-4} \text{ M}}$.

(b) Adding $CaCl_2$ provides Ca^{2+} ions that can combine with OH^- ions to form a precipitate of $Ca(OH)_2$. The equation for the dissolution of $Ca(OH)_2$ in water is:

$$Ca(OH)_2(s) \rightleftharpoons Ca^{2+}(aq) + 2OH^-(aq) \qquad\qquad K_{sp} = [Ca^{2+}][OH^-]^2$$

To determine whether a precipitate will form, calculate the initial concentration of Ca^{2+} and use that along with $[OH^-]$ from part (a) to calculate Q, and compare the value of Q with K_{sp}. (From Table 15.2, $K_{sp} = 4.0 \times 10^{-6}$.) Again, assume that the volume of solution is constant.

$$[Ca^{2+}] = \frac{4.8 \text{ g } CaCl_2}{2.0 \text{ L}} \times \frac{1 \text{ mol } CaCl_2}{110.98 \text{ g } CaCl_2} \times \frac{1 \text{ mol } Ca^{2+}}{1 \text{ mol } CaCl_2} = 2.2 \times 10^{-2} \text{ M}$$

$$Q = [Ca^{2+}][OH^-]^2 = (2.2 \times 10^{-2})(3.8 \times 10^{-4})^2 = 3.2 \times 10^{-9}$$

$Q < K_{sp}$, so **NO, a precipitate will not form**

(c)   Since no precipitate of $Ca(OH)_2$ will form, $[Ca^{2+}]$ calculated in part (b) above will be the equilibrium concentration. **$[Ca^{2+}] = 2.2 \times 10^{-2}$ M**

64. This problem is related to Example 15.4(a); see also the discussion in the text immediately above Example 15.4. Let s = the molar solubility of CaC_2O_4. The equation for the compound dissolving in water is:

$$CaC_2O_4(s) \rightleftharpoons Ca^{2+}(aq) + C_2O_4^{2-}(aq) \qquad\qquad K_{sp} = [Ca^{2+}][C_2O_4^{2-}]$$

and by analogy with problem 14(a) above, the relationship between s and K_{sp} is:

$$K_{sp} = [Ca^{2+}][C_2O_4^{2-}] = (s)(s) = s^2 \qquad\qquad OR \quad s = K_{sp}^{1/2}$$

This will apply at any temperature. Thus, at 95 °C:

$$s = K_{sp}^{1/2} = (1 \times 10^{-8})^{1/2} = 1 \times 10^{-4} \ M$$

In other words, 1×10^{-4} mol of the compound will dissolve in 1 liter at that temperature. Therefore the mass of solute in 500 mL of this saturated solution is:

$$mass \ CaC_2O_4 = \frac{1 \times 10^{-4} \ mol \ CaC_2O_4}{1L} \times 0.500 \ L \times \frac{128.10 \ g \ CaC_2O_4}{1 \, mol \ CaC_2O_4} = 6 \times 10^{-3} \ g$$

At 25 °C, the same calculations give:

$$s = K_{sp}^{1/2} = (4 \times 10^{-9})^{1/2} = 6 \times 10^{-5} \ M$$

and the mass of solute in 500 mL of this saturated solution is:

$$mass \ CaC_2O_4 = \frac{6 \times 10^{-5} \ mol \ CaC_2O_4}{1L} \times 0.500 \ L \times \frac{128.10 \ g \ CaC_2O_4}{1 \, mol \ CaC_2O_4} = 4 \times 10^{-3} \ g$$

At the higher temperature, a greater quantity of solute can be accommodated by the solution. Upon cooling, less CaC_2O_4 can remain in solution and so the excess must precipitate from solution. Thus the mass of solute that will precipitate upon cooling is the difference between these two masses:

mass of precipitate = $(6 \times 10^{-3}$ g$) - (4 \times 10^{-3}$ g$)$ = **2×10^{-3} g**   or   **2 mg**

CONCEPTUAL PROBLEMS

66. Since the first box in the problem represents a saturated solution, we know that four MX_2 units will dissolve in one liter. For the subsequent boxes, dissolve up to four MX_2 units and leave any remainder undissolved.

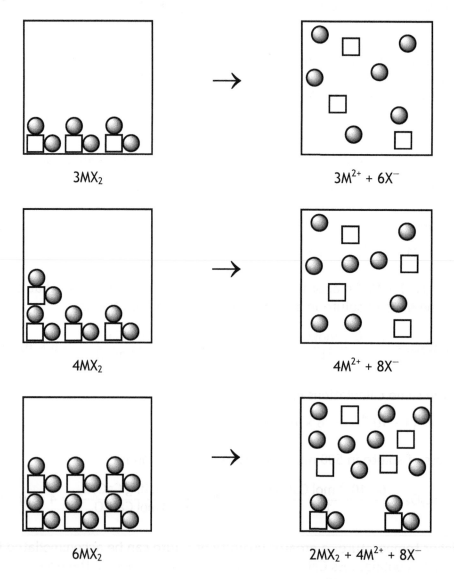

3MX$_2$ → 3M^{2+} + 6X$^-$

4MX$_2$ → 4M^{2+} + 8X$^-$

6MX$_2$ → 2MX$_2$ + 4M^{2+} + 8X$^-$

68. (a) **True.** The expression for K_{sp} is always the product of concentrations of ions. Thus if the substance is insoluble the concentrations of the ions in solution will be very low and hence K_{sp} will be much less than one.

(b) **False.** Solubility is related to K_{sp} and therefore an increase in solubility corresponds to an increase in the value of K_{sp}. From Section 12-5, K increases with temperature for **endothermic** processes, so it must be concluded that dissolving $PbCl_2$ is an endothermic process.

(c) **False.** A saturated solution of $Cu(OH)_2$ involves the following equilibrium:

$$Cu(OH)_2(s) \rightleftharpoons Cu^{2+}(aq) + 2OH^-(aq)$$

whereas strips of copper metal will provide Cu(s) which is not involved in this equilibrium reaction and therefore will not affect it.

70. Since the plot of solubility against temperature yields a straight line with a negative slope it can be concluded that the solubility of this compound **decreases** with temperature. Solubility is related to K_{sp} and therefore an increase in solubility corresponds to an increase in the value of K_{sp}. From Section 12-5 (page 327), K decreases with temperature for exothermic processes, so: **YES, dissolving compound X is an exothermic process**.

72. (a) Adding HCl (a strong acid) should increase the solubility of the carbonate salt $CaCO_3$ (see page 400): **graph (3)**.

   (b) Increasing temperature will increase K (= K_{sp}) for an endothermic process (Section 12-5) and therefore increase the solubility: **graph (5)**.

   (c) Adding $CaCl_2$ introduces a common ion (Ca^{2+}) and therefore will decrease the solubility of $CaCO_3$ (see pages 393–394): **graph (2)**.

   (d) Adding NaCl introduces Na^+ and Cl^-, neither of which are involved in the dissolution of $CaCO_3$; nor will they form precipitates since the other possible cation/anion combinations (Na_2CO_3 and $CaCl_2$) are soluble species. Thus there will be no effect upon the solubility of $CaCO_3$: **graph (1)**.

CHALLENGE PROBLEMS

73. To determine K for the reaction in the question, proceed similarly to Example 15.9(a).

   The equation for the dissolution of $Zn(OH)_2$ in water is below; the equilibrium constant for this reaction, K_1, is simply K_{sp} for dissolution of $Zn(OH)_2$ (see Table 15.2; K_{sp} for $Zn(OH)_2$ is 4×10^{-17}):

   $$Zn(OH)_2(s) \rightleftharpoons Zn^{2+}(aq) + 2OH^-(aq) \qquad \ldots \text{①} \quad K_1 = K_{sp} \text{ (Zn(OH}_2\text{)}$$

   The OH^- formed in reaction ① must react with the weak acid HCN:

   $$HCN(aq) + OH^-(aq) \rightleftharpoons CN^-(aq) + H_2O \qquad \ldots \text{②} \quad K_2 = 1/K_b \text{ (NH}_3\text{)}$$

   Reaction ② is the reverse of the reaction $CN^-(aq) + H_2O \rightleftharpoons HCN(aq) + OH^-(aq)$, for which the equilibrium constant is K_b for CN^- (see Section 13-5; K_b for CN^- is 1.7×10^{-5}). Thus, to obtain K_2, the equilibrium constant for reaction ②, apply the Reciprocal Rule (Table 12.3): $K_2 = 1/K_b$. Since there are two mol OH^- in equation ①, we require 2 mol of HCN and hence 2 mol of reaction ②; apply the Coefficient Rule (Table 12.3) to get the equilibrium constant K_3 for this reaction:

   $$2HCN(aq) + 2OH^-(aq) \rightleftharpoons 2CN^-(aq) + 2H_2O \qquad \ldots \text{③} \quad K_3 = (K_2)^2 = (1/K_b)^2$$

Finally, we obtain the overall equation and hence the net ionic equation by adding together ① + ③, so we must apply the Product Rule (Table 12.3) to get K for the overall reaction:

①: $\qquad$ $Zn(OH)_2(s) \rightleftharpoons Zn^{2+}(aq) + 2OH^-(aq)$ $\qquad$ $K_1 = K_{sp} = 4 \times 10^{-17}$

③: $\qquad$ $2HCN(aq) + 2OH^-(aq) \rightleftharpoons 2CN^-(aq) + 2H_2O$ $\qquad$ $K_3 = (1/K_b)^2 = 3.5 \times 10^9$

$$Zn(OH)_2(s) + 2HCN(aq) + 2OH^-(aq)$$
$$\rightleftharpoons Zn^{2+}(aq) + 2OH^-(aq) + 2CN^-(aq) + 2H_2O$$

$$Zn(OH)_2(s) + 2HCN(aq) \rightleftharpoons Zn^{2+}(aq) + 2CN^-(aq) + 2H_2O$$

$$K = K_1 \times K_3 = K_{sp} \times (1/K_b)^2 = (4 \times 10^{-17}) \times (3.5 \times 10^9) = 1 \times 10^{-7}$$

The expression for K is:

$$K = \frac{[Zn^{2+}][CN^-]^2}{[HCN]^2}$$

and note from the stoichiometry of the reaction that $[CN^-] = 2[Zn^{2+}]$. Let s = solubility of $Zn(OH)_2$ in mol/L. Thus:

$$[Zn^{2+}] = \frac{s \text{ mol } Zn(OH)_2}{1L} \times \frac{1 \text{ mol } Zn^{2+}}{1 \text{ mol } Zn(OH)_2} = s \text{ mol/L} \quad \text{and} \quad [CN^-] = 2s \text{ mol/L}$$

Substitute these two, along with [HCN] given in the question, into the expression for K:

$$K = \frac{s(2s)^2}{(0.15)^2} = 1 \times 10^{-7} \quad \rightarrow \quad s^3 = 6 \times 10^{-10} \quad \rightarrow \quad s = \underline{\mathbf{8 \times 10^{-4} M}}$$

To obtain the sdolubility of $Zn(OH)_2$ in pure H_2O, proceed as in Example 15.4; see also the discussion in the text immediately before that example. From Table 15.2, $K_{sp} = 4 \times 10^{-17}$. Let s = the molar solubility of $Zn(OH)_2$. For the compound dissolving in water:

$$Zn(OH)_2(s) \rightleftharpoons Zn^{2+}(aq) + 2OH^-(aq) \qquad\qquad K_{sp} = [Zn^{2+}][OH^-]^2$$
$$\quad s \qquad\qquad\quad s \qquad\quad 2s$$

The equilibrium concentration of Zn^{2+} is equal to the solubility of $Zn(OH)_2$ and the equilibrium concentration of OH^- is equal to twice the solubility of $Zn(OH)_2$:

$$[Zn^{2+}] = \frac{s \text{ mol } Zn(OH)_2}{1L} \times \frac{1 \text{ mol } Zn^{2+}}{1 \text{ mol } Zn(OH)_2} = s \text{ mol/L}$$

$$[OH^-] = \frac{s \text{ mol } Zn(OH)_2}{1L} \times \frac{2 \text{ mol } OH^-}{1 \text{ mol } Zn(OH)_2} = 2s \text{ mol/L}$$

Thus:

$$K_{sp} = [Zn^{2+}][OH^-]^2 = (s)(2s)^2 = 4s^3 \rightarrow s = \left(\frac{K_{sp}}{4}\right)^{\frac{1}{3}} = \left(\frac{4 \times 10^{-17}}{4}\right)^{\frac{1}{3}} = \underline{\mathbf{2 \times 10^{-6}\ M}}$$

The solubility of $Zn(OH)_2$ in pure water is **2×10^{-6} M**, whereas in 0.15 M HCN its solubility is **8×10^{-4} M**; so **the solubility is improved by a factor of 400 in the acid solution**.

74. For the given buffer components, use Equation 14.1 to determine $[H^+]$. From Table 13.2, K_a for $HCHO_2$ is 1.9×10^{-4}.

$$[H^+] = K_a \times \frac{[HCHO_2]}{[CHO_2^-]} = (1.9 \times 10^{-4}) \times \frac{0.30}{0.20} = 2.8 \times 10^{-4}\ M$$

For the equation given in the question, refer to problem 40(a) above. The overall reaction is related to the following two reactions:

$$CaF_2(s) \rightleftharpoons Ca^{2+}(aq) + 2F^-(aq) \qquad \ldots \text{①} \qquad K_1 = K_{sp(CaF_2)}$$
$$H^+(aq) + F^-(aq) \rightleftharpoons HF(aq) \qquad \ldots \text{②} \qquad K_2 = 1/K_{a(HF)}$$

Reaction ② is the reverse of the reaction $HF(aq) \rightleftharpoons H^+(aq) + F^-(aq)$ for which the equilibrium constant is K_a for HF. To obtain K_2, apply the Reciprocal Rule (Table 12.3). Since there are 2 mol F^- in equation ①, 2 mol of reaction ② will be required to provide 2 mol H^+; to obtain the equilibrium constant, K_3, for this reaction apply the Coefficient Rule (Table 12.3):

$$2H^+(aq) + 2F^-(aq) \rightleftharpoons 2HF(aq) \qquad \ldots \text{③} \qquad K_3 = (K_2)^2 = [1/K_{a(HF)}]^2$$

The overall equation is obtained by adding reactions ① and ③; to obtain the equilibrium constant, K, for the overall reaction apply the Product Rule (Table 12.3):

$$\text{① + ③:} \qquad CaF_2(s) + 2H^+(aq) + 2F^-(aq) \rightleftharpoons Ca^{2+}(aq) + 2F^-(aq) + 2HF(aq)$$
$$CaF_2(s) + 2H^+(aq) \rightleftharpoons Ca^{2+}(aq) + 2HF(aq)$$

$$K = K_1 \times K_3 = K_{sp(CaF_2)} \times [1/K_{a(HF)}]^2 = (1.5 \times 10^{-10}) \times [1/(6.9 \times 10^{-4})]^2 = 3.2 \times 10^{-4}$$

The expression for K is:

$$K = \frac{[Ca^{2+}][HF]^2}{[H^+]^2}$$

and note from the stoichiometry of the reaction that $[HF] = 2[Ca^{2+}]$. Let s = solubility of CaF_2 in mol/L.
Thus:

$$[Ca^{2+}] = \frac{s\ \text{mol}\ CaF_2}{1\,L} \times \frac{1\,\text{mol}\ Ca^{2+}}{1\,\text{mol}\ CaF_2} = s\ \text{mol/L} \quad \text{and} \quad [HF] = 2s\ \text{mol/L}$$

Substitute the two concentration expressions, and [H⁺] calculated above, into the expression for K:

$$K = \frac{s(2s)^2}{(2.8 \times 10^{-4})^2} = 3.2 \times 10^{-4} \quad \rightarrow \quad s^3 = 6.3 \times 10^{-12} \quad \rightarrow \quad s = \underline{1.8 \times 10^{-4}\ M}$$

Thus the solubility of CaF_2 in this buffer is $\underline{1.8 \times 10^{-4}\ M}$.

75. This problem is similar to Example 15.8. From Table 15.2, K_{sp} for AgCl is 1.8×10^{-10}, and K_{sp} for AgBr is 5×10^{-13}. For each salt, precipitation begins when $Q = K_{sp}$; therefore substitute the given concentrations of Cl⁻ and Br⁻ in the K_{sp} expression for AgCl and AgBr, respectively, and solve to find [Ag⁺] at which precipitation begins. (Assume that concentrations do not change as the concentrated $AgNO_3$ solution is added.)

> for AgCl:
> $$K_{sp} = [Ag^+][Cl^-] = [Ag^+](0.050) = 1.8 \times 10^{-10} \qquad \rightarrow \quad [Ag^+] = 3.6 \times 10^{-9}\ M$$
> for AgBr:
> $$K_{sp} = [Ag^+][Br^-] = [Ag^+](0.050) = 5 \times 10^{-13} \qquad \rightarrow \quad [Ag^+] = 1 \times 10^{-11}\ M$$

A lower concentration of Ag⁺ is required to precipitate AgBr, so AgBr will precipitate first. The concentration of Ag⁺ must be increased significantly in order to precipitate AgCl. At that point, a large amount of AgBr will already have precipitated so [Br⁻] will have been significantly reduced. (Compare Figure 15.5: as [Ag⁺] increases, [Br⁻] decreases.) To determine [Br⁻] at this point, substitute the value of [Ag⁺] required to precipitate AgCl into the K_{sp} expression for AgBr:

$$K_{sp} = [Ag^+][Br^-] = (3.6 \times 10^{-9})[Br^-] = 5 \times 10^{-13} \quad \rightarrow \quad [Br^-] = \underline{1 \times 10^{-4}\ M}$$

76. This problem is related to Example 15.8(a). (Assume that concentrations do not change as the solution providing OH⁻ ions is added.)

(a) Precipitation begins when $Q = K_{sp}$; therefore substitute the given concentration of Mg^{2+} into the K_{sp} expression for $Mg(OH)_2$, and solve to find [OH⁻] at which precipitation begins.

$$K_{sp} = [Mg^{2+}][OH^-]^2 = (0.056)[OH^-]^2 = 6 \times 10^{-12} \quad \rightarrow \quad [OH^-] = \underline{1 \times 10^{-5}\ M}$$

(b) For the other ions, determine Q using [OH⁻] from part (a) and compare Q with K_{sp} to determine whether precipitation occurs.

> For Na⁺: NaOH is highly soluble, so **no precipitate of NaOH will form.**
>
> For Ca²⁺: $Q = [Ca^{2+}][OH^-]^2 = (0.01)(1 \times 10^{-5})^2 = 1 \times 10^{-12} < 1.0 \times 10^{-10}$
>
> $Q < K_{sp}$ so **no precipitate of Ca(OH)₂ will form.**
>
> For Al³⁺: $Q = [Al^{3+}][OH^-]^3 = (4 \times 10^{-7})(1 \times 10^{-5})^3 = 4 \times 10^{-22} > 2 \times 10^{-31}$
>
> $Q > K_{sp}$ so **a precipitate of Al(OH)₃ will form.**

For Fe^{3+}: $Q = [Fe^{3+}][OH^-]^3 = (2 \times 10^{-7})(1 \times 10^{-5})^3 = 2 \times 10^{-22} > 3 \times 10^{-39}$

$Q > K_{sp}$ so **a precipitate of $Fe(OH)_3$ will form.**

(c) When 50 % of the Mg^{2+} has precipitated, $[Mg^{2+}]$ must have fallen to 0.028 M. Repeat the calculation in part (a) to determine the $[OH^-]$ required to achieve this $[Mg^{2+}]$. For the other ions, repeat part (b) with this new $[OH^-]$. For those ions that precipitate calculate their concentration at equilibrium in the presence of this $[OH^-]$, and hence how much of each ion has precipitated.

$$K_{sp} = [Mg^{2+}][OH^-]^2 = (0.028)[OH^-]^2 = 6 \times 10^{-12} \quad \rightarrow \quad [OH^-] = 1 \times 10^{-5}\ M$$

Unfortunately (to one significant figure) this is the same answer as in part (a)! Therefore, repeating part (b) we will again conclude that only $Al(OH)_3$ and $Fe(OH)_3$ will also precipitate. Now calculate the values of $[Al^{3+}]$ and $[Fe^{3+}]$ that will correspond to this $[OH^-]$ and hence what percentage of each ion has precipitated:

For Al^{3+}:

$$K_{sp} = [Al^{3+}][OH^-]^3 = [Al^{3+}](1 \times 10^{-5})^3 = 2 \times 10^{-31} \quad \rightarrow \quad [Al^{3+}] = 2 \times 10^{-16}\ M$$

% of ion remaining in solution $= \dfrac{2 \times 10^{-16}}{4 \times 10^{-7}} \times 100 = 5 \times 10^{-8}\ \%$

For Fe^{3+}:

$$K_{sp} = [Fe^{3+}][OH^-]^3 = [Fe^{3+}](1 \times 10^{-5})^3 = 3 \times 10^{-39} \quad \rightarrow \quad [Fe^{3+}] = 3 \times 10^{-24}\ M$$

% of ion remaining in solution $= \dfrac{3 \times 10^{-24}}{2 \times 10^{-7}} \times 100 = 2 \times 10^{-15}\ \%$

Thus, **essentially all of the Al^{3+} and Fe^{3+} ions are precipitated from solution.**

(d) For the three precipitates formed, use the following strategy:
change in [cation] → mol cation precipitated → mol precipitate
→ mass precipitate

For Mg^{2+}, $[Mg^{2+}]$ falls by 0.028 M.

mass $Mg(OH)_2 = \dfrac{0.028\ \text{mol } Mg^{2+}}{1L} \times 1\ L \times \dfrac{1\ \text{mol } Mg(OH)_2}{1\ \text{mol } Mg^{2+}} \times \dfrac{58.32\ \text{g } Mg(OH)_2}{1\ \text{mol } Mg(OH)_2} = 1.6\ \text{g}$

For Al^{3+}, $[Al^{3+}]$ falls by $4 \times 10^{-7}\ M$ (assuming complete precipitation - see part (c)).

mass $Al(OH)_3 = \dfrac{4 \times 10^{-7}\ \text{mol } Al^{3+}}{1L} \times 1\ L \times \dfrac{1\ \text{mol } Al(OH)_3}{1\ \text{mol } Al^{3+}} \times \dfrac{78.00\ \text{g } Al(OH)_3}{1\ \text{mol } Al(OH)_3}$

$= 3 \times 10^{-5}\ \text{g}$

For Fe^{3+}, $[Fe^{3+}]$ falls by 2×10^{-7} M (assuming complete precipitation - see part (c)).

$$\text{mass Fe(OH)}_3 = \frac{2 \times 10^{-7} \text{ mol Fe}^{3+}}{1 \text{L}} \times 1 \text{ L} \times \frac{1 \text{ mol Fe(OH)}_3}{1 \text{ mol Fe}^{3+}} \times \frac{106.87 \text{ g Fe(OH)}_3}{1 \text{ mol Fe(OH)}_3}$$

$$= 2 \times 10^{-5} \text{ g}$$

Thus, **the total mass of precipitate is 1.6 g**.

77. For the two complex ions present in the reaction given in the question, the equations for their formation, and the corresponding equilibrium constants are:

$$Zn^{2+}(aq) + 4NH_3(aq) \rightleftharpoons Zn(NH_3)_4^{2+}(aq) \quad \ldots① \qquad\qquad K_1 = K_f \, (Zn(NH_3)_4^{2+})$$

$$Zn^{2+}(aq) + 4OH^-(aq) \rightleftharpoons Zn(OH)_4^{2-}(aq) \quad \ldots② \qquad\qquad K_2 = K_f \, (Zn(OH)_4^{2-})$$

(a) In the reaction given in the question the complex ion $Zn(NH_3)_4^{2+}$ is a reactant; therefore reaction ① must be reversed (apply the Reciprocal Rule, Table 12.3):

$$Zn(NH_3)_4^{2+}(aq) \rightleftharpoons Zn^{2+}(aq) + 4NH_3(aq) \quad \ldots③ \qquad\qquad K_3 = 1 / K_f \, (Zn(NH_3)_4^{2+})$$

Adding together reactions ② and ③ gives the desired reaction; apply the Product Rule (Table 12.3) to obtain K for this reaction.

② + ③:
$$Zn^{2+}(aq) + 4OH^-(aq) + Zn(NH_3)_4^{2+}(aq)$$
$$\rightleftharpoons Zn(OH)_4^{2-}(aq) + Zn^{2+}(aq) + 4NH_3(aq)$$
$$Zn(NH_3)_4^{2+}(aq) + 4OH^-(aq) \rightleftharpoons Zn(OH)_4^{2-}(aq) + 4NH_3(aq)$$

$$K = K_2 \times K_3 = [K_f \, (Zn(OH)_4^{2-})] \times [1 / K_f \, (Zn(NH_3)_4^{2+})] = (3 \times 10^{14}) \times [1 / (3.6 \times 10^8)]$$

$$= \underline{8 \times 10^5}$$

(b) First calculate $[OH^-]$ in a 1.0 M solution of NH_3. Proceed as in Example 13.11(a); set up an equilibrium table and determine $[OH^-]$:

Use $[NH_3]_0 = 1.0$ M (given) and initially we assume there is no ionization, so $[NH_4^+]_0 = [OH^-]_0 = 0$. Let $\Delta[OH^-] = x$:

	$NH_3(aq) + H_2O \rightleftharpoons$	$NH_4^+(aq) +$	$OH^-(aq)$
$[\ ]_0$ (M)	1.0	0	0
$\Delta[\ ]$ (M)	$-x$	$+x$	$+x$
$[\ ]_{eq}$ (M)	$1.0 - x$	x	x

Now substitute the equilibrium concentrations of each species into the K_b expression:

$$K_b = 1.8 \times 10^{-5} = \frac{[NH_4^+][OH^-]}{[NH_3]} = \frac{(x)(x)}{(1.0 - x)}$$

Assume $x \ll 1.0$:

$$1.8 \times 10^{-5} = \frac{x^2}{1.0} \quad \text{so } x^2 = 1.8 \times 10^{-5} \rightarrow \quad x = 4.2 \times 10^{-3} = [OH^-]$$

Check the assumption that the value of x is indeed much less that 1.0: the % ionization = $[OH^-]_{eq}/[NH_3]_0 \times 100$ = $(4.2 \times 10^{-3})/(1) \times 100 = 0.42\% < 5\%$ (see page 345), so the assumption made in solving the above equation was valid.

Now substitute $[NH_3]$ and $[OH^-]$ into the expression for K:

$$K = 8 \times 10^5 = \frac{[Zn(OH)_4^{2-}][NH_3]^4}{[Zn(NH_3)_4^{2+}][OH^-]^4} = \frac{[Zn(OH)_4^{2-}](1.0)^4}{[Zn(NH_3)_4^{2+}](0.0042)^4}$$

$$\frac{[Zn(OH)_4^{2-}]}{[Zn(NH_3)_4^{2+}]} = 2 \times 10^{-4} \quad \rightarrow \quad \frac{[Zn(NH_3)_4^{2+}]}{[Zn(OH)_4^{2-}]} = \underline{4 \times 10^3}$$

78. For the species present in the reaction given in the question, three reactions are relevant:

$$Cd^{2+}(aq) + 4NH_3(aq) \rightleftharpoons Cd(NH_3)_4^{2+}(aq) \quad \ldots \text{①} \qquad K_1 = K_f \; (Cd(NH_3)_4^{2+})$$

$$CdF_2(s) \rightleftharpoons Cd^{2+}(aq) + 2F^-(aq) \quad \ldots \text{②} \qquad K_2 = K_{sp} \; (CdF_2)$$

$$NH_4^+(aq) \rightleftharpoons H^+(aq) + NH_3(aq) \quad \ldots \text{③} \qquad K_3 = K_a \; (NH_4^+)$$

In manipulating these equations to ultimately obtain the reaction in the question, the rules in Table 12.3 will be used. $K_1 = 2.8 \times 10^7$ (Table 15.1); $K_2 = 6.4 \times 10^{-3}$ (given); $K_3 = 5.6 \times 10^{-10}$ (Table 13.4).

Reverse ①:

$$Cd(NH_3)_4^{2+}(aq) \rightleftharpoons Cd^{2+}(aq) + 4NH_3(aq) \quad \ldots \text{④} \qquad K_4 = 1/K_1$$

Reverse ②:

$$Cd^{2+}(aq) + 2F^-(aq) \rightleftharpoons CdF_2(s) \quad \ldots \text{⑤} \qquad K_5 = 1/K_2$$

Reverse ③ and multiply coefficients by 4:

$$4H^+(aq) + 4NH_3(aq) \rightleftharpoons 4NH_4^+(aq) \quad \ldots \text{⑥} \qquad K_6 = (1/K_3)^4$$

Add ④ + ⑤ + ⑥:

$$Cd(NH_3)_4^{2+}(aq) + Cd^{2+}(aq) + 2F^-(aq) + 4H^+(aq) + 4NH_3(aq)$$
$$\rightleftharpoons Cd^{2+}(aq) + 4NH_3(aq) + CdF_2(s) + 4NH_4^+(aq)$$

Simplify (cancel duplications):

$$Cd(NH_3)_4^{2+}(aq) + 2F^-(aq) + 4H^+(aq) \rightleftharpoons CdF_2(s) + 4NH_4^+(aq)$$

This is the required equation, for which the corresponding equilibrium constant is:

$$K = K_4 \times K_5 \times K_6 = [1/K_1] \times [1/K_2] \times [1/K_3]^4$$

$$= [1/(2.8 \times 10^7)] \times [1/(6.4 \times 10^{-3})] \times [1/(5.6 \times 10^{-10})]^4$$

$$= \underline{\mathbf{5.7 \times 10^{31}}}$$

SPONTANEITY OF REACTION

16-1 SPONTANEOUS PROCESSES

2. Recall that a spontaneous process is one that can happen by itself; no external action is necessary to make it happen. Based on everyday experience we can say that:

    (a) is **spontaneous**.

    (b) is **nonspontaneous**.

    (c) is **nonspontaneous**.

4. This problem is similar to question 2 above. Recall that a spontaneous process is one that can happen by itself; no external action is necessary to make it happen. Based on everyday experience we can say that:

    (a) is **nonspontaneous**.

    (b) is **nonspontaneous**.

    (c) is **spontaneous**.

    Thus, **only process (c) is spontaneous**.

6. This problem is similar to questions 2 and 4 above. Recall that a spontaneous process is one that can happen by itself; no external action is necessary to make it happen. Consider what is happening physically and/or chemically in the equation given and decide whether that normally happens by itself. Based on everyday experience we can say that:

    (a) Solid CO_2 ("dry ice") does sublime to form $CO_2(g)$ by itself (without intervention) at 25 °C, so this reaction is **spontaneous**.

    (b) Solid NaCl (table salt) does not melt at 25 °C without some external assistance, so this process is **nonspontaneous**.

    (c) Solid NaCl (table salt) does not decompose into the elements Na and Cl_2 at 25 °C without some external assistance, so this process is **nonspontaneous**.

    (d) The gas CO_2 does not decompose into the elements C and O_2 without some external assistance (if it did, no-one would be concerned about CO_2 emissions!), so this process is **nonspontaneous**.

    Thus, **the only spontaneous process is (a)**.

16-2 ENTROPY, S: Entropy, $\Delta S°$

8. This problem is related to Example 16.1. For each pair, consider the relative disorder of each substance; the more ordered substance has the lower entropy.

(a) Both samples of O_2 contain the same number of molecules (1 mol of each) at the same temperature, but the sample at higher pressure must occupy a smaller volume (apply the Ideal Gas Law, Chapter 5). Molecules that are confined to a smaller volume have less opportunity for disorder (fewer microstates — see page 409). Therefore, **the mole of $O_2(g)$ with a pressure of 758 mm Hg has the lower entropy.**

(b) The solid glucose will be ordered and of relatively low entropy (as are all solids). When the glucose is dissolved in H_2O to form the aqueous solution, the molecules are randomly distributed, and can move randomly, throughout the whole volume of the solution. This mixture can therefore be seen to have higher entropy than the solid glucose, so **glucose (s) has the lower entropy.**

(c) Recall that gases have more entropy than liquids and therefore **Hg(*l*) has lower entropy than Hg(*g*).**

10. This problem is related to Example 16.1. In each case, consider the relative disorder of the initial and final states; the more ordered state has the lower entropy.

(a) The Ag^+ and Cl^- ions, when in solution, are free to move randomly through the entire volume of the solution; whereas, when they are precipitated as solid AgCl they are confined to the ionic solid and may be restricted to a lattice of alternating positive and negative ions (compare Figure 9.21). Thus the ions in solution can be seen to have higher entropy than in the precipitate, so the final state has lower entropy than the initial state: **ΔS is negative.**

(b) The solid sugar will be relatively ordered and of relatively low entropy (as are all solids). When it is dissolved in the coffee, the sugar molecules become randomly distributed, and can move randomly, throughout the whole volume of the coffee. This mixture can therefore be seen to have higher entropy than the isolated sugar, so the final state has higher entropy than the initial state: **ΔS is positive.**

(c) Although there is no phase change when glass turns into sand, the sand would have higher entropy than the glass: In the rigid solid glass, each atom is confined to one site in the whole material; whereas each individual grain of sand can move freely relative to the other grains of sand and in that sense the sand is more random. The final state has higher entropy than the initial state, so **ΔS is positive.**

12. This problem is similar to Example 16.1. In each case, consider the relative disorder of the reactants and products; the more ordered state has the lower entropy.

(a) The reactants consist of a liquid and five mol of gas whereas the products consist of only five mol of gas. On this basis, it is difficult to say whether reactants or products has more entropy. **The sign of ΔS cannot be determined.**

(b) The reactants consist of a liquid and a solid whereas the products consist of only gases. Gases have higher entropy than solids or liquids, so the final state has higher entropy than the initial state. **ΔS is positive.**

(c) Solid Br_2 has lower entropy than liquid Br_2 (liquids have higher entropy than solids): the final state has lower entropy than the initial state, so **ΔS is negative.**

(d) The reactants consist of **two** mol of gas whereas the products consist of **four** mol of gas. Since gases have high entropy, more moles of gas will correspond to higher entropy: the final state has higher entropy than the initial state, so **ΔS is positive.**

14. This problem is similar to Example 16.1. In each case, consider the number of moles of gas among the reactants and products; the side with fewer moles of gas has lower entropy. (Since gases have much higher entropy than solids or liquids, they will tend to dominate in terms of the entropy of the reactants and products.)

    (a) The total number of moles of gas decreases from 1 mol on the reactant side to zero mol on the product side of the equation, so entropy decreases. **ΔS is negative.**

    (b) The total number of moles of gas increases from zero mol on the reactant side to 1 mol on the product side of the equation, so entropy increases. **ΔS is positive.**

    (c) The total number of moles of gas increases from 1 mol on the reactant side to 2 mol on the product side of the equation, so entropy increases. **ΔS is positive.**

16. This problem is related to Example 16.1. In each case, consider the number of moles of gas among the reactants and products. (Since gases have much higher entropy than solids or liquids, they will tend to dominate in terms of the entropy of the reactants and products.) For the four reactions given, the change in mol of gas as each reaction proceeds is as follows:

    (a) 4 mol → 2 mol   (the net change is −2 mol of gas)

    (b) 3 mol → 0 mol   (the net change is −3 mol of gas)

    (c) 2 mol → 0 mol   (the net change is −2 mol of gas)

    (d) 2 mol → 2 mol   (no net change in mol of gas)

    On this basis, the reaction with the greatest decrease in the number of moles of gas should have the lowest value for ΔS, and so on. Thus we can place the reactions in the following order of increasing ΔS:

    (b) < (a, c) < (d)

    This does not allow us to discriminate between reactions (a) and (c) so we must also consider the nature of the products in each case:
    For reaction (a), gas is being formed, whereas in reaction (c) the product is a solid. Since solids have much lower entropy than gases, it would be expected that entropy decreases more significantly as reaction (c) proceeds than it does for reaction (a); that is, ΔS for reaction (c) is lower than that for reaction (a). We can now place all four reactions in the following order of increasing ΔS:

    (b) < (c) < (a) < (d)

18. This problem is similar to Example 16.2. Substitute $S°$ values from Table 16.1 into Equation 16.1 to obtain $\Delta S°$ for each reaction.

    (a) $\Delta S° = \Sigma S°_{products} - \Sigma S°_{reactants}$

$$= 4S° \text{ NO}(g) + 6S° \text{ H}_2\text{O}(g) - [4S° \text{ NH}_3(g) + 5S° \text{ O}_2(g)]$$

$$= 4 \text{ mol}\left(\frac{210.7 \text{ J}}{\text{mol·K}}\right) + 6 \text{ mol}\left(\frac{188.7 \text{ J}}{\text{mol·K}}\right) - 4 \text{ mol}\left(\frac{192.3 \text{ J}}{\text{mol·K}}\right) - 5 \text{ mol}\left(\frac{205.0 \text{ J}}{\text{mol·K}}\right)$$

$$= \underline{\text{+180.8 J/K}}$$

(b) $\Delta S° = \Sigma S°_{\text{products}} - \Sigma S°_{\text{reactants}}$

$$= S° \text{ N}_2(g) + 4S° \text{ H}_2\text{O}(g) - [2S° \text{ H}_2\text{O}_2(l) + S° \text{ N}_2\text{H}_4(l)]$$

$$= 1 \text{ mol}\left(\frac{191.5 \text{ J}}{\text{mol·K}}\right) + 4 \text{ mol}\left(\frac{188.7 \text{ J}}{\text{mol·K}}\right) - 2 \text{ mol}\left(\frac{109.6 \text{ J}}{\text{mol·K}}\right) - 1 \text{ mol}\left(\frac{121.2 \text{ J}}{\text{mol·K}}\right)$$

$$= \underline{\text{+605.9 J/K}}$$

(c) $\Delta S° = \Sigma S°_{\text{products}} - \Sigma S°_{\text{reactants}}$

$$= S° \text{ CO}_2(g) - [S° \text{ C}(s) + S° \text{ O}_2(g)]$$

$$= 1 \text{ mol}\left(\frac{213.6 \text{ J}}{\text{mol·K}}\right) - 1 \text{ mol}\left(\frac{5.7 \text{ J}}{\text{mol·K}}\right) - 1 \text{ mol}\left(\frac{205.0 \text{ J}}{\text{mol·K}}\right)$$

$$= \underline{\text{+2.9 J/K}}$$

(d) $\Delta S° = \Sigma S°_{\text{products}} - \Sigma S°_{\text{reactants}}$

$$= S° \text{ CHCl}_3(l) + 3S° \text{ HCl}(g) - [S° \text{ CH}_4(g) + 3S° \text{ Cl}_2(g)]$$

$$= 1 \text{ mol}\left(\frac{201.7 \text{ J}}{\text{mol·K}}\right) + 3 \text{ mol}\left(\frac{186.8 \text{ J}}{\text{mol·K}}\right) - 1 \text{ mol}\left(\frac{186.2 \text{ J}}{\text{mol·K}}\right) - 3 \text{ mol}\left(\frac{223.0 \text{ J}}{\text{mol·K}}\right)$$

$$= \underline{\text{−93.1 J/K}}$$

20. This problem is similar to Example 16.2. Substitute $S°$ values from Table 16.1 into Equation 16.1 to obtain $\Delta S°$ for each reaction.

(a) $\Delta S° = \Sigma S°_{\text{products}} - \Sigma S°_{\text{reactants}}$

$$= 5S° \text{ NO}_3^-(aq) + 3S° \text{ Mn}^{2+}(aq) + 2S° \text{ H}_2\text{O}(l)$$

$$- [5S° \text{ NO}(g) + 3S° \text{ MnO}_4^-(aq) + 4S° \text{ H}^+(aq)]$$

$$= 5 \text{ mol}\left(\frac{146.4 \text{ J}}{\text{mol·K}}\right) + 3 \text{ mol}\left(\frac{-73.6 \text{ J}}{\text{mol·K}}\right) + 2 \text{ mol}\left(\frac{69.9 \text{ J}}{\text{mol·K}}\right)$$

$$- 5 \text{ mol}\left(\frac{210.7 \text{ J}}{\text{mol·K}}\right) - 3 \text{ mol}\left(\frac{191.2 \text{ J}}{\text{mol·K}}\right) - 4 \text{ mol}\left(\frac{0.0 \text{ J}}{\text{mol·K}}\right)$$

$$= \underline{\text{−976.1 J/K}}$$

(b) $\quad \Delta S° = \Sigma\, S°_{products} - \Sigma\, S°_{reactants}$

$\quad = S°\ Cr(s) + 6S°\ Fe^{3+}(aq) + 4S°\ H_2O(l)$

$\qquad\qquad\qquad - [6S°\ Fe^{2+}(aq) + S°\ CrO_4^{2-}(aq) + 8S°\ H^+(aq)]$

$\quad = 1\ mol\left(\dfrac{23.8\ J}{mol\cdot K}\right) + 6\ mol\left(\dfrac{-315.9\ J}{mol\cdot K}\right) + 4\ mol\left(\dfrac{69.9\ J}{mol\cdot K}\right)$

$\qquad\qquad\qquad - 6\ mol\left(\dfrac{-137.7\ J}{mol\cdot K}\right) - 1\ mol\left(\dfrac{50.2\ J}{mol\cdot K}\right) - 8\ mol\left(\dfrac{0.0\ J}{mol\cdot K}\right)$

$\quad = \underline{\mathbf{-816.0\ J/K}}$

(c) $\quad \Delta S° = \Sigma\, S°_{products} - \Sigma\, S°_{reactants}$

$\quad = 2S°\ CO_2(g) + S°\ H_2O(g) - [S°\ C_2H_2(g) + \tfrac{5}{2} S°\ O_2(g)]$

$\quad = 2\ mol\left(\dfrac{213.6\ J}{mol\cdot K}\right) + 1\ mol\left(\dfrac{188.7\ J}{mol\cdot K}\right) - 1\ mol\left(\dfrac{200.8\ J}{mol\cdot K}\right) - \tfrac{5}{2}\ mol\left(\dfrac{205.0\ J}{mol\cdot K}\right)$

$\quad = \underline{\mathbf{-97.4\ J/K}}$

22. This problem is similar to Example 16.2. Substitute $S°$ values from Table 16.1 into Equation 16.1 to obtain $\Delta S°$ for each reaction.

(a) $\quad \Delta S° = \Sigma\, S°_{products} - \Sigma\, S°_{reactants}$

$\quad = 4S°\ H_2O(l) + 2S°\ NO(g) + 3S°\ S(s) - [2S°\ HNO_3(l) + 3S°\ H_2S(g)]$

$\quad = 4\ mol\left(\dfrac{69.9\ J}{mol\cdot K}\right) + 2\ mol\left(\dfrac{210.7\ J}{mol\cdot K}\right) + 3\ mol\left(\dfrac{31.8\ J}{mol\cdot K}\right)$

$\qquad\qquad\qquad - 2\ mol\left(\dfrac{155.6\ J}{mol\cdot K}\right) - 3\ mol\left(\dfrac{205.7\ J}{mol\cdot K}\right)$

$\quad = \underline{\mathbf{-131.9\ J/K}} \qquad or \qquad \underline{\mathbf{-0.1319\ kJ/K}}$

(b) $\quad \Delta S° = \Sigma\, S°_{products} - \Sigma\, S°_{reactants}$

$\quad = 5S°\ Cl^-(aq) + S°\ H_2PO_4^-(aq) + 6S°\ H^+(aq) - [S°\ PCl_5(g) + 4S°\ H_2O(l)]$

$\quad = 5\ mol\left(\dfrac{56.5\ J}{mol\cdot K}\right) + 1\ mol\left(\dfrac{90.4\ J}{mol\cdot K}\right) + 6\ mol\left(\dfrac{0.0\ J}{mol\cdot K}\right)$

$\qquad\qquad\qquad - 1\ mol\left(\dfrac{364.5\ J}{mol\cdot K}\right) - 4\ mol\left(\dfrac{69.9\ J}{mol\cdot K}\right)$

$\quad = \underline{\mathbf{-271.2\ J/K}} \qquad or \qquad \underline{\mathbf{-0.2712\ kJ/K}}$

(c) $\quad \Delta S° = \Sigma\, S°_{products} - \Sigma\, S°_{reactants}$

$\quad\quad = 3S°\ Fe^{3+}(aq) + S°\ MnO_2(s) + 2S°\ H_2O(l)$

$\quad\quad\quad\quad\quad - [S°\ MnO_4^-(aq) + 3S°\ Fe^{2+}(aq) + 4S°\ H^+(aq)]$

$\quad\quad = 3\ mol\left(\dfrac{-315.9\ J}{mol\cdot K}\right) + 1\ mol\left(\dfrac{53.0\ J}{mol\cdot K}\right) + 2\ mol\left(\dfrac{69.9\ J}{mol\cdot K}\right)$

$\quad\quad\quad\quad\quad - 1\ mol\left(\dfrac{191.2\ J}{mol\cdot K}\right) - 3\ mol\left(\dfrac{-137.7\ J}{mol\cdot K}\right) - 4\ mol\left(\dfrac{0.0\ J}{mol\cdot K}\right)$

$\quad\quad = \underline{\mathbf{-533.0\ J/K}} \quad\quad \text{or} \quad\quad \underline{\mathbf{-0.5330\ kJ/K}}$

16-4 STANDARD FREE ENERGY CHANGE, $\Delta G°$: $\Delta G°$ and the Gibbs-Helmholtz Equation

24. These questions are similar to Example 16.3(c). In each case, use temperature in K (72 °C = 345 K) and (where necessary) convert $\Delta S°$ to kJ/K; then substitute these and $\Delta H°$ into the Gibbs-Helmholtz equation (Equation 16.2) and evaluate $\Delta G°$.

(a) $\Delta S° = +457\ J/K = +0.457\ kJ/K$

$\quad \Delta G° = \Delta H° - T\Delta S° = -136\ kJ - (345\ K)(+0.457\ kJ/K) = \underline{\mathbf{-294\ kJ}}$

(b) $\Delta G° = \Delta H° - T\Delta S° = 41.5\ kJ - (345\ K)(-0.288\ kJ/K) = \underline{\mathbf{+140.9\ kJ}}$

(c) $\Delta S° = -861\ J/K = -0.861\ kJ/K$

$\quad \Delta G° = \Delta H° - T\Delta S° = -795\ kJ - (345\ K)(-0.861\ kJ/K) = \underline{\mathbf{-498\ kJ}}$

26. These questions are similar to Example 16.3. For each reaction, calculate $\Delta H°$ using $\Delta H_f°$ values (Appendix 1) as in Example 16.3(a); and convert $\Delta S°$ values (calculated in Question 18) to kJ/K. Finally substitute these and $T = 415\ K$ (given) into Equation 16.2 and evaluate $\Delta G°$. Recall that a reaction is spontaneous if $\Delta G° < 0$ (page 416).

(a) $\quad \Delta H° = \Sigma\, \Delta H_f°_{(products)} - \Sigma\, \Delta H_f°_{(reactants)}$

$\quad\quad = 4\Delta H_f°\ NO(g) + 6\Delta H_f°\ H_2O(g) - [4\Delta H_f°\ NH_3(g) + 5\Delta H_f°\ O_2(g)]$

$\quad\quad = 4\ mol\left(\dfrac{90.2\ kJ}{mol}\right) + 6\ mol\left(\dfrac{-241.8\ kJ}{mol}\right) - 4\ mol\left(\dfrac{-46.1\ kJ}{mol}\right) - 5\ mol\left(\dfrac{0\ kJ}{mol}\right)$

$\quad\quad = -905.6\ kJ$

$\Delta S° = +180.8\ J/K = +0.1808\ kJ/K$

$\Delta G° = \Delta H° - T\Delta S° = -905.6\ kJ - (415\ K)(+0.1808\ kJ/K) = \underline{\mathbf{-980.6\ kJ}}$

$\Delta G° < 0$, so the reaction is **spontaneous**.

(b) $\Delta H° = \Sigma \Delta H_f°_{\text{(products)}} - \Sigma \Delta H_f°_{\text{(reactants)}}$

$= \Delta H_f° \ N_2(g) + 4\Delta H_f° \ H_2O(g) - [2\Delta H_f° \ H_2O_2(l) + \Delta H_f° \ N_2H_4(l)]$

$= 1 \ \text{mol}\left(\dfrac{0 \ \text{kJ}}{\text{mol}}\right) + 4 \ \text{mol}\left(\dfrac{-241.8 \ \text{kJ}}{\text{mol}}\right) - 2 \ \text{mol}\left(\dfrac{-187.8 \ \text{kJ}}{\text{mol}}\right) - 1 \ \text{mol}\left(\dfrac{50.6 \ \text{kJ}}{\text{mol}}\right)$

$= -642.2 \ \text{kJ}$

$\Delta S° = +605.9 \ \text{J/K} = +0.6059 \ \text{kJ/K}$

$\Delta G° = \Delta H° - T\Delta S° = -642.2 \ \text{kJ} - (415 \ \text{K})(+0.6059 \ \text{kJ/K}) = \underline{\mathbf{-893.6 \ kJ}}$

$\Delta G° < 0$, so the reaction is **spontaneous**.

(c) $\Delta H° = \Sigma \Delta H_f°_{\text{(products)}} - \Sigma \Delta H_f°_{\text{(reactants)}}$

$= \Delta H_f° \ CO_2(g) - [\Delta H_f° \ C(s) + \Delta H_f° \ O_2(g)]$

$= 1 \ \text{mol}\left(\dfrac{-393.5 \ \text{kJ}}{\text{mol}}\right) - 1 \ \text{mol}\left(\dfrac{0 \ \text{kJ}}{\text{mol}}\right) - 1 \ \text{mol}\left(\dfrac{0 \ \text{kJ}}{\text{mol}}\right)$

$= -393.5 \ \text{kJ}$

$\Delta S° = +2.9 \ \text{J/K} = +0.0029 \ \text{kJ/K}$

$\Delta G° = \Delta H° - T\Delta S° = -393.5 \ \text{kJ} - (415 \ \text{K})(+0.0029 \ \text{kJ/K}) = \underline{\mathbf{-394.7 \ kJ}}$

$\Delta G° < 0$, so the reaction is **spontaneous**.

(d) $\Delta H° = \Sigma \Delta H_f°_{\text{(products)}} - \Sigma \Delta H_f°_{\text{(reactants)}}$

$= \Delta H_f° \ CHCl_3(l) + 3\Delta H_f° \ HCl(g) - [\Delta H_f° \ CH_4(g) + 3\Delta H_f° \ Cl_2(g)]$

$= 1 \ \text{mol}\left(\dfrac{-134.5 \ \text{kJ}}{\text{mol}}\right) + 3 \ \text{mol}\left(\dfrac{-92.3 \ \text{kJ}}{\text{mol}}\right) - 1 \ \text{mol}\left(\dfrac{-74.8 \ \text{kJ}}{\text{mol}}\right) - 3 \ \text{mol}\left(\dfrac{0 \ \text{kJ}}{\text{mol}}\right)$

$= -336.6 \ \text{kJ}$

$\Delta S° = -93.1 \ \text{J/K} = -0.0931 \text{kJ/K}$

$\Delta G° = \Delta H° - T\Delta S° = -336.6 \ \text{kJ} - (415 \ \text{K})(-0.0931 \ \text{kJ/K}) = \underline{\mathbf{-298.0 \ kJ}}$

$\Delta G° < 0$, so the reaction is **spontaneous**.

28. This problem is similar to Example 16.4. Substitute $\Delta G_f°$ values from Appendix 1 into Equation 16.3 to obtain $\Delta G°$ for each reaction.

(a) $\Delta G° = \Sigma \Delta G_f°_{\text{(products)}} - \Sigma \Delta G_f°_{\text{(reactants)}}$

$$= 5\Delta G_f^\circ\ NO_3^-(aq) + 3\Delta G_f^\circ\ Mn^{2+}(aq) + 2\Delta G_f^\circ\ H_2O(l)$$
$$- [5\Delta G_f^\circ\ NO(g) + 3\Delta G_f^\circ\ MnO_4^-(aq) + 4\Delta G_f^\circ\ H^+(aq)]$$

$$= 5\ mol\left(\frac{-108.7\ kJ}{mol}\right) + 3\ mol\left(\frac{-228.1 kJ}{mol}\right) + 2\ mol\left(\frac{-237.2\ kJ}{mol}\right)$$
$$- 5\ mol\left(\frac{86.6\ kJ}{mol}\right) - 3\ mol\left(\frac{-447.2\ kJ}{mol}\right) - 4\ mol\left(\frac{0\ kJ}{mol}\right)$$

$$= \underline{-793.6\ kJ}$$

(b) $\quad \Delta G^\circ = \Sigma\ \Delta G_f^\circ{}_{(products)} - \Sigma\ \Delta G_f^\circ{}_{(reactants)}$

$$= \Delta G_f^\circ\ Cr(s) + 6\Delta G_f^\circ\ Fe^{3+}(aq) + 4\Delta G_f^\circ\ H_2O(l)$$
$$- [6\Delta G_f^\circ\ Fe^{2+}(aq) + \Delta G_f^\circ\ CrO_4^{2-}(aq) + 8\Delta G_f^\circ\ H^+(aq)]$$

$$= 1\ mol\left(\frac{0\ kJ}{mol}\right) + 6\ mol\left(\frac{-4.7\ kJ}{mol}\right) + 4\ mol\left(\frac{-237.2\ kJ}{mol}\right)$$
$$- 6\ mol\left(\frac{-78.9\ kJ}{mol}\right) - 1\ mol\left(\frac{-727.8\ kJ}{mol}\right) - 8\ mol\left(\frac{0\ kJ}{mol}\right)$$

$$= \underline{224.2\ kJ}$$

(c) $\quad \Delta G^\circ = \Sigma\ \Delta G_f^\circ{}_{(products)} - \Sigma\ \Delta G_f^\circ{}_{(reactants)}$

$$= 2\Delta G_f^\circ\ CO_2(g) + \Delta G_f^\circ\ H_2O(g) - [\Delta G_f^\circ\ C_2H_2(g) + \tfrac{5}{2}\Delta G_f^\circ\ O_2(g)]$$

$$= 2\ mol\left(\frac{-394.4\ kJ}{mol}\right) + 1\ mol\left(\frac{-228.6\ kJ}{mol}\right) - 1\ mol\left(\frac{209.2\ kJ}{mol}\right) - \tfrac{5}{2}\ mol\left(\frac{0\ kJ}{mol}\right)$$

$$= \underline{-1226.6\ kJ}$$

30. Recall the definition of ΔG_f° from page 417: it is the free energy change when **one mole** of a compound is formed from the elements in their stable states. Thus, for each compound write a reaction (with correct stoichiometry) for the formation of that compound from its elements. By definition the value of ΔH_f° for the compound (from Appendix 1) equals ΔH° for this reaction (see Section 8-5); and we can use S° values from Appendix 1 to calculate ΔS° for the reaction using Equation 16.1. Finally, combine ΔH° and ΔS° using Equation 16.2 (with $T = 298$ K) to obtain ΔG° for each reaction, which equals ΔG_f° for the given compound.

(a) solid ammonium nitrate, $NH_4NO_3(s)$

formation from its elements: $\quad 2H_2(g) + N_2(g) + \tfrac{3}{2}O_2(g) \rightarrow NH_4NO_3(s)$

$\Delta H_f^\circ\ NH_4NO_3(s) = -365.6\ kJ = \Delta H^\circ$

$\Delta S° = \Sigma \, S°_{(products)} - \Sigma \, S°_{(reactants)}$

$= S° \; NH_4NO_3(s) - [2S° \; H_2(g) + S° \; N_2(g) + \frac{3}{2} S° \; O_2(g)]$

$= 1 \; mol\left(\dfrac{0.1511 \, kJ}{mol \cdot K}\right) - 2 \; mol\left(\dfrac{0.1306 \, kJ}{mol \cdot K}\right)$

$\qquad - 1 \; mol\left(\dfrac{0.1915 \, kJ}{mol \cdot K}\right) - \frac{3}{2} \; mol\left(\dfrac{0.2050 \, kJ}{mol \cdot K}\right)$

$= -0.6091 \; kJ/K$

$\Delta G_f° \text{ (per mol)} = \Delta H° - T\Delta S° = -365.6 \; kJ - (298 \; K)(-0.6091 \; kJ/K) = \underline{\mathbf{-184.1 \; kJ}}$

(b) liquid methyl alchohol, $CH_3OH(l)$

formation from its elements: $\quad C(s) + 2H_2(g) + \frac{1}{2} O_2(g) \rightarrow CH_3OH(l)$

$\Delta H_f° \; CH_3OH(l) = -238.7 \; kJ = \Delta H°$

$\Delta S° = \Sigma \, S°_{(products)} - \Sigma \, S°_{(reactants)}$

$= S° \; CH_3OH(l) - [S° \; C(s) + 2S° \; H_2(g) + \frac{1}{2} S° \; O_2(g)]$

$= 1 \; mol\left(\dfrac{0.1268 \, kJ}{mol \cdot K}\right) - 1 \; mol\left(\dfrac{0.0057 \, kJ}{mol \cdot K}\right)$

$\qquad\qquad - 2 \; mol\left(\dfrac{0.1306 \, kJ}{mol \cdot K}\right) - \frac{1}{2} \; mol\left(\dfrac{0.2050 \, kJ}{mol \cdot K}\right)$

$= -0.2426 \; kJ/K$

$\Delta G_f° \text{ (per mol)} = \Delta H° - T\Delta S° = -238.7 \; kJ - (298 \; K)(-0.2426 \; kJ/K) = \underline{\mathbf{-166.4 \; kJ}}$

(c) solid copper(II) sulfide, $CuS(s)$

formation from its elements: $\quad Cu(s) + S(s) \rightarrow CuS(s)$

$\Delta H_f° \; CuS(s) = -53.1 \; kJ = \Delta H°$

$\Delta S° = \Sigma \, S°_{(products)} - \Sigma \, S°_{(reactants)}$

$= S° \; CuS(s) - [S° \; Cu(s) + S° \; S(s)]$

$= 1 \; mol\left(\dfrac{0.0665 \, kJ}{mol \cdot K}\right) - 1 \; mol\left(\dfrac{0.0332 \, kJ}{mol \cdot K}\right) - 1 \; mol\left(\dfrac{0.0318 \, kJ}{mol \cdot K}\right)$

$= 0.0015 \; kJ/K$

$\Delta G_f° \text{ (per mol)} = \Delta H° - T\Delta S° = -53.1 \; kJ - (298 \; K)(0.0015 \; kJ/K) = \underline{\mathbf{-53.5 \; kJ}}$

32. This problem is related to Example 16.4. Use the $\Delta G_f°$ values given in the question (plus others as needed from Appendix 1) and use Equation 16.3 (with $T = 298$ K) to obtain $\Delta G°$ for the reaction.

$$\Delta G° = \Sigma \Delta G_f°_{(products)} - \Sigma \Delta G_f°_{(reactants)}$$

$$= \Delta G_f° \text{ H}_2\text{O}_2(aq) + \Delta G_f° \text{ O}_2(g) - [\Delta G_f° \text{ O}_3(g) + \Delta G_f° \text{ H}_2\text{O}(l)]$$

$$= 1 \text{ mol}\left(\frac{-134 \text{ kJ}}{\text{mol}}\right) + 1 \text{ mol}\left(\frac{0 \text{ kJ}}{\text{mol}}\right) - 1 \text{ mol}\left(\frac{163.2 \text{ kJ}}{\text{mol}}\right) - 1 \text{ mol}\left(\frac{-237.2 \text{ kJ}}{\text{mol}}\right)$$

$$= -6.0 \times 10^1 \text{ kJ}$$

The value of $\Delta G° < 0$, so the reaction is spontaneous under these conditions and therefore **YES, the claim is plausible.**

34. This problem is similar to Example 16.5. Write a balanced equation for the reaction, for forming **one mole** of Mg(OH)_2. To calculate $\Delta G°$ for the reaction at 27°C and 39°C recall that $\Delta G_f°$ values cannot be used since they apply only at 25°C; instead calculate $\Delta H°$ and $\Delta S°$ for the reaction using tabulated $\Delta H_f°$ and $S°$ values (Appendix 1) and then use Equation 16.2 with the appropriate value of T to obtain the value of $\Delta G°$ at the two temperatures.

Reaction equation: $\quad \text{Mg}(s) + 2\text{H}_2\text{O}(l) \rightarrow \text{Mg(OH)}_2(s) + \text{H}_2(g)$

$$\Delta H° = \Sigma \Delta H_f°_{(products)} - \Sigma \Delta H_f°_{(reactants)}$$

$$= \Delta H_f° \text{ Mg(OH)}_2(s) + \Delta H_f° \text{ H}_2(g) - [\Delta H_f° \text{ Mg}(s) + 2\Delta H_f° \text{ H}_2\text{O}(l)]$$

$$= 1 \text{ mol}\left(\frac{-924.5 \text{ kJ}}{\text{mol}}\right) + 1 \text{ mol}\left(\frac{0 \text{ kJ}}{\text{mol}}\right)$$

$$- 1 \text{ mol}\left(\frac{0 \text{ kJ}}{\text{mol}}\right) - 2 \text{ mol}\left(\frac{-285.8 \text{ kJ}}{\text{mol}}\right)$$

$$= -352.9 \text{ kJ}$$

$$\Delta S° = \Sigma S°_{(products)} - \Sigma S°_{(reactants)}$$

$$= S° \text{ Mg(OH)}_2(s) + S° \text{ H}_2(g) - [S° \text{ Mg}(s) + 2S° \text{ H}_2\text{O}(l)]$$

$$= 1 \text{ mol}\left(\frac{0.0632 \text{ kJ}}{\text{mol·K}}\right) + 1 \text{ mol}\left(\frac{0.1306 \text{ kJ}}{\text{mol·K}}\right)$$

$$- 1 \text{ mol}\left(\frac{0.0327 \text{ kJ}}{\text{mol·K}}\right) - 2 \text{ mol}\left(\frac{0.0699 \text{ kJ}}{\text{mol·K}}\right)$$

$$= 0.0213 \text{ kJ/K}$$

At 27°C (300 K):

$$\Delta G° = \Delta H° - T\Delta S° = -352.9 \text{ kJ} - (300 \text{ K})(0.0213 \text{ kJ/K}) = \underline{\mathbf{-359.3 \text{ kJ}}}$$

At 39°C (312 K):

$\Delta G° = \Delta H° - T\Delta S° = -352.9 \text{ kJ} - (312 \text{ K})(0.0213 \text{ kJ/K}) = \underline{\mathbf{-359.6 \text{ kJ}}}$

36. This problem is related to Examples 16.2 and 16.3.

    (a) Use Equation 16.2 and the given values for $\Delta G°$ and $\Delta H°$ to calculate $\Delta S°$:

    $\Delta G° = \Delta H° - T\Delta S°$

    $-148.4 \text{ kJ} = -109.0 \text{ kJ} - (298 \text{ K} \times \Delta S°)$

    $\Delta S° = \dfrac{-148.4 \text{ kJ} + 109.0 \text{ kJ}}{-298 \text{ K}} = \underline{\mathbf{0.132 \text{ kJ/K}}}$

    YES, the sign of $\Delta S°$ is reasonable because the reaction involves an increase in the number of moles of gas.

    (b) Use Equation 16.1 and data from Appendix 1 to calculate $S°$ for $Na_2O_2(s)$. The value of $\Delta S°$ was calculated in part (a) above:

    $\Delta S° = \Sigma S°_{(products)} - \Sigma S°_{(reactants)}$

    $0.132 \text{ kJ/K} = 4S° \text{ NaOH}(s) + S° \text{ O}_2(g) - [2S° \text{ Na}_2\text{O}_2(s) + 2S° \text{ H}_2\text{O}(l)]$

    $0.132 \text{ kJ/K} = 4 \text{ mol}\left(\dfrac{0.0645 \text{ kJ}}{\text{mol}\cdot\text{K}}\right) + 1 \text{ mol}\left(\dfrac{0.2050 \text{ kJ}}{\text{mol}\cdot\text{K}}\right)$

    $\qquad\qquad\qquad - 2S° \text{ Na}_2\text{O}_2(s) - 2 \text{ mol}\left(\dfrac{0.0699 \text{ kJ}}{\text{mol}\cdot\text{K}}\right)$

    $2S° \text{ Na}_2\text{O}_2(s) = 0.191 \text{ kJ/K} \qquad\qquad \rightarrow \qquad S° \text{ Na}_2\text{O}_2(s) = \underline{\mathbf{0.0955 \text{ kJ/K}}}$

    (c) Similarly to part (b) above, use Equation 8.4 and data from Appendix 1 to calculate $\Delta H_f°$ for $Na_2O_2(s)$; the value of $\Delta H°$ for the reaction is given in the question:

    $\Delta H° = \Sigma \Delta H_f°_{(products)} - \Sigma \Delta H_f°_{(reactants)}$

    $-109.0 \text{ kJ} = 4\Delta H_f° \text{ NaOH}(s) + \Delta H_f° \text{ O}_2(g) - [2\Delta H_f° \text{ Na}_2\text{O}_2(s) + 2\Delta H_f° \text{ H}_2\text{O}(l)]$

    $-109.0 \text{ kJ} = 4 \text{ mol}\left(\dfrac{-425.6 \text{ kJ}}{\text{mol}}\right) + 1 \text{ mol}\left(\dfrac{0 \text{ kJ}}{\text{mol}}\right)$

    $\qquad\qquad\qquad - 2\Delta H_f° \text{ Na}_2\text{O}_2(s) - 2 \text{ mol}\left(\dfrac{-285.8 \text{ kJ}}{\text{mol}}\right)$

    $2\Delta H_f° \text{ Na}_2\text{O}_2(s) = -1021.8 \text{ kJ} \qquad\qquad \rightarrow \qquad \Delta H_f° \text{ Na}_2\text{O}_2(s) = \underline{\mathbf{-510.9 \text{ kJ}}}$

38. This problem is related to Examples 16.2 and 16.3.

    (a) Use Equation 16.2 and the given values for $\Delta G°$ and $\Delta H°$ to calculate $\Delta S°$:

    $\Delta G° = \Delta H° - T\Delta S°$

    $-1750.9 \text{ kJ} = -2024.6 \text{ kJ} - (298 \text{ K} \times \Delta S°)$

$$\Delta S° = \frac{-1750.9 \text{ kJ} + 2024.6 \text{ kJ}}{-298 \text{ K}} = \underline{-0.918 \text{ kJ/K}}$$

(b) Use Equation 16.1 to calculate $S°$ for $Co^{2+}(aq)$. The value of $\Delta S°$ was calculated in part (a) above; use $S°$ data from Appendix 1 and $S°$ $Co(s)$ given in the question:

$$\Delta S° = \Sigma\, S°_{(products)} - \Sigma\, S°_{(reactants)}$$

$$-0.918 \text{ kJ/K} = 2S°\, Mn^{2+}(aq) + 5S°\, Co^{2+}(aq) + 8S°\, H_2O(l)$$
$$- [2S°\, MnO_4^-(aq) + 16S°\, H^+(aq) + 5S°\, Co(s)]$$

$$-0.918 \text{ kJ/K} = 2 \text{ mol}\left(\frac{-0.0736 \text{ kJ}}{\text{mol}\cdot\text{K}}\right) + 5S°\, Co^{2+}(aq) + 8 \text{ mol}\left(\frac{0.0699 \text{ kJ}}{\text{mol}\cdot\text{K}}\right)$$

$$- 2 \text{ mol}\left(\frac{0.1912 \text{ kJ}}{\text{mol}\cdot\text{K}}\right) - 16 \text{ mol}\left(\frac{0 \text{ kJ}}{\text{mol}\cdot\text{K}}\right) - 5 \text{ mol}\left(\frac{0.03004 \text{ kJ}}{\text{mol}\cdot\text{K}}\right)$$

$$5S°\, Co^{2+}(aq) = -0.797 \text{ kJ/K} \qquad \rightarrow \qquad S°\, Co^{2+}(aq) = \underline{-0.159 \text{ kJ/K}}$$

16-5 EFFECT OF TEMPERATURE, PRESSURE AND CONCENTRATION ON REACTION SPONTANEITY: Temperature Dependence of Spontaneity

40. Refer to Table 16.2 and answer based on the signs of $\Delta H°$ and $\Delta S°$.

(a) $\Delta H° > 0$ and $\Delta S° > 0$ (case III in Table 16.2), so **the reaction is nonspontaneous at low temperature and spontaneous at high temperature**.

(b) $\Delta H° < 0$ and $\Delta S° < 0$ (case IV in Table 16.2), so **the reaction is spontaneous at low temperature and nonspontaneous at high temperature**.

(c) $\Delta H° < 0$ and $\Delta S° > 0$ (case I in Table 16.2), so **the reaction is spontaneous at ALL temperatures; changing temperature has no effect on spontaneity**.

42. (a) In a case such as this, where the spontaneity changes with temperature, we can use Equation 16.2 to calculate at what temperature this change takes place. At low temperature $\Delta G° > 0$ (nonspontaneous) and at high temperature $\Delta G° < 0$ (spontaneous); therefore the temperature where $\Delta G° = 0$ corresponds to the point at which the change in spontaneity occurs. Use Equation 16.2 with this value of $\Delta G°$ and the given values of $\Delta H°$ and $\Delta S°$:

$$\Delta G° = \Delta H° - T\Delta S°$$
$$0 = 128 \text{ kJ} - (T \times 0.0895 \text{ kJ/K})$$

$$T = \frac{-128 \text{ kJ}}{-0.0895 \text{ kJ/K}} = 1430 \text{ K}$$

We know that the reaction is spontaneous at high temperature, so we can say that **the reaction is spontaneous above 1430 K (1157°C)**.

(b) As in part (a) above we can use Equation 16.2 with $\Delta G° = 0$ and the given values of $\Delta H°$ and $\Delta S°$ to calculate at what temperature the change in spontaneity takes place:

$$\Delta G° = \Delta H° - T\Delta S°$$

$$0 = -20.4 \text{ kJ} - (T \times -0.1563 \text{ kJ/K})$$

$$-T = \frac{20.4 \text{ kJ}}{-0.1563 \text{ kJ/K}} \qquad \rightarrow T = 131 \text{ K}$$

We know that the reaction is spontaneous at low temperature, so we can say that **the reaction is spontaneous below 131 K (−142°C)**.

(c) As noted in Question 40(c) above, **the reaction is spontaneous at ALL temperatures**.

44. This problem is similar to Example 16.6. (As in Example 16.6, we require values of $\Delta H°$ and $\Delta S°$, which must be calculated as in Example 16.5 using data from Appendix 1.)

$$\Delta H° = \Sigma \Delta H_f°_{(products)} - \Sigma \Delta H_f°_{(reactants)}$$

$$= 2\Delta H_f° \text{ CO}_2(g) + \Delta H_f° \text{ Sn}(s) - [\Delta H_f° \text{ SnO}_2(s) + 2\Delta H_f° \text{ CO}(g)]$$

$$= 2 \text{ mol}\left(\frac{-393.5 \text{ kJ}}{\text{mol}}\right) + 1 \text{ mol}\left(\frac{0 \text{ kJ}}{\text{mol}}\right)$$

$$\qquad - 1 \text{ mol}\left(\frac{-580.7 \text{ kJ}}{\text{mol}}\right) - 2 \text{ mol}\left(\frac{-110.5 \text{ kJ}}{\text{mol}}\right)$$

$$= 14.7 \text{ kJ}$$

$$\Delta S° = \Sigma S°_{(products)} - \Sigma S°_{(reactants)}$$

$$= 2S° \text{ CO}_2(g) + S° \text{ Sn}(s) - [S° \text{ SnO}_2(s) + 2S° \text{ CO}(g)]$$

$$= 2 \text{ mol}\left(\frac{0.2136 \text{ kJ}}{\text{mol·K}}\right) + 1 \text{ mol}\left(\frac{0.0516 \text{ kJ}}{\text{mol·K}}\right)$$

$$\qquad - 1 \text{ mol}\left(\frac{0.0523 \text{ kJ}}{\text{mol·K}}\right) - 2 \text{ mol}\left(\frac{0.1976 \text{ kJ}}{\text{mol·K}}\right)$$

$$= 0.0313 \text{ kJ/K}$$

To calculate the temperature where $\Delta G° = 0$, use Equation 16.2 with this value of $\Delta G°$ and the values of $\Delta H°$ and $\Delta S°$ calculated above:

$$\Delta G° = \Delta H° - T\Delta S°$$

$$0 = 14.7 \text{ kJ} - (T \times 0.0313 \text{ kJ/K})$$

$$T = \frac{-14.7 \text{ kJ}}{-0.0313 \text{ kJ/K}} = \textbf{470 K} \text{ or } \textbf{197 °C}$$

46. (a) To obtain an expression for $\Delta G°$ as a function of temperature, simply use Equation 16.2 with the values of $\Delta H°$ and $\Delta S°$ given in the question; for consistency of units, express $\Delta S°$ in kJ/K:

$$\Delta G° = \Delta H° - T\Delta S°$$

$$\mathbf{\Delta G° = 492 \text{ kJ} - T(0.327 \text{ kJ/K})}$$

For each value of T in the range $T = 100$ K to $T = 500$ K, substitute for T in the above equation to calculate $\Delta G°$ at that temperature:

At $T = 100$ K: $\quad \Delta G° = 492 \text{ kJ} - (100 \text{ K})(0.327 \text{ kJ/K}) = \underline{\mathbf{459 \text{ kJ}}}$

At $T = 200$ K: $\quad \Delta G° = 492 \text{ kJ} - (200 \text{ K})(0.327 \text{ kJ/K}) = \underline{\mathbf{427 \text{ kJ}}}$

At $T = 300$ K: $\quad \Delta G° = 492 \text{ kJ} - (300 \text{ K})(0.327 \text{ kJ/K}) = \underline{\mathbf{394 \text{ kJ}}}$

At $T = 400$ K: $\quad \Delta G° = 492 \text{ kJ} - (400 \text{ K})(0.327 \text{ kJ/K}) = \underline{\mathbf{361 \text{ kJ}}}$

At $T = 500$ K: $\quad \Delta G° = 492 \text{ kJ} - (500 \text{ K})(0.327 \text{ kJ/K}) = \underline{\mathbf{328 \text{ kJ}}}$

T (K)	100	200	300	400	500
$\Delta G°$ (kJ)	459	427	394	361	328

(b) To determine the temperature at which $\Delta G°$ becomes zero, substitute this value of $\Delta G°$ into the equation in part (a) and solve for T:

$$\Delta G° = 492 \text{ kJ} - T(0.327 \text{ kJ/K})$$

$$0 = 492 \text{ kJ} - T(0.327 \text{ kJ/K})$$

$$T = \frac{492 \text{ kJ}}{0.327 \text{ kJ/K}} = 1505 \text{ K}$$

Thus, $\underline{\mathbf{\Delta G° \text{ becomes zero at } 1505 \text{ K}}}$; at higher temperatures the reaction becomes spontaneous ($\Delta G° < 0$).

48. This problem is related to Examples 16.5 and 16.6. For each method, write a balanced equation and calculate values of $\Delta H°$ and $\Delta S°$ using data from Appendix 1; then determine the lowest temperature for which each method is spontaneous using Equation 16.2 with $\Delta G° = 0$.

(a) Balanced equation: $MnO_2(s) \rightarrow Mn(s) + O_2(g)$

$$\Delta H° = \Sigma \, \Delta H_f°_{\text{(products)}} - \Sigma \, \Delta H_f°_{\text{(reactants)}}$$

$$= \Delta H_f° \, Mn(s) + \Delta H_f° \, O_2(g) - \Delta H_f° \, MnO_2(s)$$

$$= 1 \text{ mol}\left(\frac{0 \text{ kJ}}{\text{mol}}\right) + 1 \text{ mol}\left(\frac{0 \text{ kJ}}{\text{mol}}\right) - 1 \text{ mol}\left(\frac{-520.0 \text{ kJ}}{\text{mol}}\right)$$

$$= 520.0 \text{ kJ}$$

$$\Delta S° = \Sigma \, S°_{\text{(products)}} - \Sigma \, S°_{\text{(reactants)}}$$

$$= S° \, Mn(s) + S° \, O_2(g) - S° \, MnO_2(s)$$

$$= 1 \text{ mol}\left(\frac{0.0320 \text{ kJ}}{\text{mol} \cdot \text{K}}\right) + 1 \text{ mol}\left(\frac{0.2050 \text{ kJ}}{\text{mol} \cdot \text{K}}\right) - 1 \text{ mol}\left(\frac{0.0530 \text{ kJ}}{\text{mol} \cdot \text{K}}\right)$$

$$= 0.1840 \text{ kJ/K}$$

$$\Delta G° = \Delta H° - T\Delta S°$$

$$0 = 520.0 \text{ kJ} - (T \times 0.1840 \text{ kJ/K}) \qquad \rightarrow \qquad T = \frac{520.0 \text{ kJ}}{0.1840 \text{ kJ/K}} = 2826 \text{ K}$$

(b) Balanced equation: $MnO_2(s) + 2H_2(g) \rightarrow Mn(s) + 2H_2O(g)$

$$\Delta H° = \Sigma \Delta H_f°_{(products)} - \Sigma \Delta H_f°_{(reactants)}$$

$$= \Delta H_f° \text{ Mn}(s) + 2\Delta H_f° \text{ H}_2\text{O}(g) - [\Delta H_f° \text{ MnO}_2(s) + 2\Delta H_f° \text{ H}_2(g)]$$

$$= 1 \text{ mol}\left(\frac{0 \text{ kJ}}{\text{mol}}\right) + 2 \text{ mol}\left(\frac{-241.8 \text{ kJ}}{\text{mol}}\right) - 1 \text{ mol}\left(\frac{-520.0 \text{ kJ}}{\text{mol}}\right) - 2 \text{ mol}\left(\frac{0 \text{ kJ}}{\text{mol}}\right)$$

$$= 36.4 \text{ kJ}$$

$$\Delta S° = \Sigma S°_{(products)} - \Sigma S°_{(reactants)}$$

$$= S° \text{ Mn}(s) + 2S° \text{ H}_2\text{O}(g) - [S° \text{ MnO}_2(s) + 2S° \text{ H}_2(g)]$$

$$= 1 \text{ mol}\left(\frac{0.0320 \text{ kJ}}{\text{mol} \cdot \text{K}}\right) + 2 \text{ mol}\left(\frac{0.1887 \text{ kJ}}{\text{mol} \cdot \text{K}}\right)$$

$$- 1 \text{ mol}\left(\frac{0.0530 \text{ kJ}}{\text{mol} \cdot \text{K}}\right) - 2 \text{ mol}\left(\frac{0.1306 \text{ kJ}}{\text{mol} \cdot \text{K}}\right)$$

$$= 0.0952 \text{ kJ/K}$$

$$\Delta G° = \Delta H° - T\Delta S°$$

$$0 = 36.4 \text{ kJ} - (T \times 0.0952 \text{ kJ/K}) \qquad \rightarrow \qquad T = \frac{36.4 \text{ kJ}}{0.0952 \text{ kJ/K}} = 382 \text{ K}$$

(c) Balanced equation: $MnO_2(s) + C(s) \rightarrow Mn(s) + CO_2(g)$

$$\Delta H° = \Sigma \Delta H_f°_{(products)} - \Sigma \Delta H_f°_{(reactants)}$$

$$= \Delta H_f° \text{ Mn}(s) + \Delta H_f° \text{ CO}_2(g) - [\Delta H_f° \text{ MnO}_2(s) + \Delta H_f° \text{ C}(s)]$$

$$= 1 \text{ mol}\left(\frac{0 \text{ kJ}}{\text{mol}}\right) + 1 \text{ mol}\left(\frac{-393.5 \text{ kJ}}{\text{mol}}\right) - 1 \text{ mol}\left(\frac{-520.0 \text{ kJ}}{\text{mol}}\right) - 1 \text{ mol}\left(\frac{0 \text{ kJ}}{\text{mol}}\right)$$

$$= 126.5 \text{ kJ}$$

$$\Delta S° = \Sigma S°_{(products)} - \Sigma S°_{(reactants)}$$

$$= S° \text{ Mn}(s) + S° \text{ CO}_2(g) - [S° \text{ MnO}_2(s) + S° \text{ C}(s)]$$

$$= 1 \text{ mol}\left(\frac{0.0320 \text{ kJ}}{\text{mol·K}}\right) + 1 \text{ mol}\left(\frac{0.2136 \text{ kJ}}{\text{mol·K}}\right)$$

$$- 1 \text{ mol}\left(\frac{0.0530 \text{ kJ}}{\text{mol·K}}\right) - 1 \text{ mol}\left(\frac{0.0057 \text{ kJ}}{\text{mol·K}}\right)$$

$$= 0.1869 \text{ kJ/K}$$

$$\Delta G° = \Delta H° - T\Delta S°$$

$$0 = 126.5 \text{ kJ} - (T \times 0.1869 \text{kJ/K}) \quad \rightarrow \quad T = \frac{126.5 \text{ kJ}}{0.1869 \text{ kJ/K}} = 676.8 \text{ K}$$

For each proposed method, $\Delta H° > 0$ and $\Delta S° > 0$ (case III in Table 16.2), so the reaction is nonspontaneous at low temperature and spontaneous at high temperature.
Thus, **method (b) is the preferred method since it becomes spontaneous at the lowest temperature**.

50. Refer to the discussion below Example 16.6. The temperature at which the two phases are in equilibrium can be obtained from $T = \Delta H°/\Delta S°$. Thus we must calculate $\Delta H°$ and $\Delta S°$ for the process using the information given in the question. The process occurring can be represented as:

Sn(s, white) → Sn(s, gray)

and $\Delta H°$ and $\Delta S°$ can be calculated as for any other reaction, using $\Delta H_f°$ and $S°$ values.

$$\Delta H° = \Sigma \Delta H_f°_{(products)} - \Sigma \Delta H_f°_{(reactants)}$$

$$= \Delta H_f° \text{ Sn(s, gray)} - \Delta H_f° \text{ Sn(s, white)}$$

$$= 1 \text{ mol}\left(\frac{-2.09 \text{ kJ}}{\text{mol}}\right) - 1 \text{ mol}\left(\frac{0 \text{ kJ}}{\text{mol}}\right)$$

$$= -2.09 \text{ kJ}$$

$$\Delta S° = \Sigma S°_{(products)} - \Sigma S°_{(reactants)}$$

$$= S° \text{ Sn(s, gray)} - S° \text{ Sn(s, white)}$$

$$= 1 \text{ mol}\left(\frac{0.04414 \text{ kJ}}{\text{mol·K}}\right) - 1 \text{ mol}\left(\frac{0.05155 \text{ kJ}}{\text{mol·K}}\right)$$

$$= -0.00741 \text{ kJ/K}$$

Thus the equilibrium temperature for the transition is:

$$T = \Delta H°/\Delta S° = (-2.09 \text{ kJ})/(-0.00741 \text{ kJ/K}) = \underline{\textbf{282 K}}$$

52. This problem is very similar to Question 50 above. Refer to the discussion below Example 16.6. The temperature at which the two phases are in equilibrium can be obtained from $T = \Delta H° / \Delta S°$. Thus we must calculate $\Delta H°$ and $\Delta S°$ for the process shown in the question. In addition to the $\Delta H_f°$ and $S°$ values for diamond given in the question, we are told that graphite is the stable elemental form of carbon and therefore $\Delta H_f°$ and $S°$ values listed for carbon in Appendix 1 are those for graphite. Calculate $\Delta H°$ and $\Delta S°$ as for any other reaction, then use them to determine the equilibrium temperature.

$$\Delta H° = \Sigma \Delta H_f°_{(products)} - \Sigma \Delta H_f°_{(reactants)}$$

$$= \Delta H_f° \text{ C(diamond)} - \Delta H_f° \text{ C(graphite)}$$

$$= 1 \text{ mol}\left(\frac{1.9 \text{ kJ}}{\text{mol}}\right) - 1 \text{ mol}\left(\frac{0 \text{ kJ}}{\text{mol}}\right)$$

$$= 1.9 \text{ kJ}$$

$$\Delta S° = \Sigma S°_{(products)} - \Sigma S°_{(reactants)}$$

$$= S° \text{ C(diamond)} - S° \text{ C(graphite)}$$

$$= 1 \text{ mol}\left(\frac{0.0024 \text{ kJ}}{\text{mol·K}}\right) - 1 \text{ mol}\left(\frac{0.0057 \text{ kJ}}{\text{mol·K}}\right)$$

$$= -0.0033 \text{ kJ/K}$$

Thus the equilibrium temperature for the transition is:

$$T = \Delta H° / \Delta S° = (1.9 \text{ kJ})/(-0.0033 \text{ kJ/K}) = \underline{\mathbf{-5.8 \times 10^2 \text{ K}}}$$

The value of the equilibrium temperature is below 0 K, which is impossible; in other words **the two forms of carbon are not in equilibrium at any temperature when $P = 1$ atm.**

54. This problem is very similar to Questions 50 and 52 above. Refer to the discussion below Example 16.6. When bromine is at its normal boiling point, two phases (liquid and gas) are in equilibrium, and the temperature at which this equilibrium occurs can be obtained from $T = \Delta H° / \Delta S°$. The process can be represented as:

$$\text{Br}_2(l) \rightleftharpoons \text{Br}_2(g)$$

for which we must calculate $\Delta H°$ and ΔS. In addition to the $\Delta H_f°$ and $S°$ values given in the question, we need $\Delta H_f°$ for $\text{Br}_2(l)$: since this is the stable elemental form of bromine, by definition $\Delta H_f°$ is zero. Calculate $\Delta H°$ and $\Delta S°$ as for any other reaction (remember to convert the given $S°$ values into kJ/mol·K), then use them to determine the equilibrium temperature, that is, the normal boiling point of Br_2.

$$\Delta H° = \Sigma \Delta H_f°_{(products)} - \Sigma \Delta H_f°_{(reactants)}$$

$$= \Delta H_f° \text{ Br}_2(g) - \Delta H_f° \text{ Br}_2(l)$$

$$= 1 \text{ mol}\left(\frac{30.91 \text{ kJ}}{\text{mol}}\right) - 1 \text{ mol}\left(\frac{0 \text{ kJ}}{\text{mol}}\right)$$

$$= 30.91 \text{ kJ}$$

$$\Delta S° = \Sigma S°_{(products)} - \Sigma S°_{(reactants)}$$

$$= S° \text{ Br}_2(g) - S° \text{ Br}_2(l)$$

$$= 1 \text{ mol}\left(\frac{0.2454 \text{ kJ}}{\text{mol·K}}\right) - 1 \text{ mol}\left(\frac{0.1522 \text{ kJ}}{\text{mol·K}}\right)$$

$$= 0.0932 \text{ kJ/K}$$

Thus the equilibrium temperature for the transition is:

$$T = \Delta H°/\Delta S° = (30.91 \text{ kJ})/(0.0932 \text{ kJ/K}) = 332 \text{ K}$$

and hence **the normal boiling point of bromine is 332 K**.

16-5   EFFECT OF TEMPERATURE, PRESSURE AND CONCENTRATION ON REACTION SPONTANEITY: Effect of Concentration/Pressure on Spontaneity

56. This problem is very similar to Example 16.7(b). Use Equation 16.4 in order to calculate ΔG under the given conditions; the reaction is spontaneous if $\Delta G < 0$. For the reaction in the question, $\Delta G°$ is 18.0 kJ (given) and the expression for Q is:

$$Q = \frac{[H^+][F^-]}{[HF]}$$

(a) For the concentrations given, Equation 16.4 becomes:

$$\Delta G = 18.0 \text{ kJ} + (0.00831 \text{ kJ/K})(298 \text{ K}) \ln\frac{(0.78)(0.78)}{0.24}$$

$$\underline{\Delta G = 20.3 \text{ kJ}}$$

$\Delta G > 0$, so **the reaction is nonspontaneous under these conditions**.

(b) For the concentrations given, Equation 16.4 becomes:

$$\Delta G = 18.0 \text{ kJ} + (0.00831 \text{ kJ/K})(298 \text{ K}) \ln\frac{(0.0030)(0.0030)}{1.85}$$

$$\underline{\Delta G = -12.3 \text{ kJ}}$$

$\Delta G < 0$, so **the reaction is spontaneous under these conditions**.

58. This problem is very similar to Example 16.7(a,b).

(a) Calculate $\Delta G°$ using Equation 16.3 and $\Delta G_f°$ values from Appendix 1.

$$\Delta G° = \Sigma \Delta G_f°_{(products)} - \Sigma \Delta G_f°_{(reactants)}$$

$$\Delta G° = \Delta G_f° \text{ H}_2(g) + \Delta G_f° \text{ Br}_2(l) + 2\Delta G_f° \text{ OH}^-(aq) - [2\Delta G_f° \text{ H}_2\text{O}(l) + 2\Delta G_f° \text{ Br}^-(aq)]$$

$$= 1 \text{ mol}\left(\frac{0 \text{ kJ}}{\text{mol}}\right) + 1 \text{ mol}\left(\frac{0 \text{ kJ}}{\text{mol}}\right) + 2 \text{ mol}\left(\frac{-157.2 \text{ kJ}}{\text{mol}}\right)$$

$$- 2 \text{ mol}\left(\frac{-237.2 \text{ kJ}}{\text{mol}}\right) - 2 \text{ mol}\left(\frac{-104.0 \text{ kJ}}{\text{mol}}\right)$$

$$= \underline{\textbf{368.0 kJ}}$$

(b) Use Equation 16.4 to calculate ΔG under the given conditions. For this reaction the expression for Q is:

$$Q = \frac{(P_{H_2})[OH^-]^2}{[Br^-]^2}$$

The expression for Q requires $[OH^-]$, which is obtained from the pH (Equation 13.3):

$$pOH = 14.00 - pH = 4.28 \qquad [OH^-] = 10^{-pOH} = 10^{-4.28} = 5.2 \times 10^{-5} \text{ M}$$

Using this and the other conditions stated in the question, along with $\Delta G°$ from part (a) above, Equation 16.4 becomes:

$$\Delta G = 368.0 \text{ kJ} + (0.00831 \text{ kJ/K})(298 \text{ K}) \ln\frac{(0.878)(5.2 \times 10^{-5})^2}{(0.423)^2}$$

$$\underline{\textbf{\Delta G = 323.1 kJ}}$$

60. This problem is very similar to Example 16.7(a,c).

(a) Calculate $\Delta G°$ using Equation 16.3 and $\Delta G_f°$ values from Appendix 1.

$$\Delta G° = \Sigma \Delta G_f°_{(products)} - \Sigma \Delta G_f°_{(reactants)}$$
$$\Delta G° = \Delta G_f° \text{ Ag}^+(aq) + \Delta G_f° \text{ Cl}^-(aq) - \Delta G_f° \text{ AgCl}(s)$$

$$= 1 \text{ mol}\left(\frac{77.1 \text{ kJ}}{\text{mol}}\right) + 1 \text{ mol}\left(\frac{-131.2 \text{ kJ}}{\text{mol}}\right) - 1 \text{ mol}\left(\frac{-109.8 \text{ kJ}}{\text{mol}}\right)$$

$$= \underline{\textbf{55.7 kJ}}$$

(b) Let $[Ag^+] = [Cl^-] = x$. Substitute these into Equation 16.4 with $\Delta G = -1.0$ kJ and $\Delta G°$ calculated in part (a) above, and solve for x. For this reaction the expression for Q is:

$$Q = [Ag^+][Cl^-] = (x)(x) = x^2$$

Thus, Equation 16.4 becomes:

$$-1.0 \text{ kJ} = 55.7 \text{ kJ} + (0.00831 \text{ kJ/K})(298 \text{ K}) \ln(x^2)$$

$$\ln x^2 = 2\ln x = \frac{-1.0\,\text{kJ} - 55.7\,\text{kJ}}{(0.00831\,\text{kJ/K})(298\,\text{K})} \quad \rightarrow \quad \ln x = \frac{-1.0\,\text{kJ} - 55.7\,\text{kJ}}{(2)(0.00831\,\text{kJ/K})(298\,\text{K})} = -11.4$$

$$x = e^{-11.4} = 1.1 \times 10^{-5}$$

Therefore, the concentrations required such that the reaction is just spontaneous are **$[Ag^+] = [Cl^-] = 1.1 \times 10^{-5}$ M.**

(c) (See Section 15-3.) For the values of $[Ag^+]$ and $[Cl^-]$ calculated in part (b) above, the product $[Ag^+][Cl^-] = (1.1 \times 10^{-5})(1.1 \times 10^{-5}) = 1.2 \times 10^{-10}$, which is very close to the value of K_{sp} for AgCl (1.8×10^{-10}). Since $Q = [Ag^+][Cl^-]$ is slightly less than K_{sp} it follows that the solution is (slightly) below saturation and thus dissolving AgCl should indeed occur spontaneously (albeit only slightly) as the equation in the problem shows. **Yes, the answer to (b) is reasonable.**

16-6   THE FREE ENERGY CHANGE AND THE EQUILIBRIUM CONSTANT

62. This problem is similar to Example 16.8.

(a) Use Equation 16.3 to calculate ΔG°; use ΔG_f° data given in the question, along with values from Appendix 1.

$$\Delta G^\circ = \Sigma\,\Delta G_f^\circ{}_{(\text{products})} - \Sigma\,\Delta G_f^\circ{}_{(\text{reactants})}$$

$$\Delta G^\circ = \Delta G_f^\circ\,H^+(aq) + \Delta G_f^\circ\,NH_3(aq) - \Delta G_f^\circ\,NH_4^+(aq)$$

$$= 1\,\text{mol}\left(\frac{0\,\text{kJ}}{\text{mol}}\right) + 1\,\text{mol}\left(\frac{-26.7\,\text{kJ}}{\text{mol}}\right) - 1\,\text{mol}\left(\frac{-79.3\,\text{kJ}}{\text{mol}}\right)$$

$$= \underline{\textbf{52.6 kJ}}$$

(b) See Section 13-4. The reaction given in the question represents the ionization of the weak acid NH_4^+ for which the corresponding equilibrium constant is K_a. Use Equation 16.5 to calculate K, which is K_a, for the reaction.

$$\Delta G^\circ = -RT\ln K$$

$$52.6\,\text{kJ} = -(0.00831\,\text{kJ/K})(298\,\text{K})\ln K_a$$

$$\ln K_a = \left(\frac{52.6\,\text{kJ}}{-(0.00831\,\text{kJ/K})(298\,\text{K})}\right) = -21.2 \qquad \rightarrow \qquad K_a = e^{-21.2} = \underline{\textbf{6} \times \textbf{10}^{-10}}$$

64. This problem is related to Example 16.8.

(a) Use Equation 16.5 to calculate ΔG° for the reaction.

$$\Delta G^\circ = -RT\ln K = -(0.00831\,\text{kJ/K})(298\,\text{K})\ln(4.4 \times 10^{-19}) = \underline{\textbf{105 kJ}}$$

(b) Knowing ΔG° for this reaction (from part (a) above), we can use Equation 16.3 to calculate ΔG_f° for $N_2O(g)$; other necessary ΔG_f° values are given in Appendix 1.

$$\Delta G^\circ = \Sigma\, \Delta G_f^\circ{}_{(products)} - \Sigma\, \Delta G_f^\circ{}_{(reactants)}$$

$$\Delta G^\circ = 3\Delta G_f^\circ\, NO(g) - [\Delta G_f^\circ\, N_2O(g) + \Delta G_f^\circ\, NO_2(g)]$$

$$105\ kJ = 3\ mol\left(\frac{86.6\ kJ}{mol}\right) - 1\ mol[\Delta G_f^\circ\, N_2O(g)] - 1\ mol\left(\frac{51.3\ kJ}{mol}\right)$$

$$\Delta G_f^\circ\, N_2O(g) = \underline{\mathbf{104\ kJ/mol}}$$

66. This problem is similar to Example 16.8.

(a) Use Equation 16.5 to calculate ΔG° for the reaction at $T = 373\ K$ (100 °C).

$$\Delta G^\circ = -RT\ \ln K = -(0.00831\ kJ/K)(373\ K)\ln(11) = \underline{\mathbf{-7.43\ kJ}}$$

(b) Use Equation 16.3 to calculate ΔG° using ΔG_f° for $N_2O_4(g)$ given in the question, and other necessary ΔG_f° data from Appendix 1 (since we seek ΔG° at 25 °C, ΔG_f° values can be used); then use Equation 16.5 to determine K.

$$\Delta G^\circ = \Sigma\, \Delta G_f^\circ{}_{(products)} - \Sigma\, \Delta G_f^\circ{}_{(reactants)}$$

$$\Delta G^\circ = 2\Delta G_f^\circ\, NO_2(g) - \Delta G_f^\circ\, N_2O_4(g)$$

$$\Delta G^\circ = 2\ mol\left(\frac{51.3\ kJ}{mol}\right) - 1\ mol\left(\frac{97.9\ kJ}{mol}\right) = 4.7\ kJ$$

$$\Delta G^\circ = -RT\ \ln K$$

$$4.7\ kJ = -(0.00831\ kJ/K)(298\ K)\ln K$$

$$\ln K = \left(\frac{4.7\ kJ}{-(0.00831\ kJ/K)(298\ K)}\right) = -1.9 \qquad \rightarrow \quad K = e^{-1.9} = \underline{\mathbf{0.15}}$$

68. This problem is similar to Example 16.8. See also Section 13-4: Recall that the equilibrium constant K_a refers to the ionization of a weak acid, in this case HF.

$$HF(aq) \rightleftharpoons H^+(aq) + F^-(aq)$$

The value of K (i.e., K_a) for this reaction can be calculated using Equation 16.5, for which a value for ΔG° is required. This may be obtained similarly to Example 16.5, using Equation 13.2 and ΔH° and ΔS° calculated using data from Appendix 1 and given in the problem. Therefore: calculate ΔH° and ΔS° and use them to calculate ΔG° and hence K_a.

$$\Delta H^\circ = \Sigma\, \Delta H_f^\circ{}_{(products)} - \Sigma\, \Delta H_f^\circ{}_{(reactants)}$$

$$\Delta H^\circ = \Delta H_f^\circ\, H^+(aq) + \Delta H_f^\circ\, F^-(aq) - \Delta H_f^\circ\, HF(aq)$$

$$\Delta H^\circ = 1\ mol\left(\frac{0\ kJ}{mol}\right) + 1\ mol\left(\frac{-332.6\ kJ}{mol}\right) - 1\ mol\left(\frac{-320.1\ kJ}{mol}\right) = -12.5\ kJ$$

$$\Delta S° = \Sigma\, S°_{(products)} - \Sigma\, S°_{(reactants)}$$

$$\Delta S° = S°\, H^+(aq) + S°\, F^-(aq) - S°\, HF(aq)$$

$$\Delta S° = 1\ mol\left(\frac{0\ kJ}{mol\cdot K}\right) + 1\ mol\left(\frac{-0.0138\ kJ}{mol\cdot K}\right) - 1\ mol\left(\frac{0.0887\ kJ}{mol\cdot K}\right) = -0.1025\ kJ/K$$

$$\Delta G° = \Delta H° - T\Delta S° = -12.5\ kJ - (298\ K)(-0.1025\ kJ/K) = 18.0\ kJ$$

$$\Delta G° = -RT\ \ln K$$

$$18.0\ kJ = -(0.00831\ kJ/K)(298\ K)\ln K_a$$

$$\ln K_a = \left(\frac{18.0\ kJ}{-(0.00831\ kJ/K)(298\ K)}\right) = -7.27 \qquad \rightarrow \qquad K = e^{-7.27} = \underline{\mathbf{7.0 \times 10^{-4}}}$$

70. See Section 13-5 and Example 13.11: Recall that the equilibrium constant K_b for ionization of a weak base, in this case R_2NH, refers to the following equation:

$$R_2NH(aq) + H_2O \rightleftharpoons R_2NH_2^+(aq) + OH^-(aq) \qquad K_b = \frac{[R_2NH_2^+][OH^-]}{[R_2NH]}$$

Following the procedure in Example 13.11, the expression for K_b becomes $K_b = \dfrac{[OH^-]^2}{[R_2NH]_0}$

where $[R_2NH]_0$ is the initial concentration of the base, in this case 0.250 M. The value of $[OH^-]$ is obtained from the pH and K_w:

For a solution with pH = 10.60, $[H^+] = 10^{-10.60}$ (Equation 13.3) and hence from Equations 13.1 and 13.2:

$$[OH^-] = \frac{K_w}{[H^+]} = \frac{1.0 \times 10^{-14}}{10^{-10.60}} = 4.0 \times 10^{-4}\ M$$

Substitute this value for $[OH^-]$ and $[R_2NH]_0$ into the expression for K_b above:

$$K_b = \frac{[OH^-]^2}{[R_2NH]_0} = \frac{(4.0 \times 10^{-4})^2}{0.250} = 6.4 \times 10^{-7}$$

Finally, knowing K_b for the reaction, use Equation 16.5 to calculate the corresponding $\Delta G°$:

$$\Delta G° = -RT\ \ln K_b = -(0.00831\ kJ/K)(298\ K)\ln(6.4 \times 10^{-7}) = \underline{\mathbf{35.3\ kJ}}$$

16-7   ADDITIVITY OF FREE ENERGY CHANGES; COUPLED REACTIONS

72. Refer to the discussion at the start fof Section 16-7. Label the two reactions given in the question as (1) and (2), respectively. The required coupled reaction can then be obtained by adding together [the reverse of (1)] + 2 × [the reverse of (2)], respectively:

$$2NO(g) + \tfrac{3}{2}O_2(g) \rightarrow N_2O_5(g) \qquad \Delta G° = -\Delta G°(1) = 59.2 \text{ kJ}$$

$$\underline{2NO_2(g) \rightarrow 2NO(g) + O_2(g)} \qquad \Delta G° = -2 \times \Delta G°(2) = 71.2 \text{ kJ}$$

$$2NO(g) + \tfrac{3}{2}O_2(g) + 2NO_2(g)$$
$$\rightarrow N_2O_5(g) + 2NO(g) + O_2(g)$$

$$\mathbf{2NO_2(g) + \tfrac{1}{2}O_2(g) \rightarrow N_2O_5(g)}$$

Note that $\Delta G°$ for the two new reactions above (given the the right of each reaction) are obtained by the same manipulations that were applied to the reactions themselves. When two reactions are added together their free energy changes are additive. Thus, to obtain $\Delta G°$ for the coupled reaction, add together the $\Delta G°$ values of the two contributing reactions:

$$\Delta G° = 59.2 \text{ kJ} + 71.2 \text{ kJ} = \underline{\textbf{130.4 kJ}}$$

74. (a) Calculate $\Delta G°$ for the given reaction using Equation 16.3:

$$\Delta G° = \Sigma \Delta G_f°_{(products)} - \Sigma \Delta G_f°_{(reactants)}$$
$$\Delta G° = \Delta G_f° \text{ Zn}(s) + \Delta G_f° \text{ S}(s) - \Delta G_f° \text{ ZnS}(s)$$
$$\Delta G° = 1 \text{ mol}\left(\frac{0 \text{ kJ}}{\text{mol}}\right) + 1 \text{ mol}\left(\frac{0 \text{ kJ}}{\text{mol}}\right) - 1 \text{ mol}\left(\frac{-201.3 \text{ kJ}}{\text{mol}}\right) = \underline{\textbf{201.3 kJ}}$$

$\Delta G°$ for the reaction is positive so the reaction is not feasible at 25°C (it is nonspontaneous).

(b) For the reaction given in this part, calculate $\Delta G°$ using Equation 16.3:

$$\Delta G° = \Sigma \Delta G_f°_{(products)} - \Sigma \Delta G_f°_{(reactants)}$$
$$\Delta G° = \Delta G_f° \text{ SO}_2(g) - [\Delta G_f° \text{ S}(s) + \Delta G_f° \text{ O}_2(g)]$$
$$\Delta G° = 1 \text{ mol}\left(\frac{-300.2 \text{ kJ}}{\text{mol}}\right) - 1 \text{ mol}\left(\frac{0 \text{ kJ}}{\text{mol}}\right) - 1 \text{ mol}\left(\frac{0 \text{ kJ}}{\text{mol}}\right) = -300.2 \text{ kJ}$$

Adding together the two reactions in the question gives the required coupled reaction:

$$\text{ZnS}(s) \rightarrow \text{Zn}(s) + \text{S}(s)$$
$$\underline{\text{S}(s) + \text{O}_2(g) \rightarrow \text{SO}_2(g)}$$
$$\text{ZnS}(s) + \text{S}(s) + \text{O}_2(g)$$
$$\rightarrow \text{Zn}(s) + \text{S}(s) + \text{SO}_2(g)$$

$$\mathbf{ZnS(s) + O_2(g) \rightarrow Zn(s) + SO_2(g)}$$

When two reactions are added together their free energy changes are additive. Thus, to obtain $\Delta G°$ for the coupled reaction, add together the $\Delta G°$ values of the two contributing reactions:

$$\Delta G° = 201.3 \text{ kJ} + (-300.2 \text{ kJ}) = \underline{-98.9 \text{ kJ}}$$

$\Delta G°$ for the coupled reaction is negative so the reaction is feasible at 25°C (it is spontaneous).

76. Refer to the discussion above Figure 16.10; see also Example 16.9. Note that $\Delta G°$ for the second equation given in the question is *per mol* of ADP consumed. Since free energy changes are additive, it follows that for x mol of ADP being consumed, the free energy change is $31x$ kJ. To determine how many mol of ADP must be consumed in order to give $\Delta G° = -390$ kJ for the coupled reaction, solve the following equation:

$$\Delta G°_{(coupled\ reaction)} = \Delta G°_{(glucose\ reaction)} + x\ \Delta G°_{(ADP\ reaction)}$$

In other words, the glucose reaction is to be coupled with x mol of ADP reaction. Thus:

$$-390 \text{ kJ} = -2870 \text{ kJ} + 31x \text{ kJ} \quad \rightarrow \quad 31x \text{ kJ} = 2480 \text{ kJ} \quad \rightarrow \quad x = 80$$

To obtain the overall coupled reaction multiply the ADP reaction by 80 and add that to the glucose reaction:

$C_6H_{12}O_6(aq) + 6O_2(g) \rightarrow 6CO_2(g) + 6H_2O$	$\Delta G° = -2870$ kJ
$80\ [ADP(aq) + HPO_4^{2-}(aq) + 2H^+(aq) \rightarrow ATP(aq) + H_2O]$	$\Delta G° = 80(31)$ kJ
$C_6H_{12}O_6(aq) + 6O_2(g) + 80ADP(aq) + 80HPO_4^{2-}(aq) + 160H^+(aq)$	$\Delta G° = -390$ kJ
$\quad\quad \rightarrow 6CO_2(g) + 6H_2O + 80ATP(aq) + 80H_2O$	

The overall coupled reaction is:

$$C_6H_{12}O_6(aq) + 6O_2(g) + 80ADP(aq) + 80HPO_4^{2-}(aq) + 160H^+(aq)$$
$$\rightarrow 6CO_2(g) + 86H_2O + 80ATP(aq)$$

UNCLASSIFIED PROBLEMS

78. This problem is related to Examples 16.5 and 16.6. For the protein denaturation process, we are given values of $\Delta H° = 142$ kJ and $\Delta S° = 278$ J/K = 0.278 kJ/K. (Note that $\Delta H° > 0$ and $\Delta S° > 0$ (case III in Table 16.2), so the reaction is nonspontaneous at low temperature and spontaneous at high temperature.) To determine the lowest temperature at which the process becomes spontaneous, use Equation 16.2 with $\Delta G° = 0$.

$$\Delta G° = \Delta H° - T\Delta S°$$

$$0 = 142 \text{ kJ/mol} - (T \times 0.278 \text{ kJ/K}) \quad \rightarrow \quad T = \frac{142 \text{ kJ}}{0.278 \text{ kJ/K}} = 511 \text{ K}$$

Thus, the **protein begins to denature at 511 K.**

80. This problem is related to the discussion above Figure 16.10; see also Example 16.9. Before we can couple the two reactions, $\Delta G°$ for the second reaction (H_2S oxidation) must be calculated using Equation 16.3 and data from Appendix 1.

$$\Delta G° = \Sigma \Delta G_f°_{(products)} - \Sigma \Delta G_f°_{(reactants)}$$

$$\Delta G° = \Delta G_f° \, H_2O(l) + \Delta G_f° \, S(s) - [\Delta G_f° \, H_2S(g) + \tfrac{1}{2}\Delta G_f° \, O_2(g)]$$

$$\Delta G° = 1 \text{ mol}\left(\frac{-237.2 \text{ kJ}}{\text{mol}}\right) + 1 \text{ mol}\left(\frac{0 \text{ kJ}}{\text{mol}}\right) - 1 \text{ mol}\left(\frac{-33.6 \text{ kJ}}{\text{mol}}\right) - \tfrac{1}{2} \text{ mol}\left(\frac{0 \text{ kJ}}{\text{mol}}\right)$$

$$\Delta G° = -203.6 \text{ kJ}$$

The stoichiometry of the coupled reaction given in the question indicates 24 mol of H_2S are required; thus 24 mol of the above reaction must be coupled with the glucose synthesis reaction:

$$6CO_2(g) + 6H_2O(l) \rightarrow C_6H_{12}O_6(aq) + 6O_2(g) \qquad\qquad \Delta G° = 2870 \text{ kJ}$$
$$\underline{24 \, [H_2S(g) + \tfrac{1}{2}O_2(g) \rightarrow H_2O(l) + S(s)]} \qquad\qquad \underline{\Delta G° = 24(-203.6) \text{ kJ}}$$
$$6CO_2(g) + 6H_2O(l) + 24H_2S(g) + 12O_2(g) \qquad\qquad \Delta G° = 2870 \text{ kJ} + 24(-203.6\text{kJ})$$
$$\rightarrow C_6H_{12}O_6(aq) + 6O_2(g) + 24H_2O(l) + 24S(s)$$

The overall coupled reaction is:

$$6CO_2(g) + 24H_2S(g) + 6O_2(g) \rightarrow C_6H_{12}O_6(aq) + 18H_2O(l) + 24S(s)$$

for which $\Delta G° = 2870 \text{ kJ} + 24(-203.6\text{kJ}) = \underline{\mathbf{-2016 \text{ kJ}}}$

$\Delta G° < 0$ so the reaction is spontaneous at 25 °C.

82. For this reaction, the expression for K is

$$K = \frac{(P_{O_2})[\text{Hb·CO}]}{(P_{CO})[\text{Hb·O}_2]}$$

and when the pressure of CO is the same as that of O_2, the expression for K becomes:

$$K = \frac{[\text{Hb·CO}]}{[\text{Hb·O}_2]}$$

We are told that the value of $\Delta G°$ is -14 kJ at a temperature of 37°C (310 K); substituting these values and the expression for K into Equation 16.5 gives:

$$\Delta G° = -RT \ln K$$

$$-14 \text{ kJ} = -(0.00831 \text{ kJ/K})(310 \text{ K})\ln\frac{[\text{Hb·CO}]}{[\text{Hb·O}_2]}$$

$$\ln\frac{[\text{Hb·CO}]}{[\text{Hb·O}_2]} = \frac{-14 \text{ kJ}}{-(0.00831 \text{ kJ/K})(310 \text{ K})} = 5.4$$

$$\frac{[\text{Hb·CO}]}{[\text{Hb·O}_2]} = e^{5.4} \qquad \rightarrow \qquad \frac{[\text{Hb·O}_2]}{[\text{Hb·CO}]} = e^{-5.4} = \underline{\mathbf{4.5 \times 10^{-3}}}$$

CONCEPTUAL PROBLEMS

84. (a) **False**. See pages 407–408: It is true that **many** exothermic reactions **are spontaneous**, but not all reactions: there are many endothermic processes that also are spontaneous and many exothermic processes that are nonspontaneous. The enthalpy change must be considered alongside the entropy change since one might oppose the other: a favorable entropy change might make an endothermic reaction spontaneous and an unfavorable entropy change might make an exothermic reaction nonspontaneous.

    (b) **False**. See pages 419–422: A positive value for $\Delta G°$ simply means that the reaction is nonspontaneous **under standard conditions**. However, as Equation 16.4 shows, an adjustment in pressure or concentration away from standard conditions (so that $Q \neq 1$) can make the non-standard ΔG value negative (and therefore spontaneous under those conditions) even if $\Delta G°$ is positive.

    (c) **False**. See page 412: In general, $\Delta S°$ is positive for a reaction that results in an **increase in the number of moles _of gas_**. In other cases, $\Delta S°$ may be either positive or negative.

    (d) **False**. See Table 16.2: When both $\Delta H°$ and $\Delta S°$ are negative, $\Delta G°$ is negative only at low temperature; at high temperature $\Delta G°$ becomes positive.

86. (a) Using Equation 16.2, if $\Delta H° = \Delta G°$, then it follows that $T\Delta S° = 0$ and hence $T = 0$. Thus:

    $\underline{\Delta H° \text{ and } \Delta G° \text{ become equal at 0 K}}$.

    (b) Using Equation 16.4, if $\Delta G° = \Delta G$, then it follows that $RT\ln Q = 0$, so $\ln Q = 0$ and hence $Q = e^0 = 1$. Thus:

    $\underline{\Delta G° \text{ and } \Delta G \text{ are equal when } Q = 1}$.

    (c) See page 409: The entropy of a gas is always greater than the entropy of the liquid from which it is formed. (This can be confirmed by comparing $S°$ values for $H_2O(l)$ and $H_2O(g)$ in Table 16.1.) Thus:

    $\underline{S° \text{ for steam is greater than } S° \text{ for water}}$.

88. (a) See Section 16-2. Gases have much higher entropy (per mole of substance) than either solids or liquids, and therefore if the number of moles of gas decreases in a reaction then the total entropy should decrease significantly; that is, $\Delta S°$ will be negative.

    (b) See Section 16-2. The entropy of a given substance increases with temperature. However, when calculating $\Delta S°$ we are comparing the entropies of two (or more) substances — in other words, we look at the **difference in entropy** between reactants and products — and therefore $\Delta S°$ may be either positive or negative. As temperature increases, the entropy of **both** reactants **and** products will increase at approximately the same rate and hence the **difference** between them remains essentially constant.

(c) See Sections 16-1 and 16-2. In a solid, the particles have restricted movement: each particle essentially has a fixed location and the whole assembly is ordered. Therefore, since the number of ways of arranging the particles in the solid is very limited, the solid has low entropy. In the corresponding liquid, the particles are free to move and can access the whole volume of the liquid. In this way there are many different ways in which the particles can be distributed, and hence the entropy of the liquid is larger.

90. (a) The graph shows that $\Delta G°$ is negative (the reaction is spontaneous) at low temperatures ($T < 350$ K) and $\Delta G°$ becomes increasingly positive (the reaction is nonspontaneous) at higher temperatures ($T > 350$ K). The relationship between $\Delta G°$ and temperature is linear and **the spontaneity decreases with increasing temperature**.

(b) Referring to Table 16.2 (case III), it can be concluded that a reaction for which $\Delta G°$ is negative at low temperatures and positive at high temperatures has both $\Delta H°$ and $\Delta S°$ negative. Therefore: **YES, the reaction is exothermic**.

(c) See part (b) above: **NO, $\Delta S° < 0$**.

(d) See page 414: Recall that the reaction at standard conditions is at equilibrium when $\Delta G° = 0$. Reading from the graph, the temperature at which this occurs is approximately **350 K**.

(e) At 97°C (370 K), reading from the graph $\Delta G°$ is approximately 10 kJ. Substitute these values of T and $\Delta G°$ into Equation 16.5:

$$\Delta G° = -RT \ln K$$

$$10 \text{ kJ} = -(0.00831 \text{ kJ/K})(370 \text{ K})\ln K$$

$$\ln K = \left(\frac{10 \text{ kJ}}{-(0.00831 \text{ kJ/K})(370 \text{ K})} \right) = -3.3 \qquad \rightarrow \qquad K = e^{-3.3} = \mathbf{4 \times 10^{-2}}$$

92. (a) The reaction involves 2 mol of gas forming 2 mol of solid; since solids have less entropy than gases, the entropy change therefore should always be negative:

At all temperatures, $\Delta S°$ **LT** 0

(b) We are told that the reaction vessel feels warm to the touch after the reaction; in other words, heat is given out during the reaction: the enthalpy change is negative:

At all temperatures, $\Delta H°$ **LT** 0

(c) Refer to Table 16.2: a reaction for which both $\Delta H°$ and $\Delta S°$ are negative is spontaneous at low temperature ($\Delta G° < 0$) and nonspontaneous ($\Delta G° > 0$) at higher temperature. Thus the required statement cannot be answered in the given form; more information must be supplied:

At all temperatures, $\Delta G°$ **MI** 0

CHALLENGE PROBLEMS

93. See Sections 9-2 and 16-5. In considering the vaporization of C_2H_5OH and the entropy of C_2H_5OH as a vapor, let's focus on the liquid/vapor phase change for C_2H_5OH:

$$C_2H_5OH(l) \rightleftharpoons C_2H_5OH(g)$$

Recall (page 421) that at a phase change $T = \Delta H° / \Delta S°$. Knowing ΔH_{vap} and the temperature at which this phase change occurs, we can obtain ΔS_{vap}. We are told that at 63 °C (336 K) the vapor pressure of C_2H_5OH is 392 mm Hg and we are given ΔH_{vap}: use the Clausius-Clapeyron equation (Equation 9.1) to calculate the normal boiling point of C_2H_5OH — it is, by definition, the temperature at which the vapor pressure reaches 760 mm Hg.

The Clausius-Clapeyron equation is:

$$\ln \frac{P_2}{P_1} = \frac{\Delta H_{vap}}{R} \left[\frac{1}{T_1} - \frac{1}{T_2} \right]$$

We know $P_1 = 392$ mm Hg at $T_1 = 336$ K and $\Delta H_{vap} = 42.9$ kJ $= 42.9 \times 10^3$ J; we seek T_1 such that $P_2 = 760$ mm Hg.

$$\ln \frac{760 \text{ mm Hg}}{392 \text{ mm Hg}} = \frac{42900 \text{ J}}{8.3145 \text{ J/K}} \left[\frac{1}{336 \text{ K}} - \frac{1}{T_2} \right]$$

$$\left[\frac{1}{336 \text{ K}} - \frac{1}{T_2} \right] = \frac{8.3145 \text{ J/K}}{42900 \text{ J}} \ln \frac{760 \text{ mm Hg}}{392 \text{ mm Hg}} = 1.28 \times 10^{-4} \text{ /K}$$

$$\frac{1}{T_2} = \frac{1}{336 \text{ K}} - 1.28 \times 10^{-4} \text{ /K} = 2.85 \times 10^{-3} \text{ /K} \qquad \rightarrow T_2 = 351 \text{ K}$$

At the phase change $T = \Delta H° / \Delta S° \quad \rightarrow \quad \Delta S° = \Delta S_{vap} = \Delta H_{vap}/T$

$$\Delta S_{vap} = (42.9 \times 10^3 \text{ J})/(351 \text{ K}) = 122 \text{ J/K}$$

Thus we know ΔS for the liquid/vapor equilibrium:

$$C_2H_5OH(l) \rightleftharpoons C_2H_5OH(g)$$

and since $S°$ $C_2H_5OH(l)$ is given in Table 16.1, we can obtain $S°$ $C_2H_5OH(g)$ using Equation 16.1:

$$\Delta S° = \Delta S_{vap} = \Sigma S°_{(products)} - \Sigma S°_{(reactants)}$$

$$122 \text{ J/K} = S° \text{ } C_2H_5OH(g) - S° \text{ } C_2H_5OH(l)$$

$$122 \text{ J/K} = (1 \text{ mol})(S° \text{ } C_2H_5OH(g)) - 1 \text{ mol} \left(\frac{160.7 \text{ J}}{\text{mol·K}} \right)$$

$S°$ $C_2H_5OH(g)$ = __283 J/mol·K__

94. In order to calculate equilibrium partial pressures, the value of K must first be obtained. Determine $\Delta G°$ as in Example 16.5, using $\Delta H°$ and $\Delta S°$ calculated from data in the question and in Appendix 1; then use $\Delta G°$ to calculate K using Equation 16.5.

$$\Delta H° = \Sigma \Delta H_f°_{(products)} - \Sigma \Delta H_f°_{(reactants)}$$

$$= [\Delta H_f° \, H_2(g) + \Delta H_f° \, I_2(g)] - 2\Delta H_f° \, HI(g)$$

$$= 1 \text{ mol}\left(\frac{0 \text{ kJ}}{\text{mol}}\right) + 1 \text{ mol}\left(\frac{62.4 \text{ kJ}}{\text{mol}}\right) - 2 \text{ mol}\left(\frac{26.5 \text{ kJ}}{\text{mol}}\right) = 9.4 \text{ kJ}$$

$$\Delta S° = \Sigma S°_{(products)} - \Sigma S°_{(reactants)}$$

$$= [S° \, H_2(g) + S° \, I_2(g)] - 2S° \, HI(g)$$

$$= 1 \text{ mol}\left(\frac{0.1306 \text{ kJ}}{\text{mol·K}}\right) + 1 \text{ mol}\left(\frac{0.2607 \text{ kJ}}{\text{mol·K}}\right) - 2 \text{ mol}\left(\frac{0.2065 \text{ kJ}}{\text{mol·K}}\right) = -0.0217 \text{ kJ/K}$$

$$\Delta G° = \Delta H° - T\Delta S° = 9.4 \text{ kJ} - (773 \text{ K})(-0.0217 \text{ kJ/K}) = 26.2 \text{ kJ}$$

$$\Delta G° = -RT \ln K$$

$$26.2 \text{ kJ} = -(0.00831 \text{ kJ/K})(773 \text{ K})\ln K$$

$$\ln K = \frac{26.2 \text{ kJ}}{-(0.00831 \text{ kJ/K})(773 \text{ K})} = -4.08 \qquad \rightarrow \qquad K = e^{-4.08} = 1.7 \times 10^{-2}$$

Now proceed similarly to Example 12.6. Set up an equilibrium table and insert the given information. Let ΔP for H_2 be $+x$ and complete the table:

	2HI(g)	$\rightleftharpoons$	H₂(g)	+	I₂(g)
P_0 (atm)	0.200		0.200		0.200
ΔP (atm)	$-2x$		$+x$		$+x$
P_{eq} (atm)	$0.200 - 2x$		$0.200 + x$		$0.200 + x$

Substitute into the expression for K and solve for x:

$$K = \frac{(P_{H_2})(P_{I_2})}{(P_{HI})^2} = \frac{(0.200 + x)(0.200 + x)}{(0.200 - 2x)^2} = \frac{(0.200 + x)^2}{(0.200 - 2x)^2} = 1.7 \times 10^{-2}$$

Hence, taking square roots: $\qquad \dfrac{(0.200 + x)}{(0.200 - 2x)} = (1.7 \times 10^{-2})^{\frac{1}{2}} = \pm 0.13$

Thus: $\qquad 0.200 + x = \pm 0.13(0.200 - 2x) \qquad\qquad \rightarrow \quad x = -0.138 \quad \text{or} \quad -0.31$

The root $x = -0.31$ is not acceptable since it will give $P_{H_2} < 0$; thus $x = -0.138$ and the equilibrium partial pressures of all three gasses are:

$$P_{H_2} = P_{I_2} = 0.200 + x = \underline{\mathbf{0.062 \text{ atm}}} \qquad\qquad P_{HI} = 0.200 - 2x = \underline{\mathbf{0.476 \text{ atm}}}$$

95. (a) (See also Section 8-4.) The heat of fusion corresponds to $\Delta H°$ for the process

$$H_2O(s) \rightleftharpoons H_2O(l)$$

but $\Delta H°$ should be expressed in kJ/mol. Thus:

$$\Delta H° = \Delta H_{fusion} = \frac{333\ J}{1\,g\,H_2O} \times \frac{18.02\ g\,H_2O}{1\,mol\,H_2O} \times \frac{1\,kJ}{1000\ J} = \mathbf{\underline{6.00\ kJ/mol}}$$

(b) At 0 °C, and assuming a pressure of 1 atm, $H_2O(s)$ and $H_2O(l)$ are at equilibrium because this is the temperature and pressure at which this phase change occurs. Since the system is at equilibrium and under standard conditions, $\mathbf{\underline{\Delta G° = 0}}$.

(c) See page 421: At a phase change such as here, $T = \Delta H°/\Delta S°$. Since we know $T = 0$ °C (or $T = 273.15$ K) for this phase change we can solve for $\Delta S°$. (Assume 1 mol of reaction.)

$$\Delta S° = \Delta H°/T = \frac{6.00\ kJ}{273.15\ K} = \mathbf{\underline{0.0220\ kJ/K}}$$

(d) (See Example 16.5.) Knowing $\Delta H°$ and $\Delta S°$ for the reaction we can use Equation 16.2 to calculate $\Delta G°$ at any T, making the assumption that $\Delta H°$ and $\Delta S°$ are constant. Thus, at 253 K:

$$\Delta G° = \Delta H° - T\Delta S° = 6.00\ kJ - (253\ K)(0.0220\,kJ/K) = \mathbf{\underline{0.43\ kJ}}$$

(e) This is similar to part (d) above: use Equation 16.2 to calculate $\Delta G°$ at $T = 293$ K:

$$\Delta G° = \Delta H° - T\Delta S° = 6.00\ kJ - (293\ K)(0.0220\,kJ/K) = \mathbf{\underline{-0.45\ kJ}}$$

Note that, as is to be expected, the answers to parts (d) and (e) show that ice does NOT melt spontaneously at −20°C but it DOES melt spontaneously at +20°C.

96. (a) We are told that the useful work produced by the sugar metabolism is 25% of the free energy change for the reaction. Therefore, to determine the work obtained at 37°C, we need to know $\Delta G°$ at that temperature. This will be similar to Example 16.5. Given $\Delta H°$ and $\Delta G°$ at 25°C, we can use Equation 16.2 to calculate $\Delta S°$:

$$\Delta G° = \Delta H° - T\Delta S°$$

$$\rightarrow \Delta S° = (\Delta H° - \Delta G°)/T = \frac{-5650\ kJ - (-5790\ kJ)}{298\ K} = 0.470\ kJ/K$$

and then recall that $\Delta H°$ and $\Delta S°$ are essentially constant so we can now use Equation 16.2 to calculate $\Delta G°$ at $T = 310$ K (37°C):

$$\Delta G° = \Delta H° - T\Delta S° = -5650\ kJ - (310\ K)(0.470\ kJ/K) = -5796\ kJ$$

This is the free energy change produced at 37°C per mol of reaction, that is, *per mol of sugar*, of which 25% is available as useful work. Use this $\Delta G°$ value to determine the useful work obtained per gram of sugar.

Strategy: $\Delta G°$ (per mol) → $\Delta G°$ (per g) → useful work (= 25% of $\Delta G°$ per g)

$$\text{useful work} = \frac{-5796 \text{ kJ}}{1 \text{ mol } C_{12}H_{22}O_{11}} \times \frac{1 \text{ mol } C_{12}H_{22}O_{11}}{342.3 \text{ g } C_{12}H_{22}O_{11}} \times 0.25 = -4.2 \text{ kJ/g } C_{12}H_{22}O_{11}$$

Thus, one gram of sugar provides **4.2 kJ of useful work**.

(b) To use the equation for w given in the question, the body mass m must first be expressed in kg. Use conversion factors from Table 1.3:

$$m = 120 \text{ lb} \times \frac{453.6 \text{ g}}{1 \text{ lb}} \times \frac{1 \text{ kg}}{1000 \text{ g}} = 54.4 \text{ kg}$$

Now use the given equation to determine w for this mass and $h = 4158$ m (given):

$$w = 9.79 \times 10^{-3} \, mh = (9.79 \times 10^{-3})(54.4 \text{ kg})(4158 \text{ m}) = 2.21 \times 10^3 \text{ kJ}$$

Thus, the mass of sugar required to provide this energy is:

$$2.21 \times 10^3 \text{ kJ} \times \frac{1 \text{ g sugar}}{4.2 \text{ kJ}} = \mathbf{5.3 \times 10^2 \text{ g}}$$

97. The reaction described in the question, the decomposition of CaH_2 into Ca and H_2, is:

$$CaH_2(s) \rightleftharpoons Ca(s) + H_2(g)$$

Note that the expression for K for this reaction is simply $K = P_{H_2}$ since CaH_2 and Ca are both solids and so do not appear in the expression for K. Thus, the value of K equals the partial pressure of H_2 at equilibrium; and since the question asks for a partial pressure of H_2 of 1 atm, we must determine the temperature where $K = P_{H_2} = 1$. This value of K also implies the value of $\Delta G°$ is zero:

$$\Delta G° = -RT \ln K = -RT \ln 1 = 0$$

In effect, the question now becomes similar to Example 16.6. Calculate $\Delta H°$ and $\Delta S°$, then substitute them (along with $\Delta G° = 0$) into Equation 16.2 and solve for T.

$$\begin{aligned}
\Delta H° &= \Sigma \, \Delta H_f°_{(products)} - \Sigma \, \Delta H_f°_{(reactants)} \\
&= \Delta H_f° \, Ca(s) + \Delta H_f° \, H_2(g) - \Delta H_f° \, CaH_2(s) \\
&= 1 \text{ mol}\left(\frac{0 \text{ kJ}}{\text{mol}}\right) + 1 \text{ mol}\left(\frac{0 \text{ kJ}}{\text{mol}}\right) - 1 \text{ mol}\left(\frac{-186.2 \text{ kJ}}{\text{mol}}\right) = 186.2 \text{ kJ}
\end{aligned}$$

$$\begin{aligned}
\Delta S° &= \Sigma \, S°_{(products)} - \Sigma \, S°_{(reactants)} \\
&= S° \, Ca(s) + S° \, H_2(g) - S° \, CaH_2(s)
\end{aligned}$$

$$= 1 \text{ mol}\left(\frac{0.0414 \text{ kJ}}{\text{mol·K}}\right) + 1 \text{ mol}\left(\frac{0.1306 \text{ kJ}}{\text{mol·K}}\right) - 1 \text{ mol}\left(\frac{0.0420 \text{ kJ}}{\text{mol·K}}\right) = 0.1300 \text{ kJ/K}$$

$\Delta G° = \Delta H° - T\Delta S°$

$0 = 186.2 \text{ kJ} - T(0.1300 \text{ kJ/K})$ $\qquad \rightarrow \qquad$ $T =$ **1432 K**

A temperature of 1432 K (1159 °C) is required to produce H_2 at a partial pressure of 1 atm.

98. The initial conversion of Cu to CuO proceeds according to the following reaction, for which $\Delta G°$ may be calculated using Equation 16.3 and data from Appendix 1:

$Cu(s) + \frac{1}{2}O_2(g) \rightarrow CuO(s)$

$\begin{aligned} \Delta G° \quad &= \Sigma \Delta G_f°_{(products)} - \Sigma \Delta G_f°_{(reactants)} \\ &= \Delta G_f° \text{ CuO}(s) - [\Delta G_f° \text{ Cu}(s) + \frac{1}{2}\Delta G_f° \text{ O}_2(g)] \\ &= 1 \text{ mol}\left(\frac{-129.7 \text{ kJ}}{\text{mol}}\right) - 1 \text{ mol}\left(\frac{0 \text{ kJ}}{\text{mol}}\right) - \frac{1}{2} \text{ mol}\left(\frac{0 \text{ kJ}}{\text{mol}}\right) = -129.7 \text{ kJ} \end{aligned}$

For this reaction, $\Delta G° < 0$, so the black CuO forms spontaneously at room temperature.

In considering the other two reactions given in the question, we must determine the temperature at which they become spontaneous: see Section 16-5 and Example 16.6.

For the conversion of CuO to Cu_2O:

$2CuO(s) \rightarrow Cu_2O(s) + \frac{1}{2}O_2(g)$

$\begin{aligned} \Delta H° \quad &= \Sigma \Delta H_f°_{(products)} - \Sigma \Delta H_f°_{(reactants)} \\ &= \Delta H_f° \text{ Cu}_2\text{O}(s) + \frac{1}{2}\Delta H_f° \text{ O}_2(g) - 2\Delta H_f° \text{ CuO}(s) \\ &= 1 \text{ mol}\left(\frac{-168.6 \text{ kJ}}{\text{mol}}\right) + \frac{1}{2} \text{ mol}\left(\frac{0 \text{ kJ}}{\text{mol}}\right) - 2 \text{ mol}\left(\frac{-157.3 \text{ kJ}}{\text{mol}}\right) = 146.0 \text{ kJ} \end{aligned}$

$\begin{aligned} \Delta S° \quad &= \Sigma S°_{(products)} - \Sigma S°_{(reactants)} \\ &= S° \text{ Cu}_2\text{O}(s) + \frac{1}{2}S° \text{ O}_2(g) - 2S° \text{ CuO}(s) \\ &= 1 \text{ mol}\left(\frac{0.0931 \text{ kJ}}{\text{mol·K}}\right) + \frac{1}{2} \text{ mol}\left(\frac{0.2050 \text{ kJ}}{\text{mol·K}}\right) - 2 \text{ mol}\left(\frac{0.0426 \text{ kJ}}{\text{mol·K}}\right) = 0.1104 \text{ kJ/K} \end{aligned}$

Both $\Delta H°$ and $\Delta S°$ are positive, so this reaction is expected (see Table 16.3) to become spontaneous at higher temperature. The value of T above which the reaction is spontaneous is obtained from Equation 16.2 with $\Delta G° = 0$. (See also problem 42 above.)

$\Delta G° = \Delta H° - T\Delta S°$

$0 = 146.0 \text{ kJ} - T(0.1104 \text{ kJ/K})$ $\qquad \rightarrow \qquad$ $T = 1322 \text{ K}$

Thus, **black CuO spontaneously converts to red Cu₂O at temperatures above 1322 K (1049 °C).**

For the conversion of Cu₂O to Cu:

$$Cu_2O(s) \rightarrow 2Cu(s) + \tfrac{1}{2}O_2(g)$$

$$\Delta H° = \Sigma \Delta H_f°_{(products)} - \Sigma \Delta H_f°_{(reactants)}$$

$$= 2\Delta H_f° \; Cu(s) + \tfrac{1}{2}\Delta H_f° \; O_2(g) - \Delta H_f° \; Cu_2O(s)$$

$$= 2 \; mol\left(\frac{0\,kJ}{mol}\right) + \tfrac{1}{2} \; mol\left(\frac{0\,kJ}{mol}\right) - 1 \; mol\left(\frac{-168.6\,kJ}{mol}\right)$$

$$= 168.6 \; kJ$$

$$\Delta S° = \Sigma S°_{(products)} - \Sigma S°_{(reactants)}$$

$$= 2S° \; Cu(s) + \tfrac{1}{2}S° \; O_2(g) - S° \; Cu_2O(s)$$

$$= 2 \; mol\left(\frac{0.0332\,kJ}{mol \cdot K}\right) + \tfrac{1}{2} \; mol\left(\frac{0.2050\,kJ}{mol \cdot K}\right) - 1 \; mol\left(\frac{0.0931\,kJ}{mol \cdot K}\right)$$

$$= 0.0758 \; kJ/K$$

Both $\Delta H°$ and $\Delta S°$ are positive, so this reaction is expected (see Table 16.3) to become spontaneous at higher temperature. The value of T above which the reaction is spontaneous is obtained from Equation 16.2 with $\Delta G° = 0$. (See also problem 42 above.)

$$\Delta G° = \Delta H° - T\Delta S°$$

$$0 = 168.6 \; kJ - T(0.0758 \; kJ/K) \qquad \rightarrow \qquad T = 2224 \; K$$

Thus, **red Cu₂O spontaneously converts to Cu metal at temperatures above 2224 K (1951 °C).**

99. Given values of K_a at the two different temperatures, we can use Equation 16.5 to calculate the corresponding value of $\Delta G°$ at these two temperatures:

At 25 °C (298 K): $\qquad \Delta G° = -RT\ln K_a = -(0.00831 \; kJ/K)(298 \; K)\ln(1.754 \times 10^{-5}) = 27.1 \; kJ$

At 50 °C (323 K): $\qquad \Delta G° = -RT\ln K_a = -(0.00831 \; kJ/K)(323K)\ln(1.633 \times 10^{-5}) = 29.6 \; kJ$

In addition, we also know that the Gibbs-Helmholtz equation (Equation 16.2) relate $\Delta G°$, $\Delta H°$ and $\Delta S°$ at any temperature, so we can write expressions relating these three quantities using Equation 16.2:

At 25 °C (298 K): $\qquad 27.1 \; kJ = \Delta H° - (298 \; K)\Delta S°$

At 50 °C (323 K): $\qquad 29.6 \; kJ = \Delta H° - (323 \; K)\Delta S°$

Now assuming that $\Delta H°$ is constant these give us two simultaneous equations that allow us to determine $\Delta S°$. Subtracting the second equation from the first gives:

$$27.1 \text{ kJ} = \Delta H° - (298 \text{ K})\Delta S°$$

$$-[\, 29.6 \text{ kJ} = \Delta H° - (323 \text{ K})\Delta S°\,]$$

$$27.1 \text{ kJ} - 29.6 \text{ kJ} = \Delta H° - (298 \text{ K})\Delta S° - [\Delta H° - (323 \text{ K})\Delta S°]$$

$$-2.5 \text{ kJ} = - (298 \text{ K})\Delta S° + (323 \text{ K})\Delta S° = (25 \text{ K})\Delta S°$$

$$\underline{\Delta S° = -0.10 \text{ kJ/K}}$$

100. This is related to Example 12.4. Set up an equilibrium table and insert the given information. Let ΔP for H_2 be $+x$ and complete the table:

	$2HI(g)$	$\rightleftharpoons$	$H_2(g)$	$+$	$I_2(g)$
P_0 (atm)	0.200		0.200		0.200
ΔP (atm)	$-2x$		$+x$		$+x$
P_{eq} (atm)	$0.200 - 2x$		$0.200 + x$		$0.200 + x$

However, we are told that P_{HI} at equilibrium is 0.48 atm. Thus:

$$0.48 = 0.200 - 2x \qquad \rightarrow \qquad x = -0.14$$

And the other equilibrium partial pressures are $P_{H_2} = P_{I_2} = 0.200 + x = 0.06$ atm. Substitute into the expression for K and evaluate:

$$K = \frac{(P_{H_2})(P_{I_2})}{(P_{HI})^2} = \frac{(0.06)(0.06)}{(0.48)^2} = 0.0156$$

Now use Equation 16.5 to calculate the corresponding value of $\Delta G°$ at this temperature:

$$\Delta G° = -RT\ln K = -(0.00831 \text{ kJ/K})(773 \text{ K})\ln(0.0156) = \underline{\textbf{26.7 kJ}}$$

101. Recall that when $K = 1$, $\Delta G° = 0$ (using Equation 16.5). Thus we must find T such that $\Delta G° = 0$: this can be done using Equation 16.2 and therefore we must calculate the corresponding values of $\Delta H°$ and $\Delta S°$ for this reaction using data from Appendix 1.

$$\Delta H° = \Sigma \Delta H_f°_{(products)} - \Sigma \Delta H_f°_{(reactants)}$$

$$= 2\Delta H_f° \; NH_3(g) - [\Delta H_f° \; N_2(g) + 3\Delta H_f° \; H_2(g)]$$

$$= 2 \text{ mol}\left(\frac{-46.1 \text{ kJ}}{\text{mol}}\right) - 1 \text{ mol}\left(\frac{0 \text{ kJ}}{\text{mol}}\right) - 3 \text{ mol}\left(\frac{0 \text{ kJ}}{\text{mol}}\right) = -92.2 \text{ kJ}$$

$$\Delta S° = \Sigma S°_{(products)} - \Sigma S°_{(reactants)}$$

$$= 2S° \; NH_3(g) - [S° \; N_2(g) + 3S° \; H_2(g)]$$

$$= 2 \text{ mol}\left(\frac{0.1923 \text{ kJ}}{\text{mol·K}}\right) - 1 \text{ mol}\left(\frac{0.1915 \text{ kJ}}{\text{mol·K}}\right) - 3 \text{ mol}\left(\frac{0.1306 \text{ kJ}}{\text{mol·K}}\right) = -0.1987 \text{ kJ/K}$$

$$\Delta G° = \Delta H° - T\Delta S°$$

$$0 = -92.2 \text{ kJ} - T(-0.1987 \text{ kJ/K}) \qquad \rightarrow \qquad \underline{T = 464 \text{ K}}$$

Thus, **the value of K = 1.00 at 464 K (191 °C).**

102. We are given values of K at two different temperatures: the variation of K with temperature is given by the van't Hoff equation (Equation 12.5) , from which we can obtain the value of $\Delta H°$ for the given reaction. Thus, using Equation 12.5 with K_1 = 98 at a temperature of 333 °C (T_1 = 606 K) and K_2 = 62.5 at 527 °C (T_2 = 800 K):

$$\ln\frac{K_2}{K_1} = \frac{\Delta H°}{R}\left[\frac{1}{T_1} - \frac{1}{T_2}\right]$$

$$\ln\frac{62.5}{98} = \frac{\Delta H°}{8.3145 \text{ J/mol}\cdot\text{K}}\left[\frac{1}{606 \text{ K}} - \frac{1}{800 \text{ K}}\right]$$

$$\ln(0.6\underline{38}) = -0.45 = \Delta H°(4.81 \times 10^{-5} \text{ mol/J})$$

$$\Delta H° = -9.35 \times 10^3 \text{ J/mol} = -9.35 \text{ kJ/mol}$$

In addition, given the value of K at 527 °C (800 K) we can obtain $\Delta G°$ at the same temperature using Equation 16.5 with the appropriate value of T:

$$\Delta G° = -RT \ln K = -(0.0083145 \text{ kJ/mol}\cdot\text{K})(800 \text{ K})\ln(62.5) = -27.5 \text{ kJ/mol}$$

Now, knowing $\Delta H°$ and $\Delta G°$, we can use Equation 16.2 to obtain $\Delta S°$ for the given reaction at T = 800 K (527 °C):

$$\Delta G° = \Delta H° - T\Delta S°$$

$$-27.5 \text{ kJ/mol} = -9.35 \text{ kJ/mol} -(800 \text{ K} \times \Delta S°)$$

$$\Delta S° = \frac{-27.5 \text{ kJ/mol} + 9.35 \text{ kJ/mol}}{-800 \text{ K}} = 0.0227 \text{ kJ/mol}\cdot\text{K} = 22.7 \text{ J/mol}\cdot\text{K}$$

Finally, use Equation 16.1 along with data from Appendix I to obtain $S°$ for $I_2(g)$ at this temperature:

$$\Delta S° = \Sigma S°_{products} - \Sigma S°_{reactants}$$

$$22.7 \text{ J/mol}\cdot\text{K} = 2S° \text{ HI}(g) - [S° \text{ H}_2(g) + S° \text{ I}_2(g)]$$

$$22.7 \text{ J/mol}\cdot\text{K} = 2 \text{ mol}\left(\frac{206.5 \text{ J}}{\text{mol}\cdot\text{K}}\right) - 1 \text{ mol}\left(\frac{130.6 \text{ J}}{\text{mol}\cdot\text{K}}\right) - 1 \text{ mol } S° \text{ I}_2(g)$$

$$S° \text{ I}_2(g) = +259.7 \text{ J/mol}\cdot\text{K} = \underline{\textbf{0.2597 kJ/mol}\cdot\textbf{K}}$$

ELECTROCHEMISTRY

17-1 OXIDATION-REDUCTION REACTIONS REVISITED: Oxidation-Reduction Reactions

2. These problems are similar to Example 17.1. (See also Figure 17.1.) Begin by assigning oxidation numbers to each element. Balance the element oxidized or reduced. Find the total oxidation number in each side of the equation then balance the total oxidation number by adding electrons. (Note that in reduction half equations, the electrons are added to the reactant side while in oxidation half equations, the electrons are added to the product side.) Balance the charge by adding H^+ (if acidic) or OH^- (if basic). Balance H by adding H_2O. Check that all elements (e.g. O) are balanced.

(a) $CH_3OH(aq) \rightarrow CO_2(g)$ [acidic]

oxidation numbers	C: $-2 \rightarrow +4$; C is **oxidized**
atom balance	1 C on each side, no adjustment is required
total oxidation number	C: $1(-2)= -2 \rightarrow$ C: $1(+4)= +4$
add electrons	The total oxidation number for C goes from -2 to $+4$. The total oxidation number increased by 6. Add 6 electrons to the product side. $$CH_3OH(aq) \rightarrow CO_2(g) + 6e^-$$
balance charge	reactant: 0 product: $6(-1)= -6$ acidic medium: add H^+ To balance the charge, add 6 H^+ to the product side. $$CH_3OH(aq) \rightarrow CO_2(g) + 6H^+(aq) + 6e^-$$ reactant: 0 product: $6(+1) + 6(-1)= 0$
balance H	reactant: 4 H product: 6 H To balance H, add 1 H_2O to the reactant side. $$CH_3OH(aq) + H_2O \rightarrow CO_2(g) + 6H^+(aq) + 6e^-$$ reactant: 6 H product: 6 H
check O	reactant: 2 O product: 2 O
balanced half-equation:	$CH_3OH(aq) + H_2O \rightarrow CO_2(g) + 6H^+(aq) + 6e^-$

(b) $NO_3^-(aq) \rightarrow NH_4^+(aq)$ [acidic]

oxidation numbers	N: $+5 \rightarrow -3$; N is **reduced**
atom balance	1 N on each side, no adjustment is required
total oxidation number	N: $1(+5)=+5 \rightarrow$ N: $1(-3)= -3$

add electrons	The total oxidation number for N goes from +5 to −3. It is reduced by 8. Add 8 electrons to the reactant side. <center>$NO_3^-(aq) + 8e^- \rightarrow NH_4^+(aq)$</center>
balance charge	reactant: −1 + 8(−1)= −9; product: +1 acidic medium: add H^+ To balance the charge, add 10 H^+ to the reactant side. $NO_3^-(aq) + 10H^+(aq) + 8e^- \rightarrow NH_4^+(aq)$ reactant: −1 + 10(+1) + 8(−1)= +1; product: +1
balance H	reactant: 10 H product: 4 H To balance H, add 3 H_2O to the product side. $NO_3^-(aq) + 10H^+(aq) + 8e^- \rightarrow NH_4^+(aq) + 3H_2O$ reactant: 10 H product: 10 H
check O	reactant: 3 O product: 3 O
balanced half-equation:	$NO_3^-(aq) + 10H^+(aq) + 8e^- \rightarrow NH_4^+(aq) + 3H_2O$

(c) $Fe^{3+}(aq) \rightarrow Fe(s)$ [basic]

oxidation numbers	Fe: +3 $\rightarrow$ 0; Fe is **reduced**
atom balance	1 Fe on each side, no adjustment is required
total oxidation number	Fe: 1(+3)=+3 $\rightarrow$ Fe: 0
add electrons	The total oxidation number for Fe goes from +3 to 0. It is reduced by 3. Add 3 electrons to the reactant side. <center>$Fe^{3+}(aq) + 3e^- \rightarrow Fe(s)$</center>
balance charge	reactant: +3 + 3(−1)= 0 product: 0
balance H	reactant: 0 H product: 0 H
check O	reactant: 0 O product: 0 O
balanced half-equation:	$Fe^{3+}(aq) + 3e^- \rightarrow Fe(s)$

(d) $V^{2+}(aq) \rightarrow VO_3^-(aq)$ [basic]

oxidation numbers	V: +2 $\rightarrow$ +5; V is **oxidized**
atom balance	1 V on each side, no adjustment is required
total oxidation number	V: 1(+2)=+2 $\rightarrow$ V: 1(+5)= +5
add electrons	The total oxidation number for V goes from +2 to +5. The total oxidation number increased by 3. Add 3 electrons to the product side.

	$V^{2+}(aq) \rightarrow VO_3^-(aq) + 3e^-$
balance charge	reactant: $1(+2)= +2$; product: $-1 + 3(-1)= -4$ basic medium: add OH^-. To balance the charge, add 6 OH^- to the reactant side. $\quad V^{2+}(aq) + 6OH^-(aq) \rightarrow VO_3^-(aq) + 3e^-$ reactant: $1(+2) + 6(-1) = -4$ product: $-1 + 3(-1)= -4$
balance H	reactant: 6 H product: 0 H To balance H, add 3 H_2O to the product side. $\quad V^{2+}(aq) + 6OH^-(aq) \rightarrow VO_3^-(aq) + 3H_2O + 3e^-$ reactant: 6 H product: 6 H
check O	reactant: 6 O product: 6 O
balanced half-equation:	$V^{2+}(aq) + 6OH^-(aq) \rightarrow VO_3^-(aq) + 3H_2O + 3e^-$

4. These problems are similar to Example 17.2.   To balance the equation, follow the steps
    outlined in the text for balancing redox equations. Begin by splitting the reaction into
    two half-equations. Balance each of the half-equations separately with respect to both
    number of atoms and charges. Eliminate the electrons. If the number of electrons in the
    two balanced half-equations is not the same, multiply the equations by an appropriate
    number to make the number of electrons equal. Combine the two half-equations and
    cancel those species which are duplicated or appear on both sides of the reaction.

    (a) $H_2O_2(aq) + Ni^{2+}(aq) \rightarrow Ni^{3+}(aq) + H_2O$ [acidic]

a.	Split into two half-equations.	$H_2O_2(aq) \rightarrow H_2O$
		$Ni^{2+}(aq) \rightarrow Ni^{3+}(aq)$

b.	Balance one of the half-equations: $H_2O_2(aq) \rightarrow H_2O$	
	oxidation numbers	O (oxidation no.): $-1 \rightarrow -2$; O is reduced
	atom balance	reactant: 2 O product: 1 O Multiply H_2O by 2: $H_2O_2(aq) \rightarrow 2H_2O$
	total oxidation number	O: $2(-1)= -2 \rightarrow$ O: $2(-2)= -4$
	add electrons	The total oxidation number for O goes from -2 to -4. It is reduced by 2. Add 2 electrons to the reactant side. $\quad H_2O_2(aq) + 2e^- \rightarrow 2H_2O$
	balance charge	reactant: $2(-1)= -2$; product: 0 acidic medium: add H^+ To balance the charge, add 2 H^+ to the reactant side. $\quad H_2O_2(aq) + 2e^- + 2H^+(aq) \rightarrow 2H_2O$ reactant: $2(-1) + 2(+1)= 0$; product: 0
	balance H	reactant: 4 H product: 4 H
	check O	reactant: 2 O product: 2 O

	Balanced Reduction Half-Equation	$H_2O_2(aq) + 2e^- + 2H^+(aq) \rightarrow 2H_2O$
c.	Balance the other half-equation: $Ni^{2+}(aq) \rightarrow Ni^{3+}(aq)$	
	oxidation numbers	Ni: $+2 \rightarrow +3$; Ni is oxidized
	atom balance	1 Ni on each side, no adjustment is required
	total oxidation number	Ni: $1(+2) = +2 \rightarrow$ Ni: $1(+3) = +3$
	add electrons	The total oxidation number for Ni goes from +2 to +3. The total oxidation number increased by 1. Add 1 electron to the product side. $Ni^{2+}(aq) \rightarrow Ni^{3+}(aq) + e^-$
	balance charge	reactant: $+2$ product: $1(+3) + 1(-1) = +2$
	balance H	reactant: 0 H product: 0 H
	check O	reactant: 0 O product: 0 O
	Balanced Oxidation Half-Equation	$Ni^{2+}(aq) \rightarrow Ni^{3+}(aq) + e^-$
d.	Make the number of electrons in the two half-equations equal (multiply by an appropriate number if necessary).	Balanced half-equations:

reduction	$H_2O_2(aq) + 2e^- + 2H^+(aq) \rightarrow 2H_2O$	
oxidation: multiply by 2	$2\,[Ni^{2+}(aq) \rightarrow Ni^{3+}(aq) + e^-]$	
	$2Ni^{2+}(aq) \rightarrow 2Ni^{3+}(aq) + 2e^-$	

e.	Combine the two half-equations.	$2Ni^{2+}(aq) + H_2O_2(aq) + 2H^+(aq) + 2e^-$ $\rightarrow 2Ni^{3+}(aq) + 2H_2O + 2e^-$
	Cancel duplication.	$2Ni^{2+}(aq) + H_2O_2(aq) + 2H^+(aq) \rightarrow 2Ni^{3+}(aq) + 2H_2O$
Balanced Equation:		$\mathbf{2Ni^{2+}(aq) + H_2O_2(aq) + 2H^+(aq) \rightarrow 2Ni^{3+}(aq) + 2H_2O}$

(b) $Cr_2O_7{}^{2-}(aq) + Sn^{2+}(aq) \rightarrow Cr^{3+}(aq) + Sn^{4+}(aq)$ [acidic]

a.	Split into two half-equations.	$Cr_2O_7{}^{2-}(aq) \rightarrow Cr^{3+}(aq)$
		$Sn^{2+}(aq) \rightarrow Sn^{4+}(aq)$
b.	Balance one of the half-equations: $Cr_2O_7{}^{2-}(aq) \rightarrow Cr^{3+}(aq)$	
	oxidation numbers	Cr (oxidation no.): $+6 \rightarrow +3$; Cr is reduced
	atom balance	reactant: 2 Cr product: 1 Cr Multiply Cr^{3+} by 2: $Cr_2O_7{}^{2-}(aq) \rightarrow 2Cr^{3+}(aq)$
	total oxidation number	Cr: $2(+6) = +12 \rightarrow$ Cr: $2(+3) = +6$
	add electrons	The total oxidation number for Cr goes from +12 to +6. It is reduced by 6. Add 6 electrons to the reactant side. $Cr_2O_7{}^{2-}(aq) + 6e^- \rightarrow 2Cr^{3+}(aq)$

balance charge	reactant: $-2 + 6(-1) = -8$; product: $2(+3) = +6$
	acidic medium: add H^+
	To balance the charge, add 14 H^+ to the reactant side.
	$Cr_2O_7^{2-}(aq) + 14H^+(aq) + 6e^- \rightarrow 2Cr^{3+}(aq)$
	reactant: $-2 + 14(+1) + 6(-1) = +6$
	product: $2(+3) = +6$
balance H	reactant: 14 H product: 0 H
	To balance H, add 7 H_2O to the product side.
	$Cr_2O_7^{2-}(aq) + 14H^+(aq) + 6e^- \rightarrow 2Cr^{3+}(aq) + 7H_2O$
	reactant: 14 H product: 14 H
check O	reactant: 7 O product: 7 O
Balanced Reduction Half-Equation	$Cr_2O_7^{2-}(aq) + 14H^+(aq) + 6e^- \rightarrow 2Cr^{3+}(aq) + 7H_2O$

c.   Balance the other half-equation: $Sn^{2+}(aq) \rightarrow Sn^{4+}(aq)$

oxidation numbers	Sn (oxidation no.): $+2 \rightarrow +4$; Sn is oxidized
atom balance	1 Sn on each side, no adjustment is required
total oxidation number	Sn: $1(+2) = +2 \rightarrow$ Sn: $1(+4) = +4$
add electrons	The total oxidation number for Sn goes from +2 to +4.
	The total oxidation number increased by 2.
	Add 2 electrons to the product side.
	$Sn^{2+}(aq) \rightarrow Sn^{4+}(aq) + 2e^-$
balance charge	reactant: +2 product: $1(+4) + 1(-2) = +2$
balance H	reactant: 0 H product: 0 H
check O	reactant: 0 O product: 0 O
Balanced Oxidation Half-Equation	$Sn^{2+}(aq) \rightarrow Sn^{4+}(aq) + 2e^-$

d. Make the number of electrons in the two half-equations equal (multiply by an appropriate number if necessary).	Balanced half-equations:	
	reduction	$Cr_2O_7^{2-}(aq) + 14H^+(aq) + \mathbf{6e^-}$ $\rightarrow 2Cr^{3+}(aq) + 7H_2O$
	oxidation: multiply by 3	$3 [Sn^{2+}(aq) \rightarrow Sn^{4+}(aq) + 2e^-]$
		$3Sn^{2+}(aq) \rightarrow 3Sn^{4+}(aq) + \mathbf{6e^-}$
e. Combine the two half-equations.	$Cr_2O_7^{2-}(aq) + 14H^+(aq) + 3Sn^{2+}(aq) + \mathbf{6e^-}$ $\rightarrow 2Cr^{3+}(aq) + 7H_2O + 3Sn^{4+}(aq) + \mathbf{6e^-}$	
Cancel duplication.	$Cr_2O_7^{2-}(aq) + 14H^+(aq) + 3Sn^{2+}(aq)$ $\rightarrow 2Cr^{3+}(aq) + 7H_2O + 3Sn^{4+}(aq)$	

Balanced Equation:
$$Cr_2O_7^{2-}(aq) + 14H^+(aq) + 3Sn^{2+}(aq) \rightarrow 2Cr^{3+}(aq) + 7H_2O + 3Sn^{4+}(aq)$$

6. These problems are similar to Example 17.2. To balance the equation, follow the steps outlined in the text for balancing redox equations. Begin by splitting the reaction into two half-equations. Balance each of the half-equations separately with respect to both number of atoms and charges. Eliminate the electrons. If the number of electrons in the two balanced half-equations is not the same, multiply the equations by an appropriate number to make the number of electrons equal. Combine the two half-equations and cancel those species which are duplicated or appear on both sides of the reaction.

(a) $P_4(s) + Cl^-(aq) \rightarrow PH_3(g) + Cl_2(g)$ [acidic]

a.	Split into two half-equations.	$P_4(s) \rightarrow PH_3(g)$
		$Cl^-(aq) \rightarrow Cl_2(g)$
b.	Balance one of the half-equations:	$P_4(s) \rightarrow PH_3(g)$
	oxidation numbers	P (oxidation no.): $0 \rightarrow -3$; P is reduced
	atom balance	reactant: 4 P product: 1 P
		Multiply PH_3 by 4: $P_4(s) \rightarrow 4PH_3(g)$
	total oxidation number	P: 0 $\rightarrow$ P: $4(-3) = -12$
	add electrons	The total oxidation number for P goes from 0 to -12. It is reduced by 12. Add 12 electrons to the reactant side.
		$P_4(s) + 12e^- \rightarrow 4PH_3(g)$
	balance charge	reactant: $12(-1) = -12$; product: 0
		acidic medium: add H^+
		To balance the charge, add 12 H^+ to the reactant side.
		$P_4(s) + 12H^+(aq) + 12e^- \rightarrow 4PH_3(g)$
		reactant: $12(+1) + 12(-1) = 0$; product: 0
	balance H	reactant: 12 H product: 12 H
	check O	reactant: 0 O product: 0 O
	Balanced Reduction Half-Equation	$P_4(s) + 12H^+(aq) + 12e^- \rightarrow 4PH_3(g)$
c.	Balance the other half-equation:	$Cl^-(aq) \rightarrow Cl_2(g)$
	oxidation numbers	Cl (oxidation no.): $-1 \rightarrow 0$; Cl is oxidized
	atom balance	reactant: 1 Cl product: 2 Cl
		Multiply Cl^- by 2: $2Cl^-(aq) \rightarrow Cl_2(g)$
	total oxidation number	Cl: $2(-1) = -2 \rightarrow$ Cl: $2(0) = 0$
	add electrons	The total oxidation number for Cl goes from -2 to 0. The total oxidation number increased by 2. Add 2 electrons to the product side.
		$2Cl^-(aq) \rightarrow Cl_2(g) + 2e^-$
	balance charge	reactant: -2 product: -2
	balance H	reactant: 0 H product: 0 H

check O	reactant: 0 O product: 0 O
Balanced Oxidation Half-Equation	$2Cl^-(aq) \rightarrow Cl_2(g) + 2e^-$

d. Make the number of electrons in the two half-equations equal (multiply by an appropriate number if necessary).	Balanced half-equations:	
	reduction	$P_4(s) + 12H^+(aq) + \mathbf{12e^-} \rightarrow 4PH_3(g)$
	oxidation: multiply by 6	$6\ [2Cl^-(aq) \rightarrow Cl_2(g) + 2e^-]$
		$12Cl^-(aq) \rightarrow 6Cl_2(g) + \mathbf{12e^-}$

e. Combine the two half-equations.	$P_4(s) + 12H^+(aq) + 12Cl^-(aq) + \mathbf{12e^-}$ $\rightarrow 4PH_3(g) + 6Cl_2(g) + \mathbf{12e^-}$
Cancel duplication.	$P_4(s) + 12H^+(aq) + 12Cl^-(aq) \rightarrow 4PH_3(g) + 6Cl_2(g)$

Balanced Equation:	$\mathbf{P_4(s) + 12H^+(aq) + 12Cl^-(aq) \rightarrow 4PH_3(g) + 6Cl_2(g)}$

(b) $MnO_4^-(aq) + NO_2^-(aq) \rightarrow Mn^{2+}(aq) + NO_3^-(aq)$ [acidic]

a. Split into two half-equations.	$MnO_4^-(aq) \rightarrow Mn^{2+}(aq)$
	$NO_2^-(aq) \rightarrow NO_3^-(aq)$

b. Balance one of the half-equations: $MnO_4^-(aq) \rightarrow Mn^{2+}(aq)$	
oxidation numbers	Mn (oxidation no.): $+7 \rightarrow +2$; Mn is reduced
atom balance	1 Mn on each side, no adjustment is required
total oxidation number	Mn: $1(+7) = +7 \rightarrow$ Mn: $1(+2) = +2$
add electrons	The total oxidation number for Mn goes from +7 to +2. It is reduced by 5. Add 5 electrons to the reactant side. $MnO_4^-(aq) + 5e^- \rightarrow Mn^{2+}(aq)$
balance charge	reactant: $-1 + 5(-1) = -6$; product: $+2$ acidic medium: add H^+ To balance the charge, add 8 H^+ to the reactant side. $MnO_4^-(aq) + 5e^- + 8H^+(aq) \rightarrow Mn^{2+}(aq)$ reactant: $-1 + 5(-1) + 8(+1) = +2$; product: $+2$
balance H	reactant: 8 H product: 0 H To balance H, add 4 H_2O to the product side. $MnO_4^-(aq) + 5e^- + 8H^+(aq) \rightarrow Mn^{2+}(aq) + 4H_2O$ reactant: 8 H product: 8 H
check O	reactant: 4 O product: 4 O
Balanced Reduction Half-Equation	$MnO_4^-(aq) + 5e^- + 8H^+(aq) \rightarrow Mn^{2+}(aq) + 4H_2O$

c. Balance the other half-equation: $NO_2^-(aq) \rightarrow NO_3^-(aq)$	
oxidation numbers	N (oxidation no.): $+3 \rightarrow +5$; N is oxidized
atom balance	1 N on each side, no adjustment is required

total oxidation number	N: 1(+3)= +3 → N: 1(+5)=+5
add electrons	The total oxidation number for N goes from +3 to +5. The total oxidation number increased by 2. Add 2 electrons to the product side. $NO_2^-(aq) \rightarrow NO_3^-(aq) + 2e^-$
balance charge	reactant: −1 product: −1 + 2(−1) = −3 acidic medium: add H^+ To balance the charge, add 2 H^+ to the product side. $NO_2^-(aq) \rightarrow NO_3^-(aq) + 2e^- + 2H^+(aq)$ reactant: −1 product: −1 + 2(−1) + 2(+1) = −1
balance H	reactant: 0 H product: 2 H To balance H, add H_2O to the reactant side. $NO_2^-(aq) + H_2O \rightarrow NO_3^-(aq) + 2e^- + 2H^+(aq)$ reactant: 2 H product: 2 H
check O	reactant: 3 O product: 3 O
Balanced Oxidation Half-Equation	$NO_2^-(aq) + H_2O \rightarrow NO_3^-(aq) + 2e^- + 2H^+(aq)$

d. Make the number of electrons in the two half-equations equal (multiply by an appropriate number if necessary).	Balanced half-equations:	
	reduction: multiply by 2	$2[MnO_4^-(aq) + 5e^- + 8H^+(aq)$ $\rightarrow Mn^{2+}(aq) + 4H_2O]$
		$2MnO_4^-(aq) + 10e^- + 16H^+(aq)$ $\rightarrow 2Mn^{2+}(aq) + 8H_2O$
	oxidation: multiply by 5	$5[NO_2^-(aq) + H_2O$ $\rightarrow NO_3^-(aq) + 2e^- + 2H^+(aq)]$
		$5NO_2^-(aq) + 5H_2O$ $\rightarrow 5NO_3^-(aq) + 10e^- + 10H^+(aq)$
e. Combine the two half-equations.	$2MnO_4^-(aq) + 16H^+(aq) + 5NO_2^-(aq) + 5H_2O + 10e^-$ $\rightarrow 2Mn^{2+}(aq) + 8H_2O + 5NO_3^-(aq) + 10H^+(aq) + 10e^-$	
Cancel duplication.	$2MnO_4^-(aq) + 6H^+(aq) + 5NO_2^-(aq)$ $\rightarrow 2Mn^{2+}(aq) + 3H_2O + 5NO_3^-(aq)$	

Balanced Equation:
$2MnO_4^-(aq) + 6H^+(aq) + 5NO_2^-(aq) \rightarrow 2Mn^{2+}(aq) + 3H_2O + 5NO_3^-(aq)$

(c) $HBrO_3(aq) + Bi(s) \rightarrow HBrO_2(aq) + Bi_2O_3(s)$ [acidic]

a. Split into two half-equations.	$HBrO_3(aq) \rightarrow HBrO_2(aq)$
	$Bi(s) \rightarrow Bi_2O_3(s)$
b. Balance one of the half-equations: $HBrO_3(aq) \rightarrow HBrO_2(aq)$	
oxidation numbers	Br (oxidation no.): +5 → +3; Br is reduced
atom balance	1 Br on each side, no adjustment is required
total oxidation number	Br: 1(+5)= +5 → Br: 1(+3)= +3

add electrons	The total oxidation number for Br goes from +5 to +3. It is reduced by 2. Add 2 electrons to the reactant side. $HBrO_3(aq) + 2e^- \rightarrow HBrO_2(aq)$
balance charge	reactant: $2(-1) = -2$; product: 0 acidic medium: add H^+ To balance the charge, add 2 H^+ to the reactant side. $HBrO_3(aq) + 2e^- + 2H^+(aq) \rightarrow HBrO_2(aq)$ reactant: $2(-1) + 2(+1) = 0$; product: 0
balance H	reactant: 3 H product: 1 H To balance H, add 1 H_2O to the product side. $HBrO_3(aq) + 2e^- + 2H^+(aq) \rightarrow HBrO_2(aq) + H_2O$ reactant: 3 H product: 3 H
check O	reactant: 3 O product: 3 O
Balanced Reduction Half-Equation	$HBrO_3(aq) + 2e^- + 2H^+(aq) \rightarrow HBrO_2(aq) + H_2O$

c.   Balance the other half-equation: $Bi(s) \rightarrow Bi_2O_3(s)$

oxidation numbers	Bi (oxidation no.): $0 \rightarrow +3$; Bi is oxidized
atom balance	reactant: 1 Bi product: 2 Bi Multiply Bi by 2: $2Bi(s) \rightarrow Bi_2O_3(s)$
total oxidation number	Bi: $2(0) = 0 \rightarrow$ Bi: $2(+3) = +6$
add electrons	The total oxidation number for Bi goes from 0 to +6. The total oxidation number increased by 6. Add 6 electrons to the product side. $2Bi(s) \rightarrow Bi_2O_3(s) + 6e^-$
balance charge	reactant: 0 product: $6(-1) = -6$ acidic medium: add H^+ To balance the charge, add 6 H^+ to the product side. $2Bi(s) \rightarrow Bi_2O_3(s) + 6e^- + 6H^+(aq)$ reactant: 0 product: 0
balance H	reactant: 0 H product: 6 H To balance H, add 3 H_2O to the reactant side. $2Bi(s) + 3H_2O \rightarrow Bi_2O_3(s) + 6e^- + 6H^+(aq)$ reactant: 6 H product: 6 H
check O	reactant: 3 O product: 3 O
Balanced Oxidation Half-Equation	$2Bi(s) + 3H_2O \rightarrow Bi_2O_3(s) + 6e^- + 6H^+(aq)$

d. Make the number of electrons in the two half-equations equal (multiply by an appropriate number if necessary).	Balanced half-equations:	
	reduction: multiply by 3	$3[HBrO_3(aq) + 2e^- + 2H^+(aq)$ $\rightarrow HBrO_2(aq) + H_2O]$
		$3HBrO_3(aq) + 6e^- + 6H^+(aq)$ $\rightarrow 3HBrO_2(aq) + 3H_2O$
	oxidation	$2Bi(s) + 3H_2O$ $\rightarrow Bi_2O_3(s) + 6e^- + 6H^+(aq)$

e. Combine the two half-equations.	$2Bi(s) + \mathbf{3H_2O} + 3HBrO_3(aq) + \mathbf{6H^+(aq)} + \mathbf{6e^-}$ $\rightarrow Bi_2O_3(s) + \mathbf{6H^+(aq)} + 3HBrO_2(aq) + \mathbf{3H_2O} + \mathbf{6e^-}$
Cancel duplication.	$2Bi(s) + 3HBrO_3(aq) \rightarrow Bi_2O_3(s) + 3HBrO_2(aq)$
Balanced Equation:	$\mathbf{2Bi(s) + 3HBrO_3(aq) \rightarrow Bi_2O_3(s) + 3HBrO_2(aq)}$

(d) $CrO_4^{2-}(aq) + SO_3^{2-}(aq) \rightarrow Cr^{3+}(aq) + SO_4^{2-}(aq)$ [acidic]

a. Split into two half-equations.	$CrO_4^{2-}(aq) \rightarrow Cr^{3+}(aq)$
	$SO_3^{2-}(aq) \rightarrow SO_4^{2-}(aq)$
b. Balance one of the half-equations: $CrO_4^{2-}(aq) \rightarrow Cr^{3+}(aq)$	
oxidation numbers	Cr (oxidation no.): $+6 \rightarrow +3$; Cr is reduced
atom balance	1 Cr on each side, no adjustment is required
total oxidation number	Cr: $1(+6)=+6 \rightarrow$ Cr: $1(+3)=+3$
add electrons	The total oxidation number for Cr goes from +6 to +3. It is reduced by 3. Add 3 electrons to the reactant side. $CrO_4^{2-}(aq) + 3e^- \rightarrow Cr^{3+}(aq)$
balance charge	reactant: $-2 + 3(-1) = -5$; product: $+3$ acidic medium: add H^+ To balance the charge, add 8 H^+ to the reactant side. $CrO_4^{2-}(aq) + 3e^- + 8H^+(aq) \rightarrow Cr^{3+}(aq)$ reactant: $-2 + 3(-1) + 8(+1) = +3$; product: $+3$
balance H	reactant: 8 H product: 0 H To balance H, add 4 H_2O to the product side. $CrO_4^{2-}(aq) + 3e^- + 8H^+(aq) \rightarrow Cr^{3+}(aq) + 4H_2O$ reactant: 8 H product: 8 H
check O	reactant: 4 O product: 4 O
Balanced Reduction Half-Equation	$CrO_4^{2-}(aq) + 3e^- + 8H^+(aq) \rightarrow Cr^{3+}(aq) + 4H_2O$
c. Balance the other half-equation: $SO_3^{2-}(aq) \rightarrow SO_4^{2-}(aq)$	
oxidation numbers	S (oxidation no.): $+4 \rightarrow +6$; S is oxidized
atom balance	1 S on each side, no adjustment is required
total oxidation number	S: $+4 \rightarrow$ S: $+6$
add electrons	The total oxidation number for S goes from +4 to +6. The total oxidation number increased by 2. Add 2 electrons to the product side. $SO_3^{2-}(aq) \rightarrow SO_4^{2-}(aq) + 2e^-$
balance charge	reactant: -2 product: $-2 + 2(-1) = -4$ acidic medium: add H^+ To balance the charge, add 2 H^+ to the product side.

	$SO_3^{2-}(aq) \rightarrow SO_4^{2-}(aq) + 2e^- + 2H^+(aq)$
	reactant: -2 product: $-2 + 2(-1) + 2(+1) = -2$
balance H	reactant: 0 H product: 2 H
	To balance H, add 1 H_2O to the reactant side.
	$SO_3^{2-}(aq) + H_2O \rightarrow SO_4^{2-}(aq) + 2e^- + 2H^+(aq)$
	reactant: 2 H product: 2 H
check O	reactant: 4 O product: 4 O
Balanced Oxidation Half-Equation	$SO_3^{2-}(aq) + H_2O \rightarrow SO_4^{2-}(aq) + 2e^- + 2H^+(aq)$

d. Make the number of electrons in the two half-equations equal (multiply by an appropriate number if necessary).	Balanced half-equations:	
	reduction: multiply by 2	$2\,[CrO_4^{2-}(aq) + 3e^- + 8H^+(aq)$ $\rightarrow Cr^{3+}(aq) + 4H_2O]$
		$2CrO_4^{2-}(aq) + 6e^- + 16H^+(aq)$ $\rightarrow 2Cr^{3+}(aq) + 8H_2O$
	oxidation: multiply by 3	$3\,[SO_3^{2-}(aq) + H_2O$ $\rightarrow SO_4^{2-}(aq) + 2e^- + 2H^+(aq)]$
		$3SO_3^{2-}(aq) + 3H_2O$ $\rightarrow 3SO_4^{2-}(aq) + 6e^- + 6H^+(aq)$

e. Combine the two half-equations.	$2CrO_4^{2-}(aq) + \mathbf{16H^+}(aq) + 3SO_3^{2-}(aq) + \mathbf{3H_2O} + 6e^-$ $\rightarrow 2Cr^{3+}(aq) + \mathbf{8H_2O} + 3SO_4^{2-}(aq) + \mathbf{6H^+}(aq) + \mathbf{6e^-}$
Cancel duplication.	$2CrO_4^{2-}(aq) + 10H^+(aq) + 3SO_3^{2-}(aq)$ $\rightarrow 2Cr^{3+}(aq) + 5H_2O + 3SO_4^{2-}(aq)$

Balanced Equation:
$$2CrO_4^{2-}(aq) + 10H^+(aq) + 3SO_3^{2-}(aq) \rightarrow 2Cr^{3+}(aq) + 5H_2O + 3SO_4^{2-}(aq)$$

8. These problems are similar to Example 17.2. To balance the equation, follow the steps outlined in the text for balancing redox equations. Begin by splitting the reaction into two half-equations. Balance each of the half-equations separately with respect to both number of atoms and charges. Eliminate the electrons. If the number of electrons in the two balanced half-equations is not the same, multiply the equations by an appropriate number to make the number of electrons equal. Combine the two half-equations and cancel those species which are duplicated or appear on both sides of the reaction.

(a) $Ca(s) + VO_4^{3-}(aq) \rightarrow Ca^{2+}(aq) + V^{2+}(aq)$ [basic]

a. Split into two half-equations.	$VO_4^{3-}(aq) \rightarrow V^{2+}(aq)$
	$Ca(s) \rightarrow Ca^{2+}(aq)$

b. Balance one of the half-equations: $VO_4^{3-}(aq) \rightarrow V^{2+}(aq)$	
oxidation numbers	V (oxidation no.): $+5 \rightarrow +2$; V is reduced
atom balance	1 V on each side, no adjustment is required
total oxidation number	V: $1(+5) = +5 \rightarrow$ V: $1(+2) = +2$

add electrons	The total oxidation number for V goes from +5 to +2. It is reduced by 3. Add 3 electrons to the reactant side. $$VO_4{}^{3-}(aq) + 3e^- \rightarrow V^{2+}(aq)$$
balance charge	reactant: $-3 + 3(-1) = -6$; product: +2 basic medium: add OH^-. To balance the charge, add 8 OH^- to the product side. $$VO_4{}^{3-}(aq) + 3e^- \rightarrow V^{2+}(aq) + 8OH^-(aq)$$ reactant: $-3 + 3(-1) = -6$; product: $+2 + 8(-1) = -6$
balance H	reactant: 0 H product: 8 H To balance H, add 4 H_2O to the reactant side. $$VO_4{}^{3-}(aq) + 4H_2O + 3e^- \rightarrow V^{2+}(aq) + 8OH^-(aq)$$ reactant: 8 H product: 8 H
check O	reactant: 8 O product: 8 O
Balanced Reduction Half-Equation	$VO_4{}^{3-}(aq) + 4H_2O + 3e^- \rightarrow V^{2+}(aq) + 8OH^-(aq)$

c.   Balance the other half-equation: $Ca(s) \rightarrow Ca^{2+}(aq)$

oxidation numbers	Ca (oxidation no.): $0 \rightarrow +2$; Ca is oxidized
atom balance	1 Ca on each side, no adjustment is required
total oxidation number	Ca: $0 \rightarrow$ Ca: +2
add electrons	The total oxidation number for Ca goes from 0 to +2. It is oxidized by 2. Add 2 electrons to the product side. $$Ca(s) \rightarrow Ca^{2+}(aq) + 2e^-$$
balance charge	reactant: 0 product: $+2 + 2(-1) = 0$
balance H	reactant: 0 H product: 0 H
check O	reactant: 0 O product: 0 O
Balanced Oxidation Half-Equation	$Ca(s) \rightarrow Ca^{2+}(aq) + 2e^-$

d. Make the number of electrons in the two half-equations equal (multiply by an appropriate number if necessary).	Balanced half-equations:	
	reduction: multiply by 2	$2 [VO_4{}^{3-}(aq) + 4H_2O + 3e^-$ $\rightarrow V^{2+}(aq) + 8OH^-(aq)]$
		$2VO_4{}^{3-}(aq) + 8H_2O + \mathbf{6e^-}$ $\rightarrow 2V^{2+}(aq) + 16OH^-(aq)$
	oxidation: multiply by 3	$3 [Ca(s) \rightarrow Ca^{2+}(aq) + 2e^-]$
		$3Ca(s) \rightarrow 3Ca^{2+}(aq) + \mathbf{6e^-}$

e. Combine the two half-equations.	$3Ca(s) + 2VO_4{}^{3-}(aq) + 8H_2O + \mathbf{6e^-}$ $\rightarrow 3Ca^{2+}(aq) + 2V^{2+}(aq) + 16OH^-(aq) + \mathbf{6e^-}$
Cancel duplication.	$3Ca(s) + 2VO_4{}^{3-}(aq) + 8H_2O$ $\rightarrow 3Ca^{2+}(aq) + 2V^{2+}(aq) + 16OH^-(aq)$

Balanced Equation:
$$3Ca(s) + 2VO_4{}^{3-}(aq) + 8H_2O \rightarrow 3Ca^{2+}(aq) + 2V^{2+}(aq) + 16OH^-(aq)$$

(b) $C_2H_4(g) + BiO_3^-(aq) \rightarrow CO_2(g) + Bi^{3+}(aq)$ [basic]

a.	Split into two half-equations.	$C_2H_4(g) \rightarrow CO_2(g)$
		$BiO_3^-(aq) \rightarrow Bi^{3+}(aq)$
b.	Balance one of the half-equations: $BiO_3^-(aq) \rightarrow Bi^{3+}(aq)$	
	oxidation numbers	Bi (oxidation no.): +5 → +3; Bi is reduced
	atom balance	1 Bi on each side, no adjustment is required
	total oxidation number	Bi: 1(+5) = +5 → Bi: 1(+3) = +3
	add electrons	The total oxidation number for Bi goes from +5 to +3. It is reduced by 2. Add 2 electrons to the reactant side. $BiO_3^-(aq) + 2e^- \rightarrow Bi^{3+}(aq)$
	balance charge	reactant: −1 + 2(−1) = −3; product: +3 basic medium: add OH^-. To balance the charge, add 6 OH^- to the product side. $BiO_3^-(aq) + 2e^- \rightarrow Bi^{3+}(aq) + 6OH^-(aq)$ reactant: −1 + 2(−1) = −3; product: +3 + 6(−1) = −3
	balance H	reactant: 0 H product: 6 H To balance H, add 3 H_2O to the reactant side. $BiO_3^-(aq) + 3H_2O + 2e^- \rightarrow Bi^{3+}(aq) + 6OH^-(aq)$ reactant: 6 H product: 6 H
	check O	reactant: 6 O product: 6 O
	Balanced Reduction Half-Equation	$BiO_3^-(aq) + 3H_2O + 2e^- \rightarrow Bi^{3+}(aq) + 6OH^-(aq)$
c.	Balance the other half-equation: $C_2H_4(g) \rightarrow CO_2(g)$	
	oxidation numbers	C (oxidation no.): −2 → +4; C is oxidized
	atom balance	reactant: 2 C product: 1 C Multiply CO_2 by 2: $C_2H_4(g) \rightarrow 2CO_2(g)$
	total oxidation number	C: 2(−2) = −4 → C: 2(+4) = +8
	add electrons	The total oxidation number for C goes from −4 to +8. It is oxidized by 12. Add 12 electrons to the product side. $C_2H_4(g) \rightarrow 2CO_2(g) + 12e^-$
	balance charge	reactant: 0; product: 12(−1) = −12 basic medium: add OH^-. To balance the charge, add 12 OH^- to the reactant side. $C_2H_4(g) + 12OH^-(aq) \rightarrow 2CO_2(g) + 12e^-$ reactant: 12(−1) = −12; product: 12(−1) = −12
	balance H	reactant: 16 H product: 0 H To balance H, add 8 H_2O to the product side. $C_2H_4(g) + 12OH^-(aq) \rightarrow 2CO_2(g) + 12e^- + 8H_2O$ reactant: 16 H product: 16 H
	check O	reactant: 12 O product: 12 O

Balanced Oxidation Half-Equation	$C_2H_4(g) + 12OH^-(aq) \rightarrow 2CO_2(g) + 12e^- + 8H_2O$	
d. Make the number of electrons in the two half-equations equal (multiply by an appropriate number if necessary).	Balanced half-equations:	
	reduction: multiply by 6	$6\,[BiO_3^-(aq) + 3H_2O + 2e^- \rightarrow$ $Bi^{3+}(aq) + 6\,OH^-(aq)]$
		$6BiO_3^-(aq) + 18H_2O + 12e^-$ $\rightarrow 6Bi^{3+}(aq) + 36OH^-(aq)$
	oxidation	$C_2H_4(g) + 12OH^-(aq)$ $\rightarrow 2CO_2(g) + 12e^- + 8H_2O$
e. Combine the two half-equations.	$C_2H_4(g) + 12OH^-(aq) + 6BiO_3^-(aq) + 18H_2O + 12e^-$ $\rightarrow 2CO_2(g) + 12e^- + 8H_2O + 6Bi^{3+}(aq) + 36OH^-(aq)$	
Cancel duplication.	$C_2H_4(g) + 6BiO_3^-(aq) + 10H_2O$ $\rightarrow 2CO_2(g) + 6Bi^{3+}(aq) + 24OH^-(aq)$	

Balanced Equation:
$$C_2H_4(g) + 6BiO_3^-(aq) + 10H_2O \rightarrow 2CO_2(g) + 6Bi^{3+}(aq) + 24OH^-(aq)$$

(c) $PbO_2(s) + H_2O \rightarrow O_2(g) + Pb^{2+}(aq)$ [basic]

a. Split into two half-equations.	$PbO_2(s) \rightarrow Pb^{2+}(aq)$
	$H_2O \rightarrow O_2(g)$
b. Balance one of the half-equations: $PbO_2(s) \rightarrow Pb^{2+}(aq)$	
oxidation numbers	Pb (oxidation no.): $+4 \rightarrow +2$; Pb is reduced
atom balance	1 Pb on each side, no adjustment is required
total oxidation number	Pb: $1(+4) = +4 \rightarrow$ Pb: $1(+2) = +2$
add electrons	The total oxidation number for Pb goes from +4 to +2. It is reduced by 2. Add 2 electrons to the reactant side. $PbO_2(s) + 2e^- \rightarrow Pb^{2+}(aq)$
balance charge	reactant: $2(-1) = -2$; product: $+2$ basic medium: add OH^-. To balance the charge, add 4 OH^- to the product side. $PbO_2(s) + 2e^- \rightarrow Pb^{2+}(aq) + 4OH^-(aq)$ reactant: $2(-1) = -2$; product: $+2 + 4(-1) = -2$
balance H	reactant: 0 H product: 4 H To balance H, add 2 H_2O to the reactant side. $PbO_2(s) + 2H_2O + 2e^- \rightarrow Pb^{2+}(aq) + 4OH^-(aq)$ reactant: 4 H product: 4 H
check O	reactant: 4 O product: 4 O
Balanced Reduction Half-Equation	$PbO_2(s) + 2H_2O + 2e^- \rightarrow Pb^{2+}(aq) + 4OH^-(aq)$
c. Balance the other half-equation: $H_2O \rightarrow O_2(g)$	

oxidation numbers	O (oxidation no.): $-2 \rightarrow 0$; O is oxidized
atom balance	reactant: 1 O $\qquad$ product: 2 O Multiply H_2O by 2: $\qquad$ $2H_2O \rightarrow O_2(g)$
total oxidation number	O: $2(-2) = -4 \rightarrow$ O: $2(0) = 0$
add electrons	The total oxidation number for O goes from -4 to 0. It is oxidized by 4. Add 4 electrons to the product side. $\qquad$ $2H_2O \rightarrow O_2(g) + 4e^-$
balance charge	reactant: 0; $\qquad$ product: $4(-1) = -4$ basic medium: add OH^-. To balance the charge, add 4 OH^- to the reactant side. $\qquad$ $2H_2O + 4OH^-(aq) \rightarrow O_2(g) + 4e^-$ reactant: $4(-1) = -4$; $\qquad$ product: $4(-1) = -4$
balance H	reactant: 8 H $\qquad$ product: 0 H To balance H, add 4 H_2O to the product side. $\qquad$ $2H_2O + 4OH^-(aq) \rightarrow O_2(g) + 4H_2O + 4e^-$ reactant: 8 H $\qquad$ product: 8 H
check O	reactant: 6 O $\qquad$ product: 6 O
Balanced Oxidation Half-Equation	$2H_2O + 4OH^-(aq) \rightarrow O_2(g) + 4H_2O + 4e^-$
Cancel duplication	$4OH^-(aq) \rightarrow O_2(g) + 2H_2O + 4e^-$

d. Make the number of electrons in the two half-equations equal (multiply by an appropriate number if necessary).	Balanced half-equations:	
	reduction: $\qquad$ multiply by 2	$2 [PbO_2(s) + 2H_2O + 2e^-$ $\qquad \rightarrow Pb^{2+}(aq) + 4OH^-(aq)]$
		$2PbO_2(s) + 4H_2O + 4e^-$ $\qquad \rightarrow 2Pb^{2+}(aq) + 8OH^-(aq)$
	oxidation	$4OH^-(aq) \rightarrow O_2(g) + 2H_2O + 4e^-$

e. Combine the two half-equations.	$2PbO_2(s) + 4H_2O + 4OH^-(aq) + 4e^-$ $\qquad \rightarrow 2Pb^{2+}(aq) + 8OH^-(aq) + O_2(g) + 2H_2O + 4e^-$
Cancel duplication.	$2PbO_2(s) + 2H_2O \rightarrow 2Pb^{2+}(aq) + 4OH^-(aq) + O_2(g)$
Balanced Equation:	$2PbO_2(s) + 2H_2O \rightarrow 2Pb^{2+}(aq) + 4OH^-(aq) + O_2(g)$

(d) $IO_3^-(aq) + Cl^-(aq) \rightarrow Cl_2(g) + I_3^-(aq)$ $\qquad$ [basic]

a. Split into two half-equations.	$IO_3^-(aq) \rightarrow I_3^-(aq)$
	$Cl^-(aq) \rightarrow Cl_2(g)$

b. Balance one of the half-equations: $\qquad IO_3^-(aq) \rightarrow I_3^-(aq)$	
oxidation numbers	I (oxidation no.): $+5 \rightarrow -\frac{1}{3}$; I is reduced
atom balance	reactant: 1 I $\qquad$ product: 3 I Multiply IO_3^- by 3: $\qquad$ $3IO_3^-(aq) \rightarrow I_3^-(aq)$
total oxidation number	I: $3(+5) = +15 \rightarrow$ I: $3(-\frac{1}{3}) = -1$

add electrons	The total oxidation number for I goes from +15 to −1. It is reduced by 16. Add 16 electrons to the reactant side. $$3IO_3^-(aq) + 16e^- \rightarrow I_3^-(aq)$$
balance charge	reactant: $3(-1) + 16(-1) = -19$; product: −1 basic medium: add OH^-. To balance the charge, add 18 OH^- to the product side. $$3IO_3^-(aq) + 16e^- \rightarrow I_3^-(aq) + 18OH^-(aq)$$ reactant: $3(-1) + 16(-1) = -19$ product: $-1 + 18(-1) = -19$
balance H	reactant: 0 H product: 18 H To balance H, add 9 H_2O to the reactant side. $$3IO_3^-(aq) + 9H_2O + 16e^- \rightarrow I_3^-(aq) + 18OH^-(aq)$$ reactant: 18 H product: 18 H
check O	reactant: 18 O product: 18 O
Balanced Reduction Half-Equation	$3IO_3^-(aq) + 9H_2O + 16e^- \rightarrow I_3^-(aq) + 18OH^-(aq)$

c.   Balance the other half-equation: $Cl^-(aq) \rightarrow Cl_2(g)$

oxidation numbers	Cl (oxidation no.): $-1 \rightarrow 0$; Cl is oxidized
atom balance	reactant: 1 Cl product: 2 Cl Multiply Cl^- by 2: $2Cl^-(aq) \rightarrow Cl_2(g)$
total oxidation number	Cl: $2(-1)= -2 \rightarrow$ Cl: $2(0)=0$
add electrons	The total oxidation number for Cl goes from −2 to 0. The total oxidation number increased by 2. Add 2 electrons to the product side. $$2Cl^-(aq) \rightarrow Cl_2(g) + 2e^-$$
balance charge	reactant: −2 product: −2
balance H	reactant: 0 H product: 0 H
check O	reactant: 0 O product: 0 O
Balanced Oxidation Half-Equation	$2Cl^-(aq) \rightarrow Cl_2(g) + 2e^-$

d. Make the number of electrons in the two half-equations equal (multiply by an appropriate number if necessary).

Balanced half-equations:		
reduction		$3IO_3^-(aq) + 9H_2O + \mathbf{16e^-}$ $\rightarrow I_3^-(aq) + 18OH^-(aq)$
oxidation: multiply by 8	$8 [2Cl^-(aq) \rightarrow Cl_2(g) + 2e^-]$	
	$16Cl^-(aq) \rightarrow 8Cl_2(g) + \mathbf{16e^-}$	

e. Combine the two half-equations.

$3IO_3^-(aq) + 9H_2O + 16Cl^-(aq) + \mathbf{16e^-}$
$\rightarrow I_3^-(aq) + 18OH^-(aq) + 8Cl_2(g) + \mathbf{16e^-}$

Cancel duplication.

$3IO_3^-(aq) + 9H_2O + 16Cl^-(aq) \rightarrow I_3^-(aq) + 18OH^-(aq) + 8Cl_2(g)$

Balanced Equation:
$$3IO_3^-(aq) + 9H_2O + 16Cl^-(aq) \rightarrow I_3^-(aq) + 18OH^-(aq) + 8Cl_2(g)$$

10. These problems are similar to Example 17.2. To balance the equation, follow the steps outlined in the text for balancing redox equations. Begin by writing the given word equation as a reaction using chemical symbols, then split the reaction into two half-equations. Balance each of the half-equations separately with respect to both number of atoms and charges. Eliminate the electrons. If the number of electrons in the two balanced half-equations is not the same, multiply the equations by an appropriate number to make the number of electrons equal. Combine the two half-equations and cancel those species which are duplicated or appear on both sides of the reaction.

(a) Unbalanced net ionic equation: $NO(g) + H_2(g) \rightarrow NH_3(g) + H_2O(g)$ [acidic]

a.	Split into two half-equations.	$NO(g) \rightarrow NH_3(g)$
		$H_2(g) \rightarrow H_2O(g)$

b.	Balance one of the half-equations: $NO(g) \rightarrow NH_3(g)$	
	oxidation numbers	N (oxidation no.): $+2 \rightarrow -3$; N is reduced
	atom balance	1 N on each side, no adjustment is required
	total oxidation number	N: $1(+2) = +2 \rightarrow$ N: $1(-3) = -3$
	add electrons	The total oxidation number for N goes from $+2$ to -3. It is reduced by 5. Add 5 electrons to the reactant side. $\qquad NO(g) + 5e^- \rightarrow NH_3(g)$
	balance charge	reactant: $5(-1) = -5$; product: 0 acidic medium: add H^+ To balance the charge, add 5 H^+ to the reactant side. $\qquad NO(g) + 5H^+(aq) + 5e^- \rightarrow NH_3(g)$ reactant: $5(+1) + 5(-1) = 0$; product: 0
	balance H	reactant: 5 H product: 3 H To balance H, add 1 H_2O to the product side. $\qquad NO(g) + 5H^+(aq) + 5e^- \rightarrow NH_3(g) + H_2O(g)$ reactant: 5 H product: 5 H
	check O	reactant: 1 O product: 1 O
	Balanced Reduction Half-Equation	$NO(g) + 5H^+(aq) + 5e^- \rightarrow NH_3(g) + H_2O(g)$

c.	Balance the other half-equation: $H_2(g) \rightarrow H_2O(g)$	
	oxidation numbers	H (oxidation no.): $0 \rightarrow +1$; H is oxidized
	atom balance	2 H on each side, no adjustment is required
	total oxidation number	H: $2(0) = 0 \rightarrow$ H: $2(+1) = +2$
	add electrons	The total oxidation number for H goes from 0 to $+2$. The total oxidation number increased by 2. Add 2 electrons to the product side. $\qquad H_2(g) \rightarrow H_2O(g) + 2e^-$
	balance charge	reactant: 0 product: $2(-1) = -2$

---

		acidic medium: add H⁺
		To balance the charge, add 2 H⁺ to the product side.
		$H_2(g) \rightarrow H_2O(g) + 2e^- + 2H^+(aq)$
		reactant: 0 product: 0
	balance H	reactant: 2 H product: 4 H
		To balance H, add 1 H_2O to the reactant side.
		$H_2(g) + H_2O \rightarrow H_2O + 2e^- + 2H^+(aq)$
		reactant: 4 H product: 4 H
	check O	reactant: 1 O product: 1 O
	Balanced Oxidation Half-Equation	$H_2(g) + H_2O \rightarrow H_2O + 2e^- + 2H^+(aq)$
		Cancel duplication: $H_2(g) \rightarrow 2H^+(aq) + 2e^-$
d.	Make the number of electrons in the two half-equations equal (multiply by an appropriate number if necessary).	Balanced half-equations:

reduction: multiply by 2	$2 [NO(g) + 5H^+(aq) + 5e^-$ $\rightarrow NH_3(g) + H_2O(g)]$
	$2NO(g) + 10H^+(aq) + 10e^-$ $\rightarrow 2NH_3(g) + 2H_2O(g)$
oxidation: multiply by 5	$5 [H_2(g) \rightarrow 2e^- + 2H^+(aq)]$
	$5H_2(g) \rightarrow 10e^- + 10H^+(aq)$

e.	Combine the two half-equations.	$2NO(g) + 10H^+(aq) + 5H_2(g) + 10e^-$ $\rightarrow 2NH_3(g) + 2H_2O(g) + 10H^+(aq) + 10e^-$
	Cancel duplication.	$2NO(g) + 5H_2(g) \rightarrow 2NH_3(g) + 2H_2O(g)$

Balanced Equation: $2NO(g) + 5H_2(g) \rightarrow 2NH_3(g) + 2H_2O(g)$

(b) Unbalanced net ionic equation: $H_2O_2(aq) + ClO^-(aq) \rightarrow O_2(g) + Cl_2(g)$ [acidic]

a.	Split into two half-equations.	$H_2O_2(aq) \rightarrow O_2(g)$
		$ClO^-(aq) \rightarrow Cl_2(g)$
b.	Balance one of the half-equations: $ClO^-(aq) \rightarrow Cl_2(g)$	
	oxidation numbers	Cl (oxidation no.): +1 → 0; Cl is reduced
	atom balance	reactant: 1 Cl product: 2 Cl
		Multiply ClO^- by 2: $2ClO^-(aq) \rightarrow Cl_2(g)$
	total oxidation number	Cl: 2(+1)=+2 → Cl: 2(0)= 0
	add electrons	The total oxidation number for Cl goes from +2 to 0. It is reduced by 2. Add 2 electrons to the reactant side.
		$2ClO^-(aq) + 2e^- \rightarrow Cl_2(g)$
	balance charge	reactant: 2(−1) + 2(−1) = −4; product: 0
		acidic medium: add H⁺
		To balance the charge, add 4 H⁺ to the reactant side.
		$2ClO^-(aq) + 2e^- + 4H^+(aq) \rightarrow Cl_2(g)$
		reactant: 2(−1) + 2(−1) + 4(+1) = 0; product: 0

balance H	reactant: 4 H product: 0 H
	To balance H, add 2 H_2O to the product side.
	$2ClO^-(aq) + 2e^- + 4H^+(aq) \rightarrow Cl_2(g) + 2H_2O$
	reactant: 4 H product: 4 H
check O	reactant: 2 O product: 2 O
Balanced Reduction Half-Equation	$2ClO^-(aq) + 2e^- + 4H^+(aq) \rightarrow Cl_2(g) + 2H_2O$

c. Balance the other half-equation: $H_2O_2(aq) \rightarrow O_2(g)$

oxidation numbers	O (oxidation no.): $-1 \rightarrow 0$; O is oxidized
atom balance	2 O on each side, no adjustment is required
total oxidation number	O: $2(-1) = -2 \rightarrow$ O: $2(0) = 0$
add electrons	The total oxidation number for O goes from -2 to 0.
	The total oxidation number increased by 2.
	Add 2 electrons to the product side.
	$H_2O_2(aq) \rightarrow O_2(g) + 2e^-$
balance charge	reactant: 0 product: $2(-1) = -2$
	acidic medium: add H^+
	To balance the charge, add 2 H^+ to the product side.
	$H_2O_2(aq) \rightarrow O_2(g) + 2H^+(aq) + 2e^-$
	reactant: 0 product: $2(-1) + 2(+1) = 0$
balance H	reactant: 2 H product: 2 H
check O	reactant: 2 O product: 2 O
Balanced Oxidation Half-Equation	$H_2O_2(aq) \rightarrow O_2(g) + 2H^+(aq) + 2e^-$

d. Make the number of electrons in the two half-equations equal (multiply by an appropriate number if necessary).	Balanced half-equations:	
	reduction	$2ClO^-(aq) + 2e^- + 4H^+(aq) \rightarrow Cl_2(g) + 2H_2O$
	oxidation	$H_2O_2(aq) \rightarrow O_2(g) + 2H^+(aq) + 2e^-$

e. Combine the two half-equations.	$H_2O_2(aq) + 2ClO^-(aq) + 2e^- + 4H^+(aq)$
	$\rightarrow O_2(g) + 2e^- + 2H^+(aq) + Cl_2(g) + 2H_2O$
Cancel duplication.	$H_2O_2(aq) + 2ClO^-(aq) + 2H^+(aq) \rightarrow O_2(g) + Cl_2(g) + 2H_2O$

Balanced Equation:
$$H_2O_2(aq) + 2ClO^-(aq) + 2H^+(aq) \rightarrow O_2(g) + Cl_2(g) + 2H_2O$$

(c) Unbalanced net ionic equation: $Zn(s) + VO^{2+}(aq) \rightarrow V^{3+}(aq) + Zn^{2+}(aq)$ [acidic]

a. Split into two half-equations.	$Zn(s) \rightarrow Zn^{2+}(aq)$
	$VO^{2+}(aq) \rightarrow V^{3+}(aq)$

b. Balance one of the half-equations: $VO^{2+}(aq) \rightarrow V^{3+}(aq)$

oxidation numbers	V (oxidation no.): +4 → +3; V is reduced
atom balance	1 V on each side, no adjustment is required
total oxidation number	V: 1(+4)=+4 → V: 1(+3)= +3
add electrons	The total oxidation number for V goes from +4 to +3. It is reduced by 1. Add 1 electron to the reactant side. $VO^{2+}(aq) + e^- \rightarrow V^{3+}(aq)$
balance charge	reactant: 1(+2) + 1(−1) = +1; product: +3 acidic medium: add H^+ To balance the charge, add 2 H^+ to the reactant side. $VO^{2+}(aq) + 2H^+(aq) + e^- \rightarrow V^{3+}(aq)$ reactant: 1(+2) + 2(+1) + 1(−1) = +3; product: +3
balance H	reactant: 2 H product: 0 H To balance H, add 1 H_2O to the product side. $VO^{2+}(aq) + 2H^+(aq) + e^- \rightarrow V^{3+}(aq) + H_2O$ reactant: 2 H product: 2 H
check O	reactant: 1 O product: 1 O
Balanced Reduction Half-Equation	$VO^{2+}(aq) + 2H^+(aq) + e^- \rightarrow V^{3+}(aq) + H_2O$

c.   Balance the other half-equation: $Zn(s) \rightarrow Zn^{2+}(aq)$

oxidation numbers	Zn (oxidation no.): 0 → +2; Zn is oxidized
atom balance	1 Zn on each side, no adjustment is required
total oxidation number	Zn: 0 → Zn: +2
add electrons	The total oxidation number for Zn goes from 0 to +2. The total oxidation number increased by 2. Add 2 electrons to the product side. $Zn(s) \rightarrow Zn^{2+}(aq) + 2e^-$
balance charge	reactant: 0 product: 0
balance H	reactant: 0 H product: 0 H
check O	reactant: 0 O product: 0 O
Balanced Oxidation Half-Equation	$Zn(s) \rightarrow Zn^{2+}(aq) + 2e^-$

d. Make the number of electrons in the two half-equations equal (multiply by an appropriate number if necessary).	Balanced half-equations:	
	reduction: multiply by 2	$2 [VO^{2+}(aq) + 2H^+(aq) + e^-$ $\rightarrow V^{3+}(aq) + H_2O]$
		$2VO^{2+}(aq) + 4H^+(aq) + 2e^-$ $\rightarrow 2V^{3+}(aq) + 2H_2O$
	oxidation	$Zn(s) \rightarrow Zn^{2+}(aq) + 2e^-$

e. Combine the two half-equations.	$2VO^{2+}(aq) + 4H^+(aq) + Zn(s) + \mathbf{2e^-}$ $\rightarrow 2V^{3+}(aq) + 2H_2O + Zn^{2+}(aq) + \mathbf{2e^-}$
Cancel duplication.	$2VO^{2+}(aq) + 4H^+(aq) + Zn(s) \rightarrow 2V^{3+}(aq) + 2H_2O + Zn^{2+}(aq)$

Balanced Equation:
$$2VO^{2+}(aq) + 4H^+(aq) + Zn(s) \rightarrow 2V^{3+}(aq) + 2H_2O + Zn^{2+}(aq)$$

17-2 VOLTAIC CELLS

12. See Example 17.3. Recall that the oxidation reaction is always on the left side of the cell notation (to the left of the double vertical line) and the reduction reaction is always on the right side. Elements are in their standard forms and ions are in aqueous solution. Note that the Pt electrodes in parts (b) and (c) are inert and do not take part in the reaction (see also Figure 17.5). The half-equations can be balanced and combined into the overall redox reaction as described in Section 17-1 (see Example 17.2), although all that is required to balance the half-equations in parts (a) and (b) is simply the addition of electrons to the appropriate side of each half-equation.

(a) The two unbalanced half-equations are:

at the anode (oxidation): $\qquad$ $Cd(s) \rightarrow Cd^{2+}(aq)$

at the cathode (reduction): $\qquad$ $Sb^{3+}(aq) \rightarrow Sb(s)$

Balance the oxidation half-equation by adding electrons to the product side:

$Cd(s) \rightarrow Cd^{2+}(aq) + 2e^-$

Balance the reduction half-equation by adding electrons to the reactant side:

$Sb^{3+}(aq) + 3e^- \rightarrow Sb(s)$

Multiply the oxidation half-equation by 3 and the reduction half-equation by 2 to make the number of electrons in each half-equation equal 6:

$3Cd(s) \rightarrow 3Cd^{2+}(aq) + 6e^-$

$2Sb^{3+}(aq) + 2e^- \rightarrow 2Sb(s)$

Finally add the two half-equations together and cancel any duplication:

$3Cd(s) + 2Sb^{3+}(aq) + 6e^- \rightarrow 3Cd^{2+}(aq) + 6e^- + 2Sb(s)$

$\mathbf{3Cd(s) + 2Sb^{3+}(aq) \rightarrow 3Cd^{2+}(aq) + 2Sb(s)}$

(b) The two unbalanced half-equations are:

at the anode (oxidation): $\qquad$ $Cu^+(aq) \rightarrow Cu^{2+}(aq)$

at the cathode (reduction): $\qquad$ $Mg^{2+}(aq) \rightarrow Mg(s)$

Balance the oxidation half-equation by adding electrons to the product side:

$Cu^+(aq) \rightarrow Cu^{2+}(aq) + e^-$

Balance the reduction half-equation by adding electrons to the reactant side:

$Mg^{2+}(aq) + 2e^- \rightarrow Mg(s)$

Multiply the oxidation half-equation by 2 to make the number of electrons in each half-equation equal:

$$2Cu^+(aq) \rightarrow 2Cu^{2+}(aq) + 2e^-$$

Finally add the two half-equations together and cancel any duplication:

$$2Cu^+(aq) + Mg^{2+}(aq) + 2e^- \rightarrow 2Cu^{2+}(aq) + 2e^- + Mg(s)$$

$$\mathbf{2Cu^+(aq) + Mg^{2+}(aq) \rightarrow 2Cu^{2+}(aq) + Mg(s)}$$

(c) The two unbalanced half-equations are:

at the anode (oxidation):	$Cr^{3+}(aq) \rightarrow Cr_2O_7^{2-}(aq)$	[acid]
at the cathode (reduction):	$ClO_3^-(aq) \rightarrow Cl^-(aq)$	[acid]

(i) Balance one of the half-equations: $Cr^{3+}(aq) \rightarrow Cr_2O_7^{2-}(aq)$	
oxidation numbers	Cr (oxid. no.): $+3 \rightarrow +6$; Cr is oxidized
atom balance	reactant: 1 Cr product: 2 Cr Multiply Cr^{3+} by 2: $2Cr^{3+}(aq) \rightarrow Cr_2O_7^{2-}(aq)$
total oxidation number	Cr: $2(+3) = +6 \rightarrow$ Cr: $2(+6) = +12$
add electrons	The total oxidation number for Cr goes from +6 to +12. It is oxidized by 6. Add 6 electrons to the product side. $\qquad 2Cr^{3+}(aq) \rightarrow Cr_2O_7^{2-}(aq) + 6e^-$
balance charge	reactant: $2(+3) = +6$; product: $1(-2) + (-6) = -8$ acidic medium: add H^+. To balance the charge, add 14 H^+ to the product side. $\qquad 2Cr^{3+}(aq) \rightarrow Cr_2O_7^{2-}(aq) + 6e^- + 14H^+(aq)$ reactant: $2(+3) = +6$; product: $1(-2) + (-6) + 14(+1) = +6$
balance H	reactant: 0 H product: 14 H To balance H, add 7 H_2O to the reactant side $2Cr^{3+}(aq) + 7H_2O \rightarrow Cr_2O_7^{2-}(aq) + 6e^- + 14H^+(aq)$ reactant: 14 H product: 14 H
check O	reactant: 7 O product: 7 O
Balanced Oxidation Half-Equation	$2Cr^{3+}(aq) + 7H_2O \rightarrow Cr_2O_7^{2-}(aq) + 6e^- + 14H^+(aq)$
(ii) Balance the other half-equation: $ClO_3^-(aq) \rightarrow Cl^-(aq)$	
oxidation numbers	Cl: $+5 \rightarrow -1$; Cl is reduced
atom balance	1 Cl on each side, no adjustment is required
total oxidation number	Cl: $1(+5) = +5 \rightarrow$ Cl: $1(-1) = -1$
add electrons	The total oxidation number for Cl goes from +5 to -1: The total oxidation number decreased by 6.

	Add 6 electrons to the reactant side. $$ClO_3^-(aq) + 6e^- \rightarrow Cl^-(aq)$$
balance charge	reactant: $-1 + 6(-1) = -7$ product: -1 acidic medium: add H^+. To balance the charge, add 6 H^+ to the reactant side. $$ClO_3^-(aq) + 6H^+(aq) + 6e^- \rightarrow Cl^-(aq)$$ reactant: $-1 + 6(+1) + 6(-1) = -1$ product: -1
balance H	reactant: 6 H product: 0 H To balance H, add 3 H_2O to the product side $$ClO_3^-(aq) + 6H^+(aq) + 6e^- \rightarrow Cl^-(aq) + 3H_2O$$
check O	reactant: 3 O product: 3 O
Balanced Reduction Half-Equation	$ClO_3^-(aq) + 6H^+(aq) + 6e^- \rightarrow Cl^-(aq) + 3H_2O$
(iii) Make the number of electrons in the two half-equations equal (multiply by an appropriate number if necessary).	Balanced half-equations:

	oxidation	$2Cr^{3+}(aq) + 7H_2O$ $\rightarrow Cr_2O_7^{2-}(aq) + 6e^- + 14H^+(aq)$
	reduction	$ClO_3^-(aq) + 6H^+(aq) + 6e^-$ $\rightarrow Cl^-(aq) + 3H_2O$

(iv) Combine the two half-equations.	$2Cr^{3+}(aq) + \mathbf{7H_2O} + ClO_3^-(aq) + \mathbf{6H^+(aq)} + \mathbf{6e^-}$ $\rightarrow Cr_2O_7^{2-}(aq) + \mathbf{6e^-} + \mathbf{14H^+(aq)} + Cl^-(aq) + \mathbf{3H_2O}$
Cancel duplication.	$2Cr^{3+}(aq) + 4H_2O + ClO_3^-(aq)$ $\rightarrow Cr_2O_7^{2-}(aq) + 8H^+(aq) + Cl^-(aq)$

Balanced Equation: $2Cr^{3+}(aq) + 4H_2O + ClO_3^-(aq)$
$$\rightarrow Cr_2O_7^{2-}(aq) + 8H^+(aq) + Cl^-(aq)$$

14. See Example 17.3 and Figure 17.5.

(a) Sn(s) is *oxidized* to $Sn^{2+}(aq)$ at the **anode**; $Ag^+(aq)$ is *reduced* to Ag(s) at the **cathode**, so electrons will move from Sn to Ag. Anions will move toward the Sn anode and cations will move towards the Ag cathode.

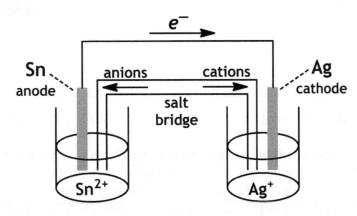

(b) $H_2(g)$ is _oxidized_ to $H^+(aq)$ at the __anode__; $Hg_2^{2+}(aq)$ is _reduced_ to $Hg(l)$ at the __cathode__, so electrons will move from H to Hg. Anions will move toward the H anode and cations will move towards the Hg cathode. Note that both anode and cathode require inert Pt electrodes (see also Figure 17.4) since a solid metal is not part of the cell reaction.

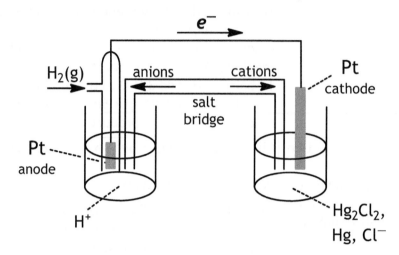

(c) $Pb(s)$ is _oxidized_ to Pb^{2+} (as solid $PbSO_4$) at the __anode__; Pb^{4+} (as solid PbO_2) is _reduced_ to Pb^{2+} (as solid $PbSO_4$) at the __cathode__, so electrons will move from Pb to PbO_2. Anions will move toward the anode and cations will move towards the cathode.

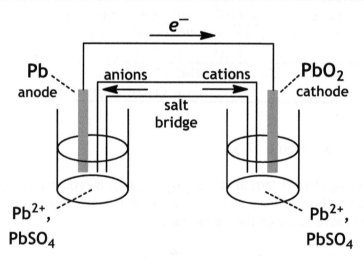

16. See Example 17.3.

(a)   Separate the given redox reaction into two half reactions:

$$Fe^{2+}(aq) \rightarrow Fe^{3+}(aq) \qquad \text{and} \qquad MnO_4^-(aq) \rightarrow Mn^{2+}(aq)$$

The oxidation number of Fe changes from +2 to +3, so it is oxidized; and therefore the Fe^{2+}/Fe^{3+} half-cell is the anode. The oxidation number of Mn changes from +7 to +2, so it is reduced; and therefore the MnO_4^-/Mn^{2+} half-cell is the cathode. __Electrons flow from anode to cathode__ in the external circuit, which is __from the Fe half-cell to the Mn half-cell__.

(b) A solid metal is not part of the half-reaction taking place at the anode and so a **Pt electrode** (or other inert electrode) must be used.

(c) Balance the half equation as described in Section 17-1 (see Example 17.1). Note that the presence of $H^+(aq)$ in the given reaction indicates that the reaction takes place under acidic conditions

$$MnO_4^-(aq) \rightarrow Mn^{2+}(aq) \qquad \text{[acidic]}$$

oxidation numbers	Mn: $+7 \rightarrow +2$; Mn is **reduced**
atom balance	1 Mn on each side, no adjustment is required
total oxidation number	Mn: $1(+7)=+7 \rightarrow$ Mn: $1(+2)= +2$
add electrons	The total oxidation number for Mn goes from +7 to +2. It is reduced by 5. Add 5 electrons to the reactant side. $MnO_4^-(aq) + 5e^- \rightarrow Mn^{2+}(aq)$
balance charge	reactant: $-1 + 5(-1)= -6$; product: +2 acidic medium: add H^+. To balance the charge, add 8 H^+ to the reactant side. $8H^+(aq) + MnO_4^-(aq) + 5e^- \rightarrow Mn^{2+}(aq)$ reactant: $8(+1) + (-1) + 5(-1)= +2$ product: +2
balance H	reactant: 8 H product: 0 H To balance H, add 4 H_2O to the product side. $8H^+(aq) + MnO_4^-(aq) + 5e^- \rightarrow Mn^{2+}(aq) + 4H_2O$ reactant: 8 H product: 8 H
check O	reactant: 4 O product: 4 O
balanced half equation	$8H^+(aq) + MnO_4^-(aq) + 5e^- \rightarrow Mn^{2+}(aq) + 4H_2O$

The reaction occurring at the cathode is:

$$8H^+(aq) + MnO_4^-(aq) + 5e^- \rightarrow Mn^{2+}(aq) + 4H_2O$$

18. This problem is similar to Example 17.3; see also Figure 17.5. Recall that oxidation takes place at the anode and reduction takes place at the cathode.

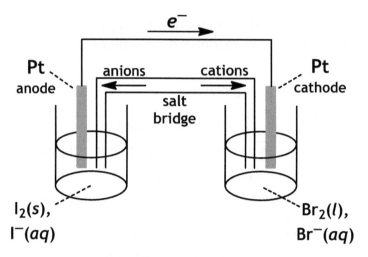

The two half-reactions are:

anode (oxidation): $2I^-(aq) \rightarrow I_2(s) + 2e^-$

cathode (reduction): $Br_2(l) + 2e^- \rightarrow 2Br^-(aq)$

and the two half-reactions may be added together directly to give the overall reaction; $2e^-$ on each side will cancel out:

$2I^-(aq) + Br_2(l) + 2e^- \rightarrow I_2(s) + 2e^- + 2Br^-(aq)$

$2I^-(aq) + Br_2(l) \rightarrow I_2(s) + 2Br^-(aq)$

The abbreviated cell notation is:

$Pt \, | \, I^- \, | \, I_2 \, \| \, Br_2 \, | \, Br^- \, | \, Pt$

17-3 STANDARD VOLTAGES: Strength of Oxidizing and Reducing Species

20. See Example 17.4b and Table 17.1. Recall that a reducing agent causes another species to be reduced and is itself oxidized in the process. In Table 17.1, going *up* the right column corresponds to increasing strength as a reducing agent, so for the three given pairs find which species is higher in the right column.

(a) Br^- is above Cl^-, so **Br^- is the stronger reducing agent.**

(b) Ni is above Cu, so **Ni is the stronger reducing agent.**

(c) Hg_2^{2+} is above $NO(g)$, **so Hg_2^{2+} is the stronger reducing agent.**

22. See Example 17.4b and Table 17.1. Recall that an oxidizing agent causes another species to be oxidized and is itself reduced in the process. In Table 17.1, going *down* the left column corresponds to increasing strength as an oxidizing agent, so find the five given

oxidizing agents and arrange them according to their position in the left column of the table. The order of increasing oxidizing strength is:

$$K^+ < Mn^{2+} < S < Cl_2 < Co^{3+}$$

24. This problem is similar to Example 17.4. Recall that oxidizing agents are located in the left column of Table 17.1 and reducing agents are found in the right column of the same table.

Species	Column	Oxidizing Agent or Reducing Agent
Cu^+	left and right	both oxidizing agent *and* reducing agent
Zn	right	reducing agent
Ni^{2+}	left	oxidizing agent
Fe^{2+}	left and right	both oxidizing agent *and* reducing agent
H^+ (acidic)	left	oxidizing agent

In order to compare the relative oxidizing and reducing strengths of all these species, the relevant parts of Table 17.1 have been selected and summarized below:

reactant		product	E°_{red} (V)
Zn^{2+}	$\rightarrow$	**Zn**	-0.762
Fe^{2+}	$\rightarrow$	Fe	-0.409
Ni^{2+}	$\rightarrow$	Ni	-0.236
H$^+$ (acidic)	$\rightarrow$	H$_2$	0.000
Cu^{2+}	$\rightarrow$	**Cu$^+$**	$+0.161$
Cu$^+$	$\rightarrow$	Cu	$+0.518$
Fe^{3+}	$\rightarrow$	**Fe^{2+}**	$+0.769$

For the oxidizing agents, going **down** the left column corresponds to increasing strength as an oxidizing agent, so the order of increasing oxidizing strength is:

$$Fe^{2+} < Ni^{2+} < H^+ \text{(acidic)} < Cu^+$$

For the reducing agents, going **up** the right column corresponds to increasing strength as a reducing agent, so the order of increasing reducing strength is:

$$Fe^{2+} < Cu^+ < Zn$$

26. This Problem is similar in part to Example 17.4. Note that the reactions listed in the problem are arranged by **decreasing** E°_{red} value, which is the opposite of Table 17.1; this must be remembered when comparing the relative oxidizing or reducing strength of the species listed. Thus, going **up** the left column of the given reactions will correspond to increasing oxidizing strength and going **down** the right column corresponds to increasing reducing strength.

(a) **Co²⁺(aq)** is highest up in the right column, so it is the weakest reducing agent.

(b) **Mn(s)** is lowest down in the right column, so it is the strongest reducing agent.

(c) **Co³⁺(aq)** is highest up in the left column, so it is the strongest oxidizing agent.

(d) **Mn²⁺(aq)** is lowest down in the left column, so it is the weakest oxidizing agent.

(e) **Yes:** Pb(s) is a stronger reducing agent than Fe^{2+}(aq), and Fe^{3+}(aq) is a stronger oxidizing agent than Pb^{2+}(aq).

(f) **No:** I⁻(aq) is a weaker reducing agent than Pb(s), and Pb^{2+}(aq) is a weaker oxidizing agent than I_2(aq), so the reverse reaction would happen.

(g) **Co³⁺(aq) and Fe³⁺(aq):** Pb(s) will be able to reduce those ions which, when reduced, give products that are weaker reducing agents than Pb(s). In this table, Co^{2+}(aq) and Fe^{2+}(aq) are weaker reducing agents than Pb^{2+}(aq), so Co^{3+}(aq) and Fe^{3+}(aq) can be reduced by Pb(s). (Another way of saying this is to note that Co^{3+}(aq) and Fe^{3+}(aq) are stronger oxidizing agents than Pb^{2+}(aq). Note that I_2(aq) is not an ion, so it is not a possible answer here.)

(h) **Pb(s), Cd(s) and Mn(s):** Fe^{3+}(aq) will be able to oxidize those metals which are stronger reducing agents than Fe^{2+}(aq). In this table, Pb(s), Cd(s) and Mn(s) are stronger reducing agents than Fe^{2+}(aq), so they will be oxidized. (Note that I⁻(aq) is not a metal, so it is not a possible answer here.)

28. (a) Use Table 17.1, under Basic Solution, and locate the two reactions which, when reversed, will correspond to the oxidations Cl^-(aq) → ClO_3^-(aq) and ClO_3^-(aq) → ClO_4^-(aq). Next, locate the oxidizing agents in the left column that are stronger than ClO_4^-(aq) but weaker than ClO_3^-(aq). (In other words, find oxidizing agents that are lower than ClO_4^- but higher than ClO_3^- in Table 17.1.) In this case, the only possible answer is **O₂(g)**.

    (b) Use Table 17.1, under Acidic Solution, and locate the two reactions which correspond to the reductions Mg^{2+}(aq) → Mg(s) and Ba^{2+}(aq) → Ba(s). Next, locate reducing agents in the right column that are stronger than Mg(s) but weaker than Ba(s). (In other words, find reducing agents that are higher than Mg but lower than Ba in Table 17.1.) In this case, possible answers are **Na(s)** and **Ca(s)**.

    (c) Use Table 17.1, under Acidic Solution, and locate the two reactions which correspond to the reductions Na^+(aq) → Na(s) and Li^+(aq) → Li(s). Next, locate reducing agents in the right column that are stronger than Na(s) but weaker than Li(s). (In other words, find reducing agents that are higher than Na but lower than Li in Table 17.1.) In this case, possible answers are **Ca(s)**, **Ba(s)** and **K(s)**.

17-3 STANDARD VOLTAGES: Calculation of E^o

30. This problem is similar to Example 17.5(a). For each cell: (i) assign oxidation numbers to each element, to decide which elements were oxidized/reduced; (ii) write the oxidation and reduction half-reactions with their E^o_{ox} and E^o_{red} values using Table 17.1; (iii) calculate E^o using $E^o = E^o_{ox} + E^o_{red}$.

(a) Na: $0 \rightarrow +1$ (oxidation)   Fe: $+2 \rightarrow 0$ (reduction)

half reactions: $Na(s) \rightarrow Na^+(aq) + e^-$ $E^o_{ox} = -E^o_{red} = +2.714$ V

$Fe^{2+}(aq) + 2e^- \rightarrow Fe(s)$ $E^o_{red} = -0.409$ V

$E^o = E^o_{ox} + E^o_{red} = +2.714$ V $+ (-0.409$ V$) = \underline{\textbf{+2.305 V}}$

(b) Al: $0 \rightarrow +3$ (oxidation)   S: $+6 \rightarrow +4$ (reduction)

half reactions: $Al(s) \rightarrow Al^{3+}(aq) + 3e^-$ $E^o_{ox} = -E^o_{red} = +1.68$ V

$SO_4{}^{2-}(aq) + 4H^+(aq) + 2e^-$

$\rightarrow SO_2(g) + 2H_2O$ $E^o_{red} = +0.155$ V

$E^o = E^o_{ox} + E^o_{red} = +1.68$ V $+ 0.155$ V $= \underline{\textbf{+1.84 V}}$

(c) O: $-2 \rightarrow 0$ (oxidation)   S: $0 \rightarrow -2$ (reduction)

half reactions: $4OH^-(aq) \rightarrow$

$O_2(g) + 2H_2O + 4e^-$ $E^o_{ox} = -E^o_{red} = -0.401$ V

$S(s) + 2e^- \rightarrow S^{2-}(aq)$ $E^o_{red} = -0.445$ V

$E^o = E^o_{ox} + E^o_{red} = -0.401$ V $+ (-0.445$ V$) = \underline{\textbf{-0.846 V}}$

32. This problem is similar to Example 17.5(a). For each reaction: (i) assign oxidation numbers to each element, to decide which elements were oxidized/reduced; (ii) write the oxidation and reduction half-reactions with their E^o_{ox} and E^o_{red} values using Table 17.1; (iii) calculate E^o using $E^o = E^o_{ox} + E^o_{red}$. (Refer to Section 2-7 for a review of the oxidation number notation using Roman numerals.)

(a) Cr: $+2 \rightarrow +3$ (oxidation)   Sn: $+4 \rightarrow +2$ (reduction)

half reactions: $Cr^{2+}(aq) \rightarrow Cr^{3+}(aq) + e^-$ $E^o_{ox} = -E^o_{red} = +0.408$ V

$Sn^{4+}(aq) + 2e^- \rightarrow Sn^{2+}(aq)$ $E^o_{red} = +0.154$ V

$E^o = E^o_{ox} + E^o_{red} = +0.408$ V $+ 0.154$ V $= \underline{\textbf{+0.562 V}}$

(b) Mn: $+2 \rightarrow +4$ (oxidation)   O: $-1 \rightarrow -2$ (reduction)

half reactions: $Mn^{2+}(aq) + 2H_2O \rightarrow$

$MnO_2(s) + 4H^+(aq) + 2e^-$ $E^o_{ox} = -E^o_{red} = -1.229$ V

$H_2O_2(aq) + 2H^+(aq) + 2e^-$

$\rightarrow 2H_2O$ $E^o_{red} = +1.763$ V

$$E^\circ = E^\circ_{ox} + E^\circ_{red} = -1.229 \text{ V} + 1.763 \text{ V} = \underline{\textbf{+0.534 V}}$$

34. This problem is related to Example 17.5(a); see also Example 17.3. For each cell, recall that the oxidation reaction takes place on the left side of the cell notation and the reduction reaction is on the right side. Write the oxidation and reduction half-reactions taking place, along with their E°_{ox} and E°_{red} values from Table 17.1 and then calculate E° using $E^\circ = E^\circ_{ox} + E^\circ_{red}$.

(a) half reactions: $\text{Ag}(s) \rightarrow \text{Ag}^+(aq) + e^-$ $E^\circ_{ox} = -E^\circ_{red} = -0.799 \text{ V}$

           $\text{Sn}^{4+}(aq) + 2e^- \rightarrow \text{Sn}^{2+}(aq)$ $E^\circ_{red} = +0.154 \text{ V}$

$$E^\circ = E^\circ_{ox} + E^\circ_{red} = -0.799 \text{ V} + 0.154 \text{ V} = \underline{\textbf{−0.645 V}}$$

(b) half reactions: $\text{Al}(s) \rightarrow \text{Al}^{3+}(aq) + 3e^-$ $E^\circ_{ox} = -E^\circ_{red} = +1.68 \text{ V}$

           $\text{Cu}^{2+}(aq) + 2e^- \rightarrow \text{Cu}(s)$ $E^\circ_{red} = +0.339 \text{ V}$

$$E^\circ = E^\circ_{ox} + E^\circ_{red} = +1.68 \text{ V} + 0.339 \text{ V} = \underline{\textbf{+2.02 V}}$$

(c) half reactions: $\text{Fe}^{2+}(aq) \rightarrow \text{Fe}^{3+}(aq) + e^-$ $E^\circ_{ox} = -E^\circ_{red} = -0.769 \text{ V}$

           $\text{MnO}_4^-(aq) + 8\text{H}^+(aq) + 5e^-$

                    $\rightarrow \text{Mn}^{2+}(aq) + 4\text{H}_2\text{O}$ $E^\circ_{red} = +1.512 \text{ V}$

$$E^\circ = E^\circ_{ox} + E^\circ_{red} = -0.769 \text{ V} + 1.512 \text{ V} = \underline{\textbf{+0.743 V}}$$

36. (a) Note that this reaction is the reverse of the standard hydrogen reduction reaction and recall that $E^\circ_{ox} = -E^\circ_{red}$: therefore, given:

         $\text{H}^+ \rightarrow \text{H}_2$ $E^\circ_{red} = 0.300 \text{ V}$

         $\text{H}_2 \rightarrow \text{H}^+$ $E^\circ_{ox} = -E^\circ_{red} = \underline{\textbf{−0.300 V}}$

(b) By definition, E°_{red} values such as those listed in Table 17.1 indicate the difference in reduction potential between the reaction of interest and the "reference reaction", namely $\text{H}^+ \rightarrow \text{H}_2$, which is normally assigned a value of $E^\circ_{red} = 0.000 \text{ V}$. Thus, in Table 17.1 the value of E°_{red} for $\text{Br}_2 \rightarrow \text{Br}^-$ is given as 1.077 V or, more specifically, it is **1.077 V higher than** E°_{red} for $\text{H}^+ \rightarrow \text{H}_2$. Therefore, if E°_{red} for $\text{H}^+ \rightarrow \text{H}_2$ is arbitrarily defined as 0.300 V, then E°_{red} for $\text{Br}_2 \rightarrow \text{Br}^-$ must be 1.077 V higher than that, or $(0.300 + 1.077) \text{ V} = \underline{\textbf{+1.377 V}}$.

(c) As seen in part (b) above, when E_{red}^o for $H^+ \rightarrow H_2$ is 0.300 V, E_{red}^o for other reactions will be the value given in Table 17.1 plus 0.300 V. Thus, for the cell given in Problem 34(c) the half reactions and corresponding potentials become:

$$Fe^{2+} \rightarrow Fe^{3+} \qquad E_{ox}^o = -E_{red}^o = -(0.769 \text{ V} + 0.300 \text{ V}) = -1.069 \text{ V}$$

$$MnO_4^- \rightarrow Mn^{2+} \qquad E_{red}^o = +1.512 \text{ V} + 0.300 \text{ V} = +1.812 \text{ V}$$

(Note that the 0.300 V 'adjustment' is added to E_{red}^o, **before** calculating E_{ox}^o.) The cell potential is then:

$$E^o = E_{ox}^o + E_{red}^o = -1.069 \text{ V} + 1.812 \text{ V} = \underline{\textbf{+0.743 V}}$$

This is the same as the value calculated in Problem 34(c). Since E^o measures the **difference** between E_{ox}^o and E_{red}^o, this will always be true. Assigning a value of <u>zero</u> to E_{red}^o for the $H^+ \rightarrow H_2$ reduction is arbitrary: any number could be used because it will always cancel out when E^o is calculated.

17-3 STANDARD VOLTAGES: Spontaneity and E^o

38. This problem is similar to Example 17.6. For each reaction: (i) assign oxidation numbers to each element, to decide which elements were oxidized/reduced; (ii) write the oxidation and reduction half-reactions with their E_{ox}^o and E_{red}^o values using Table 17.1; (iii) calculate E^o using $E^o = E_{ox}^o + E_{red}^o$. The reaction is spontaneous if $E^o > 0$.

(a) Ba: $0 \rightarrow +2$ (oxidation)   Cr: $+3 \rightarrow +2$ (reduction)

half reactions: $Ba(s) \rightarrow Ba^{2+}(aq) + 2e^-$ $E_{ox}^o = -E_{red}^o = +2.906 \text{ V}$

$\qquad\qquad\qquad\quad Cr^{3+}(aq) + e^- \rightarrow Cr^{2+}(aq)$ $E_{red}^o = -0.408 \text{ V}$

$$E^o = E_{ox}^o + E_{red}^o = +2.906 \text{ V} + (-0.408 \text{ V}) = +2.498 \text{ V}$$

$E^o > 0$ so the reaction is **spontaneous**.

(b) Cu: $0 \rightarrow +2$ (oxidation)   Co: $+2 \rightarrow 0$ (reduction)

half reactions: $Cu(s) \rightarrow Cu^{2+}(aq) + 2e^-$ $E_{ox}^o = -E_{red}^o = -0.339 \text{ V}$

$\qquad\qquad\qquad\quad Co^{2+}(aq) + 2e^- \rightarrow Co(s)$ $E_{red}^o = -0.282 \text{ V}$

$$E^o = E_{ox}^o + E_{red}^o = -0.339 \text{ V} + (-0.282 \text{ V}) = -0.621 \text{ V}$$

$E^\circ < 0$ so the reaction is **non-spontaneous**.

(c) O: $-2 \to 0$ (oxidation)   S: $0 \to -2$ (reduction)

half reactions:4OH$^-$(aq) $\to$

$$O_2(g) + 2H_2O + 4e^-   \qquad E^\circ_{ox} = -E^\circ_{red} = -0.401 \text{ V}$$

$$S(s) + 2e^- \to S^{2-}(aq) \qquad E^\circ_{red} = -0.445 \text{ V}$$

$$E^\circ = E^\circ_{ox} + E^\circ_{red} = -0.401 \text{ V} + (-0.445 \text{ V}) = -0.846 \text{ V}$$

$E^\circ < 0$ so the reaction is **non-spontaneous**.

40. The three given equations must be combined in such a way that $E^\circ = (E^\circ_{ox} + E^\circ_{red}) > 0$, since this is the requirement for a spontaneous reaction. From Table 17.1, the E°_{red} values for the three given reactions are:

1. $I_2(s) + 2e^- \to 2I^-(aq)$ $\qquad\qquad\qquad\qquad E^\circ_{red} = 0.534 \text{ V}$

2. $Co^{3+}(aq) + e^- \to Co^{2+}(aq)$ $\qquad\qquad\qquad E^\circ_{red} = 1.953 \text{ V}$

3. $Cr_2O_7^{2-}(aq) + 14H^+(aq) + 6e^- \to 2Cr^{3+}(aq) + 7H_2O$ $\qquad E^\circ_{red} = 1.33 \text{ V}$

Recall that $E^\circ_{ox} = -E^\circ_{red}$, so each time we calculate E° we are, in effect, calculating the difference between two E°_{red} values. Therefore, since all three E°_{red} values shown above are positive numbers, we must simply ensure that for each pair (oxidation and reduction) the larger E°_{red} number refers to the reduction half-equation and the smaller E°_{red} value refers to the oxidation half-equation. When combining the two half-equations, remember that electrons must cancel out.

($2 \times $ **2**) $-$ **1**: $2Co^{3+}(aq) + 2e^- + 2I^-(aq) \to 2Co^{2+}(aq) + I_2(s) + 2e^-$
simplified: $2Co^{3+}(aq) + 2I^-(aq) \to 2Co^{2+}(aq) + I_2(s)$

the oxidation half reaction is:
$2I^-(aq) \to I_2(s) + 2e^-$ $\qquad\qquad\qquad E^\circ_{ox} = -0.534 \text{ V}$

the reduction half reaction is:
$Co^{3+}(aq) + e^- \to Co^{2+}(aq)$ $\qquad\qquad\qquad E^\circ_{red} = +1.953 \text{ V}$

$$E^\circ = E^\circ_{ox} + E^\circ_{red} = -0.534 \text{ V} + 1.953 \text{ V} = \underline{+1.419 \text{ V}}$$

($6 \times $ **2**) $-$ **3**: $6Co^{3+}(aq) + 6e^- + 2Cr^{3+}(aq) + 7H_2O$
$\to 6Co^{2+}(aq) + Cr_2O_7^{2-}(aq) + 14H^+(aq) + 6e^-$

simplified: **6Co³⁺(aq) + 2Cr³⁺(aq) + 7H₂O → 6Co²⁺(aq) + Cr₂O₇²⁻(aq) + 14H⁺(aq)**

the oxidation half reaction is:

$$2Cr^{3+}(aq) + 7H_2O \rightarrow Cr_2O_7^{2-}(aq) + 14H^+(aq) + 6e^- \qquad E^0_{ox} = -1.33 \text{ V}$$

the reduction half reaction is:

$$Co^{3+}(aq) + e^- \rightarrow Co^{2+}(aq) \qquad\qquad\qquad E^0_{red} = +1.953 \text{ V}$$

$$E^0 = E^0_{ox} + E^0_{red} = -1.33 \text{ V} + 1.953 \text{ V} = \underline{\textbf{+0.62 V}}$$

3 − (3 × 1): $Cr_2O_7^{2-}(aq) + 14H^+(aq) + 6e^- + 6I^-(aq) \rightarrow 2Cr^{3+}(aq) + 7H_2O + 3I_2(s) + 6e^-$
simplified: **Cr₂O₇²⁻(aq) + 14H⁺(aq) + 6I⁻(aq) → 2Cr³⁺(aq) + 7H₂O + 3I₂(s)**

the oxidation half reaction is:
$$2I^-(aq) \rightarrow I_2(s) + 2e^- \qquad\qquad\qquad\qquad E^0_{ox} = -0.534 \text{ V}$$

the reduction half reaction is:

$$Cr_2O_7^{2-}(aq) + 14H^+(aq) + 6e^- \rightarrow 2Cr^{3+}(aq) + 7H_2O \quad E^0_{red} = +1.33 \text{ V}$$

$$E^0 = E^0_{ox} + E^0_{red} = -0.534 \text{ V} + 1.33 \text{ V} = \underline{\textbf{+0.80 V}}$$

42. This problem is similar to Example 17.6. For each possible reaction: (i) write the oxidation and reduction half-reactions with their E^0_{ox} and E^0_{red} values using Table 17.1; (ii) calculate E^0 using $E^0 = E^0_{ox} + E^0_{red}$. The reaction will occur (will be spontaneous) if $E^0 > 0$.

(a) possible half reactions:

$$\underline{\text{oxidations:}} \quad Cu(s) \rightarrow Cu^{2+}(aq) + 2e^- \qquad E^0_{ox} = -E^0_{red} = -0.339 \text{ V}$$

$$Au(s) \rightarrow Au^{3+}(aq) + 3e^- \qquad E^0_{ox} = -E^0_{red} = -1.498 \text{ V}$$

$$\underline{\text{reduction:}} \quad NO_3^-(aq) + 4H^+(aq) + 3e^-$$
$$\rightarrow NO(g) + 2H_2O \qquad E^0_{red} = +0.964 \text{ V}$$

for the oxidation of copper, $E^0 > 0$ (spontaneous):

$$E^0 = E^0_{ox} + E^0_{red} = -0.339 \text{ V} + 0.964 \text{ V} = +0.625 \text{ V}$$

for the oxidation of gold, $E^0 < 0$ (non-spontaneous):

$$E^0 = E^0_{ox} + E^0_{red} = -1.498 \text{ V} + 0.964 \text{ V} = -0.534 \text{ V}$$

Therefore, **acidic nitrate will oxidize copper to copper(II) ions, but will not oxidize gold to gold(III) ions.**

(b) possible half reactions:

$$\underline{\text{oxidations:}} \quad Cu(s) \rightarrow Cu^+(aq) + e^- \qquad E^0_{ox} = -E^0_{red} = -0.518 \text{ V}$$

$$Cu(s) \rightarrow Cu^{2+}(aq) + 2e^- \qquad E^o_{ox} = -E^o_{red} = -0.339 \text{ V}$$

reduction: $\qquad Fe^{2+}(aq) + 2e^- \rightarrow Fe(s) \qquad E^o_{red} = -0.409 \text{ V}$

for the formation of copper(I) ions, $E^o < 0$ (non-spontaneous):

$$E^o = E^o_{ox} + E^o_{red} = -0.518 \text{ V} + (-0.409 \text{ V}) = -0.927 \text{ V}$$

for the formation of copper(II) ions, $E^o < 0$ (non-spontaneous):

$$E^o = E^o_{ox} + E^o_{red} = -0.339 \text{ V} + (-0.409 \text{ V}) = -0.748 \text{ V}$$

Therefore, **iron(II) ions cannot oxidize copper to copper(I) or copper(II) ions: no reaction will occur.**

(c) possible half reactions:

oxidations: $\quad ClO_3^-(aq) + H_2O$

$\qquad\qquad \rightarrow ClO_4^-(aq) + 2H^+(aq) + 2e^- \qquad E^o_{ox} = -E^o_{red} = -1.19 \text{ V}$

$\qquad\qquad ClO_3^-(aq) + 2OH^-(aq)$

$\qquad\qquad \rightarrow ClO_4^-(aq) + H_2O + 2e^- \qquad E^o_{ox} = -E^o_{red} = -0.398 \text{ V}$

reductions: $\quad O_2(g) + 4H^+(aq) + 4e^- \rightarrow 2H_2O \qquad E^o_{red} = +1.229 \text{ V}$

$\qquad\qquad O_2(g) + 2H_2O + 4e^- \rightarrow 4OH^-(aq) \quad E^o_{red} = +0.401 \text{ V}$

(Note that the $ClO_4^- \rightarrow ClO_3^-$ reaction in basic solution (Table 17.1) can be converted to the corresponding reaction in acidic solution simply by adding $2H^+$ to each side of the reaction and cancelling duplications.)

for the oxidation of O_2 by ClO_3^- in acidic solution, $E^o > 0$ (spontaneous):

$$E^o = E^o_{ox} + E^o_{red} = -1.19 \text{ V} + 1.229 = +0.04 \text{ V}$$

for the oxidation of O_2 by ClO_3^- in basic solution, $E^o > 0$ (spontaneous):

$$E^o = E^o_{ox} + E^o_{red} = -0.398 \text{ V} + 0.401 = +0.003 \text{ V}$$

Therefore, **ClO_3^- ions can reduce O_2 in both acidic and basic solutions.**

44. This problem is similar to Example 17.6. Consider the species present and their possible half-reactions. (Find these species in Table 17.1 to see what reactions are possible.) Then, for each possible reaction: (i) write the oxidation and reduction half-reactions with their E^o_{ox} and E^o_{red} values using Table 17.1; (ii) calculate E^o using $E^o = E^o_{ox} + E^o_{red}$. A reaction will occur (will be spontaneous) under standard conditions if $E^o > 0$.

(a) possible half reactions:

oxidation: $\quad 2Hg(l) \rightarrow Hg_2^{2+}(aq) + 2e^- \qquad E^o_{ox} = -E^o_{red} = -0.796 \text{ V}$

reductions: $\quad S(s) + 2H^+(aq) + 2e^-$

$\qquad\qquad \rightarrow H_2S(aq) \qquad E^o_{red} = +0.144 \text{ V}$

$$S(s) + 2e^- \rightarrow S^{2-}(aq) \qquad E^o_{red} = -0.445 \text{ V}$$

for the reaction in acidic solution, $E^o < 0$ (non-spontaneous):

$$E^o = E^o_{ox} + E^o_{red} = -0.796 \text{ V} + 0.144 \text{ V} = -0.652 \text{ V}$$

for the reaction in basic solution, $E^o < 0$ (non-spontaneous):

$$E^o = E^o_{ox} + E^o_{red} = -0.796 \text{ V} + (-0.445 \text{ V}) = -1.241 \text{ V}$$

Therefore, **no reaction occurs between mercury and sulfur.**

(b) possible half reactions:

<u>oxidation:</u> $\quad 2Hg(l) \rightarrow Hg_2^{2+}(aq) + 2e^- \qquad E^o_{ox} = -E^o_{red} = -0.796 \text{ V}$

<u>reduction:</u> $\quad MnO_2(s) + 4H^+(aq) + 2e^-$
$$\rightarrow Mn^{2+}(aq) + 2H_2O \qquad E^o_{red} = +1.229 \text{ V}$$

for these two half-reactions, $E^o > 0$ (spontaneous):

$$E^o = E^o_{ox} + E^o_{red} = -0.796 \text{ V} + 1.229 \text{ V} = +0.433 \text{ V}$$

Therefore, **a reaction WILL occur between mercury and MnO$_2$ in acidic solution.** To obtain the overall reaction, simply add together the two half-reactions; electrons will cancel out:

$2Hg(l) + MnO_2(s) + 4H^+(aq) \rightarrow Hg_2^{2+}(aq) + Mn^{2+}(aq) + 2H_2O \qquad E^o = +0.433 \text{ V}$

(c) possible half reactions:

<u>oxidation:</u> $\quad Al(s) \rightarrow Al^{3+}(aq) + 3e^- \qquad E^o_{ox} = -E^o_{red} = +1.68 \text{ V}$

<u>reductions:</u> $\quad K^+(aq) + e^- \rightarrow K(s) \qquad E^o_{red} = -2.936 \text{ V}$

for these two half-reactions, $E^o < 0$ (non-spontaneous):

$$E^o = E^o_{ox} + E^o_{red} = +1.68 \text{ V} + (-2.936 \text{ V}) = -1.26 \text{ V}$$

Therefore, **no reaction occurs between aluminum metal and potassium ions.**

46. This problem is related to Example 17.6. The question implies that the 1 *M* solution of HBr will act as an oxidizing agent and therefore the HBr itself will be reduced. This HBr(*aq*) solution contains H$^+$ and Br$^-$ ions, and since Br$^-$ is already in a reduced state H$^+$(*aq*) is the oxidizing agent that will itself be reduced.

This reduction half-reaction is: $\qquad 2H^+(aq) + 2e^- \rightarrow H_2(g) \qquad E^o_{red} = 0.000 \text{ V}$

In deciding which species can be oxidized by the HBr: (i) write the oxidation half-reaction along with its E^o_{ox} value using Table 17.1; then (ii) calculate E^o using $E^o = E^o_{ox} + E^o_{red}$. A reaction will occur (will be spontaneous) if $E^o > 0$.

(a) oxidation half-reaction: $Na(s) \rightarrow Na^+(aq) + e^-$ $E^o_{ox} = -E^o_{red} = +2.714$ V

   for the reaction between HBr and Na, $E^o > 0$ (spontaneous):

   $E^o = E^o_{ox} + E^o_{red} = +2.714$ V $+ 0.000$ V $= +2.714$ V

   Therefore, **Na will be oxidized by 1 *M* HBr.**

(b) oxidation half-reaction: $2Hg(l) \rightarrow Hg_2^{2+}(aq) + 2e^-$ $E^o_{ox} = -E^o_{red} = -0.796$ V

   for the reaction between HBr and Hg, $E^o < 0$ (non-spontaneous):

   $E^o = E^o_{ox} + E^o_{red} = -0.796$ V $+ 0.000$ V $= -0.796$ V

   Therefore, **Hg will not be oxidized by 1 *M* HBr.**

(c) oxidation half-reaction: $Pb(s) \rightarrow Pb^{2+}(aq) + 2e^-$ $E^o_{ox} = -E^o_{red} = +0.127$ V

   for the reaction between HBr and Pb, $E^o > 0$ (spontaneous):

   $E^o = E^o_{ox} + E^o_{red} = +0.127$ V $+ 0.000$ V $= +0.127$ V

   Therefore, **Pb will be oxidized by 1 *M* HBr.**

(d) oxidation half-reactions:
   $Mn^{2+}(aq) + 4H_2O \rightarrow MnO_4^-(aq) + 8H^+(aq) + 5e^-$ $E^o_{ox} = -E^o_{red} = -1.512$ V
   $Mn^{2+}(aq) + 2H_2O \rightarrow MnO_2(s) + 4H^+(aq) + 2e^-$ $E^o_{ox} = -E^o_{red} = -1.229$ V

   for the reaction between HBr and Mn^{2+} giving MnO_4^-, $E^o < 0$ (non-spontaneous):

   $E^o = E^o_{ox} + E^o_{red} = -1.512$ V $+ 0.000$ V $= -1.512$ V

   for the reaction between HBr and Mn^{2+} giving MnO_2, $E^o < 0$ (non-spontaneous):

   $E^o = E^o_{ox} + E^o_{red} = -1.229$ V $+ 0.000$ V $= -1.229$ V

   Therefore, **Mn^{2+} will not be oxidized by 1 *M* HBr.**

48. This problem is similar to Example 17.6; see also Problem 46 above. From Table 17.1 we can see that in acidic aqueous solution, sulfur will undergo the following reduction reaction:

   $S(s) + 2H^+(aq) + 2e^- \rightarrow H_2S(aq)$ $E^o_{red} = +0.144$ V

The sulfur, therefore, will act as an oxidizing agent. In deciding which species can be oxidized by the sulfur, note that the three given compounds (a)–(c) are ionic and that both cation and anion might be oxidized. For each cation/anion combination (i) write any possible oxidation half-reaction along with its E^o_{ox} value using Table 17.1; then (ii) calculate E^o using $E^o = E^o_{ox} + E^o_{red}$. A reaction will occur (will be spontaneous) if $E^o > 0$.

(a) species present: $Mg^{2+}(aq)$, $Br^-(aq)$

   possible oxidation half-reactions:

   $2Br^-(aq) \rightarrow Br_2(aq) + 2e^-$ $\qquad\qquad$ $E^o_{ox} = -E^o_{red} = -1.077$ V

   (Mg^{2+} is already in an oxidized state and will not oxidize further; it can only be reduced.)

   for the reaction between S and Br^-, $E^o < 0$ (non-spontaneous):

   $E^o = E^o_{ox} + E^o_{red} = -1.077$ V $+ 0.144$ V $= -0.933$ V

   Therefore, **no reaction occurs between MgBr₂ and sulfur.**

(b) species present: $Sn^{2+}(aq)$, $NO_3^-(aq)$

   possible oxidation half-reactions:

   $Sn^{2+}(aq) \rightarrow Sn^{4+}(aq) + 2e^-$ $\qquad\qquad$ $E^o_{ox} = -E^o_{red} = -0.154$ V

   (NO_3^- is already in an oxidized state and will not oxidize further; it can only be reduced.)

   for the reaction between S and Sn^{2+}, $E^o < 0$ (non-spontaneous):

   $E^o = E^o_{ox} + E^o_{red} = -0.154$ V $+ 0.144$ V $= -0.010$ V

   Therefore, **no reaction occurs between Sn(NO₃)₂ and sulfur.**

(c) species present: $Cr^{2+}(aq)$, $ClO_3^-(aq)$

   possible oxidation half-reactions:

   $Cr^{2+}(aq) \rightarrow Cr^{3+}(aq) + e^-$ $\qquad\qquad$ $E^o_{ox} = -E^o_{red} = +0.408$ V

   $ClO_3^-(aq) + H_2O$
   $\qquad\qquad \rightarrow ClO_4^-(aq) + 2H^+(aq) + 2e^-$ $\qquad$ $E^o_{ox} = -E^o_{red} = -1.19$ V

   (The latter reaction comes from Problem 32(c).)

   for the reaction between S and Cr^{2+}, $E^o > 0$ (spontaneous):

   $E^o = E^o_{ox} + E^o_{red} = 0.408$ V $+ 0.144$ V $= +0.552$ V

   for the reaction between S and ClO_3^-, $E^o < 0$ (non-spontaneous):

   $E^o = E^o_{ox} + E^o_{red} = -1.19$ V $+ 0.144$ V $= -1.05$ V

Therefore, **a reaction will occur between Cr(ClO$_3$)$_2$ and sulfur,** in which the Cr^{2+} ion is oxidized to Cr^{3+}. To obtain the balanced equation, multiply the coefficients of the oxidation half-reaction by two and add it to the reduction half-reaction; electrons will cancel out. The reaction occurring is:

$$2Cr^{2+}(aq) + S(s) + 2H^+(aq) \rightarrow 2Cr^{3+}(aq) + H_2S(aq) \qquad E° = +0.552 \text{ V}$$

17-4 RELATIONS BETWEEN *E°*, Δ*G°*, AND *K*

50. This problem is related to Example 17.7. The following three equations provide relationships between *E°*, Δ*G°*, and *K*:

Equation 17.2: $\quad \Delta G° = -nFE°$

Equation 17.3: $\quad E° = \dfrac{(0.0257 \text{ V})}{n} \ln K$

Equation 16.5: $\quad \Delta G° = -RT \ln K$

Thus, if one quantity is known the other two can be calculated. Note that *T* = 298 K and *n* = 4 mol.

(a) Given: *K* = 1.6 × 10^{-3}; use Equation 16.5 to calculate Δ*G°* and Equation 17.3 to calculate *E°*.

$$\Delta G° = -RT \ln K = -(8.3145 \text{ J/mol·K})(298 \text{ K}) \times \ln(1.6 \times 10^{-3})$$

$$= 16 \times 10^3 \text{ J or } \textbf{16 kJ}$$

$$E° = \dfrac{(0.0257 \text{ V})}{n} \ln K = \dfrac{(0.0257 \text{ V})}{4} \ln(1.6 \times 10^{-3}) = \textbf{−0.041 V}$$

(b) Given: *E°* = 0.117 V; use Equation 17.2 to calculate Δ*G°* and Equation 17.3 to calculate *K*.

$$\Delta G° = -nFE° = -(4 \text{ mol})(9.648 \times 10^4 \frac{\text{J}}{\text{mol·V}})(0.117 \text{ V})$$

$$= -45.2 \times 10^3 \text{ J or } \textbf{−45.2 kJ}$$

$$E° = \dfrac{(0.0257 \text{ V})}{n} \ln K \qquad \text{so} \qquad +0.117 \text{ V} = \dfrac{(0.0257 \text{ V})}{4} \ln K$$

$$\ln K = \frac{4(+0.117 \text{ V})}{(0.0257 \text{ V})} = 18.2 \qquad \text{so} \qquad K = e^{18.2} = \textbf{8.0} \times \textbf{10}^7$$

(c) Given: $\Delta G^\circ = -5.8$ kJ $= -5.8 \times 10^3$ J; use Equation 17.2 to calculate E° and Equation 16.5 to calculate K.

$$\Delta G^\circ = -nFE^\circ \qquad \text{so} \qquad E^\circ = -\frac{\Delta G^\circ}{nF} = -\frac{-5.8 \times 10^3 \text{ J}}{(4 \text{ mol})(9.648 \times 10^4 \frac{\text{J}}{\text{mol} \cdot \text{V}})} = \textbf{0.015 V}$$

$$\Delta G^\circ = -RT \ln K \qquad \text{so} \qquad -5.8 \times 10^3 \text{ J} = -(8.3145 \text{ J/mol·K})(298 \text{ K})\ln K$$

$$\ln K = 2.3 \qquad \text{so} \qquad K = e^{2.3} = \underline{\textbf{10}}$$

52. This problem is related to Example 17.7(a). Use Equation 17.2 with the given E° and the appropriate values of n.

(a) $\Delta G^\circ = -nFE^\circ = -(1 \text{ mol})(9.648 \times 10^4 \frac{\text{J}}{\text{mol} \cdot \text{V}})(1.08 \text{ V})$

$$= -104 \times 10^3 \text{ J} \quad \text{or} \quad \underline{\textbf{-104 kJ}}$$

(b) $\Delta G^\circ = -nFE^\circ = -(2 \text{ mol})(9.648 \times 10^4 \frac{\text{J}}{\text{mol} \cdot \text{V}})(1.08 \text{ V})$

$$= -208 \times 10^3 \text{ J} \quad \text{or} \quad \underline{\textbf{-208 kJ}}$$

(c) $\Delta G^\circ = -nFE^\circ = -(3 \text{ mol})(9.648 \times 10^4 \frac{\text{J}}{\text{mol} \cdot \text{V}})(1.08 \text{ V})$

$$= -313 \times 10^3 \text{ J} \quad \text{or} \quad \underline{\textbf{-313 kJ}}$$

The number of electrons exchanged, n, does not affect the spontaneity of the reaction since a change in n does not change the sign of ΔG°. Note also that the position of equilibrium also does not change.

54. This problem is similar to Example 17.7. Assign oxidation numbers to Mn and Cl and split the redox reaction into two half-reactions along with the corresponding E°_{ox} and E°_{red} values (from Table 17.1). Calculate E° using $E^\circ = E^\circ_{ox} + E^\circ_{red}$ and determine the value of n from the number of electrons cancelled out when balancing the overall redox reaction. Finally, use Equations 17.2 and 17.3 to obtain ΔG° and K. (Note that the presence of H^+ in the redox equation indicates that the reaction takes place in acidic solution.)

Oxidation numbers:　　Cl:　$0 \to +5$ (oxidation)
　　　　　　　　　　　Mn:　$+7 \to +2$ (reduction)

Half reactions (from Table 17.1):
$$\tfrac{1}{2}Cl_2(g) + 3H_2O \;\to\; ClO_3^-(aq) + 6H^+(aq) + 5e^- \qquad E^o_{ox} = -E^o_{red} = -1.458 \text{ V}$$
$$MnO_4^-(aq) + 8H^+(aq) + 5e^- \;\to\; Mn^{2+}(aq) + 4H_2O \qquad E^o_{red} = +1.512 \text{ V}$$

$$E^o = E^o_{ox} + E^o_{red} = -1.458 \text{ V} + +1.512 \text{ V} = \underline{\mathbf{+0.054 \text{ V}}}$$

The balanced equation is obtained by multiplying the coefficients of both half-reactions by two and then adding them together:

$$2[\; \tfrac{1}{2}Cl_2(g) + 3H_2O \;\to\; ClO_3^-(aq) + 6H^+(aq) + 5e^- \;]$$
$$2[\; MnO_4^-(aq) + 8H^+(aq) + 5e^- \;\to\; Mn^{2+}(aq) + 4H_2O \;]$$

$$Cl_2(g) + 6H_2O + 2MnO_4^-(aq) + 16H^+(aq) + 10e^-$$
$$\to\; 2ClO_3^-(aq) + 12H^+(aq) + 10e^- + 2Mn^{2+}(aq) + 8H_2O$$

Cancelling duplications then yields the equation given in the problem. Note that 10 electrons are transferred in the equation above; therefore $n = 10$. Thus:

$$\Delta G^o = -nFE^o \;\; = -(10 \text{ mol})(9.648 \times 10^4 \, \frac{J}{mol \cdot V})(+0.054 \text{ V})$$

$$= -52 \times 10^3 \text{ J} \quad \text{or} \quad \underline{\mathbf{-52 \text{ kJ}}}$$

$$\Delta G^o = -RT \ln K \qquad \text{so} \qquad K = e^{-\Delta G^o/RT}$$

$$K = e^{-(-52 \times 10^3 \text{ J})/(8.3145 \text{ J/mol·K})(298 \text{ K})} = \underline{\mathbf{1.3 \times 10^9}}$$

56. This problem is similar to Example 17.7(a). Use Equation 17.2 along with the value of E^o calculated in Problem 30, and the appropriate value of n. Determine the value of n from the number of electrons cancelled out when balancing the overall redox reaction.

(a) From Problem 30, $E^o = +2.305$ V and the half-reactions are:

$$Na(s) \to Na^+(aq) + e^- \qquad \text{(oxidation)}$$
$$Fe^{2+}(aq) + 2e^- \to Fe(s) \qquad \text{(reduction)}$$

The overall balanced equation is obtained by doubling the coefficients of the oxidation half-reaction and then adding the reduction half-reaction:
$$2Na(s) + Fe^{2+}(aq) + 2e^- \to 2Na^+(aq) + 2e^- + Fe(s)$$

$$2Na(s) + Fe^{2+}(aq) \to 2Na^+(aq) + Fe(s)$$

2 electrons were cancelled out, so $n = 2$.

$$\Delta G^\circ = -nFE^\circ = -(2 \text{ mol})\left(9.648 \times 10^4 \frac{\text{J}}{\text{mol} \cdot \text{V}}\right)(+2.305 \text{ V})$$

$$= -444.8 \times 10^3 \text{ J} \quad \text{or} \quad \underline{\mathbf{-444.8 \text{ kJ}}}$$

(b) From Problem 30, $E^\circ = +1.84$ V and the half-reactions are:

$$\text{Al}(s) \rightarrow \text{Al}^{3+}(aq) + 3e^- \qquad\qquad \text{(oxidation)}$$

$$\text{SO}_4^{2-}(aq) + 4\text{H}^+(aq) + 2e^- \rightarrow \text{SO}_2(g) + 2\text{H}_2\text{O} \qquad \text{(reduction)}$$

The overall balanced equation is obtained by doubling the coefficients of the oxidation half-reaction and tripling the coefficients of the reduction half-reaction, and then adding the two together:

$$3\text{SO}_4^{2-}(aq) + 12\text{H}^+(aq) + 2\text{Al}(s) + 6e^- \rightarrow 3\text{SO}_2(g) + 2\text{Al}^{3+}(aq) + 6e^- + 6\text{H}_2\text{O}$$

$$3\text{SO}_4^{2-}(aq) + 12\text{H}^+(aq) + 2\text{Al}(s) \rightarrow 3\text{SO}_2(g) + 2\text{Al}^{3+}(aq) + 6\text{H}_2\text{O}$$

6 electrons were cancelled out, so $n = 6$.

$$\Delta G^\circ = -nFE^\circ = -(6 \text{ mol})\left(9.648 \times 10^4 \frac{\text{J}}{\text{mol} \cdot \text{V}}\right)(+1.84 \text{ V})$$

$$= -1070 \times 10^3 \text{ J} \quad \text{or} \quad \underline{\mathbf{-1.07 \times 10^3 \text{ kJ}}}$$

(c) From Problem 30, $E^\circ = -0.846$ V and the half-reactions are:

$$4\text{OH}^-(aq) \rightarrow \text{O}_2(g) + 2\text{H}_2\text{O} + 4e^- \qquad\qquad \text{(oxidation)}$$

$$\text{S}(s) + 2e^- \rightarrow \text{S}^{2-}(aq) \qquad\qquad\qquad\qquad \text{(reduction)}$$

The overall balanced equation is obtained by doubling the coefficients of the reduction half-reaction and then adding the oxidation half-reaction:

$$2\text{S}(s) + 4e^- + 4\text{OH}^-(aq) \rightarrow \text{O}_2(g) + 2\text{H}_2\text{O} + 4e^- + 2\text{S}^{2-}(aq)$$

$$2\text{S}(s) + 4\text{OH}^-(aq) \rightarrow \text{O}_2(g) + 2\text{H}_2\text{O} + 2\text{S}^{2-}(aq)$$

4 electrons then cancel out, so $n = 4$.

$$\Delta G^\circ = -nFE^\circ = -(4 \text{ mol})\left(9.648 \times 10^4 \frac{\text{J}}{\text{mol} \cdot \text{V}}\right)(-0.846 \text{ V})$$

$$= +326 \times 10^3 \text{ J} \quad \text{or} \quad \underline{\mathbf{+326 \text{ kJ}}}$$

58. This problem is similar to Example 17.7(b) and Problem 56 above. Use Equation 17.3, along with the value of E° calculated in Problem 32 and the appropriate value of n. Determine the value of n from the number of electrons cancelled out when balancing the overall redox reaction.

(a) From Problem 32, E° = +0.562 V and the half-reactions are:

$$Cr^{2+}(aq) \rightarrow Cr^{3+}(aq) + e^- \qquad \text{(oxidation)}$$
$$Sn^{4+}(aq) + 2e^- \rightarrow Sn^{2+}(aq) \qquad \text{(reduction)}$$

The overall balanced equation is obtained by doubling the coefficients of the oxidation half-reaction and then adding the reduction half-reaction:

$$2Cr^{2+}(aq) + Sn^{4+}(aq) + \mathbf{2e^-} \rightarrow 2Cr^{3+}(aq) + \mathbf{2e^-} + Sn^{2+}(aq)$$

$$2Cr^{2+}(aq) + Sn^{4+}(aq) \rightarrow 2Cr^{3+}(aq) + Sn^{2+}(aq)$$

2 electrons were cancelled out, so $n = 2$.

$$E^\circ = \frac{(0.0257 \text{ V})}{n} \ln K \qquad \text{so} \qquad +0.562 \text{ V} = \frac{(0.0257 \text{ V})}{2} \ln K$$

$$\ln K = 43.7 \qquad \text{so} \qquad K = e^{43.7} = \underline{1 \times 10^{19}}$$

(b) From Problem 32, E° = +0.534 V and the half-reactions are:

$$Mn^{2+}(aq) + 2H_2O \rightarrow MnO_2(s) + 4H^+(aq) + 2e^- \qquad \text{(oxidation)}$$
$$H_2O_2(aq) + 2H^+(aq) + 2e^- \rightarrow 2H_2O \qquad \text{(reduction)}$$

The overall balanced equation is obtained by adding together the oxidation and reduction half-reactions, then simplifying:

$$Mn^{2+}(aq) + 2H_2O + H_2O_2(aq) + 2H^+(aq) + \mathbf{2e^-} \rightarrow MnO_2(s) + 4H^+(aq) + \mathbf{2e^-} + 2H_2O$$

$$Mn^{2+}(aq) + H_2O_2(aq) \rightarrow MnO_2(s) + 2H^+(aq)$$

2 electrons were cancelled out, so $n = 2$.

$$E^\circ = \frac{(0.0257 \text{ V})}{n} \ln K \qquad \text{so} \qquad +0.534 \text{ V} = \frac{(0.0257 \text{ V})}{2} \ln K$$

$$\ln K = 41.6 \qquad \text{so} \qquad K = e^{41.6} = \underline{1 \times 10^{18}}$$

60. This problem is related to Example 17.7(b); see also Section 15-1 and Example 15-1. Recall that the formation constant K_f is the concentration equilibrium constant for formation of the complex ion (in this case $AuCl_4^-$) from the Lewis acidic metal cation (here Au^{3+}) and Lewis bases (here Cl^-). Therefore we require K for the following reaction:

$$Au^{3+}(aq) + 4Cl^-(aq) \rightleftharpoons AuCl_4^-(aq)$$

This, of course, is not a redox reaction, but consulting Table 17.1, it may be seen that the following reaction involves the same complex ion:

$$AuCl_4^-(aq) + 3e^- \rightarrow Au(s) + 4Cl^-(aq) \qquad E_{red}^o = +1.001\ V \qquad ......(1)$$

To form the complex ion, $Au^{3+}(aq)$ is required rather than $Au(s)$, so the following reaction may also be useful:

$$Au^{3+}(aq) + 3e^- \rightarrow Au(s) \qquad E_{red}^o = +1.498\ V \qquad ......(2)$$

To obtain the required reaction, reduction half-reaction **(1)** above must be reversed so that it becomes an oxidation:

$$Au(s) + 4Cl^-(aq) \rightarrow AuCl_4^-(aq) + 3e^- \qquad E_{ox}^o = -E_{red}^o = -1.001\ V \qquad ......(3)$$

then adding together reactions **(2)** and **(3)**, and cancelling duplications, gives the desired reaction:

$$\mathbf{Au(s) + 4Cl^-(aq) + Au^{3+}(aq) + 3e^- \rightarrow AuCl_4^-(aq) + 3e^- + Au(s)}$$

$$Au^{3+}(aq) + 4Cl^-(aq) \rightarrow AuCl_4^-(aq)$$

For this reaction, $E^o = E_{ox}^o + E_{red}^o = -1.001\ V + 1.498\ V = +0.497\ V$. Using this value of E^o and $n = 3$ (3 electrons were cancelled out) in Equation 17.3 allows calculation of the corresponding equilibrium constant for this reaction, which is K_f.

$$E^o = \frac{(0.0257\ V)}{n} \ln K \qquad \text{so} \qquad +0.497\ V = \frac{(0.0257\ V)}{3} \ln K$$

$$\ln K = 58.0 \qquad \text{so} \qquad K = K_f = e^{58.0} = \underline{\mathbf{2 \times 10^{25}}}$$

62. This problem is related to Example 17.7; see also Section 15-2. Recall that K_{sp} is the concentration equilibrium constant for dissolving of an ionic solid, such as $BaCO_3$. Thus the value of K_{sp} for $BaCO_3$ (given in the Problem) applies to the following reaction:

$$BaCO_3(s) \rightleftharpoons Ba^{2+}(aq) + CO_3^{2-}(aq) \qquad K_{sp} = 1.1 \times 10^{-10}$$

The above equilibrium reaction is the reverse of the one in the Problem: to obtain K for the given reaction, apply the reciprocal rule (see Section 12-2 and Table 12.3):

$$Ba^{2+}(aq) + CO_3^{2-}(aq) \rightleftharpoons BaCO_3(s) \qquad K = 1/K_{sp(BaCO_3)}$$

$$= 1/(1.1 \times 10^{-10}) = 9.1 \times 10^9$$

Since K for the given reaction is known, therefore, we can use Equation 17.3 to obtain E^o. The other variable remaining to be deduced is the value of n, which should be 2: Ba^{2+} ions are involved in this reaction and Ba only exists as Ba atoms or Ba^{2+} ions, as in the following reaction: $Ba^{2+}(aq) + 2e^- \rightarrow Ba(s)$.

$$E^o = \frac{(0.0257 \text{ V})}{n} \ln K = \frac{(0.0257 \text{ V})}{2} \ln(9.1 \times 10^9) = \underline{\textbf{+0.29 V}}$$

17-5 EFFECT OF CONCENTRATION ON VOLTAGE: Nernst Equation

64. This problem is similar to Example 17.8.

(a) This part is similar to Example 17.5(a).
First assign oxidation numbers to determine which element was oxidized/reduced, write the oxidation and reduction half-reactions with their E^o_{ox} and E^o_{red} values from Table 17.1; then calculate E^o using $E^o = E^o_{ox} + E^o_{red}$.

Mn: $+4 \rightarrow +2$ (reduction)   Br: $-1 \rightarrow 0$ (oxidation)

Half-reactions: $MnO_2(s) + 4H^+(aq) + 2e^-$
$\rightarrow Mn^{2+}(aq) + 2H_2O$ $E^o_{red} = +1.229$ V
$2Br^-(aq) \rightarrow Br_2(l) + 2e^-$ $E^o_{ox} = -E^o_{red} = -1.077$ V

$E^o = E^o_{ox} + E^o_{red} = -1.077$ V $+ 1.229$ V $= \underline{\textbf{+0.152 V}}$

(b) The Nernst Equation is:

$$E = E^o - \frac{(0.0257 \text{ V})}{n} \ln Q$$

for which values of E^o and n, and an expression for Q, are required. We can use E^o from part (a) above. To find n: the given redox reaction is obtained from the two half-reactions by adding together the reduction and oxidation half-equations:

$MnO_2(s) + 4H^+(aq) + 2e^- \rightarrow Mn^{2+}(aq) + 2H_2O$
$2Br^-(aq) \rightarrow Br_2(l) + 2e^-$
_____
$MnO_2(s) + 4H^+(aq) + 2e^- + 2Br^-(aq)$
$\rightarrow Mn^{2+}(aq) + 2H_2O + Br_2(l) + 2e^-$

Cancelling duplications then yields the equation given in the problem. Note that 2 electrons are transferred in the equation above; therefore $n = 2$.

The expression for the reaction quotient Q (see Section 12-4) is:

$$Q = \frac{[Mn^{2+}]}{[H^+]^4[Br^-]^2}$$

Thus the Nernst Equation for this reaction is:

$$E = 0.152 \text{ V} - \frac{(0.0257 \text{ V})}{2} \ln \frac{[Mn^{2+}]}{[H^+]^4[Br^-]^2}$$

(c) The expression for Q requires $[H^+]$, but in this case pH is given. To obtain $[H^+]$, use $[H^+]$ $= 10^{-pH} = 10^{-3.17} = 0.00068$ *M*. (See Section 13-3 and Equation 13.3.)

Under the given conditions, the Nernst Equation for this reaction becomes:

$$E = 0.152 \text{ V} - \frac{(0.0257 \text{ V})}{2} \ln \frac{(0.60)}{(0.00068)^4(0.83)^2} = \underline{-0.221 \text{ V}}$$

66. This problem is similar to Example 17.8.

(a) This part is similar to Example 17.5(a).
First assign oxidation numbers to determine which element was oxidized/reduced, write the oxidation and reduction half-reactions with their E^o_{ox} and E^o_{red} values from Table 17.1; and then calculate E^o using $E^o = E^o_{ox} + E^o_{red}$.

S: $-2 \to 0$ (oxidation)   N: $+5 \to +2$ (reduction)

half-reactions: $S^{2-}(aq) \to S(s) + 2e^-$ $E^o_{ox} = -E^o_{red} = +0.445 \text{ V}$

$NO_3^-(aq) + 2H_2O + 3e^-$
        $\to NO(g) + 4OH^-(aq)$ $E^o_{red} = -0.140 \text{ V}$

$E^o = E^o_{ox} + E^o_{red} = +0.445 + (-0.140 \text{ V}) = \underline{+0.305 \text{ V}}$

(b) To use the Nernst Equation values of E^o and *n*, and an expression for Q, are required. To find *n*: the given redox reaction is obtained from the two half-reactions by adding together 3 × (oxidation half-reaction) and 2 × (reduction half-reaction):

$3S^{2-}(aq) \to 3S(s) + 6e^-$
$\underline{2NO_3^-(aq) + 4H_2O + 6e^- \to 2NO(g) + 8OH^-(aq)}$

$2NO_3^-(aq) + 4H_2O + \mathbf{6e^-} + 3S^{2-}(aq)$
        $\to 2NO(g) + 8OH^-(aq) + 3S(s) + \mathbf{6e^-}$

Cancelling electrons then yields the equation given in the problem. Note that 6 electrons are transferred in the equation above; therefore *n* = 6.

The expression for the reaction quotient Q (see Section 12-4) is:

$$Q = \frac{(P_{NO})^2[OH^-]^8}{[NO_3^-]^2[S^{2-}]^3}$$

Thus the Nernst Equation for this reaction is:

$$E = 0.305 \text{ V} - \frac{(0.0257 \text{ V})}{6} \ln \frac{(P_{NO})^2[OH^-]^8}{[NO_3^-]^2[S^{2-}]^3}$$

(c) The expression for Q requires $[OH^-]$, but in this case pH is given. To obtain $[OH^-]$, recall that pH + pOH = 14.00; therefore pOH = 14.00 − pH = 14.00 − 13.7 = 0.3. Then use $[OH^-] = 10^{-pOH} = 10^{-0.3} = 0.5$ M. (See Section 13-3 and Equation 13.4.)

Under the given conditions, the Nernst Equation for this reaction becomes:

$$E = 0.305 \text{ V} - \frac{(0.0257 \text{ V})}{6} \ln \frac{(0.994)^2(0.5)^8}{(0.472)^2(0.154)^3} = \underline{+0.298 \text{ V}}$$

---

68. For both parts of this problem: first obtain the balanced redox reaction and calculate $E°$ as in Example 17.5(a) and Problem 34 above; then calculate E as in Example 17.8 and Problem 64 above.

(a) Recall that the oxidation reaction takes place on the left side of the cell notation and the reduction reaction is on the right side.

Half-reactions: $Zn(s) \rightarrow Zn^{2+}(aq) + 2e^-$ $E_{ox}^o = -E_{red}^o = +0.762$ V

$Cd^{2+}(aq) + 2e^- \rightarrow Cd(s)$ $E_{red}^o = -0.402$ V

$E° = E_{ox}^o + E_{red}^o = +0.762$ V + (−0.402 V)= +0.360 V

The balanced equation for the reaction is obtained by adding together the two half-reactions; 2 electrons cancel out:

$Zn(s) + Cd^{2+}(aq) \rightarrow Zn^{2+}(aq) + Cd(s)$

Two electrons are transferred in this reaction (two electrons were cancelled) so $n = 2$; and the expression for Q for this reaction (see Section 12-4) is:

$$Q = \frac{[Zn^{2+}]}{[Cd^{2+}]}$$

Thus the Nernst Equation for this reaction is:

$$E = 0.360 \text{ V} - \frac{(0.0257 \text{ V})}{2} \ln\frac{[Zn^{2+}]}{[Cd^{2+}]}$$

Under the given conditions, the Nernst Equation for this reaction becomes:

$$E = 0.360 \text{ V} - \frac{(0.0257 \text{ V})}{2} \ln\frac{0.50}{0.020} = \underline{+0.319 \text{ V}}$$

(b) Recall that the Pt at the cathode is inert and takes no place in the reaction.

half-reactions: $\quad Cu(s) \rightarrow Cu^{2+}(aq) + 2e^- \qquad E^o_{ox} = -E^o_{red} = -0.339 \text{ V}$

$$2H^+(aq) + 2e^- \rightarrow H_2(g) \; E^o_{red} = 0.000 \text{ V}$$

$$E^o = E^o_{ox} + E^o_{red} = -0.339 \text{ V} + 0.000 \text{ V} = -0.339 \text{ V}$$

The balanced equation for the reaction is obtained by adding together the two half-reactions; electrons cancel out:

$$Cu(s) + 2H^+(aq) \rightarrow Cu^{2+}(aq) + H_2(g)$$

Two electrons are transferred in this reaction (two electrons were cancelled) so $n = 2$; and the expression for Q for this reaction is:

$$Q = \frac{[Cu^{2+}](P_{H_2})}{[H^+]^2}$$

Thus the Nernst Equation for this reaction is:

$$E = -0.339 \text{ V} - \frac{(0.0257 \text{ V})}{2} \ln\frac{[Cu^{2+}](P_{H_2})}{[H^+]^2}$$

Under the given conditions, the Nernst Equation for this reaction becomes:

$$E = -0.339 \text{ V} - \frac{(0.0257 \text{ V})}{2} \ln\frac{(0.0010)(1.00)}{(0.010)^2} = \underline{-0.369 \text{ V}}$$

70. This problem is related to Example 17.8.

First, obtain the balanced equation as in Example 17.5(a): assign oxidation numbers to determine which element was oxidized/reduced, write the oxidation and reduction half-reactions with their E^o_{ox} and E^o_{red} values from Table 17.1; and then calculate E^o using

$$E^o = E^o_{ox} + E^o_{red}.$$

S: 0 → −2 (reduction)   Ag: 0 → +1 (oxidation)

half-reactions: Ag(s) + Br⁻(aq) → AgBr(s) + e⁻ E^o_{ox} = −E^o_{red} = −0.073 V

S(s) + 2H⁺(aq) + 2e⁻ → H₂S(aq) E^o_{red} = +0.144 V

E^o = E^o_{ox} + E^o_{red} = −0.073 V + 0.144 V = 0.071 V

Next, use the Nernst Equation, for which values of E^o and n, and an expression for Q, are required. To find n: the given redox reaction is obtained from the two half-reactions by adding together 2 × (oxidation half-reaction) and 1 × (reduction half-reaction):

2Ag(s) + 2Br⁻(aq) → 2AgBr(s) + 2e⁻
S(s) + 2H⁺(aq) + 2e⁻ → H₂S(aq)
_____

2Ag(s) + 2Br⁻(aq) + S(s) + 2H⁺(aq) + **2e⁻**
→ 2AgBr(s) + **2e⁻** + H₂S(aq)

Cancelling electrons then yields the equation given in the problem. Note that 2 electrons are transferred in the equation above; therefore $n = 2$.

The expression for the reaction quotient Q (see Section 12-4) is:

$$Q = \frac{[H_2S]}{[H^+]^2[Br^-]^2}$$

Thus the Nernst Equation for this reaction, using E^o determined above, is:

$$E = 0.071\ V - \frac{(0.0257\ V)}{2} \ln \frac{[H_2S]}{[H^+]^2[Br^-]^2}$$

We are told to find the pH such that E is zero when all other species are at standard conditions, which means the concentrations of the solutes H₂S and Br⁻ are both 1.00 M; the other species present are already in their standard states. Substitute these two concentrations into the Nernst Equation above, and solve for [H⁺]; then use [H⁺] to obtain the corresponding pH using Equation 13.3.

Doing this, the Nernst Equation for this reaction becomes:

$$0 = 0.071\ V - \frac{(0.0257\ V)}{2} \ln \frac{(1.00)}{[H^+]^2(1.00)^2}$$

$$-0.071\ V = -\frac{(0.0257\ V)}{2} \ln \frac{1.00}{[H^+]^2}$$

$$+0.071\ V = \frac{(0.0257\ V)}{2} \times \{\ln(1.00) - \ln[H^+]^2\} = \frac{(0.0257\ V)}{2} \times \{-2\ln[H^+]\}$$

$$\ln[H^+] = \frac{2(0.071\,\text{V})}{-2(0.0257\,\text{V})} = -2.76 \quad \text{so} \quad [H^+] = e^{-2.76} = 0.063\,M$$

$$pH = -\log_{10}[H^+] = -\log_{10}(0.063) = \underline{\textbf{1.20}}$$

72. First obtain the balanced redox reaction and calculate E° as in Example 17.5(a) and Problem 34 above; then obtain the Nernst Equation for the reaction as in Example 17.8 and Problem 64 above; finally solve for the required $[H^+]$.

Recall that the oxidation reaction takes place on the left side of the cell notation and the reduction reaction is on the right side, and that the Pt at the cathode is inert and takes no part in the reaction.

half-reactions:
$$Zn(s) \rightarrow Zn^{2+}(aq) + 2e^- \qquad E^\circ_{ox} = -E^\circ_{red} = +0.762\,\text{V}$$
$$2H^+(aq) + 2e^- \rightarrow H_2(g) \quad E^\circ_{red} = 0.000\,\text{V}$$

$$E^\circ = E^\circ_{ox} + E^\circ_{red} = +0.762\,\text{V} + 0.000\,\text{V} = +0.762\,\text{V}$$

We obtain the balanced equation for the reaction by adding together the two half-reactions:

$$Zn(s) + 2H^+(aq) \rightarrow Zn^{2+}(aq) + H_2(g)$$

Two electrons are transferred in this reaction (two electrons were cancelled) so $n = 2$; and the expression for Q for this reaction (see Section 12-4) is:

$$Q = \frac{[Zn^{2+}](P_{H_2})}{[H^+]^2}$$

Thus the Nernst Equation for this reaction is:

$$E = 0.762\,\text{V} - \frac{(0.0257\,\text{V})}{2} \ln \frac{[Zn^{2+}](P_{H_2})}{[H^+]^2}$$

Under the given conditions, with $E = +0.40\,\text{V}$, the Nernst Equation becomes:

$$+0.40\,\text{V} = 0.762\,\text{V} - \frac{(0.0257\,\text{V})}{2} \ln \frac{(1.00)(1.0)}{[H^+]^2}$$

$$+0.40\,\text{V} - 0.762\,\text{V} = -\frac{(0.0257\,\text{V})}{2} \ln \frac{1.0}{[H^+]^2}$$

$$-0.36\,\text{V} = -\frac{(0.0257\,\text{V})}{2} \times (-2\ln[H^+])$$

$$\ln[\text{H}^+] \; = \; \frac{-2(-0.36 \text{ V})}{-2(0.0257 \text{ V})} \; = \; -14.0 \qquad \text{so} \quad [\text{H}^+] \; = \; e^{-14.0} \; = \; \underline{\textbf{8} \times \textbf{10}^{-7} \textbf{\textit{M}}}$$

Thus the complete cell notation is:

$$\text{Zn} \,|\, \text{Zn}^{2+}(1.00 \; M) \,\|\, \text{H}^+(8 \times 10^{-7} \; M) \,|\, \text{H}_2(1.00 \text{ atm}) \,|\, \text{Pt}$$

74. This problem is related to Example 17.8. First, proceed as in Example 17.5(a) by assigning oxidation numbers to determine which element was oxidized/reduced, write the oxidation and reduction half-reactions with their E^o_{ox} and E^o_{red} values from Table 17.1, and then calculate E^o using $E^o = E^o_{ox} + E^o_{red}$.

Cr: 0 → +3 (oxidation) \qquad\qquad S: +6 → +4 (reduction)

Half-reactions: $\text{Cr}(s) \rightarrow \text{Cr}^{3+}(aq) + 3e^-$ \qquad $E^o_{ox} \; = \; -E^o_{red} \; = \; +0.744 \text{ V}$

$\text{SO}_4{}^{2-}(aq) + 4\text{H}^+(aq) + 2e^-$

$\rightarrow \text{SO}_2(g) + 2\text{H}_2\text{O}$ \qquad $E^o_{red} \; = \; +0.155 \text{ V}$

$$E^o \; = \; E^o_{ox} + E^o_{red} \; = \; 0.744 \text{ V} + 0.155 \text{ V} \; = \; \underline{\textbf{+0.899 V}}$$

(a) Recall that the reaction is spontaneous if $E > 0$. At standard conditions, $E = E^o = +0.899$ V, so **the reaction is spontaneous.**

(b) Under non-standard conditions, we must use the Nernst Equation to determine E; we also need values of E^o (from part (a) above) and n, and an expression for Q. To find n: the given redox reaction is obtained from the two half-reactions in part (a) by adding together 2 × (oxidation half-reaction) and 3 × (reduction half-reaction):

$2\text{Cr}(s) \rightarrow 2\text{Cr}^{3+}(aq) + 6e^-$

$3\text{SO}_4{}^{2-}(aq) + 12\text{H}^+(aq) + 6e^- \rightarrow 3\text{SO}_2(g) + 6\text{H}_2\text{O}$

$2\text{Cr}(s) + 3\text{SO}_4{}^{2-}(aq) + 12\text{H}^+(aq) + \textbf{6}e^-$

$\rightarrow 2\text{Cr}^{3+}(aq) + \textbf{6}e^- + 3\text{SO}_2(g) + 6\text{H}_2\text{O}$

Cancelling electrons then yields the equation given in the problem. Note that <u>6 electrons</u> are transferred in the equation above; therefore $n = 6$.

The expression for the reaction quotient Q (see Section 12-4) is:

$$Q \; = \; \frac{[\text{Cr}^{3+}]^2 (P_{\text{SO}_2})^3}{[\text{SO}_4{}^{2-}]^3 [\text{H}^+]^{12}}$$

Thus the Nernst Equation for this reaction is:

$$E = 0.899 \text{ V} - \frac{(0.0257 \text{ V})}{6} \ln \frac{[Cr^{3+}]^2 (P_{SO_2})^3}{[SO_4^{2-}]^3 [H^+]^{12}}$$

The expression for Q requires $[H^+]$, but in this case pH is given. To obtain $[H^+]$, use Equation 13.3: $[H^+] = 10^{-pH} = 10^{-3.00} = 0.0010 \text{ M}$. Under the given conditions, the Nernst Equation for this reaction becomes:

$$E = 0.899 \text{ V} - \frac{(0.0257 \text{ V})}{6} \ln \frac{(0.100)^2 (1.00)^3}{(0.100)^3 (0.0010)^{12}} = +0.534 \text{ V}$$

Under these conditions, $E = +0.534$ V, so **the reaction is spontaneous.**

(c) Again, pH is given and must be converted to $[H^+]$ as in part (b) above: $[H^+] = 10^{-pH} = 10^{-8.00} = 1.0 \times 10^{-8} \text{ M}$. Under the given conditions, the Nernst Equation for this reaction then becomes:

$$E = 0.899 \text{ V} - \frac{(0.0257 \text{ V})}{6} \ln \frac{(0.100)^2 (1.00)^3}{(0.100)^3 (1.0 \times 10^{-8})^{12}} = -0.058 \text{ V}$$

Under these conditions, $E = -0.058$ V, so **the reaction is non-spontaneous.**

(d) Recall that the reaction is at equilibrium if $E = 0$. Substitute this value of E, along with the given concentrations and pressure, into the Nernst Equation for the reaction, and solve for $[H^+]$; then use $[H^+]$ to obtain the corresponding pH using Equation 13.3.

$$E = 0.899 \text{ V} - \frac{(0.0257 \text{ V})}{6} \ln \frac{[Cr^{3+}]^2 (P_{SO_2})^3}{[SO_4^{2-}]^3 [H^+]^{12}}$$

$$0 = 0.899 \text{ V} - \frac{(0.0257 \text{ V})}{6} \ln \frac{(0.100)^2 (1.00)^3}{(0.100)^3 [H^+]^{12}}$$

$$0.899 \text{ V} = \frac{(0.0257 \text{ V})}{6} \ln \frac{1}{(0.100)[H^+]^{12}}$$

$$\frac{6(0.899 \text{ V})}{(0.0257 \text{ V})} = -\ln(0.100) - 12\ln[H^+]$$

$$2.10 \times 10^2 = -\ln(0.100) - 12\ln[H^+]$$

$$\ln[H^+] = \frac{1}{12}[-\ln(0.100) - 2.10 \times 10^2] = -17.3 \qquad \text{so} \qquad [H^+] = e^{-17.3} = 3.1 \times 10^{-8} \text{ M}$$

$$\text{pH} = -\log[H^+] = -\log(3.1 \times 10^{-8}) = \underline{\textbf{7.51}}$$

76. (a) This part of the problem is similar to Example 17.5(a): First assign oxidation numbers to determine which element was oxidized/reduced, write the oxidation and reduction half-reactions with their E_{ox}^o and E_{red}^o values from Table 17.1, and then calculate E^o using $E^o = E_{ox}^o + E_{red}^o$.

Pb: 0 → +2 (oxidation)   H: +1 → 0 (reduction)

half-reactions: $Pb(s) → Pb^{2+}(aq) + 2e^-$ $E_{ox}^o = -E_{red}^o = +0.127$ V
                   $2H^+(aq) + 2e^- → H_2(g)$ $E_{red}^o = +0.000$ V

$E^o = E_{ox}^o + E_{red}^o = +0.127$ V $+ 0.000$ V $= \underline{\textbf{+0.127 V}}$

(b) This part of the problem is related to Example 17.8. Use the Nernst Equation, for which values of E^o and n, and an expression for Q, are required. From part (a), $E^o = +0.127$ V. Since the given redox reaction is obtained simply by adding together the two half-reactions in part (a) and cancelling electrons, it can be seen that two electrons are transferred in the given reaction, so $n = 2$.
The expression for the reaction quotient Q (see Section 12-4) is:

$$Q = \frac{[Pb^{2+}](P_{H_2})}{[H^+]^2}$$

Thus the Nernst Equation for this reaction is:

$$E = 0.127 \text{ V} - \frac{(0.0257 \text{ V})}{2} \ln \frac{[Pb^{2+}](P_{H_2})}{[H^+]^2}$$

Substitute the given values of E, $[H^+]$ and P_{H_2} and solve for $[Pb^{2+}]$:

$$+0.210 \text{ V} = 0.127 \text{ V} - \frac{(0.0257 \text{ V})}{2} \ln \frac{[Pb^{2+}](1.0)}{(1.0)^2}$$

$$+0.210 \text{ V} - 0.127 \text{ V} = -\frac{(0.0257 \text{ V})}{2} \ln[Pb^{2+}]$$

$$+0.083 \text{ V} = -\frac{(0.0257 \text{ V})}{2} \ln[Pb^{2+}]$$

$$\ln[Pb^{2+}] = -\frac{2(0.083 \text{ V})}{(0.0257 \text{ V})} = -6.46 \quad \text{so} \quad [Pb^{2+}] = e^{-6.46} = \underline{\textbf{1.6 × 10}^{-3} \textbf{ M}}$$

(c) Recall that K_{sp} is the concentration equilibrium constant for dissolving an ionic solid, in this case $PbCl_2$ (see Section 15-2):

$$PbCl_2(s) \rightleftharpoons Pb^{2+}(aq) + 2Cl^-(aq) \qquad K_{sp} = [Pb^{2+}][Cl^-]^2$$

We are told that Pb^{2+} ions are added until $PbCl_2$ just begins to precipitate; in other words, when $Q = K_{sp}$. Thus, using $[Pb^{2+}]$ from part (b) and $[Cl^-]$ given:

$$K_{sp} = [Pb^{2+}][Cl^-]^2 = (1.6 \times 10^{-3})(0.10)^2 = \mathbf{1.6 \times 10^{-5}}$$

17-6 ELECTROLYTIC CELLS

78. (a) Use Avogadro's number to convert mol $e^- \to$ number of e^-:

$$\text{number of } e^- = 0.228 \text{ mol } e^- \times \frac{6.022 \times 10^{23} \, e^-}{1 \text{ mol } e^-} = \mathbf{1.37 \times 10^{23} \, e^-}$$

(b) To convert mol e^- to coulombs, recall that 1 mol $e^- = 9.648 \times 10^4$ C (see Table 17.3):

$$\text{charge (C)} = 0.228 \text{ mol } e^- \times \frac{9.648 \times 10^4 \, C}{1 \text{ mol } e^-} = \mathbf{2.20 \times 10^4 \, C}$$

(c) This part of the problem is related to Example 17.10. First split the balanced equation into two half-reactions and balance those. Next, for each half-reaction, use stoichiometry to obtain mol of each product; and then convert mol to mass product using molar mass. Note that the balanced equation given indicates that the reaction takes place in basic solution.

For H_2: $H_2O \to H_2(g)$ oxidation number $+1 \to 0$ (reduction) and from Table 17.1, the half-reaction for reduction of H_2O to H_2 in basic solution is:

$$2H_2O + 2e^- \to H_2(g) + 2OH^-(aq)$$

Strategy: mol $e^- \to$ mol $H_2 \to$ mass H_2

$$\text{mol } H_2 \text{ produced} = 0.228 \text{ mol } e^- \times \frac{1 \text{ mol } H_2}{2 \text{ mol } e^-} = 0.114 \text{ mol } H_2$$

$$\text{mass } H_2 = 0.114 \text{ mol } H_2 \times \frac{2.016 \text{ g } H_2}{1 \text{ mol } H_2} = \mathbf{0.230 \text{ g } H_2}$$

For Cl_2: $Cl^-(aq) \to Cl_2(g)$ oxidation number $-1 \to 0$ (oxidation) and from Table 17.1, the half-reaction for oxidation of Cl^- to Cl_2 is:

$$2Cl^-(aq) \to Cl_2(g) + 2e^-$$

Strategy: mol $e^- \to$ mol $Cl_2 \to$ mass Cl_2

$$\text{mol } Cl_2 \text{ produced} = 0.228 \text{ mol } e^- \times \frac{1 \text{ mol } Cl_2}{2 \text{ mol } e^-} = 0.114 \text{ mol } Cl_2$$

$$\text{mass } Cl_2 = 0.114 \text{ mol } Cl_2 \times \frac{70.9054 \text{ g } Cl_2}{1 \text{ mol } Cl_2} = \mathbf{8.08 \text{ g } Cl_2}$$

80. This problem is related to Example 17.10. Note that the metal ion M^{2+} produces atoms of the metal M upon electrolysis and so the reduction half-reaction will be:

$$M^{2+}(aq) + 2e^- \rightarrow M(s)$$

(a) To obtain the charge supplied by the battery, recall from Table 17.3 that 1 A = 1 C/s, so the current of 7.8 A corresponds to 7.8 C/s:

$$\text{charge} = \frac{7.8\,C}{1\,s} \times 15.5\,\text{min} \times \frac{60\,s}{1\,\text{min}} = \underline{\mathbf{7.3 \times 10^3\,C}}$$

(b) To identify the metal M we can use its molar mass, for which need a mass and a number of moles. Mass is given (2.39 g) and therefore the number of moles must be obtained from the electrolysis data.
First convert the charge from part (a) above to moles of e^- and then use the stoichiometry of the reduction half-reaction to obtain moles of M produced.

To convert charge to mol e^- recall that 1 mol e^- = 9.648 × 10^4 C (Table 17.3):

$$\text{mol } e^- = 7.3 \times 10^3\,C \times \frac{1\,\text{mol } e^-}{9.648 \times 10^4\,C} = 7.6 \times 10^{-2}\,\text{mol } e^-$$

$$\text{mol M produced} = 7.6 \times 10^{-2}\,\text{mol } e^- \times \frac{1\,\text{mol M}}{2\,\text{mol } e^-} = 3.8 \times 10^{-2}\,\text{mol M}$$

Finally use the number of moles of M and the mass to calculate molar mass:

$$\text{molar mass of M} = \frac{\text{mass of M}}{\text{mol of M}} = \frac{2.39\,g}{3.8 \times 10^{-2}\,\text{mol}} = 63\,g/\text{mol}$$

Based on its molar mass, 63 g/mol, the metal M is **copper**.

82. (a) First calculate the volume of the gold plating using its surface area and thickness (assume the plating is of uniform thickness); then convert volume to mass using the density as conversion factor. Recall from Table 1.3 that 1 in = 2.54 cm.

$$\begin{aligned}
\text{vol Au plating} &= \text{surface area} \times \text{thickness} \\
&= 17.21\,\text{in}^2 \times 0.00200\,\text{in} \\
&= 0.0344\,\text{in}^3
\end{aligned}$$

$$\text{mass of Au required} = 0.0344\,\text{in}^3 \times \left(\frac{2.54\,\text{cm}}{1\,\text{in}}\right)^3 \times \frac{10.5\,g}{1\,\text{cm}^3} = \underline{\mathbf{5.92\,g}}$$

(b) This part of the problem is related to Example 17.10. Since a solution of AuCN is used as gold source, the gold species being electrolyzed is $Au^+(aq)$ for which the reduction half-reaction is

$$Au^+(aq) + e^- \rightarrow Au(s)$$

from which it is deduced that 1 mol e^- will produce 1 mol Au.
From part (a), 5.92 g Au are required; convert this to mol Au and hence to mol e^- using the stoichiometry of the reaction:

$$\text{mass Au} \xrightarrow{\text{molar mass Au}} \text{mol Au} \xrightarrow{1\,\text{mol } e^- \,/\, 1\,\text{mol Au}} \text{mol } e^-$$

Finally, mol e^- may be converted to charge (coulombs) and then to time (s) using conversion factors from Table 17.3:

$$\text{mol } e^- \xrightarrow{1\,\text{mol } e^- = 9.648 \times 10^4 \text{ C}} \text{charge (C)} \xrightarrow{7.00 \text{ A} = 7.00 \text{ C/s}} \text{time (s)}$$

$$\text{mol } e^- = 5.92 \text{ g Au} \times \frac{1\,\text{mol Au}}{196.967 \text{ g Au}} \times \frac{1\,\text{mol } e^-}{1\,\text{mol Au}} = 0.0301 \text{ mol } e^-$$

$$\text{time } = 0.0301 \text{ mol } e^- \times \frac{9.648 \times 10^4 \text{ C}}{1\,\text{mol } e^-} \times \frac{1\,\text{s}}{7.00 \text{ C}} \times \frac{1\,\text{min}}{60 \text{ s}} = \underline{\textbf{6.91 min}}$$

84. (a) This problem is similar to Example 17.10(c). Since molten $CaCl_2$ is used as the calcium source, the calcium species being electrolyzed is $Ca^{2+}(l)$ for which the reduction half-reaction is:

$$Ca^{2+}(l) + 2e^- \rightarrow Ca(l)$$

from which it is deduced that 2 mol e^- will produce 1 mol Ca.
First convert the mass of Ca to mol e^- using stoichiometry and recalling (Table 1.3) that 1 lb = 453.6 g:

$$\text{lb of Ca} \rightarrow \text{g of Ca} \xrightarrow{\text{molar mass Ca}} \text{mol Ca} \xrightarrow{2\,\text{mol } e^- \,/\, 1\,\text{mol Ca}} \text{mol } e^-$$

$$\text{mol } e^- = 12.0 \text{ lb Ca} \times \frac{453.6 \text{ g Ca}}{1\,\text{lb Ca}} \times \frac{1\,\text{mol Ca}}{40.078 \text{ g Ca}} \times \frac{2\,\text{mol } e^-}{1\,\text{mol Ca}} = 272 \text{ mol } e^-$$

Then convert mol e^- to charge (coulombs) and then to energy using conversion factors from Table 17.3:

$$\text{mol } e^- \xrightarrow{1\,\text{mol } e^- = 9.648 \times 10^4 \text{ C}} \text{charge (C)} \xrightarrow{V} \text{C·V} \xrightarrow{1\,\text{J} = 1\,\text{V·C}} \text{energy (J)}$$

$$\text{energy } = 272 \text{ mol } e^- \times \frac{9.648 \times 10^4 \text{ C}}{1\,\text{mol } e^-} \times 3.2 \text{ V} \times \frac{1\,\text{J}}{1\,\text{V·C}} = \underline{\textbf{8.4} \times \textbf{10}^7 \textbf{ J}}$$

(b) Use the conversion factor 1 kWh = 3.600×10^6 J (from Table 17.3) to convert J $\rightarrow$ kWh; then use 1 kWh = 9 ¢ to obtain the cost of the electrical energy:

$$\text{cost} = 8.4 \times 10^7 \text{ J} \times \frac{1 \text{ kWh}}{3.600 \times 10^6 \text{ J}} \times \frac{9 \cancel{c}}{1 \text{ kWh}} = \underline{\underline{2.1 \times 10^2 \cancel{c}}} \quad \text{or} \quad \underline{\underline{\$2.1}}$$

UNCLASSIFIED PROBLEMS

86. The half-reactions taking place in the battery are given on page 460:

$$Cd(s) + 2OH^-(aq) \rightarrow Cd(OH)_2(s) + 2e^-$$

$$2NiO(OH)(s) + 2H_2O + 2e^- \rightarrow 2Ni(OH)_2(s) + 2OH^-(aq)$$

Note that in the latter reaction 2NiO(OH) is equivalent to $Ni_2O_3 + H_2O$, so the reaction could be rewritten as:

$$Ni_2O_3(s) + 3H_2O + 2e^- \rightarrow 2Ni(OH)_2(s) + 2OH^-(aq)$$

Thus from these reactions it is deduced than 2 mol e^- are produced by the consumption of 1 mol of Cd, and 2 mol e^- are used when 1 mol of Ni_2O_3 is consumed.

Use the given current (0.175 A = 0.175 C/s) and time (1.5 h) along with conversion factors from Table 17.3 to calculate mol e^-; then, for each species, use stoichiometry to convert mol e^- to mol Cd or Ni_2O_3 consumed and hence to the mass of each species (using their molar masses):

$$\text{mol } e^- = \frac{0.175 \text{ C}}{1 \text{ s}} \times 1.5 \text{ h} \times \frac{3600 \text{ s}}{1 \text{ h}} \times \frac{1 \text{ mol } e^-}{9.648 \times 10^4 \text{ C}} = 9.79 \times 10^{-3} \text{ mol } e^-$$

$$\text{mass Cd consumed} = 9.79 \times 10^{-3} \text{ mol } e^- \times \frac{1 \text{ mol Cd}}{2 \text{ mol } e^-} \times \frac{112.41 \text{ g Cd}}{1 \text{ mol Cd}}$$

$$= \underline{\underline{0.550 \text{ g Cd}}}$$

$$\text{mass Ni}_2\text{O}_3 \text{ consumed} = 9.79 \times 10^{-3} \text{ mol } e^- \times \frac{1 \text{ mol Ni}_2\text{O}_3}{2 \text{ mol } e^-} \times \frac{165.38 \text{ g Ni}_2\text{O}_3}{1 \text{ mol Ni}_2\text{O}_3}$$

$$= \underline{\underline{0.810 \text{ g Ni}_2\text{O}_3}}$$

88. (a) During the electrolysis of $NiCl_2(aq)$ — which contains $Ni^{2+}(aq)$ and $Cl^-(aq)$ — the half-reactions taking place, along with their E^o_{ox} and E^o_{red} values (from Table 17.1), are:

$$2Cl^-(aq) \rightarrow Cl_2(g) + 2e^- \qquad\qquad E^o_{ox} = -E^o_{red} = -1.360 \text{ V}$$

$$Ni^{2+}(aq) + 2e^- \rightarrow Ni(s) \qquad\qquad E^o_{red} = -0.236 \text{ V}$$

Thus, for the overall reaction

$$Ni^{2+}(aq) + 2Cl^-(aq) \rightarrow Ni(s) + Cl_2(g)$$

$$E^o = E^o_{ox} + E^o_{red} = -1.360 \text{ V} - 0.236 \text{ V} = -1.596 \text{ V}$$

and therefore a voltage of *at least* **1.596 V** must be applied in order to carry out the reaction.

(b) This part of the problem is related to Example 17.10(c). We are told that 10.00 kJ of energy are consumed while operating at a voltage of 3.0 V. Using conversion factors from Table 17.3, we can use this information to determine the total chage (coulombs) supplied and hence mol electrons.

$$\text{energy (kJ)} \xrightarrow{1J=1V\cdot C} C\cdot V \xrightarrow{V} \text{charge (C)} \xrightarrow{9.648\times 10^4 \, C = 1mol \, e^-} \text{mol } e^-$$

$$\text{mol } e^- = 10.00 \text{ kJ} \times \frac{1000 \text{ J}}{1 \text{ kJ}} \times \frac{1 \text{ V} \cdot \text{C}}{1 \text{ J}} \times \frac{1}{3.0 \text{ V}} \times \frac{1 \text{mol } e^-}{9.648 \times 10^4 \text{ C}} = 0.35 \text{ mol } e^-$$

From the reduction half-reaction in part (a) above, we know that 2 mol e^- are required to produce 1 mol Ni, so we can related mol e^- consumed to mol Ni produced and hence (using molar mass Ni) the mass of Ni produced.

$$\text{mol } e^- \xrightarrow{2 \, mol \, e^- = 1 mol \, Ni} \text{mol Ni} \xrightarrow{MM \, Ni} \text{mass Ni}$$

$$\text{mass Ni} = 0.035 \text{ mol } e^- \times \frac{1 \text{mol Ni}}{2 \text{mol } e^-} \times \frac{58.6934 \text{ g Ni}}{1 \text{mol Ni}} = \underline{\textbf{1.0 g}}$$

90. The given reaction indicates that Cu is being oxidized: $Cu(s) \rightarrow Cu^{2+}(aq) + 2e^-$ and therefore we expect the mass of the Cu electrode to <u>decrease</u> since Cu atoms are leaving the electrode and moving into solution. Note also that $2e^-$ are produced per mol Cu consumed.

Determine the mass of Cu consumed following the procedure in Example 17.10(a). The current is 3 A = 3 C/s, passed for 70 min 20 s = 4220 s.

Strategy: current (A) $\rightarrow$ charge (C) $\rightarrow$ mol $e^- \rightarrow$ mol Cu $\rightarrow$ mass Cu

$$\text{mol } e^- = \frac{3 \text{ C}}{1 \text{ s}} \times 4220 \text{ s} \times \frac{1 \text{mol } e^-}{9.648 \times 10^4 \text{ C}} = 0.131 \text{ mol } e^-$$

$$\text{mass Cu consumed} = 0.131 \text{ mol } e^- \times \frac{1 \text{mol Cu}}{2 \text{mol } e^-} \times \frac{63.546 \text{ g Cu}}{1 \text{mol Cu}} = 4 \text{ g Cu}$$

Thus the final mass of the Cu electrode is $(100.0 - 4)$g = **96 g**.

92. The thermodynamic data given in Appendix 1 can be used to calculate ΔH° and ΔS° for the reaction and these values combined to calculate ΔG° for the reaction at 1000°C (1273 K) using $\Delta G^\circ = \Delta H^\circ - T\Delta S^\circ$ as in Example 16.5. Finally, Equation 17.2 can be used to convert ΔG° to E° for the given reaction and conditions. To obtain the value of n needed for the latter calculation, note that the oxidation number of C in the given reaction increases from +2 (in CO) to +4 (in CO_2) and therefore $2e^-$ are transferred in the reaction, so $n = 2$.

$$\Delta H° = \Sigma \Delta H_f°_{(products)} - \Sigma \Delta H_f°_{(reactants)}$$

$$= [(1 \text{ mol})(-393.5 \text{ kJ/mol})] - [(1 \text{ mol})(-110.5 \text{ kJ/mol}) + (½ \text{ mol})(0 \text{ kJ/mol})]$$

$$= -283.0 \text{ kJ}$$

$$\Delta S° = \Sigma S°_{(products)} - \Sigma S°_{(reactants)}$$

$$= [(1 \text{ mol})(0.2136 \text{ kJ/mol·K})]$$

$$\qquad - [(1 \text{ mol})(0.1976 \text{ kJ/mol·K}) + (½ \text{ mol})(0.2050 \text{ kJ/mol·K})]$$

$$= -0.0865 \text{ kJ/K}$$

$$\Delta G° = -283.0 \text{ kJ} - (1273 \text{ K})(-0.0865 \text{ kJ/K}) = -172.9 \text{ kJ}$$

$$\Delta G° = -nFE° \qquad \text{so} \qquad E° = -\frac{\Delta G°}{nF} = -\frac{-172.9 \times 10^3 \text{ J}}{(2 \text{ mol})(9.648 \times 10^4 \frac{\text{J}}{\text{mol·V}})} = \underline{\textbf{+0.896 V}}$$

94. This problem is related to Example 17.8. First assign oxidation numbers to determine which element was oxidized/reduced, write the oxidation and reduction half-reactions with their $E°_{ox}$ and $E°_{red}$ values from Table 17.1, then calculate $E°$.

Cl: $-1 \to 0$ (oxidation)   O: $0 \to -2$ (reduction)

half-reactions: $2Cl^-(aq) \to Cl_2(g) + 2e^-$ $E°_{ox} = -E°_{red} = -1.360 \text{ V}$

$\qquad\qquad\qquad\quad O_2(g) + 4H^+(aq) + 4e^- \to 2H_2O \qquad E°_{red} = +1.229 \text{ V}$

$$E° = E°_{ox} + E°_{red} = -1.360 \text{ V} + 1.229 \text{ V} = \underline{\textbf{-0.131 V}}$$

To calculate E for the given non-standard conditions, use the Nernst Equation:

$$E = E° - \frac{(0.0257 \text{ V})}{n} \ln Q$$

for which values of $E°$ (above) and n, and an expression for Q, are required.
To find n: the given redox reaction is obtained from the two half-reactions by adding together 2 × (oxidation half-reaction) and the reduction half-reaction:

$$4Cl^-(aq) \to 2Cl_2(g) + 4e^-$$
$$\underline{O_2(g) + 4H^+(aq) + 4e^- \to 2H_2O}$$
$$O_2(g) + 4H^+(aq) + 4Cl^-(aq) \to 2H_2O + 2Cl_2(g)$$

Cancelling duplications then yields the equation given in the problem. Note that <u>4 electrons</u> are transferred in the equation above; therefore $n = 4$.

The expression for the reaction quotient Q (see Section 12-4) is:

$$Q = \frac{(P_{Cl_2})^2}{(P_{O_2})[H^+]^4[Cl^-]^4}$$

Thus the Nernst Equation for this reaction is:

$$E = -0.131 \text{ V} - \frac{(0.0257 \text{ V})}{4} \ln \frac{(P_{Cl_2})^2}{(P_{O_2})[H^+]^4[Cl^-]^4}$$

The expression for Q requires $[H^+]$, which in this case can be obtained from the given concentrations of buffer components and K_a value using Equation 14.1:

$$[H^+] = K_a \times \frac{[HB]}{[B^-]} = (1.4 \times 10^{-4}) \times \frac{0.125}{0.125} = 1.4 \times 10^{-4} \text{ M}$$

Thus, under the given conditions, the cell voltage for this reaction is:

$$E = -0.131 \text{ V} - \frac{(0.0257 \text{ V})}{4} \ln \frac{(1.00)^2}{(1.00)(1.4 \times 10^{-4})^4 (0.200)^4} = \underline{-0.400 \text{ V}}$$

Since a voltaic cell employs a **spontaneous** redox reaction ($E > 0$) we can say that **the cell will not function as a voltaic cell under the conditions given.**

CONCEPTUAL PROBLEMS

96. Electrolysis of water produces H_2 and O_2 (see pages 445–446):

    $$2H_2O + 2e^- \rightarrow H_2(g) + 2OH^-(aq) \qquad \text{and} \qquad 2H_2O \rightarrow O_2(g) + 4H^+(aq) + 4e^-$$

    for which the net reaction is: $\quad 2H_2O \rightarrow 2H_2(g) + O_2(g)$.

    Look for the box containing two types of molecule: one type consisting of two circles (H_2 molecules) and one consisting of two squares (O_2 molecules); there should also be twice as many H_2 molecules as there are O_2 molecules.

    Box (c) represents the products of electrolysis of H_2O.

98. First obtain the equation for the cell reaction and then use the Nernst Equation to examine the effect upon E of the given changes.
    Recall that the oxidation reaction takes place on the left side (anode) of the cell notation and the reduction reaction is on the right side (cathode); the Pt at the cathode is inert and takes no place in the reaction.

    half-reactions: $\quad$ $Co(s) \rightarrow Co^{2+}(aq, 0.010 \text{ M}) + 2e^-$
    $\qquad\qquad\qquad\quad$ $2H^+(aq, 0.010 \text{ M}) + 2e^- \rightarrow H_2(g, 0.500 \text{ atm})$

The balanced equation for the reaction is obtained by adding together the two half-reactions; electrons cancel out:

$$Co(s) + 2H^+(aq, 0.010\ M) \rightarrow Co^{2+}(aq, 0.010\ M) + H_2(g, 0.500\ atm)$$

Two electrons are transferred in this reaction (two electrons were cancelled) so $n = 2$; and the expression for Q for this reaction (see Section 12-4) is:

$$Q = \frac{[Co^{2+}](P_{H_2})}{[H^+]^2}$$

Thus the Nernst Equation for this reaction is:

$$E = E^\circ - \frac{(0.0257\ V)}{2} \ln \frac{[Co^{2+}](P_{H_2})}{[H^+]^2}$$

(a) The voltage depends on $[Co^{2+}]$, not on the volume of the solution, so increasing the volume has <u>no effect</u>.

(b) The voltage depends on $[H^+]$, which appears on the denominator of Q. Increasing $[H^+]$ will decrease Q and therefore will decrease $\ln Q$: the size of the $\frac{(0.0257\ V)}{2} \ln Q$ term decreases, and since that is being subtracted from E°, <u>E will increase</u>.

(c) The voltage depends on H_2 pressure, which appears on the numerator of Q. Increasing H_2 pressure will increase Q and therefore will increase $\ln Q$: the size of the $\frac{(0.0257\ V)}{2} \ln Q$ term increases, and since that is being subtracted from E°, <u>E will decrease</u>.

(d) The mass of the Co electrode does not affect Q so changing its mass has <u>no effect</u>.

(e) The voltage depends on $[Co^{2+}]$, which appears on the numerator of Q. Increasing $[Co^{2+}]$ will increase Q and hence increases $\ln Q$: the size of $\frac{(0.0257\ V)}{2} \ln Q$ increases, and since that quantity is being subtracted from E°, <u>E will decrease</u>.

Thus, **only (b) will increase the cell voltage.**

100. For each cell, recall that the oxidation reaction takes place on the left side of the cell notation and the reduction reaction is on the right side; the platinum electrode in cells (1) and (2) is inert and is not involved in the reaction. First, write the oxidation and reduction half-reactions taking place, and combine them to answer part (a). Next use the given E°_{red} values to calculate E° [part (b)] using $E^\circ = E^\circ_{ox} + E^\circ_{red}$. Then use Equation 17.2 to calculate ΔG° for part (c).

<u>Cell (1)</u>

(a) half-reactions: $\quad\quad Tl(s) \rightarrow Tl^+(aq) + e^- \quad\quad\quad E^\circ_{ox} = -E^\circ_{red} = +0.34\ V$

$$Tl^{3+}(aq) + 2e^- \rightarrow Tl^+(aq) \qquad E^o_{red} = +1.28 \text{ V}$$

The overall reaction is obtained by doubling the coefficients of the oxidation half-reaction and then adding the reduction half-reaction and cancelling duplications:

$$2Tl(s) \rightarrow 2Tl^+(aq) + \mathbf{2e^-}$$
$$\underline{Tl^{3+}(aq) + \mathbf{2e^-} \rightarrow Tl^+(aq)}$$
$$\mathbf{2Tl(s) + Tl^{3+}(aq) \rightarrow 3Tl^+(aq)}$$

(b) $E^o = E^o_{ox} + E^o_{red} = +0.34 \text{ V} + 1.28 \text{ V} = \underline{\mathbf{+1.62 \text{ V}}}$

(c) Since $2e^-$ were cancelled out while obtaining the overall equation in part (a), $n = 2$.

$$\Delta G^o = -nFE^o = -(2 \text{ mol})(9.648 \times 10^4 \frac{J}{mol \cdot V})(+1.62 \text{ V})$$
$$= -313 \times 10^3 \text{ J} \quad \text{or} \quad \underline{\mathbf{-313 \text{ kJ}}}$$

Cell (2)

(a) half-reactions:

$$Tl(s) \rightarrow Tl^{3+}(aq) + 3e^- \qquad E^o_{ox} = -E^o_{red} = -0.74 \text{ V}$$
$$Tl^{3+}(aq) + 2e^- \rightarrow Tl^+(aq) \qquad E^o_{red} = +1.28 \text{ V}$$

The overall reaction is obtained by adding together 2 × (oxidation half-reaction) and 3 × (reduction half-reaction) and cancelling duplications:

$$2Tl(s) \rightarrow 2Tl^{3+}(aq) + \mathbf{6e^-}$$
$$\underline{3Tl^{3+}(aq) + \mathbf{6e^-} \rightarrow 3Tl^+(aq)}$$
$$\mathbf{2Tl(s) + Tl^{3+}(aq) \rightarrow 3Tl^+(aq)}$$

(b) $E^o = E^o_{ox} + E^o_{red} = -0.74 \text{ V} + 1.28 \text{ V} = \underline{\mathbf{+0.54 \text{ V}}}$

(c) Since $6e^-$ were cancelled out while obtaining the overall equation in part (a), $n = 6$.

$$\Delta G^o = -nFE^o = -(6 \text{ mol})(9.648 \times 10^4 \frac{J}{mol \cdot V})(+0.54 \text{ V})$$
$$= -313 \times 10^3 \text{ J} \quad \text{or} \quad \underline{\mathbf{-313 \text{ kJ}}}$$

Cell (3)

(a) half-reactions:

$$Tl(s) \rightarrow Tl^+(aq) + e^- \qquad E^o_{ox} = -E^o_{red} = +0.34 \text{ V}$$
$$Tl^{3+}(aq) + 3e^- \rightarrow Tl(s) \qquad E^o_{red} = +0.74 \text{ V}$$

The overall reaction is obtained by tripling the coefficients of the oxidation half-reaction and then adding the reduction half-reaction and cancelling duplications:

$$3Tl(s) \rightarrow 3Tl^+(aq) + 3e^-$$

$$Tl^{3+}(aq) + 3e^- \rightarrow Tl(s)$$

$$\overline{2Tl(s) + Tl^{3+}(aq) \rightarrow 3Tl^+(aq)}$$

(b) $E^\circ = E^\circ_{ox} + E^\circ_{red} = +0.34\ V + 0.74\ V = \underline{+1.08\ V}$

(c) Since $3e^-$ were cancelled out while obtaining the overall equation in part (a), $n = 3$.

$$\Delta G^\circ = -nFE^\circ \quad = -(3\ mol)(9.648 \times 10^4\ \frac{J}{mol \cdot V})(+1.08\ V)$$

$$= -313 \times 10^3\ J \quad or \quad \underline{-313\ kJ}$$

(d) The **overall** chemical reaction taking place in each cell is the same, but the path to each of them is different (the electrodes differ in each cell). ΔG° is the same for all three cells and therefore it is a state property because state properties are path-independent. However, E° values will differ because E° depends on the path.

102. This first experiment tells us nothing about the comparative reducing strength of X, Y, Z. However, the second experiment tells us that Y is a stronger reducing agent than X because X cannot displace Y from its salt YA. The third experiment tells us that X is a stronger reducing agent than Z because it can displace Z from its salt ZA. (Since we know that X, Y, and Z are metals, this reaction corresponds to *reduction* of cations of Z to atoms of Z. The discoloration of X is likely due to some metallic Z forming on the surface of X and/or the formation of XA or some other salt of X.) Thus the order of decreasing strength as reducing agents is:

   Y > X > Z

CHALLENGE PROBLEMS

103. See Section 17-6. The electroplating process will involve two reduction reactions:

   $$Sn^{2+}(aq) + 2e^- \rightarrow Sn(s) \quad\quad and \quad\quad Cu^{2+}(aq) + 2e^- \rightarrow Cu(s)$$

   from which it is seen that 2 mol e^- are required to plate each mol of Sn or Cu. To simplify the calculation, assume that the current is passed for a time such that 1 mol e^- is supplied to the solution, and recall (page 453) that the number of mol of electrons supplied is proportional to the current. Thus, if 1 mol e^- is supplied and 20.0% of the current plates Sn while 80.0% of the current plates Cu, then 0.200 mol of e^- are used to plate Sn and 0.800 mol e^- are used to plate Cu.

   For each metal, convert mol $e^- \rightarrow$ mol metal $\rightarrow$ mass metal; then determine the mass % of each metal present.

$$\text{mass Sn plated} = 0.200 \text{ mol } e^- \times \frac{1 \text{ mol Sn}}{2 \text{ mol } e^-} \times \frac{118.71 \text{ g Sn}}{1 \text{ mol Sn}} = 11.9 \text{ g Sn}$$

$$\text{mass Cu plated} = 0.800 \text{ mol } e^- \times \frac{1 \text{ mol Cu}}{2 \text{ mol } e^-} \times \frac{63.546 \text{ g Cu}}{1 \text{ mol Cu}} = 25.4 \text{ g Cu}$$

total mass of alloy plated = 11.9 g + 25.4 g = 37.3 g

$$\text{mass \% Sn} = \frac{11.9 \text{ g Sn}}{37.3 \text{ g alloy}} \times 100 = \mathbf{31.9 \text{ \% Sn}}$$

mass % Cu = $(100 - 31.9)$ % = **68.1 % Cu**

104. See Sections 17-5 and 17-7. The redox reaction taking place in the battery is given on page 459:

$$Pb(s) + PbO_2(s) + 2H^+(aq) + 2HSO_4^-(aq) \rightarrow 2PbSO_4(s) + 2H_2O$$

and note from the discussion on the same page that $2e^-$ are transferred during the reaction ($n = 2$), and that $\Delta G°$ is -371.4 kJ. Hence, using Equation 17.2

$$\Delta G° = -371.4 \times 10^3 \text{ J} = -(2 \text{ mol})(9.648 \times 10^4 \frac{J}{\text{mol} \cdot V}) E°$$

$$E° = +1.925 \text{ V}$$

The solution is 38 % H_2SO_4 by mass (38 g H_2SO_4 in 100 g solution); use this, along with the density, to determine the nominal $[H_2SO_4]_0$ in the solution.
Strategy: % H_2SO_4 → mass H_2SO_4 → mol H_2SO_4; mass solution → vol solution; then combine to obtain $[H_2SO_4]_0$.

$$\text{mol } H_2SO_4 = 38 \text{ g } H_2SO_4 \times \frac{1 \text{ mol } H_2SO_4}{98.08 \text{ g } H_2SO_4} = 0.39 \text{ mol } H_2SO_4$$

$$\text{vol solution} = 100 \text{ g solution} \times \frac{1 \text{ cm}^3}{1.286 \text{ g}} \times \frac{1 L}{1000 \text{ cm}^3} = 0.0778 \text{ L}$$

$$[H_2SO_4]_0 = \frac{0.39 \text{ mol } H_2SO_4}{0.0778 \text{ L solution}} = 5.0 \text{ M}$$

We assume that the H_2SO_4 completely dissociates and that only the first dissociation of H_2SO_4 is significant (further dissociation of HSO_4^- is ignored):

$$H_2SO_4(aq) \rightarrow H^+(aq) + HSO_4^-(aq)$$

so $\quad [H^+] = \dfrac{5.0\,\text{mol}\,H_2SO_4}{1\,\text{L solution}} \times \dfrac{1\,\text{mol}\,H^+}{1\,\text{mol}\,H_2SO_4} = 5.0\,M$

and $\quad [HSO_4^-] = \dfrac{5.0\,\text{mol}\,H_2SO_4}{1\,\text{L solution}} \times \dfrac{1\,\text{mol}\,HSO_4^-}{1\,\text{mol}\,H_2SO_4} = 5.0\,M$

To obtain the value of E under these non-standard conditions, use the Nernst Equation. To do this, we also need the expression for Q for the cell reaction:

$$Q = \dfrac{1}{[H^+]^2[HSO_4^-]^2}$$

Thus the Nernst Equation for this reaction is:

$$E = E° - \dfrac{(0.0257\,V)}{n}\ln Q$$

$$= +1.925\,V - \dfrac{(0.0257\,V)}{2}\ln\dfrac{1}{[H^+]^2[HSO_4^-]^2}$$

$$= +1.925\,V - \dfrac{(0.0257\,V)}{2}\ln\dfrac{1}{(5.0)^2(5.0)^2} = \underline{\textbf{+2.008 V}}$$

105. (a) See Example 17.5.

The half-reactions are:

$$Zn(s) \rightarrow Zn^{2+}(aq) + 2e^- \qquad E°_{ox} = -E°_{red} = +0.762\,V$$
$$Sn^{2+}(aq) + 2e^- \rightarrow Sn(s) \qquad E°_{red} = -0.141\,V$$

$$E° = E°_{ox} + E°_{red} = +0.762\,V + (-0.141\,V) = \underline{\textbf{+0.621 V}}$$

(b) Since $E° > 0$, the reaction is spontaneous in the forward direction. Thus when the cell operates, $Zn^{2+}(aq)$ ions are being formed and $Sn^{2+}(aq)$ ions are being consumed, so **[Zn²⁺] will increase and [Sn²⁺] will decrease.**

(c) This part is related to Example 17.9. Use the Nernst Equation, with $E° = +0.621$ V and $n = 2$ from part (a) above. The expression for Q is:

$$Q = \dfrac{[Zn^{2+}]}{[Sn^{2+}]}$$

When E falls to zero, the Nernst Equation for this reaction becomes:

$$0\,V = +0.621\,V - \dfrac{(0.0257\,V)}{2}\ln\dfrac{[Zn^{2+}]}{[Sn^{2+}]}$$

$$-0.621 \text{ V} = -\frac{(0.0257 \text{ V})}{2} \ln \frac{[\text{Zn}^{2+}]}{[\text{Sn}^{2+}]}$$

$$\ln \frac{[\text{Zn}^{2+}]}{[\text{Sn}^{2+}]} = -\frac{2(-0.621 \text{ V})}{(0.0257 \text{ V})} = 48.3 \quad \text{so} \quad \frac{[\text{Zn}^{2+}]}{[\text{Sn}^{2+}]} = e^{48.3} = \underline{\mathbf{1 \times 10^{21}}}$$

(d) Let the amount by which each concentration changes be *x*. (By stoichiometry, change in $[\text{Zn}^{2+}]$ must equal change in $[\text{Sn}^{2+}]$.) Then when *E* falls to zero:

$$[\text{Zn}^{2+}] = 1.0 + x \qquad\qquad [\text{Sn}^{2+}] = 1.0 - x$$

Substitute these two expression for concentration into the ratio obtained in part (c) above:

$$\frac{[\text{Zn}^{2+}]}{[\text{Sn}^{2+}]} = \frac{1.0 + x}{1.0 - x} = 1 \times 10^{21}$$

$$1.0 + x = (1 \times 10^{21})(1.0 - x) = 1 \times 10^{21} - (1 \times 10^{21})x$$
$$1 \times 10^{21} - 1.0 = [(1 \times 10^{21}) + 1]x \qquad\qquad \text{or } 1 \times 10^{21} \approx (1 \times 10^{21})x$$

Thus *x* = 1, so $[\text{Zn}^{2+}] = 1.0 + 1 = \underline{\mathbf{2 \, M}}$ and $[\text{Sn}^{2+}] = 1.0 - 1 = 0$ M. However, recall from part (c) that $[\text{Zn}^{2+}]/[\text{Sn}^{2+}] = 1 \times 10^{21}$, whereas if $[\text{Sn}^{2+}]$ is zero the ratio of concentrations becomes infinitely large. Instead use the relationship calculated at the end of part (c), with $[\text{Zn}^{2+}] = 2$ *M*:

$$\frac{[\text{Zn}^{2+}]}{[\text{Sn}^{2+}]} = 1 \times 10^{21} \quad \text{so} \quad [\text{Sn}^{2+}] = \frac{[\text{Zn}^{2+}]}{1 \times 10^{21}} = \frac{2 \, M}{1 \times 10^{21}} = \underline{\mathbf{2 \times 10^{-21} \, M}}$$

106. (a) Use Equation 17.2:

First step:

$$\Delta G^{\circ\prime} = -nFE^{\circ\prime} = -(2 \text{ mol})(9.648 \times 10^4 \, \frac{J}{\text{mol} \cdot \text{V}})(-0.581 \text{ V}) = \underline{\mathbf{1.12 \times 10^5 \, J}}$$

Second step:

$$\Delta G^{\circ\prime} = -nFE^{\circ\prime} = -(2 \text{ mol})(9.648 \times 10^4 \, \frac{J}{\text{mol} \cdot \text{V}})(-0.197 \text{ V}) = \underline{\mathbf{3.80 \times 10^4 \, J}}$$

Overall:
$$\Delta G^{\circ\prime} = \Delta G^{\circ\prime}{}_{(\text{step 1})} + \Delta G^{\circ\prime}{}_{(\text{step 2})} = (1.12 \times 10^5 \text{ J}) + (3.80 \times 10^4 \text{ J}) = \underline{\mathbf{1.50 \times 10^5 \, J}}$$

(b) Use Equation 17.2. Note that $4e^-$ are transferred in the overall reaction (*n* = 4):

$$\Delta G^{\circ\prime} = -nFE^{\circ\prime} \qquad \text{so} \quad 1.50 \times 10^5 \text{ J} = -(4 \text{ mol})(9.648 \times 10^4 \frac{\text{J}}{\text{mol} \cdot \text{V}}) E^{\circ\prime}$$

$$E^{\circ\prime} = \underline{-0.389 \text{ V}}$$

107. Recall that the anode (oxidation) is on the left of the cell notation and the cathode (reduction) is on the right, and that the Pt electrodes are inert and play no part in the redox reaction. The two half-reactions are:

anode: $H_2(g) \rightarrow 2H^+(aq) + 2e^-$ $E^{\circ}_{ox} = -E^{\circ}_{red} = 0.000 \text{ V}$

cathode: $2H^+(aq) + 2e^- \rightarrow H_2(g)$ $E^{\circ}_{red} = 0.000 \text{ V}$

Adding the two half-reactions together and eliminating $2e^-$ gives the cell reaction:

$$H_2(g)_{anode} + 2H^+(aq)_{cathode} \rightarrow H_2(g)_{cathode} + 2H^+(aq)_{anode}$$

$$E^{\circ} = E^{\circ}_{ox} + E^{\circ}_{red} = 0.000V$$

At the anode, pH = 7.0, so (using Equation 13.3) $[H^+] = 10^{-7.0} = 1.0 \times 10^{-7}$ M; at the cathode pH = 0.0 so $[H^+] = 10^{-0.0} = 1.0$ M. Use the Nernst Equation to calculate the non-standard cell voltage, with $n = 2$, $E^{\circ} = 0.000$ V; the expression for Q will be:

$$Q = \frac{(P_{H_2})_{cathode}[H^+]^2_{anode}}{(P_{H_2})_{anode}[H^+]^2_{cathode}}$$

Thus, the Nernst Equation for this reaction becomes:

$$E = 0.000 \text{ V} - \frac{(0.0257 \text{ V})}{2} \ln \frac{(P_{H_2})_{cathode}[H^+]^2_{anode}}{(P_{H_2})_{anode}[H^+]^2_{cathode}}$$

$$= 0.000 \text{ V} - \frac{(0.0257 \text{ V})}{2} \ln \frac{(1.0)(1.0)^2}{(1.0)(1.0 \times 10^{-7})^2} = \underline{+0.414 \text{ V}}$$

108. See Chemistry Beyond the Classroom, page 461−463: the half-reaction that consumes H_2 in the fuel cell is:

$$H_2(g) \rightarrow 2H^+(aq) + 2e^-$$

Use the Ideal Gas Law (Chapter 5) to determine how many mol H_2 are available in the tank and hence (using the stoichiometry of the above reaction) how many mol e^- it can provide.

$$\text{mol } H_2 \text{ in tank} = n = \frac{PV}{RT} = \frac{(200 \text{ atm})(1.0 \text{ L})}{(0.0821 \text{ L} \cdot \text{atm/mol} \cdot \text{K})(298 \text{ K})} = 8.17 \text{ mol } H_2$$

$$\text{mol } e^- \text{ produced} = 8.17 \text{ mol } H_2 \times \frac{2 \text{ mol } e^-}{1 \text{ mol } H_2} = 16.3 \text{ mol } e^-$$

Finally, use the current (1.5 A = 1.5 C/s) and conversion factors from Table 17.3 to relate mol e^- to time:

$$\text{time} = 16.3 \text{ mol } e^- \times \frac{9.648 \times 10^4 \text{ C}}{1 \text{ mol } e^-} \times \frac{1 \text{ s}}{1.5 \text{ C}} = \underline{\mathbf{1.0 \times 10^6 \text{ s}}} \quad \text{or} \quad \underline{\mathbf{12 \text{ d}}}$$

109. See section 17-6. The electroplating process will involve two reduction reactions:

$$Zn^{2+}(aq) + 2e^- \rightarrow Zn(s) \qquad \text{and} \qquad Sn^{2+}(aq) + 2e^- \rightarrow Sn(s)$$

from which it is seen that 2 mol e^- are required to plate each mol of Zn or Sn. We are given an alloy that is 68% Zn and 32 % Sn. Assume 100 g of alloy; so within that sample there will be 68 g Zn and 32 g Sn and the corresponding mol of each metal in the 100 g of alloy is:

$$\text{mol Zn in 100 g alloy} = 68 \text{ g Zn} \times \frac{1 \text{ mol Zn}}{65.39 \text{ g Zn}} = 1.0 \text{ mol Zn}$$

$$\text{mol Sn in 100 g alloy} = 32 \text{ g Sn} \times \frac{1 \text{ mol Sn}}{118.71 \text{ g Sn}} = 0.27 \text{ mol Sn}$$

Since it takes 2 mol e^- to plate every mol of Zn or Sn, the mol e^- required to plate each metal will be in the same proportions as the metals themselves; likewise, the total charge required to plate the 100 g of alloy will also be distributed in the same proportions as the relative numbers of moles of Zn and Sn. Thus, since current is proportional to charge passed:

% of current required to plate 1.0 mol Zn

$$= \frac{\text{mol Zn}}{\text{total mol}} \times 100\% = \frac{1.0 \text{ mol}}{(1.0 + 0.27) \text{ mol}} \times 100\% = \underline{\mathbf{79 \%}}$$

% of current required to plate 0.27 mol Sn

$$= \frac{\text{mol Sn}}{\text{total mol}} \times 100\% = \frac{0.27 \text{ mol}}{(1.0 + 0.27) \text{ mol}} \times 100\% = \underline{\mathbf{21 \%}}$$

NUCLEAR CHEMISTRY

18-1 NUCLEAR STABILITY

2. This example is similar to Example 18.1. For each part, determine the neutron:proton ratio, then check the belt of stability (Figure 18.1). For these lighter elements, n/p^+ should be close to 1 or a little above 1.

(a) $^{28}_{14}\text{Si}$: $n/p^+ = \dfrac{14}{14} = 1$

$^{29}_{14}\text{Si}$: $n/p^+ = \dfrac{15}{14} = 1.07$

$n/p^+ = 1$ is closer to the belt of stability, so $^{28}_{14}\text{Si}$ **is more stable.**

(b) $^{6}_{3}\text{Li}$: $n/p^+ = \dfrac{3}{3} = 1$

$^{8}_{3}\text{Li}$: $n/p^+ = \dfrac{5}{3} = 1.67$

$n/p^+ = 1$ is closer to the belt of stability, so $^{6}_{3}\text{Li}$ **is more stable.**

(c) $^{23}_{11}\text{Na}$: $n/p^+ = \dfrac{12}{11} = 1.09$

$^{20}_{11}\text{Na}$: $n/p^+ = \dfrac{9}{11} = 0.82$

$n/p^+ = 1.09$ is closer to the belt of stability and a value of $n/p^+ = 0.82$ is below the belt of stability, so $^{23}_{11}\text{Na}$ **is more stable.**

4. See Section 18.1 and refer to the empirical rules for nuclear stability.

(a) All isotopes of Ni have 28 protons: since 28 is an even number and one of the *magic numbers*, **isotopes of Ni will tend to be more stable** than the isotopes of Cu, which have an odd number (29) of protons.

(b) **Isotopes of selenium will tend to be more stable**, as they have an even number of protons (34), whereas isotopes of Sb contain an odd number (51) of protons.

(c) **Isotopes of cadmium will tend to be more stable**, as they have an even number of protons (48), whereas isotopes of Au contain an odd number (79) of protons.

18-2   RADIOACTIVITY: Nuclear Equations

6. This problem is similar to Example 18.2(a).

Chromium-51 ($Z = 24$, $A = 51$) may be written as $^{51}_{24}Cr$; recall that the particle emitted is the positron ($^{0}_{1}e$).

Unbalanced equation:	$^{51}_{24}Cr \rightarrow ^{0}_{1}e + ^{A}_{Z}X$
Balance mass and atomic number:	$^{51}_{24}Cr \rightarrow ^{0}_{1}e + ^{51}_{23}X$
Identify element X with $Z = 23$:	$^{51}_{24}Cr \rightarrow ^{0}_{1}e + ^{51}_{23}V$

8. This problem is similar to Example 18.2 and problem 6 above. Write unbalanced equations for the two processes, balance mass and atomic number, then identify the missing species.

(a) Particle emitted: α particle ($^{4}_{2}He$)

Unbalanced equation:	$^{235}_{92}U \rightarrow ^{4}_{2}He + ^{A}_{Z}X$
Balance mass and atomic number:	$^{235}_{92}U \rightarrow ^{4}_{2}He + ^{231}_{90}X$
Identify element X with $Z = 90$:	$^{235}_{92}U \rightarrow ^{4}_{2}He + ^{231}_{90}Th$

The daughter product, formed by alpha emission from uranium-235, is $^{231}_{90}Th$.

(b) Particle emitted: β particle ($^{0}_{-1}e$)

Unbalanced equation:	$^{231}_{90}Th \rightarrow ^{0}_{-1}e + ^{A}_{Z}X$
Balance mass and atomic number:	$^{231}_{90}Th \rightarrow ^{0}_{-1}e + ^{231}_{91}X$
Identify element X with $Z = 91$:	$^{231}_{90}Th \rightarrow ^{0}_{-1}e + ^{231}_{91}Pa$

The daughter product, formed by alpha emission from thorium-231, is $^{231}_{91}Pa$.

10. These problems are similar to Example 18.2 and problem 6 above. Write unbalanced equations for each process, balance mass and atomic number, then identify the missing species.

(a) Particle emitted: positron ($^{0}_{1}e$)

Unbalanced equation:	$^{A}_{Z}X \rightarrow ^{0}_{1}e + ^{52}_{25}Mn$
Balance mass and atomic number:	$^{52}_{26}X \rightarrow ^{0}_{1}e + ^{52}_{25}Mn$
Identify element X with $Z = 26$:	$^{52}_{26}Fe \rightarrow ^{0}_{1}e + ^{52}_{25}Mn$

(b) Particle emitted: β particle ($^{0}_{-1}e$)

Unbalanced equation:	$^{A}_{Z}X \rightarrow ^{0}_{-1}e + ^{228}_{89}Ac$
Balance mass and atomic number:	$^{228}_{88}X \rightarrow ^{0}_{-1}e + ^{228}_{89}Ac$

Identify element X with $Z = 88$: $\quad {}^{228}_{88}\text{Ra} \rightarrow {}^{0}_{-1}e + {}^{228}_{89}\text{Ac}$

(c) Particle emitted: $\qquad\qquad\qquad\quad \alpha$ particle ($ {}^{4}_{2}\text{He}$)

Unbalanced equation: $\qquad\qquad\quad {}^{A}_{Z}\text{X} \rightarrow {}^{4}_{2}\text{He} + {}^{232}_{93}\text{Np}$

Balance mass and atomic number: $\quad {}^{236}_{95}\text{X} \rightarrow {}^{4}_{2}\text{He} + {}^{232}_{93}\text{Np}$

Identify element X with $Z = 95$: $\quad {}^{236}_{95}\text{Am} \rightarrow {}^{4}_{2}\text{He} + {}^{232}_{93}\text{Np}$

12. These problems are similar to Example 18.2 and problem 6 above. Write unbalanced equations for each process, balance mass and atomic number, then identify the missing species.

(a) Particle emitted: $\qquad\qquad\qquad\quad \alpha$ particle ($ {}^{4}_{2}\text{He}$)

Unbalanced equation: $\qquad\qquad\quad {}^{230}_{90}\text{Th} \rightarrow {}^{4}_{2}\text{He} + {}^{A}_{Z}\text{X}$

Balance mass and atomic number: $\quad {}^{230}_{90}\text{Th} \rightarrow {}^{4}_{2}\text{He} + {}^{226}_{88}\text{X}$

Identify element X with $Z = 88$: $\quad {}^{230}_{90}\text{Th} \rightarrow {}^{4}_{2}\text{He} + {}^{226}_{88}\text{Ra}$

(b) Particle emitted: $\qquad\qquad\qquad\quad \beta$ particle ($ {}^{0}_{-1}e$)

Unbalanced equation: $\qquad\qquad\quad {}^{210}_{82}\text{Pb} \rightarrow {}^{0}_{-1}e + {}^{A}_{Z}\text{X}$

Balance mass and atomic number: $\quad {}^{210}_{82}\text{Pb} \rightarrow {}^{0}_{-1}e + {}^{210}_{83}\text{X}$

Identify element X with $Z = 83$: $\quad {}^{210}_{82}\text{Pb} \rightarrow {}^{0}_{-1}e + {}^{210}_{83}\text{Bi}$

(c) Fission (see section 18.5) implies that a neutron ($ {}^{1}_{0}n$) interacts with the nucleus: to produce an *excess* of 2 neutrons, a total of 3 must be produced (the initial one plus 2 more).

Unbalanced equation: $\qquad\qquad\quad {}^{235}_{92}\text{U} + {}^{1}_{0}n \rightarrow {}^{140}_{56}\text{Ba} + {}^{A}_{Z}\text{X} + 3{}^{1}_{0}n$

Balance mass and atomic number: $\quad {}^{235}_{92}\text{U} + {}^{1}_{0}n \rightarrow {}^{140}_{56}\text{Ba} + {}^{93}_{36}\text{X} + 3{}^{1}_{0}n$

Identify element X with $Z = 36$: $\quad {}^{235}_{92}\text{U} + {}^{1}_{0}n \rightarrow {}^{140}_{56}\text{Ba} + {}^{93}_{36}\text{Kr} + 3{}^{1}_{0}n$

(d) Particle *captured*: $\qquad\qquad\qquad$ electron ($ {}^{0}_{-1}e$)

Unbalanced equation: $\qquad\qquad\quad {}^{37}_{18}\text{Ar} + {}^{0}_{-1}e \rightarrow {}^{A}_{Z}\text{X}$

Balance mass and atomic number: $\quad {}^{37}_{18}\text{Ar} + {}^{0}_{-1}e \rightarrow {}^{37}_{17}\text{X}$

Identify element X with $Z = 17$: $\quad {}^{37}_{18}\text{Ar} + {}^{0}_{-1}e \rightarrow {}^{37}_{17}\text{Cl}$

14. These problems can be approached in a way similar to Example 18.2 and problem 6 above. Write unbalanced equations for each process, balance mass and atomic number, then identify the missing species.

(a) Unbalanced equation: $\qquad\qquad\quad {}^{209}_{83}\text{Bi} + {}^{64}_{28}\text{Ni} \rightarrow {}^{1}_{0}n + {}^{A}_{Z}\text{X}$

Balance mass and atomic number: $^{209}_{83}Bi + ^{64}_{28}Ni \rightarrow ^{1}_{0}n + ^{272}_{111}X$

Identify element X with Z = 111: $^{209}_{83}Bi + ^{64}_{28}Ni \rightarrow ^{1}_{0}n + ^{272}_{111}Rg$

The isotope formed is $^{272}_{111}Rg$.

(b) Particles emitted: three α particles ($3\,^{4}_{2}He$) — for solving this problem, treat them together.

Unbalanced equation: $^{272}_{111}Rg \rightarrow 3\,^{4}_{2}He + ^{A}_{Z}X$

Balance mass and atomic number: $^{272}_{111}Rg \rightarrow 3\,^{4}_{2}He + ^{260}_{105}X$

Identify element X with Z = 105: $^{272}_{111}Rg \rightarrow 3\,^{4}_{2}He + ^{260}_{105}Db$

The isotope formed is $^{260}_{105}Db$.

16. This problem is similar to Example 18.2 and problem 8 above. Write unbalanced equations for the two processes, then balance mass and atomic number to obtain the product nuclide. (Give the product an arbitrary element symbol such as Z.)

Positron emission:

    Particle emitted: positron ($^{0}_{1}e$)

    Unbalanced equation: $^{282}_{115}X \rightarrow ^{0}_{1}e + ^{A}_{Z}Z$

    Balance mass and atomic number: $^{282}_{115}X \rightarrow ^{0}_{1}e + ^{282}_{114}Z$

K-capture:

    Particle *captured*: electron ($^{0}_{-1}e$)

    Unbalanced equation: $^{282}_{115}X + ^{0}_{-1}e \rightarrow ^{A}_{Z}Z$

    Balance mass and atomic number: $^{282}_{115}X + ^{0}_{-1}e \rightarrow ^{282}_{114}Z$

The product ($^{282}_{114}Z$) is the same in each case.

18. These problems can be approached in a way similar to Example 18.2 and problem 14 above. Write unbalanced equations for each process, then balance mass and atomic number and identify the missing species.

(a) Unbalanced equation: $^{54}_{26}Fe + ^{4}_{2}He \rightarrow 2\,^{1}_{1}H + ^{A}_{Z}X$

    Balance mass and atomic number: $^{54}_{26}Fe + ^{4}_{2}He \rightarrow 2\,^{1}_{1}H + ^{56}_{26}X$

    Identify element X with Z = 26: $^{54}_{26}Fe + ^{4}_{2}He \rightarrow 2\,^{1}_{1}H + ^{56}_{26}Fe$

(b) Unbalanced equation: $^{96}_{42}Mo + ^{2}_{1}H \rightarrow ^{1}_{0}n + ^{A}_{Z}X$

    Balance mass and atomic number: $^{96}_{42}Mo + ^{2}_{1}H \rightarrow ^{1}_{0}n + ^{97}_{43}X$

    Identify element X with Z = 43: $^{96}_{42}Mo + ^{2}_{1}H \rightarrow ^{1}_{0}n + ^{97}_{43}Tc$

(c) Unbalanced equation: $^{40}_{18}Ar + ^{A}_{Z}X \rightarrow ^{43}_{19}K + ^{1}_{1}H$

Balance mass and atomic number: $^{40}_{18}Ar + ^{4}_{2}X \rightarrow ^{43}_{19}K + ^{1}_{1}H$

Identify element X with $Z = 2$: $^{40}_{18}Ar + ^{4}_{2}He \rightarrow ^{43}_{19}K + ^{1}_{1}H$

(d) Unbalanced equation: $^{A}_{Z}X + ^{1}_{0}n \rightarrow ^{1}_{1}H + ^{31}_{15}P$

Balance mass and atomic number: $^{31}_{16}X + ^{1}_{0}n \rightarrow ^{1}_{1}H + ^{31}_{15}P$

Identify element X with $Z = 16$: $^{31}_{16}S + ^{1}_{0}n \rightarrow ^{1}_{1}H + ^{31}_{15}P$

20. These problems can be approached in a way similar to Example 18.2 and problem 18 above. Write unbalanced equations for each process by converting the species in the word equations into symbols of the form $^{A}_{Z}X$, then balance mass and atomic number and identify the missing species as before.

(a) Unbalanced equation: $^{240}_{99}Es + ^{1}_{0}n \rightarrow 2^{1}_{0}n + ^{A}_{Z}X + ^{161}_{64}Gd$

Balance mass and atomic number: $^{240}_{99}Es + ^{1}_{0}n \rightarrow 2^{1}_{0}n + ^{78}_{35}X + ^{161}_{64}Gd$

Identify element X with $Z = 35$: $^{240}_{99}Es + ^{1}_{0}n \rightarrow 2^{1}_{0}n + ^{78}_{35}Br + ^{161}_{64}Gd$

The complete equation is: **Es-240 + neutron $\rightarrow$ 2 neutrons + Br-78 + Gd-161**

(b) Unbalanced equation: $^{A}_{Z}X \rightarrow ^{0}_{-1}e + ^{59}_{27}Co$

Balance mass and atomic number: $^{59}_{26}X \rightarrow ^{0}_{-1}e + ^{59}_{27}Co$

Identify element X with $Z = 26$: $^{59}_{26}Fe \rightarrow ^{0}_{-1}e + ^{59}_{27}Co$

The complete equation is: **Fe-59 $\rightarrow$ β-particle + Co-59**

(c) Unbalanced equation: $4^{1}_{1}H \rightarrow ^{A}_{Z}X + 2^{0}_{1}e$

Balance mass and atomic number: $4^{1}_{1}H \rightarrow ^{4}_{2}X + 2^{0}_{1}e$

Identify particle X with $Z = 2, A = 4$: $4^{1}_{1}H \rightarrow ^{4}_{2}He + 2^{0}_{1}e$

The complete equation is: $4^{1}_{1}H \rightarrow$ **α-particle + 2 positrons**

(d) Unbalanced equation: $^{24}_{12}Mg + ^{1}_{0}n \rightarrow ^{1}_{1}H + ^{A}_{Z}X$

Balance mass and atomic number: $^{24}_{12}Mg + ^{1}_{0}n \rightarrow ^{1}_{1}H + ^{24}_{11}X$

Identify element X with $Z = 11$: $^{24}_{12}Mg + ^{1}_{0}n \rightarrow ^{1}_{1}H + ^{24}_{11}Na$

The complete equation is: **Mg-24 + 1 neutron $\rightarrow$ proton + Na-24**

18-3 RATE OF RADIOACTIVE DECAY

22. Recall that the curie (Ci) expresses radioactive decay in terms of the rate of disintegration in atoms per second: specifically, 1 Ci = 3.700 × 10¹⁰ atoms/s.

$$\text{Thus the activity } = 15.0 \text{ mCi} \times \frac{1 \text{ Ci}}{1000 \text{ mCi}} \times \frac{3.700 \times 10^{10} \text{ atoms/s}}{1 \text{ Ci}} = 5.55 \times 10^{8} \text{ atoms/s}$$

Then during 2.63 h the number of disintegrations is

$$= 2.63 \text{ h} \times \frac{60 \text{ min}}{1 \text{ h}} \times \frac{60 \text{ s}}{1 \text{ min}} \times \frac{5.55 \times 10^{8} \text{ atoms}}{1 \text{ s}}$$

$$= \mathbf{5.25 \times 10^{12} \text{ atoms}}$$

24. In order to relate the actual activity (units: Ci) to the measured activity (19.4 × 10³ counts/min), they should both be expressed in the same units. First, therefore, convert the measured activity into Ci. Recall that 1 Ci = 3.700 × 10¹⁰ atoms/s; and note that each measured disintegration (each "count" on the Geiger counter) corresponds to one atom decaying.

$$\text{measured activity} = \frac{19.4 \times 10^{3} \text{ atoms}}{1 \text{ min}} \times \frac{1 \text{ min}}{60 \text{ s}} \times \frac{1 \text{ Ci}}{3.700 \times 10^{10} \text{ atoms/s}} = 8.74 \times 10^{-9} \text{ Ci}$$

We are also told that the measured activity is 0.070% of the actual activity; in other words:

$$\text{\% emitted particles counted } = \frac{\text{measured activity}}{\text{actual activity}} \times 100\% = 0.070\%$$

From this last equation, we can rearrange to solve for the actual activity:

$$\text{actual activity} = \frac{\text{measured activity} \times 100\%}{0.070 \%} = \frac{(8.74 \times 10^{-9} \text{ Ci}) \times 100\%}{0.070 \%} = \mathbf{1.25 \times 10^{-5} \text{ Ci}}$$

26. This is similar to Example 18.3(b).

Strategy: 1. Calculate the number of nuclei N in 2.00 mg of ⁸⁷Kr (assume its molar mass is 87.0 g/mol).
   2. Use N to calculate the activity A in atoms/s using A = kN.
   3. Convert the activity to Ci.

$$N = 2.00 \text{ mg} \times \frac{1 \text{ g}}{1000 \text{ mg}} \times \frac{6.022 \times 10^{23} \text{ atoms}}{87.0 \text{ g }^{87}\text{Kr}} = 1.38 \times 10^{19} \text{ atoms}$$

$$A = kN \quad = (1.5 \times 10^{-4} \text{ s}^{-1})(1.38 \times 10^{19} \text{ atoms})$$

$$= 2.1 \times 10^{15} \text{ atoms/s}$$

$$A \, (Ci) \quad = \frac{2.1 \times 10^{15} \text{ atoms}}{1 \text{ s}} \times \frac{1 \text{ Ci}}{3.700 \times 10^{10} \text{ atoms/s}}$$

$$= \underline{\mathbf{5.6 \times 10^4 \ Ci}}$$

28. This is similar to Example 18.3(a) and (b).

Strategy: 1. Calculate k (in s^{-1}) using $t_{1/2}$.
   2. Calculate the number of nuclei N in 1.50 mg of ^{64}Cu. (Assume its molar mass is 64.0 g/mol).
   3. Use N to calculate the activity A in atoms/s using $A = kN$.
   4. Convert the activity to Ci.

$$k = \frac{0.693}{t_{1/2}} = \frac{0.693}{12.8 \text{ h}} \times \frac{1 \text{ h}}{60 \text{ min}} \times \frac{1 \text{ min}}{60 \text{ s}} = 1.50 \times 10^{-5} \ s^{-1}$$

$$N = 1.50 \text{ mg} \times \frac{1 \text{ g}}{1000 \text{ mg}} \times \frac{6.022 \times 10^{23} \text{ atoms}}{64.0 \text{ g } ^{64}Cu} = 1.41 \times 10^{19} \text{ atoms}$$

$$A = kN = (1.50 \times 10^{-5} \ s^{-1})(1.41 \times 10^{19} \text{ atoms}) = 2.12 \times 10^{14} \text{ atoms/s}$$

$$A \, (Ci) = \frac{2.12 \times 10^{14} \text{ atoms}}{1 \text{ s}} \times \frac{1 \text{ Ci}}{3.700 \times 10^{10} \text{ atoms/s}} = \underline{\mathbf{5.73 \times 10^3 \ Ci}}$$

30. This problem is related to Example 18.3. However, in this case the activity is known and is used to calculate the total number of nuclei of Br-82 present and hence the mass of this isotope.

Strategy: 1. Calculate k (in min^{-1}) using $t_{1/2}$.
   2. Use k and the given activity A to calculate the number of nuclei N in the sample. Since $A = kN$, $N = A/k$.
   3. Use Avogadro's number to convert N to mass of ^{82}Br (assume its molar mass is 82.0 g/mol).

$$k = \frac{0.693}{t_{1/2}} = \frac{0.693}{36 \text{ h}} \times \frac{1 \text{ h}}{60 \text{ min}} = 3.2 \times 10^{-4} \ min^{-1}$$

$$A = \frac{1.2 \times 10^5 \text{ disintegrations}}{1 \text{ min}} \times \frac{1 \text{ atom}}{1 \text{ disintegration}} = 1.2 \times 10^5 \text{ atoms min}^{-1}$$

$$N = \frac{A}{k} = \frac{1.2 \times 10^5 \text{ atoms min}^{-1}}{3.2 \times 10^{-4} \text{ min}^{-1}} = 3.8 \times 10^8 \text{ atoms}$$

$$\text{mass of } ^{82}\text{Br} \ = 3.8 \times 10^8 \text{ atoms} \times \frac{82.0 \text{ g } ^{82}\text{Br}}{6.022 \times 10^{23} \text{ atoms}}$$

$$= \underline{\mathbf{5.2 \times 10^{-14} \text{ g}}}$$

32. This is similar to Example 18.3(b) and (c).

   Strategy: 1. Calculate the number of nuclei N in 1.000 mg of ^{36}Cl (assume its molar mass is 36.0 g/mol).
   2. Calculate the activity A in atoms/min using $A = kN$: since each atom disintegrating produces 1 β particle, this activity will equal the number of β particles produced per minute.
   3. Convert the activity to Ci.

$$N \ = 1.000 \text{ mg} \times \frac{1 \text{ g}}{1000 \text{ mg}} \times \frac{6.022 \times 10^{23} \text{ atoms}}{36.0 \text{ g } ^{36}\text{Cl}} = 1.67 \times 10^{19} \text{ atoms}$$

$$A \ = kN = \frac{2.3 \times 10^{-6}}{1 \text{ y}} \times \frac{1 \text{ y}}{365 \text{ d}} \times \frac{1 \text{ d}}{24 \text{ h}} \times \frac{1 \text{ h}}{60 \text{ min}} \times (1.67 \times 10^{19} \text{ atoms})$$

$$= 7.3 \times 10^7 \text{ atoms/min} = \text{ number of disintegrations/min}$$

Hence **7.3×10^7 β-particles are produced per minute.**

$$\text{Activity, } A \text{ (Ci)} \ = \frac{7.3 \times 10^7 \text{ atoms}}{1 \text{ min}} \times \frac{1 \text{ min}}{60 \text{ s}} \times \frac{1 \text{ Ci}}{3.700 \times 10^{10} \text{ atoms/s}} = \underline{\mathbf{3.3 \times 10^{-5} \text{ Ci}}}$$

34. This is related to Example 18.4; note that in this case, since a time variable is involved, we must use equation **18.2**, which requires a value for k.

   (a) For the sample to exhibit 68% of its original activity, $A = 0.68A_0$.

   Strategy: 1. Calculate k (in d^{-1}) using $t_{\frac{1}{2}}$.
   2. Use k and the "integrated" rate equation (**18.2**) to determine a value for t.

$$k \ = \frac{0.693}{t_{\frac{1}{2}}} = \frac{0.693}{14.3 \text{ d}} = 0.0485 \text{ d}^{-1}$$

$$kt = \ln\frac{A_0}{A} = \ln\frac{A_0}{0.68 A_0} = \ln\frac{1}{0.68}$$

so the time t taken for the activity to fall to 68% of its original value is

$$t = \frac{1}{k}\ln\frac{1}{0.68} = \frac{1}{0.0485 \text{ d}^{-1}}\ln\frac{1}{0.68} = \underline{\mathbf{8.0 \text{ d}}}$$

(b) In this part we are dealing with numbers of atoms, N, and since N is proportional to activity A ($A = kN$ or $N = A/k$) equation **18.2** can be rewritten as

$$\ln\frac{N_0}{N} = kt$$

To determine N, the number of atoms remaining after 755 d, we first need to calculate N_0.

Strategy: 1. Calculate the number of nuclei N_0 in 10.0 μg of ^{32}P (assume its molar mass is 32.0 g/mol).
2. Use the above equation to calculate N for $t = 755$ d.

$$N_0 = 10.0 \times 10^{-6} \text{ g } ^{32}P \times \frac{6.022 \times 10^{23} \text{ atoms}}{32.0 \text{ g } ^{32}P} = 1.88 \times 10^{17} \text{ atoms}$$

$$\ln\frac{N_0}{N} = \ln\frac{1.88 \times 10^{17} \text{ atoms}}{N} = kt \qquad \text{so:} \qquad \frac{1.88 \times 10^{17} \text{ atoms}}{N} = e^{kt}$$

$$N = \frac{1.88 \times 10^{17} \text{ atoms}}{e^{kt}} = \frac{1.88 \times 10^{17} \text{ atoms}}{e^{(0.0485 \text{ d}^{-1})(755 \text{ d})}} = \underline{\textbf{24 atoms}}$$

(remember that N must be a whole number!)

36. This problem is similar to Example 18.4. We know the present quantity of ^{14}C is 0.972 times the ^{14}C content in living material – in other words, $X = 0.972X_0$, since X_0 is constant while the material is living and up to $t = 0$, the time that the organism dies.

Strategy: 1. Calculate k (in y^{-1}) using $t_{1/2}$.
2. Use k and the "integrated" rate equation to determine t, the age of the canvas.

$$k = \frac{0.693}{t_{1/2}} = \frac{0.693}{5730 \text{ y}} = 1.21 \times 10^{-4} \text{ y}^{-1}$$

$$\ln\frac{X_0}{X} = kt \qquad \text{so:} \qquad t = \frac{1}{k}\ln\frac{X_0}{X}$$

$$t = \frac{1}{1.21 \times 10^{-4} \text{ y}^{-1}}\ln\frac{X_0}{0.972X_0} = \frac{1}{1.21 \times 10^{-4} \text{ y}^{-1}}\ln\frac{1}{0.972} = \underline{\textbf{235 y}}$$

The painting is dated to be only 235 years old, so **it does not come from the time of Michelangelo.**

38. This problem is similar to Example 18.4: we know $t_{1/2}$, A and A_0.

Strategy: 1. Calculate k (in y^{-1}) using $t_{1/2}$.
2. Use k and Equation **18.2** to determine t, the age of the artifact.

$$k = \frac{0.693}{t_{1/2}} = \frac{0.693}{5730 \text{ y}} = 1.21 \times 10^{-4} \text{ y}^{-1}$$

$$\ln\frac{A_0}{A} = kt \qquad\qquad \text{so:} \qquad t = \frac{1}{k}\ln\frac{A_0}{A}$$

$$t = \frac{1}{1.21 \times 10^{-4} \text{ y}^{-1}}\ln\frac{15.3 \text{ disintegrations/min/ g C}}{5.0 \text{ disintegrations/min/ g C}} = \underline{\textbf{9.2} \times \textbf{10}^3 \text{ \textbf{y}}}$$

40. This problem is similar to Problem 36 above. We are told the quantity of ^3H in the whiskey is 62% of the ^3H content in fresh water — in other words, $X = 0.62X_0$, and X_0 is constant up to $t = 0$, the time that the water is bottled as whiskey.

Strategy: 1. Calculate k (in y^{-1}) using $t_{1/2}$ given in Problem 39
2. Use k and Equation 18.2 to determine t, the age of the whiskey.

$$k = \frac{0.693}{t_{1/2}} = \frac{0.693}{12.3 \text{ y}} = 0.0563 \text{ y}^{-1}$$

$$\ln\frac{X_0}{X} = kt \qquad\qquad \text{and} \qquad t = \frac{1}{k}\ln\frac{X_0}{X}$$

$$t = \frac{1}{0.0563 \text{ y}^{-1}}\ln\frac{X_0}{0.62X_0} = \frac{1}{0.0563 \text{ y}^{-1}}\ln\frac{1}{0.62} = \underline{\textbf{8.5 y}}$$

42. This problem can be approached in a manner similar to Example 18.4. The decay of U-238 to Pb-206 can be represented as

$$^{238}U \rightarrow {}^{206}Pb$$

and hence simple stoichiometry can be used to relate the mass of Pb-206 now present to the mass of U-238 that decayed to form it. We can therefore calculate the total mass of U-238 originally present per gram of U-238 now found in the meteorite. These two masses are X_0 and X, respectively.

Strategy: 1. Calculate k (in y^{-1}) using $t_{1/2}$.
2. Calculate mass of U-238 required to form 0.813 g of Pb-206.
3. Use k and the integrated rate equation to determine t, the age of the meteorite.

$$k = \frac{0.693}{t_{\frac{1}{2}}} = \frac{0.693}{4.5 \times 10^9 \text{ y}} = 1.54 \times 10^{-10} \text{ y}^{-1}$$

mass ^{238}U required to form 0.813 g of ^{206}Pb =

$$0.813 \text{ g } ^{206}Pb \times \frac{1 \text{ mol } ^{206}Pb}{206 \text{ g } ^{206}Pb} \times \frac{1 \text{ mol } ^{238}U}{1 \text{ mol } ^{206}Pb} \times \frac{238 \text{ g } ^{238}U}{1 \text{ mol } ^{238}U} = 0.939 \text{ g } ^{238}U$$

In the meteorite sample, the observed mass of ^{238}U now present is X = 1.00 g and the corresponding mass of ^{238}U present at the time of its formation is X_0 = (1.00 + 0.939) g = 1.939 g

$$\ln\frac{X_0}{X} = kt \qquad\qquad \text{so:} \quad t = \frac{1}{k}\ln\frac{X_0}{X}$$

$$t = \frac{1}{1.54 \times 10^{-10} \text{ y}^{-1}}\ln\frac{1.939 \text{ g } ^{238}U}{1.00 \text{ g } ^{238}U} = \underline{\underline{4.30 \times 10^9 \text{ y}}}$$

18-4   MASS–ENERGY RELATIONS
18-5   NUCLEAR FISSION
18-6   NUCLEAR FUSION

44. (a) This part of the problem is somewhat similar to Example 18.2(a) and to Problem 6 above. Write an unbalanced equation for the process, balance mass and atomic number, then confirm the identity of product.

Particle emitted:	2β particles ($2\ _{-1}^{0}e$)
Unbalanced equation:	$_{38}^{90}Sr \rightarrow 2\ _{-1}^{0}e + _{Z}^{A}X$
Balance mass and atomic number:	$_{38}^{90}Sr \rightarrow 2\ _{-1}^{0}e + _{40}^{90}X$
Identify element X with Z = 40:	$_{38}^{90}Sr \rightarrow 2\ _{-1}^{0}e + _{40}^{90}Zr$

Parts (b) and (c) are similar to Example 18.5.

(b) Use Table 18.3 to find the nuclear mass of Sr-90 and the mass of an e and use them, along with the given mass of Zr-90, to calculate Δm for the decay of 1 mol of Sr-90.

Δm (per mol) = nuclear masses of products − nuclear masses of reactants

= (mass of 1 mol of Zr-90 + mass of 2 mol of $_{-1}^{0}e$)

− (mass of 1 mol of Sr-90)

= (89.8824 g + (2 × 0.00055 g)) − (89.8869 g)

= $\underline{\underline{-0.0034 \text{ g/mol Sr-90}}}$

(c) Find Δm for 6.50 mg of Sr-90 and then use equation **18.3** to determine ΔE.

Δm (for 6.50 mg)

$$= 6.50 \text{ mg } {}^{90}\text{Sr} \times \frac{1 \text{g } {}^{90}\text{Sr}}{1000 \text{ mg } {}^{90}\text{Sr}} \times \frac{1 \text{mol } {}^{90}\text{Sr}}{89.8869 \text{ g } {}^{90}\text{Sr}} \times \frac{-0.0034 \text{ g}}{1 \text{mol } {}^{90}\text{Sr}} = -2.5 \times 10^{-7} \text{ g}$$

$$\Delta E = 9.00 \times 10^{10} \frac{\text{kJ}}{\text{g}} (\Delta m) = 9.00 \times 10^{10} \frac{\text{kJ}}{\text{g}} (-2.5 \times 10^{-7} \text{ g}) = \underline{-2.2 \times 10^{4} \text{ kJ}}$$

Thus, **2.2×10^{4} kJ of energy are released by the decay of 6.50 mg ^{90}Sr.**

46. This problem is similar to Example 18.6. The Al-28 nucleus contains 13 protons and 15 neutrons:

$$\frac{28}{13}\text{Al} \rightarrow 13 {}^{1}_{1}\text{H} + 15 {}^{1}_{0}n$$

(a) Use Table 18.3 to find Δm

Δm (per mol) = 13(mass of 1 mol of ${}^{1}_{1}\text{H}$) + 15(mass of 1 mol of ${}^{1}_{0}n$)

$$- \text{ (mass of 1 mol of } {}^{28}_{13}\text{Al)}$$

$$= 13(1.00728 \text{ g}) + 15(1.00867 \text{ g}) - (27.97477 \text{ g})$$

$$= \underline{0.24992 \text{ g/mol of Al-28}}$$

Thus, **the mass defect of ^{28}Al is $\underline{0.24992}$ g/mol.**

(b) Use equation **18.3** to find ΔE in kJ/mol

$$\Delta E = 9.00 \times 10^{10} \frac{\text{kJ}}{\text{g}} (\Delta m) = 9.00 \times 10^{10} \frac{\text{kJ}}{\text{g}} (0.24922 \frac{\text{g}}{\text{mol}}) = \underline{2.25 \times 10^{10} \text{ kJ/mol}}$$

Thus, **the binding energy of ^{28}Al is $\underline{2.25 \times 10^{10}}$ kJ/mol.**

48. Calculate the mass defect for each nuclide: since binding energy is proportional to mass defect, the nuclide with the higher mass defect will also have the higher binding energy. Calculate the mass defects as in the first part of Example 18.6, using nuclear masses from Table 18.3.

K-40 contains 19 protons and 21 neutrons:

$$\frac{40}{19}\text{K} \rightarrow 19 {}^{1}_{1}\text{H} + 21 {}^{1}_{0}n$$

Δm (per mol) = 19(mass of 1 mol of ${}^{1}_{1}\text{H}$) + 21(mass of 1 mol of ${}^{1}_{0}n$)

$$- \text{ (mass of 1 mol of } {}^{40}_{19}\text{K)}$$

$$= 19(1.00728 \text{ g}) + 21(1.00867 \text{ g}) - (39.95358 \text{ g})$$

$$= \underline{0.36681 \text{ g/mol K-40}}$$

Ca-40 contains 20 protons and 20 neutrons:

$$_{20}^{40}\text{Ca} \rightarrow 20\,_1^1\text{H} + 20\,_0^1 n$$

Δm (per mol) $\quad = 20(\text{mass of 1 mol of } _1^1\text{H}) + 20(\text{mass of 1 mol of } _0^1 n)$

$$- \,(\text{mass of 1 mol of } _{20}^{40}\text{Ca})$$

$$= 20(1.00728 \text{ g}) + 20(1.00867 \text{ g}) - (39.95162 \text{ g})$$

$$= \underline{0.36738 \text{ g/mol Ca-40}}$$

Ca-40 has the higher mass defect, so **Ca-40 will have the higher binding energy.**

50. This problem is similar to Example 18.7. Use Table 18.3 to find any required nuclear masses and use them to calculate Δm for each reaction per mol of reactants. Convert Δm to a value per gram of reactants and then use Equation **18.3** to determine ΔE per gram of reactants.

Recall that for each reaction

$\quad \Delta m$ (per mol) $=$ nuclear masses of products $-$ nuclear masses of reactants

fission reaction:

$\Delta m/\text{mol U-235} \quad = [(\text{mass Zr-94}) + (\text{mass Ce-140}) + 6(\text{mass } _{-1}^{0}e) + 2(\text{mass } _0^1 n)]$

$$- [(\text{mass U-235}) + (\text{mass } _0^1 n)]$$

$$= [(93.8841) + (139.8734) + 6(0.00055) + 2(1.00867)] \text{ g}$$
$$- [(234.9934) + (1.00867)] \text{ g}$$

$$= -0.2239 \text{ g}$$

$\Delta m/\text{g U-235} \quad = \dfrac{-0.2239 \text{ g}}{\text{mol U-235}} \times \dfrac{1 \text{ mol U-235}}{234.9934 \text{ g U-235}}$

$$= -9.528 \times 10^{-4} \text{ g/g U-235}$$

$\Delta E/\text{g U-235} \quad = 9.00 \times 10^{10} \dfrac{\text{kJ}}{\text{g}} (\Delta m)$

$$= 9.00 \times 10^{10} \dfrac{\text{kJ}}{\text{g}} (-9.528 \times 10^{-4} \text{ g/g U-235})$$

$$= \underline{-8.58 \times 10^7 \text{ kJ/g U-235}}$$

fusion reaction:

$\Delta m/\text{mol H-2} \quad = [(\text{mass H-3}) + (\text{mass H-1})] - [2(\text{mass H-2})]$

$$= [(3.01550) + (1.00728)] \text{ g} - [2(2.01355)] \text{ g}$$

$$= -0.00432 \text{ g}$$

$$\Delta m/\text{g H-2} \quad = \frac{-0.00432 \text{ g}}{\text{mol H-2}} \times \frac{1 \text{ mol H-2}}{2.01355 \text{ g H-2}}$$

$$= -2.14 \times 10^{-3} \text{ g/g H-2}$$

$$\Delta E/\text{g H-2} \quad = 9.00 \times 10^{10} \frac{\text{kJ}}{\text{g}} (\Delta m)$$

$$= 9.00 \times 10^{10} \frac{\text{kJ}}{\text{g}} (-2.14 \times 10^{-3} \text{ g/g H-2})$$

$$= \underline{-1.93 \times 10^{8} \text{ kJ/g H-2}}$$

The fusion of deuterium produces more energy per gram of reacting material.

52. This problem is similar in part to Example 18.7.

   Strategy: 1. Obtain a balanced nuclear equation for the fission reaction

   2. Use Table 18.3 to find any required nuclear masses and use them to calculate Δm per mol of U-235.

   3. Convert Δm to a value per microgram of U-235.

   4. Use Equation **18.3** to determine ΔE per microgram of U-235.

   5. Calculate the mass of PETN required to produce the same ΔE.

Unbalanced equation: $\quad {}^{235}_{92}\text{U} + {}^{1}_{0}n \rightarrow {}^{144}_{58}\text{Ce} + {}^{89}_{37}\text{Rb} + \text{X} {}^{0}_{-1}e + \text{Y} {}^{1}_{0}n$

Balance atomic number (X = 3): ${}^{235}_{92}\text{U} + {}^{1}_{0}n \rightarrow {}^{144}_{58}\text{Ce} + {}^{89}_{37}\text{Rb} + 3 {}^{0}_{-1}e + \text{Y} {}^{1}_{0}n$

Balance mass number (Y = 3): $\quad {}^{235}_{92}\text{U} + {}^{1}_{0}n \rightarrow {}^{144}_{58}\text{Ce} + {}^{89}_{37}\text{Rb} + 3 {}^{0}_{-1}e + 3 {}^{1}_{0}n$

$$\Delta m/\text{mol U-235} = [(\text{mass Ce-144}) + (\text{mass Rb-89}) + 3(\text{mass } {}^{0}_{-1}e) + 3(\text{mass } {}^{1}_{0}n)]$$
$$- [(\text{mass U-235}) + (\text{mass } {}^{1}_{0}n)]$$

$$= [(143.8817) + (88.8913) + 3(0.00055) + 3(1.00867)] \text{ g}$$
$$- [(234.9934) + (1.00867)] \text{ g}$$

$$= -0.2014 \text{ g}$$

$$\Delta m/\mu\text{g U-235} \quad = \frac{-0.2014 \text{ g}}{\text{mol U-235}} \times \frac{1 \text{ mol U-235}}{234.9934 \text{ g U-235}} \times \frac{1 \text{ g U-235}}{10^{6} \mu\text{g U-235}}$$

$$= -8.570 \times 10^{-10} \text{ g/}\mu\text{g U-235}$$

$$\Delta E/\mu\text{g U-235} \quad = 9.00 \times 10^{10} \frac{\text{kJ}}{\text{g}} (\Delta m)$$

$$= 9.00 \times 10^{10} \frac{\text{kJ}}{\text{g}} (-8.570 \times 10^{-10} \text{ g/}\mu\text{g U-235})$$

$$= \mathbf{-77.1 \text{ kJ/}\mu\text{g U-235}}$$

Hence 77.1 kJ of energy are released when 1 μg U-235 undergoes fission. We are told that 1 g of PETN evolves 6.47 kJ of energy, so the mass of PETN (in kg) required to produce 77.1 kJ is:

$$77.1 \text{ kJ} \times \frac{1 \text{ g } C_5H_8N_4O_{12}}{6.47 \text{ kJ}} \times \frac{1 \text{ kg } C_5H_8N_4O_{12}}{1000 \text{ g } C_5H_8N_4O_{12}} = \mathbf{1.19 \times 10^{-2} \text{ kg } C_5H_8H_4O_{12}}$$

UNCLASSIFIED PROBLEMS

54. See Section 18.3 and Example 18.3.

(a) Strategy: 1. Calculate k (in d^{-1}) using $t_{1/2}$.
2. Use the integrated form of the decay law to calculate the ratio X_0/X at t = 2.0 d.
3. Use this to determine the ratio X/X_0, which is the fraction remaining after 2.0 d, and hence the percentage disintegrated.

$$k = \frac{0.693}{t_{1/2}} = \frac{0.693}{8.1 \text{ d}} = 0.086 \text{ d}^{-1}$$

$$\ln\frac{X_0}{X} = kt = (0.086 \text{ d}^{-1})(2.0 \text{ d})$$

$$\frac{X_0}{X} = e^{(0.086 \text{ d}^{-1})(2.0 \text{ d})} = 1.2 \qquad \text{so} \quad \frac{X}{X_0} = \frac{1}{1.2} = 0.84$$

The fraction of I-131 remaining after 2.0 d (X/X_0) is 0.84, or 84 %, so __16 % had disintegrated__.

(b) The patient must receive a sample with the same activity as 15.0 mg of fresh iodine-131. Since the fraction of I-131 remaining in the 2-day-old sample is only 84 %, we deduce that each milligram of sample contains only 0.84 mg of I-131. Hence the mass of sample required is:

$$15.0 \text{ mg I-131} \times \frac{1 \text{ mg sample}}{0.84 \text{ mg I-131}} = \underline{\textbf{18 mg sample}}$$

56. See Section 18.3 and Example 18.3. The sample in the smoke detector shows a decay rate of 10 disintegrations per second or 10 atoms/s.

(a) Strategy: 1. Use equation **18.1** to obtain a value for N, the number of atoms of Am-241 that would correspond to an activity of 10 atoms/s. ($A = kN$, so $N = A/k$)
2. Calculate the mass of N atoms of Am-241, assuming a molar mass of 241 g/mol.

$$N = \frac{A}{k} = \frac{\left(\dfrac{10 \text{ atoms}}{1 \text{ s}}\right)}{\left(\dfrac{1.51 \times 10^{-3}}{1 \text{ y}}\right)} = \frac{10 \text{ atoms}}{1 \text{ s}} \times \left(\frac{1 \text{ y}}{1.51 \times 10^{-3}} \times \frac{365 \text{ d}}{1 \text{ y}} \times \frac{24 \text{ h}}{1 \text{ d}} \times \frac{3600 \text{ s}}{1 \text{ h}}\right)$$

$$= 2.1 \times 10^{11} \text{ atoms}$$

$$= 2.1 \times 10^{11} \text{ atoms} \times \frac{241 \text{ g Am-241}}{6.022 \times 10^{23} \text{ atoms}} = \underline{\mathbf{8.4 \times 10^{-11} \text{ g Am-241}}}$$

58. See Sections 4.1, 15.2, and 18.3. Note that the two solutions being mixed contain equal numbers of moles of $Ag^+(aq)$ and $I^-(aq)$, and the given precipitation reaction removes equal numbers of each ion from solution. Hence, after the precipitation has taken place the concentrations of each ion will be the same. Any $I^-(aq)$ in solution can be considered to have come from the precipitate dissolving; the filtrate is, in effect, a saturated solution of AgI.

For AgI: $K_{sp} = [Ag^+][I^-] = 1.0 \times 10^{-16}$

and since $[Ag^+] = [I^-]$, we can substitute into the K_{sp} expression to obtain

   $K_{sp} = [I^-]^2 = 1.0 \times 10^{-16}$

from which we can calculate:

   $[I^-] = [K_{sp}]^{1/2} = (1.0 \times 10^{-16})^{1/2} = 1.0 \times 10^{-8} \ M$

We are told that the activity of a 0.050 M solution of $NaI(aq)$ is 1.25×10^{10} counts/min·mL. (The concentration of $I^-(aq)$ in this solution is also 0.050 M.) Recall that the activity of the solution will be proportional to the number of atoms of radioactive iodine present; and since the activity is given in units of counts per minute **per mL** (counts/min·mL) we can take the *concentration* as being proportional to the activity.

From above, concentration of I^- in the filtrate = $1.0 \times 10^{-8} \ M$

The activity of the filtrate will therefore be

$$= 1.0 \times 10^{-8} \ M \ I^- \ \times \ \frac{1.25 \times 10^{10} \text{ counts/min} \cdot \text{mL}}{0.050 \ M \ I^-} = \underline{\mathbf{2.5 \times 10^3 \text{ counts/min·mL}}}$$

60. This is similar to Example 18.3(a) and (b), and Problem 28 above.

   Strategy: 1.   Calculate k (in s^{-1}) using $t_{1/2}$.
   2.   Use the Ideal Gas Law (Chapter 5) to calculate the number of moles of $^3H_2$ in 1.00 mL at STP.
   2.   Calculate the number of nuclei N present in the sample of $^3H_2$.
   3.   Use N to calculate the activity A in atoms/s using $A = kN$.
   4.   Convert the activity to Ci.

$$k = \frac{0.693}{t_{1/2}} = \frac{0.693}{12.3\,y} \times \frac{1y}{365\,d} \times \frac{1d}{24\,h} \times \frac{1h}{60\,min} \times \frac{1min}{60\,s} = 1.79 \times 10^{-9}\,s^{-1}$$

$$n = \frac{PV}{RT} = \frac{(1\,atm)(1.00\,mL \times \frac{1L}{1000\,mL})}{(0.0821\,L \cdot atm/mol \cdot K)(273\,K)} = 4.46 \times 10^{-5}\,mol\ {}^3H_2$$

$$N = 4.46 \times 10^{-5}\,mol\ {}^3H_2 \times \frac{6.022 \times 10^{23}\ molecules\ {}^3H_2}{1\,mol\ {}^3H_2} \times \frac{2\ atoms\ {}^3H}{1\,molecule\ {}^3H_2}$$

$$= 5.37 \times 10^{19}\ atoms\ {}^3H$$

$$A = kN = (1.79 \times 10^{-9}\,s^{-1})(5.37 \times 10^{19}\ atoms)$$

$$= 9.61 \times 10^{10}\ atoms/s$$

$$A\ (Ci) = \frac{9.61 \times 10^{10}\ atoms}{1\,s} \times \frac{1\,Ci}{3.700 \times 10^{10}\ atoms/s}$$

$$= \underline{\mathbf{2.60\ Ci}}$$

62. See Section 18.3. (We assume that the injected sample does not change the animal's total circulatory volume.) The injected sample contains a total activity of 1.7×10^5 cpm that becomes distributed throughout the animal's circulatory system; when 5 mL is analyzed it has an activity of 1.3×10^3 cpm. Since the activity of any sample is proportional to the quantity of that sample we can write:

$$\frac{total\ activity}{total\ blood\ volume} = \frac{activity\ of\ 5\,mL\ sample}{volume\ of\ 5\,mL\ sample}$$

Hence the total blood volume

$$= total\ activity \times \frac{volume\ of\ 5\,mL\ sample}{activity\ of\ 5\,mL\ sample}$$

$$= 1.7 \times 10^5\ cpm \times \frac{5\,mL}{1.3 \times 10^3\ cpm} = \underline{\mathbf{6.5 \times 10^2\ mL}}$$

64. See Section 18.3. Since we are seeking the stoichiometric ratio of $C_2O_4^{2-}:Cr^{3+}$, begin by converting the activity of each reactant to cpm per mol of each species. Use these to determine the number of moles of each in the product and hence the ratio.

$$Activity\ of\ Cr^{3+}\ source = \frac{765\ cpm}{g\ Na_2CrO_4} \times \frac{162.0\ g\ Na_2CrO_4}{1\,mol\ Na_2CrO_4} \times \frac{1\,mol\ Na_2CrO_4}{1\,mol\ Cr^{3+}}$$

$$= 1.24 \times 10^5\ cpm/mol\ Cr^{3+}$$

$$\text{Activity of } C_2O_4{}^{2-} \text{ source} = \frac{512 \text{ cpm}}{\text{g } H_2C_2O_4} \times \frac{90.04 \text{ g } H_2C_2O_4}{1 \text{ mol } H_2C_2O_4} \times \frac{1 \text{ mol } H_2C_2O_4}{1 \text{ mol } C_2O_4{}^{2-}}$$

$$= 4.61 \times 10^4 \text{ cpm/mol } C_2O_4{}^{2-}$$

$$\text{mol } Cr^{3+} \text{ in product} = 314 \text{ cpm} \times \frac{1 \text{ mol } Cr^{3+}}{1.24 \times 10^5 \text{ cpm}} = 2.53 \times 10^{-3} \text{ mol } Cr^{3+}$$

$$\text{mol } C_2O_4{}^{2-} \text{ in product} = 235 \text{ cpm} \times \frac{1 \text{ mol } C_2O_4{}^{2-}}{4.61 \times 10^4 \text{ cpm}} = 5.10 \times 10^{-3} \text{ mol } C_2O_4{}^{2-}$$

The ratio mol $C_2O_4{}^{2-}$:mol Cr^{3+} is 5.10×10^{-3} mol$/2.53 \times 10^{-3}$ mol $= 2.02{:}1 \approx 2{:}1$. Hence, **there are 2 $C_2O_4{}^{2-}$ ions bound to every Cr^{3+} ion.**

66. See Sections 18.2 and 18.3. Recall that alpha emission produces He-4 nuclei (α particles) and hence the quantity of He gas produced will be directly proportional to the quantity of Po-210 that decays: one Po-210 nucleus produces one He-4.

Strategy: 1. Calculate k (in h^{-1}) using $t_{1/2}$.

2. Calculate the number of nuclei N present in 25.00 g of Po-210. Its molar mass is 209.9368 g/mol (Table 18.3).

3. Use N to calculate the activity A in atoms/h using $A = kN$, and hence how many atoms decay in 75 h.

4. Convert the number of atoms decaying into mol He-4 produced.

5. Use the Ideal Gas Law (Chapter 5) to calculate the volume of He at 25 °C and 1.20 atm.

The alpha emission is: $\quad {}^{210}_{84}\text{Po} \rightarrow {}^{4}_{2}\text{He} + {}^{206}_{82}\text{Pb}$

$$k = \frac{0.693}{t_{1/2}} = \frac{0.693}{138 \text{ d}} \times \frac{1 \text{ d}}{24 \text{ h}} = 2.09 \times 10^{-4} \text{ h}^{-1}$$

$$N = 25.00 \text{ g} \times \frac{6.022 \times 10^{23} \text{ atoms } {}^{210}\text{Po}}{209.9386 \text{ g } {}^{210}\text{Po}} = 7.171 \times 10^{22} \text{ atoms } {}^{210}\text{Po}$$

$$A = kN = \frac{2.09 \times 10^{-4}}{1 \text{ h}} \times 7.171 \times 10^{22} \text{ atoms } {}^{210}\text{Po} = 1.50 \times 10^{19} \text{ atoms } {}^{210}\text{Po h}^{-1}$$

$$\text{atoms decaying in 75 h} = \frac{1.50 \times 10^{19} \text{ atoms } {}^{210}\text{Po}}{1 \text{ h}} \times 75 \text{ h} = 1.1 \times 10^{21} \text{ atoms } {}^{210}\text{Po}$$

$$\text{mol He produced} = 1.1 \times 10^{21} \text{ atoms } {}^{210}\text{Po} \times \frac{1 \text{ mol } {}^{210}\text{Po}}{6.022 \times 10^{23} \text{ atoms } {}^{210}\text{Po}} \times \frac{1 \text{ mol } {}^{4}\text{He}}{1 \text{ mol } {}^{210}\text{Po}}$$

$$= 1.8 \times 10^{-3} \text{ mol He}$$

$$\text{vol He produced} \; = V = \frac{nRT}{P} = \frac{(1.8 \times 10^{-3}\,\text{mol})(0.0821\,\text{L} \cdot \text{atm/mol} \cdot \text{K})(298\,\text{K})}{1.20\,\text{atm}}$$

$$= \underline{\mathbf{3.7 \times 10^{-2}\,L}} \quad \text{or} \quad \underline{\mathbf{37\,mL}}$$

68. See Sections 6.1 and 18.4.

Strategy: 1. Calculate Δm, the mass lost when the two particles annihilate each other. Find the electron mass in Table 18.3; recall that the positron has an identical mass.
2. Determine the total energy produced, ΔE, using $\Delta E = c^2 \Delta m$, and hence the energy per photon.
3. Use $\Delta E = hc/\lambda$ or $\lambda = hc/\Delta E$ to calculate the photon's wavelength.

$$\Delta m \text{ (per mol)} = \text{(electron mass)} + \text{(positron mass)} = 0.00055\,\text{g} + 0.00055\,\text{g}$$
$$= 0.00011\,\text{g}$$

$$\Delta E \text{ (per mol)} = c^2 \Delta m = (3.00 \times 10^8\,\text{ms}^{-1})^2 \times 0.00011\,\text{g} \times \frac{1\,\text{kg}}{1000\,\text{g}} = 9.9 \times 10^{10}\,\text{J}$$

This is the energy produced per mol of reaction, which produces 2 mol of photons.

$$\Delta E \text{ (per photon)} = \frac{9.9 \times 10^{10}\,\text{J}}{2\,\text{mol photons}} \times \frac{1\,\text{mol photons}}{6.022 \times 10^{23}\,\text{photons}} = 8.2 \times 10^{-14}\,\text{J}$$

$$\text{wavelength, } \lambda = \frac{hc}{\Delta E} = \frac{(6.626 \times 10^{-34}\,\text{J s})(3.00 \times 10^8\,\text{m s}^{-1})}{8.2 \times 10^{-14}\,\text{J}} = 2.4 \times 10^{-12}\,\text{m}$$

$$= 2.4 \times 10^{-12}\,\text{m} \times \frac{10^9\,\text{nm}}{1\,\text{m}} = \underline{\mathbf{2.4 \times 10^{-3}\,nm}}$$

CONCEPTUAL QUESTIONS

70. (a) See Section 18.2 and Figure 18.2. Alpha radiation consists of positively charged He-4 nuclei, which are attracted to the negative pole of an electric field; whereas beta radiation consists of negatively charged electrons, which are attracted to the positive pole of the electric field.

(b) See Section 18.2, Applications. The patient receives a dose of a substance containing C-11, a positron emitter, and the labeled compound can localize in different parts of the brain. Positron Emission Tomography (PET) scans of the patient's brain will establish the distribution of the labeled material and hence give a picture of its uptake and metabolism.

(c) See Section 18.5. A nuclear fission reaction is initiated by a neutron, but the process also produces neutrons which, in turn, can initiate further fission reactions. Since the process (once started) can continue in this way, it is self-sustaining.

72. See Sections 11.3 and 18.3. Radioactive decay follows first-order kinetics, for which a plot of ln of concentration against time gives a straight line of slope −*k*.
   In this case, the "concentration" data available are the activities (which are proportional to concentration of nuclide present); a plot of ln(activity) against time will give a straight line whose slope is −*k*, with units h⁻¹. Determine the slope and use it to calculate $t_{1/2}$.

Time, *t* (h)	0.00	0.50	1.00	1.50	2.00	2.50
Activity, *A* (disintegrations/h)	14472	13095	11731	10615	9605	8504
ln(*A*)	9.580	9.480	9.370	9.270	9.170	9.048

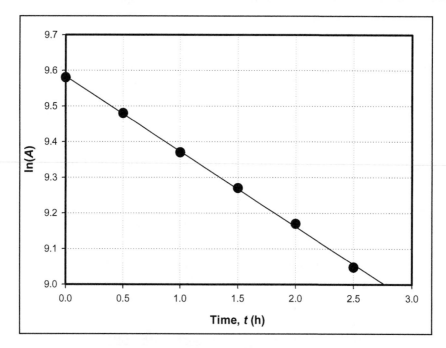

(See Figure 11.6 and associated discussion, pages 283–285.)
Taking the values of ln(*A*) as 9.58 and 9.00 at times *t* = 0.0 h and 2.7 h, respectively, the slope is

$$\text{slope} = -k = \frac{\Delta \ln(A)}{\Delta t} = \frac{9.00 - 9.58}{(2.7 - 0.0)\text{h}} = -0.21 \text{ h}^{-1}; \quad \text{so } k = 0.21 \text{ h}^{-1}.$$

Now use *k* to calculate the half-life:

$$t_{1/2} = \frac{0.693}{k} = \frac{0.693}{0.21/\text{h}} = 3.3 \text{ h}$$

74. See Sections 18.2 and 18.3. Use the "integrated" decay law to determine the proportion of Am-241 that remains in a sample after 1 year:

$$\ln\frac{X_0}{X} = kt \quad = (1.51 \times 10^{-3} \text{ y}^{-1})(1 \text{ y})$$

$$\frac{X_0}{X} = e^{(1.51\times10^{-3}\text{ y}^{-1})(1\text{y})} \qquad \text{so } \frac{X}{X_0} = e^{-(1.51\times10^{-3}\text{ y}^{-1})(1\text{y})} = 0.998$$

The quantity remaining, X, is 0.998 times the initial quantity, X_0; in other words, 99.8 % remains after 1 y. The decay of Am-241 is so slow that the sample will last many years — only around 14% decays in 100 y.

CHALLENGE PROBLEMS

76. See Section 18.3.

    Strategy: 1. Calculate k (in s^{-1}) using $t_{\frac{1}{2}}$.
                2. Convert the given activity (20 pCi) into atoms/s.
                3. Determine N using $N = A/k$.
                4. Use N to calculate how many mol Rn-222 are present in the liter of air.

$$k = \frac{0.693}{t_{\frac{1}{2}}} = \frac{0.693}{3.82 \text{ d}} = \frac{0.693}{3.82 \text{ d}} \times \frac{1\text{d}}{24\text{ h}} \times \frac{1\text{h}}{3600 \text{ s}} = 2.10 \times 10^{-6} \text{ s}^{-1}$$

$$A = 20 \times 10^{-12} \text{ Ci} \times \frac{3.700 \times 10^{10} \text{ atoms/s}}{1\text{Ci}} = 0.74 \text{ atoms/s}$$

$$N = \frac{A}{k} = \frac{0.74 \text{ atoms s}^{-1}}{2.10 \times 10^{-6} \text{ s}^{-1}} = 3.5 \times 10^5 \text{ atoms}$$

This is the number of atoms Rn-222 per liter of air; hence, the concentration of Rn-222 is

$$\frac{3.5 \times 10^5 \text{ atoms}}{1\text{L}} \times \frac{1\text{mol}}{6.022 \times 10^{23} \text{ atoms}} = \textbf{5.8} \times \textbf{10}^{-19} \textbf{ mol/L}$$

77. See (a) Section 18.3; (b) Section 18.4; and (c) "Chemistry Beyond the Classroom" (page 485).

    (a) Using $A = kN$ (equation **18.1**), one can obtain the activity of the sample in atoms/min where N is the number of atoms present. However, since mass is proportional to the number of atoms present, the same equation can be used to express the activity in units of (g Pu-239)/min. Thus, the activity is:

$$A = kN = 5.5 \times 10^{-11} \text{ /min} \times (1.00 \text{ g Pu-239}) = 5.5 \times 10^{-11} \text{ (g Pu-239)/min}$$

    so in 45 min, the mass that decomposes is:

$$\frac{5.5 \times 10^{-11}\,\text{g Pu-239}}{1\,\text{min}} \times 45\,\text{min} = \textbf{2.5} \times \textbf{10}^{-9}\,\textbf{g Pu-239}$$

(b) Use Table 18.3 to find nuclear masses and use them to calculate Δm for the decay of 1 mol of Pu-239, and hence Δm for 2.5×10^{-9} g Pu-239; then use equation **18.3** to determine ΔE.

Δm (per mol) = nuclear masses of products $-$ nuclear masses of reactants

$\qquad\qquad$ = (mass of U-235 + mass of He-4) $-$ (mass of Pu-239)

$\qquad\qquad$ = (234.9934 g + 4.00150 g) $-$ (239.0006 g)

$\qquad\qquad$ = **−0.0057 g/mol Pu-239**

Δm (for 2.5×10^{-9} g Pu-239)

$$= 2.5 \times 10^{-9}\,\text{g Pu-239} \times \frac{1\,\text{mol Pu-239}}{239.0006\,\text{g Pu-239}} \times \frac{-0.0057\,\text{g}}{1\,\text{mol Pu-239}}$$

$$= -6.0 \times 10^{-14}\,\text{g}$$

$\Delta E \quad = 9.00 \times 10^{10}\,\dfrac{\text{kJ}}{\text{g}}\,(\Delta m)$

$\qquad\quad = 9.00 \times 10^{10}\,\dfrac{\text{kJ}}{\text{g}}\,(-6.0 \times 10^{-14}\,\text{g})$

$\qquad\quad = \textbf{−5.4} \times \textbf{10}^{-3}\,\textbf{kJ}$

(c) Recall (p. 485) that dosage in rems is given by $n \times$ **(number of rads)**, where 1 rad = 10^{-2} J/kg and n = 10—20 for α particles. The energy (part (b) above) emitted by the Pu-239 source in 45 min is 5.4×10^{-3} kJ.

Exposure in rads for 75 kg man $= \dfrac{5.4 \times 10^{-3}\,\text{kJ}}{75\,\text{kg}} \times \dfrac{1000\,\text{J}}{1\,\text{kJ}} \times \dfrac{1\,\text{rad}}{10^{-2}\,\text{J/kg}} = 7.2\,\text{rad}$

Exposure in rems (using n = 10) $= 7.2\,\text{rad} \times \dfrac{10\,\text{rem}}{1\,\text{rad}} = \underline{\textbf{72 rems}}$

78. (a) Use the equation given, along with the values provided:

$$E = 8.99 \times 10^9\,\frac{q_1 q_2}{r} = 8.99 \times 10^9 \times \frac{(1.60 \times 10^{-19}\,\text{C})^2}{2 \times 10^{-15}\,\text{m}} = \underline{\textbf{1} \times \textbf{10}^{-13}\,\textbf{J}}$$

(b) From Table 18.3, the mass of a deuteron is 2.01355 amu: this must be converted to kilograms; then use the given equation, solve for v, and substitute values for E and m.

$$\text{deuteron mass (kg)} = \frac{2.01355\,\text{g}}{1\,\text{mol}} \times \frac{1\,\text{mol}}{6.022 \times 10^{23}} \times \frac{1\,\text{kg}}{1000\,\text{g}} = 3.344 \times 10^{-27}\,\text{kg}$$

Since two deuterons are present, $m = 2(3.344 \times 10^{-27}$ kg$) = 6.688 \times 10^{-27}$ kg

$$E = mv^2/2 \qquad\qquad \text{so } v = [2E/m]^{1/2}$$

$$v = \left[\frac{2(1 \times 10^{-13}\, \text{J})}{6.688 \times 10^{-27}\, \text{kg}}\right]^{1/2} = \underline{\textbf{5} \times \textbf{10}^6\ \textbf{m/s}}$$

79. (a) This problem is identical to Example 18.7(a). Use Table 18.3 to find the required nuclear masses and use them to calculate Δm for each reaction per mol of reactants. Convert Δm to a value per gram of reactants and then use equation **18.3** to determine ΔE per gram of reactants.

Recall that Δm (per mol of reaction)
= nuclear masses of products − nuclear masses of reactants

$$\begin{aligned}\Delta m/\text{mol reaction} &= (\text{mass He-4}) - [2(\text{mass H-2})] \\ &= 4.00150\ \text{g} - [2(2.01355)]\ \text{g} \\ &= -0.02560\ \text{g}\end{aligned}$$

$$\begin{aligned}\Delta m/\text{g H-2} &= \frac{-0.02560\ \text{g}}{\text{mol reaction}} \times \frac{1\,\text{mol reaction}}{2\ \text{mol H-2}} \times \frac{1\,\text{mol H-2}}{2.01355\ \text{g H-2}} \\ &= -6.36 \times 10^{-3}\ \text{g/g H-2}\end{aligned}$$

$$\begin{aligned}\Delta E/\text{g H-2} &= 9.00 \times 10^{10}\ \frac{\text{kJ}}{\text{g}}\ (\Delta m) \\[4pt] &= 9.00 \times 10^{10}\ \frac{\text{kJ}}{\text{g}}\ (-6.36 \times 10^{-3}\ \text{g/g H-2}) \\[4pt] &= \underline{\textbf{−5.72} \times \textbf{10}^8\ \textbf{kJ/g H-2}}\end{aligned}$$

(b) Determine the mass of deuterium in seawater and use that, along with the result from part (a), to calculate the energy available.

$$\begin{aligned}\text{mass H-2 present in seawater} &= 1.3 \times 10^{24}\ \text{g seawater} \times \frac{0.0017\ \text{g H-2}}{100\ \text{g seawater}} \\ &= 2.2 \times 10^{19}\ \text{g H-2}\end{aligned}$$

$$\begin{aligned}\text{energy available from seawater} &= 2.2 \times 10^{19}\ \text{g H-2} \times \frac{-5.72 \times 10^8\ \text{kJ}}{1\,\text{g H-2}} \\ &= \underline{\textbf{−1.3} \times \textbf{10}^{28}\ \textbf{kJ}}\end{aligned}$$

(c) The fraction of seawater that would need to be consumed is

$$\frac{\text{total global energy requirement}}{\text{total energy available from seawater}} = \frac{2.3 \times 10^{17} \text{ kJ}}{1.3 \times 10^{28} \text{ kJ}} = \underline{\mathbf{1.8 \times 10^{-11}}}$$

80. We must first determine the quantity of NO_2 formed by decay of the ^{14}C-labeled CO_2 (See also Example 18.4.)

Strategy: 1. Calculate k (in y^{-1}) using $t_{1/2}$.
2. Use k and the "integrated" rate equation to determine the ratio X_0/X. Recall that X_0 and X need only be proportional to the number of nuclei present. Since the Ideal Gas Law (Chapter 5) tells us that n is proportional to P (if V and T are constant) we can use partial pressure in atmospheres as a measure of the number of nuclei present.

$$k = \frac{0.693}{t_{1/2}} = \frac{0.693}{5730 \text{ y}} = 1.21 \times 10^{-4} \text{ y}^{-1}$$

$$\ln\frac{X_0}{X} = kt \qquad \text{so:} \qquad \frac{X_0}{X} = e^{kt} = e^{(1.21 \times 10^{-4} \text{ y}^{-1})(20000 \text{ y})} = 11.2$$

$$X = \frac{X_0}{11.2} = \frac{1.00 \text{ atm}}{11.2} = 0.0893 \text{ atm}$$

This is the partial pressure of CO_2 remaining after 20000 y; hence the pressure of NO_2 is $(1 - 0.0893)$ atm $= 0.911$ atm

Now set up an equilibrium table (see Chapters 12 and 13, particularly Examples 12.4 and 12.6), taking 0.911 atm as P_0 for NO_2 and 0 atm as P_0 for N_2O_4. Use this, plus the expression for K to solve for the pressure of N_2O_4 present after 20000 y.

	$NO_2(g)$	$\rightleftharpoons$	$N_2O_4(g)$
P_0 (atm)	0.911		0
ΔP (atm)	$-2x$		$+x$
P_{eq} (atm)	$0.911-2x$		x

$$K = \frac{P_{N_2O_4}}{(P_{NO_2})^2}$$

$$1 \times 10^{-5} = \frac{x}{(0.911 - 2x)^2} \approx \frac{x}{(0.911)^2} \qquad \text{(assuming that } 2x \ll 0.911)$$

$$x = (1 \times 10^{-5})(0.911)^2 = 8.3 \times 10^{-6} \qquad \text{(so the assumption was valid)}$$

$$P_{N_2O_4} = x = \underline{\mathbf{8.3 \times 10^{-6} \text{ atm}}}$$

81. See Sections 13.3, 13.4, and 18.3, and Examples 13.7 and 18.3.
    We are required to form a solution of pH = 6; in other words, $[H^+] = 10^{-pH} = 10^{-6}$ *M*. It can be shown (Section 13.4, pages 416–417) that for weak acid ionization

$$K_a = \frac{x^2}{a} \qquad \text{or} \qquad a = \frac{x^2}{K_a}$$

where $x = [H^+]$ and a = original concentration of the weak acid. In this case, therefore,

$$[Zn(H_2O)_4{}^{2+}] = a = \frac{x^2}{K_a} = \frac{[1\times10^{-6}]^2}{3.3\times10^{-10}} = 3.03 \times 10^{-3}\ M$$

Each U atom that fissions is assumed to form 1 atom Zn, which will form one complex ion $Zn(H_2O)_4{}^{2+}$, so the stoichiometry of the process is

$$1\ \text{mol U} \rightarrow 1\ \text{mol Zn} \rightarrow 1\ \text{mol Zn(H}_2\text{O)}_4{}^{2+}$$

and hence $[Zn(H_2O)_4{}^{2+}]$ formed = [U] decayed. Thus we know the initial concentration of uranium = X_0 = 1.00 *M*; and the concentration of U remaining = X = $(1.00 - 3.03 \times 10^{-3})$ *M* = 0.997 *M*. Now calculate k (in y^{-1}) using $t_{1/2}$ and then finally use k and the "integrated" rate equation to determine a value for t.

$$k = \frac{0.693}{t_{1/2}} = \frac{0.693}{7\times10^8\ y} = 1 \times 10^{-9}\ y^{-1}$$

$$kt = \ln\frac{X_0}{X} \qquad \text{so}\quad t = \frac{1}{k}\ln\frac{X_0}{X} = \frac{1}{1\times10^{-9}\ y^{-1}}\ln\frac{1.00}{0.997} = \underline{3 \times 10^6\ y}$$

82. See Sections 15.2 and 18.3, and Example 15.4.
    For Hg_2I_2 dissolving in water the net ionic equation is:

$$Hg_2I_2(s) \rightleftharpoons Hg_2{}^{2+}(aq) + 2I^-(aq)$$

and if its molar solubility is denoted s, then at equilibrium $[Hg_2{}^{2+}] = s$ and $[I^-] = 2s$; and $K_{sp} = [Hg_2{}^{2+}][I^-]^2 = s \times (2s)^2 = 4s^3$. In order to determine K_{sp}, we must use the information given to obtain the solubility of Hg_2I_2. Note that the measured activity of the solution will be proportional to the number of mol of Hg_2I_2 present in the solution.

$$\text{mol } Hg_2I_2 \text{ present} = 33\ \text{counts/min} \times \frac{1\ \text{mol I}}{5.0\times10^{11}\ \text{counts/min}} \times \frac{1\ \text{mol } Hg_2I_2}{2\ \text{mol I}}$$

$$= 3.3 \times 10^{-11}\ \text{mol } Hg_2I_2$$

$$\text{concentration of } Hg_2I_2 \text{ present in solution} = s = \frac{3.3\times10^{-11}\ \text{mol } Hg_2I_2}{0.150\ \text{L solution}} = 2.2 \times 10^{-10}\ \text{mol/L}$$

Hence: $\quad K_{sp} = 4s^3 = 4(2.2 \times 10^{-10})^3 = \underline{4.3 \times 10^{-29}}$

COMPLEX IONS

19-1 COMPOSITION OF COMPLEX IONS

2. This problem is similar to Example 19.1.

   (a) Recall that ligands are those species that are bonded to the metal atom (in this case Cd). Thus, the ligands present in the complex ion [Cd(en)(SCN)$_2$(OH)$_2$$^-$] are **$en$, SCN$^-$, and OH$^-$**:

      en is the neutral molecule ethylenediamine, H$_2$N—(CH$_2$)$_2$—NH$_2$, which has **no charge**.
      SCN$^-$ is the thiocyanate anion, which has a charge of **−1**.
      OH$^-$ is the hydroxide anion, which has a charge of **−1**.

   (b) Apply the following relation (page 489):

      charge of complex = oxidation number of central metal + charges of ligands

      −1 = oxidation number + [(2 × −1) + (2 × −1)]
      oxidation number = **+3**

   (c) Recall that magnesium ions have a charge of +2, Mg^{2+}. In order to form an electrically neutral compound with this complex ion, therefore, two of the complex anions must combine with each Mg^{2+} cation.
      Thus, the formula of the compound formed is **Mg[Cd(en)(SCN)$_2$(OH)$_2$]$_2$**.

4. Recall that "platinum(II)" refers to the Pt^{2+} ion. For each set of ligands, combine the Pt^{2+} ion with the given ligands and add up the charges of the individual components to determine the overall charge. Represent ethylenediamine by en.

   (a) Pt^{2+} + 2NH$_3$ + C$_2$O$_4$$^{2-}$:   total charge = (+2) + 2(0) + (−2) = **0**

      formula of complex: **Pt(NH$_3$)$_2$(C$_2$O$_4$)**

   (b) Pt^{2+} + 2NH$_3$ + SCN$^-$ + Br$^-$:   total charge = (+2) + 2(0) + (−1) + (−1) = **0**

      formula of complex: **Pt(NH$_3$)$_2$(SCN)(Br)**

   (c) Pt^{2+} + en + 2NO$_2$$^-$:   total charge = (+2) + (0) + 2(−1) = **0**

      formula of complex: **Pt(en)(NO$_2$)$_2$**

6. This problem is related to Example 19.2; see also page 491. Recall that the coordination number refers to the number of bonds made with the central metal atom. For each complex, determine how many bonds each ligand of each type makes and hence the total number of bonds.

   (a) [Fe(H$_2$O)$_6$$^{3+}$]:   each H$_2$O molecule makes 1 bond to the Fe atom

      coordination number = 6 × 1 = **6**

   (b) [Pt(NH$_3$)Br$_3$$^-$]: the NH$_3$ molecule makes 1 bond to the Pt atom and each Br$^-$ ion makes 1 bond to the Pt atom

      coordination number = 1 + (3 × 1) = **4**

    (c) [V(*en*)Cl$_4^{2-}$]:   the *en* molecule makes 2 bonds to the V atom and each Cl$^-$ ion makes 1 bond to the V atom

        coordination number = 2 + (4 × 1) = **6**

    (d) [Au(CN)$_2^+$]:   each CN$^-$ ion makes 1 bond to the Au atom

        coordination number = 2 × 1 = **2**

8. This problem is similar to Example 19.1(a). In each case apply the following relationship, given on page 489:

    charge of complex = oxidation number of central metal + charges of ligands

    (a) [Fe(H$_2$O)$_6^{3+}$]:
        +3 = oxidation number + (6 × 0)   →   oxidation number = **+3**

    (b) [Pt(NH$_3$)Br$_3^-$]:
        −1 = oxidation number + [(1 × 0) + (3 × −1)] → oxidation number = **+2**

    (c) [V(*en*)Cl$_4^{2-}$]:
        −2 = oxidation number + [(1 × 0) + (4 × −1)] → oxidation number = **+2**

    (d) [Au(CN)$_2^+$]:
        +1 = oxidation number + (2 × −1)   →   oxidation number = **+3**

10. This problem is similar to Example 19.1(b). Note that the carbonate ion, CO$_3^{2-}$ has a charge of −2 and the barium ion, Ba^{2+}, has a charge of +2.

    (a) [Fe(H$_2$O)$_6^{3+}$]:
        The complex ion has a charge of +3, so it must combine with carbonate. In order to form an electrically neutral compound, two complex ions must combine with three carbonate ions.
        Thus, the formula of the compound formed is **[Fe(H$_2$O)$_6$]$_2$(CO$_3$)$_3$**.

    (b) [Pt(NH$_3$)Br$_3^-$]:
        The complex ion has a charge of −1, so it must combine with barium. In order to form an electrically neutral compound, two complex ions must combine with one barium ion.
        Thus, the formula of the compound formed is **Ba[Pt(NH$_3$)Br$_3$]$_2$**.

    (c) [V(*en*)Cl$_4^{2-}$]:
        The complex ion has a charge of −2, so it must combine with barium. In order to form an electrically neutral compound, one complex ion must combine with one barium ion.
        Thus, the formula of the compound formed is **Ba[V(*en*)Cl$_4$]**.

    (d) [Au(CN)$_2^+$]:
        The complex ion has a charge of +1, so it must combine with carbonate. In order to form an electrically neutral compound, two complex ions must combine with one carbonate ion.

Thus, the formula of the compound formed is **[Au(CN)₂]₂CO₃**.

12. This problem is related to Example 19.2. For each metal–ligand pair: first determine the coordination number of the metal (Table 19.2) and the number of bonds that each ligand forms with the metal, and combine these to predict how many ligands combine with each metal center. Finally combine the metal with the ligands and add up the charges of the individual components to determine the overall charge of the complex formed.

(a) Pt^{4+} has a coordination number of 6, so 6 bonds with ligands are needed.
   Each NH_3 ligand makes one bond with the metal, so 6 molecules of NH_3 are needed.

   Pt^{4+} + $6NH_3$:   total charge = (+4) + 6(0) = +4
   formula of complex: **$Pt(NH_3)_6{}^{4+}$**

(b) Ag^+ has a coordination number of 2, so 2 bonds with ligands are needed.
   Each CN^- ligand makes one bond with the metal, so 2 CN^- ions are needed.

   Ag^+ + $2CN^-$:   total charge = (+1) + 2(−1) = −1
   formula of complex: **$Ag(CN)_2{}^-$**

(c) Zn^{2+} has a coordination number of 4, so 4 bonds with ligands are needed.
   Each $C_2O_4{}^{2-}$ ligand makes two bonds with the metal, so 2 $C_2O_4{}^{2-}$ ions are needed.

   Zn^{2+} + $2C_2O_4{}^{2-}$:   total charge = (+2) + 2(−2) = −2
   formula of complex: **$Zn(C_2O_4)_2{}^{2-}$**

(d) Cd^{2+} has a coordination number of 4, so 4 bonds with ligands are needed.
   Each CN^- ligand makes one bond with the metal, so 4 CN^- ions are needed.

   Cd^{2+} + $4CN^-$:   total charge = (+2) + 4(−1) = −2
   formula of complex: **$Cd(CN)_4{}^{2-}$**

14. This problem is similar to Example 3.4(a). Since the complex ion has a charge of +1 and the sulfate ion, $SO_4{}^{2-}$, has a charge of −2, two complex ions must combine with each sulfate ion in order to form an electrically neutral compound. Thus the formula of the sulfate salt must be $[Ni(H_2O)_4Cl_2]_2SO_4$, which has a molar mass of 499.38 g/mol.

To find the mass percent of Cl, start with 1 mol of the salt and follow this strategy:
   mol salt → mol Cl → mass Cl → mass % Cl

   mass Cl in 1 mol salt =

$$1 \text{ mol } [Ni(H_2O)_4Cl_2]_2SO_4 \times \frac{4 \text{ mol Cl}}{1 \text{ mol } [Ni(H_2O)_4Cl_2]_2SO_4} \times \frac{35.453 \text{ g Cl}}{1 \text{ mol Cl}} = 141.81 \text{ g}$$

This is contained in 499.38 g of the salt (i.e., 1 mol) and therefore the mass percent Cl is:

$$\frac{141.81 \text{ g Cl}}{499.38 \text{ g } [Ni(H_2O)_4Cl_2]_2SO_4} \times 100 = \underline{\textbf{28.4 \%}}$$

16. This problem is related to Example 3.4(a), in reverse. Recall that 4.4 % cobalt by mass is equivalent to 4.4 g cobalt in 100 g of vitamin B_{12}. Given the mass percent of Co and the molar mass of the compound, start with 1 mol of the compound and follow this strategy:

    mass % Co → mass Co → mol Co

mass Co in 1 mol vitamin B_{12} =

$$1 \text{ mol vitamin } B_{12} \times \frac{1.3 \times 10^3 \text{ g vitamin } B_{12}}{1 \text{ mol vitamin } B_{12}} \times \frac{4.4 \text{ g Co}}{100 \text{ g vitamin } B_{12}} = 57 \text{ g}$$

Thus, mol Co in 1 mol vitamin B_{12} is:

$$\frac{57 \text{ g Co}}{1 \text{ mol vitamin } B_{12}} \times \frac{1 \text{ mol Co}}{58.93 \text{ g Co}} = 0.97 \text{ mol Co/mol vitamin } B_{12}$$

This is approximately 1 mol Co per mol vitamin B_{12}, so **there is <u>one</u> atom of cobalt per molecule of vitamin B_{12}.**

19-2   NAMING COMPLEX IONS AND COORDINATION COMPOUNDS

18. This problem is related to Example 19.3. For each compound, identify the cation and/or anion. For the complex ion, identify each ligand and use its prefix to determine the number of each ligand present (if there is no prefix, there is only one such ligand present). Identify the metal center and note its oxidation number is given in Roman numerals in parentheses after the metal's name. Next combine the metal with the ligands and add up the charges of the individual components to determine the overall charge of the complex formed. Finally, use the charge on the cation and anion to decide how many of each must combine in order to form a neutral compound.

    (a) <u>diamminetetraquachromium(II) chloride</u>:

    cation: <u>diamminetetraquachromium(II)</u>   *<this is the complex ion>*

    anion: chloride, Cl^-

    components of the complex ion:

diammine	$2 \times NH_3$
tetraqua	$4 \times H_2O$
chromium(II)	$1 \times Cr^{2+}$

    charge of the complex ion:

    $Cr^{2+} + 2NH_3 + 4H_2O$ total charge $= (+2) + 2(0) + 4(0) = +2$

    formula of the complex ion: $Cr(NH_3)_2(H_2O)_4^{2+}$

    The complex ion has a charge of +2 and the chloride anion has a charge of −1, so each complex ion must combine with two chloride ions in order to form an electrically neutral compound. Thus the formula of the compound is **$[Cr(NH_3)_2(H_2O)_4]Cl_2$.**

(b) tetrachlorosulfatocadmium(III):

components of the complex ion:

tetrachloro	$4 \times Cl^-$
sulfato	$1 \times SO_4^{2-}$
cadmium(III)	$1 \times Cd^{3+}$

charge of the complex ion:

$Cd^{3+} + 4Cl^- + SO_4^{2-}$ total charge $= (+3) + 4(-1) + 1(-2) = -3$

formula of the complex ion: $CdCl_4(SO_4)^{3-}$

No counteranion is specified, so the formula of the complex ion is **$CdCl_4(SO_4)^{3-}$**.

(c) sodium tetrahydroxonickelate(II):

cation: sodium, Na^+

anion: tetrahydroxonickelate(II) *<this is the complex ion>*

components of the complex ion:

tetrahydroxo	$4 \times OH^-$
nickelate(II)	$1 \times Ni^{2+}$

charge of the complex ion:

$Ni^{2+} + 4OH^-$ total charge $= (+2) + 4(-1) = -2$

formula of the complex ion: $Ni(OH)_4^{2-}$
The complex ion has a charge of -2 and the sodium cation has a charge of $+1$, so each complex ion must combine with two cations in order to form an electrically neutral compound. Thus the formula of the compound is **$Na_2[Ni(OH)_4]$**.

(d) dibromo*bis*(ethylenediamine)iron(III):

components of the complex ion (represent the ethylenediamine ligand by *en*):

dibromo	$2 \times Br^-$
bis(ethylenediamine)	$2 \times en$
iron(III)	$1 \times Fe^{3+}$

charge of the complex ion:

$Fe^{3+} + 2Br^- + 2en$ total charge $= (+3) + 2(-1) + 2(0) = +1$

formula of the complex ion: $FeBr_2(en)_2^+$
No counteranion is specified, so the formula of the ion is **$FeBr_2(en)_2^{\pm}$**.

20. This problem is similar to Example 19.3. For each compound, identify the complex ion and determine its charge and hence find the oxidation number of the metal. (See Example 19.1(a).) Write the name of each ligand and use the appropriate prefix where more than one such ligand is present, then assemble all the ligands in alphabetical order. Place the metal center after the ligands, with its oxidation number in parentheses after its name; if

the complex ion is an anion, add —ate to the metal's name. Finally combine the complex ion with the counterion.

(a) Na[Al(OH)$_4$]:

cation (listed first):   Na$^+$ sodium

anion (listed second): Al(OH)$_4^-$ (Since each complex ion combines with one Na$^+$ cation, the charge on the complex ion must be −1.) Note that in the name of this complex **_anion_**, the metal center must carry the —ate suffix.

oxidation number of Al: −1 = (oxid. no.) + 4(−1) → oxid. no. = +3

ligands with prefix:

(OH)$_4$ tetrahydroxo

name of the compound: **sodium tetrahydroxoaluminate(III)**

(b) [Co(C$_2$O$_4$)$_2$(H$_2$O)$_2^-$]:

anion: Co(C$_2$O$_4$)$_2$(H$_2$O)$_2^-$ Note that in the name of this complex **_anion_**, the metal center must carry the —ate suffix.

oxidation number of Co: −1 = (oxid. no.) + 2(−2) + 2(0) → oxid. no. = +3

ligands with prefix:

(C$_2$O$_4$)$_2$ dioxalato

(H$_2$O)$_2$ diaqua

ligands in alphabetical order: diaquadioxalato

name of the complex ion: **diaquadioxalatocobaltate(III)**

(c) [Ir(NH$_3$)$_3$Cl$_3$]:

oxidation number of Ir: 0 = (oxid. no.) + 3(0) + 3(−1) → oxid. no. = +3

ligands with prefix:

(NH$_3$)$_3$ triammine

Cl$_3$ trichloro

ligands in alphabetical order: triamminetrichloro

name of the compound: **triamminetrichloroiridium(III)**

(d) [Cr(*en*)(NH$_3$)$_2$Br$_2$]$_2$SO$_4$:

cation (listed first):   [Cr(*en*)(NH$_3$)$_2$Br$_2$]$^+$ (Since two complex ions combine with one SO$_4^{2-}$ anion, the charge on the complex ion must be +1.)

anion (listed second): SO$_4^{2-}$, sulfate

oxidation number of Cr: +1 = (oxid. no.) + 1(0) + 2(0) + 2(−1) → oxid. no. = +3

ligands with prefix:

en   ethylenediamine

(NH$_3$)$_2$ diammine

Br$_2$ dibromo

ligands in alphabetical order: di<u>a</u>mminedi<u>b</u>romo<u>e</u>thylenediamine

name of the compound: <u>**diamminedibromoethylenediaminechromium(III) sulfate**</u>

22. This problem is similar to Example 19.3 and Question 20 above. For each complex ion use its charge and hence find the oxidation number of the metal. (See Example 19.1(a).) Write the name of each ligand and use the appropriate prefix where more than one such ligand is present, then assemble all the ligands in alphabetical order. Place the metal center after the ligands, with its oxidation number in parentheses after its name; if the complex ion is an anion, add —ate to the metal's name.

(a) $[Fe(H_2O)_6^{3+}]$:

oxidation number of Fe: $+3 = (oxid.\ no.) + 6(0) \rightarrow oxid.\ no. = +3$

ligands with prefix:

$(H_2O)_6$ hexaaqua

name of the complex ion: <u>**hexaaquairon(III)**</u>

(b) $[Pt(NH_3)Br_3^-]$:

(Note that this is a complex ___anion___, so the metal center must carry the —ate suffix.)

oxidation number of Pt: $-1 = (oxid.\ no.) + 1(0) + 3(-1) \rightarrow oxid.\ no. = +2$

ligands with prefix:

NH_3 ammine

Br_3 tribromo

ligands in alphabetical order: <u>a</u>mminetri<u>b</u>romo

name of the complex ion: <u>**amminetribromoplatinate(II)**</u>

(c) $[V(en)Cl_4^{2-}]$:

(Note that this is a complex ___anion___, so the metal center must carry the —ate suffix.)

oxidation number of V: $-2 = (oxid.\ no.) + 1(0) + 4(-1) \rightarrow oxid.\ no. = +2$

ligands with prefix:

en   ethylenediamine

Cl_4 tetrachloro

ligands in alphabetical order: tetra<u>c</u>hloro<u>e</u>thylenediamine

name of the complex ion: <u>**tetrachloroethylenediaminevanadate(II)**</u>

(d) $[Au(CN)_2^+]$:

oxidation number of Au: $+1 = (oxid.\ no.) + 2(-1) \rightarrow oxid.\ no. = +3$

ligands with prefix:

$(CN)_2$ dicyano

name of the complex ion: <u>**dicyanogold(III)**</u>

19-3   GEOMETRY OF COMPLEX IONS

24. Refer to Table 19.1 and the discussion in Section 19.3 to confirm the geometric arrangement of ligands around the metal center for a particular metal ion and coordination number. Recall that when a polyatomic ligand such as water bonds to a metal center, it is the oxygen at that bonds to the metal (it is this atom that supplies the electron pair when H_2O acts as a Lewis base). Note, therefore, that H_2O is often written OH_2 to emphasize that it is the oxygen atom that is bonded to the metal center; and similarly for other ligands.

(a)   The Cu center bears six ligands and therefore will have an octahedral geometry. The *trans* prefix indicates that the two Br^- ligands should be as far apart as possible. (Compare Figure 19.6.)

(b)   We are told that the metal is tetrahedral; there is only one way for the four ligands to be arranged around the Zn center.

(c)   The Pt center bears four identical ligands and has a square planar geometry. There is only one way to arrange the ligands around the Pt center.

(d)   The Ni center in this anionic complex ion has two neutral H_2O ligands plus two $C_2O_4^{2-}$ ligands, meaning that the metal is Ni^{2+}. The coordinating O atoms of the chelating *ox* ligand are constrained geometrically to occupy adjacent (*cis*) coordination sites, and the *cis-* prefix in the formula indicates that the two H_2O ligands must also be in adjacent positions.

simplified representation:

(e)   The Au center bears one CN^- ligand and one NH_3 ligand, so the metal center must be Au^+ which prefers a linear geometry (see Table 19.2). This is consistent with there being only two ligands.

$$NC\!\!-\!\!Au\!\!-\!\!NH_3$$

26. The *acac⁻* anion is a chelating ligand like ethylenediamine or oxalate: it will bind to the Fe^{3+} cation using both of its oxygen atoms; thus three *acac* ligands will occupy six coordination sites at the Fe center, giving an octahedral complex.

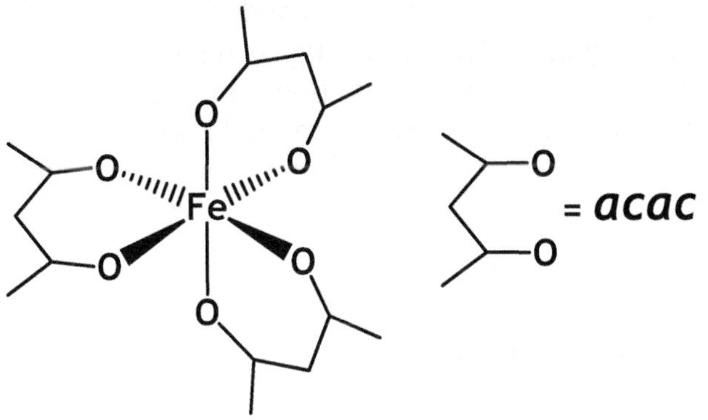

simplified representation:

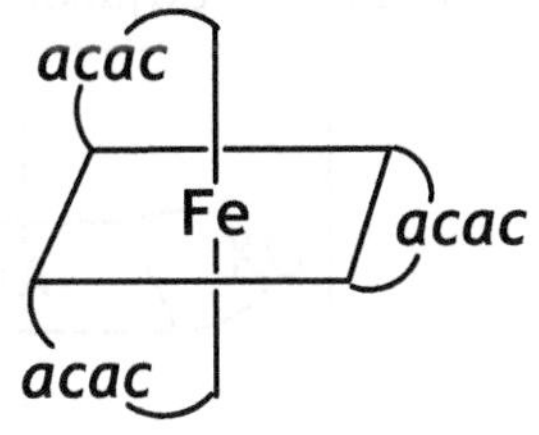

28. **All three complex ions show geometric isomerism.**

(a) $[Mn(H_2O)_2(NO_2)_4{}^{2-}]$:

The complex is similar to the one in Figure 19.6. It is an octahedral complex with two ligands of one type (H_2O) and four of another type (NO_2^-). Thus the two H_2O ligands may be either *cis* or *trans* to one another, with the NO_2^- ligands occupying the four remaining sites.

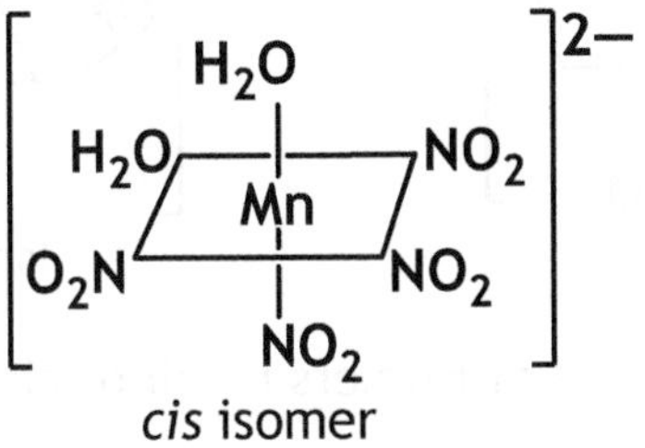

cis isomer 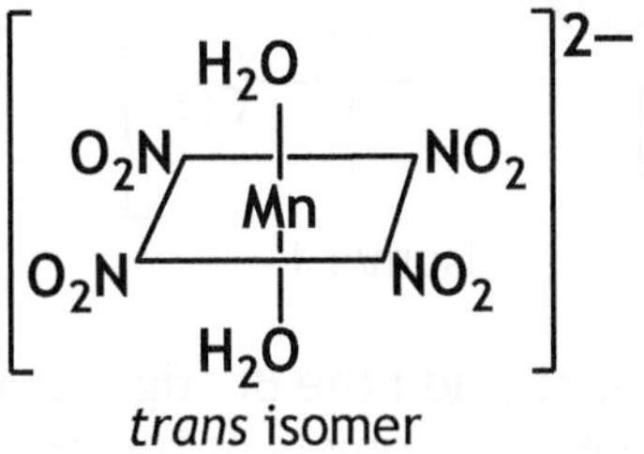   *trans* isomer

(b) $[Pt(NH_3)_3Cl_3{}^+]$:

The complex is very similar to the one in Example 19.4. It is an octahedral complex with three ligands of one type (NH_3) and three of another type (Cl^-). Two isomers are possible, analogous to Isomers I and II in Example 19.4: arrange the NH_3 ligands as in that example and then fill the remaining sites with the three Cl^- ligands.

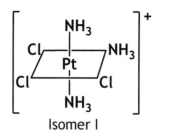

Isomer I   Isomer II

(c) [Al(C$_2$O$_4$)(CO)$_2$Br$_2^-$]:

This is an octahedral complex with one bidentate oxalate ligand, C$_2$O$_4^{2-}$ (abbreviated as *ox*) and two each of the CO and Br$^-$ ligands. The chelating *ox* ligand is constrained geometrically to occupy adjacent (*cis*) coordination sites. Begin by placing the *ox* ligand at two adjacent corners of the square (compare Example 19.4):

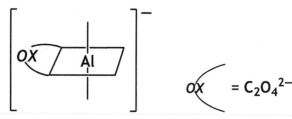

Next attach the two CO ligands; there are three different ways to do this. The two CO ligands may be opposite the *ox* ligand on the square (Isomer I below), "above and below" the square (Isomer II below), or having one CO on the square and one above the square (Isomer III below; note that the two sites on the square are equivalent and that the "above" and "below" sites are equivalent).

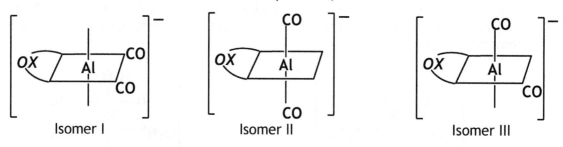

Isomer I   Isomer II   Isomer III

Finally add the Br$^-$ ligands in the remaining sites of Isomers I – III to obtain the three possible isomers of this compound:

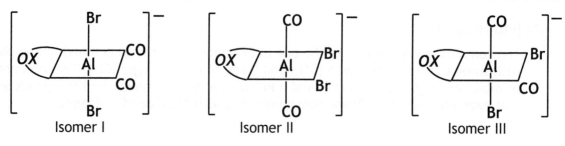

Isomer I   Isomer II   Isomer III

30. This compound is related to the complex in Question 28(b). Begin by arranging the three H$_2$O ligands like the NH$_3$ ligands are arranged in Isomers I and II in Example 19.4:

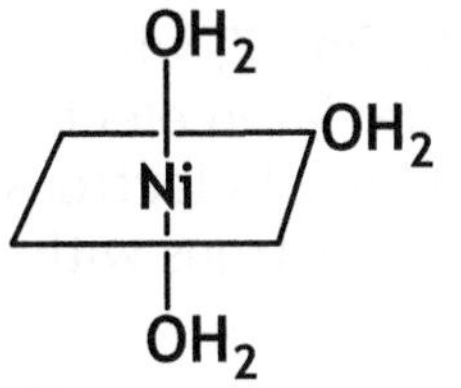

"Isomer I" 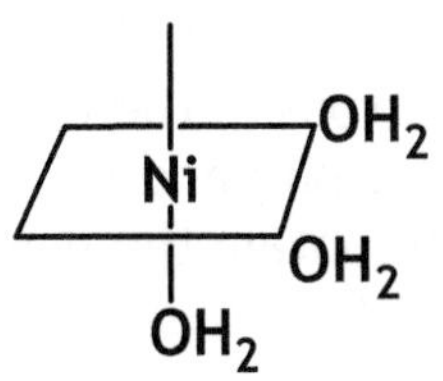 "Isomer II"

Next attach the OH$^-$ ligands: from "Isomer I" above, the two OH$^-$ ligands may be either adjacent (*cis*) or non-adjacent (*trans*):

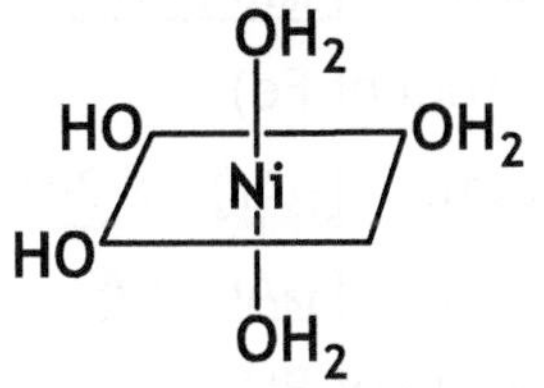

"Isomer I + *cis*-(OH)$_2$" 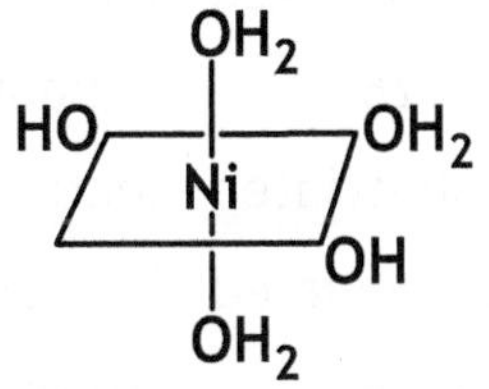 "Isomer I + *trans*-(OH)$_2$"

and from "Isomer II" above, all three of the remaining sites are equivalent so there is only one way of attaching the two OH$^-$ ligands:

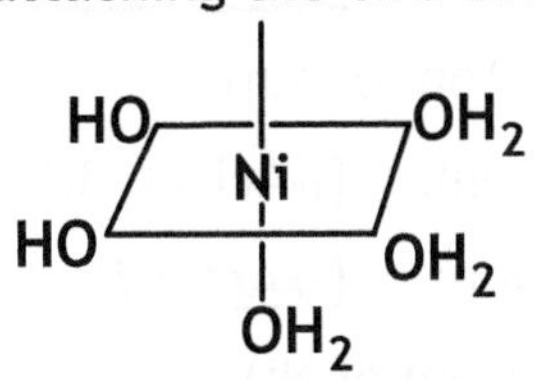

"Isomer II + (OH)$_2$"

Finally add the Cl$^-$ ligand to each of the above intermediate structures to obtain the three possible isomers of this compound:

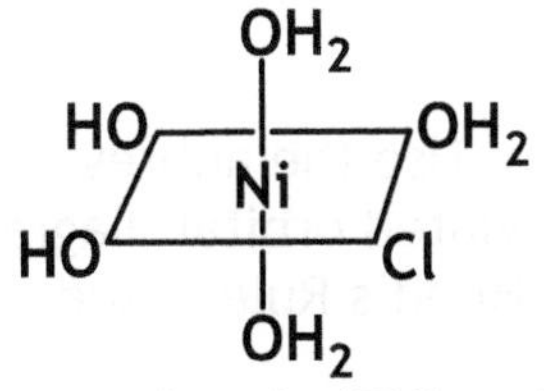

"Isomer I + *cis*-(OH)$_2$ + Cl" 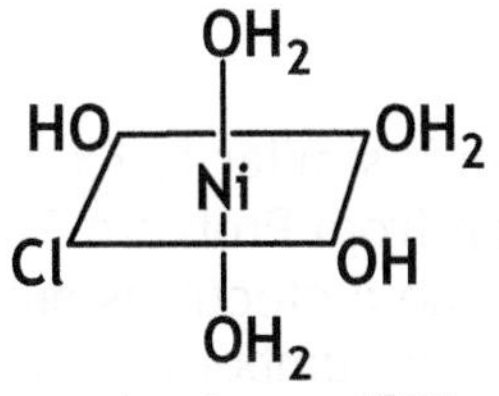 "Isomer I + *trans*-(OH)$_2$ + Cl"

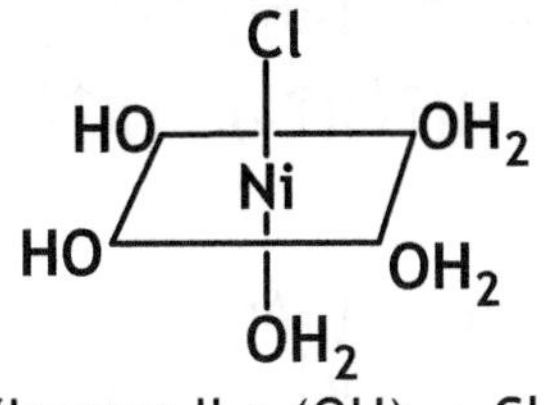

"Isomer II + (OH)$_2$ + Cl"

19-4 ELECTRONIC STRUCTURE OF COMPLEX IONS: Electronic Structure of Metal Ions

32. This problem is similar to Example 19.5(a). For each ion: (i) find Z, which gives the number of electrons in the neutral atom; (ii) write the abbreviated electron configuration for the neutral atom; and (iii) remove the appropriate number of electrons from the neutral atom to obtain the given cation, remembering that electrons with the highest n are removed first.

   (a) Cd^{2+}: $Z = 48$ (there are 48 electrons in an atom of Cd)

          abbreviated electron configuration for Cd atom: $[_{36}Kr]\ 5s^2\ 4d^{10}$

          abbreviated electron configuration for Cd^{2+} ion: **$[_{36}Kr]\ 4d^{10}$**

   (b) Fe^{2+}: $Z = 26$ (there are 26 electrons in an atom of Fe)

          abbreviated electron configuration for Fe atom: $[_{18}Ar]\ 4s^2\ 3d^6$

          abbreviated electron configuration for Fe^{2+} ion: **$[_{18}Ar]\ 3d^6$**

   (c) Pt^{2+}: $Z = 78$ (there are 78 electrons in an atom of Pt)

          abbreviated electron configuration for Pt atom: $[_{54}Xe]\ 6s^2\ 4f^{14}\ 5d^8$

          abbreviated electron configuration for Pt^{2+} ion: **$[_{54}Xe]\ 4f^{14}\ 5d^8$**

   (d) Mn^{2+}: $Z = 25$ (there are 25 electrons in an atom of Mn)

          abbreviated electron configuration for Mn atom: $[_{18}Ar]\ 4s^2\ 3d^5$

          abbreviated electron configuration for Mn^{2+} ion: **$[_{18}Ar]\ 3d^5$**

   (e) Ni^{3+}: $Z = 28$ (there are 28 electrons in an atom of Ni)

          abbreviated electron configuration for Ni atom: $[_{18}Ar]\ 4s^2\ 3d^8$

          abbreviated electron configuration for Ni^{3+} ion: **$[_{18}Ar]\ 3d^7$**

34. This problem is similar to Example 19.5(b). For each ion: Use the abbreviated electron configuration from Problem 32 above to write an abbreviated orbital diagram for the ion, focusing on the d electrons only. (Remember to follow Hund's Rule — see pages 146–147.) Then count the number of unpaired electrons.

   (a) Cd^{2+}:   abbreviated electron configuration for Cd^{2+} ion: $[_{36}Kr]\ 4d^{10}$

          abbreviated orbital diagram: **$[_{36}Kr]$ (↑↓)(↑↓)(↑↓)(↑↓)(↑↓)**

          There are **zero** unpaired electrons.

   (b) Fe^{2+}:   abbreviated electron configuration for Fe^{2+} ion: $[_{18}Ar]\ 3d^6$
          abbreviated orbital diagram: **$[_{18}Ar]$ (↑↓)(↑)(↑)(↑)(↑)**

          There are **four** unpaired electrons.

   (c) Pt^{2+}:   abbreviated electron configuration for Pt^{2+} ion: $[_{54}Xe]\ 4f^{14}\ 5d^8$

          abbreviated orbital diagram: **$[_{54}Xe]\ 4f^{14}$ (↑↓)(↑↓)(↑↓)(↑)(↑)**

          There are **two** unpaired electrons.

   (d) Mn^{2+}:   abbreviated electron configuration for Mn^{2+} ion: $[_{18}Ar]\ 3d^5$

abbreviated orbital diagram: [$_{18}$Ar] (↑)(↑)(↑)(↑)(↑)

There are <u>**five**</u> unpaired electrons.

(e) Ni^{3+}:   abbreviated electron configuration for Ni^{3+} ion: [$_{18}$Ar] 3d^7

abbreviated orbital diagram: [$_{18}$Ar] (↑↓)(↑↓)(↑)(↑)(↑)

There are <u>**three**</u> unpaired electrons.

**19-4   ELECTRONIC STRUCTURE OF COMPLEX IONS:
Electron Distributions and Crystal Field Theory**

36. This problem is similar to Example 19.6. For each ion: Obtain the abbreviated electron configuration for the neutral atom and then for the ion, as in Example 19.5. In drawing the electron distribution diagrams: for the low-spin complex, first distribute electrons to the *lower* energy orbitals and if any remain place those in the *upper* orbitals (remember to follow Hund's Rule (pages 146–147) at each step); for the high-spin complex distribute the electrons to *all* of the orbitals. (Again, remember to follow Hund's Rule.)

(a) Zn^{2+}:   Z = 30

electron configurations: Zn, [$_{18}$Ar] 4s^2 3d^{10};   Zn^{2+}, [$_{18}$Ar] 3d^{10}; **10 *d* electrons**

(↑ ↓)(↑ ↓)   (↑ ↓)(↑ ↓)

(↑ ↓)(↑ ↓)(↑ ↓)   (↑ ↓)(↑ ↓)(↑ ↓)

low spin   high spin

Since Zn^{2+} has 10 *d* electrons, the low spin and high spin configurations are identical.

(b) Ni^{3+}:   Z = 28

electron configurations: Ni, [$_{18}$Ar] 4s^2 3d^8;   Ni^{3+}, [$_{18}$Ar] 3d^7; **7 *d* electrons**

(↑)( )   (↑)(↑)

(↑ ↓)(↑ ↓)(↑ ↓)   (↑ ↓)(↑ ↓)(↑)

low spin   high spin

38. This problem is related to Example 19.6. For each ion: Obtain the abbreviated electron configuration for the neutral atom and then for the ion, as in Example 19.5. In drawing the electron distribution diagrams: for the low-spin complex, first distribute electrons to the *lower* energy orbitals and if any remain place those in the *upper* orbitals (remember to follow Hund's Rule (pages 146–147) at each step); for the high-spin complex distribute the electrons to *all* of the orbitals. (Again, remember to follow Hund's Rule.)

Mn^{3+}:   Z = 25

---

electron configurations: Mn, [$_{18}$Ar] $4s^2$ $3d^5$;   Mn^{3+}, [$_{18}$Ar] $3d^4$; **4 d electrons**

$\quad\quad$ ( )( ) $\quad\quad\quad\quad\quad\quad\quad\quad$ (↑)( )

$\quad$ (↑ ↓)(↑)(↑) $\quad\quad\quad\quad\quad\quad$ (↑)(↑)(↑)

$\quad\quad$ low spin $\quad\quad\quad\quad\quad\quad\quad\quad\quad$ high spin

Mn^{4+}: $Z = 25$

$\quad$ electron configurations: Mn, [$_{18}$Ar] $4s^2$ $3d^5$;   Mn^{4+}, [$_{18}$Ar] $3d^3$; **3 d electrons**

$\quad\quad$ ( )( ) $\quad\quad\quad\quad\quad\quad\quad\quad$ ( )( )

$\quad$ (↑)(↑)(↑) $\quad\quad\quad\quad\quad\quad$ (↑)(↑)(↑)

$\quad\quad$ "low spin" $\quad\quad\quad\quad\quad\quad\quad\quad$ "high spin"

As can be seen, the two electron distribution diagrams for Mn^{4+} are identical. With a d^3 electron configuration, only one electron distribution is possible — indeed (see page501), the terms low-spin and high-spin are only meaningful for complexes with d^4, d^5, d^5 or d^7 configurations. Hence Mn^{3+} (d^4) can have both low- and high-spin configurations but Mn^{4+} (d^3) cannot.

40. First, apply the relation below (from page 489) to identify the metal ion present in each complex:

$\quad$ charge of complex ion = oxidation number of central metal + charges of ligands

For [Co(NH$_3$)$_6^{3+}$] $\quad\quad\quad$ $+3$ = oxid. no. + (6 × 0) $\quad$ → oxid. no. = +3: Co^{3+}
For [CoF$_6^{3-}$] $\quad\quad\quad\quad\quad$ -3 = oxid. no. + (6 × −1) $\quad$ → oxid. no. = +3: Co^{3+}

Both complex ions contain a Co^{3+} ion. Now proceed as in Example 19.6 to derive the electron distribution diagrams for low- and high-spin Co^{3+} and the number of unpaired electrons for each distribution:

Co^{3+}: $Z = 27$

$\quad$ electron configurations: Co, [$_{18}$Ar] $4s^2$ $3d^7$;   Co^{3+}, [$_{18}$Ar] $3d^6$; **6 d electrons**

$\quad\quad$ ( )( ) $\quad\quad\quad\quad\quad\quad\quad\quad$ (↑)(↑)

$\quad$ (↑ ↓)(↑ ↓)(↑ ↓) $\quad\quad\quad\quad\quad$ (↑ ↓)(↑)(↑)

$\quad\quad$ low spin $\quad\quad\quad\quad\quad\quad\quad\quad\quad$ high spin
$\quad$ *zero* unpaired electrons $\quad\quad\quad\quad$ *four* unpaired electrons

Referring to the spectrochemical series on page 503, it can be seen that NH$_3$ is a stronger field ligand than F$^-$ and therefore the magnitude of the d splitting, Δ_o, for the complex [Co(NH$_3$)$_6^{3+}$] is larger than for [CoF$_6^{3-}$]. In addition, recall from pages 500—501 that larger Δ_o values favor low-spin complexes and smaller Δ_o values favor high-spin complexes.

Thus we expect [Co(NH₃)₆³⁺] to be a low-spin complex, with **zero unpaired electrons** and therefore **diamagnetic**; whereas [CoF₆³⁻] is a high-spin complex, with **four unpaired electrons** and therefore **paramagnetic**.

42. Recall from pages 500−501 that **weak**-field ligands form complexes with **smaller** Δ_o values which therefore have **high-spin** electron distributions. For each ion, proceed as in Example 19.6 to obtain the electron configuration of the ion and the high-spin electron distribution diagram for that ion; then count the number of unpaired electrons. (Note that for the Ag and Au atoms a filled *d* subshell is preferred; see pages 146−147.)

(a) Rh³⁺, Z = 45
electron configurations:
Rh, [₃₆Kr] 5s² 4d⁷
Rh³⁺, [₃₆Kr] 4d⁶

(↑)(↑)

(↑ ↓)(↑)(↑)

four unpaired electrons

(b) Mn³⁺, Z = 25
electron configurations:
Mn, [₁₈Ar] 4s² 3d⁵
Mn³⁺, [₁₈Ar] 3d⁴

(↑)( )

(↑)(↑)(↑)

four unpaired electrons

(c) Ag⁺, Z = 47
electron configurations:
Ag, [₃₆Kr] 5s¹ 4d¹⁰
Ag⁺, [₃₆Kr] 4d¹⁰

(↑ ↓)(↑ ↓)

(↑ ↓)(↑ ↓)(↑ ↓)

zero unpaired electrons

(d) Pt⁴⁺, Z = 78
electron configurations:
Pt, [₅₄Xe] 6s² 4f¹⁴ 5d⁸
Pt⁴⁺, [₅₄Xe] 4f¹⁴ 5d⁶

(↑)(↑)

(↑ ↓)(↑)(↑)

four unpaired electrons

(e) Au³⁺, Z = 79
electron configurations:
Au, [₅₄Xe] 6s¹ 4f¹⁴ 5d¹⁰
Au³⁺, [₅₄Xe] 4f¹⁴ 5d⁸

(↑)(↑)

(↑ ↓)(↑ ↓)(↑ ↓)

two unpaired electrons

44. This problem requires a calculation that is the reverse of the one shown on page 502; see also Example 6.2. First convert the units of the Δ_o value from kJ/mol to J (per photon):

$$E \text{ (per photon)} = \frac{2.60 \times 10^2 \text{ kJ}}{1 \text{ mol}} \times \frac{1 \text{ mol}}{6.022 \times 10^{23}} \times \frac{1000 \text{ J}}{1 \text{ kJ}} = 4.32 \times 10^{-19} \text{ J}$$

Now use Equation 6.2 to convert this energy into the corresponding wavelength:

$$\lambda = \frac{hc}{E} = \frac{(6.626 \times 10^{-34} \text{ J·s})(2.998 \times 10^{8} \text{ m/s})}{4.32 \times 10^{-19} \text{ J}} = \underline{4.60 \times 10^{-7} \text{ m}} \text{ or } \underline{4.60 \times 10^{2} \text{ nm}}$$

46. As stated in Figure 19.13 (caption), the color of a complex is complementary to the color of the light absorbed. For red light, the complementary color is green-blue. Referring to Figure 6.3 and Table 6.1, the wavelength of green-blue light is around **500 nm**.

UNCLASSIFIED PROBLEMS

48. (a) See Section 3.2. This problem is similar to Example 3.5. For simplicity, assume 100.0 g of compound, then the % masses of each element will be masses in grams. Based on these, calculate the number of moles of each element present then follow the flowchart in Figure 3.7.

Co: $\dfrac{22.0 \text{ g Co}}{58.93 \text{ g/mol}} = 0.373$ mol

N: $\dfrac{31.4 \text{ g N}}{14.01 \text{ g/mol}} = 2.24$ mol

H: $\dfrac{6.78 \text{ g H}}{1.008 \text{ g/mol}} = 6.73$ mol

Cl: $\dfrac{39.8 \text{ g Cl}}{34.45 \text{ g/mol}} = 1.16$ mol

Now divide by the smallest number of moles (0.373) to get the mol ratios:
Co: 0.373 mol/0.373 = 1 mol
N: 2.24 mol/0.373 = 6 mol
H: 6.73 mol/0.373 = 18 mol
Cl: 1.16 mol/.373 = 3 mol

Thus, the simplest formula for the compound is $\underline{CoN_6H_{18}Cl_3}$.

(b) One mole of the compound dissociates in solution to give 4 mol ions: considering the formula of the compound, it seems likely that the three Cl atoms are chloride anions and the remaining Co, N, and H atoms would correspond to a cation of formula $CoN_6H_{18}^{3+}$ (since it combines with $3Cl^-$). The $\{N_6H_{18}\}$ grouping probably corresponds to 6 NH_3 ligands; thus the formula of the complex would be $[Co(NH_3)_6]Cl_3$. Upon dissolving in water, the reaction taking place would be:

$$[Co(NH_3)_6]Cl_3(s) \rightleftharpoons Co(NH_3)_6^{3+}(aq) + 3Cl^-(aq)$$

CONCEPTUAL PROBLEMS

50. (a) **False**. While it is true that there are five ligands in the complex, the *en* ligand is a chelating ligand, and will occupy *two* coordination sites. Thus **the coordination number is 6**.

(b) **False**. (See pages 500–501 and the spectrochemical series on page 503.) The CN^- ligand is a stronger field ligand than NH_3, and therefore Δ_o for $[Ni(CN)_6^{4-}]$ will be larger than Δ_o for $[Ni(NH_3)_6^{2+}]$. This means that $[Ni(CN)_6^{4-}]$ will absorb light of higer energy, and therefore **shorter** wavelength, than $[Ni(NH_3)_6^{2+}]$.

(c) **True**. (See pages 500–501.) The Cr^{3+} ion has a d^3 configuration and therefore it has a partially filled d sublevel. Thus it will form complexes in which electrons in the split d orbitals can absorb a portion of the visible spectrum and hence cause the complex to appear colored. Ions such as Zn^{2+}, which has a d^{10} configuration and therefore a completely filled d sublevel, cannot do this and so do not appear colored.

(d) **False**. (See pages 500–501.) Ions with **seven** d electrons **can** form both high- and low-spin complexes. However, **ions with eight or more d electrons** cannot.

CHALLENGE PROBLEMS

51. Refer to the box "Chemistry Beyond the Classroom – Chelates: Natural and Synthetic" on pages 503–504. The EDTA ion is a powerful chelating ligand that can occupy all six of the coordination sites around an octahedral central metal cation, and in so doing forms very stable 1:1 complexes with the metal ion. With Pb, therefore, Pb·EDTA will form. From Figure B on page 504, the formula of the EDTA ion is $C_{10}H_{12}N_2O_8^{4-}$ and in this case it is supplied as the sodium salt $Na_4(C_{10}H_{12}N_2O_8)$. Thus the reaction between Na_4(EDTA) and Pb can be represented as follows (ignoring the charge on Pb):

$$Na_4(EDTA)(aq) + Pb(aq) \rightarrow Pb \cdot (EDTA)(aq) + 4Na^+(aq)$$

Refer to the flowchart for solution stoichiometry (Figure 4.6) and use the following strategy:

mass paint → mass Pb → mol Pb → mol EDTA → mol Na_4EDTA → mass Na_4EDTA

mol Pb consumed =

$$10.0 \text{ g paint} \times \frac{5.0 \text{ g Pb}}{100 \text{ g paint}} \times \frac{1 \text{ mol Pb}}{207.2 \text{ g Pb}} = 2.4 \times 10^{-3} \text{ mol Pb}$$

mass Na_4EDTA required =

$$2.4 \times 10^{-3} \text{ mol Pb} \times \frac{1 \text{ mol EDTA}}{1 \text{ mol Pb}} \times \frac{1 \text{ mol Na}_4\text{EDTA}}{1 \text{ mol EDTA}} \times \frac{380.2 \text{ g Na}_4\text{EDTA}}{1 \text{ mol Na}_4\text{EDTA}} = \underline{\mathbf{0.91 \text{ g}}}$$

52. See Section 3.2 and in particular pages 62–63. This problem is similar to Example 3.7. First determine the molar mass of the simplest formula $PtN_2H_6Cl_2$, then calculate the ratio

$$\frac{\text{actual molar mass}}{\text{simplest formula molar mass}}$$

and then multiply the subscripts in the simplest formula to get the formula for the compound.

For $PtN_2H_6Cl_2$ the molar mass =

195.1 g Pt + 2(14.01 g N) + 6(1.008 g H) + 2(35.45 g Cl) = 300.1 g/mol

$$\frac{\text{actual molar mass}}{\text{simplest formula molar mass}} = \frac{600 \text{ g/mol}}{300.1 \text{ g/mol}} = 2.00$$

Thus, the formula for the compound is $Pt_{(1 \times 2)}N_{(2 \times 2)}H_{(6 \times 2)}Cl_{(2 \times 2)}$; that is $\mathbf{Pt_2N_4H_{12}Cl_4}$.

Since the compound contains two Pt atoms and we are told that it contains both a complex cation and a complex anion (each of which require a metal atom), there must be one Pt in the cation and one Pt in the anion. Referring to Table 19.2, we see that Pt can either be Pt^{2+} (square planar, CN = 4) or Pt^{4+} (octahedral, CN = 6). Now notice that in the compound's formula that there are 4 Cl atoms (of which up to 4 could be Cl^- ligands) and 4 N atoms (of which up to 4 could be the Lewis basic atom of a ligand such as NH_3). Thus it appears that we have up to 8 ligands available, implying that the two Pt atoms in our compound are Pt^{2+}, each with a coordination number of 4.

Assuming that the N_4H_{12} part of the compound's formula corresponds to 4 NH_3 molecules, we then have two Pt^{2+} centers, and $4Cl^-$ and $4NH_3$ ligands which have to be combined in such a way as to give a complex cation and a complex anion. The complexes will have the formulas $Pt(NH_3)_m(Cl)_{4-m}$ and $Pt(NH_3)_n(Cl)_{4-n}$ (where m and n could be 0, 1, 2, 3, 4 and $m + n = 4$). Note that $m = n = 2$ gives a neutral complex, which is not permitted. Thus we are left with two possible candidates for the formula of our compound:

$[Pt(NH_3)_4^{2+}][PtCl_4^{2-}]$ ($m = 4$, $n = 0$ *or* $m = 0$, $n = 4$)

$[Pt(NH_3)_3Cl^+][Pt(NH_3)Cl_3^-]$ ($m = 3$, $n = 1$ *or* $m = 1$, $n = 3$)

Thus the compound is either $\mathbf{[Pt(NH_3)_4][PtCl_4]}$ or $\mathbf{[Pt(NH_3)_3Cl][Pt(NH_3)Cl_3]}$.

53. (a) See Section 3.2. This problem is similar to Example 3.5. For simplicity, assume 100.0 g of compound, then the % masses of each element will equal masses in grams. Based on these, calculate the number of moles of each element present then follow the flowchart in Figure 3.7.

Cu: $\dfrac{20.25 \text{ g Cu}}{63.55 \text{ g/mol}} = 0.3186 \text{ mol}$

C: $\dfrac{15.29 \text{ g C}}{12.01 \text{ g/mol}} = 1.273 \text{ mol}$

H: $\dfrac{7.07 \text{ g H}}{1.008 \text{ g/mol}} = 7.01 \text{ mol}$

N: $\dfrac{26.86 \text{ g N}}{14.01 \text{ g/mol}} = 1.917 \text{ mol}$

S: $\dfrac{10.23 \text{ g S}}{32.07 \text{ g/mol}} = 0.3190 \text{ mol}$

O: $\dfrac{20.39 \text{ g O}}{16.00 \text{ g/mol}} = 1.274 \text{ mol}$

Now divide by the smallest number of moles (0.3186) to get the mol ratios:

Cu: 0.3186 mol/0.3186 = 1 mol
C: 1.273 mol/0.3186 = 4 mol
H: 7.01 mol/0.3186 = 22 mol
N: 1.917 mol/0.3186 = 6 mol
S: 0.3190 mol/0.3186 = 1 mol
O: 1.274 mol/0.3186 = 4 mol

Thus, the simplest formula for the compound is <u>$CuC_4H_{22}N_6SO_4$</u>.

(b) The question tells us that the compound contains a complex *ion* and that the metal center is Cu^{2+}. The fact that a complex ion is present requires the presence of a counterion and the "SO_4" at the end of the formula suggests the presence of the sulfate anion SO_4^{2-}; and since the anion carries a 2− charge and the copper carries a 2+ charge, the net charge on the remaining ligands must be neutral.

Referring to Table 19.2, we see that Cu^{2+} favors octahedral geometry (coordination number = 4) and the 6 N atoms that remain in the simplest formula are likely to be the 6 Lewis basic N atoms of ligands such as NH_3 that will occupy the six bonding sites at the Cu^{2+} center. However, the remaining unaccounted-for atoms ($C_4H_{22}N_6$) cannot only be NH_3 molecules as there are C atoms present. A ligand mentioned in this chapter that contains C and N atoms is ethylenediamine (*en*) which has the formula $C_2H_8N_2$: subtracting two of these from the simplest formula, we are left with H_6N_2, which probably corresponds to 2 NH_3 ligands.

Thus the formula of our compound is proposed to be **[Cu(NH₃)₂(*en*)₂]SO₄**. The complex cation in this compound can have two geometric isomers where the NH_3 ligands are either *cis* or *trans* with respect to each other (like the two Cl ligands in the Co complexes in Figure 19.6):

cis   *trans*

54. When white light strikes the complex, it absorbs a portion of the visible spectrum, with the wavelength of the light absorbed corresponding to the energy required to promote an electron from the lower *d* orbitals to the upper *d* orbitals: the magnitude of this energy jump equals Δ_o for the complex.
Thus, this problem is similar (in part) to Question 44 above. Perform a calculation that is the reverse of the one shown on page 502; see also Example 6.2. First convert the units of the Δ_o value from kcal/mol to J (per photon), recalling (page 189) that 1 cal = 4.184 J:

$$E \text{ (per photon)} = \frac{55\,\text{kcal}}{1\,\text{mol}} \times \frac{1000\,\text{cal}}{1\,\text{kcal}} \times \frac{4.184\,\text{J}}{1\,\text{cal}} \times \frac{1\,\text{mol}}{6.022 \times 10^{23}} = 3.8 \times 10^{-19}\,\text{J}$$

Now use Equation 6.2 to convert this energy into the corresponding wavelength:

$$\lambda = \frac{hc}{E} = \frac{(6.626 \times 10^{-34} \text{ J} \cdot \text{s})(2.998 \times 10^8 \text{ m/s})}{3.8 \times 10^{-19} \text{ J}} = 5.2 \times 10^{-7} \text{ m} \quad \text{or} \quad 5.2 \times 10^2 \text{ nm}$$

Thus the Δ_o value corresponds to a photon of wavelength around 520 nm. Referring to Figure 6.2 and Table 6.1, this wavelength corresponds to green light; this is the color of light that the complex **absorbs**. Finally, recall (pages 502−503) that the observed color of the complex is the complement of the color of light that it absorbs. Using Figure 19.13 to predict the color of the complex: its color is **<u>red-violet</u>**.

CHEMISTRY OF THE METALS

2. The equation is given on page 509:

$$2Al_2O_3(l) \rightarrow 4Al(l) + 3O_2(g)$$

Strategy: Use the Ideal Gas Law (see Chapter 5) to obtain mol O_2;

then: mol $O_2 \rightarrow$ mol Al $\rightarrow$ mass Al

$$\text{mol } O_2 = n = \frac{PV}{RT} = \frac{\left(751\,\text{mm Hg} \times \frac{1\,\text{atm}}{760\,\text{mm Hg}}\right)(2.00\,\text{L})}{(0.0821\,\text{L}\cdot\text{atm/mol}\cdot\text{K})(298\,\text{K})} = 0.0808\,\text{mol } O_2$$

$$\text{mass Al} = 0.0808\,\text{mol } O_2 \times \frac{4\,\text{mol Al}}{3\,\text{mol } O_2} \times \frac{26.98\,\text{g Al}}{1\,\text{mol Al}} = \underline{\textbf{2.91 g Al}}$$

4. The equation is given on page 511:

$$Cu_2S(s) + O_2(g) \rightarrow 2Cu(s) + SO_2(g)$$

6. This problem is similar to Example 20.1(a). In this case, however, the values of $\Delta H°$ and $\Delta S°$ must first be calculated using tabulated thermodynamic data (Appendix 1). (See also Example 16.3.)

Strategy: Calculate $\Delta H°$ and $\Delta S°$;
then use $\Delta G° = \Delta H° - T\Delta S°$

$\Delta H° = \Sigma\, \Delta H_f°_{\text{(products)}} - \Sigma\, \Delta H_f°_{\text{(reactants)}}$
$\Delta H° = [(2\,\text{mol})(0\,\text{kJ/mol}) + (1\,\text{mol})(-296.8\,\text{kJ/mol})]$
$\qquad - [(1\,\text{mol})(-79.5\,\text{kJ/mol}) + (1\,\text{mol})(0\,\text{kJ/mol})]$
$\Delta H° = -217.3\,\text{kJ}$

$\Delta S° = \Sigma\, S°_{\text{(products)}} - \Sigma\, S°_{\text{(reactants)}}$
$\Delta S° = [(2\,\text{mol})(0.0332\,\text{kJ/mol}\cdot\text{K}) + (1\,\text{mol})(0.2481\,\text{kJ/mol}\cdot\text{K})]$
$\qquad - [(1\,\text{mol})(0.1209\,\text{kJ/mol}\cdot\text{K}) + (1\,\text{mol})(0.2050\,\text{kJ/mol}\cdot\text{K})]$
$\Delta S° = -0.0114\,\text{kJ/K}$

$\Delta G° = -217.3$ kJ $- (473$ K$)(-0.0114$ kJ/K$) = \underline{\mathbf{-211.9\ kJ}}$

8. (a) This problem is similar to Example 20.2(a). This reaction is given on page 510:

$$Fe_2O_3(s) + 3CO(g) \longrightarrow 2Fe(l) + 3CO_2(g)$$

(b) This reaction is given on page 510:

$$C(s) + O_2(g) \longrightarrow CO_2(g)$$

10. This problem is similar to Example 17.8. Recall that 1 metric ton = 10^3 kg (Table 1.3); 1 J = 1 V·C and 1 kWh = 3.600×10^6 J (Table 17.3)

Strategy: mass Zn $\longrightarrow$ mol Zn $\longrightarrow$ mol e^- $\longrightarrow$ charge (C) $\longrightarrow$ energy (J)
 $\longrightarrow$ energy (kWh)

$$mol\ Zn\ =\ 10^3\ kg\ Zn \times \frac{1000\ g\ Zn}{1\ kg\ Zn} \times \frac{1\ mol\ Zn}{65.39\ g\ Zn}\ =\ 1.529 \times 10^4\ mol\ Zn$$

$$charge\ =\ 1.529 \times 10^4\ mol\ Zn \times \frac{2\ mol\ e^-}{1\ mol\ Zn} \times \frac{9.648 \times 10^4\ C}{1\ mol\ e^-}\ =\ 2.950 \times 10^9\ C$$

$$energy\ =\ V \cdot C\ =\ (3.0\ V)(2.950 \times 10^9\ C)\ =\ 8.9 \times 10^9\ J$$

$$=\ 8.9 \times 10^9\ J \times \frac{1\ kWh}{3.600 \times 10^6\ J}\ =\ \underline{\mathbf{2.5 \times 10^3\ kWh}}$$

12. (i) In chalcopyrite, $CuFeS_2$, the ratio of Cu:S is 1:2; thus for each mole of Cu present, two moles of SO_2 are produced.

(ii) Chalcopyrite ore is not pure chalcopyrite, but contains $CuFeS_2$ and other minerals. Since the ore contains 0.75% Cu, 100 g ore contains 0.75 g Cu.

Strategy: volume ore $\longrightarrow$ mass ore $\longrightarrow$ mass Cu $\longrightarrow$ mol Cu $\longrightarrow$ mol S
 $\longrightarrow$ mol $SO_2(g)$ $\longrightarrow$ vol $SO_2(g)$

$$mass\ ore\ =\ 4.00 \times 10^3\ ft^3 \times \left(\frac{12\ in}{1\ ft}\right)^3 \times \left(\frac{2.54\ cm}{1\ in}\right)^3 \times \frac{2.6\ g\ ore}{1\ cm^3}\ =\ 2.9 \times 10^8\ g\ ore$$

$$mol\ Cu\ =\ 2.9 \times 10^8\ g\ ore \times \frac{0.75\ g\ Cu}{100\ g\ ore} \times \frac{1\ mol\ Cu}{63.55\ g\ Cu}\ =\ 3.4 \times 10^4\ mol\ Cu$$

$$mol\ SO_2\ =\ 3.4 \times 10^4\ mol\ Cu \times \frac{2\ mol\ S}{1\ mol\ Cu} \times \frac{1\ mol\ SO_2}{1\ mol\ S}\ =\ 6.8 \times 10^4\ mol\ SO_2$$

$$\text{vol } SO_2 \ = \ V \ = \ \frac{nRT}{P} = \frac{(6.8 \times 10^4 \text{ mol})(0.0821 \text{ L} \cdot \text{atm/mol} \cdot \text{K})(298 \text{ K})}{1 \text{ atm}} \ = \ \underline{\textbf{1.7} \times \textbf{10}^6 \textbf{ L } SO_2}$$

20-2 REACTIONS OF THE ALKALI METALS AND ALKALINE EARTH METALS

14. See Table 20.1. Review also the nomenclature of binary ionic compounds (Section 2-7).

   (a) **potassium nitride** K_3N
   (b) **potassium iodide** KI
   (c) **potassium hydroxide** KOH
   (d) **potassium hydride** KH
   (e) **potassium sulfide** K_2S

16. See Table 20.1 and Example 20.5.

   (a) $Na_2O_2(s) \ + \ 2H_2O(l) \ \rightarrow \ 2Na^+(aq) \ + \ 2OH^-(aq) \ + \ H_2O_2(aq)$

                                       sodium ion   **hydroxide ion**   **hydrogen peroxide**

   (b) $2Ca(s) \ + \ O_2(g) \ \rightarrow \ 2CaO(s)$

                                          calcium oxide

   (c) $Rb(s) \ + \ O_2(g) \ \rightarrow \ RbO_2(s)$

                                     rubidium superoxide

   (d) $SrH_2(s) \ + \ 2H_2O(l) \ \rightarrow \ Sr^{2+}(aq) \ + \ 2OH^-(aq) \ + \ 2H_2(g)$

                                  strontium ion   **hydroxide ion**   **hydrogen gas**

18. The relevant reactions (given on page 515) are:

   $4KO_2(s) + 2H_2O(g) \ \rightarrow \ 3O_2(g) + 4KOH(s)$
   $KOH(s) + CO_2(g) \ \rightarrow \ KHCO_3(g)$

   Strategy: Use the Ideal Gas Law (Chapter 5) to obtain mol air; then:
                   mol air $\rightarrow$ mol CO_2 present $\rightarrow$ mol CO_2 to be removed $\rightarrow$ mass KO_2

   $$\text{mol air} = n = \frac{PV}{RT} = \frac{(1.00 \text{ atm})(1.00 \text{ L})}{(0.0821 \text{ L} \cdot \text{atm/mol} \cdot \text{K})(310 \text{ K})} = 0.0393 \text{ mol air}$$

$$\text{mol CO}_2 \text{ present} = 0.0393 \text{ mol air} \times \frac{5.00 \text{ mol CO}_2}{100 \text{ mol air}} = 0.00196 \text{ mol CO}_2$$

$$\text{mol CO}_2 \text{ to be removed} = 0.00196 \text{ mol CO}_2 \times 0.900 = 0.00176 \text{ mol CO}_2$$

$$\text{mass KO}_2 = 0.00176 \text{ mol CO}_2 \times \frac{1 \text{ mol KOH}}{1 \text{ mol CO}_2} \times \frac{4 \text{ mol KO}_2}{4 \text{ mol KOH}} \times \frac{71.10 \text{ g KO}_2}{1 \text{ mol KO}_2} = \underline{\mathbf{0.125 \text{ g KO}_2}}$$

20-3 REDOX CHEMISTRY OF THE TRANSITION METALS

20. (a) See Reactions of Transition Metals with Acids (pages 517–519) and Table 20.3. Cobalt is in the left column of Table 20.3, so it will be oxidized by H^+ ions to form the Co^{2+} cation, and the H^+ ions are reduced to H_2:

$$Co(s) \rightarrow Co^{2+}(aq) + 2e^- \qquad \text{oxidation half-reaction}$$
$$\underline{2H^+(aq) + 2e^- \rightarrow H_2(g)} \qquad \text{reduction half-reaction}$$
$$Co(s) + 2H^+(aq) + 2e^- \rightarrow Co^{2+}(aq) + H_2(g) + 2e^-$$
$$\mathbf{Co(s) + 2H^+(aq) \rightarrow Co^{2+}(aq) + H_2(g)}$$

(b) The reaction is given on page 518:

$$\mathbf{3Cu(s) + 8H^+(aq) + 2NO_3^-(aq) \rightarrow 3Cu^{2+}(aq) + 2NO(g) + 4H_2O}$$

(c) This reaction is given on page 521 and in Table 17.1:

$$\mathbf{Cr_2O_7^{2-}(aq) + 14H^+(aq) + 6e^- \rightarrow 2Cr^{3+}(aq) + 7H_2O}$$

22. This problem is similar to Example 17.2: balance the redox reaction in the same way; see also Section 17-1 and Problem 60 in Chapter 4.

By analogy with the reaction of gold with aqua regia (page 519), we can propose a similar <u>unbalanced</u> redox reaction:

$$Cd(s) + H^+(aq) + Cl^-(aq) + NO_3^-(aq) \rightarrow CdCl_4^{2-}(aq) + NO(g) + H_2O$$

(a) Split into two half-equations.	$Cd(s) \rightarrow CdCl_4^{2-}(aq)$
	$NO_3^-(aq) \rightarrow NO(g)$
(b) Balance one of the half-equations: $Cd(s) \rightarrow CdCl_4^{2-}(aq)$ (omit the chlorides for now): $\quad Cd(s) \rightarrow Cd^{2+}(aq)$	
oxidation numbers	Cd (oxid. no.): $0 \rightarrow +2$; Cd is oxidized
atom balance	1 Cd on each side, no adjustment is required
	$Cd(s) \rightarrow Cd^{2+}(aq)$

total oxidation number	Cd: 0 → Cd: +2
add electrons	The total oxidation number for Cd goes from 0 to +2. It is oxidized by 2. Add 2 electrons to the product side. $Cd(s) \rightarrow Cd^{2+}(aq) + 2e^-$
balance charge	reactant: 0; product: 1(2+) + 2(1−) = 0 charge is balanced
balance H	reactant: 0 H product: 0 H
check O	reactant: 0 O product: 0 O
Balanced Oxidation Half-Equation	$Cd(s) \rightarrow Cd^{2+}(aq) + 2e^-$
Include chlorides (add 4Cl⁻ to reactant side to balance 4 Cl⁻ in the complex ion)	$Cd(s) + 4Cl^- \rightarrow CdCl_4^{2-}(aq) + 2e^-$

(c) Balance the other half-equation: $NO_3^-(aq) \rightarrow NO(g)$	
Balanced Reduction Half-Equation (given in Table 17.1)	$NO_3^-(aq) + 4H^+(aq) + 3e^- \rightarrow NO(g) + 2H_2O$

(d) Make the number of electrons in the two half-equations equal (multiply by an appropriate number if necessary).	Balanced half-equations:	
	oxidation: multiply by 3	3 [$Cd(s) + 4Cl^-$ $\rightarrow CdCl_4^{2-}(aq) + 2e^-$]
		$3Cd(s) + 12Cl^-$ $\rightarrow 3CdCl_4^{2-}(aq) + 6e^-$
	reduction: multiply by 2	2 [$NO_3^-(aq) + 4H^+(aq) + 3e^-$ $\rightarrow NO(g) + 2H_2O$]
		$2NO_3^-(aq) + 8H^+(aq) + 6e^-$ $\rightarrow 2NO(g) + 4H_2O$

(e) Combine the two half-equations.	$3Cd(s) + 12Cl^- + 2NO_3^-(aq) + 8H^+(aq) + 6e^-$ $\rightarrow 3CdCl_4^{2-}(aq) + 6e^- + 2NO(g) + 4H_2O$
Cancel duplication.	$3Cd(s) + 12Cl^- + 2NO_3^-(aq) + 8H^+(aq)$ $\rightarrow 3CdCl_4^{2-}(aq) + 2NO(g) + 4H_2O$
Balanced Equation:	**$3Cd(s) + 12Cl^- + 2NO_3^-(aq) + 8H^+(aq)$** **$\rightarrow 3CdCl_4^{2-}(aq) + 2NO(g) + 4H_2O$**

24. These problems are similar to Example 17.2: balance the redox reactions in the same way.

(a) $Fe(s) + NO_3^-(aq) \rightarrow Fe^{3+}(aq) + NO_2(g)$ [acidic]

(a) Split into two half-equations.	$Fe(s) \rightarrow Fe^{3+}(aq)$
	$NO_3^-(aq) \rightarrow NO_2(g)$
(b) Balance one of the half-equations: $Fe(s) \rightarrow Fe^{3+}(aq)$	

oxidation numbers	Fe: 0 → +3; Fe is oxidized
atom balance	1 Fe on each side, no adjustment is required
total oxidation number	Fe: 0 → Fe: +3
add electrons	The total oxidation number for Fe goes from 0 to +3. The Fe is oxidized by 3. Add 3 electrons to the product side. $Fe(s) \rightarrow Fe^{3+}(aq) + 3e^-$
balance charge	reactant: 0 product: 1(+3) + 3(−1)= 0
balance H	reactant: 0 H product: 0 H
check O	reactant: 0 O product: 0 O
Balanced Oxidation Half-Equation	$Fe(s) \rightarrow Fe^{3+}(aq) + 3e^-$
(c) Balance the other half-equation: $NO_3^-(aq) \rightarrow NO_2(g)$	
oxidation numbers	N (oxid. No.): +5 → +4; N is reduced
atom balance	1 N on each side, no adjustment is required
total oxidation number	N: +5 → N: +4
add electrons	The total oxidation number for N goes from +5 to +4. It is reduced by 1. Add 1 electron to the reactant side. $NO_3^-(aq) + e^- \rightarrow NO_2(g)$
balance charge	reactant: 1(−1) + 1(−1) = −2; product: 0 acidic medium: add H^+ To balance the charge, add 2 H^+ to the reactant side. $NO_3^-(aq) + 2H^+(aq) + e^- \rightarrow NO_2(g)$ reactant: 1(−1) + 1(−1) + 2(+1) = 0; product: 0
balance H	reactant: 2 H product: 0 H To balance H, add 1 H_2O to the product side. $NO_3^-(aq) + 2H^+(aq) + e^- \rightarrow NO_2(g) + H_2O$ reactant: 2 H product: 2 H
check O	reactant: 3 O product: 3 O
Balanced Reduction Half-Equation	$NO_3^-(aq) + 2H^+(aq) + e^- \rightarrow NO_2(g) + H_2O$

(d) Make the number of electrons in the two half-equations equal (multiply by an appropriate number if necessary).	Balanced half-equations:	
	oxidation	$Fe(s) \rightarrow Fe^{3+}(aq) + 3e^-$

	reduction:
	   multiply by 3
		3 [$NO_3^-(aq) + 2H^+(aq) + e^-$
		   $\rightarrow NO_2(g) + H_2O$]

		$3NO_3^-(aq) + 6H^+(aq) + 3e^-$ $\rightarrow 3NO_2(g) + 3H_2O$
(e) Combine the two half-equations.	$Fe(s) + 3NO_3^-(aq) + 6H^+(aq) + 3e^-$ $\rightarrow Fe^{3+}(aq) + 3e^- + 3NO_2(g) + 3H_2O$	

Cancel duplication.	$Fe(s) + 3NO_3^-(aq) + 6H^+(aq)$ $\rightarrow Fe^{3+}(aq) + 3NO_2(g) + 3H_2O$
Balanced Equation:	$\mathbf{Fe(s) + 3NO_3^-(aq) + 6H^+(aq)}$ $\mathbf{\rightarrow Fe^{3+}(aq) + 3NO_2(g) + 3H_2O}$

(b) $Cr(OH)_3(s) + O_2(g) \rightarrow CrO_4^{2-}(aq)$ [basic]

Note that in basic solution, the reduction of $O_2(g)$ yields $OH^-(aq)$ (see Table 17.1):

$$O_2(g) + 2H_2O + 4e^- \rightarrow 4OH^-(aq)$$

(a) Split into two half-equations.	$Cr(OH)_3(s) \rightarrow CrO_4^{2-}(aq)$
	$O_2(g) \rightarrow H_2O$

(b) Balance one of the half-equations: $Cr(OH)_3(s) \rightarrow CrO_4^{2-}(aq)$

oxidation numbers	Cr: +3 $\rightarrow$ +6; Cr is oxidized
atom balance	1 Cr on each side, no adjustment is required
total oxidation number	Cr: +3 $\rightarrow$ Cr: +6
add electrons	The total oxidation number for Cr goes from +3 to +6; the Cr is oxidized by 3. Add 3 electrons to the product side. $\quad Cr(OH)_3(s) \rightarrow CrO_4^{2-}(aq) + 3e^-$
balance charge	reactant: 0 $\qquad$ product: 1(−2) + 3(−1)= −5 basic medium: add OH^-. To balance the charge, add 5 OH^- to the reactant side. $\quad Cr(OH)_3(s) + 5OH^- \rightarrow CrO_4^{2-}(aq) + 3e^-$ reactant: 5(−1) = −5 $\qquad$ product: 1(−2) + 3(−1)= −5
balance H	reactant: 8 H $\qquad$ product: 0 H To balance H, add 4 H_2O to the product side. $\quad Cr(OH)_3(s) + 5OH^- \rightarrow CrO_4^{2-}(aq) + 4H_2O + 3e^-$ reactant: 8 H $\qquad$ product: 8H
check O	reactant: 8 O $\qquad$ product: 8 O
Balanced Oxidation Half-Equation	$Cr(OH)_3(s) + 5OH^-(aq) \rightarrow CrO_4^{2-}(aq) + 4H_2O + 3e^-$
(c) Balanced Reduction Half-Equation	$O_2(g) + 2H_2O + 4e^- \rightarrow 4OH^-(aq)$

(d) Make the number of electrons in the two half-equations equal (multiply by an appropriate number if necessary).	Balanced half-equations:	
	oxidation: multiply by 4	$4 [Cr(OH)_3(s) + 5OH^-(aq)$ $\quad \rightarrow CrO_4^{2-}(aq) + 4H_2O + 3e^-]$
		$4Cr(OH)_3(s) + 20OH^-(aq)$ $\quad \rightarrow 4CrO_4^{2-}(aq) + 16H_2O + 12e^-$
	reduction : multiply by 3	$3 [2H_2O + O_2(g) + 4e^-$ $\quad \rightarrow 4OH^-(aq)]$

	$6H_2O + 3O_2(g) + 12e^-$ $\rightarrow 12OH^-(aq)$
(e) Combine the two half-equations.	$4Cr(OH)_3(s) + 20OH^-(aq) + 6H_2O + 3O_2(g) + 12e^-$ $\rightarrow 4CrO_4^{2-}(aq) + 16H_2O + 12e^- + 12OH^-(aq)$
Cancel duplication.	$4Cr(OH)_3(s) + 8OH^-(aq) + 3O_2(g)$ $\rightarrow 4CrO_4^{2-}(aq) + 10H_2O$
Balanced Equation:	$4Cr(OH)_3(s) + 8OH^-(aq) + 3O_2(g)$ $\rightarrow 4CrO_4^{2-}(aq) + 10H_2O$

26. This problem is related to Examples 20.6 and 17.6. (See also Sections 17-2 and 17-3.)

Strategy:

(1) The metals can only be oxidized; obtain the corresponding E°_{ox} from Table 20.3.

(2) Nitric acid solution contains $NO_3^-(aq)$ ions which may be reduced to $NO(g)$:

$$NO_3^-(aq) \rightarrow NO(g) \qquad\qquad E^\circ_{red} = +0.964 \text{ V} \qquad \text{(Table 17.1)}$$

(3) For a reaction to occur E° must be positive. Hence:

$$E^\circ = E^\circ_{ox}(\text{metal}) + E^\circ_{red} > 0$$

$$E^\circ_{ox}(\text{metal}) + 0.964 \text{ V} > 0$$

$$E^\circ_{ox}(\text{metal}) > -0.964 \text{ V}$$

Consulting Table 20.3, the metals for which $E^\circ_{ox} > -0.964$ V are **Cd, Cr, Co, Ag**.

(a) For $Cd \rightarrow Cd^{2+}$, $E^\circ_{ox} = 0.402$ V

$$E^\circ = E^\circ_{ox} + E^\circ_{red} = 0.402 \text{ V} + 0.964 \text{ V} = \underline{\textbf{1.366 V}} \qquad E^\circ > 0 \text{ so reaction occurs.}$$

(b) For $Cr \rightarrow Cr^{2+}$, $E^\circ_{ox} = 0.912$ V

$$E^\circ = E^\circ_{ox} + E^\circ_{red} = 0.912 \text{ V} + 0.964 \text{ V} = \underline{\textbf{1.876 V}} \qquad E^\circ > 0 \text{ so reaction occurs.}$$

(*NOTE:* for $Cr \rightarrow Cr^{3+}$, $E^\circ_{ox} = 0.744$ V (Table 17.1), so $E^\circ = 0.744$ V + 0.964 V = **1.708 V** and that oxidation may occur preferentially.)

(c) For $Co \rightarrow Co^{2+}$, $E^\circ_{ox} = 0.282$ V

$$E^\circ = E^\circ_{ox} + E^\circ_{red} = 0.282 \text{ V} + 0.964 \text{ V} = \underline{\textbf{1.246 V}} \qquad E^\circ > 0 \text{ so reaction occurs.}$$

(d) For $Ag \rightarrow Ag^+$, $E^\circ_{ox} = -0.799$ V

$$E^\circ = E^\circ_{ox} + E^\circ_{red} = -0.799 \text{ V} + 0.964 \text{ V} = \underline{\textbf{0.165 V}} \qquad E^\circ > 0 \text{ so reaction occurs.}$$

(e) For $Au \rightarrow Au^{3+}$, $E^\circ_{ox} = -1.498$ V

$$E° = E°_{ox} + E°_{red} = -1.498 \text{ V} + 0.964 = \underline{-0.534 \text{ V}} \qquad E° < 0 \text{ so no reaction.}$$

28. See Sections 20-3 and 17-3. This problem is similar to Example 17.5.

    Strategy: Split the equation into two half-equations and use Table 17.1 to find $E°_{ox}$ and $E°_{red}$ for the two half-equations; then calculate $E° = E°_{ox} + E°_{red}$. (Remember that $E°$ is intensive, so it has the same value regardless of how the half-equation has been written.)

    (a) $2Co^{3+}(aq) + 2e^- \rightarrow 2Co^{2+}(aq)$ $E°_{red} = +1.953 \text{ V}$

         $H_2O \rightarrow \frac{1}{2}O_2(g) + 2H^+(aq) + 2e^-$ $E°_{ox} = -E°_{red} = -1.229 \text{ V}$

         $E° = E°_{ox} + E°_{red} = -1.229 \text{ V} + 1.953 \text{ V} = \underline{+0.724 \text{ V}}$

    (b) $I_2(s) + 2e^- \rightarrow 2I^-(aq)$ $E°_{red} = +0.534 \text{ V}$

         $2Cr^{2+}(aq) \rightarrow 2Cr^{3+}(aq) + 2e^-$ $E°_{ox} = -E°_{red} = +0.408 \text{ V}$

         $E° = E°_{ox} + E°_{red} = +0.534 \text{ V} + 0.408 \text{ V} = \underline{+0.942 \text{ V}}$

30. This problem is similar to Example 17.5. (See also Section 17-3.) Recall that disproportionation means that the Au^+ ion is reduced *and* oxidized.

    In Example 20.7(b) it is shown that Au^+ will disproportionate into Au and Au^{3+}; the unbalanced redox reaction is

         $Au^+(aq) \rightarrow Au(s) + Au^{3+}(aq)$

    from which the unbalanced oxidation and reduction half-reactions (respectively) are:

         $Au^+(aq) \rightarrow Au^{3+}(aq)$ $Au^+(aq) \rightarrow Au(s)$

    Balance the oxidation by adding 2 e^- to the product side and balance the reduction by adding 1 e^- to the reactant side:

         $Au^+(aq) \rightarrow Au^{3+}(aq) + 2e^-$ $Au^+(aq) + e^- \rightarrow Au(s)$

    Finally, the balanced equation is obtained by doubling the coefficients of the reduction half-reaction and adding the oxidation half-reaction; $2e^-$ cancel out ($n = 2$):

$$2Au^+(aq) + 2e^- \rightarrow 2Au(s)$$
$$\underline{Au^+(aq) \rightarrow Au^{3+}(aq) + 2e^-}$$
$$3Au^+(aq) \rightarrow 2Au(s) + Au^{3+}(aq)$$

(a) Strategy: Determine $E°$ then use Equation **17.3** in the textbook to calculate K. Obtain E°_{red} values from Table 20.4.

$$Au^+(aq) \rightarrow Au^{3+}(aq) + 2e^- \qquad\qquad E^\circ_{ox} = -E^\circ_{red} = -1.400 \text{ V}$$
$$Au^+(aq) + e^- \rightarrow Au(s) \qquad\qquad E^\circ_{red} = 1.695 \text{ V}$$

$$E° = E^\circ_{ox} + E^\circ_{red} = -1.400 \text{ V} + 1.695 \text{ V} = +0.295 \text{ V}$$

$$E° = \frac{(0.0257 \text{ V})}{n}\ln K \qquad\text{so}\qquad \ln K = \frac{nE°}{(0.0257 \text{ V})} = \frac{2(0.295 \text{ V})}{(0.0257 \text{ V})} = 23.0$$

$$K = e^{23.0} = \underline{\mathbf{1 \times 10^{10}}}$$

(b) From the balanced equation for the redox reaction, derive an expression for K (see Section 12-2; recall that the solid Au(s) will not appear in the expression for K); then use that to solve for [Au+].

$$K = \frac{[Au^{3+}]}{[Au^+]^3}$$

$$[Au^+]^3 = \frac{[Au^{3+}]}{K} = \frac{0.10}{1 \times 10^{10}} = 1 \times 10^{-11} \text{ M}$$

$$[Au^+] = \underline{\mathbf{2 \times 10^{-4} \text{ M}}}$$

UNCLASSIFIED PROBLEMS

32. This is related to Problem 18 above.

The relevant reaction is:

$$4KO_2(s) + 2H_2O(g) \rightarrow 3O_2(g) + 4KOH(s)$$

Strategy: vol of exhaled air $\rightarrow$ vol of $H_2O(g)$ in exhaled air $\rightarrow$ mol H_2O
$\rightarrow$ mol KO_2 required $\rightarrow$ mass KO_2 required $\rightarrow$ mass KO_2 remaining

$$\text{volume } H_2O(g) \text{ present} = 116 \text{ L air} \times \frac{6.2 \text{ L } H_2O}{100 \text{ L air}} = 7.2 \text{ L } H_2O(g)$$

$$\text{mol } H_2O \text{ present} = n = \frac{PV}{RT} = \frac{\left(748 \text{ mm Hg} \times \dfrac{1 \text{ atm}}{760 \text{ mm Hg}}\right)(7.2 \text{ L})}{(0.0821 \text{ L} \cdot \text{atm/mol} \cdot \text{K})(310 \text{ K})} = 0.28 \text{ mol } H_2O$$

$$\text{mass } KO_2 \text{ required} = 0.28 \text{ mol } H_2O \times \frac{4 \text{ mol } KO_2}{2 \text{ mol } H_2O} \times \frac{71.10 \text{ g } KO_2}{1 \text{ mol } KO_2} = 40 \text{ g } KO_2$$

$$\text{mass } KO_2 \text{ remaining} = 248 \text{ g} - 40 \text{ g} = \textbf{208 g}$$

34. See Section 13-3. Obtain an expression for K for the given reaction (see Section 12-2 and recall that the solvent H_2O will not appear in the expression for K); rearrange to solve for $[H^+]$; then calculate $[H^+]$ for the given concentrations of CrO_4^{2-} and $Cr_2O_7^{2-}$. Use Equation 13.3 with this $[H^+]$ to calculate pH.

$$K = \frac{[Cr_2O_7^{2-}]}{[CrO_4^{2-}]^2[H^+]^2}$$

$$[H^+]^2 = \frac{[Cr_2O_7^{2-}]}{[CrO_4^{2-}]^2(K)} = \frac{0.10 \text{ } M}{(0.10 \text{ } M)^2(3 \times 10^{14})} = 3 \times 10^{-14}$$

$$[H^+] = 2 \times 10^{-7} \text{ } M$$

$$pH = -\log_{10}[H^+] = -\log_{10}(2 \times 10^{-7}) = \textbf{6.7}$$

36. See Section 20-3.

The reaction of the Zn in the alloy with acid (see pages 517–518) is:

$$Zn(s) + 2H^+(aq) \rightarrow Zn^{2+}(aq) + H_2(g)$$

Calculate the partial pressure of $H_2(g)$ that has been collected (see Section 5-5). Recall that the gas collected is a mixture of H_2O vapor plus the H_2 produced in the reaction.

$$P_{H_2} = P_{total} - P_{H_2O} = (755 - 26.74) \text{ mm Hg} = 728 \text{ mm Hg} \times \frac{1 \text{ atm}}{760 \text{ mmHg}} = 0.958 \text{ atm}$$

Use the following strategy to determine the % mass Zn in the alloy:
    partial pressure $H_2(g) \rightarrow$ mol $H_2(g) \rightarrow$ mol Zn $\rightarrow$ mass Zn $\rightarrow$ % mass Zn

$$\text{mol } H_2 = n = \frac{PV}{RT} = \frac{(0.958 \text{ atm})(0.1057 \text{ L})}{(0.0821 \text{ L} \cdot \text{atm/mol} \cdot \text{K})(300 \text{ K})} = 0.00411 \text{ mol } H_2$$

$$\text{mass Zn} = 0.00411 \text{ mol } H_2 \times \frac{1 \text{ mol Zn}}{1 \text{ mol } H_2} \times \frac{65.39 \text{ g Zn}}{1 \text{ mol Zn}} = 0.269 \text{ g Zn}$$

$$\text{mass \% Zn} = \frac{0.269 \text{ g Zn}}{0.500 \text{ g alloy}} \times 100\% = \underline{\textbf{53.8 \%Zn}}$$

$$\text{mass \% Cu} = 100\% - 53.8\% \text{ Zn} = \underline{\textbf{46.2\% Cu}}$$

38. See Section 20-1 and Chapter 4.

Since gold is extracted under basic conditions (pages 512–513), it may be assumed that the extraction of silver takes place under similar conditions. (Recall from Section 13-6 that solutions of NaCN are basic.) Bearing this in mind, obtain balanced redox equations for the two parts of this process: (a) extraction from Ag_2S using $CN^-(aq)$; and (b) reduction of $Ag(CN)_2^-$ using Zn. Analogies can be drawn with the extraction of gold described in the textbook. To simplify the process, consider the $O_2(g)$ to be reduced to hydroxide ion and ignore the spectator Na^+ ions.

(a) The *unbalanced* redox equation for the extraction step is:

$$Ag_2S(s) + CN^-(aq) + O_2(g) \rightarrow SO_2(g) + Ag(CN)_2^-(aq)$$

Balance this equation as in Example 17.2. For simplicity, the Ag^+ and CN^- ions can be omitted for now.

$$S^{2-}(aq) + O_2(g) \rightarrow SO_2(g) \qquad \text{[basic]}$$

(a) Split into two half-equations.	$S^{2-}(aq) \rightarrow SO_2(g)$
	$O_2(g) \rightarrow OH^-(aq)$
(b) Balance one of the half-equations: $S^{2-}(aq) \rightarrow SO_2(g)$	
oxidation numbers	S (oxid. no.): $-2 \rightarrow +4$; S is oxidized
atom balance	1 S on each side, no adjustment is required $S^{2-}(aq) \rightarrow SO_2(g)$
total oxidation number	S: $-2 \rightarrow$ S: $+4$
add electrons	The total oxidation number for S goes from -2 to $+4$. It is oxidized by 6. Add 6 electrons to the product side. $S^{2-}(aq) \rightarrow SO_2(g) + 6e^-$
balance charge	reactant: $2-$; product: $0 + 6(1-) = 6-$ basic medium: add OH^- To balance charge, add 4 OH^- to reactant side $S^{2-}(aq) + 4OH^-(aq) \rightarrow SO_2(g) + 6e^-$ reactant: $(2-) + 4(1-) = 6-$; product: $0 + 6(1-) = 6-$ charge is balanced
balance H	reactant: 4 H product: 0 H To balance H, add 2 H_2O to the product side $S^{2-}(aq) + 4OH^-(aq) \rightarrow SO_2(g) + 6e^- + 2H_2O$ reactant: 4 H product: 4 H

check O	reactant: 4 O product: 4 O
Balanced Oxidation Half-Equation	$S^{2-}(aq) + 4OH^-(aq) \rightarrow SO_2(g) + 2H_2O + 6e^-$
(c) Balance the other half-equation: $O_2(g) \rightarrow OH^-(aq)$	
Balanced Reduction Half-Equation (given in Table 17.1)	$O_2(g) + 2H_2O + 4e^- \rightarrow 4OH^-(aq)$

(d) Make the number of electrons in the two half-equations equal (multiply by an appropriate number if necessary).	Balanced half-equations:	
	oxidation: multiply by 2	$2\ [S^{2-}(aq) + 4OH^-(aq)$ $\rightarrow SO_2(g) + 2H_2O + 6e^-]$
		$2S^{2-}(aq) + 8OH^-(aq)$ $\rightarrow 2SO_2(g) + 4H_2O + 12e^-$
	reduction: multiply by 3	$3\ [O_2(g) + 2H_2O + 4e^-$ $\rightarrow 4OH^-(aq)]$
		$3O_2(g) + 6H_2O + 12e^-$ $\rightarrow 12OH^-(aq)$

(e) Combine the two half-equations.	$2S^{2-}(aq) + 8OH^-(aq) + 3O_2(g) + 6H_2O + 12e^-$ $\rightarrow 2SO_2(g) + 4H_2O + 12e^- + 12OH^-(aq)$
Cancel duplication.	$2S^{2-}(aq) + 3O_2(g) + 2H_2O \rightarrow 2SO_2(g) + 4OH^-(aq)$
Balanced Equation:	$2S^{2-}(aq) + 3O_2(g) + 2H_2O \rightarrow 2SO_2(g) + 4OH^-(aq)$

Now add the Ag^+ and CN^- ions back in, keeping the proper stoichiometric ratios. There are 2 Ag^+ per S^- in Ag_2S, so add 4 Ag^+ to each side:

$$2Ag_2S(s) + 2H_2O + 3O_2(g) \rightarrow 2SO_2(g) + 4OH^-(aq) + 4Ag^+(aq)$$

There are 2 CN^- per Ag^+ in $Ag(CN)_2{}^-$, so add 8 CN^- to each side:

$$2Ag_2S(s) + 2H_2O + 3O_2(g) + 8CN^-(aq) \rightarrow 2SO_2(g) + 4OH^-(aq) + 4Ag(CN)_2{}^-(aq)$$

(b) The unbalanced equation for the reduction step is:

$$Zn(s) + Ag(CN)_2{}^-(aq) \rightarrow Ag(s) + Zn(CN)_4{}^{2-}(aq)$$

Balance this equation similarly; omit the CN^- ions for now.

$$Zn(s) + Ag^+(aq) \rightarrow Ag(s) + Zn^{2+}(aq)$$

(a) Split into two half-equations.	$Zn(s) \rightarrow Zn^{2+}(aq)$
	$Ag^+(aq) \rightarrow Ag(s)$
(b) Balance one of the half-equations: $Zn(s) \rightarrow Zn^{2+}(aq)$	
oxidation numbers	Zn (oxid. no.): $0 \rightarrow +2$; Zn is oxidized
atom balance	1 Zn on each side, no adjustment is required $Zn(s) \rightarrow Zn^{2+}(aq)$
total oxidation	Zn: $0 \rightarrow$ Zn: $+2$

number	
add electrons	The total oxidation number for Zn goes from 0 to +2. It is oxidized by 2. Add 2 electrons to the product side. $Zn(s) \ \rightarrow \ Zn^{2+}(aq) + 2e^-$
balance charge	reactant: 0; product: (2+) + 2(1−) = 0 charge is balanced
balance H	reactant: 0 H product: 0 H
check O	reactant: 0 O product: 0 O
Balanced Oxidation Half-Equation	$Zn(s) \ \rightarrow \ Zn^{2+}(aq) + 2e^-$

(c) Balance the other half-equation: $Ag^+(aq) \ \rightarrow \ Ag(s)$	
oxidation numbers	Ag (oxid. no.): +1 → 0; Ag is oxidized
atom balance	1 Ag on each side, no adjustment is required $Ag^+(aq) \ \rightarrow \ Ag(s)$
total oxidation number	Ag: +1 → Ag: 0
add electrons	The total oxidation number for Ag goes from +1 to 0. It is reduced by 1. Add 1 electron to the reactant side. $Ag^+(aq) \ + e^- \rightarrow \ Ag(s)$
balance charge	reactant: (1+) + (1−) = 0; product: 0 charge is balanced
balance H	reactant: 0 H product: 0 H
check O	reactant: 0 O product: 0 O
Balanced Reduction Half-Equation	$Ag^+(aq) \ + e^- \rightarrow \ Ag(s)$

(d) Make the number of electrons in the two half-equations equal (multiply by an appropriate number if necessary).	Balanced half-equations:	
	oxidation:	$Zn(s) \ \rightarrow \ Zn^{2+}(aq) + 2e^-$
	reduction: multiply by 2	$2\,[Ag^+(aq) \ + e^- \rightarrow \ Ag(s)]$
		$2Ag^+(aq) \ + 2e^- \rightarrow \ 2Ag(s)$

(e) Combine the two half-equations.	$Zn(s) + 2Ag^+(aq) + 2e^- \ \rightarrow \ Zn^{2+}(aq) + 2e^- + 2Ag(s)$
Cancel duplication.	$Zn(s) + 2Ag^+(aq) \ \rightarrow \ Zn^{2+}(aq) + 2Ag(s)$
Balanced Equation:	$Zn(s) + 2Ag^+(aq) \ \rightarrow \ Zn^{2+}(aq) + 2Ag(s)$

Now add the CN^- ions back in, again observing the proper stoichiometric ratios: add 4 CN^- to each side

$$Zn(s) + 2Ag(CN)_2^-(aq) \ \rightarrow \ 2Ag(s) + Zn(CN)_4^{2-}(aq)$$

40. See Section 17-3 and Example 17.4.

Recall (page 440) that the more positive E°_{red} is, the stronger the oxidizing agent. (When a species in the center of Table 20.4 acts as an oxidizing agent, it is itself reduced. Thus the column of E°_{red} values to the __*right*__ of center in Table 20.4 should therefore be considered when comparing oxidizing strength.) In addition, the more positive E°_{ox} is, the stronger the reducing agent. (When a species in the center of Table 20.4 acts as a reducing agent it is itself oxidized. Thus the column of E°_{red} values to the __*left*__ of center in Table 20.4 should therefore be converted to E°_{ox} values (recall that $E^{\circ}_{ox} = -E^{\circ}_{red}$), and then those values considered when comparing reducing strength.)

(a) Of the choices, **Cr^{2+}** has the most positive value for E°_{ox} (+0.408 V), and is thus the strongest reducing agent.

(b) Of the choices, **Au^{+}** has the most positive E°_{red} value (+1.695 V), and is thus the strongest oxidizing agent.

(c) Of the choices, **Co^{2+}** has the least positive E°_{ox} value (−1.953 V), and is thus the weakest reducing agent.

(d) Of the choices, **Mn^{2+}** has the least positive E°_{red} value (−1.182 V), and is thus the weakest oxidizing agent.

CHALLENGE PROBLEMS

41. See Section 20-2, Reaction with Oxygen; and Table 20.1 & Example 20.4(b).

The relevant reactions are:

$$2Ba(s) + O_2(g) \rightarrow 2BaO(s)$$

$$Ba(s) + O_2(g) \rightarrow BaO_2(s)$$

The only reactants in the two reactions are barium and oxygen and the Ba:O atom ratios in the products are 1:1 and 1:2, respectively.

Calculate the amount of oxygen that reacts and then the Ba:O mole ratio. The excess moles of oxygen must be used to form BaO_2, and hence will indicate the amount of BaO_2 formed.

mass of oxygen in product mixture
= mass of mixture − initial mass of Ba
= 22.38 g − 20.00 g
= 2.38 g oxygen

$$\text{mol Ba present} \quad = 20.00 \text{ g Ba} \times \frac{1 \text{ mol Ba}}{137.3 \text{ g Ba}} = 0.1457 \text{ mol Ba}$$

$$\text{mol O reacting} \quad = 2.38 \text{ g O} \times \frac{1 \text{ mol O}}{16.00 \text{ g O}} = 0.149 \text{ mol O}$$

Thus, the mole ratio Ba:O in the product mixture is

$$0.1457 \text{ mol} : 0.149 \text{ mol} \quad = \frac{0.1457 \text{ mol}}{0.1457 \text{ mol}} : \frac{0.149 \text{ mol}}{0.1457 \text{ mol}} = 1 : 1.02$$

This indicates that there are 0.02 mol of 'excess' O atoms for every 1 mol of Ba atoms. Any barium oxide (BaO) present has a 1:1 Ba:O ratio, so the 'excess' O atoms must belong to barium peroxide (BaO_2): thus, for every mole of Ba present, 0.02 mol is as BaO_2, and the remainder (0.98 mol) is as BaO. This corresponds to a mixture whose composition is 2% BaO_2 and 98% BaO.

This can also be shown mathematically by setting x = mol Ba in BaO, and y = mol Ba in BaO_2. Hence, mol O in BaO = x, and mol O in BaO_2 = $2y$. Then, using the calculated Ba:O ratio of 1:1.02 we obtain (proportionally):

Total moles of Ba: $\quad x + y = 1$
Total moles of O: $\quad x + 2y = 1.02$

Solving gives y = moles of BaO_2 = 0.02 mol; and x = moles of BaO = 0.98 mol.

42. See Chapters 3 and 4, Chapter 13, and Chapter 19

(a) Oxalic acid ($H_2C_2O_4$) is a diprotic acid (see Section 13-4) which, after donating both of its protons, can act as a source (see page 346) of the oxalate dianion ($C_2O_4^{2-}$). The oxalate dianion is also a chelating ligand (Section 19-1) that can occupy two of the coordination sites on a metal center when it forms a complex ion. The Fe^{3+} ion present in $Fe(OH)_3$ has 6 coordination sites (Table 19.2), suggesting that it will form a complex ion with 3 oxalate ligands that would have the formula $Fe(C_2O_4)_3^{3-}$. Thus $Fe(OH)_3$ and $H_2C_2O_4$ must react in the ratio 1:3.

$$Fe(OH)_3(s) + 3H_2C_2O_4(aq) \rightarrow Fe(C_2O_4)_3^{3-}(aq) + 3H_2O + 3H^+(aq)$$

(b) Use the flowchart for solution stoichiometry (Figure 4.6)
Strategy: mass $Fe(OH)_3 \rightarrow$ mol $Fe(OH)_3 \rightarrow$ mol $H_2C_2O_4 \rightarrow$ volume $H_2C_2O_4$

Vol of 0.10 M $H_2C_2O_4$ required =

$$1.0 \text{ g Fe(OH)}_3 \times \frac{1 \text{ mol Fe(OH)}_3}{106.9 \text{ g}} \times \frac{3 \text{ mol H}_2\text{C}_2\text{O}_4}{1 \text{ mol Fe(OH)}_3} \times \frac{1 \text{ L solution}}{0.10 \text{ mol H}_2\text{C}_2\text{O}_4} = \underline{\textbf{0.28 L}}$$

43. See Section 20-3 and Chapter 4

Two reactions are involved here:

(i) excess Fe^{2+} reduces all the MnO_4^- to Mn^{2+};

and then

(ii) any Fe^{2+} that remains is determined by titration with $Cr_2O_7^{2-}$.

Balanced equations for each redox reaction are needed.

(i) An unbalanced reaction is:

$$Fe^{2+}(aq) + MnO_4^-(aq) \rightarrow Fe^{3+}(aq) + Mn^{2+}(aq)$$

For which the unbalanced half-reactions are:

$$Fe^{2+}(aq) \rightarrow Fe^{3+}(aq) \qquad \text{(Fe oxidation number: +2} \rightarrow \text{+3; oxidation)}$$
$$MnO_4^-(aq) \rightarrow Mn^{2+}(aq) \qquad \text{(Mn oxidation number: +7} \rightarrow \text{+2; reduction)}$$

The balanced reduction half-reaction is given in Table 17.1 and the oxidation half-reaction can be obtained by reversing the $Fe^{3+} \rightarrow Fe^{2+}$ reduction from the same table:

$$Fe^{2+}(aq) \rightarrow Fe^{3+}(aq) + e^- \qquad \text{(oxidation)}$$
$$MnO_4^-(aq) + 8H^+(aq) + 5e^- \rightarrow Mn^{2+}(aq) + 4H_2O \qquad \text{(reduction)}$$

and the balanced overall reaction is obtained by multiplying the coefficients of the oxidation half-reaction by 5, and then adding the reduction half-reaction and cancelling duplications:

$$5Fe^{2+}(aq) \rightarrow 5Fe^{3+}(aq) + 5e^-$$
$$\underline{MnO_4^-(aq) + 8H^+(aq) + 5e^- \rightarrow Mn^{2+}(aq) + 4H_2O}$$
$$5Fe^{2+}(aq) + MnO_4^-(aq) + 8H^+(aq) + 5e^-$$
$$\rightarrow 5Fe^{3+}(aq) + 5e^- + Mn^{2+}(aq) + 4H_2O$$

$$5Fe^{2+}(aq) + MnO_4^-(aq) + 8H^+(aq) \rightarrow 5Fe^{3+}(aq) + Mn^{2+}(aq) + 4H_2O$$

(ii) An unbalanced reaction is:

$$Fe^{2+}(aq) + Cr_2O_7^{2-}(aq) \rightarrow Fe^{3+}(aq) + Cr^{3+}(aq)$$

For which the unbalanced half-reactions are:

$$Fe^{2+}(aq) \rightarrow Fe^{3+}(aq) \qquad \text{(Fe oxidation number: +2} \rightarrow \text{+3; oxidation)}$$
$$Cr_2O_7^{2-}(aq) \rightarrow Cr^{3+}(aq) \qquad \text{(Cr oxidation number: +6} \rightarrow \text{+3; reduction)}$$

The balanced reduction half-reaction is given in Table 17.1 and the oxidation half-reaction is the same as above:

$$Fe^{2+}(aq) \rightarrow Fe^{3+}(aq) + e^- \qquad \text{(oxidation)}$$
$$Cr_2O_7^{2-}(aq) + 14H^+(aq) + 6e^- \rightarrow 2Cr^{3+}(aq) + 7H_2O \qquad \text{(reduction)}$$

The balanced redox reaction is obtained by multiplying the coefficients of the oxidation half-reaction by 6, then adding the reduction half-reaction and cancelling duplications:

$$6Fe^{2+}(aq) \rightarrow 6Fe^{3+}(aq) + 6e^-$$
$$\underline{Cr_2O_7^{2-}(aq) + 14H^+(aq) + 6e^- \rightarrow 2Cr^{3+}(aq) + 7H_2O}$$
$$6Fe^{2+}(aq) + Cr_2O_7^{2-}(aq) + 14H^+(aq) + 6e^-$$
$$\rightarrow 6Fe^{3+}(aq) + 6e^- + 2Cr^{3+}(aq) + 7H_2O$$

$$6Fe^{2+}(aq) + Cr_2O_7^{2-}(aq) + 14H^+(aq) \rightarrow 6Fe^{3+}(aq) + 2Cr^{3+}(aq) + 7H_2O$$

Use the flowchart for solution stoichiometry (Figure 4.6): Calculate the initial quantity of Fe^{2+} present, and the amount of Fe^{2+} that reacts with $Cr_2O_7^{2-}$ in reaction (ii) above. The difference between these two gives the quantity of Fe^{2+} that reacted with MnO_4^- (in reaction (i) above), and hence the amount of MnO_4^- formed when the Mn in the steel is oxidized initially. The amount of MnO_4^- can then be used to calculate the mass of Mn and hence %Mn in the original sample.

mol of Fe^{2+} present initially (in the added $FeSO_4$ solution)

$$= 75.00 \text{ mL solution} \times \frac{1L}{1000 \text{ mL}} \times \frac{0.125 \text{ mol FeSO}_4}{1L \text{ solution}} \times \frac{1 \text{ mol Fe}^{2+}}{1 \text{ mol FeSO}_4} = 0.00938 \text{ mol Fe}^{2+}$$

mol of Fe^{2+} reacting with $Cr_2O_7^{2-}$ (when $K_2Cr_2O_7$ oxidizes Fe^{2+}; reaction (ii) above)

$$= 13.50 \text{ mL solution} \times \frac{1L}{1000 \text{ mL}} \times \frac{0.100 \text{ mol K}_2\text{Cr}_2\text{O}_7}{1L \text{ solution}} \times \frac{1 \text{ mol Cr}_2\text{O}_7^{2-}}{1 \text{ mol K}_2\text{Cr}_2\text{O}_7} \times \frac{6 \text{ mol Fe}^{2+}}{1 \text{ mol Cr}_2\text{O}_7^{2-}}$$

$$= 0.00810 \text{ mol Fe}^{2+}$$

Hence, mol of Fe^{2+} that had reacted with MnO_4^- (reaction (i) above)

$$= (\text{mol of Fe}^{2+} \text{ present initially}) - (\text{mol of Fe}^{2+} \text{ reacting with Cr}_2\text{O}_7^{2-})$$

$$= 0.00938 \text{ mol} - 0.00810 \text{ mol} = 0.00128 \text{ mol Fe}^{2+}$$

Thus, the mass of Mn present in the sample

$$= 0.00128 \text{ mol Fe}^{2+} \times \frac{1 \text{ mol MnO}_4^-}{5 \text{ mol Fe}^{2+}} \times \frac{1 \text{ mol Mn}}{1 \text{ mol MnO}_4^-} \times \frac{54.94 \text{ g}}{1 \text{ mol Mn}} = 0.0141 \text{ g Mn}$$

$$\text{mass \% Mn} = \frac{\text{mass Mn}}{\text{mass sample}} \times 100\% = \frac{0.0141 \text{ g}}{0.500 \text{ g}} \times 100\% = \underline{\textbf{2.82 \%}}$$

44. See Chapters 8 and 16.

Recall that $\quad \Delta G° = \Delta H° - T\Delta S° \quad$ (Equation 16.2)

and $\quad \Delta G° = -RT \ln K \quad$ (Equation 16.5)

Since K = 1.00, substituting this value of K into Equation 16.5 gives $\Delta G°$ = 0.

Hence, we require T such that $\Delta G°$ = 0 = $\Delta H° - T\Delta S°$ or $\Delta H°$ = $T\Delta S°$.

Calculate $\Delta H°$ and $\Delta S°$ (see problem 6 above and Example 16.3) using thermodynamic data from Appendix 1, and then solve for T.

$\Delta H°$ = $\Sigma \Delta H_f°$ (products) $- \Sigma \Delta H_f°$ (reactants)

$\Delta H°$ = [(1 mol)(0 kJ/mol) + (1 mol)(0 kJ/mol)] $-$ [(1 mol)(-520.0 kJ/mol)]

$\Delta H°$ = +520.0 kJ

$\Delta S°$ = $\Sigma S°$ (products) $- \Sigma S°$ (reactants)

$\Delta S°$ = [(1 mol)(0.0320 kJ/mol·K) + (1 mol)(0.2050 kJ/mol·K)]

$\qquad - $ [(1 mol)(0.0530 kJ/mol·K)]

$\Delta S°$ = +0.184 kJ/K

$\Delta H°$ = $T\Delta S°$, so T = $\dfrac{\Delta H°}{\Delta S°}$ = $\dfrac{520.0 \text{ kJ}}{0.184 \text{ kJ/K}}$ = **2.83 × 10³ K**

45. See Section 20-3, Oxoanions of the Transition Metals; and Chapters 4, 15, and 19.

The red $Cr_2O_7^{2-}$ ion is stable in acid solution but converts to yellow CrO_4^{2-} in basic solution; see page 521:

$$Cr_2O_7^{2-}(aq) + 2OH^-(aq) \rightarrow 2CrO_4^{2-}(aq) + H_2O$$

Ag^+ ions form a red precipitate of Ag_2CrO_4 with CrO_4^{2-}; see Figure 15.6 and Example 15.8:

$$CrO_4^{2-}(aq) + 2Ag^+(aq) \rightarrow Ag_2CrO_4(s)$$

Precipitates containing Ag^+ ions form a stable complex ion with NH_3 so that the precipitate redissolves; see Section 15-4

$$Ag_2CrO_4(s) + 4NH_3(aq) \rightarrow 2Ag(NH_3)_2^+(aq) + CrO_4^{2-}(aq)$$

The molecules of NH_3 in the complex ion react with H^+ from the nitric acid in a simple acid–base reaction, forming NH_4^+ and freeing the Ag^+ ions which once again combine with CrO_4^{2-} to give the Ag_2CrO_4 precipitate:

$$2Ag(NH_3)_2^+(aq) + CrO_4^{2-}(aq) + 4H^+(aq) \rightarrow Ag_2CrO_4(s) + 4NH_4^+(aq)$$

CHEMISTRY OF THE NONMETALS

FORMULAS, EQUATIONS AND REACTIONS

2. See Sections 2-5 and 2-6 and Examples 2.8–2.10. The names of the oxyanions in these salts can be deduced by analogy with the corresponding oxoanions of chlorine.

    (a) **bromic acid**   (b) **potassium hypoiodite**

    (c) **sodium chlorite**   (d) **sodium perbromate**

4. See Sections 2-5 and 2-6. The formulas of the oxyanions in these salts can be deduced by analogy with the corresponding oxoanions of chlorine.

    (a) **$KBrO_2$**   (b) **$CaBr_2$**

    (c) **$NaIO_4$**   (d) **$Mg(ClO)_2$**

6. See Oxidizing and Reducing Strength in Section 21-4 (pages 540–541). According to the first general principle given in this section, an anion cannot act as a reducing agent if it is already in its highest oxidation state. (To be a reducing agent, the anion itself would have to be oxidized, which cannot happen.) Thus we require the highest oxidation state anions of the given elements.

    (a) **NO_3^-**   (N oxidation number +5)   (b) **SO_4^{2-}** (S oxidation number +6)
    (c) **ClO_4^-**   (Cl oxidation number +7)

8. See Reaction of Nonmetal Oxides with Water in Section 21-3; see also Example 21.6. Refer to Table 21.4 and find an acid with the same oxidation number as the given oxide. Note that in these reactions the nonmetal atoms should not change their oxidation number.

    (a) $SO_2 + H_2O \rightarrow H_2SO_3$ (S oxidation number +4)

    (b) $Cl_2O + H_2O \rightarrow 2HClO$ (Cl oxidation number +1)

    (c) $P_4O_6 + 6H_2O \rightarrow 4H_3PO_3$ (P oxidation number +3)

10. See Section 21-2 and Chapter 2.

    (a) **NaN_3**   (b) **H_2SO_3**   (c) **N_2H_4**   (d) **NaH_2PO_4**

12. See Section 21-2 (particularly Table 21.2), and Chapter 2.

   (a) H_2S (b) N_2H_4 or HN_3 (c) PH_3

14. See Sections 21-2 and 21-4; see also Chapter 13.

   (a) NH_3, N_2H_4 (b) HNO_3 (c) HNO_2, HN_3, HCN (d) HNO_3

16. See Sections 21-1 and 17-1. These problems are similar to Example 17.2: balance the redox reactions in the same way.

   (a) An unbalanced equation for the reaction would be

   $$I^-(aq) + SO_4^{2-}(aq) \rightarrow I_2(s) + SO_2(g) \qquad \text{[acidic]}$$

(a) Split into two half-equations.	$I^-(aq) \rightarrow I_2(s)$
	$SO_4^{2-}(aq) \rightarrow SO_2(g)$
(b) Balance one of the half-equations: $I^-(aq) \rightarrow I_2(s)$	
oxidation numbers	I: $-1 \rightarrow 0$; I is oxidized
atom balance	reactant: 1 I product: 2 I
	multiply I^- by 2: $2I^-(aq) \rightarrow I_2(s)$
total oxidation number	I: $-2 \rightarrow$ I: 0
add electrons	The total oxidation number for I goes from -2 to 0. The I is oxidized by 2. Add 2 electrons to the product side. $\quad 2I^-(aq) \rightarrow I_2(s) + 2e^-$
balance charge	reactant: $2(-1) = -2$ product: $0 + 2(-1) = -2$
balance H	reactant: 0 H product: 0 H
check O	reactant: 0 O product: 0 O
Balanced Oxidation Half-Equation	$2I^-(aq) \rightarrow I_2(s) + 2e^-$
(c) Balance the other half-equation: $SO_4^{2-}(aq) \rightarrow SO_2(g)$	
oxidation numbers	S (oxid. no.): $+6 \rightarrow +4$; S is reduced
atom balance	1 S on each side, no adjustment is required
total oxidation number	S: $+6 \rightarrow$ S: $+4$
add electrons	The total oxidation number for S goes from $+6$ to $+4$. It is reduced by 2. Add 2 electrons to the reactant side. $\quad SO_4^{2-}(aq) + 2e^- \rightarrow SO_2(g)$
balance charge	reactant: $1(-2) + 2(-1) = -4$; product: 0 acidic medium: add H^+

	To balance the charge, add 4 H⁺ to the reactant side.
	$$4H^+(aq) + SO_4^{2-}(aq) + 2e^- \rightarrow SO_2(g)$$
	reactant: $4(+1) + 1(-2) + 2(-1) = 0$; product: 0
balance H	reactant: 4 H product: 0 H
	To balance H, add 2 H_2O to the product side.
	$$4H^+(aq) + SO_4^{2-}(aq) + 2e^- \rightarrow SO_2(g) + 2H_2O$$
	reactant: 4 H product: 4 H
check O	reactant: 4 O product: 4 O
Balanced Reduction Half-Equation	$4H^+(aq) + SO_4^{2-}(aq) + 2e^- \rightarrow SO_2(g) + 2H_2O$

(d) Make the number of electrons in the two half-equations equal (multiply by an appropriate number if necessary).

Balanced half-equations:

oxidation	$2I^-(aq) \rightarrow I_2(s) + 2e^-$
reduction	$4H^+(aq) + SO_4^{2-}(aq) + 2e^-$ $\rightarrow SO_2(g) + 2H_2O$

(e) Combine the two half-equations.	$2I^-(aq) + SO_4^{2-}(aq) + 4H^+(aq) + \mathbf{2e^-}$ $\rightarrow I_2(s) + \mathbf{2e^-} + SO_2(g) + 2H_2O$
Cancel duplication.	$2I^-(aq) + SO_4^{2-}(aq) + 4H^+(aq) \rightarrow I_2(s) + SO_2(g) + 2H_2O$
Balanced Equation:	$\mathbf{2I^-(aq) + SO_4^{2-}(aq) + 4H^+(aq) \rightarrow I_2(s) + SO_2(g) + 2H_2O}$

(b) An unbalanced equation for the reaction would be

$$I^-(aq) + Cl_2(g) \rightarrow I_2(s) + Cl^-(aq)$$

(a) Split into two half-equations.	$I^-(aq) \rightarrow I_2(s)$
	$Cl_2(g) \rightarrow Cl^-(aq)$

(b) Balance one of the half-equations: $I^-(aq) \rightarrow I_2(s)$

oxidation numbers	I: $-1 \rightarrow 0$; I is oxidized
atom balance	reactant: 1 I product: 2 I
	multiply I^- by 2: $2I^-(aq) \rightarrow I_2(s)$
total oxidation number	I: $-2 \rightarrow$ I: 0
add electrons	The total oxidation number for I goes from -2 to 0. The I is oxidized by 2. Add 2 electrons to the product side. $$2I^-(aq) \rightarrow I_2(s) + 2e^-$$
balance charge	reactant: $2(-1) = -2$ product: $0 + 2(-1) = -2$
balance H	reactant: 0 H product: 0 H
check O	reactant: 0 O product: 0 O
Balanced Oxidation Half-Equation	$2I^-(aq) \rightarrow I_2(s) + 2e^-$

(c) Balance the other half-equation: $Cl_2(g) \rightarrow Cl^-(aq)$	
oxidation numbers	Cl (oxid. no.): $0 \rightarrow -1$; Cl is reduced
atom balance	reactant: 2 Cl product: 1 Cl multiply Cl^- by 2: $Cl_2(g) \rightarrow 2Cl^-(aq)$
total oxidation number	Cl: $0 \rightarrow$ Cl: -2
add electrons	The total oxidation number for Cl goes from 0 to -2. It is reduced by 2. Add 2 electrons to the reactant side. $Cl_2(g) + 2e^- \rightarrow 2Cl^-(aq)$
balance charge	reactant: $0 + 2(-1) = -2$; product: $2(-1) = -2$ charge is balanced
balance H	reactant: 0 H product: 0 H
check O	reactant: 0 O product: 0 O
Balanced Reduction Half-Equation	$Cl_2(g) + 2e^- \rightarrow 2Cl^-(aq)$
(d) Make the number of electrons in the two half-equations equal (multiply by an appropriate number if necessary).	Balanced half-equations: oxidation $2I^-(aq) \rightarrow I_2(s) + 2e^-$ reduction $Cl_2(g) + 2e^- \rightarrow 2Cl^-(aq)$
(e) Combine the two half-equations.	$2I^-(aq) + Cl_2(g) + 2e^- \rightarrow I_2(s) + 2Cl^-(aq) + 2e^-$
Cancel duplication.	$2I^-(aq) + Cl_2(g) \rightarrow I_2(s) + 2Cl^-(aq)$
Balanced Equation:	$\mathbf{2I^-(aq) + Cl_2(g) \rightarrow I_2(s) + 2Cl^-(aq)}$

18. See Sections 21-1 and 21-4, and Section 17-1. These problems are similar to Example 17.2: balance the redox reactions in the same way.

    (a) An unbalanced equation for the reaction would be

$$HClO(aq) \rightarrow Cl_2(g) + HClO_2(aq) \qquad \text{[acidic]}$$

or with H^+ cations removed for simplicity:

$$ClO^-(aq) \rightarrow Cl_2(g) + ClO_2^-(aq) \qquad \text{[acidic]}$$

(a) Split into two half-equations.	$ClO^-(aq) \rightarrow ClO_2^-(aq)$
	$ClO^-(aq) \rightarrow Cl_2(g)$
(b) Balance one of the half-equations: $ClO^-(aq) \rightarrow ClO_2^-(aq)$	
oxidation numbers	Cl: $+1 \rightarrow +3$; Cl is oxidized
atom balance	1 Cl on each side; no adjustment necessary

total oxidation number	Cl: +1 $\rightarrow$ Cl: +3
add electrons	The total oxidation number for Cl goes from +1 to +3. The Cl is oxidized by 2. Add 2 electrons to the product side. $ClO^-(aq) \rightarrow ClO_2^-(aq) + 2e^-$
balance charge	reactant: -1 product: $-1 + 2(-1) = -3$ acidic medium: add H^+ To balance the charge, add 2 H^+ to the product side. $ClO^-(aq) \rightarrow ClO_2^-(aq) + 2H^+(aq) + 2e^-$ reactant: -1 product: $-1 + 2(+1) + 2(-1) = -1$
balance H	reactant: 0 H product: 2 H To balance H, add H_2O to the reactant side. $ClO^-(aq) + H_2O \rightarrow ClO_2^-(aq) + 2H^+(aq) + 2e^-$ reactant: 2 H product: 2 H
check O	reactant: 2 O product: 2 O
Balanced Oxidation Half-Equation	$ClO^-(aq) + H_2O \rightarrow ClO_2^-(aq) + 2H^+(aq) + 2e^-$

(c) Balance the other half-equation: $ClO^-(aq) \rightarrow Cl_2(g)$

oxidation numbers	Cl (oxid. no.): +1 $\rightarrow$ 0; Cl is reduced
atom balance	reactant: 1 Cl product: 2 Cl multiply ClO^- by 2: $2ClO^-(aq) \rightarrow Cl_2(g)$
total oxidation number	Cl: +2 $\rightarrow$ Cl: 0
add electrons	The total oxidation number for Cl goes from +2 to 0. It is reduced by 2. Add 2 electrons to the reactant side. $2ClO^-(aq) + 2e^- \rightarrow Cl_2(g)$
balance charge	reactant: $2(-1) + 2(-1) = -4$; product: 0 acidic medium: add H^+ To balance the charge, add 4 H^+ to the reactant side. $2ClO^-(aq) + 4H^+(aq) + 2e^- \rightarrow Cl_2(g)$ reactant: $2(-1) + 4(+1) + 2(-1) = 0$; product: 0
balance H	reactant: 4 H product: 0 H To balance H, add 2 H_2O to the product side $2ClO^-(aq) + 4H^+(aq) + 2e^- \rightarrow Cl_2(g) + 2H_2O$ reactant: 4 H product: 4 H
check O	reactant: 2 O product: 2 O
Balanced Reduction Half-Equation	$2ClO^-(aq) + 4H^+(aq) + 2e^- \rightarrow Cl_2(g) + 2H_2O$

(d) Make the number of electrons in the two half-equations equal	Balanced half-equations:	
	oxidation	$ClO^-(aq) + H_2O$ $\rightarrow ClO_2^-(aq) + 2H^+(aq) + 2e^-$

(multiply by an appropriate number if necessary).	reduction	$2ClO^-(aq) + 4H^+(aq) + 2e^-$ $\rightarrow Cl_2(g) + 2H_2O$
(e) Combine the two half-equations.		$ClO^-(aq) + H_2O + 2ClO^-(aq) + 4H^+(aq) + 2e^-$ $\rightarrow ClO_2^-(aq) + 2H^+(aq) + 2e^- + Cl_2(g) + 2H_2O$
Cancel duplication.		$3ClO^-(aq) + 2H^+(aq) \rightarrow ClO_2^-(aq) + Cl_2(g) + H_2O$
Balanced Equation:		$3ClO^-(aq) + 2H^+(aq) \rightarrow ClO_2^-(aq) + Cl_2(g) + H_2O$

Finally, add H^+ to each side of the equation and make molecules of HClO and $HClO_2$. Since these are weak acids (and the reaction is taking place in acidic solution), they preferentially exist as the acid molecules rather than the anions and this should be reflected in the net ionic equation:

$$3HClO(aq) \rightarrow Cl_2(g) + HClO_2(aq) + H_2O$$

(b) An unbalanced equation for the reaction would be

$$ClO_3^-(aq) \rightarrow ClO_4^-(aq) + ClO_2^-(aq)$$

For simplicity, assume the reaction takes place in acidic solution.

(a) Split into two half-equations.	$ClO_3^-(aq) \rightarrow ClO_4^-(aq)$
	$ClO_3^-(aq) \rightarrow ClO_2^-(aq)$
(b) Balance one of the half-equations: $ClO_3^-(aq) \rightarrow ClO_4^-(aq)$	
Balanced Oxidation Half-Equation (from page 540)	$ClO_3^-(aq) + H_2O \rightarrow ClO_4^-(aq) + 2H^+(aq) + 2e^-$
(c) Balance the other half-equation: $ClO_3^-(aq) \rightarrow ClO_2^-(aq)$	
oxidation numbers	Cl (oxid. no.): +5 $\rightarrow$ +3; Cl is reduced
atom balance	1 Cl on each side; no adjustment required
total oxidation number	Cl: +5 $\rightarrow$ Cl: +3
add electrons	The total oxidation number for Cl goes from +5 to +3. It is reduced by 2. Add 2 electrons to the reactant side. $ClO_3^-(aq) + 2e^- \rightarrow ClO_2^-(aq)$
balance charge	reactant: $-1 + 2(-1) = -3$; product: -1 acidic medium: add H^+ To balance the charge, add 2 H^+ to the reactant side. $ClO_3^-(aq) + 2H^+(aq) + 2e^- \rightarrow ClO_2^-(aq)$ reactant: $-1 + 2(+1) + 2(-1) = -1$; product: -1
balance H	reactant: 2 H product: 0 H To balance H, add H_2O to the product side $ClO_3^-(aq) + 2H^+(aq) + 2e^- \rightarrow ClO_2^-(aq) + H_2O$ reactant: 2 H product: 2 H

check O	reactant: 3 O	product: 3 O
Balanced Reduction Half-Equation	$ClO_3^-(aq) + 2H^+(aq) + 2e^- \rightarrow ClO_2^-(aq) + H_2O$	
(d) Make the number of electrons in the two half-equations equal (multiply by an appropriate number if necessary).	Balanced half-equations:	
	oxidation	$ClO_3^-(aq) + H_2O$ $\rightarrow ClO_4^-(aq) + 2H^+(aq) + 2e^-$
	reduction	$ClO_3^-(aq) + 2H^+(aq) + 2e^-$ $\rightarrow ClO_2^-(aq) + H_2O$
(e) Combine the two half-equations.	$ClO_3^-(aq) + H_2O + ClO_3^-(aq) + 2H^+(aq) + 2e^-$ $\rightarrow ClO_4^-(aq) + 2H^+(aq) + 2e^- + ClO_2^-(aq) + H_2O$	
Cancel duplication.	$2ClO_3^-(aq) \rightarrow ClO_4^-(aq) + ClO_2^-(aq)$	
Balanced Equation:	$\mathbf{2ClO_3^-(aq) \rightarrow ClO_4^-(aq) + ClO_2^-(aq)}$	

20. See Sections 21-1 and 17-3.

For the four possible redox reactions, $E°$ for the reaction can be calculated using $E°_{red}$ values listed in Table 17.1. A reaction will occur (will be spontaneous) of $E° > 0$.

(a) The two half-reactions will be:

$2Br^-(aq) \rightarrow Br_2(l) + 2e^-$ $E°_{ox} = -E°_{red} = -1.077$ V

$Cl_2(g) + 2e^- \rightarrow 2Cl^-(aq)$ $E°_{red} = +1.360$ V

Adding the two half-reactions together gives the overall reaction:

$\mathbf{Cl_2(g) + 2Br^-(aq) \rightarrow 2Cl^-(aq) + Br_2(l)}$

$E° = E°_{ox} + E°_{red} = -1.077 + 1.360 = +0.283$ V

$E° > 0$ so the reaction occurs.

(b) The two half-reactions will be:

$2Cl^-(aq) \rightarrow Cl_2(g) + 2e^-$ $E°_{ox} = -E°_{red} = -1.360$ V

$I_2(s) + 2e^- \rightarrow 2I^-(aq)$ $E°_{red} = +0.534$ V

Adding the two half-reactions together gives the overall reaction:

$I_2(s) + 2Cl^-(aq) \rightarrow 2I^-(aq) + Cl_2(g)$

$E° = E°_{ox} + E°_{red} = -1.360 + 0.534 = -0.826$ V

$E° < 0$ so **NR**

(c) The two half-reactions will be:

$2Br^-(aq) \rightarrow Br_2(l) + 2e^-$ $E°_{ox} = -1.077$ V

$I_2(s) + 2e^- \rightarrow 2I^-(aq)$ $E°_{red} = +0.534$ V

Adding the two half-reactions together gives the overall reaction:

$$I_2(s) + 2Br^-(aq) \rightarrow 2I^-(aq) + Br_2(l)$$

$$E° = E°_{ox} + E°_{red} = -1.077 + 0.534 = -0.543 \text{ V}$$

$$E° < 0 \text{ so } \textbf{NR}$$

(d) The two half-reactions will be:

$$2Cl^-(aq) \rightarrow Cl_2(g) + 2e^- \qquad E°_{ox} = -E°_{red} = -1.360 \text{ V}$$

$$Br_2(l) + 2e^- \rightarrow 2Br^-(aq) \qquad E°_{red} = +1.077 \text{ V}$$

Adding the two half-reactions together gives the overall reaction:

$$Br_2(l) + 2Cl^-(aq) \rightarrow 2Br^-(aq) + Cl_2(g)$$

$$E° = E°_{ox} + E°_{red} = -1.360 + 1.077 = -0.283 \text{ V}$$

$$E° < 0 \text{ so } \textbf{NR}$$

Note also (Section 21-1) that the oxidizing power of the halogens decreases in the order $Cl_2 > Br_2 > I_2$ and on that basis it can be predicted that neither Br_2 nor I_2 could oxidize Cl^- ions and I_2 would not oxidize Br^- ions [answers (d), (b), (c), respectively], but Cl_2 *will* oxidize Br^- ions [answer (a)].

22. (a) $\textbf{Pb(N}_3\textbf{)}_2\textbf{(s)} \rightarrow \textbf{3N}_2\textbf{(g)} + \textbf{Pb(s)}$ (the equation is given on page 526)

    (b) $\textbf{2O}_3\textbf{(g)} \rightarrow \textbf{3O}_2\textbf{(g)}$

    (c) $\textbf{2H}_2\textbf{S(g)} + \textbf{O}_2\textbf{(g)} \rightarrow \textbf{2S(s)} + \textbf{2H}_2\textbf{O}$ (the equation is given on page 531)

24. See Section 21-2.

    (a) $\textbf{Cd}^{2+}\textbf{(aq)} + \textbf{H}_2\textbf{S(aq)} \rightarrow \textbf{CdS(s)} + \textbf{2H}^+\textbf{(aq)}$ (precipitation reaction; given on page 531)

    (b) $\textbf{H}_2\textbf{S(aq)} + \textbf{OH}^-\textbf{(aq)} \rightarrow \textbf{H}_2\textbf{O} + \textbf{HS}^-\textbf{(aq)}$ (acid-base reaction: H_2S is a weak Brønsted-Lowry acid; see page 531)

    (c) $\textbf{2H}_2\textbf{S(aq)} + \textbf{O}_2\textbf{(g)} \rightarrow \textbf{2H}_2\textbf{O} + \textbf{2S(s)}$ (redox reaction; given on page 531)

26. See Section 21-4 and Chapter 4.

    (a) See Example 4.2. The overall reaction is:

$$CaCO_3(s) + H_2SO_4(aq) \rightarrow CaSO_4(aq) + H_2O + CO_2(g)$$

H_2SO_4 is a strong acid, so the reacting species is H^+ and SO_4^{2-} is a spectator.

Net ionic equation:

$$\textbf{CaCO}_3\textbf{(s)} + \textbf{2H}^+\textbf{(aq)} \rightarrow \textbf{Ca}^{2+}\textbf{(aq)} + \textbf{H}_2\textbf{O} + \textbf{CO}_2\textbf{(g)}$$

(b) See Table 4.2 and Example 4.4. The overall reaction is:

$$2NaOH(aq) + H_2SO_4(aq) \rightarrow Na_2SO_4(aq) + 2H_2O$$

H_2SO_4 is a strong acid, so the reacting species is H^+ and SO_4^{2-} is a spectator; NaOH is a strong base, so the reacting species is OH^- and Na^+ is a spectator.

Net ionic equation:

$$\mathbf{H^+(aq) + OH^-(aq) \rightarrow H_2O}$$

(c) As the question implies, this is a redox reaction and is best approached using the standard procedure from Chapter 17. By analogy with the reaction of Cu with HNO_3 (page 542), assume that Cu is oxidized to Cu^{2+}.

The unbalanced redox reaction is

$$Cu(s) + SO_4^{2-}(aq) \rightarrow Cu^{2+}(aq) + SO_2(g)$$

and the balanced net ionic equation is given on page 544:

$$\mathbf{Cu(s) + 4H^+(aq) + SO_4^{2-}(aq) \rightarrow Cu^{2+}(aq) + SO_2(g) + 2H_2O}$$

MOLECULAR STRUCTURE

28. See Sections 21-3 and 7-1. The structure of N_2O is given in Figure 21.9. Note (page 536) that the phosphorus atoms in P_4 molecules form a tetrahedron.

30. See Sections 7-2 and 7-3.

(a) **Polar**. Since Cl_2O is bent (AX_2E_2), the dipoles in each O–Cl bond do not cancel.

(b) **Polar**. The molecule is not symmetric so electrons must be distributed unsymmetrically, giving a dipole.

(c) and (d) Non-polar. In each molecule, all the atoms are identical so none of the bonds can be polar; hence there cannot be a molecular dipole.

32. See Sections 21-4 and 7-2. Recall (Section 13-1) that the conjugate base of an acid is the species formed when a proton is removed from the acid. (Note that for each of the Lewis structures shown below several resonance forms are possible.)

(a) The conjugate base of HNO_3 is NO_3^-

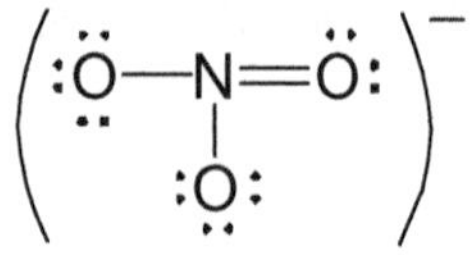

This Lewis structure has a formal charge of +1 on the N atom and formal charge of −1 on two of the O atoms.

(b) The conjugate base of H_2SO_4 is HSO_4^-

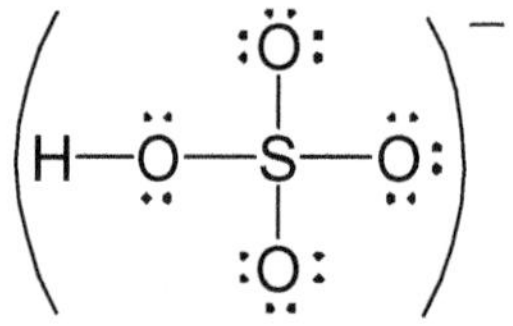

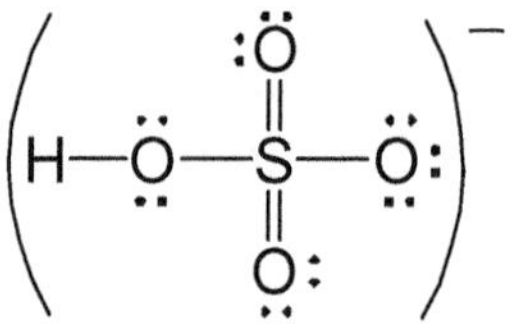

The Lewis structure on the left has a formal charge of +2 on the S atom and a formal charge of −1 on three of the O atoms; the one on the right has a formal charge of −1 on only one of the O atoms.

(c) The conjugate base of H_3PO_4 is $H_2PO_4^-$

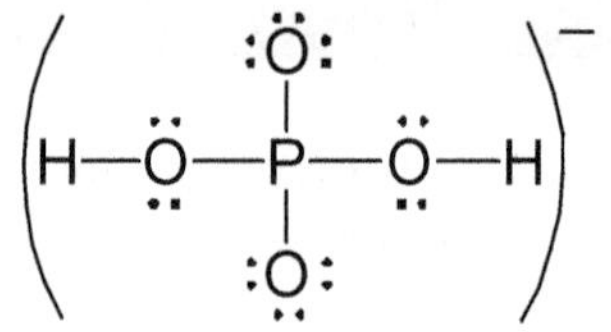

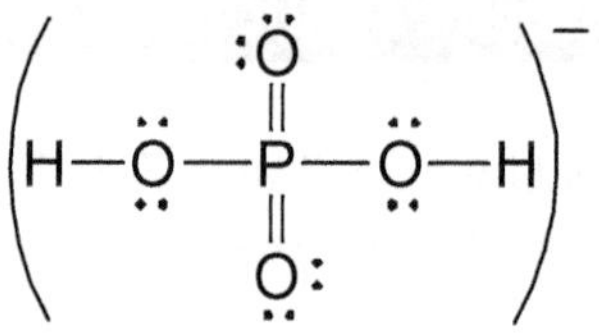

The Lewis structure on the left has a formal charge of +1 on the P atom and a formal charge of −1 on two of the O atoms; the one on the right has a formal charge of −1 on only one of the O atoms.

34. See Sections 21-4 and 7-1.

(a) N_2O_5

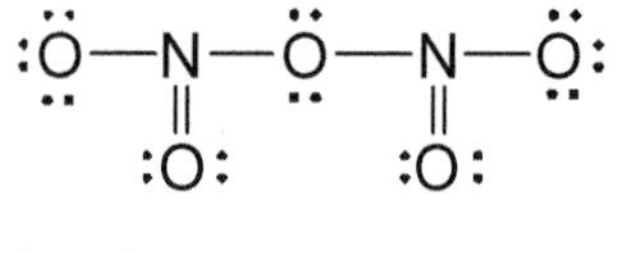

(See Figure 21.9.) This Lewis structure has a formal charge of +1 on the N atoms and a formal charge of −1 on the two terminal −O atoms; several resonance forms are possible.

(b) HNO_3

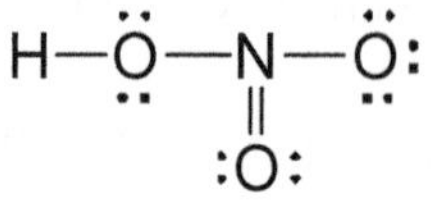

This Lewis structure has a formal charge of +1 on the N atom and a formal charge of −1 on the terminal −O atom; several resonance forms are possible.

(c) SO_4^{2-}

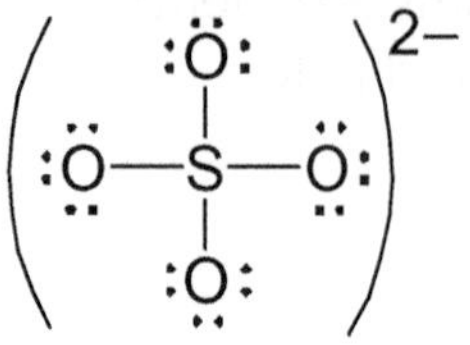

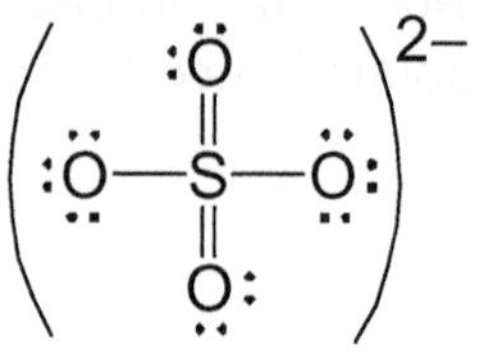

The Lewis structure on the left has a formal charge of +2 on the S atom and a formal charge of −1 on all four O atoms; the one on the right has a formal charge of −1 on only the two −O atoms, and has several possible resonance forms.

STOICHIOMETRY

36. See Section 21-1. The balanced equation for the reaction, from Problem 16(b) above, is:

$$2I^-(aq) + Cl_2(g) \rightarrow I_2(s) + 2Cl^-(aq)$$

or, with Na^+ ions included:

$$2NaI(aq) + Cl_2(g) \rightarrow I_2(s) + 2NaCl(aq)$$

Strategy: mass NaI → mol NaI → mol Cl_2 required; then use the Ideal Gas Law (Chapter 5) to obtain the corresponding vol Cl_2.

$$\text{mol } Cl_2 \text{ required } = 175 \text{ g NaI} \times \frac{1 \text{ mol NaI}}{149.89 \text{ g NaI}} \times \frac{1 \text{ mol } Cl_2}{2 \text{ mol NaI}} = 0.584 \text{ mol } Cl_2$$

$$\text{vol } Cl_2 \text{ required } = \frac{nRT}{P} = \frac{(0.584 \text{ mol})(0.0821 \text{ L} \cdot \text{atm/mol} \cdot \text{K})(298 \text{ K})}{758 \text{ mmHg} \times \dfrac{1 \text{ atm}}{760 \text{ mmHg}}} = \underline{\textbf{14.3 L}}$$

38. Strategy: use the Ideal Gas Law (see Chapter 5) to obtain mol HBr;

then use mol HBr and assume vol H_2O = vol solution → M HBr solution

$$\text{mol HBr } = n = \frac{PV}{RT} = \frac{(0.974 \text{ atm})(1.283 \text{ L})}{(0.0821 \text{ L} \cdot \text{atm/mol} \cdot \text{K})(298 \text{ K})} = 0.0511 \text{ mol HBr}$$

$$\text{molarity of HBr solution } = \frac{\text{mol HBr}}{\text{L of solution}} = \frac{0.0511 \text{ mol}}{250.0 \text{ mL} \times \dfrac{1 \text{ L}}{1000 \text{ mL}}} = \underline{\textbf{0.204 M HBr}}$$

40. First calculate the amount of S(s) present in the coal sample and how much $SO_2(g)$ will be produced when it is burned. (Recall from Table 1.3 that 1 metric ton = 10^3 kg.) The equation for this reaction is:

$$S(s) + O_2(g) \rightarrow SO_2(g)$$

Strategy: mass coal → mol S present → mol SO_2 produced

From mol of SO_2, mol H_2S needed and hence vol H_2S can be calculated, and the mass of S produced can also be calculated. Strategies:

mol SO_2 → mol H_2S → vol H_2S (use the Ideal Gas Law, from Chapter 5)

mol SO_2 → mol S → mass S

The reaction between SO_2 and H_2S, forming S, is a redox process. To obtain stoichiometric ratios, a balanced equation is needed. An unbalanced equation for the reaction is:

$SO_2(g) + H_2S(g) \rightarrow S(s) + H_2O$

| (a) Split into two half-equations. | $H_2S(g) \rightarrow S(s)$ |
| | $SO_2(g) \rightarrow S(s)$ |

(b) Balance one of the half-equations: $H_2S(g) \rightarrow S(s)$

oxidation numbers	S (oxid. no.): $-2 \rightarrow 0$; S is oxidized
atom balance	1 S on each side; no adjustment required
total oxidation number	S: $-2 \rightarrow$ S: 0
add electrons	The total oxidation number for S goes from -2 to 0. It is oxidized by 2. Add 2 electrons to the product side. $\qquad H_2S(g) \rightarrow S(s) + 2e^-$
balance charge	reactant: 0; product: $0 + 2(-1) = -2$ acidic medium: add H^+ To balance the charge, add 2 H^+ to the product side. $\qquad H_2S(g) \rightarrow S(s) + 2H^+(aq) + 2e^-$ reactant: 0; product: $0 + 2(+1) + 2(-1) = 0$
balance H	reactant: 2 H $\qquad$ product: 2 H
check O	reactant: 0 O $\qquad$ product: 0 O
Balanced Oxidation Half-Equation	$H_2S(g) \rightarrow S(s) + 2H^+(aq) + 2e^-$

(c) Balance the other half-equation: $SO_2(g) \rightarrow S(s)$

oxidation numbers	S (oxid. no.): $+4 \rightarrow 0$; S is reduced
atom balance	1 S on each side; no adjustment required
total oxidation number	S: $+4 \rightarrow$ S: 0
add electrons	The total oxidation number for S goes from $+4$ to 0. It is reduced by 4. Add 4 electrons to the reactant side. $\qquad SO_2(g) + 4e^- \rightarrow S(s)$
balance charge	reactant: $0 + 4(-1) = -4$; product: 0 acidic medium: add H^+ To balance the charge, add 4 H^+ to the reactant side. $\qquad SO_2(g) + 4H^+(aq) + 4e^- \rightarrow S(s)$

	reactant: 0 + 4(+1) + 4(−1) = 0; product: 0
balance H	reactant: 4 H product: 0 H To balance H, add 2H$_2$O to the product side $\quad$ SO$_2$(g) + 4H$^+$(aq) + 4e$^-$ → S(s) + 2H$_2$O reactant: 4 H product: 4 H
check O	reactant: 2 O product: 2 O
Balanced Reduction Half-Equation	SO$_2$(g) + 4H$^+$(aq) + 4e$^-$ → S(s) + 2H$_2$O

(d) Make the number of electrons in the two half-equations equal (multiply by an appropriate number if necessary).	Balanced half-equations:	
	oxidation: multiply by 2	2 [H$_2$S(g) → S(s) + 2H$^+$(aq) + 2e$^-$]
		2H$_2$S(g) → 2S(s) + 4H$^+$(aq) + 4e$^-$
	reduction	SO$_2$(g) + 4H$^+$(aq) + 4e$^-$ $\qquad$ → S(s) + 2H$_2$O

(e) Combine the two half-equations.	2H$_2$S(g) + SO$_2$(g) + **4H$^+$(aq)** + **4e$^-$** $\qquad$ → 2S(s) + **4H$^+$(aq)** + **4e$^-$** + S(s) + 2H$_2$O
Cancel duplication.	2H$_2$S(g) + SO$_2$(g) → 3S(s) + 2H$_2$O

Balanced Equation:	2H$_2$S(g) + SO$_2$(g) → 3S(s) + 2H$_2$O

mol SO$_2$ produced

$$= 10^3 \text{ kg coal} \times \frac{1000 \text{ g}}{1 \text{ kg}} \times \frac{5.0 \text{ g S}}{100 \text{ g coal}} \times \frac{1 \text{ mol S}}{32.066 \text{ g S}} \times \frac{1 \text{ mol SO}_2}{1 \text{ mol S}} = 1.6 \times 10^3 \text{ mol SO}_2$$

$$\text{mol H}_2\text{S required} = 1.6 \times 10^3 \text{ mol SO}_2 \times \frac{2 \text{ mol H}_2\text{S}}{1 \text{ mol SO}_2} = 3.2 \times 10^3 \text{ mol H}_2\text{S}$$

vol H$_2$S required

$$= V = \frac{nRT}{P} = \frac{(3.2 \times 10^3 \text{ mol})(0.0821 \text{ L} \cdot \text{atm/mol} \cdot \text{K})(300 \text{ K})}{755 \text{ mmHg} \times \dfrac{1 \text{ atm}}{760 \text{ mmHg}}} = \underline{7.9 \times 10^4 \text{ L H}_2\text{S}}$$

$$\text{mass S produced} = 1.6 \times 10^3 \text{ mol SO}_2 \times \frac{3 \text{ mol S}}{1 \text{ mol SO}_2} \times \frac{32.066 \text{ g S}}{1 \text{ mol S}} = \underline{1.5 \times 10^5 \text{ g S}}$$

42. First calculate how many mol of H$_2$S(aq) are present in the water and how many mol of Cl$_2$(g) will be required to remove it. (Recall (Table 1.3) that 1 gal = 4 qt and 1 L = 1.057 qt.) Then use the Ideal Gas Law (Chapter 5) to obtain the corresponding volume of Cl$_2$(g).

Strategy: vol H$_2$O (gal) → vol H$_2$O (mL) → mass H$_2$O → mass H$_2$S → mol H$_2$S → mol Cl$_2$
$\qquad$ → vol Cl$_2$

$$\text{mass } H_2O = 1.00 \times 10^3 \text{ gal} \times \frac{4 \text{ qt}}{1 \text{ gal}} \times \frac{1 L}{1.057 \text{ qt}} \times \frac{1000 \text{ mL}}{1 L} \times \frac{1.00 \text{ g}}{1 \text{ mL}} = 3.78 \times 10^6 \text{ g } H_2O$$

Since the $H_2S(aq)$ solution is very dilute (5.0 ppm H_2S), we can assume that this mass of the solvent (H_2O) is the same as the mass of the solution.

$$\text{mol } H_2S = 3.78 \times 10^6 \text{ g solution} \times \frac{5.0 \text{ g } H_2S}{10^6 \text{ g solution}} \times \frac{1 \text{ mol } H_2S}{34.08 \text{ g } H_2S} = 0.55 \text{ mol } H_2S$$

$$\text{mol } Cl_2 = 0.55 \text{ mol } H_2S \times \frac{1 \text{ mol } Cl_2}{1 \text{ mol } H_2S} = 0.55 \text{ mol } Cl_2$$

$$\text{vol } Cl_2 = V = \frac{nRT}{P} = \frac{(0.55 \text{ mol})(0.0821 L \cdot atm/mol \cdot K)(273 K)}{1 \text{ atm}} = \underline{\textbf{12 L } Cl_2}$$

To calculate the pH of the resulting solution, calculate the moles of H^+ produced and hence the concentration of $H^+(aq)$ and then use Equation 13.3 to determine the pH.

$$\text{mol } H^+ \text{ produced} = 0.55 \text{ mol } Cl_2 \times \frac{2 \text{ mol } H^+}{1 \text{ mol } Cl_2} = 1.1 \text{ mol } H^+$$

$$[H^+] = \frac{\text{mol } H^+}{L \text{ solution}} = \frac{1.1 \text{ mol } H^+}{3.78 \times 10^6 \text{ g solution} \times \frac{1 \text{ mL}}{1.00 \text{ g}} \times \frac{1 L}{1000 \text{ mL}}} = 2.9 \times 10^{-4} M$$

$$pH = -\log_{10}[H^+] = -\log(2.9 \times 10^{-4}) = \underline{\textbf{3.54}}$$

EQUILIBRIA

44. See Section 13-4 and Examples 13.7 and 13.8.

$$HClO(aq) \rightleftharpoons H^+(aq) + ClO^-(aq) \qquad K_a = 2.8 \times 10^{-8}$$

Set up an equilibrium table with $[HClO]_0 = 0.10 M$ and let $\Delta[H^+] = \Delta[ClO^-] = x \ M$.

	HClO	$\rightleftharpoons$	H^+	+	ClO^-
$[\ ]_0$ (M)	0.10		0		0
$\Delta[\ ]$ (M)	$-x$		$+x$		$+x$
$[\ ]_{eq}$ (M)	$0.10 - x$		x		x

Now substitute the equilibrium concentrations of each species into the expression for K_a:

$$K_a = 2.8 \times 10^{-8} = \frac{[H^+][ClO^-]}{[HClO]} = \frac{(x)(x)}{(0.10 - x)} \approx \frac{x^2}{0.10} \qquad \text{(assuming that } x \ll 0.10\text{)}$$

hence $x^2 = 2.8 \times 10^{-9}$ and so $x = 5.3 \times 10^{-5}$.

Note that the value of x is indeed much less that 0.10, so the assumption made in solving the above equation was valid; % ionization = $[H^+]_{eq}/[HClO]_0 \times 100$ = $(5.3 \times 10^{-5})/(0.10) \times 100$ = 0.053% < 5% (see page 345).

Thus, $x = [H^+] = [ClO^-] = 5.3 \times 10^{-5} M$ so pH = $-\log_{10}[H^+]$ = $-\log_{10}(5.3 \times 10^{-5})$ = **4.28**

and the equilibrium concentration of HClO = $[HClO]_{eq}$ = $0.10 - x$ = **0.10 M**

46. See Section 12-2 and Table 12.3. The third reaction is simply the sum of the first two reactions given. Thus K for the required reaction is simply the product of the equilibrium constants for the first two reactions (Rule of Multiple Equilibria: see Table 12.3).

$$K = (6.9 \times 10^{-4}) \times (2.7) = \mathbf{1.9 \times 10^{-3}}$$

48. See Section 15-2. This problem is similar to Example 15.6.

$$BaF_2(s) \rightleftharpoons Ba^{2+}(aq) + 2F^-(aq) \qquad K_{sp} = [Ba^{2+}][F^-]^2 = 1.8 \times 10^{-7}$$

Let s be the solubility of BaF_2 (in mol/L) = mol BaF_2 dissolved in 1L of solution. Initially, only $BaCl_2$ is present in solution: no F^- ions are present, and no Ba^{2+} ions are contributed by the BaF_2 but some *are* contributed by the $BaCl_2$. Set up an equilibrium table with $[Ba^{2+}]_0 = 0.10 M$ and $[F^-]_0 = 0$. Then $\Delta[Ba^{2+}]$ will equal s M and $\Delta[F^-] = 2s$ M. (Recall that for every s mol of BaF_2 dissolving, s mol of Ba^{2+} ions and $2s$ mol of F^- ions are provided.)

	BaF_2 $\rightleftharpoons$	Ba^{2+} +	$2F^-$
[ ]$_0$ (M)		0.10	0
Δ[ ] (M)		+ s	+ 2s
[ ]$_{eq}$ (M)		0.10 + s	2s

Now substitute the equilibrium concentrations of each species into the expression for K_{sp}:

$$K_{sp} = 1.8 \times 10^{-7} = [Ba^{2+}][F^-]^2 \qquad = (0.10 + s)(2s)^2$$
$$\approx (0.10)(4s^2) \quad \text{(assuming } s << 0.10)$$
$$s^2 = \frac{1.8 \times 10^{-7}}{(0.10)(4)} = 4.5 \times 10^{-7} \text{ so } s = 6.7 \times 10^{-4} M$$

Note that s is indeed much smaller than 0.10 so the assumption is justified.
Thus 6.7×10^{-4} moles of BaF_2 would dissolve in 1 L of 0.10 M $BaCl_2$, so the solubility in grams per 100 mL is

$$\frac{6.7 \times 10^{-4} \text{ mol } BaF_2}{1 \text{ L solution}} \times \frac{175.33 \text{ g } BaF_2}{1 \text{ mol } BaF_2} \times \frac{1 L}{1000 \text{ mL}} \times 100 \text{ mL solution} = \mathbf{0.012 \text{ g } BaF_2}$$

THERMODYNAMICS

50. See Sections 16-3—16-5. This problem is related to Examples 16.3 and 16.6.

Strategy: Use $\Delta H_f°$ and $S°$ values from Appendix 1 and given in the question to calculate $\Delta H°$ and $\Delta S°$, then use Equation 16.2 to calculate $\Delta G°$.

$\Delta H°$ = $\Sigma \Delta H_f°_{\text{(products)}}$ — $\Sigma \Delta H_f°_{\text{(reactants)}}$

$\Delta H°$ = [(2 mol)(−397.7 kJ/mol) + (1 mol)(0 kJ/mol)]

$\qquad$ − [(2 mol)(−360.2 kJ/mol) + (1 mol)(0 kJ/mol)]

$\Delta H°_{\text{(reaction)}}$ = −75.0 kJ

$\Delta S°$ = $\Sigma S°_{\text{(products)}}$ — $\Sigma S°_{\text{(reactants)}}$

$\Delta S°$ = [(2 mol)(0.1431 kJ/mol·K) + (1 mol)(0.1522 kJ/mol·K)]

$\qquad$ − [(2 mol)(0.1492 kJ/mol·K) + (1 mol)(0.2230 kJ/mol·K)]

$\Delta S°$ = −0.0830 kJ/K

$\Delta G°$ = −75.0 kJ − (298 K)(−0.0830 kJ/K) = −50.3 kJ

$\Delta G° < 0$, so **the reaction is spontaneous.**

The reaction is nonspontaneous when $\Delta G°$ becomes positive so the change in spontaneity occurs when $\Delta G° = 0$.

0 = −75.0 kJ − (T)(−0.0830 kJ/K)

Solving for $T = \dfrac{75.0\,\text{kJ}}{0.0830\,\text{kJ/K}}$ = 904 K

The reaction becomes non-spontaneous at 904 K and we know it is spontaneous at 298 K, and therefore it must be spontaneous at all temperatures below 904 K. Thus the **lowest temperature at which the reaction is spontaneous is 0 K.**

52. See Sections 16-3—16-6.

(a) This part of the problem is related to Example 16.3.

Strategy: Use $\Delta H°_f$ and $S°$ values from Appendix 1 and given in the question to calculate $\Delta H°$ and $\Delta S°$, then use Equation 16.2 to calculate $\Delta G°$.

$\Delta H°$ = $\Sigma \Delta H_f°_{\text{(products)}}$ — $\Sigma \Delta H_f°_{\text{(reactants)}}$

$\Delta H°$ = [(2 mol)(0 kJ/mol) + (2 mol)(−285.8 kJ/mol)]

$\qquad$ − [(4 mol)(−92.3 kJ/mol) + (1 mol)(0 kJ/mol)]

$\Delta H°$ = −202.4 kJ

$$\Delta S° = \Sigma S°_{(products)} - \Sigma S°_{(reactants)}$$

$$\Delta S° = [(2 \text{ mol})(0.2230 \text{ kJ/mol·K}) + (2 \text{ mol})(0.0699 \text{ kJ/mol·K})]$$
$$- [(4 \text{ mol})(0.1868 \text{ kJ/mol·K}) + (1 \text{ mol})(0.2050 \text{ kJ/mol·K})]$$

$$\Delta S° = -0.3664 \text{ kJ/K}$$

$$\Delta G° = -202.4 \text{ kJ} - (298 \text{ K})(-0.3664 \text{ kJ/K}) = \mathbf{-93.2 \text{ kJ}}$$

$\Delta G° < 0$, therefore **the reaction is spontaneous.**

(b) Use Equation 16.5 to determine K; see Example 16.8.

$$\Delta G° = -RT\ln K$$

$$-93.2 \times 10^3 \text{ J} = -(8.3145 \text{ J/mol·K})(298 \text{ K})\ln K$$

$$\ln K = 37.6 \qquad \text{so} \quad K = e^{37.6} = \underline{\mathbf{2 \times 10^{16}}}$$

54. See Sections 16-3—16-5 and Examples 16.3 and 16.6. Sublimation is the phase change from solid to gas. During a phase change, the two phases present are in equilibrium with each other; hence, we need to know the temperature for which the following process is at equilibrium ($\Delta G° = 0$):

$$P_4(s) \rightleftharpoons P_4(g)$$

Strategy: Use $\Delta H_f°$ and $S°$ values to calculate $\Delta H°_{(reaction)}$ and $\Delta S°_{(reaction)}$; then use Equation 16.2 with $\Delta G° = 0$ and solve for T.

$$\Delta H° = \Sigma \Delta H_f°_{(products)} - \Sigma \Delta H_f°_{(reactants)}$$
$$\Delta H° = [(1 \text{ mol})(58.9 \text{ kJ/mol})] - [(1 \text{ mol})(0 \text{ kJ/mol})] = 58.9 \text{ kJ}$$

$$\Delta S° = \Sigma S°_{(products)} - \Sigma S°_{(reactants)}$$
$$\Delta S° = [(1 \text{ mol})(0.2800 \text{ kJ/mol·K}) - [(1 \text{ mol})(0.1644 \text{ kJ/mol·K})] = 0.1156 \text{ kJ/K}$$

$$\Delta G° = 58.9 \text{ kJ} - (T)(0.1156 \text{ kJ/K}) = 0 \text{ at equilibrium}$$

Solving for $T = \dfrac{58.9 \text{ kJ}}{0.1156 \text{ kJ/K}} = \underline{\mathbf{510 \text{ K}}}$

ELECTROCHEMISTRY

56. See Section 17-6. Use conversion factors from Table 17.3.

Strategy: current (A) → charge (C) → mol e^- → mol F_2 → mass F_2 (theoretical) → mass F_2 (actual)

(The actual yield is less than the theoretical yield since the cell is only operating with 95% efficiency: actual yield = 95% × theoretical yield.)

The relevant oxidation reaction is:

$$2F^-(l) \rightarrow F_2(g) + 2e^-$$

mass F_2 expected

$$= \frac{7.00 \times 10^3 \text{ C}}{1 \text{ s}} \times \frac{3600 \text{ s}}{1 \text{ h}} \times \frac{24 \text{ h}}{1 \text{ d}} \times 2 \text{ d} \times \frac{1 \text{ mol } e^-}{9.648 \times 10^4 \text{ C}} \times \frac{1 \text{ mol } F_2}{2 \text{ mol } e^-} \times \frac{37.9968 \text{ g } F_2}{1 \text{ mol } F_2}$$

$$= 2.38 \times 10^5 \text{ g } F_2$$

mass F_2 obtained = 0.95 × (theoretical yield) = 0.95 × (2.38 × 10⁵ g) = **2.26 × 10⁵ g**

58. See Section 17-6. The oxidation process is given on page 648:

$$ClO_3^-(aq) + H_2O \rightarrow ClO_4^-(aq) + 2H^+(aq) + 2e^-$$

Strategy: current (A) → charge (C) → mol e^- → mol ClO_4^- → mol $NaClO_4$
→ mass $NaClO_4$

mol ClO_4^- produced $= \dfrac{1.50 \times 10^3 \text{ C}}{1 \text{ s}} \times \dfrac{3600 \text{ s}}{1 \text{ h}} \times 8 \text{ h} \times \dfrac{1 \text{ mol } e^-}{9.648 \times 10^4 \text{ C}} \times \dfrac{1 \text{ mol } ClO_4^-}{2 \text{ mol } e^-}$

$$= 2.24 \times 10^2 \text{ mol } ClO_4^-$$

mass $NaClO_4$ $= 2.24 \times 10^2 \text{ mol } ClO_4^- \times \dfrac{1 \text{ mol } NaClO_4}{1 \text{ mol } ClO_4^-} \times \dfrac{122.44 \text{ g } NaClO_4}{1 \text{ mol } NaClO_4} \times \dfrac{1 \text{ kg}}{1000 \text{ g}}$

$$= \textbf{27.4 kg } NaClO_4$$

60. See Sections 17-3 and 17-4. For each of the species given, note that H_2O_2 is to act as an oxidizing agent, so consider the value of E_{ox}° for oxidation of the given species. Then, to determine whether the species will be oxidized by H_2O_2, calculate E° for the overall reaction using $E^\circ = E_{ox}^\circ + E_{red}^\circ (H_2O_2)$. If $E^\circ > 0$ then the reaction will occur (will be spontaneous), so that species will be oxidized by H_2O_2.

(a) Co^{2+}: oxidation half-reaction will be

$$Co^{2+}(aq) \rightarrow Co^{3+}(aq) + e^- \qquad E_{ox}^\circ = -E_{red}^\circ = -1.953 \text{ V}$$

$$E^\circ = E_{ox}^\circ + E_{red}^\circ = -1.953 \text{ V} + 1.763 \text{ V} = -0.190 \text{ V} \qquad \textbf{Not oxidized}$$

(b) Cl⁻: oxidation half-reaction will be

$$2Cl^-(aq) \rightarrow Cl_2(g) + 2e^- \qquad E^\circ_{ox} = -E^\circ_{red} = -1.360 \text{ V}$$

$$E^\circ = E^\circ_{ox} + E^\circ_{red} = -1.360 \text{ V} + 1.763 \text{ V} = +0.403 \text{ V} \qquad \textbf{Oxidized}$$

(c) Fe²⁺: oxidation half-reaction will be

$$Fe^{2+}(aq) \rightarrow Fe^{3+}(aq) + e^- \qquad E^\circ_{ox} = -E^\circ_{red} = -0.769 \text{ V}$$

$$E^\circ = E^\circ_{ox} + E^\circ_{red} = -0.769 \text{ V} + 1.763 \text{ V} = +0.994 \text{ V} \qquad \textbf{Oxidized}$$

(d) Sn²⁺: oxidation half-reaction will be

$$Sn^{2+}(aq) \rightarrow Sn^{4+}(aq) + 2e^- \qquad E^\circ_{ox} = -E^\circ_{red} = -0.154 \text{ V}$$

$$E^\circ = E^\circ_{ox} + E^\circ_{red} = -0.154 \text{ V} + 1.763 \text{ V} = +1.609 \text{ V} \qquad \textbf{Oxidized}$$

62. See Sections 17-1 and 17-5. An unbalanced equation for the redox reaction taking place is:

$$NO_3^-(aq) + SO_2(g) \rightarrow NO(g) + SO_4^{2-}(aq) \qquad \text{(in acidic solution)}$$

This equation must first be balanced and E° for the reaction obtained. Then use the Nernst equation (Equation 17.4), with the non-standard $E = 1.000$ V. Solving for [H⁺] then allows us to calculate pH using Equation 13.3.

(a) Split into two half-equations.	$SO_2(g) \rightarrow SO_4^{2-}(aq)$
	$NO_3^-(aq) \rightarrow NO(g)$

(b) Balance one of the half-equations: $SO_2(g) \rightarrow SO_4^{2-}(aq)$	
Balanced Oxidation Half-Equation (from Table17.1)	$SO_2(g) + 2H_2O \rightarrow SO_4^{2-}(aq) + 4H^+(aq) + 2e^-$

(c) Balance the other half-equation: $NO_3^-(aq) \rightarrow NO(g)$	
Balanced Reduction Half-Equation (from Table17.1)	$NO_3^-(aq) + 4H^+(aq) + 3e^- \rightarrow NO(g) + 2H_2O$

(d) Make the number of electrons in the two half-equations equal (multiply by an appropriate number if necessary).	Balanced half-equations:	
	oxidation: multiply by 3	$3[SO_2(g) + 2H_2O \rightarrow SO_4^{2-}(aq) + 4H^+(aq) + 2e^-]$
		$3SO_2(g) + 6H_2O \rightarrow 3SO_4^{2-}(aq) + 12H^+(aq) + 6e^-$
	reduction: multiply by 2	$2[NO_3^-(aq) + 4H^+(aq) + 3e^- \rightarrow NO(g) + 2H_2O]$
		$2NO_3^-(aq) + 8H^+(aq) + 6e^- \rightarrow 2NO(g) + 4H_2O$

(e) Combine the two half-equations.	$3SO_2(g) + 6H_2O + 2NO_3^-(aq) + 8H^+(aq) + 6e^- \rightarrow 3SO_4^{2-}(aq) + 12H^+(aq) + 6e^- + 2NO(g) + 4H_2O$

Cancel duplication.	$3SO_2(g) + 2H_2O + 2NO_3^-(aq)$ $\rightarrow 3SO_4^{2-}(aq) + 4H^+(aq) + 2NO(g)$
Balanced Equation:	$3SO_2(g) + 2NO_3^-(aq) + 2H_2O$ $\rightarrow 3SO_4^{2-}(aq) + 2NO(g) + 4H^+(aq)$

Now use values from Table 17.1 to calculate $E°$

for the oxidation half-reaction: $\quad\quad\quad\quad E_{ox}^° = -E_{red}^° = -0.155$ V

for the reduction half-reaction: $\quad\quad\quad E_{red}^° = +0.964$ V

$$E° = E_{ox}^° + E_{red}^° = -0.155 \text{ V} + 0.964 \text{ V} = +0.809 \text{ V}$$

When combining the two half-reactions to obtain the overall reaction, 6 electrons were cancelled so $n = 6$. Now use the Nernst Equation with $E° = 0.809$ V, $E = 1.000$ V and $n = 6$.

$$E = E° - \frac{(0.0257 \text{ V})}{n} \ln Q$$

$$1.000 \text{ V} = 0.809 \text{ V} - \frac{(0.0257 \text{ V})}{6} \ln \left(\frac{(P_{NO})^2 [SO_4^{2-}]^3 [H^+]^4}{(P_{SO_2})^3 [NO_3^-]^2} \right)$$

$$\frac{6(0.191 \text{ V})}{-(0.0257 \text{ V})} = \ln \left(\frac{(1.00)^2 [0.100]^3 [H^+]^4}{(1.00)^3 [0.100]^2} \right)$$

$$-44.6 = \ln\{(0.100)[H^+]^4\}$$

$$(0.100)[H^+]^4 = e^{-44.6} = 4.27 \times 10^{-20}$$

$$[H^+]^4 = 4.27 \times 10^{-19} \quad\quad\quad \text{so } [H^+] = 2.56 \times 10^{-5} \text{ M}$$

$$pH = -\log_{10}[H^+] = -\log_{10}(2.56 \times 10^{-5}) = \underline{\textbf{4.6}}$$

UNCLASSIFIED PROBLEMS

64. See Section 9-4; this problem is very similar to Example 9.6. For each molecule, consider its Lewis structure and determine its polarity. Check for H—N, H—O, or H—F bonds, which would indicate hydrogen bonding. (Recall also that all molecules have dispersion forces.)

(a) Lewis structure: $:\ddot{C}l\!\!-\!\!\ddot{C}l:$

    Polarity: non-polar

    H—N, H—O, or H—F bonds?   no; no hydrogen bonding

    Intermolecular forces: **dispersion forces**

(b) Lewis structure:

$$H—\overset{..}{\underset{..}{Br}}:$$

    Polarity: polar

    H—N, H—O, or H—F bonds?   no; no hydrogen bonding

    Intermolecular forces: **dispersion forces and dipole forces**

(c) Lewis structure:

$$H—\overset{..}{\underset{..}{F}}:$$

    Polarity: polar

    H—N, H—O, or H—F bonds?   yes, H—F bonds; hydrogen bonding

    Intermolecular forces: **dispersion forces, dipole forces,**
                 hydrogen bonding

(d) Lewis structure:

$$H—\overset{..}{\underset{..}{O}}—\overset{\overset{:O:}{\|}}{\underset{\underset{:O:}{\|}}{Cl}}=\overset{..}{\underset{..}{O}}$$

    Polarity: polar

    H—N, H—O, or H—F bonds?   yes, H—O bonds; hydrogen bonding

    Intermolecular forces: **dispersion forces, dipole forces,**
                 hydrogen bonding

(e) Ionic solid, so **no intermolecular forces are present**

66. See Section 4-3. This problem is similar to Example 4.6. Apply the rules for oxidation numbers (pages 88–89).

    (a) NO_2^- Rule 6:   oxid. no. O = **−2**
                     Rule 5:   oxid. no. N: $x + 2(-2) = -1$;   solve for $x \Rightarrow x = +3$;
                              oxid. no. N = +3

    (b) NO_2 Rule 6:   oxid. no. O = **−2**
                    Rule 5:   oxid. no. N: $x + 2(-2) = 0$;   solve for $x \Rightarrow x = +4$;
                              oxid. no. N = +4

    (c) HNO_3 Rule 4:   oxid. no. H = **+1**
                    Rule 6:   oxid. no. O = **−2**
                    Rule 5:   oxid. no. N: $+1 + x + 3(-2) = 0$;   solve for $x \Rightarrow x = +5$;
                              oxid. no. N = +5

    (d) NH_4^+ Rule 4:   oxid. no. H = **+1**
                    Rule 5:   oxid. no. N: $x + 4(+1) = +1$;   solve for $x \Rightarrow x = -3$;
                              oxid. no. N = −3

68. (a) **HClO, ClO₂**   (b) **S, KClO₃**   (c) **NH₃, NaClO**   (d) **HF**

70. (a) Increasing oxidation number corresponds to an increase in the number of strongly electronegative O atoms bonded to the central atom. These O atoms pull electrons away from the O atom bonded to H, making that proton more readily dissociate and hence *increasing acid strength*. The increasing number of O atoms also make the corresponding conjugate base less attractive to protons (and therefore less basic) by delocalizing its negative charge (multiple resonance forms are possible), and by drawing electron density away from the atom bearing the negative charge. Thus the conjugate base becomes *weaker*, and the acid correspondingly becomes *stronger*.

     (b) NO_2 has an odd number of valence electrons ($5 + (2 \times 6) = 17$) and therefore must have an unpaired electron, which is the requirement for paramagnetism.

     (c) In general, when the oxoanions act as oxidizing agents (that is, when they are reduced), H^+ is a *reactant* (see, for example, the reactions in Table 17.1). A lower pH (higher $[H^+]$), therefore, will cause the reaction to be more spontaneous. (See Section 17-4: increasing reactant concentration increases the value of E, which corresponds to increasing spontaneity.)

     (d) The sugar is oxidized to elemental carbon, which is black.

CHALLENGE PROBLEMS

71. To determine the amount of sulfuric acid that could be produced, one first needs to know the mass of sulfur present. This would require knowing the volume of the deposit, so its *depth* is needed in order to calculate its volume. From the volume of the deposit, the *density* is needed to calculate the mass of the deposit; and then the percent by mass of sulfur in the deposit (in other words, its *purity*) is needed to calculate the mass of sulfur.

72. Start from, say, 100 g of quartz (SiO_2). Calculate the value of the Au that could be obtained, and the cost of the HF solution required; then compare the two results to see whether the cost of the HF outweighs the value of the Au obtained.

     Strategy: (i) mass SiO_2 → mass Au (g) → mass Au (troy oz) → value of Au
                  (ii) mass SiO_2 → mol SiO_2 → mol HF → mass HF → mass HF solution
                        → vol HF solution → cost of HF solution

$$\text{(i) value of Au obtained} = 100 \text{ g SiO}_2 \times \frac{1.0 \times 10^{-3} \text{ g Au}}{100 \text{ g SiO}_2} \times \frac{1 \text{ troy oz}}{31.1 \text{ g}} \times \frac{\$425}{1 \text{ troy oz}} = \$0.014$$

(ii) cost of HF required =

$$100 \text{ g SiO}_2 \times \frac{1 \text{ mol SiO}_2}{60.08 \text{ g SiO}_2} \times \frac{4 \text{ mol HF}}{1 \text{ mol SiO}_2} \times \frac{20.01 \text{ g HF}}{1 \text{ mol HF}} \times \frac{100 \text{ g solution}}{50 \text{ g HF}}$$

$$\times \frac{1 \text{ mL solution}}{1.17 \text{ g solution}} \times \frac{1 \text{ L solution}}{1000 \text{ mL solution}} \times \frac{75 \text{¢}}{1 \text{ L solution}} \times \frac{\$1}{100\text{¢}} = \$0.17$$

The costs of HF solution is far more than the value of the recovered gold. Thus the process is **not economically feasible.**

73. Unbalanced equations for the two redox reactions occurring are:

(i) $ClO^-(aq) + I^-(aq) \rightarrow Cl^-(aq) + I_2(s)$
(ii) $S_2O_3^{2-}(aq) + I_2(s) \rightarrow S_4O_6^{2-}(aq) + I^-(aq)$

Both of these must be balanced (see Example 17-2) so that the correct mole ratios can be used in stoichiometric calculations.

(i) $ClO^-(aq) + I^-(aq) \rightarrow Cl^-(aq) + I_2(s)$

(a) Split into two half-equations.	$I^-(aq) \rightarrow I_2(s)$
	$ClO^-(aq) \rightarrow Cl^-(aq)$
(b) Balance one of the half-equations: $I^-(aq) \rightarrow I_2(s)$	
oxidation numbers	I: $-1 \rightarrow 0$; I is oxidized
atom balance	reactant: 1 I product: 2 I
	multiply I$^-$ by 2: $2I^-(aq) \rightarrow I_2(s)$
total oxidation number	I: $-2 \rightarrow$ I: 0
add electrons	The total oxidation number for I goes from -2 to 0.
	The I is oxidized by 2.
	Add 2 electrons to the product side.
	$2I^-(aq) \rightarrow I_2(s) + 2e^-$
balance charge	reactant: $2(-1) = -2$ product: $0 + 2(-1) = -2$
balance H	reactant: 0 H product: 0 H
check O	reactant: 0 O product: 0 O
Balanced Oxidation Half-Equation	$2I^-(aq) \rightarrow I_2(s) + 2e^-$
(c) Balance the other half-equation: $ClO^-(aq) \rightarrow Cl^-(aq)$	
oxidation numbers	Cl (oxid. no.): $+1 \rightarrow -1$; Cl is reduced
atom balance	1 Cl on each side, no adjustment is required
total oxidation number	Cl: $+1 \rightarrow$ Cl: -1

add electrons	The total oxidation number for Cl goes from +1 to −1. It is reduced by 2. Add 2 electrons to the reactant side. $ClO^-(aq) + 2e^- \rightarrow Cl^-(aq)$
balance charge	reactant: −1 + 2(−1) = −3; product: −1 acidic medium: add H^+ To balance the charge, add 2 H^+ to the reactant side. $2H^+(aq) + ClO^-(aq) + 2e^- \rightarrow Cl^-(aq)$ reactant: 2(+1) + (−1) + 2(−1) = −1; product: −1
balance H	reactant: 2 H product: 0 H To balance H, add H_2O to the product side. $2H^+(aq) + ClO^-(aq) + 2e^- \rightarrow Cl^-(aq) + H_2O$ reactant: 2 H product: 2 H
check O	reactant: 1 O product: 1 O
Balanced Reduction Half-Equation	$2H^+(aq) + ClO^-(aq) + 2e^- \rightarrow Cl^-(aq) + H_2O$

(d) Make the number of electrons in the two half-equations equal (multiply by an appropriate number if necessary).	Balanced half-equations:	
	oxidation	$2I^-(aq) \rightarrow I_2(s) + 2e^-$
	reduction	$2H^+(aq) + ClO^-(aq) + 2e^-$ $\rightarrow Cl^-(aq) + H_2O$

(e) Combine the two half-equations.	$2I^-(aq) + 2H^+(aq) + ClO^-(aq) + 2e^-$ $\rightarrow I_2(s) + 2e^- + Cl^-(aq) + H_2O$
Cancel duplication.	$2I^-(aq) + 2H^+(aq) + ClO^-(aq) \rightarrow I_2(s) + Cl^-(aq) + H_2O$
Balanced Equation:	$\mathbf{2I^-(aq) + 2H^+(aq) + ClO^-(aq) \rightarrow I_2(s) + Cl^-(aq) + H_2O}$

(ii) $S_2O_3^{2-}(aq) + I_2(s) \rightarrow S_4O_6^{2-}(aq) + I^-(aq)$

(a) Split into two half-equations.	$S_2O_3^{2-}(aq) \rightarrow S_4O_6^{2-}(aq)$
	$I_2(s) \rightarrow I^-(aq)$
(b) Balance one of the half-equations: $S_2O_3^{2-}(aq) \rightarrow S_4O_6^{2-}(aq)$	
oxidation numbers	S: +2 → +2.5; S is oxidized
atom balance	reactant: 2 S product: 4 S multiply $S_2O_3^{2-}$ by 2: $2S_2O_3^{2-}(aq) \rightarrow S_4O_6^{2-}(aq)$
total oxidation number	S: 4(+2) = +8 → S: 4(+2.5) = +10
add electrons	The total oxidation number for S goes from +8 to +10. The S is oxidized by 2. Add 2 electrons to the product side. $2S_2O_3^{2-}(aq) \rightarrow S_4O_6^{2-}(aq) + 2e^-$
balance charge	reactant: 2(−2) = −4 product: −2 + 2(−1) = −4

balance H	reactant: 0 H product: 0 H
check O	reactant: 6 O product: 6 O
Balanced Oxidation Half-Equation	$2S_2O_3^{2-}(aq) \rightarrow S_4O_6^{2-}(aq) + 2e^-$

(c) Balance the other half-equation: $I_2(s) \rightarrow I^-(aq)$

oxidation numbers	I (oxid. no.): $0 \rightarrow -1$; I is reduced
atom balance	reactant: 2 I product: 1 I multiply I^- by 2: $I_2(s) \rightarrow 2I^-(aq)$
total oxidation number	I: $0 \rightarrow$ I: -2
add electrons	The total oxidation number for I goes from 0 to -2. It is reduced by 2. Add 2 electrons to the reactant side. $I_2(s) + 2e^- \rightarrow 2I^-(aq)$
balance charge	reactant: $0 + 2(-1) = -2$; product: $2(-1) = -2$
balance H	reactant: 0 H product: 0 H
check O	reactant: 0 O product: 0 O
Balanced Reduction Half-Equation	$I_2(s) + 2e^- \rightarrow 2I^-(aq)$
(d) Make the number of electrons in the two half-equations equal (multiply by an appropriate number if necessary).	Balanced half-equations:
	oxidation	$2S_2O_3^{2-}(aq) \rightarrow S_4O_6^{2-}(aq) + 2e^-$
	reduction	$I_2(s) + 2e^- \rightarrow 2I^-(aq)$
(e) Combine the two half-equations.	$2S_2O_3^{2-}(aq) + I_2(s) + \mathbf{2e^-} \rightarrow S_4O_6^{2-}(aq) + \mathbf{2e^-} + 2I^-(aq)$
Cancel duplication.	$2S_2O_3^{2-}(aq) + I_2(s) \rightarrow S_4O_6^{2-}(aq) + 2I^-(aq)$
Balanced Equation:	$\mathbf{2S_2O_3^{2-}(aq) + I_2(s) \rightarrow S_4O_6^{2-}(aq) + 2I^-(aq)}$

First use reaction (ii) to determine how much iodine was oxidized by thiosolufate. Use the flow chart for solution stoichiometry (Figure 4.6).

Strategy: vol $Na_2S_2O_3 \rightarrow$ mol $Na_2S_2O_3 \rightarrow$ mol $S_2O_3^{2-} \rightarrow$ mol I_2

$$\text{mol } I_2 = 25.00 \text{ mL } Na_2S_2O_3 \times \frac{1 \text{ L}}{1000 \text{ mL}} \times \frac{0.0700 \text{ mol } Na_2S_2O_3}{1 \text{ L}} \times \frac{1 \text{ mol } S_2O_3^{2-}}{1 \text{ mol } Na_2S_2O_3}$$

$$\times \frac{1 \text{ mol } I_2}{2 \text{ mol } S_2O_3^{2-}} = 8.75 \times 10^{-4} \text{ mol } I_2$$

This is also the quantity of iodine produced in reaction (i) by oxidation of iodide. Now use reaction (i) to determine how much hypochlorite was present in the bleach in order to produce this quantity of iodine, and hence the mass percent of NaClO in the bleach.

Strategy: mol I_2 → mol OCl^- → mol $NaOCl$ → mass $NaOCl$
        then: vol bleach → mass bleach → mass percent $NaOCl$

$$\text{mass NaOCl} \quad = 8.75 \times 10^{-4} \text{ mol } I_2 \times \frac{1 \text{ mol OCl}^-}{1 \text{ mol } I_2} \times \frac{1 \text{ mol NaOCl}}{1 \text{ mol OCl}^-} \times \frac{74.44 \text{ g NaOCl}}{1 \text{ mol NaOCl}}$$

$$= 0.0651 \text{ g NaOCl}$$

$$\text{mass percent NaOCl} \quad = \frac{\text{mass NaOCl}}{\text{mass bleach}} \times 100 = \frac{0.0651 \text{ g NaOCl}}{5.00 \text{ mL bleach} \times \dfrac{1.00 \text{ g bleach}}{1 \text{ mL bleach}}} \times 100$$

$$= \underline{\textbf{1.30\%}}$$

74. Assume that the NaN_3 decomposes according to the following reaction (compare the decomposition of $Pb(N_3)_2$ on page 526):

$$2NaN_3(s) \rightarrow 2Na(s) + 3N_2(g)$$

Assume also that none of the N_2 produced escapes from the bag. Then assume that conditions are 25°C and 1 atm pressure and that the gas behaves ideally. Use the Ideal Gas Law (see Chapter 5) to calculate the moles of N_2 needed to occupy 20.0 L, and hence the quantity of NaN_3 needed to produce that amount of N_2.

Strategy: P, T, V of N_2 → mol N_2 → mol NaN_3 → mass NaN_3

$$\text{mol } N_2 = n = \frac{PV}{RT} = \frac{(1 \text{ atm})(20.0 \text{ L})}{(0.0821 \text{ L} \cdot \text{atm/mol} \cdot \text{K})(298 \text{ K})} = 0.817 \text{ mol } N_2$$

$$\text{mass NaN}_3 = 0.817 \text{ mol } N_2 \times \frac{2 \text{ mol NaN}_3}{3 \text{ mol } N_2} \times \frac{65.01 \text{ g NaN}_3}{1 \text{ mol NaN}_3} = \underline{\textbf{35.4 g NaN}_3}$$

ORGANIC CHEMISTRY

22.1 SATURATED HYDROCARBONS: Alkanes
22.2 UNSATURATED HYDROCARBONS: Alkenes and Alkynes

2. See Example 22.3.

   Strategy: Assume that the given hydrocarbon is not cyclic and may or may not contain a multiple bond. If a multiple bond is present, only one is present (either a double bond or a triple bond). The following formula applies where n is the number of C atoms:

   C_nH_{2n+2} for alkanes

   C_nH_{2n} for alkenes

   C_nH_{2n-2} for alkynes

   (a) $C_{12}H_{24}$ $n=12$, number of H = 24 = 2(12)= $2n$, thus it is an **alkene**.

   (b) C_7H_{12} $n=7$, number of H = 12 = 2(7)–2 = $2n-2$ thus it is an **alkyne**.

   (c) $C_{13}H_{28}$ $n=13$, number of H = 28 = 2(13)+2 = $2n+2$, thus it is an **alkane**.

4. See Example 22.3.

   Assume that the hydrocarbon is not cyclic and contains only one multiple bond (either one double bond or one triple bond). In the following solution, let n=number of C atoms.

   (a) For alkynes the following formula applies: C_nH_{2n-2}

   Since number of H atoms = $2n-2$ = 16, then n = 9.

   Therefore there are 9 C atoms. The formula of the compound is C_9H_{16}

   (b) For alkenes the following formula applies: C_nH_{2n}

   Since number of H atoms = $2n$ = 44, then n = 22.

   Therefore there are 22 C atoms. The formula of the compound is $C_{22}H_{44}$

   (c) For alkanes the following formula applies: C_nH_{2n+2}

   Since number of C atoms = n = 10, then H = $2n+2$ = 2(10)+2 = 22.

   Therefore there are 22 H atoms. The formula of the compound is $C_{10}H_{22}$

6. See Table 22.1 and Example 22.2.

   Strategy: (i) Find the parent chain. This is the longest carbon chain.

   (ii) Use Table 22.1 to identify the alkyl groups attached to the parent chain.

   (iii) Number the C atoms in the parent chain such that the C to which the alkyl groups are attached will have the lowest number possible.

   (a) 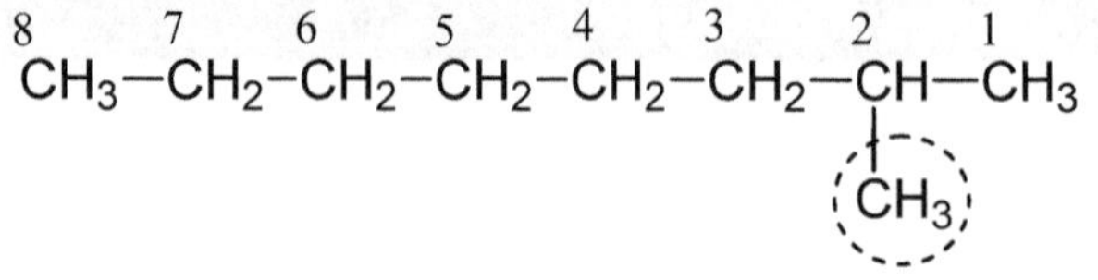
   $$\overset{8}{CH_3}-\overset{7}{CH_2}-\overset{6}{CH_2}-\overset{5}{CH_2}-\overset{4}{CH_2}-\overset{3}{CH_2}-\overset{2}{CH}-\overset{1}{CH_3}$$
   $$|$$
   $$CH_3$$

1. longest C chain	8 C atoms = octane
2. alkyl group (encircled)	1 one-carbon group = methyl
3. number/position of the C atom bonded to the alkyl group	methyl group at the C_2 position
4. name	**2-methyloctane**

   The C atoms in the parent chain are numbered in such a way that the C to which the alkyl group is attached will have the lowest number possible. In this problem, the lower number is achieved when the parent chain is numbered from right to left. Thus, the name of the alkane is **2-methyloctane** (not 7-methyloctane).

   (b) The structural formula of $(CH_3)_4C$ is shown below.

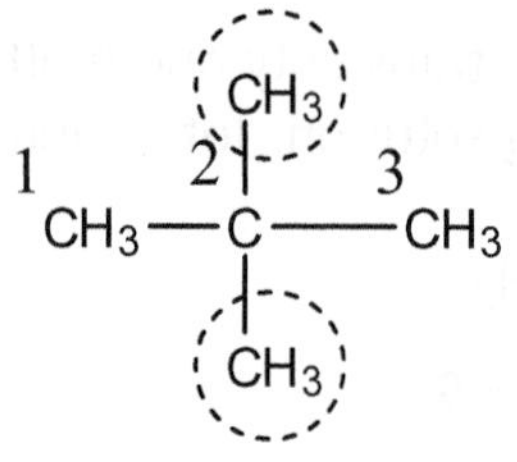

   $$CH_3$$
   $$\overset{1}{CH_3}-\overset{2}{C}-\overset{3}{CH_3}$$
   $$|$$
   $$CH_3$$

1. longest C chain	3 C atoms = propane
2. alkyl group (encircled)	2 one-carbon groups = dimethyl
3. number/position of the C atom bonded to the alkyl group	both methyl groups at the C_2 position
4. name	**2,2-dimethylpropane**

   The parent chain can be numbered either from left to right or from right to left because both will give the same number on the C where the methyl groups are attached.

(c)

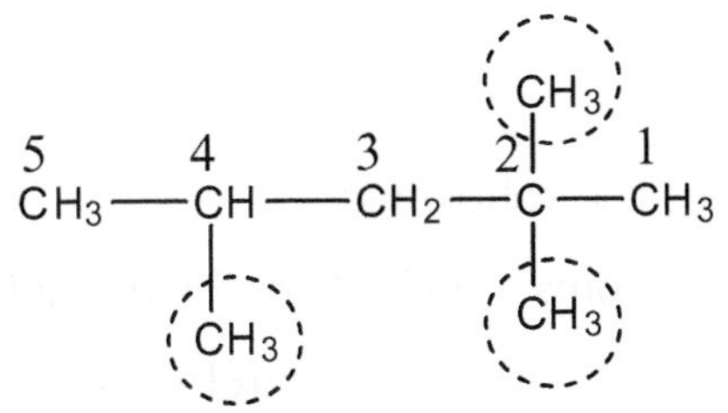

$$CH_3—CH—CH_2—C—CH_3$$

with CH₃ groups and numbering showing 5, 4, 3, 2, 1 across the chain, CH₃ encircled above C2, CH₃ encircled below C4, and CH₃ encircled below C2.

1.	longest C chain	5 C atoms = pentane
2.	alkyl group (encircled)	3 one-carbon groups = trimethyl
3.	number/position of the C atom bonded to the alkyl group	2 methyl groups at C$_2$ and one at C$_4$ position
4.	name	**2,2,4-trimethylpentane**

The parent chain is numbered from right to left because it will give the lowest number possible for the C to which the alkyl groups are attached. Thus, the name of the alkane is **2,2,4-trimethylpentane** (not 2,4,4-trimethylpentane).

(d)

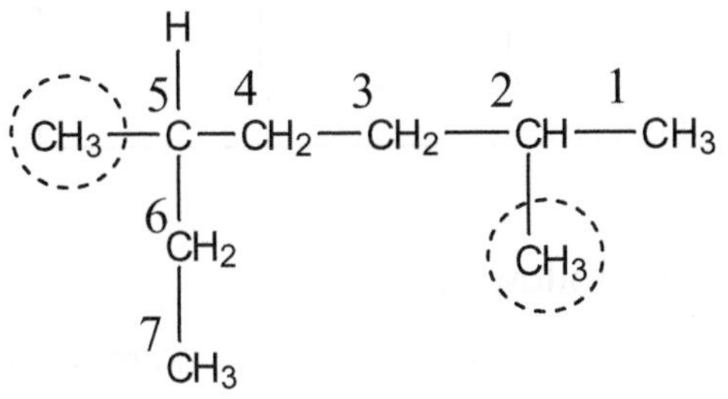

$$CH_3—C—CH_2—CH_2—CH—CH_3$$

with H above C5, numbering 5, 4, 3, 2, 1 across the chain, CH₃ encircled to left of C5, CH₂ (C6) and CH₃ (C7) below C5, and CH₃ encircled below C2.

1.	longest C chain	7 C atoms = heptane
2.	alkyl group (encircled)	2 one-carbon groups = dimethyl
3.	number/position of the C atom bonded to the alkyl group	one group at C$_2$ and another one at C$_5$
4.	name	**2,5-dimethylheptane**

The longest C chain has 7 C's and not 6. This parent chain is numbered from right to left then down because it will give the lowest number possible for the C to which the alkyl groups are attached. Thus, the name of the alkane is **2,5-dimethylheptane** (not 3,6-dimethylheptane).

8. See Table 22.1 and Examples 22.2 and 22.4.

(a) the name 2,2,4-trimethylpentane implies the following:

parent chain: pentane: "pent" = 5 C atoms; suffix "ane" = no multiple bond

alkyl branching: trimethyl = 3 methyl ($-CH_3$) groups (encircled below)

position of the alkyl groups: 2,2,4 = 2 methyl's at C_2 and 1 methyl at C_4 position

C skeleton:

Structural formula of 2,2,4-trimethylpentane (enough H is added to the C skeleton so that each C atom has 4 bonds):

(b) the name 2,2-dimethylpropane implies the following:

parent chain: propane: "prop" = 3 C atoms; suffix "ane" = no multiple bond

alkyl branching: dimethyl = 2 methyl ($-CH_3$) groups (encircled below)

position of the alkyl groups: 2,2 = 2 methyl's both at C_2 position

C skeleton:

Structural formula of 2,2-dimethylpropane (enough H is added to the C skeleton so that each C atom has 4 bonds):

(c) the name 4-isopropyloctane implies the following:

parent chain: octane: "oct"= 8 C atoms; suffix "ane" = no multiple bond

alkyl branching: isopropyl group (encircled below)

position of the alkyl groups: 4 = the isopropyl group is at C_4 position

C skeleton:

$$\overset{1}{C}-\overset{2}{C}-\overset{3}{C}-\overset{4}{C}-\overset{5}{C}-\overset{6}{C}-\overset{7}{C}-\overset{8}{C}$$

$$CH_3-C-H$$

$$CH_3$$

isopropyl group

Structural formula of 4-isopropyloctane (enough H is added to the C skeleton so that each C atom has 4 bonds):

$$CH_3-CH_2-CH_2-CH-CH_2-CH_2-CH_2-CH_3$$

$$CH_3-C-H$$

$$CH_3$$

(d) the name 2,3,4-trimethylheptane implies the following:

parent chain: heptane: "hept"= 7 C atoms; suffix "ane" = no multiple bond

alkyl branching: trimethyl = 3 methyl ($-CH_3$) groups (encircled below)

position of the alkyl groups: 2,3,4 = 1 methyl each at C_2, C_3 and C_4 position

C skeleton:

$$CH_3 \quad CH_3$$

$$\overset{1}{C}-\overset{2}{C}-\overset{3}{C}-\overset{4}{C}-\overset{5}{C}-\overset{6}{C}-\overset{7}{C}$$

$$CH_3$$

Structural formula of 2,3,4-trimethylheptane (enough H is added to the C skeleton so that each C atom has 4 bonds):

$$CH_3 \qquad CH_3$$

$$CH_3-CH-CH-CH-CH_2-CH_2-CH_3$$

$$CH_3$$

10. See Table 22.1 and Example 22.2.

(a) Reasonable structure for 2, 2-dimethylbutane:

$$\overset{\displaystyle CH_3}{\underset{\displaystyle CH_3}{\overset{4}{CH_3}-\overset{3}{C}-\overset{2}{CH_2}-\overset{1}{CH_3}}}$$

proper IUPAC name: **2,2-dimethylbutane**

The name 2, 2-dimethylbutane is incorrect because in writing the IUPAC name, the numbers are separated by commas only. There should be no space between the comma (,) and the next number.

(b) Reasonable structure for 4-methylpentane:

$$\underset{\displaystyle CH_3}{CH_3-CH_2-CH_2-CH-CH_3}$$

proper IUPAC name: **2-methylpentane**

$$\underset{\displaystyle CH_3}{\overset{1}{CH_3}-\overset{2}{CH_2}-\overset{3}{CH_2}-\overset{4}{CH}-\overset{5}{CH_3}}$$

$$\underset{\displaystyle CH_3}{\overset{5}{CH_3}-\overset{4}{CH_2}-\overset{3}{CH_2}-\overset{2}{CH}-\overset{1}{CH_3}}$$

incorrectly numbered   correctly numbered

4-methylpentane is incorrect. To properly name the compound, number the C atoms of the parent chain (pentane) from right to left rather than from left to right. This will assign a lower number to the C to which the methyl group is attached.

(c) Reasonable structure for 2-ethylpropane:

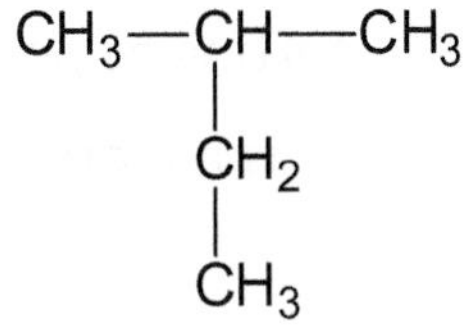

$$CH_3—CH—CH_3$$
$$|$$
$$CH_2$$
$$|$$
$$CH_3$$

proper IUPAC name: **2-methylbutane**

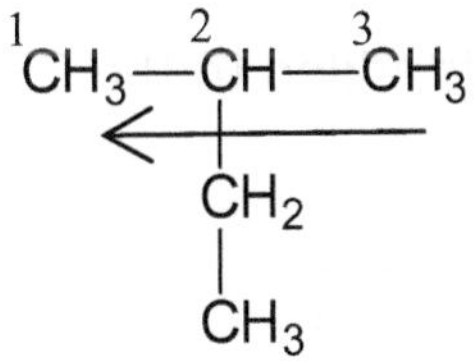

$$^1CH_3—^2CH—^3CH_3$$
$$|$$
$$CH_2$$
$$|$$
$$CH_3$$

shorter C chain (3 C atoms)
incorrect parent chain

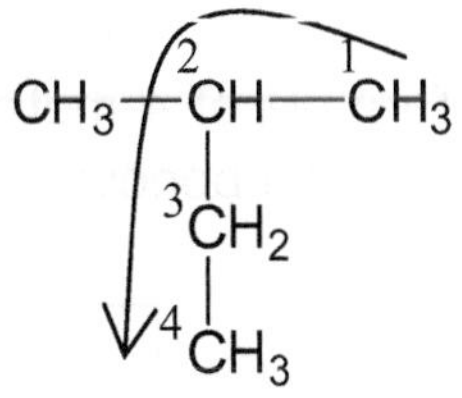

$$CH_3—^2CH—^1CH_3$$
$$|$$
$$^3CH_2$$
$$|$$
$$^4CH_3$$

longer C chain (4 C atoms)
correct parent chain

2-ethylpropane is incorrect. To name the alkane, the suffix that identifies the parent chain should have the longest continuous chain of C atoms. In this case, the chain that has 4 C atoms (butane) is longer, so the name should end in butane.

12. See Table 22.1 and Example 22.4.

(a) the name 2-pentyne implies the following:

"pent" indicates a chain of 5 C atoms

"yne" denotes the presence of a triple bond, $C \equiv C$

"2" points out that the position of the triple bond starts at C_2

C skeleton: $\underset{1}{C} — \underset{2}{C} \equiv \underset{3}{C} — \underset{4}{C} — \underset{5}{C}$

Structural formula of 2-pentyne (enough H is added to the C skeleton so that each C atom has 4 bonds):

$$CH_3—C \equiv C—CH_2—CH_3$$

(b) See Table 22.1, Example 22.4 and problem 12a above.

The name 4-methyl-2-pentyne implies the following:

"2-pentyne" indicates that the parent chain has 5 C atoms ("pent") with a triple
bond, $C \equiv C$ ("yne") starting at position C_2 (suggested by the number 2)

"4-methyl" points out that there is a methyl group attached at C_4

C skeleton:

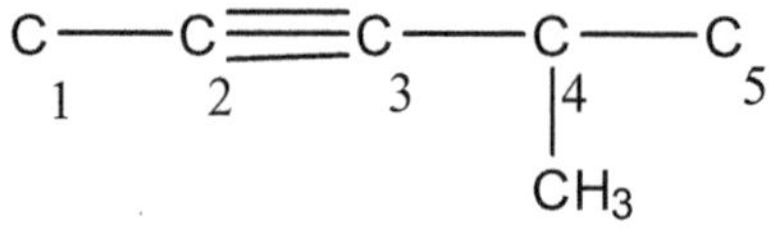

$$\underset{1}{C} - \underset{2}{C} \equiv \underset{3}{C} - \underset{4}{C} - \underset{5}{C}$$
$$\mid$$
$$CH_3$$

Structural formula of 4-methyl-2-pentyne (enough H is added to the C skeleton so that
each C atom has 4 bonds):

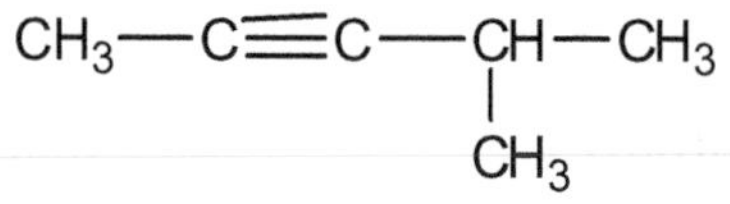

$$CH_3 - C \equiv C - \underset{\underset{CH_3}{\mid}}{CH} - CH_3$$

(c) See Table 22.1, Example 22.4 and problem 12b above.

The name 2-methyl-3-hexyne implies the following:

"3-hexyne" indicates that the parent chain has 6 C atoms ("hex") with a triple
bond, $C \equiv C$ ("yne") starting at position C_3

"2-methyl" points out that there is a methyl group attached at C_2

C skeleton:

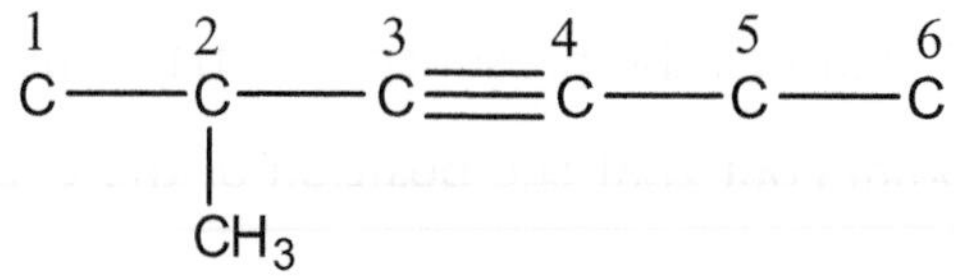

$$\underset{1}{C} - \underset{2}{C} - \underset{3}{C} \equiv \underset{4}{C} - \underset{5}{C} - \underset{6}{C}$$
$$\mid$$
$$CH_3$$

Structural formula of 2-methyl-3-hexyne:

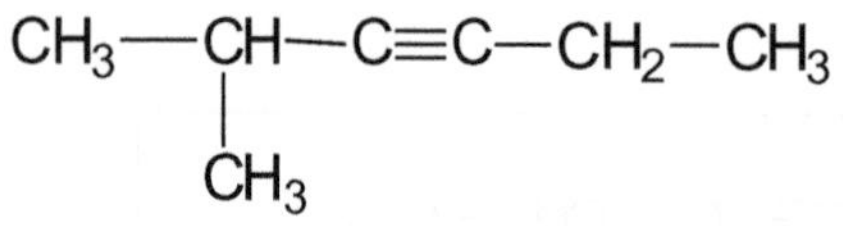

$$CH_3 - \underset{\underset{CH_3}{\mid}}{CH} - C \equiv C - CH_2 - CH_3$$

(d) See Table 22.1, Example 22.4 and problem 12b and c above.

The name 3,3-dimethyl-1-butyne implies the following:

"1-butyne" indicates that the parent chain has 4 C atoms ("but") with a triple bond, C≡C ("yne") starting at position C_1 (suggested by the number 1)

"3,3-dimethyl" the prefix "di" in dimethyl shows that there are 2 –CH_3 groups while the numbers 3,3 suggest that these 2 –CH_3 groups are both attached at C_3

C skeleton:

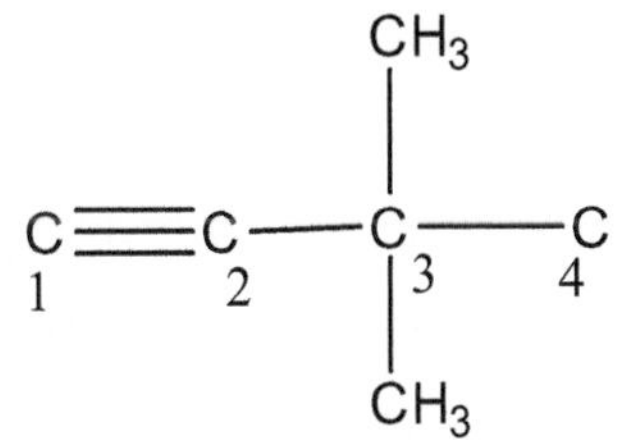

$$C \equiv C - C - C$$

Structural formula of 3,3-dimethyl-1-butyne (enough H is added to the C skeleton so that each C atom has 4 bonds):

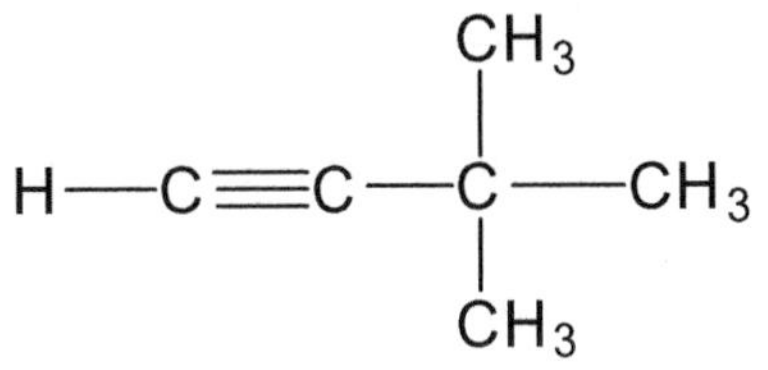

22.3 AROMATIC HYDROCARBONS AND THEIR DERIVATIVES

14. See Example 22.5.

The parent compound, toluene has the following structure:

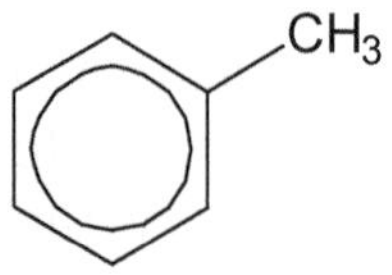

In naming compounds as derivatives of toluene, the position of the substituents can be identified in two ways:

(i) use numerical prefixes to identify the position of the substituents in the benzene ring. The C atom bearing the $-CH_3$ group is assigned position C_1.

(ii) use locator prefixes, *ortho-*, *meta-*, and *para-* using the C bearing the $-CH_3$ group as point of reference. Ortho means that the two substituents are on adjacent carbons. Meta means that substituents on the ring are one carbon away from each other. Para means that the substituents are on opposite ends of the benzene ring.

Both naming systems will be used here.

(a) *o*-chlorotoluene or 2-chlorotoluene

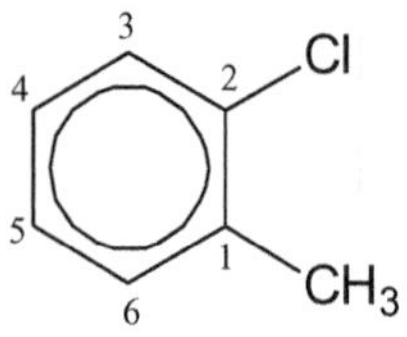

The C atoms bearing the $-CH_3$ group and the Cl substituent are adjacent to each other. Therefore, the *ortho* prefix can be used hence the name **o-chlorotoluene**. Similarly, numbers can be used. Since the C bearing the $-CH_3$ group is assigned as C_1 and the substituents of the ring should have the lowest number possible, the C atoms in the ring is numbered counterclockwise so that the C bearing Cl is C_2 (rather than C_6 if numbered clockwise). Hence, the compound is also called **2-chlorotoluene**.

(b) *m*-bromotoluene or 3-bromotoluene

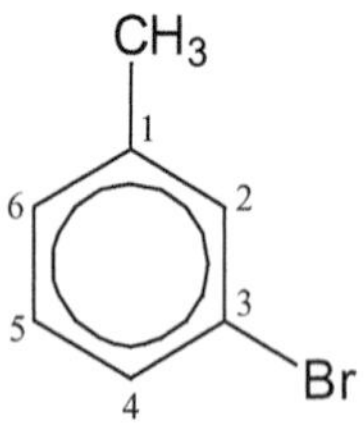

The C atoms bearing the $-CH_3$ group and the Br substituent are one C away from each other. Therefore, the *meta* prefix can be used hence the name **m-bromotoluene**. Similarly, numbers can be used. Since the C bearing the $-CH_3$ group is assigned as C_1 and the substituents of the ring should have the lowest number possible, the C atoms in the ring is numbered clockwise so that the C bearing Br is C_3 (rather than C_5 if numbered counterclockwise). Hence, the compound is also called **3-bromotoluene**.

(c) **2,3,6-tribromotoluene**

As pointed out in the textbook, the use of numbers becomes mandatory when there are three or more substituents. In this structure, there are 4 substituents on the ring. Since the compound is to be named as a derivative of toluene, the C bearing the $-CH_3$ group is assigned as C_1 and the substituents of the ring should have the lowest number possible, the C atoms in the ring is numbered clockwise so that C bearing Br is C_2, C_3 and C_6 (rather than C_2, C_5 and C_6 if numbered counterclockwise). Hence, the compound is named **2,3,6-tribromotoluene**. The prefix "tri" is used because there are 3 bromo substituents on the ring.

22.4 FUNCTIONAL GROUPS

16. See Table 22.2 and Example 22.6. The encircled group identifies the functional group.

(a) **Alcohol**. The $-OH$ group (encircled) shows that this compound is an alcohol.

$$CH_3-(CH_2)_3-OH$$

(b) **Ester**. The $-\overset{O}{\underset{}{C}}-O-$ group (encircled) suggests that this compound is an ester.

$$CH_3-CH_2-C-O-CH_2-CH_3$$

(c) **Ester and carboxylic acid**. The compound can be classified both as an ester and as a carboxylic acid because it shows the presence of these 2 functional groups:

$$-\overset{O}{\underset{}{C}}-O-C \quad \text{and} \quad -\overset{O}{\underset{}{C}}-OH \quad \text{or COOH}$$

ester   carboxylic acid

$$CH_3-CH_2-O-C-(CH_2)_6COOH$$

18. See Table 22.2 and Example 22.6. The encircled group identifies the functional group.

(a) 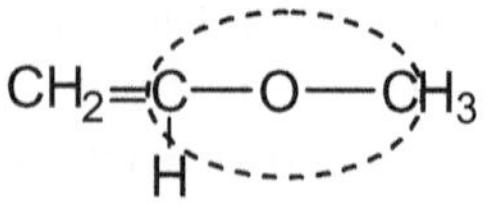 CH₂=C—O—CH₃
         |
         H

Ether. The compound is an alkene with the –C–O–C– functional group of an ether.

(b) CH₃—CH₂—C—O—CH₃
              ‖
              O

Ester. The compound has the

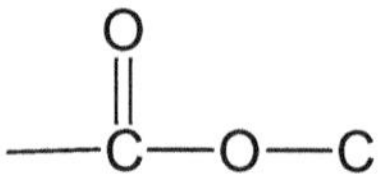

$$-\overset{\overset{\displaystyle O}{\|}}{C}-O-C$$

functional group of an ester.

(c) 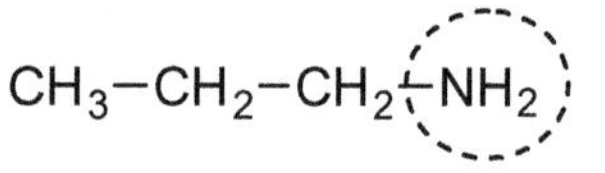 CH₃—CH₂—CH₂—NH₂

Amine. The compound has the —N— functional group of an amine.

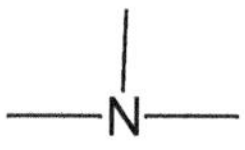

(d)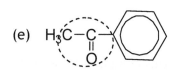

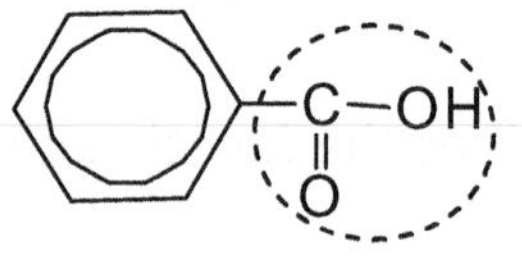

Carboxylic acid. The compound has the

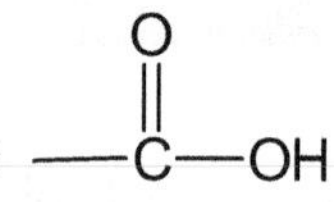

$$-\overset{\overset{\displaystyle O}{\|}}{C}-OH$$

functional group of a carboxylic acid.

(e) 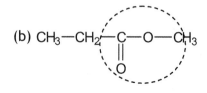 H₃C—C—
              ‖
              O

Ketone. The compound has the

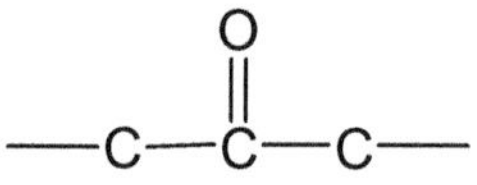

$$-C-\overset{\overset{\displaystyle O}{\|}}{C}-C-$$

functional group of a ketone.

20. See Table 22.2 and Example 22.7.

(a) i)   parent compound chain, 3 C alcohol: C–C–C

   ii)   functional group (attached to the center):   C—C—C
                                                              |
                                                              OH

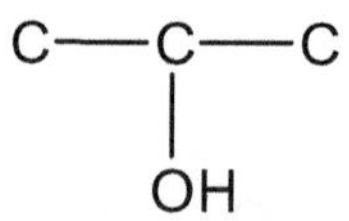

   iii) hydrogen atoms. To complete the structural formula, add enough H so that each C has 4 bonds and O has 2 bonds.

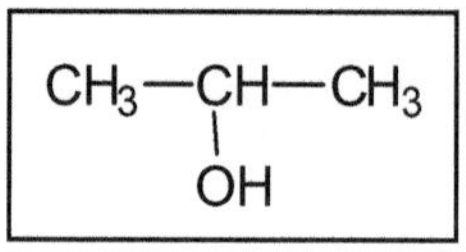

CH₃—CH—CH₃
       |
       OH

2-propanol

(b) i)   parent chain, 4-C branched carboxylic acid. Since the compound is branched, the parent chain contains less that 4 C atoms.   The methyl group must be attached to the second carbon of a three carbon chain in order for the compound to be branched. Attaching the methyl group to the terminal carbon will result in an unbranched chain.

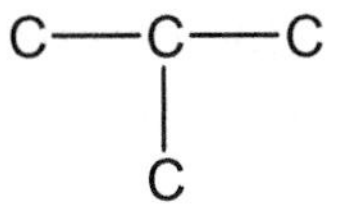

$$C-C-C$$
$$\overset{|}{C}$$

ii)   functional group: the C atom in the carboxylic acid functional group, COOH is always a terminal C.

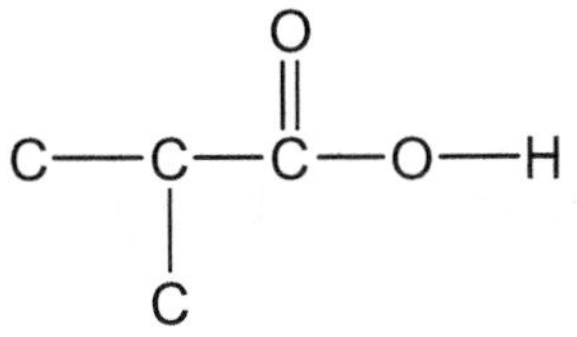

$$C-C-\overset{\overset{O}{\|}}{C}-O-H$$
$$\overset{|}{C}$$

iii) hydrogen atoms. To complete the structural formula, add enough H so that each C forms 4 bonds and each O has 2 bonds.

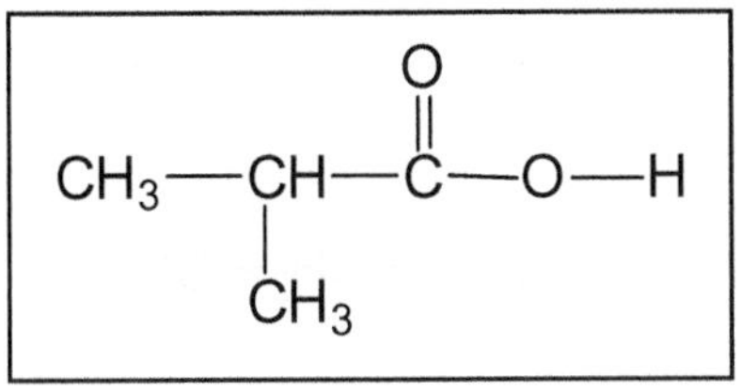

$$CH_3-\underset{\overset{|}{CH_3}}{CH}-\overset{\overset{O}{\|}}{C}-O-H$$

2-methylpropanoic acid

(c) See example 22.7b. The condensation reaction between 2-methylpropanoic acid and 2-propanol forms an ester. Water is also produced from the –OH group (removed from the carboxylic acid) and from the H atom (covalently bonded to O in the alcohol) as shown in the equation below.

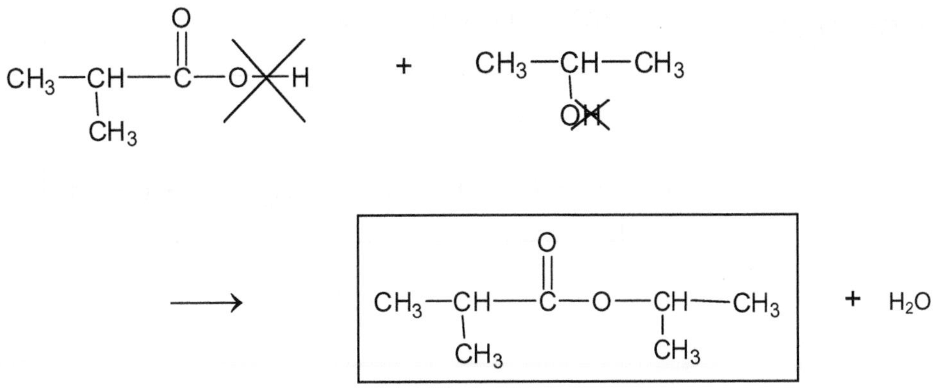

$$CH_3-\underset{\overset{|}{CH_3}}{CH}-\overset{\overset{O}{\|}}{C}-O\cancel{}H \quad + \quad CH_3-\underset{\overset{|}{O\cancel{H}}}{CH}-CH_3$$

$$\longrightarrow \quad CH_3-\underset{\overset{|}{CH_3}}{CH}-\overset{\overset{O}{\|}}{C}-O-\underset{\overset{|}{CH_3}}{CH}-CH_3 \quad + \quad H_2O$$

ester

22. See Table 22.2 and Example 22.7b.

The condensation reaction between ethylene glycol and a carboxylic acid (e.g. formic acid and acetic acid) forms an ester. Since ethylene glycol has two –OH groups and both of these groups can react with carboxylic acids, it is also possible to produce a diester. Water is also produced from the –OH group (removed from the carboxylic acid) and from the H atom (covalently bonded to O in the alcohol) as shown in the following equations.

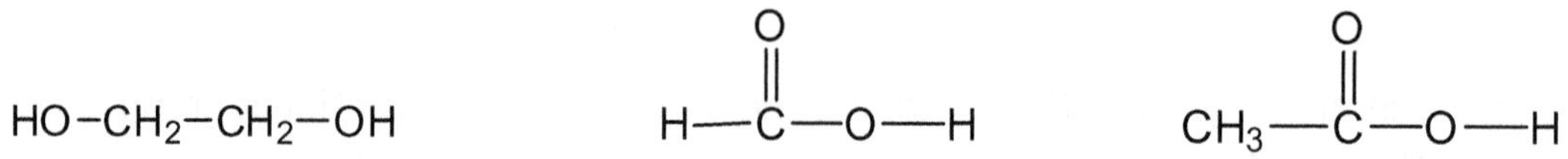

HO–CH$_2$–CH$_2$–OH

ethylene glycol

formic acid

acetic acid

Possible Reactions:

i) Reaction of formic acid and one of the –OH group of ethylene glycol.

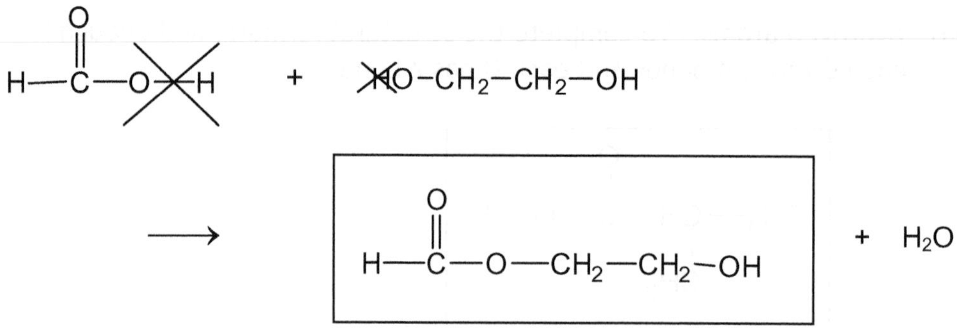

ester

ii) Reaction of formic acid with the two –OH groups of ethylene glycol to form a diester.

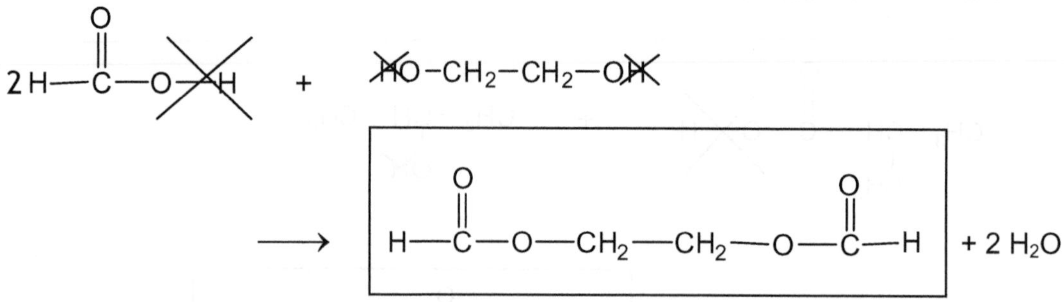

diester

iii) Reaction of acetic acid and one of the –OH group of ethylene glycol.

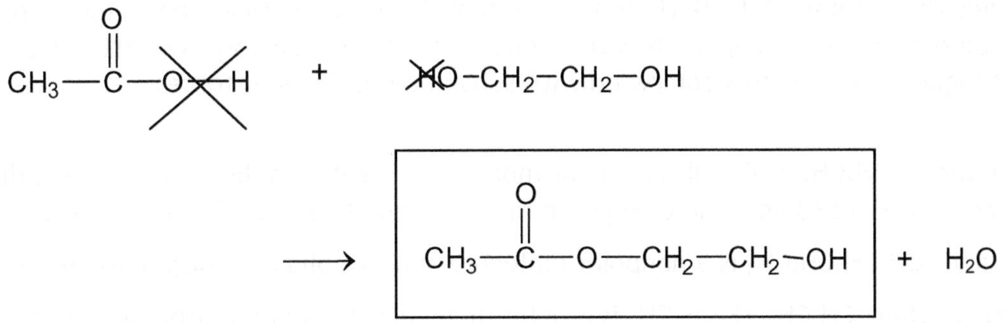

ester

iv) Reaction of acetic acid with the two –OH groups of ethylene glycol to form a diester.

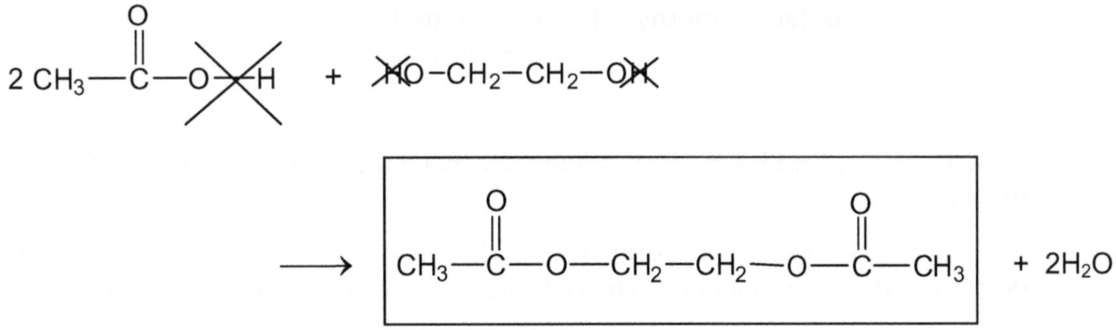

diester

v) Reaction of acetic acid and formic acid with the two –OH groups of ethylene glycol to form a diester.

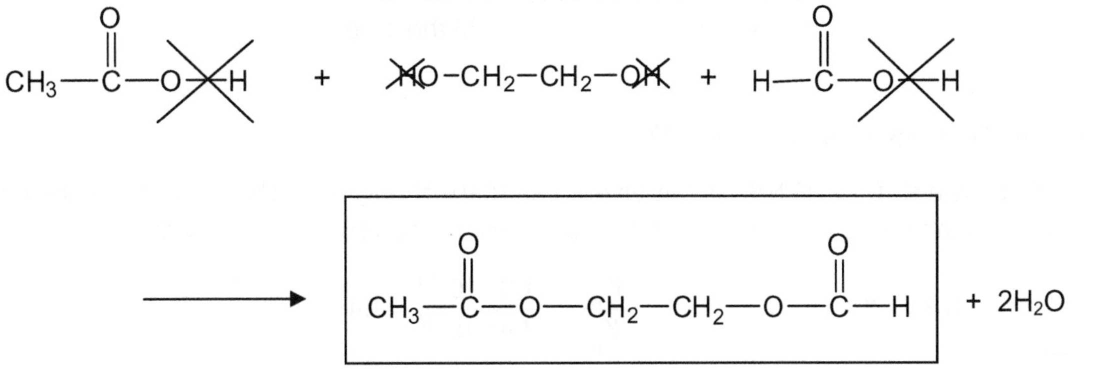

diester

24. See also Section 9.4.   All molecules exhibit dispersion forces and it is the only force among nonpolar molecules. Polar molecules have dipole-dipole and dispersion forces. Molecules that contain O–H, N–H, and F–H exhibit H–bonding aside from dispersion and dipole-dipole forces.   Compounds with stronger intermolecular forces of attraction will exhibit higher boiling points compared with substances of weaker forces.

    (a) 1-butanol, $CH_3CH_2CH_2CH_2OH$ is a polar molecule. It exhibits dispersion forces, dipole forces and H–bonding. The O–H group makes H–bonding possible in alcohols.

       butane, $CH_3CH_2CH_2CH_3$ is a nonpolar molecule and exhibits only dispersion forces.

       diethylether, $CH_3CH_2–O–CH_2CH_3$ is a polar molecule. It exhibits dispersion forces and dipole forces.

       Of the three compounds, 1-butanol has the strongest intermolecular forces of attraction while butane has the weakest.   Thus, in terms of increasing boiling point the predicted order is:

                butane < diethylether < 1-butanol
                lowest bp   highest bp

    (b) hexane, $CH_3CH_2CH_2CH_2CH_2CH_3$ is a nonpolar molecule. It exhibits only dispersion forces.

       1-hexanol, $CH_3CH_2CH_2CH_2CH_2CH_2OH$ is a polar molecule. It exhibits dispersion forces, dipole forces and H–bonding. The O–H group makes H–bonding possible in alcohols.

       dipropylether, $CH_3CH_2CH_2–O–CH_2CH_2CH_3$ is a polar molecule. It exhibits dispersion forces and dipole forces.

       Of the three compounds, 1-hexanol has the strongest intermolecular forces of attraction while hexane has the weakest.   In terms of increasing boiling point the predicted order is:

                hexane < dipropylether < 1-hexanol
                lowest bp   highest bp

26. See Sections 13.4, 13.5, and 22.4.

    Since the K_a for $C_6H_5NH_3^+$ is not given, calculate the K_a from the K_b of its corresponding (conjugate) base, $C_6H_5NH_2$ ($K_b=7.4 \times 10^{-10}$) using the following equation:

$$K_aK_b = K_w \quad \Rightarrow \quad K_a = \frac{K_w}{K_b} = \frac{1.0 \times 10^{-14}}{7.4 \times 10^{-10}} = 1.4 \times 10^{-5}$$

    Thus, the K_a for the the following reaction is 1.4×10^{-5}.

$$C_6H_5NH_3^+(aq) \rightleftharpoons C_6H_5NH_2(aq) + H^+(aq) \qquad K_a = 1.4 \times 10^{-5}$$

Set up the table:

	$C_6H_5NH_3^+$ *(aq)* $\rightleftharpoons$	$C_6H_5NH_2$*(aq)* $+$	H^+*(aq)*
$[\ ]_0$	0.100	0	0
$\Delta[\ ]$	$-x$	$+x$	$+x$
$[\ ]_{eq}$	$0.100 - x$	x	x

$$K_a = \frac{[H^+]_{eq}\,[C_6H_5NH_2]_{eq}}{[C_6H_5NH_3^+]_{eq}}$$

$$1.4 \times 10^{-5} = \frac{x^2}{0.100 - x} \qquad \text{assume } x \ll 0.100 \Rightarrow 0.100 - x \cong 0.100$$

Thus, the equation simplifies to:

$$1.4 \times 10^{-5} = \frac{x^2}{0.100}$$

$$1.4 \times 10^{-6} = x^2$$

$$x = 1.2 \times 10^{-3}\ M \qquad \text{(Note that the assumption } x \ll 0.100 \text{ is justified}$$
$$\text{because } x = 1.2 \times 10^{-3}\ M \ll 0.100\ )$$

Since $x = [H^+]_{eq}$ the pH can be calculated: $pH = -\log(1.2 \times 10^{-3}) = \underline{\textbf{2.92}}$

The pH of a 0.100 M solution of $C_6H_5NH_3^+$ is $\underline{\textbf{2.92}}$

pH of acetic acid, assume the same concentration (0.100M):

	$HC_2H_3O_2$ *(aq)* $\rightleftharpoons$	$C_2H_3O_2^-$ *(aq)* $+$	H^+*(aq)*
$[\ ]_0$	0.100	0	0
$\Delta[\ ]$	$-x$	$+x$	$+x$
$[\ ]_{eq}$	$0.100 - x$	x	x

$$K_a = 1.8 \times 10^{-5} = \frac{[H^+]_{eq}\,[C_2H_3O_2^-]_{eq}}{[HC_2H_3O_2]_{eq}} = \frac{x^2}{0.100 - x}$$

$$K_a = 1.8 \times 10^{-5} = \frac{x^2}{0.100 - x} \qquad \Rightarrow \text{assume } x \ll 0.100 \Rightarrow 0.100 - x \cong 0.100$$

Thus, the equation simplifies to:

$$1.8 \times 10^{-5} = \frac{x^2}{0.100}$$

$$1.8 \times 10^{-6} = x^2$$

$$x = 1.36 \times 10^{-3} \text{ M}$$

(Note that the assumption $x \ll 0.100$ is justified because $x = 1.36 \times 10^{-3}$ M $\ll 0.100$)

Since $x = [H^+]_{eq}$ the pH can be calculated: pH $= -\log [H^+] = -\log(1.36 \times 10^{-3}) = \underline{\textbf{2.87}}$

The pH of a 0.100 M solution of $HC_2H_3O_2$ is **2.87**.

0.100M solutions of the two acids, $C_6H_5NH_3^+$ and $HC_2H_3O_2$ have <u>comparable pH</u> values of 2.92 and 2.87, respectively.

28. See Sections 12.2 and 22.4 and Example 22.8.

Over-all Reaction: $C_6H_5NH_3^+(aq) + CH_3COO^-(aq) \rightleftharpoons C_6H_5NH_2(aq) + CH_3COOH(aq)$

Break the reaction into two equations and find the K for each reaction:

$C_6H_5NH_3^+ \rightleftharpoons C_6H_5NH_2 + H^+$ $\qquad\qquad$ $K_a = 1.35 \times 10^{-5}$

$CH_3COO^- + H^+ \rightleftharpoons CH_3COOH$ $\qquad\qquad$ $K_2 = 1/K_{a \text{ acetic acid}} = 1/(1.86 \times 10^{-5})$

$C_6H_5NH_3^+ + CH_3COO^- \rightleftharpoons C_6H_5NH_2 + CH_3COOH$ $\qquad$ $K = ?$

$$K = K_a \times K_2 = (1.35 \times 10^{-5})(1/(1.86 \times 10^{-5})) = 0.726$$

The K for the reaction is **0.726**.

22.5 ISOMERISM IN ORGANIC COMPOUNDS

30. See Example 22.10.

The alkene, C_4H_8 has one double bond. There are three structural isomers(I, II, III). The second isomer, II has two possible geometric isomers, a *cis*-isomer and a *trans*-isomer.

$$\text{CH}_3\text{—CH}_2\text{—CH}=\text{CH}_2 \qquad \text{CH}_3\text{—CH}=\text{CH}\text{—CH}_3 \qquad \overset{\displaystyle \text{CH}_3}{\underset{}{\text{CH}_3\text{—C}=\text{CH}_2}}$$

$$\text{I} \qquad\qquad\qquad \text{II} \qquad\qquad\qquad \text{III}$$

Two geometric isomers of structure II:

cis-isomer   trans-isomer

32. The textbook presents the complete solution to this problem in Example 22.9. The four structural isomers of $C_3H_6Cl_2$ are the following:

34. This type of problem is best approached by moving the 2 Cl atoms while 1 Cl atom is on fixed position.

36. See also Section 22.2 and problem 30 above.

    Use the structural formulas for the 3 butene isomers from problem 30:

    $$CH_3-CH_2-CH=CH_2 \qquad CH_3-CH=CH-CH_3 \qquad CH_3-\overset{\overset{\displaystyle CH_3}{|}}{C}=CH_2$$

    I $\qquad\qquad\qquad\qquad$ II $\qquad\qquad\qquad\qquad$ III

    Replace 2 H's of the double-bonded carbon(s), with Cl and Br. Three possible isomers, (V, W, X) can be obtained from structure I. One isomer each from structures II (Y) and III (Z) is possible.

    V $\qquad\qquad\qquad\qquad\qquad$ W $\qquad\qquad\qquad\qquad\qquad$ X

    Y $\qquad\qquad\qquad\qquad\qquad\qquad$ Z

38. See also Section 22.4.
    Since the carboxylic acid functional group (COOH) is always a terminal carbon (end of the hydrocarbon chain), there are only two possible isomers for the saturated 4-C carboxylic acid. One is a straight chain while the other is branched.

40. See Example 22.10.

All the isomers in problem #36, except the last one (structure Z), have geometric isomers. Structure Z does not have geometric isomers because it has two identical groups (–CH$_3$ on the same carbon). The geometric isomers for V, W, X and Y are shown below by a), b), c), and d), respectively.

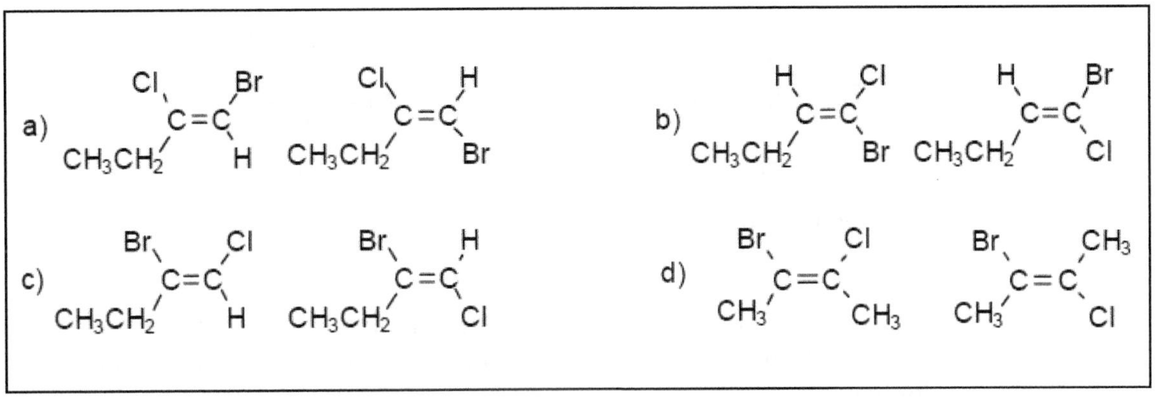

42. See Example 22.10.

(a) **No.** Geometric isomerism is possible only if the two groups attached to each of the two double-bonded carbons are different.

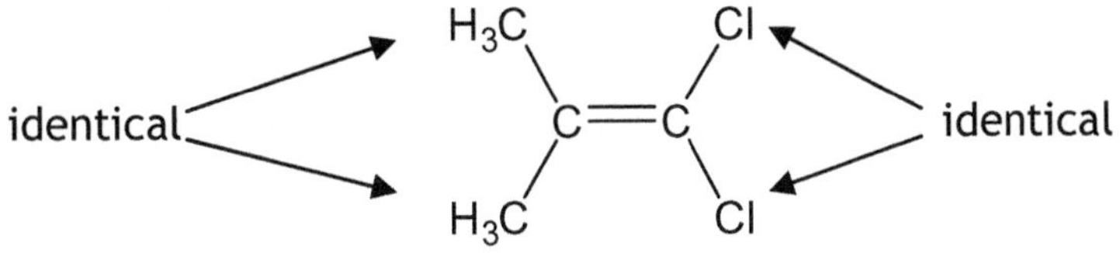

(b) **Yes**. Geometric isomerism is possible since the two groups attached to each of the two double-bonded carbons are different. The two geometric isomers are shown below.

(c) **Yes**. Geometric isomerism is possible since the two groups attached to each of the two double-bonded carbons are different. The two geometric isomers are shown below.

44. See Example 22.11.

    To show optical isomerism, the molecule must have a carbon atom attached to 4 different groups.

    (a) and (b). **No**. Both (a) dichloromethane and (b) 1,2-dichloroethane does not have a carbon atom attached to 4 different groups as shown below.

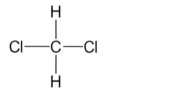

      (a)   dichloromethane 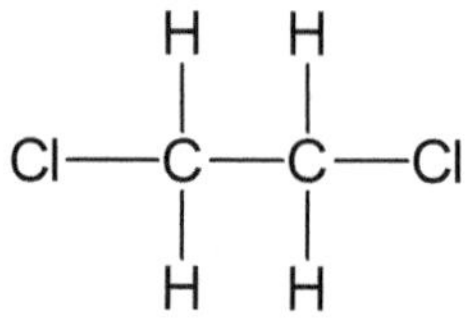   (b) 1,2-dichloroethane

    (c) and (d). **Yes**. Both (c) bromochlorofluoromethane and (d) 1-bromoethanol have a carbon atom attached to 4 different groups as shown in the structure below.

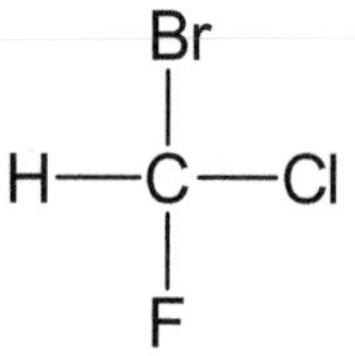

      (c)   bromochlorofluoromethane 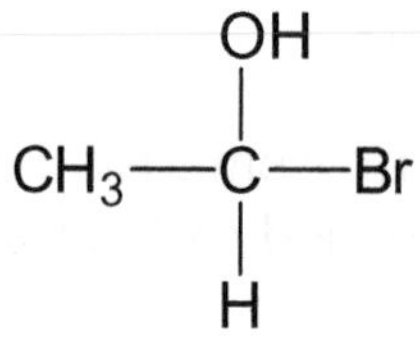   (d) 1-bromoethanol

46. See Example 22.11.

    The chiral carbons (indicated by *) are those that have four different groups attached to them.

    a. 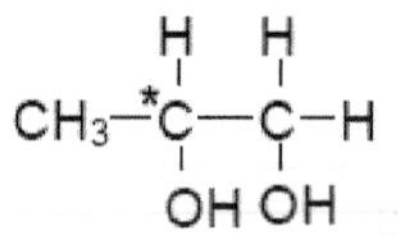 $CH_3-\overset{*}{C}-\overset{}{C}-H$ (with OH OH below)

    b. 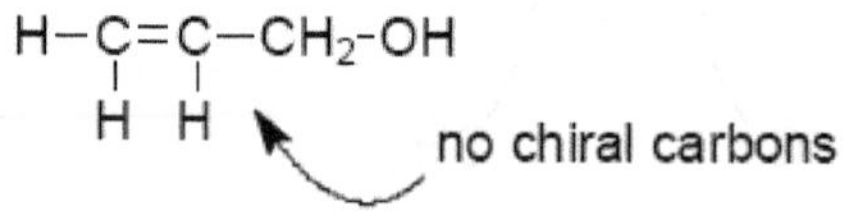 $H-C=C-CH_2-OH$ (with H H below)   no chiral carbons

    c. $CH_3-C-\overset{*}{C}-Cl$ (with Cl F above, Cl H below)

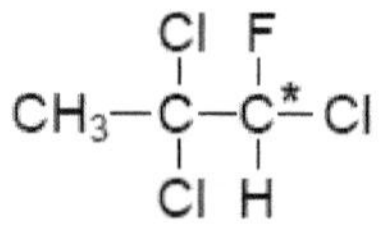

22.6 ORGANIC REACTIONS

48. See Example 22.12.

    (a) This reaction is **not an organic reaction**. Based on the definition presented in Section 22.6 it cannot be classified as in problem 47.

(b) This is an **elimination** reaction. It involved the elimination of –OH and H groups from adjacent C atoms to form an unsaturated alkene, C_3H_6 and water.

(c) This is a **condensation** reaction. The two reactants combined to form an ether and water.

50. (a)

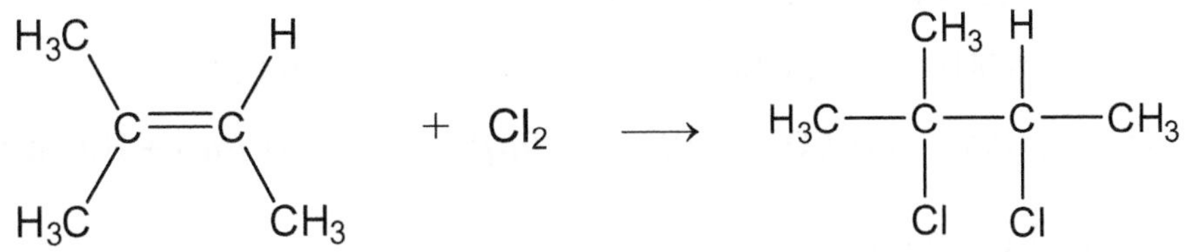

2-methyl-2-butene   **2,3-dichloro-2-methylbutane**

(b)

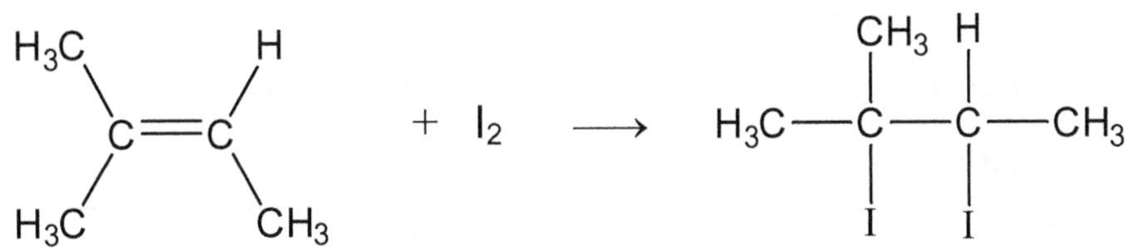

2-methyl-2-butene
2,3-diiodo-2-methylbutane

UNCLASSIFIED PROBLEMS

52. See Section 22.3. The circle in the center of the structural formula of benzene represents the six delocalized pi electrons around the ring.

54. (a) See Section 22.1. A saturated fat is one that does not have multiple bonds (C=C and C≡C). It only has C–C bonds. As quoted from the textbook, saturated means that "the ratio of hydrogen to carbon atoms is at a maximum". The maximum H:C ratio is achieved when there are no multiple bonds.

(b) See Section 22.4. A soap is a sodium (or potassium) salt of a long chain carboxylic acid.

(c) See Section 22.4. The "proof" of an alcoholic beverage is a measure of the alcohol content and is equal to twice the volume percentage of ethanol.

(d) See Section 22.4. Denatured alcohol is ethanol that has been treated with additives like methanol or benzene which impart unpleasant taste and makes the alcohol poisonous and unfit for human consumption.

56. See Sections 4.2 and 22.6.

(a) Reactants: $(CH_3)_2NH$ – weak base; does not ionize

HCl – strong acid; ionizes to give H^+ and Cl^-

Net ionic equation: $(CH_3)_2NH(aq) + H^+(aq) \rightarrow (CH_3)_2NH_2^+(aq)$

(b) Reactants: CH_3COOH – weak acid; does not ionize

$Ba(OH)_2$ – strong base; ionizes to give Ba^{2+} and OH^-

Net ionic equation: $CH_3COOH(aq) + OH^-(aq) \rightarrow CH_3COO^-(aq) + H_2O(aq)$

(c) This is a substitution reaction.

$$
\begin{array}{c}
Cl \\
| \\
CH_3 - C - CH_3 \\
| \\
H
\end{array}
$$

Reactants: 2-chloropropane

NaOH – strong base; ionizes to give Na^+ and OH^-

Net ionic equation:

$$
\begin{array}{ccccccc}
Cl & & & & OH & & \\
| & & & & | & & \\
CH_3 - C - CH_3\ (l) & + & OH^-(aq) & \rightarrow & CH_3 - C - CH_3\ (aq) & + & Cl^-\ (aq) \\
| & & & & | & & \\
H & & & & H & &
\end{array}
$$

CHALLENGED PROBLEMS

57. See "Chemistry Beyond the Classroom: Cholesterol. Figure A." In the structure, each vertex represents a carbon atom. Each carbon atom forms 4 bonds. Bonds not shown are understood to be bonded to hydrogen. Thus, the structure in Figure A can be redrawn below (right).

The number of C, H and O atoms can be counted. There are 27 C, 46 H and 1 O. Therefore, the molecular formula of cholesterol is $C_{27}H_{46}O$.

58. See Sections 22.4 and 22.5.

This type of problem is best approached systematically. Start with the linear 6-carbon alcohol molecule and move the –OH group into different C positions. Repeat the process with a branched molecule, then with a second branched molecule and so on, until all possibilities have been exhausted.

Linear 6-carbon molecules:

$$CH_3—CH_2—CH_2—CH_2—CH_2—\underset{\underset{OH}{|}}{CH_2}$$

$$CH_3—CH_2—CH_2—CH_2—\underset{\underset{OH}{|}}{CH}—CH_3$$

$$CH_3—CH_2—CH_2—\underset{\underset{OH}{|}}{CH}—CH_2—CH_3$$

Branched molecules (the longest chain has 4-C with two –CH$_3$ branch):

$$CH_3—\overset{\overset{CH_3}{|}}{CH}—\overset{\overset{CH_3}{|}}{CH}—\underset{\underset{OH}{|}}{CH_2}$$

$$CH_3—\overset{\overset{CH_3}{|}}{CH}—\underset{\underset{OH}{|}}{\overset{\overset{CH_3}{|}}{C}}—CH_3$$

$$H_3C—\underset{\underset{CH_3}{|}}{\overset{\overset{CH_3}{|}}{C}}—CH_2—\underset{\underset{OH}{|}}{CH_2}$$

$$H_3C—\underset{\underset{CH_3}{|}}{\overset{\overset{CH_3}{|}}{C}}—\underset{\underset{OH}{|}}{CH}—CH_3$$

$$H_3C—\underset{\underset{CH_2—OH}{|}}{\overset{\overset{CH_3}{|}}{C}}—CH_2—CH_3$$

Branched molecules (the longest chain has 5-C with one –CH_3 branch):

59. See Sections 22.1 and 8.5. Follow the following steps to solve this problem.

i. Calculate the amount of heat needed to raise the temperature of one quart of water at 25°C to boiling (100°C).

$$mass_{water} = one\ qt \times \frac{1L}{1.057\ qt} \times \frac{1000\ mL}{1L} \times \frac{1.0\ g}{1mL} = 9.5 \times 10^2\ g$$

$$q_{water} = mass \times c \times \Delta t$$

$$= 9.5 \times 10^2\ g \times \frac{4.18\ J}{g \cdot °C} \times (100°C - 25°C) = 3.0 \times 10^5\ J = 3.0 \times 10^2\ kJ$$

ii. Determine the amount of heat given off by the combustion of one mole of propane.

$$C_3H_8(g) + 5O_2(g) \rightarrow 3CO_2(g) + 4H_2O(l)$$

$$\Delta H° = \Sigma\Delta H_f°\,_{products} - \Sigma\Delta H_f°\,_{reactants}$$

$$\Delta H° = [3\Delta H_f°CO_2 + 4\Delta H_f°H_2O] - [\Delta H_f°C_3H_8 + 5\Delta H_f°O_2]$$

$$\Delta H° = [(3 \text{ mol})(-393.5 \text{ kJ/mol}) + (4 \text{ mol})(-285.8 \text{ kJ/mol})]$$
$$- [(1 \text{ mol})(-103.8 \text{ kJ/mol}) + (5 \text{ mol})(0 \text{ kJ/mol})]$$

$$\Delta H° = -2220 \text{ kJ/mol}$$

iii. Finally, assuming no heat lost and that all the heat released by burning propane is transferred to water, calculate the moles and grams of propane needed to raise the temperature of one quart of water from 25°C to boiling (100°C).

$$3.0 \times 10^2 \text{ kJ} \times \frac{1 \text{mol } C_3H_8}{2220 \text{ kJ}} \times \frac{44.10 \text{ g } C_3H_8}{1 \text{mol } C_3H_8} = \mathbf{6.0 \text{ g } C_3H_8}$$

60. See Sections 12.2, 13.4, 13.5, 22.4, and Example 22.8.

i. The acid–base reaction involved when chloroacetic acid reacts with trimethylamine:

$$ClCH_2COOH\ (aq) + (CH_3)_3N\ (aq) \rightleftharpoons ClCH_2COO^-\ (aq) + (CH_3)_3NH^+\ (aq)$$

ii. Determination of the equilibrium constant for the reaction in (i).

a. Ionization of the base trimethylamine and its given equilibrium constant (K_b)

$$(CH_3)_3N + H_2O \rightleftharpoons (CH_3)_3NH^+ + OH^- \qquad K_b = 5.9 \times 10^{-5}$$

b. Dissociation of the conjugate acid of trimethylamine and its equilibrium constant (K_a)

$$(CH_3)_3NH^+ \rightleftharpoons (CH_3)_3N + H^+ \qquad K_a = K_w/K_b$$
$$= (1.0 \times 10^{-14})/(5.9 \times 10^{-5})$$
$$K_a = 1.7 \times 10^{-10}$$

c. Equilibrium constant for the reverse of the reaction above (K_2):

$$(CH_3)_3N + H^+ \rightleftharpoons (CH_3)_3NH^+ \qquad K_2 = 1/K_a = 1/1.7 \times 10^{-10}$$
$$K_2 = 5.9 \times 10^9$$

d. Dissociation of chloroacetic acid and its given equilibrium constant (K_a)

$$ClCH_2COOH \rightleftharpoons ClCH_2COO^- + H^+ \qquad K_a = 1.5 \times 10^{-3}$$

Combine the information from d and c to find the equilibrium constant for the reaction between chloroacetic acid and trimethylamine (see reaction in i):

$$ClCH_2COOH \rightleftharpoons ClCH_2COO^- + H^+ \qquad K_a = 1.5 \times 10^{-3}$$

$$(CH_3)_3N + H^+ \rightleftharpoons (CH_3)_3NH^+ \qquad K_2 = 5.9 \times 10^9$$

$$ClCH_2COOH + (CH_3)_3N \rightleftharpoons ClCH_2COO^- + (CH_3)_3NH^+ \qquad K = K_a \times K_2$$

$$= (1.5 \times 10^{-3})(5.9 \times 10^9)$$

$$= 8.8 \times 10^6$$

The equilibrium constant for the reaction between chloroacetic acid and trimethylamine is **K = 8.8 × 10⁶.**

$$\mathbf{ClCH_2COOH \quad + \quad (CH_3)_3N \rightleftharpoons ClCH_2COO^- + (CH_3)_3NH^+ \qquad K = 8.8 \times 10^6}$$

iii. Equilibrium concentration of chloroacetic acid and trimethylamine when 0.10M solutions of these two species are mixed.

	$ClCH_2COOH$	+	$(CH_3)_3N$	$\rightleftharpoons$	$ClCH_2COO^-$	+	$(CH_3)_3NH^+$	$K = 8.8 \times 10^6$
[ ]$_o$	0.10		0.10		0		0	
Δ[ ]	−x		−x		+x		+x	
[ ]$_{eq}$	0.10 − x		0.10 − x		x		x	

$$8.8 \times 10^6 = \frac{x^2}{(0.10 - x)^2} \qquad \text{take the square root of both sides of the equation}$$

$$2966 = \frac{x}{0.10 - x}$$

$$x = 2966\,(0.10 - x)$$

$$x = 296.6 - 2966x$$

$$x = 0.099966$$

$$[ClCH_2COOH]_{eq} = [(CH_3)_3N]_{eq} = 0.10 - x = 0.10 - 0.099966 = \underline{\mathbf{3.4 \times 10^{-5}\ M}}$$

The equilibrium concentration of chloroacetic acid and trimethylamine is **3.4 × 10⁻⁵ *M*.**

ORGANIC POLYMERS, NATURAL AND SYNTHETIC

23.1 Synthetic Addition Polymers

2. See Section 3.2 and Examples 3.4 and 23.1.

(a) Refer to Table 23.1 for the monomer. Polymerization of tetrafluoroethylene produces Teflon where n is a large number. The double bonds in the monomer are replaced with single bonds connecting monomers together.

tetrafluoroethylene   teflon

portion of the Teflon chain (with 3 monomer units)

(b) $\dfrac{5.0 \times 10^4 \text{ CF}_2 \text{ units}}{1 \text{ mol}} \times \dfrac{50.01 \text{ g}}{1 \text{ CF}_2 \text{ unit}} = \underline{\textbf{2.5×10}^6 \textbf{ g/mol}}$

(c) Teflon is composed of C and F atoms at a ratio of 1C atom:2F atoms, or simply CF_2
molar mass of CF_2 unit:
1 mol C (12.01g/mol) + 2 mol F (19.00 g/mol) = 12.01 g C + 38.00 g F = 50.01 gCF_2

$\% \text{ C} = \dfrac{12.01 \text{ g C}}{50.01 \text{ g CF}_2} \times 100\% = \underline{\textbf{24.02\% C}}$

$\% \text{ F} = \dfrac{38.00 \text{ g F}}{50.01 \text{ g CF}_2} \times 100\% = \underline{\textbf{75.98\% F}}$

Alternatively, % F = 100% − %C = 100 − 24.02 = $\underline{\textbf{75.98\% F}}$

4. See Example 23.2.

This problem is solved the same way as Example 23.2. Styrene is similar to propene structure however instead of the CH_3 group there is a phenyl group

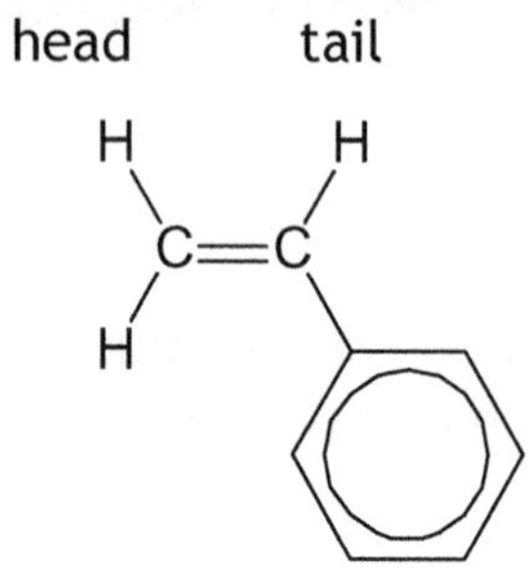

head   tail

styrene

portion of a head-to-tail addition polymer of a polystyrene molecule:

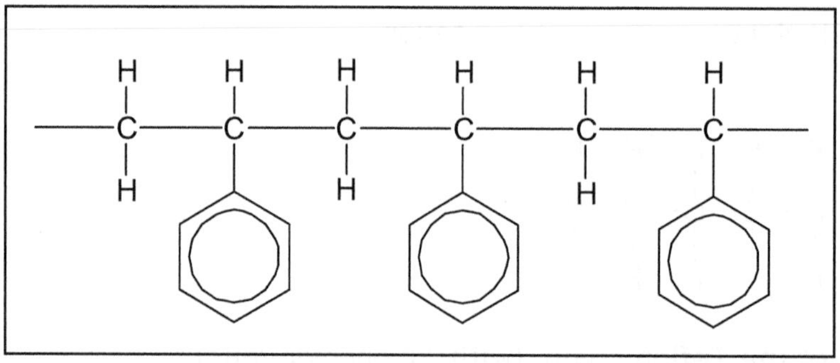

6. See Example 23.2.

(a) Based on the problem, the polymer is formed from one monomer. Addition polymers are usually formed from a monomer with a C=C bond. The identity of the monomer can be obtained from the boxed part of the polymer and replacing the C–C single bond with a C=C. The polymer is a head-to-head, tail-to-tail addition polymer. The monomer is vinyl fluoride.

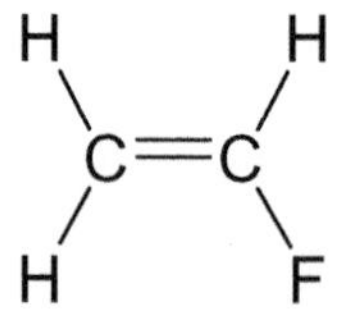

monomer: vinyl fluoride 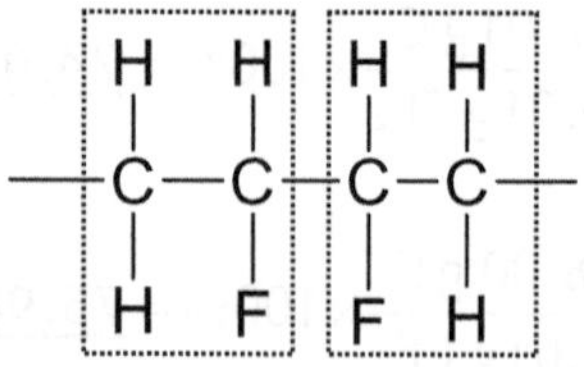   polymer

(b) Addition polymerization of 2-butene produced the given polymer. The C=C bond is
   converted to a C–C bond in the polymerization process. The monomer, 2-butene adds
   directly to another 2-butene.

monomer: 2-butene   polymer portion

8. See Section 3.2 and Examples 3.4 and 23.1.

monomer name	formula	molar mass (g)	mass C (g)	% C by mass
ethylene	C_2H_4	28.05	24.02	85.63
propylene	C_3H_6	42.08	36.03	85.63
vinyl chloride	C_2H_3Cl	62.49	24.02	38.44
acrylonitrile	C_3H_3N	53.06	36.03	67.90
styrene	C_8H_8	104.14	96.08	92.26
methyl methacrylate	$C_5H_8O_2$	100.11	60.05	59.98
tetrafluoroethylene	C_2F_4	100.02	24.02	24.02

Ethylene: $\quad$ % C by mass $= \dfrac{24.02 \text{ g C}}{28.05 \text{ g } C_2H_4} \times 100\% = 85.63\%$

Propylene: $\quad$ % C by mass $= \dfrac{36.03 \text{ g C}}{42.08 \text{ g } C_3H_6} \times 100\% = 85.63\%$

Vinyl chloride: $\quad$ % C by mass $= \dfrac{24.02 \text{ g C}}{62.49 \text{ g } C_2H_3Cl} \times 100\% = 38.44\%$

Acrylonitrile: $\quad$ % C by mass $= \dfrac{36.03 \text{ g C}}{53.06 \text{ g } C_3H_3N} \times 100\% = 67.90\%$

Styrene: $\quad$ % C by mass $= \dfrac{96.08 \text{ g C}}{104.14 \text{ g } C_8H_8} \times 100\% = 92.26\%$

Methyl methacrylate: $\quad$ % C by mass $= \dfrac{60.05 \text{ g C}}{100.11 \text{ g } C_8H_8} \times 100\% = 59.98\%$

Tetrafluoroethylene: $\quad$ % C by mass $= \dfrac{24.02 \text{ g C}}{100.02 \text{ g } C_8H_8} \times 100\% = 24.02\%$

**The polymer with the highest percent by mass of carbon is polystyrene because its
monomer, (styrene) has the highest %C by mass.**

23.2 Synthetic Condensation Polymers

10. See Example 23.3.

    Lexan chain is formed by removing the –OH group from carbonic acid and the –H from the other monomer. The two monomers are joined together at the ends where the groups were removed. (The groups removed are boxed in the monomer structure).

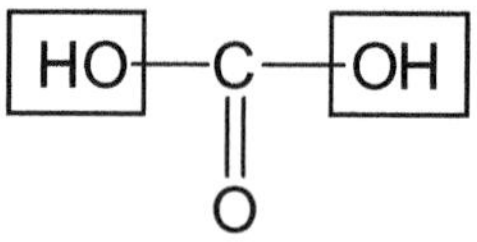

    carbonic acid (monomer 1) 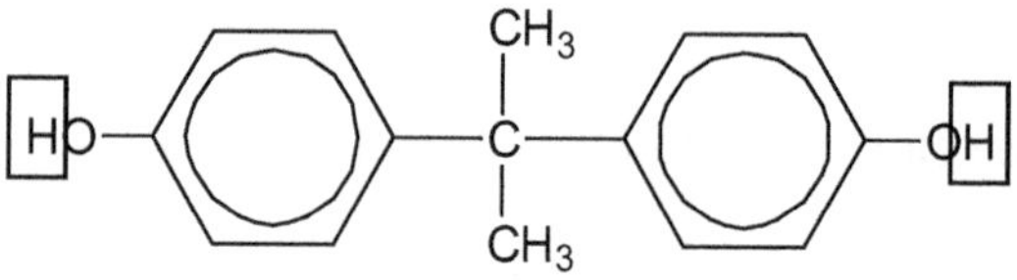   monomer 2

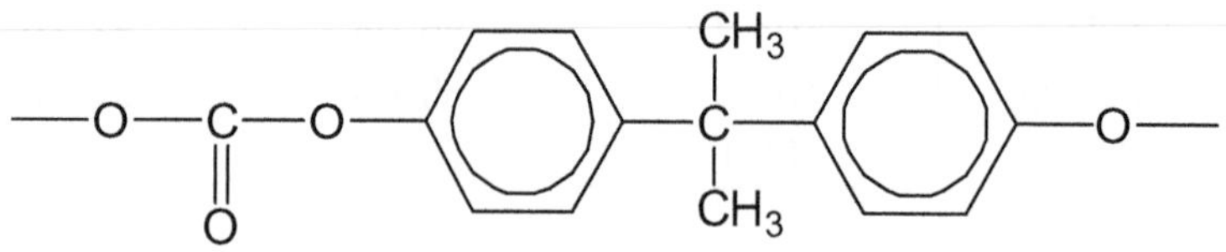

    section of the Lexan chain

12. See Example 23.4.

    Nylon–66 is formed from the following single monomer.

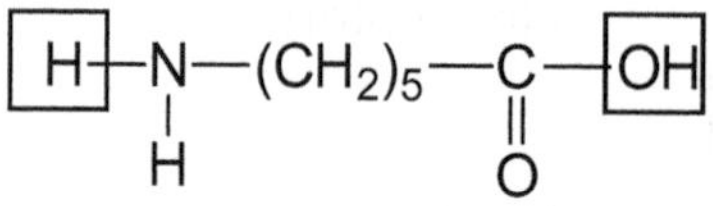

    The –OH from the COOH group and the –H from the –NH₂ group are removed. The monomer units are joined together at the ends where the groups are removed. A section of the polymer chain in Nylon-66 is shown below.

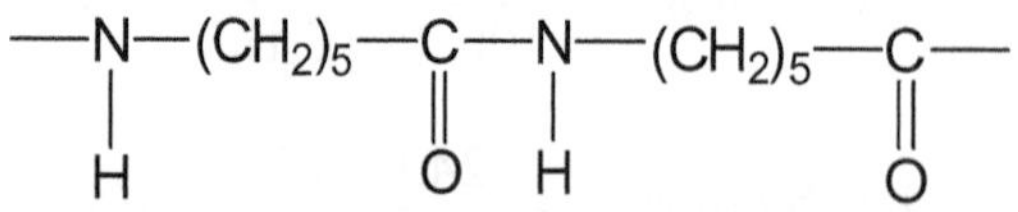

    section of the Nylon-66 chain

14. See Example 23.4.

The monomer units of a condensation polymer are identified by locating the bond forming the polymer linkage. Recall that condensation polymers are made by removing –OH from one monomer and –H from another monomer. Thus, in breaking the polymer linkage, –OH is added to one end of the bond and –H to the other. Most often than not, the –OH is from the –COOH group of one monomer while the –H group is from the amino (–NH₂) or the alcohol group (–OH) of the other monomer.

polymer linkage

(a)

$$-N-CH_2-CH_2-N-C-CH_2-C-$$

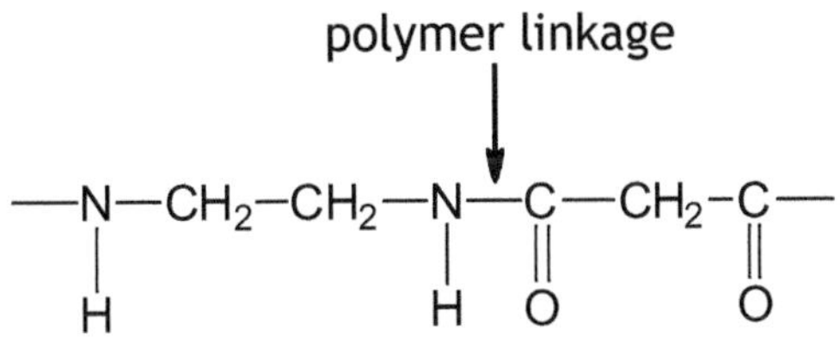

two monomers: H₂N–CH₂–CH₂–NH₂ and HOOC–CH₂–COOH

polymer linkage

(b)

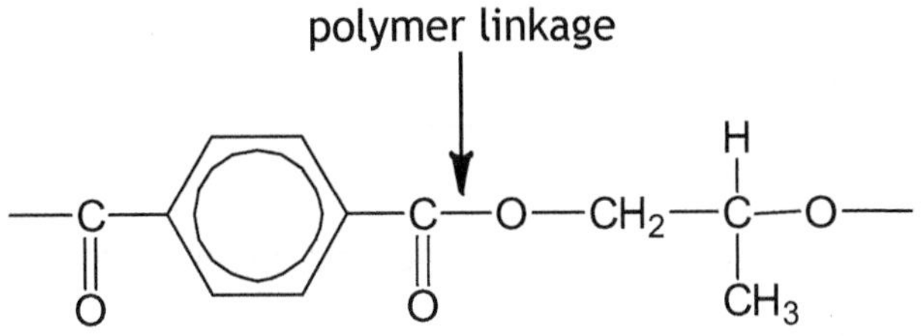

two monomers:

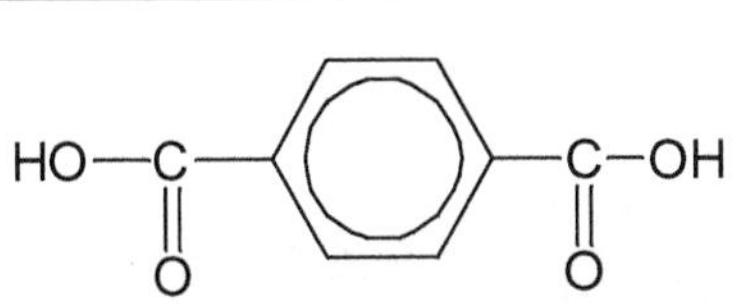

and

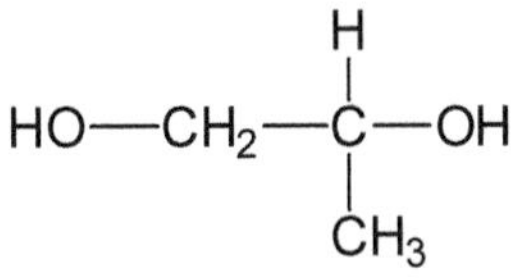

23.3 Carbohydrates

16. See Example 23.6.

The reaction of maltose with water to form glucose is just the reverse of the reaction of the formation of maltose from glucose described in the topic maltose and sucrose.

$$C_{12}H_{22}O_{11}(aq) + H_2O \rightarrow 2C_6H_{12}O_6(aq)$$

18. See Section 3.2 and Examples 3.4 and 3.7.

   The mass percent of C, H and O in starch can be determined from the simplest or empirical formula of starch. Figure 23.5 shows a portion of starch with two units of the monosaccharide. The simplest formula of the repeating unit can be obtained by counting the number of C, H and O atoms in one repeating unit of starch. Each repeating unit has 6 C atoms, 10 H atoms and 5 O atoms thus the empirical formula of starch is $C_6H_{10}O_5$.

   $C_{12}H_{20}O_{10}$ empirical formula = $C_6H_{10}O_5$

   mass of 1 mole of the empirical formula unit, $C_6H_{10}O_5$

   $$= 6(12.01) \text{ g C} + 10(1.008) \text{ g H} + 5(16.00) \text{ g O}$$

   $$= 72.06 \text{ g C} + 10.08 \text{ g H} + 80.00 \text{ g O}$$

   $$= 162.14 \text{ g/mol } C_6H_{10}O_5$$

   (a) % C by mass $= \dfrac{\text{mass C}}{\text{mass } C_6H_{10}O_5} \times 100 = \dfrac{72.06}{162.14} \times 100 = \underline{\textbf{44.44\% C}}$

   % H by mass $= \dfrac{\text{mass H}}{\text{mass } C_6H_{10}O_5} \times 100 = \dfrac{10.08}{162.14} \times 100 = \underline{\textbf{6.22\% H}}$

   % O by mass $= \dfrac{\text{mass O}}{\text{mass } C_6H_{10}O_5} \times 100 = \dfrac{80.00}{162.14} \times 100 = \underline{\textbf{49.34\% O}}$

   Alternatively, the mass percent of oxygen can be determined by difference:
   $100 - (44.44 + 6.22) = \underline{\textbf{49.34\% O}}$

   (b) number of $C_6H_{10}O_5$ units $= \dfrac{\text{molar mass of starch}}{\text{molar mass of the empirical formula}}$

   $$= \dfrac{1.0 \times 10^5 \text{g/mol}}{162.14 \text{ g/mol}} = \underline{\textbf{6.2} \times \textbf{10}^2}$$

20. See end-of-chapter question 19.

α−mannose

Assume that carbon atom 1 of the first α–mannose molecule is joined to carbon 4 of the second α–mannose molecule by an oxygen atom.

disaccharide formed by 2 units of α–mannose

22. See Section 22.5 and Examples 23.5b.

For a carbon to be chiral it has to be bonded to four different groups.

Sucrose. All the carbons which are part of the two rings of sucrose are chiral. The chiral carbons of sucrose are indicated by an asterisk (*). **Sucrose has 9 chiral carbons**.

sucrose

Maltose. All the carbons which are part of the two rings of maltose are chiral. The chiral carbons of maltose are indicated by an asterisk (*). **Maltose has 10 chiral carbons.**

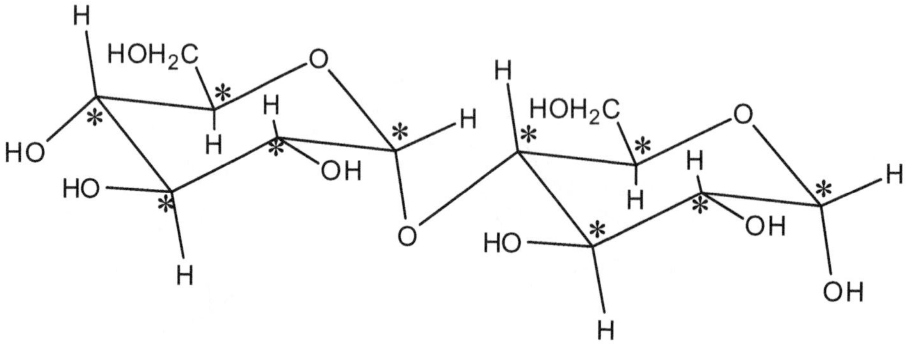

maltose

23.4 Proteins

24. See Table 23.3 and Example 23.9.

The two different dipeptides formed between leucine and lysine are produced by condensing out a water molecule made up of OH from one of the amino acids and H from the other amino acid as shown below.

First dipeptide: leu-lys: leucine with free NH_2 group and lysine with free COOH group

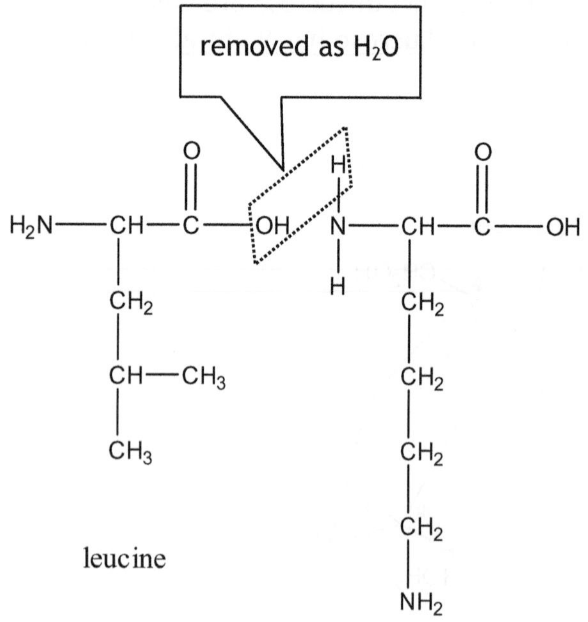

removed as H_2O

leucine

lysine

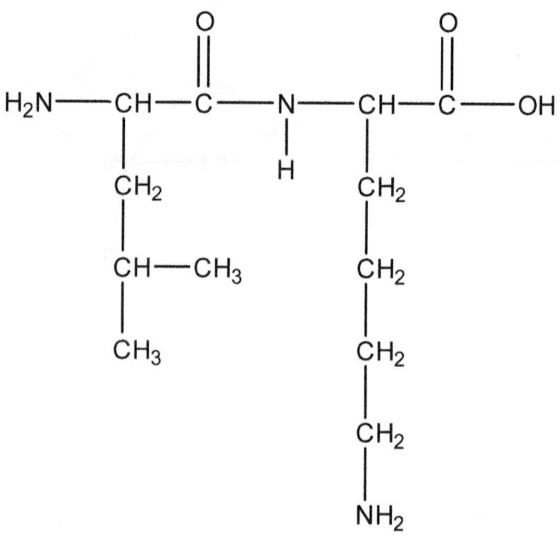

Leucyllysine (Leu-Lys)

Second dipeptide: lys-leu: lysine with free NH$_2$ group and leucine with free COOH group

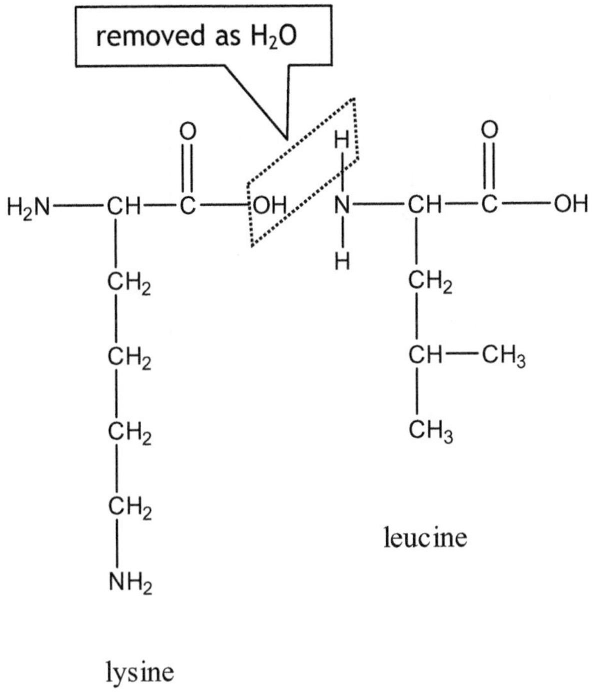

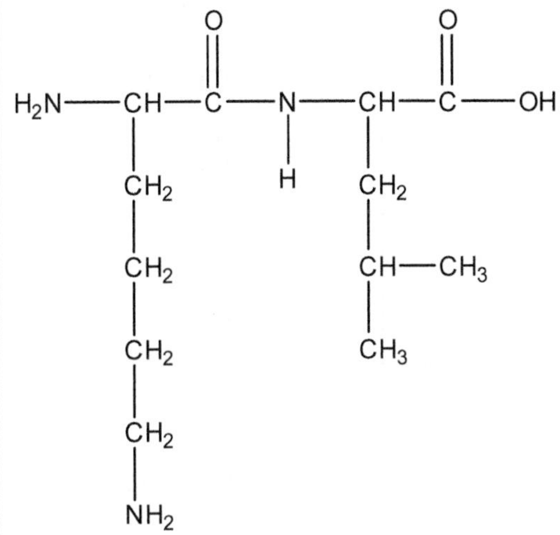

removed as H$_2$O

$$H_2N-CH-C(=O)-OH \quad N-CH-C(=O)-OH$$

lysine

leucine

Lysylleucine (Lys-Leu)

26. See Table 23.3.

    (a) There are **6 possible tripeptides** composed of valine, lysine and phenylalanine residues. These 6 tripeptides are:

        i.   **Val-Lys-Phe**

        ii.   **Val-Phe-Lys**

        iii.   **Phe-Lys-Val**

        iv.   **Phe-Val-Lys**

        v.   **Lys-Val-Phe**

        vi.   **Lys-Phe-Val**

(b) The structural formula of one of the six possible tripeptides, Val-Lys-Phe, is shown below.

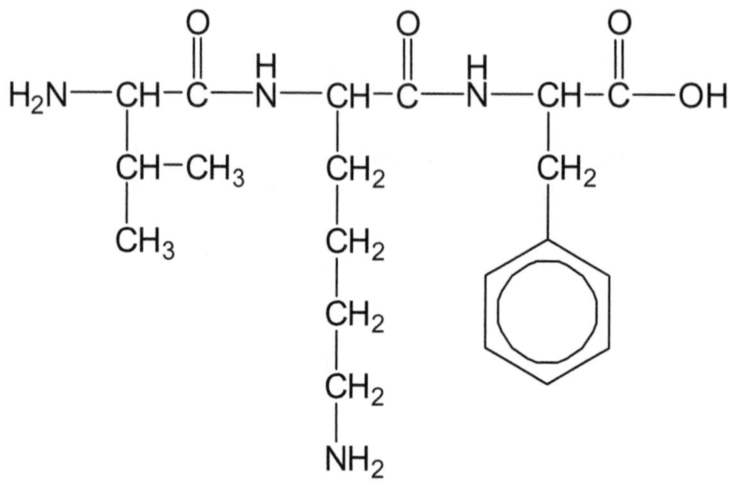

valyllysylphenylalanine (Val-Lys-Phe)

28. See Table 23.3.

(a) In an amino acid, a zwitterion is formed when the N of the NH_2 group accepts the H from the COOH group. Consequently, the zwitterions have a charge of +1 on the N of the NH_2 group and a charge of –1 in one of the O atoms in the COOH group. The overall charge however of a zwitterion is zero. The zwitterion of serine is shown.

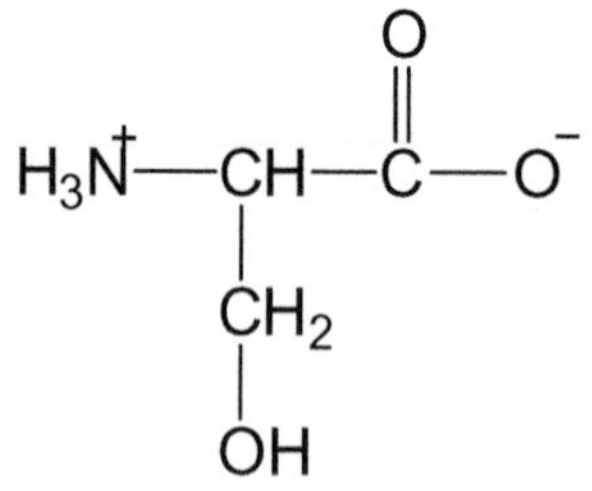

zwitterion of serine

(b) Serine is in the cation form at low pH or acidic condition. The –COO⁻ is converted back to its acid form, –COOH while the –NH₃⁺ remains protonated, resulting in a net positive charge for serine.

cation form of serine (in acid)

(c) Serine is in the anion form at high pH or basic condition. The –COOH will exist in its base form (–COO⁻) while the NH₃⁺ will be converted back to its base form, –NH₂, resulting in a net negative charge for serine.

anion form of serine (in acid)

30. See Example 23.8.

(a) $C^+(aq) \rightleftharpoons H^+(aq) + Z(aq)$ $K_{a1} = \dfrac{[H^+][Z]}{[C^+]}$ where $K_{a1} = 5.1 \times 10^{-3}$

since $[Z] = [C^+]$ then $K_{a1} = \dfrac{[H^+][\cancel{Z}]}{[\cancel{C}]}$ $\Rightarrow$ $K_{a1} = [H^+]$

thus, $[H^+] = 5.1 \times 10^{-3}$

$pH = -\log [H^+] = -\log (5.1 \times 10^{-3}) = \underline{\mathbf{2.29}}$

(b) $Z(aq) \rightleftharpoons H^+(aq) + A^-(aq)$ $K_{a2} = \dfrac{[H^+][A^-]}{[Z]}$ where $K_{a2} = 1.8 \times 10^{-10}$

since [Z] = [A]   then $K_{a2} = \dfrac{[H^+][\cancel{A}]}{[\cancel{Z}]} \Rightarrow K_{a2} = [H^+]$

$[H^+] = 1.8 \times 10^{-10}$

$pH = -\log[H^+] = -\log(1.8 \times 10^{-10}) = \underline{\textbf{9.74}}$

(c) pH at the isoelectric point = $\dfrac{pK_{a1} + pK_{a2}}{2}$

$pH = \dfrac{[-\log(5.1 \times 10^{-3})] + [-\log(1.8 \times 10^{-10})]}{2}$

$pH = \dfrac{2.29 + 9.74}{2} = \dfrac{12.03}{2} = \underline{\textbf{6.02}}$

32. See end-of-chapter question 31. The polypeptide has 6 amino acid residues:
2 alanine, 1 leucine, 1 methionine, 1 phenylalanine, 1 valine

Partial hydrolysis of the polypeptide yielded the following dipeptides:
Ala–Phe   Leu–Met   Val–Ala   Phe–Leu

Since the first amino acid in the sequence is alanine, the first dipeptide should be Ala-Phe. Align the other dipeptide fragments vertically such that they overlap one another.

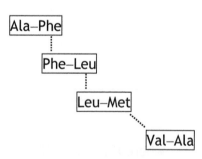

Since alanine is the last amino acid in the sequence, then Val-Ala follows Leu-Met. Thus, the complete amino acid sequence of the polypeptide is:

<u>Ala–Phe–Leu–Met–Val–Ala</u>

UNCLASSIFIED PROBLEMS

34. See Sections 23.1 and 23.2.

For a compound to be useful as a monomer for addition polymerization, it has to have a carbon–carbon double bond (C=C). This C=C bond is converted to a single bond (C–C) upon polymerization. The C=C containing monomers add to one another hence, addition polymerization.

Compounds which contain two functional groups that can form a small molecule (and eventually eliminated) can be used as monomers for condensation polymerization. The most common functional groups are: $-NH_2$, $-OH$, and $-COOH$ while the most common molecule formed and removed is water.

36. See Sections 23.1 and 23.3.

(a) Both linear and branched polyethylene is produced by addition polymerization of ethylene. Linear polyethylene or high-density polyethylene has carbon atoms bonded to each other in a linear fashion such that each carbon is bonded to two other carbons with the exception of the terminal carbons which are attached to only one C atom. On the other hand, in branched polyethylene or low-density polyethylene some carbons are bonded to three other carbons thus forming a branch from the linear carbon chain. A portion of the structure of linear and branched polyethylene is given in the textbook in 23.1.

(b) Glucose and fructose are both six-carbon containing monosaccharides. As shown in the open structure below, glucose has an aldehyde group while fructose has a ketone group. In its cyclic structure, glucose forms a 6-membered ring while fructose has a 5-membered ring.

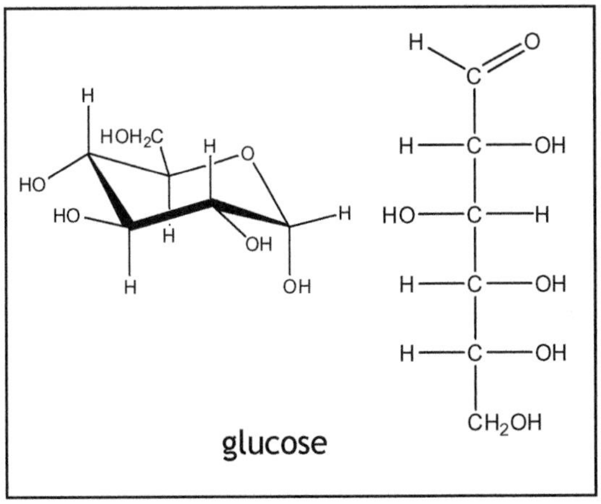

glucose

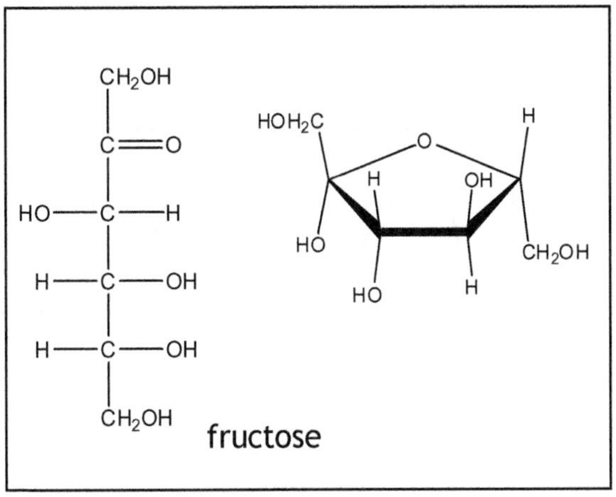

fructose

(c) Maltose and sucrose are both disaccharides which are formed by two monosaccharides. Maltose is a dimer of two glucose molecules whereas sucrose is a dimer of glucose and fructose. The structure of maltose and sucrose are presented in the carbohydrate Section 23.3.

38. See Sections 23.1 and 23.2.

(a)

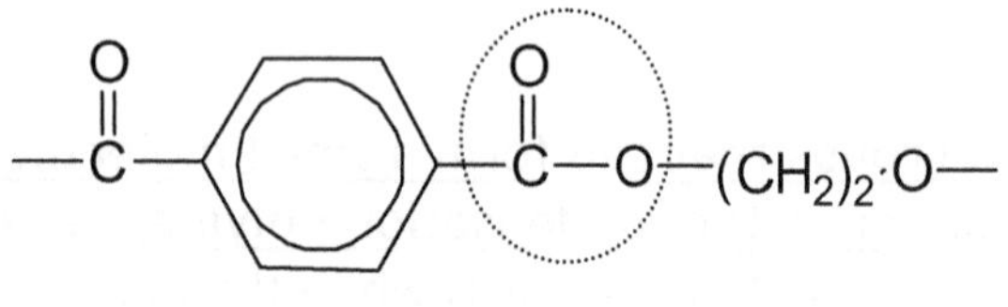

polymer

The presence of an ester group (encircled) suggests that the polymer above is a product of condensation polymerization. The two monomers are connected through this ester group. Using this functional group as reference, break the polymer into the two monomer fragments. The monomers that would be used to make the polymer are:

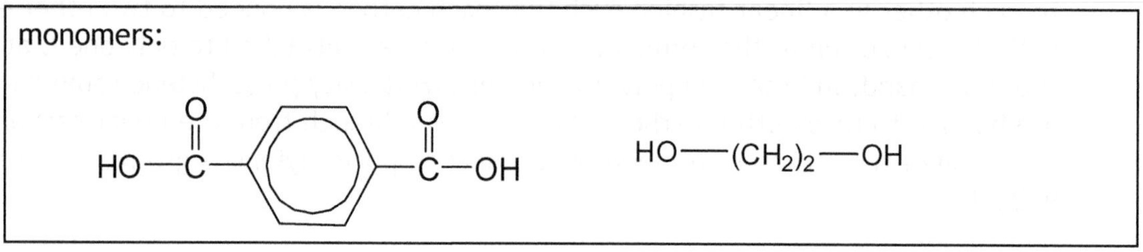

monomers:

These two monomers react to form the ester polymer with the removal of a water molecule as described in Section 23.2.

(b) As described in Section 23.1, this polymer is formed by addition polymerization of the following monomer:

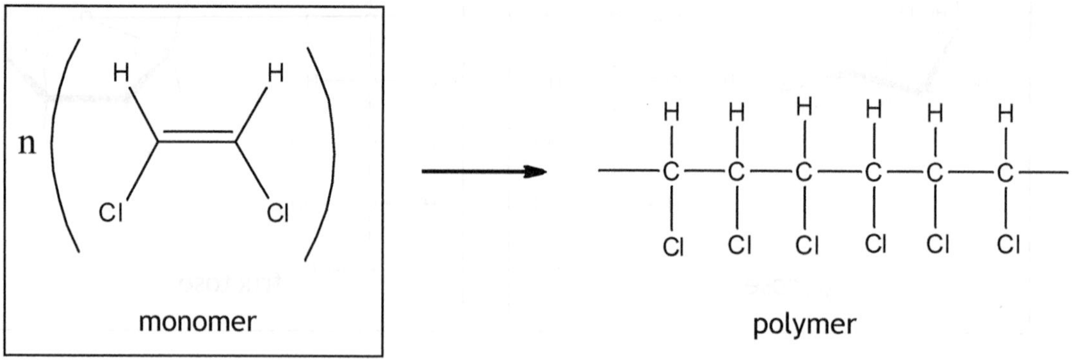

monomer   polymer

(c) The polymer is formed by the condensation polymerization of n units of the monomer.

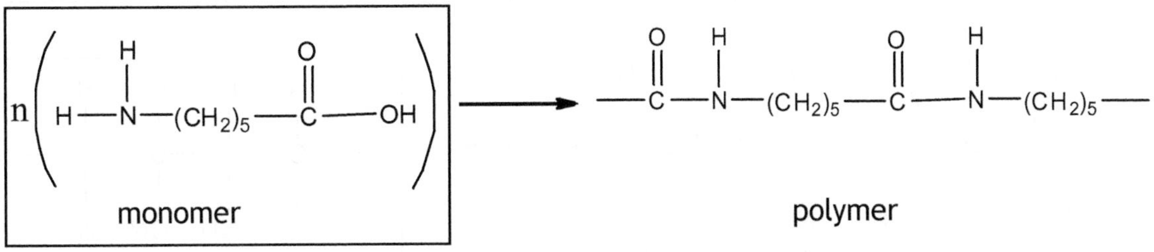
monomer
polymer

40. See Section 23.1 and Example 23.2.

monomer: acrylonitrile

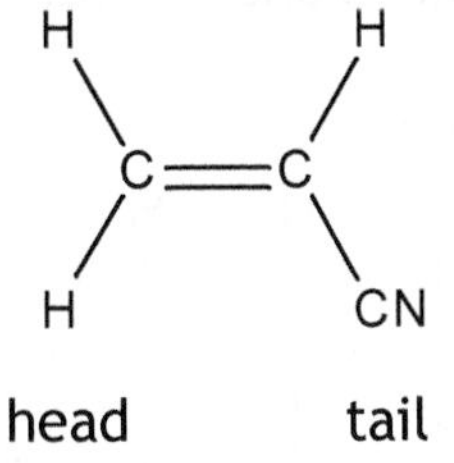
head
tail

head-to-head:

draw the monomers in a head–to–head arrangement. Convert the C=C to a C–C bond and connect the head of the first monomer to the head of the second with a C–C bond.

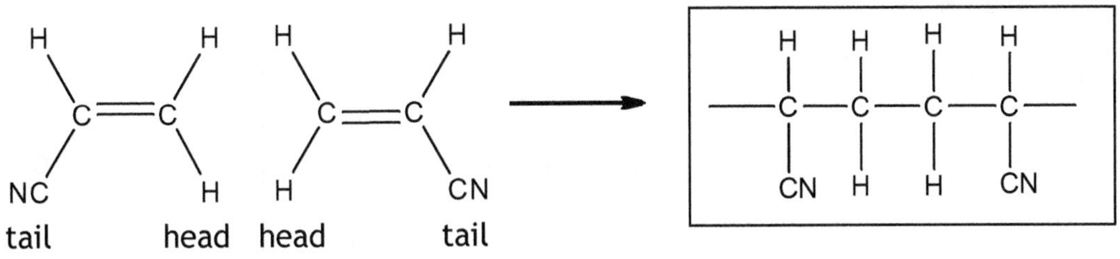
tail
head
head
tail

head-to-tail:

draw the monomers in a head–to–tail arrangement. Convert the C=C to a C–C bond and connect the head of the first monomer to the tail of the second with a C–C bond.

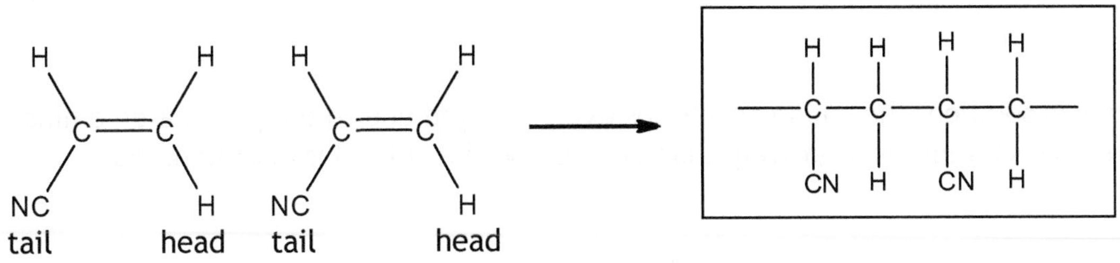
tail
head
tail
head

tail-to-tail:

draw the monomers in a tail–to–tail arrangement. Convert the C=C to a C–C bond and connect the tail of the first monomer to the tail of the second with a C–C bond.

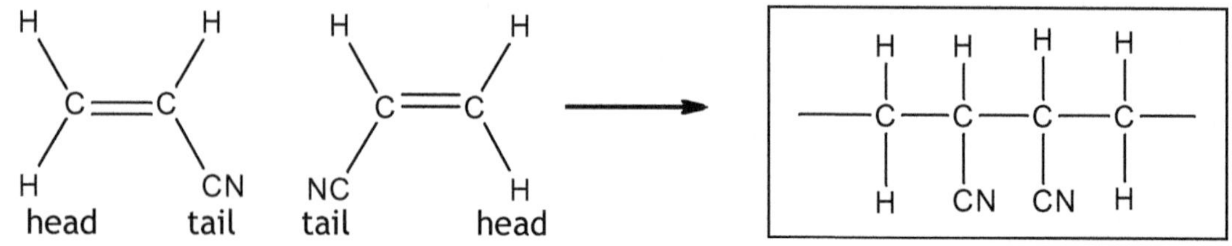

Note that if a longer portion of the head-to-head and the tail-to tail polymers is drawn, you would see that the head-to-head and the tail-to-tail polymers are identical.

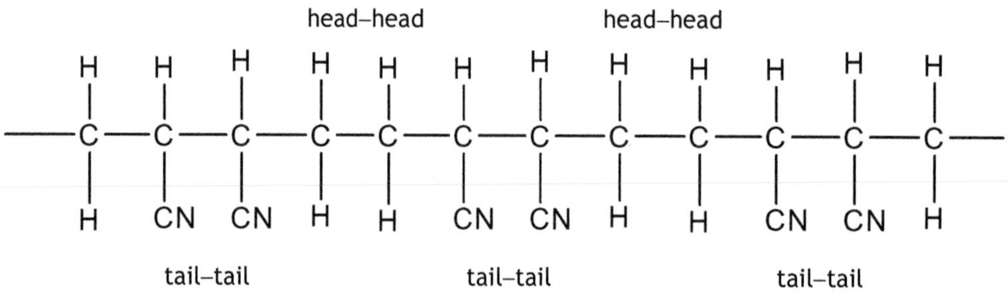

42. See Section 23.4.

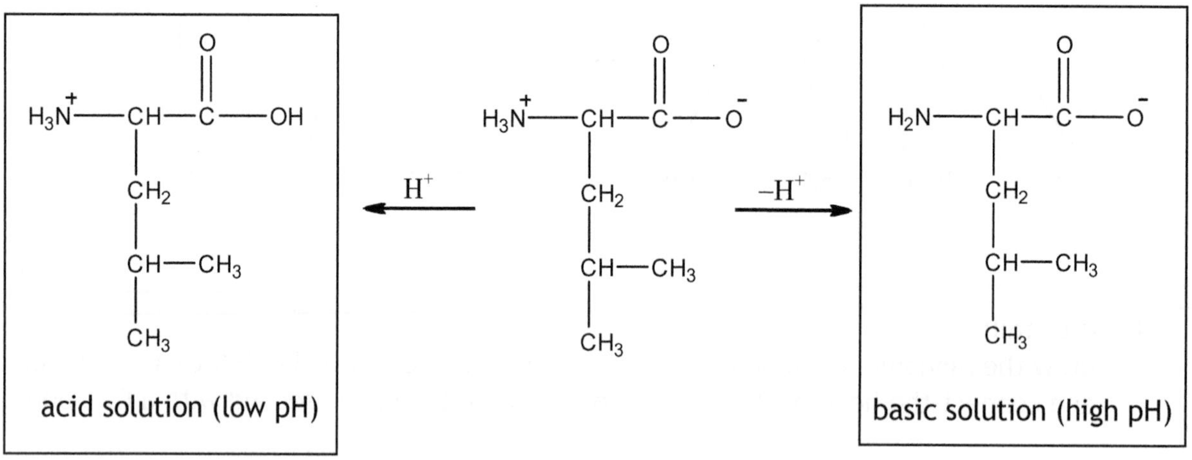

In acid solution, the amino acid is positively charged due to the protonated amino group. In basic solution, is negatively charged because the –COOH is deprotonated.

44. See Section 23.4.

Since the sample is a pure protein and hydrolysis yielded only isoleucine and leucine, then the minimum possible molar mass can be obtained by determining the mole ratio of leucine and isoleucine present in the sample.

$$\text{mol Ile} = 0.0248 \text{ mg Ile} \times \frac{1 \text{g}}{1000 \text{ mg}} \times \frac{1 \text{mol}}{131 \text{g}} = 1.89 \times 10^{-7} \text{ mol}$$

$$\text{mol Leu} = 0.0165 \text{ mg Leu} \times \frac{1 \text{g}}{1000 \text{ mg}} \times \frac{1 \text{mol}}{131 \text{g}} = 1.26 \times 10^{-7} \text{ mol}$$

total mol amino acid residue = 1.89×10^{-7} mol + 1.26×10^{-7} mol
$$= 3.15 \times 10^{-7} \text{ mol amino acids}$$

mol ratio Ile to Leu:
$$\frac{1.89 \times 10^{-7} \text{ mol Ile}}{1.26 \times 10^{-7} \text{ mol Leu}} = 1.50 \text{ mol Ile/mol Leu}$$

Therefore, the mol ratio of Ile to Leu is:
        1.50 mol Ile:1 mol Leu   or   3 mol Ile:2 mol Leu

Thus, the total mol amino acid residue, 3.15×10^{-7} corresponds to 5 amino acid residues (3 of which is isoleucine and 2 is leucine).

$$\text{mol per residue} = \frac{3.15 \times 10^{-7} \text{ mol amino acid residue}}{5 \text{ amino acids}} = 6.30 \times 10^{-8} \text{ mol residue}$$

$$\text{minimum possible molar mass of protein} = \frac{1.00 \text{ mg}}{6.30 \times 10^{-8} \text{ mol Leu}} \times \frac{1 \text{g}}{1000 \text{ mg}}$$

$$\underline{= 1.59 \times 10^4 \text{ g/mol}}$$

46. See Section 16.6 and Example 17.7

$$C_6H_{12}O_6(aq) \rightarrow 2 \ C_3H_6O_3(aq) \qquad\qquad \Delta G° = -198 \text{ kJ at pH 7.0 and 25°C}$$
   glucose   lactic acid

$$\Delta G = \Delta G° + RT \ln Q$$

$$\Delta G = -198 \text{ kJ} + \left(8.31 \times 10^{-3} \text{ kJ/K}\right)(298\text{K}) \ln \frac{(2.9 \times 10^{-3})^2}{5.0 \times 10^{-3}}$$

$$\underline{\Delta G = -214 \text{ kJ}}$$

CHALLENGED PROBLEMS

48. See Section 23.2 and Example 23.3. This is a condensation polymerization reaction.

(a)

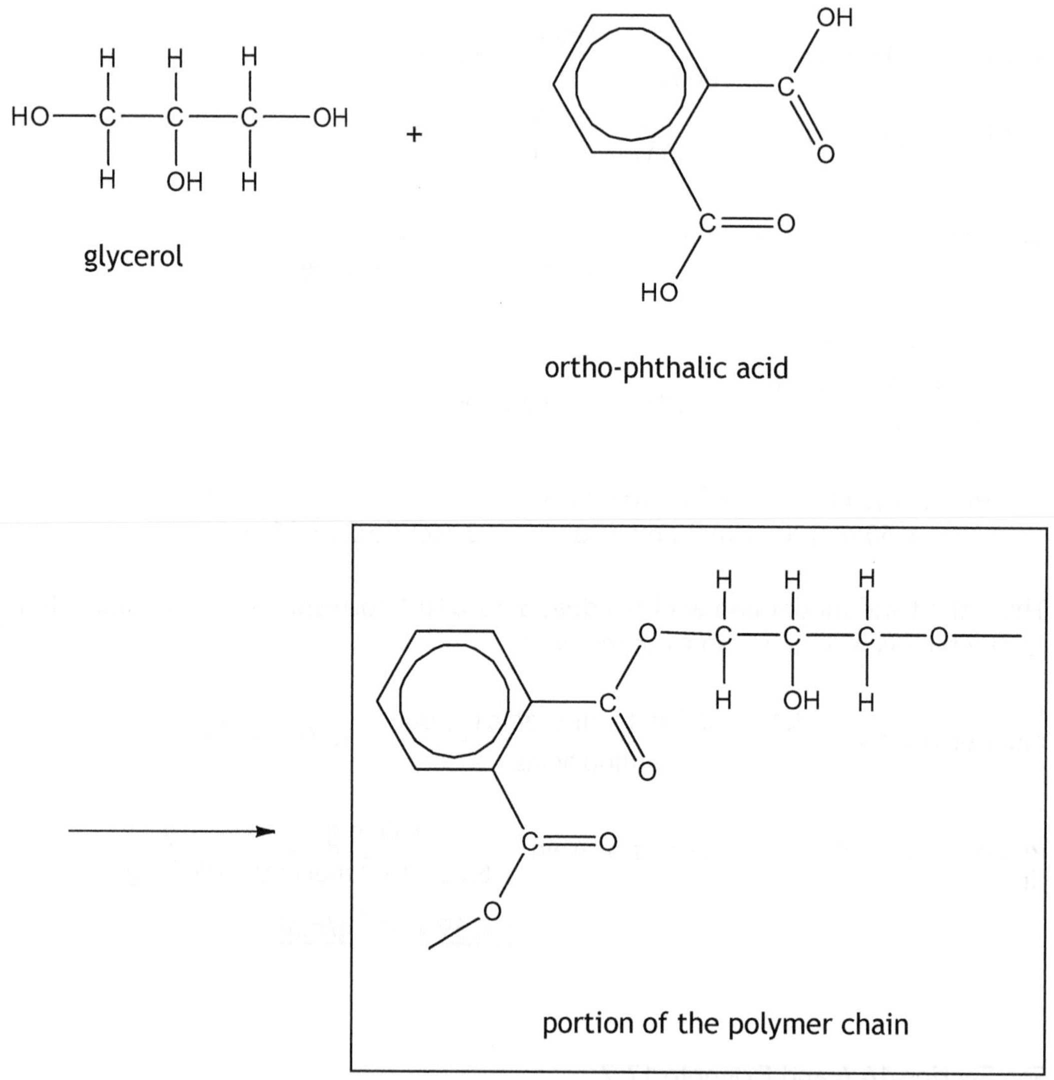

glycerol

+

ortho-phthalic acid

portion of the polymer chain

(b) Cross linking can occur in the polymer chain described in 48a above, cross-linking occurs when the –COOH groups of ortho-phthalic acid (encircled below) condenses with the –OH groups of two adjacent polymer strands.

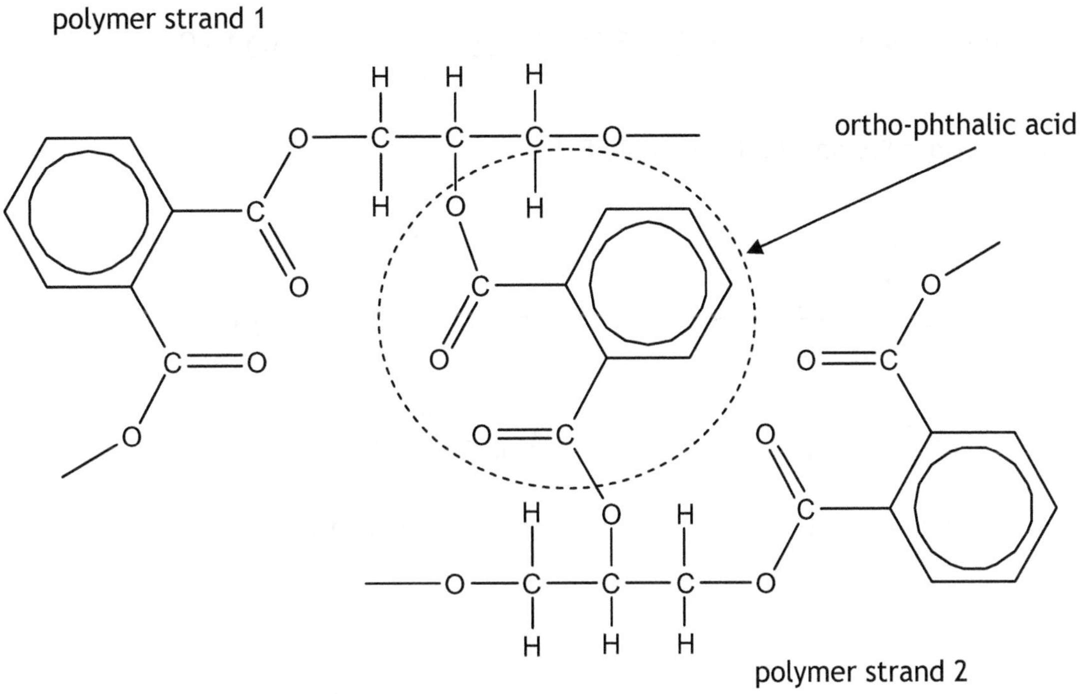

polymer strand 1

ortho-phthalic acid

polymer strand 2

49. See Sections 3.2 and 23.4 and Examples 3.4 and 23.4.

Use the formula of the repeating unit of Nylon. This formula can be obtained from the first half of the structure of the polymer portion described in Example 23.4.

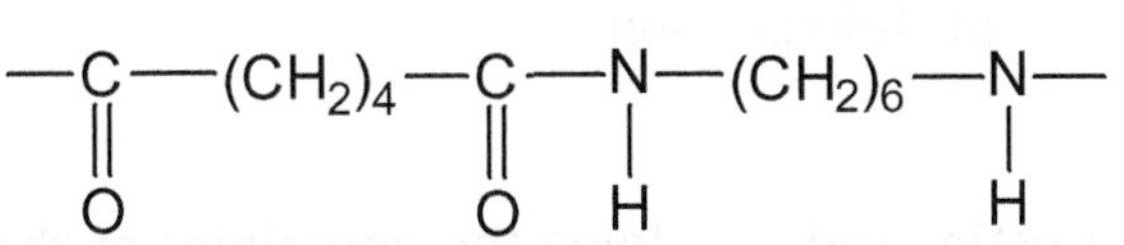

The molecular formula of the repeating unit is $C_{12}H_{22}N_2O_2$. Determine the % by mass of each element by assuming 1 mole of the repeating unit.

molar mass of the repeating unit = 12(12.01) + 22(1.008) + 2(14.01) + 2(16.00) = 226.32

$$\text{mass \% C} = \frac{12(12.01)}{226.32} \times 100 = 63.68\%$$

$$\text{mass \% H} = \frac{22(1.008)}{226.32} \times 100 = 9.80\%$$

$$\text{mass \% N} = \frac{2(14.01)}{226.32} \times 100 = 12.38\%$$

$$\text{mass \% O} = \frac{2(16.00)}{226.32} \times 100 = 14.14\%$$

Nylon is composed of 63.68% C, 9.80% H, 12.38% N, and 14.14% O.

50. See Sections 8.6 and 23.4 and Table 8.4.

amino acid 1   amino acid 2

protein portion

The change in bond enthalpy, ΔH is estimated by adding the enthalpies of the bonds that are broken and subtracting the enthalpies of the bonds that are formed. In protein formation, peptide bonds are formed by the breaking of the C–OH bond of one amino acid (encircled in amino acid 1) and the N–H bond of another amino acid (encircled in amino acid 2). The peptide bond is formed between the two amino acids by the formation of C–N, an amide bond (encircled in the protein portion). In the reaction water is also produced, thus the formation of an O–H bond.

Bond	C–O	H–N	C–N	O–H
H_{bond} (kJ/mol)	351	389	293	464

ΔH for protein formation per mole of amino acid added:

$$\Delta H = \Sigma H_{\text{bonds broken}} - \Sigma H_{\text{bonds formed}}$$

$$= (H_{C-O} + H_{H-N}) - (H_{C-N} + H_{O-H})$$

$$= (351 \text{ kJ/mol} + 389 \text{ kJ/mol}) - (293 \text{ kJ/mol} + 464 \text{ kJ/mol})$$

$$= -17 \text{ kJ/mol}$$

The bond energy involved for protein formation is −17 kJ/mol.

51. See Section 23.2.

Linear polymerization monomers:

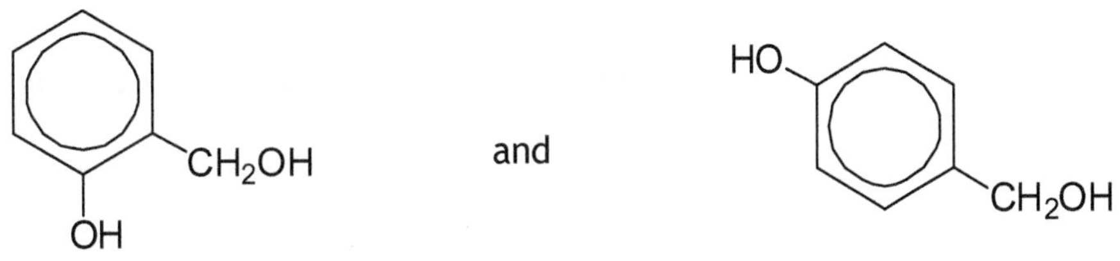

and

The linear polymer is formed when the −OH of the −CH₂OH and the −H on another benzene ring are removed as water and the benzene rings are linked by −CH₂− as shown below. The linear polymer can be composed of portions having −OH adjacent to the −CH₂ linkages, −OH opposite to the −CH₂ linkages or a combination of the two.

−OH adjacent to the −CH₂ linkages:

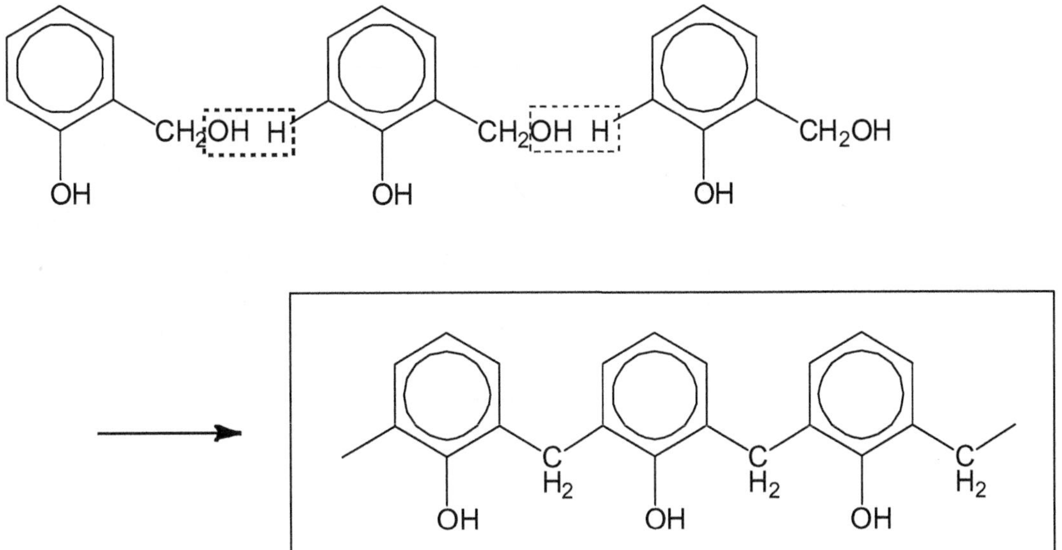

possible portion of the linear polymer chain

−OH opposite the −CH$_2$ linkages:

another possible portion of the linear polymer

combination of −OH adjacent to the −CH$_2$ linkages and −OH opposite to the −CH$_2$ linkages:

another possible portion of the linear polymer

Cross-linking monomer produced from the reaction of two $H_2C=O$ for every phenol:

monomer

The linear polymer can react with the monomer above to form a cross-linked polymer as described below:

portion of the cross-linked polymer

52. See Section 23.4.

$$pH = pK_{a1} = -\log(8.0 \times 10^{-3}) = 2.10$$

$$pH = pK_{a2} = -\log(1.4 \times 10^{-4}) = 3.85$$

$$pH = pK_{a3} = -\log(1.5 \times 10^{-10}) = 9.82$$

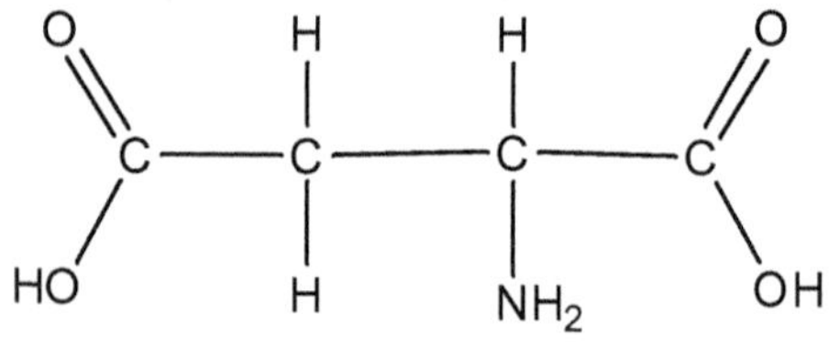

aspartic acid

Four different forms of aspartic acid in water solution:

I.   The following form predominates at pH less than 2.10

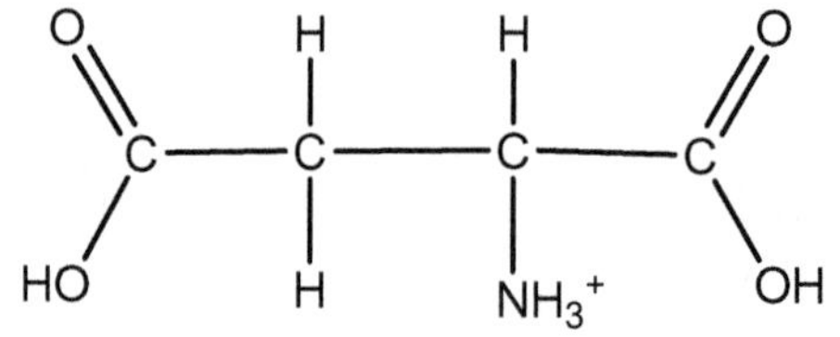

II.   The following form predominates at pH between 2.10 and 3.85.

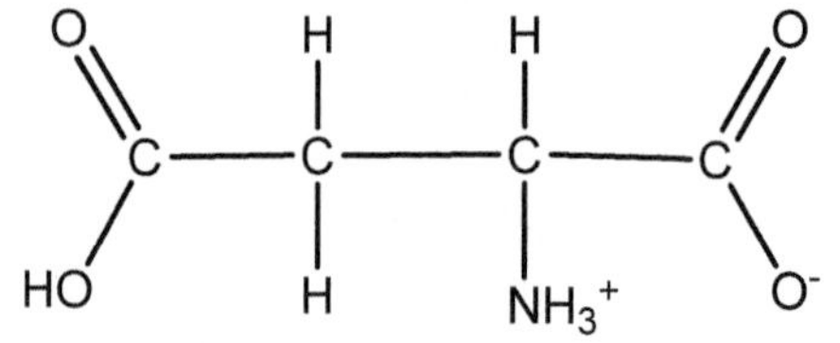

III.   The following predominates at pH between 3.85 and 9.82.

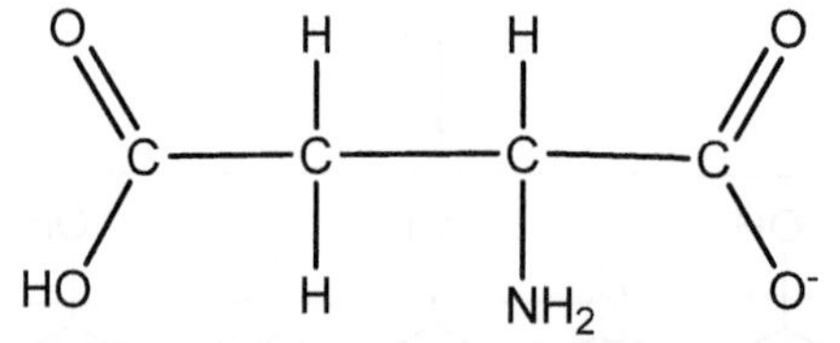

IV.   The following predominates at pH above 9.82.

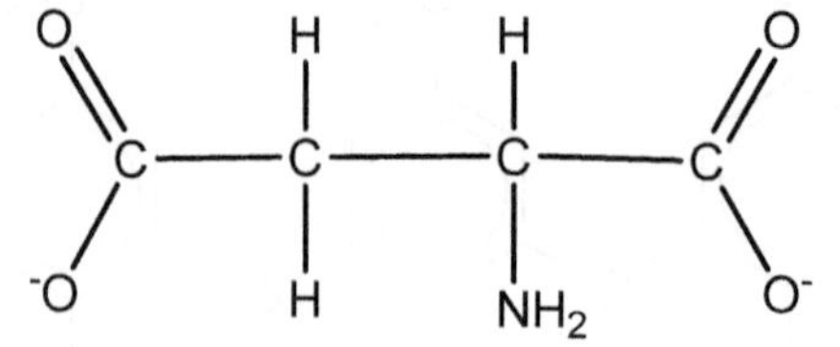